P9-CCD-984

The American City
What Works, What Doesn't

Alexander Garvin

Third Edition

New York Chicago San Francisco Athens London
Madrid Mexico City Milan New Delhi
Singapore Sydney Toronto

Cataloging-in-Publication Data is on file with the Library of Congress

1 2 3 4 5 6 7 8 9 0 DOW/DOW 1 9 8 7 6 5 4 3

ISBN 978-0-07-180162-1
MHID 0-07-180162-6

Sponsoring Editor
Michael McCabe

Editing Supervisor
Stephen M. Smith

Production Supervisor
Richard C. Ruzycka

Acquisitions Coordinator
Bridget L. Thoreson

Project Manager
Virginia E. Carroll,
North Market Street Graphics

Copy Editor
Virginia E. Carroll,
North Market Street Graphics

Proofreaders
Stewart Smith, Virginia Landis, and Sue Miller,
North Market Street Graphics

Indexer
Heidi Blough

Art Director, Cover
Jeff Weeks

Composition
Mark Righter,
North Market Street Graphics

Printed and bound by RR Donnelley.

This book is printed on acid-free paper.

• For My Students •

Contents

Preface to the Third Edition

When McGraw-Hill asked me to prepare a third edition of *The American City*, I mistakenly thought it would require updating the statistics, adding important new items like Chicago's Millennium Park or the subprime mortgage crisis, and making minor revisions. I did not imagine that American cities had, in fact, changed so much over the past decade that I would have to discuss whole neighborhoods that experienced a major revival (the Pearl District in Portland, Oregon; the Third Ward in Milwaukee; Northern Liberties in Philadelphia; and the Old Bank District in downtown Los Angeles), as well as some that were finally experiencing the revival that I claimed in past editions had eluded them (the Mexican War Streets in Pittsburgh and the Victorian District in Savannah).

Changes had also occurred to slum districts that I had discussed in previous editions. Years ago a prominent political figure announced that once you had seen one slum, "you had seen them all." He was wrong. Mott Haven, in the Bronx, which had been characterized by abandoned buildings and vacant lots, is now a thriving, if poor, Puerto Rican neighborhood. A section of Morrisania in the Bronx, where the Nehemiah Program had made major investments, is now a healthy community of homeowners. East New York in Brooklyn, which had been the target of massive investments by the Model Cities Program, on the other hand, remains a high crime location with areas of devastation. In fact, no two cities are ever the same, just as no two sections of any city are the same. Moreover, they are continually changing. That is why I could not just make minor revisions and ended up reexamining and frequently revising the wide range of places and strategies that are presented in *The American City*.

The way we perceive cities has also changed. At the time this book first appeared, many people believed that cities were in trouble, perhaps even terminally ill. The antidote that had been routinely prescribed consisted of clearance and redevelopment. Despite Jane Jacobs's devastating critique, redevelopment and suburbanization still occupied the public consciousness. Thus, despite my inclusion of chapters on housing rehabilitation, neighborhood revitalization, and historic preservation, this book concentrated on new construction. By the time I started working on the third edition, many people had noticed that young people were choosing to live in urban areas rather than suburbs. New strategies had replaced the pervasive confidence in new construction. Business improvement districts (BIDs), for example, had become major factors in almost every downtown. So I expanded the discussion of BIDs and created a separate chapter for them. Many once-pedestrianized streets had been remotorized. So, rather than isolating the discussion of pedestrianization as a separate chapter, I included it in the chapters on downtown revival.

My thinking about central business districts also has evolved greatly from the first edition. Thus, although I updated the discussion of redevelopment and placed it within the context of interstate highway development, "urban renewal" is no longer a major focus. The big change, however, is the inclusion of an entirely new chapter: "Retrofitting the City for a Modern Commercial Economy." While it includes material from previous editions, it is an entirely new take on the subject, presenting the public realm approach to planning that I set forth in my recent book, *The Planning Game: Lessons from Great Cities.*

The biggest change is not in the text, but rather in the illustrations. Now, with the exception of historical images, the book's 650 illustrations are in full color and consist nearly entirely of photographs, maps, and diagrams made especially for the third edition. As all of my students know, the photographs are particularly important because I never write about places to which I have not been. More important, though, I have selected images that illustrate these places from my point of view (or rather, that of my camera lens). The maps and diagrams are the work of Ryan Salvatore, Joshua Price, Owen Howlett, and Zakery Snider—to whom I am forever indebted for helping me to show aspects of *The American City* that would be impossible to reveal any other way. Ryan not only helped to make 40 of these illustrations, he did background research for every chapter, read the entire text, debated it with me, and made invaluable editorial suggestions. Without his input this book would be a much less persuasive volume. There is no way I can thank him enough.

Friends and acquaintances, many of whom helped me with the first and second editions, also aided me in the updating of the third edition. There are too many to enumerate all of them. You know who you are, and I am grateful to all of you. Among those who made major contributions are Miguel Angel Soler Ruiz, Con and Kathryn Howe, David and Susan Brownlee, Rick and Nancy Rubens, Alan and Leslie Beller, Bob and Hannah Kaiser, Paul and Grace Kelly, Andrés Duany and Elizabeth Plater-Zyberk, Dan Biederman, Brad Blatstein, Joe Breen, Bob Bruegmann, Anne Goulet, Matt Jacobs, Paul Levy, Dennis McClendon, John Meigs, Charles Michener, Hunter Morrison, Max Musicant, Ollie Nieuwland-Zlotnicki, Nick Peterson, Herman Pettegrove, Davis Rhorer, Inga Saffron, Jim Schroder, Danny Serviansky, Bob Stein, Scott Stone, and Elizabeth "Boo" Thomas.

I thought my education ended when I had completed 10 years at Yale and received my three degrees. It had only begun. I have continued coming to Yale for more than 46 years. On every visit I learn something new from my students. They are the most important contributors to this volume, to whom it is dedicated, and I remain eternally grateful to them.

Alexander Garvin, 2013

Preface to the Second Edition

The first edition of *The American City: What Works, What Doesn't* presented projects that were largely completed by 1990. When McGraw-Hill asked me to update it to 2000, I thought it would be an easy task. Finding contemporary statistics was not easy. Adding significant projects and programs required revising and expanding important sections of the book.

I have replaced and augmented the illustrations with several hundred new photographs and drawings. Most chapters present additional projects and programs, largely from the end of the twentieth century. In some cases, the additions have amplified the material presented in that chapter.

My arguments for market-based planning are, as before, presented in the first two chapters. In some instances, however, the examples illustrating those arguments have been changed. Chapter 5, "Retail Shopping," has been completely revised, and Chapter 9, "Reducing Housing Cost," has been substantially expanded.

Visits to Miami Beach, Scottsdale, New Albany, and other places allowed me to add material with which I had been previously unfamiliar. In other cases I revisited places one or more times to ensure that the discussion reflected current conditions. Still other subjects, such as the federal Low-Income Housing Tax Credit program, big-box retailing, or the New Urbanism, were added because of their growing role in shaping our cities and suburbs.

My professional life has deepened my understanding of what is involved in planning cities. Seven years on the New York City Planning Commission, discussing hundreds of items with my fellow commissioners and with Chairman Joseph Rose and Executive Director Lance Michaels, has forced me to reexamine a huge range of subjects.

The same is true of my consulting work in Palm Beach, Charlotte, Baton Rouge, and elsewhere. The experience in Baton Rouge was particularly important. It gave me the opportunity to work with Andrés Duany and the talented team of professionals from DPZ (described in Chapter 18). I am indebted to Mark Drennen, Raymond "Skipper" Post, Davis Rhorer, and Jeff Fluhr for their insights into the functioning of that city. Watching Elizabeth "Boo" Thomas, executive director of *Plan Baton Rouge,* convinced me that a skilled professional could make public participation in planning a gracious process that would result in desirable changes to a city.

Planning New York City's effort to hold the summer Olympics in 2012 has been the single most important experience in my professional life (prior to my current involvement with Lower Manhattan). That effort is entirely the vision of one man, Daniel Doctoroff. He provided the inspiration for many improvements to a central theme of this book: entrepreneurship. Every day I learn something else about entrepreneurship simply by observing Dan at work "getting things done."

Working on the New York City Olympics involved so much more than just planning sports facilities. The talented and extremely hardworking professionals at NYC2012 deepened my understanding of the entire planning process. For this I am indebted to George Abraham, Jay Kriegel, Wendy Hilliard, Brenda Levin, and most particularly Chris Glaisek, J. B. Clancy, and Andrew Winters. I met Chris, J.B., and Andrew as students at the Yale School of Architecture. They became teaching assistants. Together, we put together the plan for the New York City Olympic games. Nobody ever had more dedicated, loyal, or effective partners to demonstrate what planning is all about. I don't know where my thinking starts and theirs ends. Consequently, I cannot identify the many ways in which they influenced the changes in the text.

Without the help of Scott Ball, Daniel Biederman, Robert Davis, Adam Flatow, George Garvin, David Haltom, Daniel Jung, Niesen Kasdin, Edgar Lampert, Henry Lanier, Paul Levy, Dennis McClendon, Hunter Morrison, Josh Olsen, Jaquelin Robertson, the late I. D. Robbins, Marshall Rose, Robert A. M. Stern, Rob Walch, and Mtamanika Youngblood, the second edition would have been much less convincing. There is no way I can enumerate all the other people who have taken time to show me around their cities, to explain their projects, and to amplify my understanding of the many subjects covered in the text. I thank them all.

Books do not just happen. I have been blessed with the assistance of Eric Clough, who worked with me to produce so many new illustrations. Hugh Eastwood, Mark Abraham, Sam Ryerson, and Brett Rubin helped to update statistics, check facts, prepare captions, and nurse me through the year and a half it took to make all the revisions. Christine Furry, Michael E. Gonzalez, Mary Jo Fostina, and the team at North Market Street Graphics are probably the best editors I have ever worked with. More important, they found wonderful ways to showcase the photographs, maps, and drawings that are an integral part of this book.

Two people hover on every page of the second edition: Edmund Bacon and Robert Bruegmann. One of the great joys that came from the initial publication of *The American City* is my friendship with Ed Bacon. I met him the year the first edition was published. In the ensuing years he has spent hours explaining his work and telling me anything and everything that is important about planning cities. If I was in awe of his achievements when we met, I am now a thunderstruck disciple.

At about the same time I met Edmund Bacon, I met Robert Bruegmann. Nobody knows more about the American city. He has altered my thinking more than anybody, including the writers and teachers who influenced the first edition of this book. We share one prejudice. We believe you cannot understand a place that you have not visited. Our visits to various corners of the Chicago metropolitan area are among the highlights of the past few years. Our conversations have deepened my thinking and changed my mind about so very much that I cannot identify which ideas are his and which are mine. Bob was gracious enough to read large sections of the manuscript of the second edition. It is entirely due to his comments and suggestions that I can refer to this volume as the "new and improved" edition of *The American City.*

Alexander Garvin, 2002

Preface to the First Edition

Most books approach the city from one perspective (housing, zoning, historic preservation, etc.), one discipline (history, politics, law, finance, architecture, etc.), or one city (Chicago, Boston, San Francisco, etc.). Instead, *The American City: What Works, What Doesn't* presents a comprehensive, multidisciplinary review of the many attempts to fix the American city (everything from parks to shopping centers, mortgage insurance to planned new towns): what has worked, what hasn't, and why.

The book covers two centuries of activity in cities from one end of the continent to the other. More than 250 projects and programs in 100 cities are analyzed. In all cases they are projects that I have personally visited, usually many times. Because most readers will be unfamiliar with many of the places that are discussed, the text is supplemented with nearly 500 illustrations, most photographs taken during my visits.

The text examines six ingredients of project success (i.e., market, location, design, financing, time, and entrepreneurship) and discusses the ways in which those ingredients affect its outcome. Since I do not believe that project success necessarily results in any improvement to the surrounding city (indeed, some have made things worse), I also redefine successful city planning as public action that generates a desirable, widespread, and sustained private market reaction. Thus, the programs and projects discussed in the book are always examined in terms of the private market reaction they have generated. Because it is impossible to judge project impact until some years after it has been completed, I have made 1990 the cutoff point.

The book presents the logic behind a wide range of strategies for municipal improvement, the initial project or program (some dating back to the nineteenth century) that launched the strategy, the classic cases of its application, and finally a framework for predicting whether that strategy will or will not succeed. It also examines how successful that strategy has been in generating the desired private market reaction and what legislation, if any, is needed to improve its performance. The reason for this structure is to convince the reader that we have been fixing American cities for two centuries and can identify what will work and what won't.

In addition to the many secondary sources acknowledged in the footnotes, I have included information from a number of primary sources that have not until now appeared in any book, as well as observations that are at variance with often-repeated accounts of particular projects or programs. These differences should be of particular interest to specialists in the field and in some cases will cause them to revise long-held beliefs.

My purpose in presenting new information about classic programs and projects and fresh material on some that have never before been examined within a historical context is not to be original, nor to mention interesting innovations, but to evaluate the success or failure of specific strategies for fixing the American city and to recommend further action.

I hope my readers will come to believe, as I do, that we know how to fix the American city. More important, I hope that whenever their communities are considering any of the strategies presented, they will return to this book to read about what worked, what didn't, and why. Most important, I hope that this book will inspire them to get involved personally and take an active role in improving the American city.

Alexander Garvin, 1995

Acknowledgments

The number of people who have contributed to the three editions of this book far exceeds the space available to thank them. They include students who have introduced me to their hometowns; coworkers who have helped me to explore ideas, programs, and places; friends and acquaintances who have taken me around their cities; librarians and archivists who have shown me documents that I didn't know existed; public officials who have shared stories and fugitive documents about programs and cities all over America. I thank them all.

In particular, I thank my parents, who always loyally provided support and good counsel, and my brother, George, whose enthusiasm is a constant inspiration.

Everybody is indebted to his teachers. Three in particular have influenced the contents of this book. To this day I think about the ideas that I first encountered in Vincent Scully's lectures on American architecture. Christopher Tunnard insisted that I consider the broad cultural underpinnings of American city planning. Dennis Durden persuaded me always to go out and see for myself before deciding whether something worked.

Yale University has nurtured my intellectual development for nearly six decades, first while I was an undergraduate, then while I was a graduate student, and finally during the 46 years I have taught there. It has provided the best and brightest students to question my observations, talented teaching assistants to make sure I was clear and convincing in conveying my ideas, and generous colleagues with whom to discuss the widest range of subjects. I have found all sorts of unexpected treasures in its extraordinary library. For all of this, I am forever indebted to everybody at Yale (past and present) and to a series of deans of Yale College (Georges May, Horace Taft, Sidney Altman, the late Martin Griffin, Howard Lamar, Donald Kagan, Donald Engleman, Richard Brodhead, Peter Solovey, and Mary Miller) and of the Yale School of Architecture (Tom Beebe, Fred Koetter, and Robert A. M. Stern) who provided support even when they were without conventional explanations for my presence at Yale.

I learned how to operate and manage a business together with Jim Schroder and later deepened that understanding working with Nick Peterson and Joshua Price. My former partners, Alan Davies, Irwin Leimas, Fred Roth, Robert Haskell, Alvin Schein, and Larry Ackman, and former students Danny Serviansky, Matt Jacobs, Keith Rogal, and Nick Monroe provided me with countless lessons in real estate development and management. Thanks to them and to my father I understand how real estate works and am able to explain how and why public action can succeed or fail in generating any private market reaction.

Donald Elliott and Edward Robin are responsible for bringing me into New York City government. I continue to regard their achievements as the truest demonstration of the role that city planners can and should play in government. During the 10 years I spent as a full-time "bureaucrat," I came to know hundreds of dedicated men and women who worked hard and effectively to make New York City a better place in which to live and work. I was inspired by the leadership of Nathan Leventhal, Roger Starr, John Zuccotti, and the late Robert Wagner Jr. (who also helped me with the text of this book). I learned how government gets things done by watching such talented individuals as Joseph Christian, Charles Cuneo, Martha Davis, Linda Einhorn, Marilyn Gelber, Jolie Hammer, Chris Hooke, Robert Jacobson, Henry Lanier, Michael Lappin, Barbara Leeds, Jean Lerman, Barry Light, Victor Marrero, Russell McCubbin, David McGregor, Robert Milward, John Skelly, Jesse Taylor, Joe Tenga, Tupper Thomas, and Jack Toby. Fifteen years later, thanks to Joseph Rose, I reentered government, encountering a whole new generation of talented people who are working hard to make government effective. These dedicated civil servants helped me to grasp the meaning and possibilities of effective public action.

Without the help and encouragement of my friends, this book would never have been written. Many do not even know the ways in which they helped to transform me into a writer. Among those who played a critical role are Alan and Leslie Beller, Eugenie Ladner Birch, Iris and Paul Brest, David and Susan Brownlee, Bob Bruegmann, Van Burger, Sherwin Goldman, Anne Goulet, Rick Henderson, Con and Catherine Welsh Howe, Robert and Hannah Kaiser, Robert Kimball, Richard Kumro, Michael Larson, Lance and Carol Liebman, John Meigs, Charles Michener, Nick Monroe, Hunter Morrison, Richard Peiser, Herman Pettegrove, Michael Piore and Rodney Yoder, Alec and Drika Purves, Daniel and Joanna Rose, David Rose and Gail Gremse, Joe Rose, Richard Singer, Miguel Angel Soler Ruiz, Megan Tourlis, Ted Volckhausen, Tappy and Robin Wilder, and Ollie Nieuwland-Zlotnicki. Over a decade or more, Rick and Nancy Rubens have accompanied me on visits and contributed mightily to my understanding of the places we visited.

I thought I had a good understanding of American cities when the first edition of this book appeared, but, as a result of the expansion of my consulting practice, I learned a great deal more about New York City, and cities in Florida, Georgia, Louisiana, Maryland, Nebraska, Tennessee, Texas, and Utah. I am indebted to my clients for what I learned.

One client stands out among them all: Daniel Doctoroff. He hired me to plan for a Summer Olympics in New York City. Working with him and J. B. Clancy, Chris Glaisek, Andrew Winters, Nick Peterson, and a truly amazing array of great architects, I learned more about the shape and character of the city I knew best, my hometown. No professional ever had a better client or more generous colleagues to work with.

During the middle of my work on the Olympics I spent 15 months as Vice President for Planning, Design and Development of the Lower Manhattan Development Corporation (LMDC), setting the stage for the revival of the nation's fourth-largest business district and the construction of a replacement for the World Trade Center. Chris Glaisek and Andrew Winters joined me in that endeavor, along with Brett Rubin, Hugh Eastwood, and Brandon Smith. From them, from my colleagues at the LMDC, and from literally thousands of New Yorkers, I learned still more about my hometown. The sections of this book that discuss politics, planning, and development in New York have been greatly influenced by my work with these amazing people and by my experiences at the LMDC.

Arthur Klebanoff provided invaluable advice and is responsible for finding my publisher. The editorial team at McGraw-Hill—Larry Hager, Michael McCabe, and Bridget Thoreson—was of great assistance. The production team of Mark Righter, Dennis Bicksler, and Ginny Carroll transformed the text and illustrations into a beautiful volume. I cannot imagine working with more generous or supportive collaborators.

Last and most important are Ted Volckhausen and Rick Henderson, who read every word of every version of every chapter of the first edition, and Ryan Salvatore, who did the same for this edition. They patiently corrected spelling, grammar, punctuation—even footnotes. They challenged ideas that were unclear or poorly explained, questioned the logic of my theories, and were always there to provide encouragement. I can never thank them enough. My major regret is that Rick died before he could see any of the three editions of this book.

Alexander Garvin, 2013

ABOUT THE AUTHOR

ALEXANDER GARVIN has combined a career in urban planning and real estate with teaching, architecture, and public service. He is currently President and CEO of AGA Public Realm Strategists. Between 1996 and 2005 he was Managing Director for Planning of NYC2012, the committee to bring the Summer Olympics to New York in 2012. During 2002–2003, he was Vice President for Planning, Design and Development of the Lower Manhattan Development Corporation. Over the last 44 years he has held prominent positions in five New York City administrations, including Deputy Commissioner of Housing and City Planning Commissioner.

Garvin is Adjunct Professor of Urban Planning and Management at Yale University, where he has taught a wide range of subjects, including "Introduction to the Study of the City," which for more than 46 years has remained one of the most popular courses in Yale College. In addition, he teaches two courses in the School of Architecture, including a seminar on "Intermediate Planning and Development."

Among other honors, Garvin has received the 2012 Award of Merit from the New York Chapter of the American Institute of Architects (AIA) and the 2004 Distinguished Service Award from the New York City Chapter of the American Planning Association (APA). The first edition of *The American City* won the 1996 AIA Book Award in Urbanism.

Garvin is also the author of *Public Parks: The Key to Livable Communities* and *The Planning Game: Lessons from Great Cities*. He earned his B.A., M.Arch., and M.U.S. from Yale University.

Chicago, 1892 *(Chicago Historical Society)*

Chicago, 2010 *(Alexander Garvin)*

Pittsburgh, 1936 *(Carnegie Library of Pittsburgh)*

Pittsburgh, 2012 *(Alexander Garvin)*

Baltimore, 1972 *(Alexander Garvin)*

Baltimore, 2009 *(Alexander Garvin)*

1

A Realistic Approach to City and Suburban Planning

There is agreement neither on what to do to improve our cities and suburbs nor on how to get the job done. Some believe the answers are a matter of money; others believe they involve politics, or racial and ethnic conflict, or some other factor. One thing most people share, though, is disillusionment with planning as a way of fixing the American city.

This disillusionment with planning is far from justified. Dozens of projects are triumphs of American planning:

- Chicago would not have 24 miles of continuous parkland along Lake Michigan if this land had not been included in the city's comprehensive plan of 1909.
- The glorious antebellum sections of Charleston, South Carolina, would not have survived if the city had not adopted zoning in 1931.
- Pittsburgh would no longer be a major corporate headquarters center if it had not virtually rebuilt its downtown after World War II.
- Baltimore would not be a lively retail, entertainment, and employment center if it had not spent decades reconstructing its downtown waterfront.

Such triumphs are easy to overlook. Once a problem is solved it disappears and is forgotten. Other, newer concerns become paramount. During the nineteenth century, for example, Americans wanted to pave city streets, obtain a reliable supply of drinking water, dispose of sewage, and provide public parks. Once these problems were dealt with, they began confronting the onslaught of motor vehicles, deteriorating air and water quality, and the proliferation of slums. By the second half of the twentieth century, some people even thought that without massive reconstruction, cities would not be able to survive.

Many people are disillusioned with planning because so

Charleston, 2011 *(Alexander Garvin)*

many of its promises are not kept. Usually these promises are made in good faith by planners who believe that their job is to establish municipal goals and provide blueprints for a better city. Too often, the planners' efforts end before they consider how they will obtain political support for their proposals, who will execute them, or where the money to finance them will come from. Disillusionment with planning also develops when physical improvements fail to solve deep-seated social problems. This is not the fault of planning. After all, fixing cities does not fix people. The disillusionment in this case is a product of misplaced expectations. Crime, delinquency, and poverty are afflictions of city residents, not of the cities themselves. Such problems can be found in suburban and rural areas as well.

We need more realistic expectations of what planning can accomplish. While it cannot change human nature and is therefore not a panacea for all urban ills, it surely can identify opportunities to improve a city's physical plant, strategies to finance those improvements, and entities to execute them, and consequently affect the safety, utility, attractiveness, and character of city life. When Chicago began creating its waterfront parks, for example, large sections of the shoreline of Lake Michigan were being used as rail yards and garbage dumps. Simply removing these uses reduced hazards and made neighboring property more attractive.

We also need a better understanding of how effective planning is translated into a better quality of life. By themselves, planners cannot accomplish very much. Improving cities requires the active participation of property owners, bankers, developers, architects, lawyers, contractors, and all sorts of people involved with real estate. It also requires the sanction of community groups, civic organizations, elected and appointed public officials, and municipal employees. Together they provide the financial and political means of bringing plans to fruition. Without them even the best plans will remain irrelevant dreams.

Finally, the planning profession itself needs to improve its understanding of the way physical changes to a city can achieve a more smoothly functioning environment, a healthier economy, and a better quality of life. For example, the restoration of Charleston's historic district generated substantial tourist spending, and the reconstruction of the bridges and highways leading into downtown Pittsburgh reduced the cost of doing business and initiated an era of major corporate investment. These and other successful planning strategies are too frequently ignored in the search for more innovative prescriptions.

At its best, planning changes the very character of city life. During the last half of the twentieth century, Baltimore rebuilt its harbor, opened an aquarium, built baseball and football stadiums, and erected a major convention center. As a result, Baltimore became a safer, more convenient, more beautiful city. It also became a more attractive destination for the city's rapidly growing metropolitan region, drawing tens of thousands of additional weekday shoppers and weekend visitors.

Despite many remarkable successes, American planning has been plagued with continuing mistakes. These mistakes were and are avoidable. More than six decades have passed since Jane Jacobs, in her pioneering book *The Death and Life of Great American Cities,* observed that we had spent billions of dollars for

> *Housing projects that are truly marvels of dullness and regimentation. . . . Civic centers that are avoided by everyone but bums, who have fewer choices of loitering places than others. Commercial centers that are lack-luster imitations of standardized suburban chain-store shopping. Promenades that go from no place to nowhere and have no promenaders. Expressways that eviscerate great cities. This is not the rebuilding of cities. This is the sacking of cities.*[1]

Four decades and hundreds of billions of dollars later, her criticisms still ring true. Most cities continue to lack housing, civic and commercial centers, places to congregate and promenade, and traffic arteries. In too many cases, the attempt to remedy the situation constituted further "sacking of cities." These attempts may have been financially and politically feasible. However, they failed because they were conceived without proper consideration of whether they would benefit the surrounding city.

Defining the Planning Process

Much of the nation's unsuccessful planning arises from the erroneous belief that project success equals planning success. Highways that are filled with automobiles, housing projects that are fully rented, and civic centers with plenty of busy bureaucrats may be successful on their own terms. The cities around them, however, may be completely unaffected. Worse, they may be in even greater trouble than they were prior to these projects.

Only when a project also has a beneficial impact on the surrounding community can it be considered successful planning. Thus, *planning* should be defined as ***public action that generates a sustained and widespread private market reaction,*** which improves the quality of life of the affected community, thereby making it more attractive, convenient, and environmentally healthy. That is precisely what has occurred wherever planning has been successful.

- When Chicago transformed its lakeshore into a continuous park and drive, the real estate industry responded by spending billions to make it a setting for tens of thousands of new apartments. Hundreds of thousands of people responded by coming to its lovely lakeshore parks.
- When Charleston preserved its old and historic district, it

DeKalb County, Georgia, 2007 *(Alexander Garvin)*

retained an extraordinary physical asset that, decades later, would attract a growing population, draw millions of tourists, and provide the basis of a thriving economy.

- When Pittsburgh cleared its downtown of the clutter of rail yards and warehouses, reduced air and water pollution, and built new highways, bridges, and downtown garages, businesses responded by rebuilding half the central business district.
- When Baltimore invested in redeveloping its Inner Harbor, the private sector responded by erecting office buildings, retail stores, hotels, and apartment houses. Downtown Baltimore was transformed from a place that had been deserted from Friday evening to Monday morning into an area that attracted people seven days a week.

In each of these instances, the private market reaction went far beyond economic stimulus. The improvements to the environment and to the quality of life satisfied a myriad of personal desires, in the process attracting great numbers of people whose very presence brought the vitality these places had lacked.

In addition to redefining the parameters of successful planning, we need to broaden the very scope of the profession. Over the past few decades, the areas of public concern and therefore of public action have expanded both substantively and geographically. Outraged citizens have demanded action to protect the natural environment, to preserve the national heritage, to provide a range of services that had never before been considered a public responsibility, and to deal with territory outside local political jurisdictions. The country should be deeply grateful to these activists for insisting that government fill important vacuums.

Too often, we have responded to such legitimate citizen demands by creating a set of protected special interests that are excluded from competition with other equally legitimate public concerns. As a result, large geographic areas are removed from active use without consideration of the social consequences. Buildings are declared landmarks without reference to economic impact. Services are provided to socially impaired individuals without any thought of the effect on the surrounding community. Including these new areas of public concern within the scope of planning and simultaneously including a far broader range of participants in the planning process can rectify the situation.

Similarly, we must expand our understanding of who generates public action. Government agencies are not the only entities in the planning game. The public realm in many suburban areas is the creation of developers acting on the public's behalf. Nonprofit organizations are responsible for many activities that have encouraged property owners to invest in neighborhood improvements. That is why the definition of planning refers to *public* (rather than to government) action that generates a sustained and widespread private market reaction. This broad definition highlights the fact that planning is about *change:* preventing undesirable change and encouraging desirable change. It may involve a tax incentive, a zoning regulation, or some other technical prescription, but only as a mechanism for instigating change. The important element is change itself. Planners obtain changes in safety, utility, and attractiveness of city life by using three major techniques: strategic public investment, regulation, and incentives for private action.

Strategic Public Investment

Nineteenth-century planners were particularly enamored of strategic government investment. Just think of the many locations that were made more attractive for prospective developers through the installation of water mains, sewer pipes, or transit lines.

A more recent example is federal subsidization of the Interstate Highway System, created pursuant to the National Interstate and Defense Highways Act of 1956 (Public Law 84-627), which vastly increased the amount of land within commuting distance of cities and, in the process, increased the attractiveness of suburban locations. Developers eagerly purchased the newly accessible land and built houses, shopping malls, and office parks. In the process, millions of consumers were given the opportunity of owning a house in the country, close to shopping facilities and sometimes also near their jobs.

The 47,182 miles of interstate highways comprising the system as of 2010 continue to alter land use patterns

DeKalb County, Georgia, 2007 *(Alexander Garvin)*

and local economies. At many exits, "illuminated logos of McDonald's, Kentucky Fried Chicken, Waffle House, Phillips 66 . . . Super 8 and Best Western have sprung up like lollipops on 100-foot-tall sticks. . . . Just off the exits, often tucked out of sight, are its more muscular businesses—factories, meat-packing plants, warehouses and distribution centers." The effects are not just a matter of changing land use. For example, all but one of the 13 counties bordering Interstate 40 in Arkansas increased in population during the decade between 1990 and 2000, whereas 6 of the 11 counties just beyond them lost population or failed to grow.[2]

Capital investment can cause adverse impacts as well. Many interstate highways attract motorists away from traditional urban arterials, thereby reducing demand in the retail establishments that had previously catered to the large market of automobile-oriented consumers. For decades after the highways had been built, cities were plagued with blighted retail streets, unable to replace the customers who had abandoned previously active stores.

The difference between routine capital spending and strategically planned investments lies in the use of expenditures to spark further investment by private businesses, financial institutions, property owners, and developers. Governments regularly spend substantial sums of money to eliminate congestion and improve traffic flow. Smart planners also use such spending opportunities as a way of generating further private development. One of the most dramatic examples occurred in San Antonio, Texas. By the early twentieth century, flooding had become a major problem in San Antonio, where the river that looped through the business district flooded regularly. Four people died during the flood of 1913; 50 died in the flood of 1921, when 13 of the city's 27 bridges were destroyed. The first solution was straightforward: dig a bypass connecting the two ends of the loop and install locks so that, when water threatened to overflow the riverbanks, the locks could be closed and divert water past the downtown. Ordinary spending of this sort is not planning. As Walter Moody, the Chicago Plan Commission's first managing director, declared, effective planning means investing in "public improvements" with intelligence and "foresight." He called a street widening or a flood control measure, unrelated to any other improvements, "unplanning."[3]

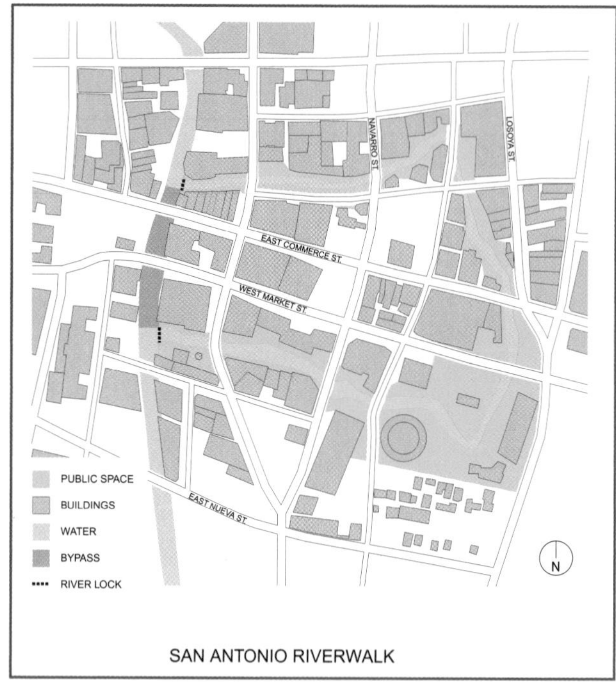

San Antonio Riverwalk *(Alexander Garvin, Joshua Price, and Ryan Salvatore)*

During the Great Depression, the Roosevelt administration convinced Congress to spend billions of dollars to stimulate the economy. Some of this money undoubtedly went into "unplanning." In Los Angeles, for example, the city spent $230,000,000 of this money diverting all but 12 miles of the Los Angeles River into a concrete culvert. All that money solved the flooding problem and disfigured 46 miles of the city's landscape, but achieved little else.

San Antonio, on the other hand, took far less money from the WPA to solve its flooding problem: $355,000. To this it added $75,000 from a bond issue to transform the downtown river loop into a public park that was completed in 1941. By

Los Angeles, 2013 *(Alexander Garvin)*

San Antonio Riverwalk, 1941 *(UTSA Institute of Texan Cultures)*

San Antonio Riverwalk, 2012 *(Alexander Garvin)*

2012, this park, known as the Paseo del Rio or Riverwalk, along with the Alamo, had become one of the top two attractions in San Antonio, the most visited city in Texas, which hosts about 26 million tourists a year.[4]

Regulation

Regulation is most often used to alter the size and character of the market and the design of the physical environment. Perhaps the single most effective example occurred during the 1930s when the federal government restructured the banking system and in the process dramatically altered the housing market. Prior to that time, few banks provided mortgage loans that covered more than half the cost of a house. These loans were extended for relatively short periods of time (two to five years) and involved little or no amortization.

The National Housing Act of 1934, which created the Federal Housing Administration (FHA), changed all that by regulating the rate of interest and the terms of every mortgage that it insured. By 1938, a house could be bought for a cash down payment equal to 10 percent of the purchase price. The other 90 percent came in the form of a 25-year, self-amortizing, FHA-insured mortgage loan. These new mortgage lending practices greatly increased the number of people who could afford a down payment on a house and monthly debt service payments on a mortgage, thereby also increasing the size of the market for single-family houses.

Not only did the FHA alter the size of the market, it also determined the design of the product. In order to be eligible for FHA mortgage insurance, a house had to conform to published minimum property standards that included regulations on structure, materials, and room sizes, the effect of which was to guarantee a minimum quality nationally. That is one of several reasons that the proportion of American households owning their homes increased from 44 percent in 1940 to 65 percent in 2010.

Ill-begotten regulations can have a deleterious effect as well. At the beginning of the twenty-first century, FHA regulations had been sufficiently watered down that institutions started making unwise loans. In the subprime mortgage crisis that followed, lenders foreclosed on tens of thousands of homeowners who were unable to pay debt service on their mortgages. Home ownership declined by 1.1 percent, the largest decline since the decade during which the FHA had been established.[5]

Regulation also can be used to alter the character of an entire area. This process usually begins with an attempt to prevent hazardous conditions. Local governments, for example, are usually interested in providing sufficient open land to permit natural drainage of rain and snow, to prevent waste from percolating through the ground to contaminate the water supply, and to ensure privacy. One way of achieving these objectives is to require a minimum lot size for any development (e.g., no more than one house per acre). The end result is the landscape of one-family houses on large lots that can be found throughout the nation.

Memphis, 2006. Routine zoning regulations produced a routine landscape of identical front and side yards. *(Alexander Garvin)*

Santa Barbara, 1880. Aerial view of the city when it was a typical wooden Western town. *(Santa Barbara Historical Society Collection)*

As with strategic government investment, a regulation such as mandating minimum lot sizes also can produce an adverse impact. Since the amount of land in any community is finite, whenever a minimum lot size is adopted the future supply of house sites is reduced. This reciprocal relationship between the degree of regulation and the size of the market for the resulting product is inevitable. It was poignantly explained by Jacob Riis, who, in 1901, was already lamenting that the minimum construction requirements of "tenement house reform . . . tended to make it impossible for anyone [not able] to pay $75 to live on Manhattan Island."[6]

Zoning regulations can be used to exclude the intrusive development that is incompatible with desired land use patterns. By eliminating the possibility of such undesirable change, they reduce the risk of future problems (e.g., traffic, pollution, and noise) and thereby increase the attractiveness of investing in real estate. In suburban communities, zoning is routinely used to provide front and side yards that are large enough for residents to park their cars, thereby leaving room for visitors to park on the street. Rear yards have to be large enough to prevent unwarranted intrusions onto neighboring properties.

Although regulations governing yard size can sustain real estate values, they rarely generate further improvements. But Santa Barbara, California, has demonstrated how land use regulation can stimulate real estate activity by reducing the risk of developing property. Civic leaders were eager to spur economic growth and decided to do so by encouraging investment in tourist-oriented facilities. Not only did they need something that would attract the growing tourist market, but they needed to induce the real estate industry to build the necessary facilities.[7]

Santa Barbara, 2013. Aerial view of the city after it had been altered to conform with the Hispano-Mediterranean esthetic required by local zoning. *(Alexander Garvin)*

At the beginning of the twentieth century, when this effort began, Santa Barbara was a dusty, wooden town, typical of those seen in Western movies. It decided to reshape itself to conform to a California-Mediterranean image and adopted building laws that required property owners to develop in compliance with that image. By mandating design requirements, Santa Barbara increased its tourist appeal. More important, since property owners were assured of compatible neighboring buildings, the risk of failure was reduced and the likelihood of capturing the customers attracted by the area's charming heritage was increased. It would be difficult to create a more auspicious climate for a tourist-based economy.

Incentives

Although the use of incentives is becoming more popular, the approach has been around a long time. One of the oldest examples is the incentive for people to own their homes. During the Civil War, Congress allowed taxpayers to deduct interest payments and local taxes from the income that formed the basis of federal tax payments. The same deduction of mortgage interest and taxes was reintroduced in 1913, when the federal income tax was adopted. In both instances the incentives triggered an increase in home ownership. Yet neither incentive was intended to encourage development or redevelopment in specific communities.

There can be no serious change in either cities or suburbs without a favorable investment climate. In many instances, government need guarantee only two things: intelligent spending on capital improvements and regulatory policies that provide stability and encourage market demand. When investment and regulation are insufficient to do the job, incentives come into play. New York City faced such a situation during the mid-1970s. The city was running a huge deficit and, as a result, the market for its bonds evaporated. Thus, it was precluded from issuing debt that provided the money for city capital investments or mortgages to private owners to improve existing or build new housing. Political gridlock prevented serious regulatory reform. At the same time the rate of housing deterioration and abandonment had reached alarming proportions. The city administration had to develop a strategy that would prevent further deterioration.[8] One technique seemed most likely to succeed: incentives that were sufficiently generous to induce private investment in the existing housing stock. Consequently, the Housing and Development Administration (HDA) proposed to revise the city's little-known J-51 Program. It provided a 12-year exemption from any increase in real estate tax assessment due to physical improvements and allowed a deduction from annual real estate tax payments of a portion of the cost of those improvements.[9]

The problem with the initial J-51 Program was that it did not apply to three-quarters of the city's housing stock. Existing apartments that were not subject to rent control (because they were in structures that had been built after 1947, or had experienced a change in occupancy after 1971, or were owner-occupied) could not obtain these benefits unless they became subject to rent control. Nonresidential structures that had been converted to residential use were completely ineligible. Without J-51 benefits, any major investment in improvements resulted in punishment: a major increase in the real estate tax assessment. This was especially burdensome to the 770,000 apartments then subject to rent stabilization, New York City's second rent regulatory system.[10]

During 1976, the administration of Mayor Abraham Beame persuaded the state legislature and the New York City Council to smash the rent-control barrier by extending eligibility to rent-stabilized apartments. J-51 benefits were also provided for cooperative and condominium apartments in newly rehabilitated residential structures and to rental apartments in buildings converted from nonresidential to residential use, provided that they would become subject to some form of rent regulation.

These tax incentives completely altered the climate for investment in existing buildings. Banks increased their lending for housing rehabilitation, building owners increased their investments in building improvements, and developers began purchasing vacant structures for conversion to residential use. In fiscal 1977–1978, the first year in which the full impact of these incentives could be measured, more than 48,000 apartments were granted J-51 benefits.

J-51 provided an incentive that was sufficiently attractive to induce major investment in housing rehabilitation. However, there was another reason that so many property owners chose to apply for benefits. The administration of the program was made user-friendly. Until 1975 the program operated subject to unpublished regulations. Specific improvements that were eligible for benefits and the maximum allowable expenditures for those improvements were listed on a typed schedule that was kept by the individual responsible for reviewing applications. Applicants had to file 26 separate forms. Program procedures were known to a few well-connected lawyers and developers, but had never been made public.[11]

Within months of enactment of the revised J-51 Program, the administration published official regulations, made public a printed schedule of all allowable costs, and reduced the required filing to three one-page forms. Even unsophisticated property owners and poorly informed mortgage officers were now able to calculate probable J-51 benefits. As a result of these efforts, hundreds of property owners who had always had an aversion to government agencies were willing to seek the assistance they needed. In the process, tens of millions of dollars were invested in improving the existing housing stock, demonstrating that properly conceived incentives can generate a desirable, sustained, and widespread market reaction.[12]

A New Approach to Planning

We need a new approach to planning that explicitly deals with both *public action* and the probable *private market reaction*. Such change-oriented planning requires general acceptance of the idea that while planners are in the change business, others will actually effect the changes: civic leaders, interest groups, community organizations, property owners, developers, bankers, lawyers, architects, engineers, elected and appointed public officials—the list is endless.

Being entirely dependent on these other players, planners must focus on increasing the chances that everybody else's agenda will be successful. They may choose to do so by targeting public investment in infrastructure and community facilities, or by shaping the regulatory system, or by introducing incentives that will encourage market activity. But whatever they select, their role must be to initiate and shepherd often-controversial expenditures and legislation. More important, the public will be able to hold them accountable by evaluating the cost-effectiveness of the private market reaction to their programs. Most important of all, to be effective, planners must satisfy and reconcile the interests of all the other players, upon whom their success is always dependent.

Only when this approach to planning takes hold will we get beyond the technical studies, needs analyses, and visions of the good city that currently masquerade as urban planning and get on with the business of fixing the American city.

Notes

1. Jane Jacobs, *The Death and Life of Great American Cities,* Random House, New York, 1961, p. 4.
2. Peter T. Kilborn, "In Rural Areas, Interstates Build Their Own Economy," *New York Times,* July 14, 2001, pp. 1, 12.
3. Walter Moody, *What of the City,* A. C. McClurg & Co., Chicago, 1919, pp. 4, 9, 10.
4. Alexander Garvin, *Public Parks: The Key to Livable Communities,* W. W. Norton & Company, New York, 2011, pp. 65–67.
5. "Housing Characteristics: 2010," U.S. Census Bureau, http://www.census.gov/prod/cen2010/briefs/c2010br-07.pdf, retrieved May 24, 2012.
6. Jacob Riis, in a letter probably to Dr. Jane Robbins, October 10, 1891, quoted by Roy Lubove in *The Progressives and the Slums,* (University of Pittsburgh Press, Pittsburgh, PA, 1962, p. 181.
7. A more detailed discussion of regulation in Santa Barbara can be found in Chapter 18.
8. The author, at that time deputy commissioner of housing in charge of J-51 and all other housing rehabilitation programs, proposed this strategy.
9. A more detailed discussion of J-51 can be found in Chapters 11, 12, and 18.
10. New York City regulates rents pursuant to two programs: rent control and rent stabilization. In 1975, 642,000 of New York City's 2,719,000 housing units were rent-controlled. Lawrence Bloomberg (with Helen Lamale), *The Rental Housing Situation in New York City 1975,* Housing & Development Administration, New York, January 1976.
11. The author was responsible for J-51 during the period in which these changes were made.
12. Between 1975 and 1990, many user-friendly characteristics of the program were eliminated. As a result, few property owners now apply for J-51 benefits without the assistance of a lawyer or expediter who specializes in agency processing.

2

Ingredients of Success

San Francisco, 2006. Ghirardelli Square. *(Alexander Garvin)*

There is no formula that guarantees a desirable private market reaction in response to public action. However, six ingredients must be intelligently dealt with for any project to succeed: market, location, design, financing, time, and entrepreneurship.

The need to consider these ingredients may seem obvious. Unfortunately, the proliferation of stillborn projects reveals how little they are understood. Otherwise, why would there be housing for which there is no *market,* commercial centers that are in the wrong *location,* civic centers for which *financing* is not available, places whose *design* makes them unpleasant and unsafe areas in which to congregate, agency-installed public works whose *time* has passed but are still under way, and economic development projects whose completion is beyond the *entrepreneurship* of the responsible public.

If any of these six ingredients is absent or if they are not combined in a mutually reinforcing fashion, the project will fail. Even properly combining all the ingredients may be insufficient to guarantee project success, because city planners, unlike chefs, cannot keep unexpected ingredients from getting into the pot. Nevertheless, an intelligent mix of market, location, design, financing, entrepreneurship, and time is the key to success. Thus, an understanding of how these elements operate and interact increases the likelihood of favorable results. And, anybody who wishes to change cities for the better needs to engage in actions to ensure that each of these ingredients separately and in combination with the others will bring about those changes.

Market

The existence of a market for any urban planning prescription is primary, for without it there is no reason even to consider action. The word *market* is not synonymous with *population.* It means a specific population's desire for something and its ability and willingness to pay for it in the face of available alternatives. Nor is *market* synonymous with *need.* Too often, what one person calls a need is really a preference for what other people ought to have.

To be successful, an urban planning prescription must reflect both market demand and supply. The demand side requires a user population with enough money to purchase what it desires and the willingness to spend it. It also requires enough spending to support both the capital cost and the operating expenses of a project. Without this critical mass of customers, any private action supported by user charges will also require subsidies; if it is a public project, the electorate will have to pay the necessary taxes.

The role that demand plays in determining the success of an urban planning prescription is illustrated by the ways in which downtown Baton Rouge has changed since World War II. As the capital of Louisiana, it offered all the opportunities of a growing market area. In 1950 the population of Baton Rouge was 126,000. By 1990 it had risen to 220,000. Employment was readily available either at the country's largest refinery or in one of the growing state agencies.

Household income had been increasing steadily. Never-

Baton Rouge, 1960s. In the 1960s, Third Street was still the shopping destination for the entire metropolitan area. *(Courtesy of* The Baton Rouge Advocate*)*

Baton Rouge, 1998. By the 1990s, shoppers had abandoned Third Street. *(Alexander Garvin)*

theless, by the 1970s downtown Baton Rouge, like so many other cities, was plagued by empty buildings. Its four department stores and three movie theaters had closed. Civic leaders tried many of the popular devices to bring customers back downtown: a performing arts center, a complex of government office buildings, a convention center, a festival marketplace, museums, even casino gambling. Third Street, the city's premier shopping street, which had been pedestrianized in 1971, continued to be plagued by empty stores. Clearly, this wasn't because the population or its spending power had declined. It was because people were spending their money at the growing number of suburban shopping centers.

At the end of the twentieth century, the city engaged in a planning process (described in Chapter 19) that revived the downtown economy. As a result of carefully selected public investments, by 2013 major hotels, retail stores, and restaurants and cafés had returned to downtown Baton Rouge. More than 40,000 people worked in the downtown business district, which attracted a million visitors annually.

Market demand often declines, as it did in downtown Baton Rouge, when consumers discover cheaper, more convenient, or more attractive places to shop. Public officials often believe erroneously that downtown redevelopment will beat out such competition. Even worse, they give insufficient attention to the additional competitors who will appear in future years. This is what happened in New Haven, Connecticut.

During the 1950s, New Haven's business district faced the usual symptoms of decline: accelerating physical deterioration, decreasing retail sales, and a diminishing tax base. The city's consultants proposed rebuilding its ostensibly obsolete physical plant and using federal urban renewal funds to pay for it. Their plan called for clearing a major portion of the central business district and creating Chapel Square: two department stores, an air-conditioned shopping mall, an office building, a hotel, and a parking garage. Since no substantial increase in demand for office space had been identified, Chapel Square was conceived as a predominantly retail center.[1]

The shopping mall did not attract many customers. In 1982, the first of the department stores closed. During the mid-1980s, in an attempt to salvage the situation, a new city administration provided substantial subsidies and brought in the Rouse Company, a well-regarded national mall operator, to renovate and remarket the project. In 1993, the second department store closed but wasn't torn down till 2007. Finally, in 2012, Gateway Community College was opened on the site once occupied by both department stores.

What went wrong? The diagnosis was faulty. New Haven was not in trouble because of an obsolete physical plant. It was in trouble because suburban competitors were doing a better job of supplying the same market. Between 1960 and 1973, seven additional shopping centers opened in the suburbs around New Haven. Nearby residents found it more convenient to buy toothbrushes and pajamas from these local retailers. Restructuring the city's business district to accommodate what proved to be unnecessary new retail structures

Church St., New Haven, 1980 *(Alexander Garvin)*

Church St., New Haven, 2005 *(Alexander Garvin)*

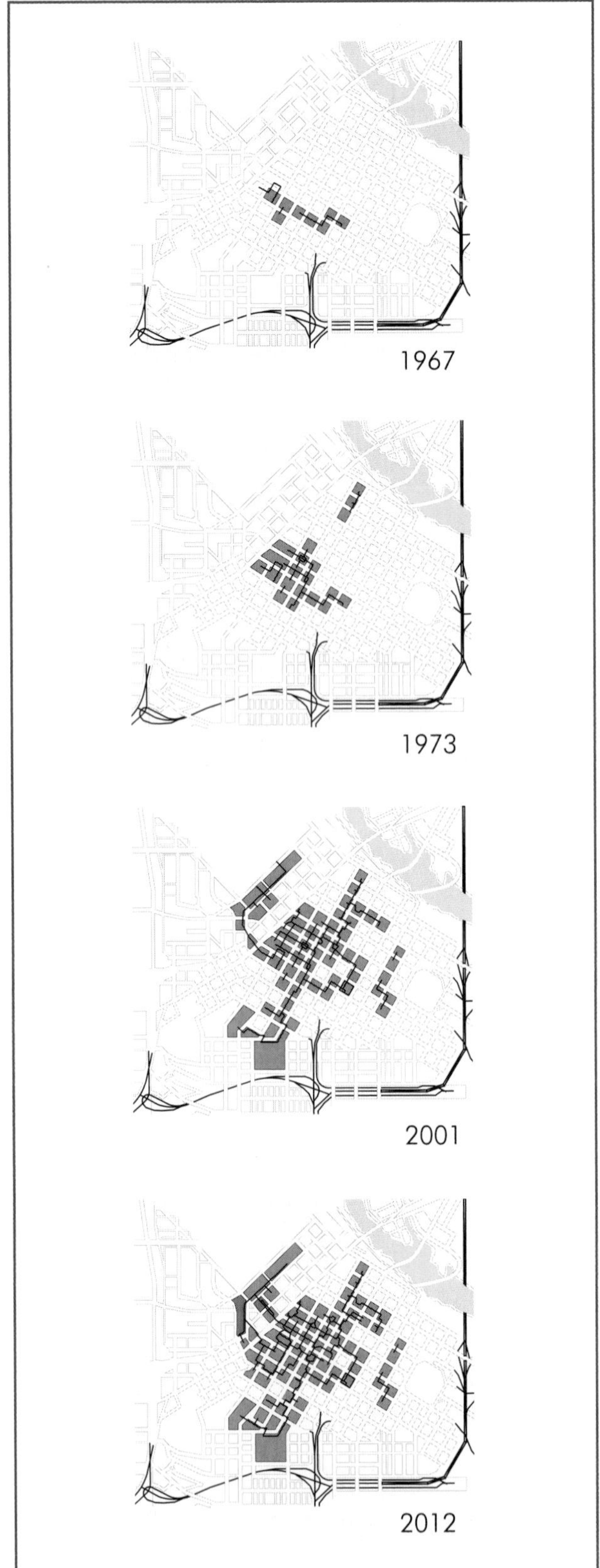

Skyways, 1967–2012 *(Alexander Garvin, Eric Clough, and Ryan Salvatore)*

Minneapolis, 2012. Skyway leading into the IDS Center. *(Alexander Garvin)*

could never be much help in retaining this large and growing market.

At the very same time, Minneapolis dealt successfully with problems that New Haven had failed to solve through downtown redevelopment. In 1956, Southdale, the world's first air-conditioned shopping mall, opened in Edina, Minnesota, just 10 miles southwest of the central business district. Downtown merchants hoped to compete with Southdale and other suburban shopping centers by creating a climate-controlled, pedestrian-friendly environment within the central business district. The scheme called for building additional downtown garages, enhancing Nicollet Avenue as the city's main shopping street, and connecting downtown buildings with a system of glass-enclosed bridges, which they called *skyways.* Its planners hoped the new skyway system would encourage pedestrians to walk from downtown building to downtown building without worrying about the weather. Together, these three components would spur office workers to do their shopping downtown rather than closer to their suburban residences and help to retain customers already attracted by the variety and quantity of shopping opportunities unavailable in the suburbs.[2]

The first skyway opened in 1962. Five years later, when Nicollet Avenue reopened, its sidewalks had been widened,

Minneapolis, 2012. The Crystal Court inside the IDS Center. *(Alexander Garvin)*

repaved, and outfitted with sculpture, fountains, and street furniture. The roadway had been reconfigured as a gently undulating right-of-way set aside exclusively for buses. Unfortunately, by 1969, only seven skyway bridges had been completed, not enough to constitute a climate-controlled business district. The real change came in 1973 when the 2.25-million-square-foot IDS Center opened. Designed by Philip Johnson and John Burgee, IDS covered an entire block with a 51-story office tower, a 279-room hotel, an 8-story office annex, and a 631-car underground garage, all built around a 121-foot-high skylit glass atrium with cafés, restaurants, and shops. Its skyways crossed all four bounding streets and extended the system into the city's two most important department stores, Dayton's and Donaldson's. Soon after opening, the skyways carried 20,000 people per day, providing the critical mass of customers needed to support the system. Within two decades, skyways connected 44 blocks and 105 buildings; by 2011, the skyway system was approximately 8 miles long and comprised 83 enclosed bridges connecting 73 blocks.[3]

Pittsburgh, 1937. Downtown traffic congestion prior to redevelopment. *(Courtesy of Carnegie Library of Pittsburgh)*

Location

Location consists of two elements: a site's inherent characteristics and its proximity and linkage to other locations. Site characteristics alone may be sufficient to make it attractive. A spectacular view is an example. Another is an architecturally distinctive housing stock, such as the one that made renovation particularly inviting in the historic districts of Charleston.

Pittsburgh, 2008. The Golden Triangle half a century after redevelopment. *(Alexander Garvin)*

Site conditions also can ruin an otherwise desirable location. During the first half of the twentieth century, air pollution in downtown Pittsburgh was so serious that streetlights often remained on 24 hours a day. Raw sewage polluted both riverfronts. Daytime traffic congestion seriously restricted both circulation and business activity. In order to alter these inhibiting site conditions, the city obtained state legislation that allowed it to regulate air and water pollution, rebuild its highways and bridges, create more than 5000 parking spaces, and clear away the tangle of downtown rail yards, dilapidated warehouses, and obsolete manufacturing lofts. Once these site conditions were eliminated, property owners invested hundreds of millions of dollars in redevelopment. Within a couple of decades, more than half of the business district had been rebuilt.[4]

Proximity involves both time and space. The temporal dimension is shaped by technology and can be understood in terms of available means of conveyance (Table 2.1). During the eighteenth century, when people were concerned with walking distances, cities had to be compact and densely built up. Now, when we measure distance in driving time, the resultant landscape is a "spread city."

Table 2.1

DISTANCE MEASURED IN TRAVELING TIME

Distances Traveled in 30 Minutes

Conveyance	Miles
Pedestrian walking leisurely	1.5
Pedestrian walking briskly	2.0
Bicycle at normal pace	5.0
Bicycle in 1-hour race	15.0
Bus (in dense city traffic)	3.0
Bus (on suburban streets)	8.0
Bus (express)	15.0
Streetcar (in mixed traffic)	4.0
Light-rail	8.0
Subway (regular service)	12.0
Train (local service)	18.0
Train (regional express)	22.0
Train (metroliner)	45.0
Train (French TGV "train a grande vitess")	80.0
Automobile (moving at normal urban speed limit)	12.0
Automobile (moving at 55 miles per hour)	27.0

Source: Sigurd Grava.

The spatial dimension of proximity involves interdependence with neighboring areas. An obvious example is the relationship between movie theaters, parking facilities, and eating establishments. On a larger scale, nineteenth-century warehouse and manufacturing districts often developed in close proximity to waterfront areas, through which they received and shipped goods and materials.

Even before the end of World War II, most mercantile districts, especially in port cities and railroad towns, had begun a slow and steady decline. There was no longer the same need for large, multistory warehouses and manufacturing structures near the traffic-congested waterfront. Now merchandise could be stored in large, prepackaged containers that were lifted by crane and shipped by truck along an increasingly convenient highway system. Container ports needed too much upland open space to be easily located along already built-up city waterfronts. Instead, they were being established along vacant shorefronts, nearer to major highways. Production was easier and cheaper in single-story, suburban factories that could provide extended horizontal production lines, easy parking for employees, and even easier highway access for trucks. Technological change had transformed proximity to the waterfront from an asset into a liability.

Recognition of changing demand for different locations is often quite slow. Most city officials became aware of the decreasing importance of waterfront shipping only from

Ghirardelli Square, San Francisco, 2006
(Alexander Garvin)

Delaware St., River Quay, Kansas City, 2011 *(Alexander Garvin)*

declining tax collections and increasing building vacancies. Recognition of the opportunities provided by declining but still attractive waterfront locations became apparent only after the success of Ghirardelli Square in San Francisco.[5]

Conceived in 1962, this project converted into an urban marketplace 2.5 acres of factory and warehouse structures that had once housed a chocolate company. The design (by architects Wurster, Bernardi & Emmons Inc. and landscape architects Lawrence Halprin & Associates) established a charming combination of fashionable retail stores and restaurants in a physical setting redolent of old San Francisco. Ghirardelli Square became an instant tourist attraction. More important, it became an inspiration for similar projects in the surrounding Fisherman's Wharf section of San Francisco and throughout the country. The imitators of Ghirardelli Square soon discovered that financial success was not guaranteed by rehabilitation and adaptive reuse of older structures or by creation of an urban marketplace with the imagery of a bygone era. River Quay in Kansas City, Missouri, is a particularly vivid example.[6]

In 1973, inspired by the success of Ghirardelli Square, enterprising planners decided to transform River Quay, the run-down district of bars, rooming houses, cheap hotels, and dilapidated buildings along Delaware Street that had been the birthplace of Kansas City, into an "old town" marketplace. Their plan called for rehabilitated buildings, restored "historic" street fronts, and decorative sidewalks with new street trees. At first these improvements brought restaurants, shops, and artists' studios. But it soon became clear that the market they attracted was too small. Retailers moved away or went out of business, and the area reverted to its former vacant and dilapidated condition. By the second decade of the twenty-first century, however, enterprising housing developers had begun renovating empty buildings and erecting apartment buildings on vacant sites. The location on the edge of the city downtown business district that had proved inappropriate for substantial retail activity half a century earlier had become attractive for city residents.[7]

At exactly the same time, Westport, another decaying commercial section of Kansas City, was transformed into a thriving urban marketplace. Westport was an intersection lined with dilapidated storefronts that in the nineteenth century had been a busy departure point for wagon caravans going west. Eventually it was overshadowed by the port of Kansas City, 3.5 miles north, and annexed. Westport then slowly declined until, years later, a developer acquired several of its run-down stores. He restored the façades, reconstructed the interior retail space, and installed new street furniture and decorative paving. The new Westport Square easily attracted middle-class clientele from the surrounding residential areas.

Like Ghirardelli Square, both Westport Square and River Quay renovated decaying, multistory commercial structures and re-created gussied-up images of nineteenth-century mercantile America. The prescription failed at River Quay because it was applied to the wrong location. The bulk of Kansas City's population lived several miles inland and was unwilling to drive to River Quay when there were more attractive alternatives (including Westport Square) closer to home. Daytime office workers were unwilling to travel a half mile from the business district, crossing a depressed multilane interstate highway, to get to River Quay. Without these customers, there was no way all the stores, restaurants, and entertainment spots could survive.

Kansas City, 2011. Westport Square was still thriving two decades after it was transformed into an "old town" marketplace. *(Alexander Garvin)*

Location is also important in terms of linkage to nearby supporting and supported land uses and activities. Half a century after completion of Ghirardelli Square, its continuing success and that of Fisherman's Wharf has been the product of their indissoluble linkage. The initial commercial failure and eventual residential revival of River Quay were similarly the result of their proximity to downtown Kansas City.

Kansas City, 2011. Country Club Plaza. *(Alexander Garvin)*

Design

The most misunderstood of the six ingredients of success is design. Too often, it is thought of as decoration that can be applied after the important decisions have been made. In fact, design is the physical manifestation of any prescription and therefore is integral to its success or failure from the time of inception.

Design is not just a matter of architectural style. Styles go in and out of fashion; successful planning has to survive for decades. Other, more enduring aspects of design are more important. They include the arrangement of project components, the relative size of those components, and their character and landscaping. Each element affects a project's utility, cost, and attractiveness. When they are organized in a mutually supportive manner, the result is an identifiable destination that provides an auspicious place for the activities occurring there. When arranged to fit the right combination of market, location, financing, time, and entrepreneurship, the result is a successful project.

America's first shopping center, Country Club Plaza in Kansas City, established the basis for the arrangement of stores in any successful retail complex.[8] In the scheme devised in 1922 by its developer, Jesse Clyde Nichols, customers drive to the plaza, park their cars, and walk to their ultimate shopping destination—generally national chain stores and popular department stores. Nichols understood that without such destinations, Country Club Plaza would not attract as many customers. Consequently, he was ready to offer the anchor stores locations at a lower rent. The stores along that route would prosper if they made enough sales to passing customers, thereby permitting them to pay a higher rent than was paid by an anchor store. The approach worked so well that Country Club Plaza remains one of Kansas City's major shopping destinations.[9]

The components of a commercial center are not always arranged as successfully as they are at Country Club Plaza. They also can be assembled in a manner that reduces their utility to retail shoppers and therefore reduces retail sales. This is what accelerated the eventual failure of New Haven's Chapel Square. Instead of being placed between the two department stores, its two-story shopping mall was built at one end of the scheme. A five-story parking garage was placed immediately adjacent to the anchor stores and provided direct access to them but did not connect to the shopping mall. As a result, none of the mall's retail tenants were able to profit from purchases made by customers stopping in or on their way from the garage to either of the two department stores. This only decreased cash flow from the already inadequate market demand.[10]

If Chapel Square illustrates how the inept arrangement of the components of a design can exacerbate already-poor market conditions, Ghirardelli Square illustrates how an in-

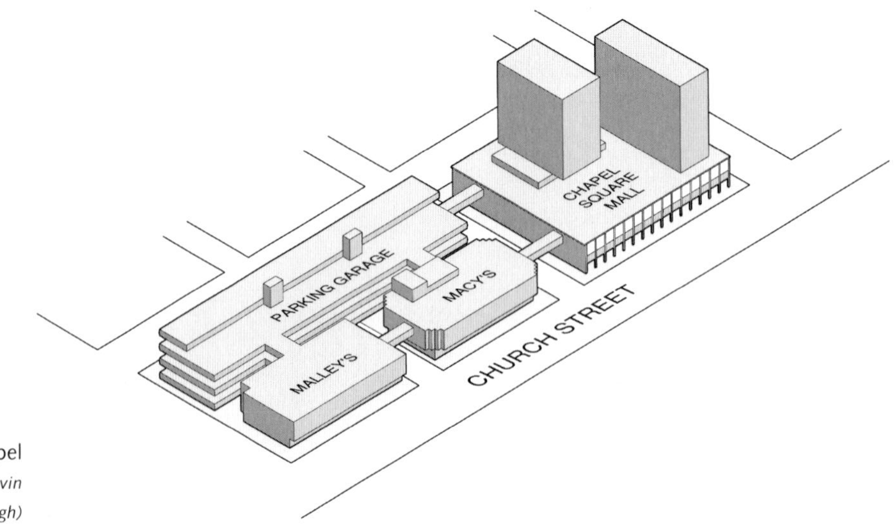

New Haven, 1970. Chapel Square Mall. *(Alexander Garvin and Eric Clough)*

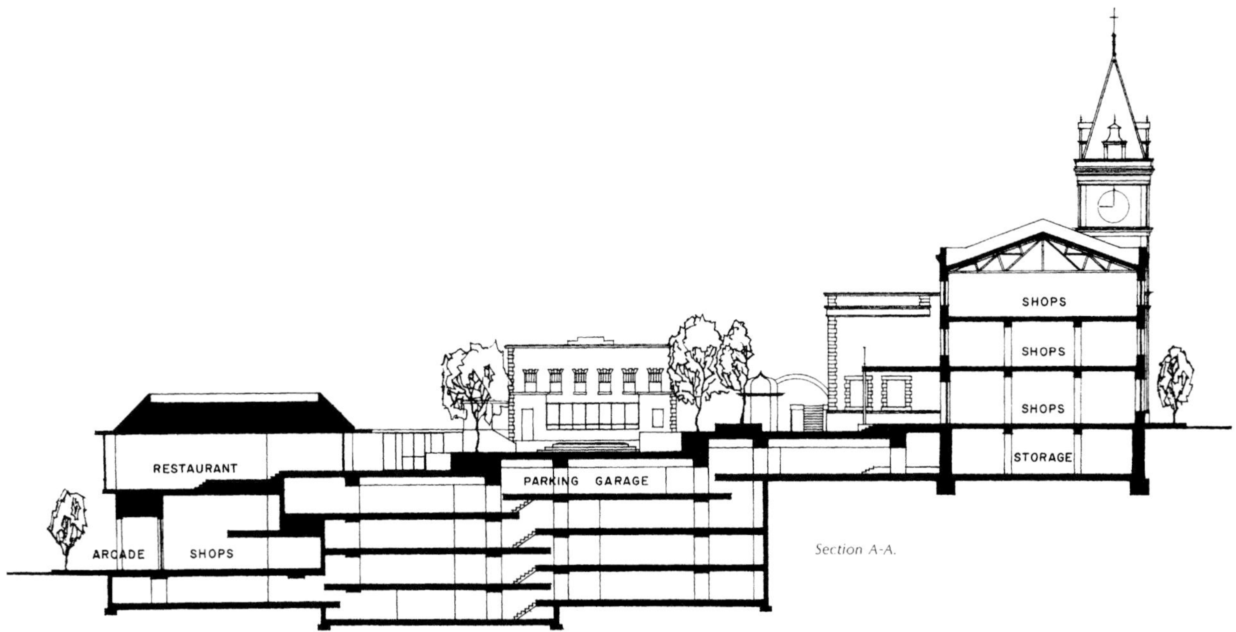

San Francisco, 1972. Ghirardelli Square—section drawing showing the arrangement of parking, pedestrian levels, shops, and restaurants. *(From L. Redstone,* New Dimensions in Shopping Centers, *McGraw-Hill, New York, 1973)*

telligent arrangement can enhance a potentially wonderful location. At Ghirardelli Square the components are terraced in a manner that increases the utility of the site, reduces costs, and attracts customers. On this steeply sloping site, parking is fitted in under several levels of shopping without taking up otherwise rentable floor area. At the higher end of the site, the parking structure provides the foundation for retail stores. In the middle, its roof provides an outdoor pedestrian level in which retail shoppers can freely circulate among the stores. Only at the lowest end of the site is parking fully underground.

By including in the design formerly obsolete buildings (especially the factory building that now includes a display about the chocolate company that was its initial occupant) and by reserving for public use spots with panoramic views of the waterfront, the design attracts additional tourists. It is a profitable combination of utility, economy, and picturesque features. Today this arrangement seems obvious, but when Ghirardelli Square was conceived, nothing like it had ever been designed.

Dimensions are as important as the arrangement of components. They have to be correct from the beginning. For example, traffic engineers suggest a width of 12 feet for every lane of traffic. That may not be enough on busy streets where trucks keep stopping to unload merchandise. Similarly, building codes mandate a minimum height for every habitable room (usually 8 feet). How much higher should one build? The answer should vary with the type and floor area of each room. But, whether the product is a traffic artery or a residence, the quality of the results will depend on dimensional appropriateness.

Southdale, the world's first shopping mall with skylit and air-conditioned common spaces, was a sensation when it opened in 1956. Even after half a century of changes in marketing and increasing competition from larger, more up-

Edina, Minnesota, 2009. Southdale—the world's first air-conditioned shopping mall. *(Alexander Garvin)*

Boston, 1968. Quincy Market prior to its transformation into a "festival marketplace." *(Alexander Garvin)*

Boston, 2010. Quincy Market as a functioning "festival marketplace." *(Alexander Garvin)*

to-date retail centers, it remains one of the more successful malls in the Midwest. Victor Gruen, its architect, conceived the scheme that continues to provide the basis of climate-controlled shopping mall design: an anchor department store at each end of a multilevel, skylit, pedestrian concourse enclosed on both sides by smaller retail stores. While Gruen and his client, the Dayton Hudson Corporation, had a great deal of experience with shopping centers, they had no experience with air-conditioned malls. They could only guess about the proper dimensions, and they guessed wrong. The central space was too wide. It was 105 feet from one side to the other, too far for customers to get a good look at the merchandise on the other side and much too far to read price tags.[11]

When Southdale's shopping concourse was extended in 1989–1991, the new common spaces were limited to 58 feet wide, a more reasonable dimension to accommodate window-shopping. To reduce the apparent distances across the older section of the concourse, its owners added a line of pushcarts whose merchandise could attract the attention of shoppers as they zigzagged between shop windows and pushcarts.

If some sections of Southdale were too big, the buildings that became Boston's Quincy Market were too small to accommodate the anchor stores that were thought essential components of any successful shopping center. The buildings that now constitute Quincy Market were initially built in 1826. In 1961, when they were acquired as part of an urban renewal project, the buildings were significant architectural landmarks without an obvious reuse. Their 50-foot width was believed to be too narrow for major retailers. To deal with this problem, architects Benjamin and Jane Thompson, working with their client, the Rouse Company, came up with a scheme that consisted largely of small food and specialty retailers. When Quincy Market reopened in 1976, James Rouse called it a "festival marketplace." Its tenants attracted enough customers to have a high volume of sales per square foot. Consequently, they could pay high rents per square foot. However, they neither needed nor could afford large establishments. Rouse and the Thompsons had transformed dimensions that were too small for large national retailers into a moneymaking asset.[12]

Character is the product of style, color, materials, and scale. The attractiveness of Charleston's Old and Historic District is largely a matter of architectural character. The redbrick paving highlights at Westport Square identify it as a distinctive retail destination, different from the ordinary gray sidewalks of that part of Kansas City. Similarly, the painted wood and brick buildings of Charleston's historic district provide qualities that are different from the nondescript materials of the city's post–World War II suburban subdivisions.

During the nineteenth and early twentieth centuries the character of subdivisions was determined by the way their designers handled the public realm. For example, in creating the distinctive character of Pinehurst, North Carolina, Frederick Law Olmsted lined public rights-of-way exclusively with evergreens (conifers, holly, magnolias, and rhododendrons) that afforded residents a degree of privacy throughout the year. More recently, designers have preferred to determine the character by regulating what property owners can build. At Sea Ranch, a planned community in northern California begun in 1964, property owners are required to build in open fields and use unpainted, natural materials; roofs are not permitted to hang out beyond building walls.[13]

The character of both of these distinctive communities is a product of their landscaping and context. Both were created on land that had been logged of all its tree cover. The conifers of Pinehurst are a distinctive note within the rolling sand hills of North Carolina. There are stands of cypress trees at Sea Ranch that act as windbreaks along the treeless coast of this stretch of northern California. Its character and context, however, are quite different from Pinehurst. Sea Ranch is an integral part of the splendid isolation of much of the Pacific Ocean in much of northern California. That context is enhanced, however, by the carefully nurtured landscape of free-growing grass meadows in which the wooden houses of Sea Ranch are clustered. Remove the regional context and the landscaping and both communities would be quite similar to any of the characterless suburban subdivisions that can be found throughout the United States. The critical component of their success proved to be excellence in design.

Financing

Every prescription for fixing cities requires financing. When this involves governmental action, as is usually the case with parks, the financing comes from taxes that are used to cover maintenance and operations. When the income stream from taxes is insufficient, parks begin to deteriorate. Thus, financing is the key to a decent park system. Major projects, however, are usually too expensive to be paid for out of tax revenues. They can be paid for in the same way that people pay for their houses: by borrowing the money in the form of a home mortgage. City governments operate in the same way when they issue bonds whose debt service is paid by using a portion of their tax revenue to cover debt service on the bonds.

Among the reasons that Minneapolis has the best-designed and best-maintained park system in America is that its elected Park and Recreation Board can levy taxes and issue bonds. As a result, it has money to pay for acquisition, design, development, program delivery, and maintenance. Elsewhere, whenever cities face a period of budget stringency, they transfer money from the parks to other "more pressing priorities." Without that money, not only is additional park development impossible, so are repair, rehabilitation, and replacement of worn-out facilities.

Parks, however, need not be dependent on government funding. Boston's 1.7-acre Post Office Square was financed with parking garage revenue. The site had been a post office until 1954, when a 950-car public parking garage replaced it. In 1987, citizen activists persuaded the city to replace it with a public park. They were successful because they had proposed a scheme that required no public money. It called for creating 1400 parking spaces beneath the park and was paid for by allocating $12 million in annual garage revenues to cover maintenance, operating costs, taxes, and payments on a $60 million loan and $30 million stock offering.[14]

Financing also is essential to private-sector activity. Privately financed projects need *capital* to cover start-up costs, a short-term *development loan* to pay expenses until it is operational, and a *permanent mortgage* to replace the other two when the project is complete and tenanted. The obvious place to obtain financing is a bank. Banks lend their depositors' money to developers, whose projects pay a large enough return to keep depositors happy and contribute toward covering the costs of bank operations. In other words, developers pay banks for the use of their money. The price depends on the bank's assessment of the risk involved. If the deal looks too risky, it will not lend a penny.

Boston, 2008. Post Office Square. *(Alexander Garvin)*

Most banks will not lend enough to cover project cost. The rest of the money, the *equity* investment, usually comes from the developer and from investors who have confidence in the venture. Investors know that the bank has not lent enough money to complete the project. They know that if the venture fails, the bank may recoup its investment but they may not. They also know how much the bank is getting for its money. Consequently, developers have to pay investors a higher price for equity funds than they are paying for bank money. Developers will put up their own money if bank mortgages and investor equity do not cover all development costs. Typically, if something goes wrong and the investment has to be liquidated, the bank mortgage will be repaid first, then the equity investors, and finally the developer. Since the developer is taking the greatest risk, he or she will not go into the venture unless the return is better than that of the bank and the equity investors.

In other words, money is obtained at different prices, depending on risk and availability. Mortgage money is usually the least expensive. Equity money is more expensive. The developer's money is the most expensive. The greater the proportion of project costs that comes from other sources (the greater the *leverage* of the developer's cash investment), the more attractive the venture will appear to the developer. Government can increase the likelihood of project success by creating an investment climate in which bank financing is readily available and developers are able to maximize leverage.

Congress has consistently tried to ensure adequate financing for housing construction by increasing the safety of residential mortgages. During the Great Depression, it restructured the banking industry and established mortgage insurance programs that eventually led to construction of millions of suburban houses. Title I of the Housing Act of 1949 (popularly known as the urban renewal program) provided two-thirds of the money needed to subsidize planning, start-up costs, property acquisition, demolition, and relocation for federally approved urban renewal projects. Local governments had to pay the remaining one-third. In 1954, Congress added federal insurance on mortgages for new or rehabilitated housing in urban renewal areas. This was followed by a series of mortgage subsidy programs that reduced housing costs to a level that was affordable for low- and moderate-income families.

These programs reduced the cost of money and assured older cities of the financing needed to pay for major neighborhood reconstruction. Some cities (in particular, Philadelphia, New Haven, and New York) used the money for more than just wholesale clearance. They recognized that cutting away scattered pockets of blight and rehabilitating structures that contributed to the area's historic character also could revive deteriorating neighborhoods. Philadelphia's Washington Square East Urban Renewal Project (better known as Society Hill) is one of the earliest and most successful examples. The money to pay for planning, property acquisition, clearance, and site preparation came from the urban renewal program. Banks provided mortgage money for housing construction and rehabilitation because it was federally insured. Payments to cover ongoing operations came from the middle-income residents of the new and renovated housing. Where necessary, financing was supplemented with further federal subsidies.[15]

Edmund Bacon, executive director of the Philadelphia City Planning Commission, planned the revitalization of Society Hill in the mid-1950s. He proposed selective clearance of only those structures that were beyond repair or that were incompatible with the rest of the neighborhood. These sites were to be filled in either with residential buildings sensitively fitted in between their older neighbors or with small greenway parks intended as both landscaped pedestrian paths and as small-scale recreation areas.

It was a perfect strategy for Society Hill. The bulk of the area's buildings were charming eighteenth- and nineteenth-century row houses. Once the blighting influence of neighboring properties had been eliminated, these row houses became extremely attractive to middle-class residents who wanted to live downtown. By 1970, owners had rehabilitated more than 600 of Society Hill's historic structures, property

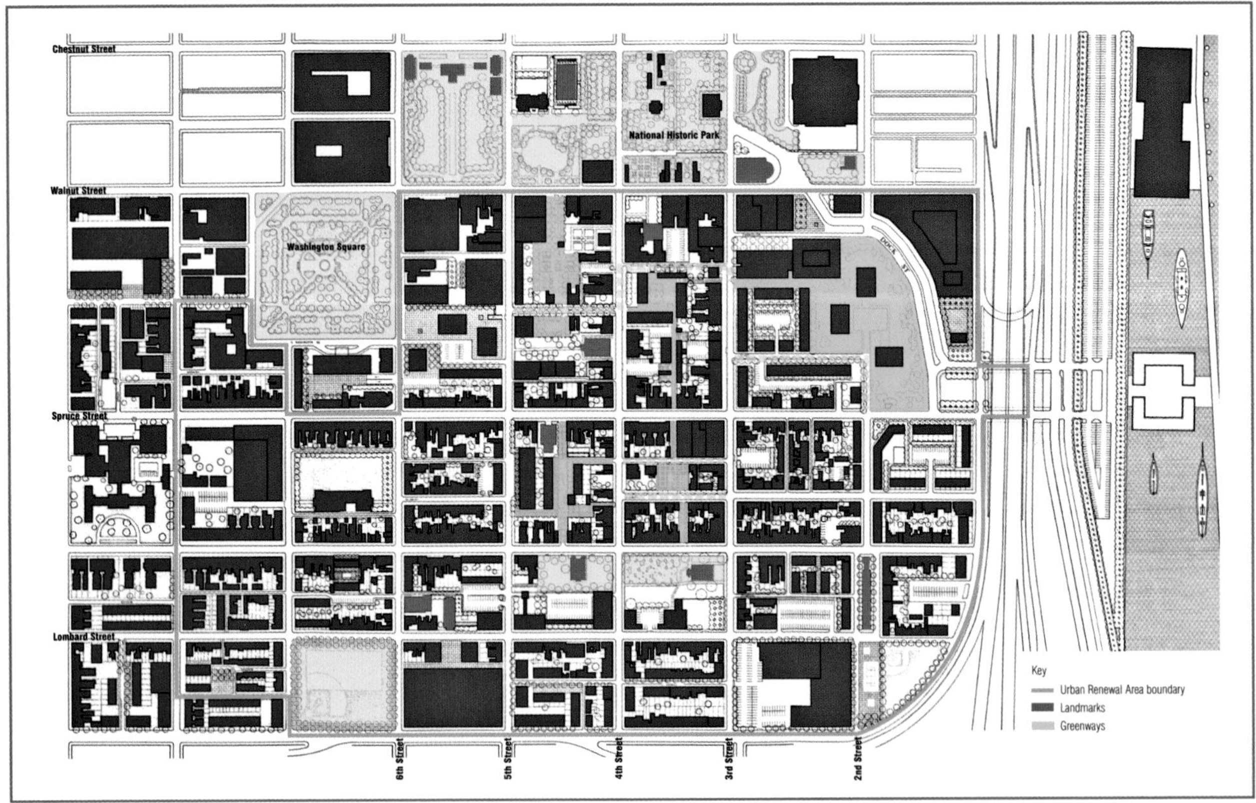

Philadelphia, c. 1967. The revitalization of Society Hill involved preserving and renovating eighteenth- and nineteenth-century row houses and selectively replacing vacant lots and nonconforming land uses with new residential buildings. *(Alexander Garvin and Zakery Snider)*

Bronx, 2012. In Mott Haven, selective filling in of vacant lots and dilapidated buildings with new residential buildings failed to revitalize surrounding areas because financing for further development was not available. *(Alexander Garvin)*

values had more than doubled, and the population had increased by a third.

More than a decade after the Society Hill project was under way, New York City adopted a similar infill strategy. Known as the *vest-pocket redevelopment program,* it was launched by Mayor John Lindsay a few months after his election in 1965. The idea was to use local (rather than federal) funds to acquire vacant lots and dilapidated buildings. These sites would then be resold to developers who would receive federally subsidized mortgages to either rehabilitate salvageable abandoned structures or build new apartment buildings, in both cases incurring little or no need to relocate existing residents.

In the Mott Haven section of the Bronx, these new and rehabilitated structures were to be the first phase of a much more ambitious community-renewal effort that included a second, third, and fourth round of housing, parks and play lots, schools, repaved streets and sidewalks, street trees, and a variety of other community facilities. From the start, most Mott Haven residents had been poor. They lived in walk-up tenements built to accommodate the overflow of immigrants from Manhattan. Over the years, little or no money had been spent on necessary repairs and maintenance, so most of the buildings were seriously run down. Proper restoration now required substantial capital investment. Owners could not recoup such investments by increasing rents because most residents could not afford to pay more rent and, those who could were prevented from doing so by rent control.[16]

Realizing that private financing was not available, planners put their trust in the federal government. They applied for and were accepted into the recently enacted Demonstration Cities and Metropolitan Development Act of 1966 (popularly known as the Model Cities Program), thereby becoming part of what promised to be an extraordinary demonstration of national financial commitment to the renewal of 150 "model" neighborhoods. Model Cities paid for comprehensive planning, involving not just rebuilding but also rehabilitation, social service delivery, citizen participation, and the non–federally subsidized portion of any federal program that required local matching funds.

City-funded early acquisition of property in Mott Haven was supposed to provide a relocation resource for residents that would be displaced by later Model Cities projects. Institutional lenders were willing to provide mortgage financing because the federal government subsidized the difference between the cost of supplying new housing and the amount of money its occupants could afford to spend. Thus, the market was directly proportional to the subsidies that were available.

Mott Haven, 2012 *(Alexander Garvin and Owen Howlett)*

Bronx, 2012. The Bronx Nehemiah Project erected affordable row houses on abandoned property and sold them to area residents at affordable prices. *(Alexander Garvin)*

The construction on the first sites in Mott Haven had just started in 1973 when President Nixon unilaterally declared a moratorium on all federal housing assistance, and the program itself was terminated in 1974. Since residents were too poor to pay for building improvements and the federal government had terminated its subsidy programs, banks withdrew the necessary mortgage financing. At that time, the New York City government was experiencing a serious fiscal crisis. It was neither willing nor able to replace federal commitments. As a result, nothing further could take place. So, for at least a decade, the first stage of Mott Haven's vest-pocket redevelopment seemed to be its last.

More properties were taken by the City of New York for failure to pay real estate taxes, but, given the market conditions, nothing could be built on the growing inventory of vacant lots without subsidies, nor could the abandoned buildings be renovated without them. As federal subsidies became available after Nixon resigned from the presidency, the city transferred ownership of abandoned properties for a nominal price to developers who were able to obtain federal subsidies. Despite the experience in Mott Haven, two decades later, a nonprofit developer demonstrated that infill housing could be financed in a low-income section of the South Bronx. In 1978, I. D. Robbins, a retired builder, wrote a column for the New York *Daily News* entitled "Blueprint for a One-Family House: For the $12,000-a-Year Income." Robbins argued for the construction of single-family owner-occupied homes rather than more expensive apartment houses. He called his proposal the Nehemiah Plan in recollection of the Old Testament prophet who led the rebuilding of Jerusalem in the fifth century B.C. In 1983, with money provided by a group of East Brooklyn churches, he began building affordable two-story row houses on vacant sites. A group of South Bronx churches asked him to do the same for them, and in 1991 Robbins broke ground a few blocks north of Mott Haven on what would become the South Bronx Nehemiah Project.[17]

Robbins's strategy was to reduce the cost to the homeowner by reducing every component of that cost. He reduced the costs of development by standardizing and repeating the same design, using prefabricated components, and obtaining vacant sites from the city at a nominal price. He reduced the price paid to contractors by paying their vouchers every week. Those payments came from a $3.5 million interest-free revolving fund established by a group of Protestant and Catholic Churches. Since one-family houses could be built in a few months, house sales replenished the fund equally quickly. Robbins operated on extremely low overhead, which eliminated most of the legal and financial charges normally added to the price paid by home buyers. With an equity investment of $5000, home buyers obtained a below-market interest rate on a $40,000 first mortgage from the State of New York Mortgage Agency and an interest-free $15,000 second mortgage from the City of New York. In the early 1990s, this was a deal that hundreds of neighborhood residents could afford.

While I. D. Robbins was starting the Brooklyn Nehemiah, the NYC Partnership (Chamber of Commerce) started its Partnership Housing Program. The city sold abandoned properties to the Partnership at the same price ($500/"affordable" dwelling unit) at which it sold them to the Nehemiah Project. The 512 households that participated in Bronx Nehemiah took pride in their houses and took care of their street, which now occupied a formerly dangerous, abandoned section of the South Bronx. Over the next two decades the

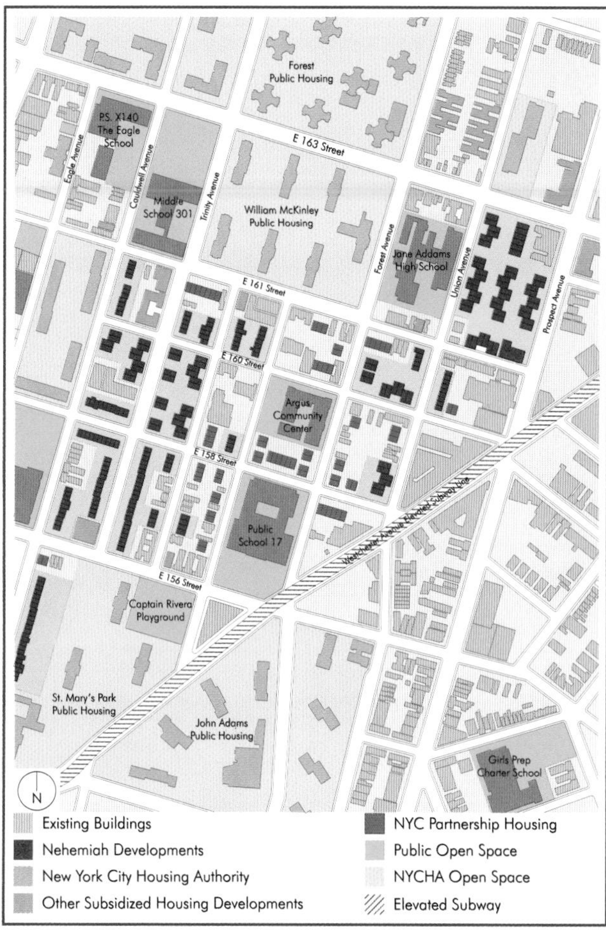

Bronx Nehemiah, 2012. Map. *(Alexander Garvin and Ryan Salvatore)*

Partnership Housing Program filled in other properties. Together, they were enough to generate private development of another eight properties. Similar unsubsidized private-sector activity has as yet not returned to Mott Haven.

Some abandoned sites in Mott Haven were developed as part of the Partnership Housing Program. Other abandoned sites were developed by Bronx Green-Up (a community outreach program of the New York Botanical Garden) for use as community gardens. While the Mott Haven Vest Pocket Project itself may have been a failure, enough other properties were put back into occupancy to keep the neighborhood from going down the drain. In time, the rest will be developed. The Society Hill and Bronx Nehemiah projects provided enough money to remove pockets of blight and to subsidize strategically located housing construction. While this stabilized both neighborhoods, in Society Hill it triggered substantial further private investment because banks had evidence of a growing market and readily provided mortgage financing to other property owners once initial housing rehabilitation and new construction proved successful. The same prescription failed to work in Mott Haven because the project remained dependent on continuing public subsidies for nearly every property.

Time

Four time sequences affect success. The first is relatively brief: the period during which a person passes through an area. The second takes into account what will occur 24 hours a day, 7 days a week. The third is the development period, which may take decades. The last will continue to affect any project during its entire existence: changes in consumer taste, and in the political and financial climates, all three of which will continuously change.

Developers of retail shopping facilities are perhaps the most skilled in predicting a person's activity pattern within an area. They have to be skilled in dealing with this brief time period because their tenants' profits are dependent on transient customer activity and their own profits are dependent on tenant success.

In any shopping center, whether visitors come by foot or by motor vehicle, they are quickly faced with a wide variety of attractions when they arrive. Passing from one to the next, these visitors invariably stop to look at or purchase something. The result is plenty of activity, a high volume of sales, and therefore high rents per square foot.

The movements of a single individual, on the other hand, are irrelevant in planning for a 24-hour day and a 7-day week. Such planning requires providing a suitable environment for a wide variety of users on a continuing basis. Thus, the crucial questions revolve around the people who are likely to be in an area over a 7-day period, what they will want to do, how many people are needed to support those activities, and in what ways the environment should be organized to accommodate satisfactorily those people and activities.

Starting in the 1940s, Detroit succeeded in using federal urban renewal funds to create a project that provided its residents all the basic components of daily life.[18] Finally in the mid-1950s, at Lafayette Park, architect Mies van der Rohe, developer Herbert Greenwald, and their associates provided simple but effective answers to these questions. Rather than creating a housing project that was devoid of people during substantial periods of the week, they created a superblock that included a school, convenience shopping, recreation facilities, and substantial open spaces. Although the population of the superblock was insufficient to support all these facilities, they were located along its bounding streets and thus were easily available to residents of the surrounding community. By planning for both residents who lived in the project and those who lived in the surrounding area, Lafayette Park was able to offer services that otherwise would not have been justifiable. The project that emerged provided the community with a suitable living environment every day of the year.[19]

Jane Jacobs calls for districts with a "diversity of uses

Detroit, 2008. Lafayette Park. *(Alexander Garvin)*

Manhattan, 2004. Lincoln Center patrons and workers spend time and money in neighborhood restaurants and stores. *(Alexander Garvin)*

that give each other constant mutual support both economically and socially."[20] But her reasoning is far more complex. She recommends districts that contain apartment houses with residents who leave for work every day, office buildings with daytime workers, performance halls that accommodate primarily nighttime customers, as well as bars, restaurants, retail stores, and all manner of service establishments. Together, they constitute a district that is alive with people 24 hours a day, 7 days a week. Such a district surrounds and includes New York's Lincoln Center. It attracts people for different purposes at different times of the day, 52 weeks a year.

Successful planning also requires a strategy that will remain appropriate over long periods of time. Lincoln Center completely altered the perception of Manhattan's West Side. Consequently, during every upturn in the real estate market, developers invested in new buildings for the area. At the start, the impact was not obvious. Only after the early 1980s, when the city had emerged from a financial downturn, did the potential scope of the private market reaction become

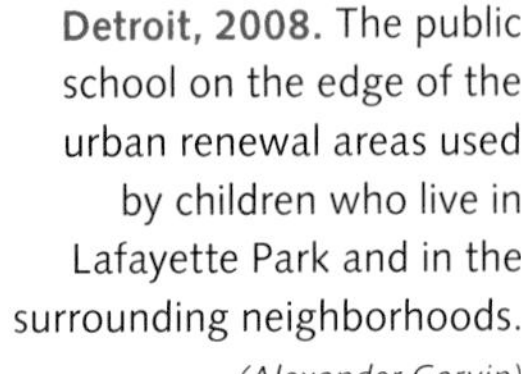

Detroit, 2008. The public school on the edge of the urban renewal areas used by children who live in Lafayette Park and in the surrounding neighborhoods. *(Alexander Garvin)*

clear. Similarly, the impact of regulating Charleston's Old and Historic District could not be measured in 1931, when the district was created. The country had to emerge from the Great Depression, World War II, and the early stages of postwar suburbanization for market demand for historic structures in older city districts to become apparent. Only then did the effectiveness of regulating Charleston's historic structures become clear.

Of all the strategies for fixing urban/suburban America, the planned "new community" is among the most sensitive to business cycles. During the decades required to plan, build, and market a new community, it will experience continually changing economic conditions, political trends, migration patterns, and consumer demand. Because of these inevitably changing market pressures, cash flow can vary substantially from year to year. However, to survive to completion, every planned new community must continue making debt-service payments on a massive, front-loaded investment in land, streets, sidewalks, sewers, water mains, and all the required infrastructure and community facilities. This requires access to plenty of capital and investors who are willing to wait for years before seeing profits. By itself, however, financing is not enough; all six ingredients of success need to come together for the project itself to succeed.

Radburn, New Jersey, perhaps the best-designed and most influential planned community in America, was never completed because it could not ride out these pressures. Radburn was developed by the City Housing Corporation, a limited-dividend company expressly created to demonstrate the efficacy of developing carefully planned new communities. In 1927, it purchased 1300 acres in Fair Lawn, New Jersey, 10 miles from the George Washington Bridge, where it intended to create "a new town for the motor age" with a projected population of 25,000.[21]

Clarence Stein and Henry Wright, Radburn's architects, devised a unique plan in which you drove to your home, parked, and entered the rear of the house. The house itself was turned around so that it faced a private yard that fronted on a landscaped pedestrian walk. These pedestrian walks opened onto beautifully landscaped common open spaces, large enough for children to play ball. They were, in turn, connected by an underpass to Radburn's school, swimming pool, and community facilities.

Radburn quickly became famous among city planners. Photographs of its underpass were printed in books and articles all over the world. Architects and planners, particularly in Europe, began copying what they called "the Radburn idea." Ironically, while giving new life to the idea of building planned new communities, Radburn itself failed. During the Great Depression, few families could afford to purchase a new house. Sales were insufficient for the City Housing Corporation to service the debt it had incurred to pay for land, infrastructure, and community facilities. Its financial backers were not willing to continue the venture without receiving a return on their investment. In 1935, after completing about 300 houses, the City Housing Corporation declared bankruptcy. Nevertheless, more than half a century since its financial collapse, Radburn remains one of the world's most beautiful and important planned new communities.

Other planned communities, like Palos Verdes Estates, California, succeeded because they could withstand constantly changing market conditions. Palos Verdes Estates occupies one of the country's loveliest sites: a hilly peninsula

Radburn, 2006. The houses at Radburn are turned around so that the backyard faces a pedestrian walk leading to a large green common open space. *(Alexander Garvin)*

Radburn, 2006. The green is used, in common by the residents, for recreational activities that are not disturbed by vehicular traffic. *(Alexander Garvin)*

jutting into the Pacific Ocean, 23 miles southwest of downtown Los Angeles. This beautiful landscape is enhanced by an extraordinary town design by Frederick Law Olmsted Jr. and Charles Cheney. Its streets, carefully fitted into the spectacular promontory, were laid out to provide building sites with even more spectacular views. Because of this sensitive planning, Palos Verdes's quasi-Mediterranean buildings seem to have been there for centuries. In fact, Palos Verdes Estates is a splendid twentieth-century oasis in the urban congestion and suburban sprawl of Los Angeles County.[22]

The site that was to become Palos Verdes Estates was first sold for development in 1913. Its buyer could not finance the purchase price and had to be bailed out by a syndicate controlled by the president of a New York bank. The project was resurrected by another developer in 1921, only to fall apart again. Finally, in 1923, the syndicate that had purchased the property more than a decade earlier initiated development of a 3200-acre planned community. Within two years, it had built and paid for 20 miles of landscaped boulevards and avenues, 60 miles of water mains, a shopping plaza, a country club and golf course, a public school, and 2500 prepared homesites.

The economic downturn of the late 1920s reduced demand for building sites, and Palos Verdes Estates had to be refinanced for a third time. Then, during the late 1930s, the community faced yet another financial hurdle: unpaid county taxes. This was overcome with state legislation that allowed Palos Verdes Estates to become an incorporated city with an independent park and recreation district, thereby eliminating further county tax payments on community-owned public open space. From that point on, there were no further financial difficulties.

By 2010, Palos Verdes Estates had a population of nearly 14,000 living in 4800 detached single-family houses whose total market value was approximately $4.1 billion. The project had taken a decade to get started and several decades more to come to fruition. However, because its developers had not sought immediate profits and had the resources to patiently withstand a series of reverses, Palos Verdes Estates was successfully carried through to completion.[23]

Palos Verdes Estates succeeded not only because it was able to withstand the vagaries of time. All six ingredients of success played a part. It was able to attract the rapidly expanding market provided by the Los Angeles metropolitan region. It had a location with spectacular site characteristics near downtown Los Angeles, a location that became even more convenient as additional traffic arteries spread through the region. Olmsted and Cheney's design, which exploited the topography and views, reinforced the attractiveness of the location to its expanding market. Its developer, a bank president, was able to obtain the necessary financing. He also had the vision and entrepreneurial skills needed to see the project past its critical early years and carry it forward to the point at which its future was assured. Without any one of these ingredients, the results would have been less than satisfactory.

Entrepreneurship

No prescription is self-implementing. Each requires talented public and private entrepreneurs. Without them, a perfectly appropriate prescription will not get off the ground. Entrepreneurs conceive projects, often when others are unaware that there are any opportunities available. They assemble and coordinate the various players who will execute whatever needs to be done. Without the extra drive that entrepreneurs supply, the other players would be overwhelmed by the uncertainties of the marketplace.

Entrepreneurs do not appear automatically whenever there is unfulfilled demand for something. They have to believe that the risk of failure is minimal and that the rewards for success will be generous. Unless such favorable conditions

Palos Verdes Estates, 1979. The streets and houses of this planned community are fitted to the topography. *(From R. Cameron,* Above Los Angeles, *Cameron & Company, San Francisco, 1978)*

are prevalent, entrepreneurs will exploit other, more attractive opportunities.

Public projects often fail because public officials ignore the role of entrepreneurship. They mistakenly believe that once a project has been assigned to a government agency, its role is purely administrative. In fact, public entrepreneurs are needed to assemble, coordinate, and inspire all the participants in the development process. Edmund Bacon performed that role in Society Hill. He successfully combined the activities of the bankers, bureaucrats, property owners, developers, architects, engineers, contractors, and countless other actors needed for the revitalization of the neighborhood. He also maintained public approval and bureaucratic momentum despite the uncertainty of acquiescence by property owners. He obtained timely approval by federal agencies and mortgage commitments from financial institutions. He stimulated developer interest in the project and political acceptance by Philadelphia's disparate civic and community groups. He sought and discovered opportunities for participation, funding, and implementation by previously uninvolved public agencies, nonprofit organizations, and private developers. Most important, Bacon implemented a strategy that had never been tried before: eliminating scattered pockets of blight and filling the resulting holes in the fabric of the neighborhood with new housing and parks.

While it is easier to understand the role of an entrepreneur in the private sector, it is essentially the same as that performed by public officials like Edmund Bacon. The role includes coordinating a plethora of participants, dealing with uncertainty, recognizing available opportunities that

Philadelphia, Society Hill, 2006
(Alexander Garvin)

have not yet been exploited, and frequently accomplishing things in ways that have never been tried before. The difference between private and public entrepreneurial activity is only in the form of payment. The private entrepreneur is paid in hard currency, the public entrepreneur in power. The sort of people capable of getting things done, however, will have to be extremely well paid in their respective coin.

In many cases, private and public entrepreneurs work side by side. This is especially true for urban renewal projects like Society Hill, where implementation is dependent on individual property owners and developers. When it enacted the Housing Act of 1949, Congress hoped to attract private developers into the business of redeveloping federally approved urban renewal areas by sharply reducing the risk of failure. This was accomplished by requiring the clearance of any blighted property that might affect the area. Local officials had to prepare a redevelopment plan that provided developers and financial institutions with certainty regarding the future of every property within the area. Most important, Congress provided the subsidies needed to reduce land prices and site development costs to a marketable level.

Despite this reduced level of risk, few of the early renewal projects went into construction very quickly. Developers either were not willing to acquire approved urban renewal sites or, if they did acquire them, were unable to persuade financial institutions to provide the necessary financing. As a result, most cities initially generated government-subsidized clearance but were then unable to find the proper combination of developers and financing to get very much built.

Not just developers but banks and insurance companies were afraid of investing in officially designated "blighted areas." Without institutional financing, developers would have had to invest substantial amounts of equity capital. Initially, neither lenders nor equity investors perceived a return commensurate with their risk.

In 1954, Congress made the changes that were needed to interest private entrepreneurs in carrying out approved redevelopment projects. The vehicle it chose was federal mortgage insurance that covered up to 95 percent of the cost of new and rehabilitated housing in urban renewal areas. Since financing now could be insured, banks were ready to issue mortgages on most approved urban renewal projects. For the first time, risk was minimal, equity capital requirements extremely low, and profits entirely a matter of entrepreneurial skill. Naturally, all sorts of businesspeople were eager to get involved.

Detroit's Gratiot Urban Renewal Project (better known as Lafayette Park) illustrates the importance of this entrepreneurial element to any redevelopment effort. The project was initially conceived in 1946. At that time, two out of three dwellings in this 129-acre residential neighborhood were considered substandard. They lacked running water, central heating, private baths, indoor toilets, or some other feature considered necessary to the health, comfort, and safety of its residents. The best way to eliminate such "slums" was thought to be clearance and redevelopment.[24]

Detroit, 2008. Lafayette Park. *(Alexander Garvin)*

Before the project could proceed, however, Lafayette Park had to overcome community opposition, then a taxpayer's lawsuit, then the high cost of acquisition. In 1952, when the site was put up for auction, there were no bidders. The following year, a developer finally agreed to buy the site but had to withdraw because he was unable to obtain construction financing.

The project languished until 1955, when Herbert Greenwald, a successful Chicago developer, assembled a development package that could be financed. Greenwald, his partners, and his architect, Mies van der Rohe, had been responsible for a series of glass apartment towers that revolutionized residential development in Chicago. He brought this successful team to Detroit, where they began building one of America's most beautiful residential redevelopment projects. It was conceived as a 78-acre self-contained neighborhood (in the jargon of the period, a *superblock*), entirely closed to through traffic, containing many of the elements of a healthy community (an elementary school, a small shopping center, a clubhouse, and a swimming pool), all organized around a 19-acre park. Within the superblock they proposed to build 2000 apartments in six towers surrounded by clusters of one- and two-story row houses. Greenwald had the vision to create something truly extraordinary and to commission an inspired design team; he also had the development company that could make the vision a reality and access to the money without which nothing could have happened.

In 1959, a year after the first residents moved into Lafayette Park, Greenwald died in a plane crash. His firm dropped out of the project. Others who lacked his vision and know-how replaced him. Although they followed the original site plan, they produced the same mediocre buildings and inadequate public spaces that characterized most federal urban renewal projects.

Since the Greenwald-Mies portion of Lafayette Park was completed, Detroit has lost half its population and experienced one of the nation's worst rates of housing deterioration and abandonment. But the Greenwald-Mies portion of the renewal project has remained fully occupied, integrated racially and ethnically, and consistently well maintained. That record is unmatched by most other housing redevelopment projects in Detroit or anywhere else. Entrepreneurship is probably the single most important of the six ingredients of success, because without somebody to implement a project, nothing will happen.

Manipulating the Ingredients of Success to Obtain Desirable Private Market Reaction

Palos Verdes Estates and Society Hill are real estate ventures that may be evaluated in terms of their profitability. They also are examples of city and suburban planning that must be evaluated in terms of the cost-effectiveness of the induced private market reaction and the civic gain (whose value is harder to assess). The same ingredients that determine the community impact of profit-motivated projects determine that reaction.

While private developers rarely seek to generate and sustain a widespread private market reaction, some of their projects make profound changes to surrounding communities. Society Hill, for example, altered the character of downtown Philadelphia by attracting a substantial amount of new residents. It was able to sustain this widespread private market reaction because, unlike Palos Verdes Estates, the ingredients of project success were manipulated in a manner that fostered the spillover of its residents into the surrounding downtown and triggered the revival of private residential construction within the business district.

The only way to ensure that market demand will spill over into the surrounding area is to plan *not* to satisfy that market within the project. Then there will be a reason for people to go elsewhere. The new apartments created in Society Hill did not satisfy all the demands of the people who chose to move to Philadelphia. Nor was Society Hill conceived as a neighborhood that would supply everything of interest to its residents. In fact, Edmund Bacon and the planners of the project hoped to attract residents who would work and shop in downtown Philadelphia.

Unlike the planners of Society Hill, the developers of Palos Verdes Estates hoped to absorb market demand without its customers interacting in competing areas. They consciously tried to satisfy consumer needs within Palos Verdes so that there would be no reason to go elsewhere. During the 1920s, when sales first began, the plains to the northeast were largely undeveloped and remained so for the next two decades, during which time the project succeeded in capturing the lion's share of the market. Only after World War II, when the project had sold out and millions of people had moved into nearby sections of suburban Los Angeles County, did the spillover of that market result in increased prices for Palos Verdes property.

For a project to generate a sustained market reaction in surrounding areas, it must exploit linkages to those areas. The IDS Center is connected to 73 blocks of downtown Minneapolis by an 8-mile-long skyway system. Customers have to pass other retailers on their way from any of those 73 blocks to the IDS Center, often making purchases along the way. The same linkage applies to automobile-oriented visitors.

The IDS Center is designed to attract office occupants from every part of downtown and customers from the many attractions along the way. It can be entered on foot from any of its four bounding streets, as well. Palos Verdes, on the other hand, is designed in a way that separates residents from surrounding areas and minimizes market spillover. Its design-

Minneapolis, 2009. The IDS Center together with the skyway system and the transformation of Nicollet Avenue into a transitway created a public realm that has helped downtown Minneapolis to become a magnet attracting customers from the entire metropolitan district. *(Alexander Garvin)*

ers chose to make the project initially accessible only along three widely separated routes.[25] Residents have to drive along one of these routes to get anywhere outside Palos Verdes Estates, usually bypassing nearby areas and continuing on to major shopping and entertainment centers 10 or 15 minutes away. It is located in a manner that minimizes linkages with surrounding communities. Because it is built on a hilly peninsula extending into the Pacific Ocean, there is nothing to influence on the ocean side. In an attempt to compensate for this isolation, the project included a school, a country club and golf course, a charming retail complex inspired by Italian piazzas, and beachfront recreation facilities. During the early years, residents left Palos Verdes when they drove to work and spent most of the rest of their time away from home at facilities provided within the community. Consequently, this growing body of consumers had little impact on the rest of Los Angeles County.

Private real estate ventures like the IDS Center and Palos Verdes do not provide financing or entrepreneurs for other projects. At best, they demonstrate the potential of further real estate activity, perhaps attracting other developers and reducing the wariness of previously skeptical lending institutions.

Government programs, on the other hand, can manipulate financing and entrepreneurship in a manner that affects market activity. The renewal program for Society Hill included mortgage insurance for banks that financed rehabilitation and new construction. Because mortgage insurance provisions also reduced cash equity to as little as 5 percent of project cost, homeowners and developers were more likely to be able to afford the equity payments needed to acquire, renovate, and build. Equally important, by eliminating all incompatible land uses from the area, the program also reduced the risk of failure, thereby attracting people who would not otherwise have been willing to get involved.

The only period of time during which a project can affect surrounding market activity is the period during which it is in operation. Its impact, however, is particularly important when it supplies neighboring businesses with additional customers during slack periods. The increased consumer spending may support neighboring businesses whose market would not otherwise be large enough. For example, the customers attracted to Ghirardelli Square during the day, at night, and on weekends bring enough spillover business to be of real help to shops and restaurants in less convenient waterfront locations.

The Role of Government

Public action is increasingly difficult to justify if it achieves only a single purpose without providing additional benefits to the surrounding community. Too many cities want a plentiful supply of electricity but are unwilling to permit power plants within their boundaries. Too many neighborhoods want clean streets but bitterly resist sanitation garages in their neighborhood. Too many homeowners want convenient neighborhood schools for their children but oppose locating them across the street. Citizens regularly defeat such single-function public projects. As a result, controversial decisions are blocked by the not-in-my-back-yard (NIMBY) faction, by the not-over-there-either (NOTE) crowd, by the build-absolutely-nothing-anywhere-near-anybody (BANANA) enthusiasts, by the not-on-planet-earth (NOPE) advocates, or by citizens-against-virtually-everything (CAVE) people.

Three questions need to be answered positively to justify public action. Can it be implemented? Will it generate a private market reaction? Is that private market reaction worth it? This chapter demonstrates that when all six ingredients of success are properly combined, a project can be successfully implemented and can generate a private market reaction. As a result of Philadelphia's urban renewal strategy for Society Hill, for example, between 1960 and 2000 the number of housing units in the neighborhood nearly doubled, and their value increased 78 times at a time when values citywide increased only 16 times. Moreover, by subsidizing housing construction in Society Hill, the city government increased the number of customers in walking distance of the downtown stores and restaurants. The additional consumer traffic allowed shops and restaurants to remain in operation for longer hours and in the process increased the safety and attractiveness of downtown streets during the early evening. The revitalization of Society Hill and the accompanying improvements to downtown Philadelphia were purchased at bargain-basement prices: the city's share of urban renewal subsidies was approximately $9.5 million.

The preceding examples involve the use of *investment* (housing subsidies), *regulation* (parking requirements), or *incentives* (a zoning bonus) to alter four of the ingredients of success (market, location, design, time of operation). Success in generating further market activity may also require the other two ingredients: financing and entrepreneurship. When Detroit finally launched Lafayette Park, mortgage insurance became available within the boundaries of the urban renewal area. Because insured mortgages were not available for surrounding areas, owners and developers in those areas found it difficult to raise money to improve their properties. As a result, Lafayette Park was unable to generate the desirable, widespread, and sustained private market reaction that could be expected from the construction of hundreds of new apartments.

We can overcome citizen opposition and ensure project feasibility if we stop thinking solely in terms of individual projects. Instead, we must make decisions that are *also* based on the probability of a desired private market reaction. Then the public dialogue will shift from consideration of the project itself to the ways in which its market, location, design, financing, times of operation, and entrepreneurs will benefit the surrounding community. More important, we will increase financial and political feasibility while simultaneously increasing the likelihood of the desirable, sustained, and widespread market reaction that is characteristic of good city and suburban planning by creating places whose magnetic attraction is not only sustainable, but must spill over into surrounding communities because the amount of market demand it attracts can never be completely satisfied on-site.

Notes

1. A more detailed discussion of downtown redevelopment in New Haven can be found in Chapter 7.
2. A more detailed discussion of downtown Minneapolis can be found in Chapter 6.
3. Lawrence M. Irwin and Jeffrey B. Groy, *The Minneapolis Skyway System,* City of Minneapolis City Planning Department, January 1982, www.ids-center.com/pages/architecture.html; Leif Pettersen, "Take the Skyway," February 17, 2011, http://www.vita.mn/story.php?id=116353724.
4. A more detailed discussion of downtown redevelopment in Pittsburgh can be found in Chapter 7.
5. A more detailed discussion of Ghirardelli Square can be found in Chapter 4.
6. Historical and statistical material on Kansas City's River Quay and Westport Square is derived from Carla C. Sabala (editor), *Kansas City Today,* Urban Land Institute, Washington, DC, 1974; Patricia Cleary Miller, *Westport: Missouri's Port of Many Returns,* Lowell Press, Kansas City,1983, pp. 104–105; and George Ehrlich, *Kansas City Missouri: An Architectural History 1826–1976,* Historic Kansas City Foundation, Kansas City, MO, 1979.
7. During the later 1980s the city successfully expanded nearby City Market, which attracts thousands of shoppers on Saturdays and Sundays. Despite all the additional traffic that passes through River Quay to get to City Market, in October 1994 only 5 of the 20 storefronts along Delaware Street between Third and Fifth Streets were occupied.
8. It was common for property owners to group retail businesses within a unified architectural design. Country Club Plaza was the first to provide enough off-street parking for a large number of stores selling a wide variety of merchandise.
9. A more detailed discussion of Country Club Plaza can be found in Chapter 4.
10. A more detailed discussion of Chapel Square Mall can be found in Chapters 4 and 7.
11. A more detailed discussion of Southdale can be found in Chapter 4.
12. A more detailed discussion of Quincy Market can be found in Chapter 4.
13. A more detailed discussion of Seaside and Sea Ranch can be found in Chapter 16.
14. Peter Harnick, "The Park at Post Office Square," pp. 146–157, in *Urban Parks and Open Space,* Alexander Garvin and Gayle Berens (principal authors), The Urban Land Institute, Washington, DC, 1997.
15. A more detailed discussion of Society Hill can be found in Chapter 12.
16. A more detailed discussion of New York City's Vest-Pocket Redevelopment Program can be found in Chapter 12
17. A more detailed discussion of the Nehemiah Project can be found in Chapter 10.
18. A more detailed discussion of Lafayette Park can be found in Chapter 12.
19. When Herbert Greenwald died, only half the apartments that had been initially planned for were completed or under way, which makes it impossible to evaluate fully the success of its planning.
20. Jane Jacobs, *The Death and Life of Great American Cities,* Random House, New York, 1961, p. 14.
21. A more detailed discussion of Radburn can be found in Chapter 14.
22. A more detailed discussion of Palos Verdes Estates can be found in Chapter 16.
23. U.S. Bureau of the Census, 2000, www.palosverdes.com/pve/.
24. A more detailed discussion of Lafayette Park can be found in Chapter 12.
25. The design provided access from the Pacific Coast Highway via Palos Verdes Boulevard, from Hawthorne Avenue via Palos Verdes Drive, and from Long Beach via Paseo del Mar.

3

Parks, Playgrounds, and Open Space

Minneapolis, 2009. Lake Calhoun. *(Alexander Garvin)*

Traditionally, parks are conceived as either restorative (an idyllic counterpoint to the congestion of the city) or therapeutic (a place to relieve the tension of urban living). There is great wisdom in these traditional conceptions of park and playground. But parks and playgrounds are more than just places for city dwellers to relieve their tensions through communion with nature or in active, organized play. In most cities, parks and playgrounds take up large amounts of land. For that reason alone, they have a major impact on the development and character of every city.[1]

The idea that parks can spur the improvement of the surrounding city has at times provided an equally compelling rationale for public investment. While this strategic approach to investment in parks has taken different forms depending on the city in which it has been pursued, it usually has been for one of three purposes: initiating urbanization at specific locations, altering land use patterns in surrounding locations, or establishing a comprehensive system that could shape the very character of city life.

To achieve these purposes, the quantity and quality of public investment have to be adequate to alter private market conditions for neighboring properties. Only then will property owners react in the desired fashion. In St. Louis, for example, the combination of 1293-acre Forest Park and a trolley system that connected it with the city's Missouri River landings 3 miles away increased demand for residences in the surrounding area. In San Antonio, transforming an unsightly riverbed into a park created a tourist attraction that could be exploited by neighboring property owners. In Minneapolis, a superbly designed and maintained 6732-acre system of parks, parkways, playgrounds, athletic fields, jogging trails, bicycles, paths, lakes, and swimming pools contributed to living conditions that were attractive enough for the city to retain a sizable middle-class population at a time when other cities were losing that market to the suburbs.

Not all public open space consists of parkland. Nature preserves, bike lanes, traffic interchanges, parkway shoulders, and all sorts of other publicly owned property are open and accessible. By the twentieth century there were active movements to convert rail corridors to trails for jogging and cycling, reclaim abandoned lots, create tree cover for parking areas, and even convert water retention basins into wading ponds. Our conception of public open space is now quite different from what it was a century ago.

During the second half of the twentieth century parks were not thought of as central to the planning of cities. Their very existence has eliminated pressure for additional park development and made it possible for other issues to dominate the urban planning agenda. Nor was there the same certainty that existed during the nineteenth century that exposure to nature or active recreation can alleviate the effects of slums and poverty. Public open space is again regarded as an effective tool for shaping the American city, and today, as never before, conditions are ripe for parks to reenter the urban planning agenda. This opportunity exists because so much inner-city land that was once actively used now lies fallow and can be reused for intelligently planned parks, because so much suburban land has been developed without adequate public open space that there is now a huge suburban constituency to support park development, and because so much undeveloped land is now subject to recently enacted legislation intended to protect the environment.

Nature as Restorative

The Twenty-third Psalm suggests that green pastures restore health and still waters restore the soul. Ralph Waldo Emerson, writing in 1836, asserts,

> *To the body and mind which have been cramped by noxious work or company, nature is medicinal and restores their tone. The tradesman, the attorney comes out of the din and craft of the street and sees the sky and the woods, and is a man again. In their eternal calm, he finds himself.*[2]

Even the root of our contemporary word *paradise* is derived from the old Persian word *pairidaeza,* which translates to "garden." Living in paradise is a universal goal. It is not surprising that the remains of gardens can be found at most excavations of ancient civilizations.

In Europe, the nobility surrounded their residences with elaborate gardens as a way of maintaining the restorative powers of nature near at hand. The design of these gardens for the aristocracy became high art in eighteenth-century England. Nature was not permitted to grow wild, and efforts to tame it were inspired by the living versions of Arcadian landscape that the French painter Claude Lorrain had produced several centuries earlier. The premier practitioner of this emerging eighteenth-century art of landscape design was an English gardener known as "Capability" Brown.[3] In designing the grounds of such estates as Blenheim, Petworth, and Sherborne, Brown evolved an approach that stripped the landscape of discordant blemishes and exploited the natural beauty of pure landscape—pure landscape as filtered through the images of Claude Lorrain.[4]

The pre-Revolutionary American landscape, unlike that of England, contained few sophisticated large gardens. A town might have a common or a few small squares, but with a continent of wilderness just beyond the edge of town, there was little demand for large open spaces. Even in the more populous cities of Europe, public parks were almost unknown during the eighteenth century. The large parks of London, Paris, and Vienna were royal estates that, in some cases (e.g., London's Hyde and Richmond parks), were open to the public, even if they were privately owned.

Birkenhead Park, in Liverpool, England, was one of the world's first truly *public parks*. Sir Joseph Paxton designed Birkenhead Park in 1843 and created a 125-acre landscape inspired by Capability Brown's gardens for the aristocracy. Like them, it had areas of meadow, clumps of trees, serpentine lakes, and pedestrian paths weaving through the landscape to create different views. To these, Paxton added circumferential carriage roads for pleasure driving, a single transverse street, and boundary roads for town traffic outside the park. Birkenhead Park inspired similar efforts throughout England. However, its most important influence was on the American Frederick Law Olmsted, who, in 1852 and 1859, published his observations on Birkenhead Park in *Walks and Talks of an American Farmer in England,* an account of his 13-week tour of Europe. What impressed him was that,

> *In democratic America there was nothing to be thought comparable with this People's Garden.... Winding paths, over acres and acres, with a constant varying surface, where on all sides were growing every variety of shrubs and flowers.*
>
> *...A party of boys in one part, and a party of gentlemen in another, were playing cricket. Beyond this was a large meadow with groups of young trees, under which... girls and women with children were playing.... I was glad to*

Manhattan, 1863. Aerial view of Central Park soon after it was opened for public use. *(Courtesy of the Museum of the City of New York, J. Clarence Davies Collection)*

observe that the privileges of the garden were enjoyed about equally by all classes.[5]

When Olmsted visited Birkenhead Park, he was an unsuccessful farmer who was trying his hand at journalism. He would go on to design many of the nation's first large city parks and to become the country's premier landscape architect and arguably its most talented urban planner (see Chapters 14 and 16).

Central Park, New York City

By the middle of the nineteenth century, vast tides of immigrants were pouring into America's cities. The more congested these cities became, the more the public demanded parks. The first American city to respond to this demand was New York.[6] In 1850, when the city's population had reached 654,000, public clamor for parks spilled over into the mayoral election campaign. The following year the newly elected mayor, Ambrose Kingsland, obtained approval from the state legislature for the creation of a major public park. As he put it,

> *There are thousands who pass the day of rest among the idle and dissolute in porter houses or in places more objectionable, who would rejoice in being able to breathe pure air in such a place, while they ride and drive through its avenues free from . . . noise, dust, and confusion.*[7]

In 1857, after much discussion, the commissioners for the new Central Park decided to hold a design competition. Thirty-three designs were submitted. A proposal entitled "Greensward" by Frederick Law Olmsted and his architect partner Calvert Vaux won the competition.

Olmsted and Vaux's design was a brilliant adaptation of English eighteenth-century private garden design to the activities that they thought would take place in a public park. Just as the gardens created for an aristocratic elite in England were intended to be a work of art, so, too, was the new Central Park, except it was to be created for an entire city population. More than 2500 men were employed for more than 10 years to transform 843 acres of rough, largely undeveloped land into Olmsted and Vaux's populist vision of Arcadia. Tons of earth were excavated and moved to soften and improve the landscape; acres of land were dredged to create artificial lakes; miles of sewer and water mains were buried to artificially drain the park and supply it with water; tens of thousands of trees, shrubs, and flowers were planted to create the illusion of an Arcadian pleasure ground; and miles of gently curving paths and roads were introduced to permit easy access and circulation by the tens of thousands of pedestrians, horses, and vehicles that were expected to use the new park every day. The result appeared to be just another picturesque English garden with rolling lawns, serpentine lakes, clumps of trees,

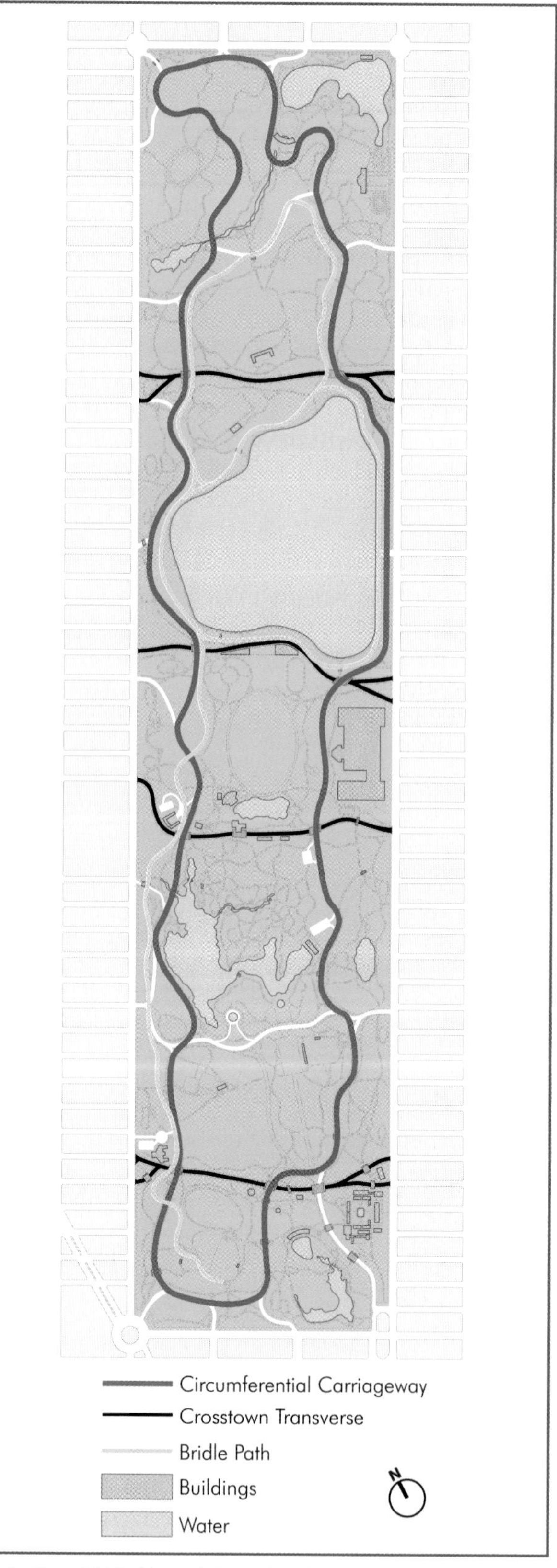

Manhattan, 2010. Map of Central Park. *(Alexander Garvin and Joshua Price)*

and meandering paths designed for the leisurely delight of the elite. In reality, though, it was a durable public park designed for active use in the rapidly growing city of New York.

When Olmsted and Vaux submitted their design, the bulk of the city's population lay miles to the south. Nevertheless, they understood that Central Park had to accommodate tens of thousands of users every day. For it to do this, they designed a unique circulation system that enabled masses of people to get to a series of very different places of recreation. Coming by carriage (or today by car), one enters at selected spots along the park's perimeter and rides along a beautifully landscaped roadway that encircles the park. Millions of people have enjoyed this pleasant drive either in the flesh or at the movies. Coming on foot, one enters at different spots and strolls to one's destination. There is also a separate bridle path for horseback riding. These three routes seldom intersect. When they do, there is often a pedestrian underpass that minimizes the chance of accidents.

Since Central Park extends north-south for 2.5 miles and divides the east and west sides of Manhattan, there is a need for traffic arteries to connect both sides of the city. To accommodate this traffic, Olmsted and Vaux designed four crosstown roads that traverse the park, providing through-traffic routes never intersected by circulation within the park. Nowhere within Central Park is one ever aware that the transverse roadways exist, because they are depressed and thereby separated from all other traffic and because they are cleverly masked by slopes, shrubbery, and trees. This allows other traffic on the carriageway and paths to pass over the transverse roads without an awareness of the vehicles below.

Manhattan, 2008. Central Park, where Olmsted intended people to escape from the hustle and bustle of the city. *(Alexander Garvin)*

Not only was Central Park conceived for an entire city of users as opposed to a much more circumscribed and elite population, but it also provided a wide variety of settings for mass recreation. Among them are meadows for playing ball, slopes for sledding, lakes for boating and ice-skating, a rustic ramble for wandering, a mall for promenading, and a variety of playing grounds. Today, on a typical weekend, at least 250,000 people can be found using the park.

After New York City created Central Park, citizens in almost every other city pressed for something similar. The ensuing demand for public parks is exemplified by a *Minneapolis Tribune* editorial printed in 1880:

> *Public parks have come to be recognized as institutions essential to the health as well as the happiness of thickly settled communities. Children suffer from privations, pine, and sicken and fall into untimely graves because [they are] cut off from the healthful provision of God's light and the pure atmosphere that circulates among the trees and through broad expanses. To keep down the death rate, and to be rid of wasting disease, a plentiful supply of parks is needed in every large town.*[8]

Similar editorials appeared in nearly every city. Fairmount Park in Philadelphia, Rock Creek Park in Washing-

Manhattan, 2005. Central Park is still actively used by hundreds of thousands of New Yorkers a century and a half after completion. *(Alexander Garvin)*

ton, DC, Forest Park in St. Louis, Golden Gate Park in San Francisco, and City Park in Denver are only a few of the more famous results.

Recreation as Therapy

At about the same time that this notion that public parks would cure disease was taking shape, a second idea was also gaining acceptance: that recreation would divert the city dweller from a life of crime. It is beautifully expressed by Jane Addams, the guiding spirit of Chicago's Hull House, a pioneering settlement house. She wrote, "To fail to provide for the recreation of youth is not only to deprive all of them of their natural form of expression, but is certainly to subject some of them to the overwhelming temptation of illicit and soul-destroying pleasures."[9]

The formulation of Addams's philosophy coincided with the rise of organized sports and was derived from the notion that city dwellers, who no longer engaged in physical activities such as farming and hunting, needed an outlet for their accumulated tensions and thus required a place for physical exercise.

In the 1820s, physical education became part of the curriculum of Harvard and Yale. Gymnastic societies, patterned after German *turnvereins,* were founded in most big cities. New field sports such as football and baseball became popular. Since there were no public parks, however, there were no public playing fields. It was not until 1871 that ball fields were included in a public park, and it was Olmsted and Vaux, who, for the first time, made playing fields a central feature of the design of Chicago's new Washington Park.

Municipal playgrounds as we know them today did not exist. In 1872, Brookline, Massachusetts, became the first

Manhattan, 1889. Jacob Riis's photograph of the slum known as "Mulberry Bend." *(Courtesy of the Museum of the City of New York, Jacob A. Riis Collection)*

Manhattan, 1900. Jacob Riis's photograph of Columbus Park, which replaced Mulberry Bend. *(Courtesy of the Museum of the City of New York, Jacob A. Riis Collection)*

city to vote that funds be allocated to the establishment of playgrounds, and Hull House began experimenting with playgrounds as early as 1884. However, it was not until the New York state legislature enacted the Small Parks Act three years later that the provision of public playgrounds became accepted as a legitimate function of government.

The Small Parks Act established the principle that local government could condemn privately owned land for the purpose of creating public playgrounds. Legislation was only the first step. Mulberry Bend Park (today Columbus Park) on Manhattan's squalid Lower East Side was the first park to result from the legislation, and reformers were eager to demonstrate that playgrounds were effective in the battle against delinquency, crime, and slums. One of the most important of them, Jacob Riis, a major figure behind both the legislation and the park, wrote books and articles to demonstrate

Manhattan, 2010. Columbus Park. *(Alexander Garvin)*

the case for public playgrounds. His writings remain a persuasively eloquent argument for their creation. In discussing Mulberry Bend Park, he wrote,

> *I do not believe that there was a week in all the twenty years I had to do with . . . [Mulberry Bend], as a police reporter, in which I was not called to record there a stabbing or shooting affair, some act of violence. It is now five years since the Bend became a park and the police reporter has not had business there during that time; not once has a shot been fired or a knife been drawn.*[10]

His promotion of Mulberry Park set the stage for a huge campaign for playgrounds on the basis that they offered social cures as much as they offered outlets for physical activity.

Boston, 1886. Charlesbank prior to becoming a public park (currently Storrow Memorial Drive and Embankment Road). *(Courtesy of the Boston Public Library, Print Department)*

Charlesbank, Boston

When the new playground at Mulberry Bend opened in 1897, play equipment was only just being invented. Just as Frederick Law Olmsted provided the paradigm for the park as a restorative Arcadia, he also provided the paradigm for the playground as recreation therapy. In the late 1880s, he presented his first designs for Charlesbank, a 10-acre site along Boston's Charles River opposite Massachusetts General Hospital. Like Mulberry Bend Park, it was intended as a nostrum for the slums, in this case Boston's West End. Among the reasons for the new playground that Olmsted enumerated were the provision of open-air facilities for an increasingly sedentary population and an attempt at the reduction of the death rate from cholera among the children of the West End slums.[11]

The final plan for Charlesbank, completed in 1892, included a riverfront promenade; a playground for little girls with sandboxes, swings, and ladders; a small women's outdoor gymnastic area with a tenth-of-a-mile running track and small areas for jumping and shot-putting; and a somewhat larger men's outdoor gymnastic field with a sixth-of-a-mile running track, a trapeze, flying rings, horizontal bars, and areas for shot-putting, pole-vaulting, and jumping. Olmsted consulted recreation experts in designing Charlesbank, as did the Department of Parks in administering it. For the design of the men's playground, Olmsted selected equipment designed by Professor Dudley Sargent of Harvard. For the supervision of the children's and women's playground, the Department of Parks employed staff trained by the Massachusetts Emergency and Hygiene Association.[12]

Now that generations of children and adults have grown up thinking of sandboxes, swings, slides, running tracks, and

Boston, 1889. Charlesbank men's gymnasium (replaced by Storrow Memorial Drive and Embankment Road). *(Courtesy of the Boston Public Library, Print Department)*

playing fields as common fixtures of the urban landscape, it is hard to imagine how revolutionary Olmsted's design for Charlesbank was. Its success is easier to visualize. In its first year of operation, daily attendance at the women's and children's gymnastic field averaged 840.[13] As a result of that success, Boston had by 1898 approved legislation calling for one playground to be established in each of the city's wards.

Chicago followed suit in 1903, authorizing a bond issue of $1 million for the creation of "small parks or pleasure grounds not more than 10 acres each." Olmsted's firm, now managed by his sons, was hired to design the new parks. Within two years, Chicago had 10 new playgrounds and San Francisco had three. The increasing proliferation of playgrounds led, in 1906, to the establishment of the Playground Association of America.

Today, Charlesbank has been replaced by urban arterials. However, its importance cannot be overestimated. Olmsted's pioneering efforts in providing facilities for urban recreation influenced the creation of tens of thousands of American playgrounds with swings, slides, seesaws, and sandboxes.

TABLE 3.1

APPROPRIATE DISTANCES APART, IN MILES, OF PARKS AND PLAYGROUNDS

Facility	Chicago	Denver	Minneapolis
Playground	¼	½	¼–½
Neighborhood park	⅜	½	—
Playfield	—	1½	½–1
Community park	¾	—	—
District park	2½–3	3	3
Regional park	—	10	30

SOURCE: Chicago Recreation Commission, *Suggested Goals in Park and Recreation Planning*, 1959; Denver Inter-County Regional Planning Commission, Standards for New Urban Development, Denver, 1962; Minneapolis Park Board, Planning Division, *Park and Recreation Facilities and Standards*, Minneapolis, 1964.

Standardization and Mass Production

Olmsted supplied the paradigm for both the park and the playground. Reformers made them part of the public agenda. Now government had to standardize the product and make it generally available. What was the proper number and location for parks and playgrounds? Committees, organizations, and agencies studied the problem. Experts differed. Some proposed a standard of 30 square feet per child. Others proposed setting aside 10 acres per 1000 for playgrounds and 40 acres per 1000 for large parks. Eventually, parks advocates settled on a standard of 10 acres for every 1000 people.[14]

Accordingly, cities prepared master plans measuring current population, estimating the probable growth in population, cataloging current park acreage, tracing each facility's service radius, projecting the deficit in public open space, and proposing projects to fill the gaps (Table 3.1). San Francisco, in 1942, established guidelines for the distances between libraries, schools, and recreation places. Philadelphia, in 1968, published a document entitled *Comprehensive Plan for Swimming Pools,* which advocated pools serving from 20,000 to 30,000 people in every neighborhood.[15]

Since most standards failed to take into account differences in scale, the National Park Service in 1938 proposed differential standards for park acreage based on the size of a community (Table 3.2). The irony of recommending more park acreage for communities that were less densely settled and therefore had more private open space apparently eluded these single-function planners.

In 1943, the American Society of Planning Officials (ASPO) proposed lowering the standard to 10 acres for every 3000 city residents in cities with populations above 1 million, because higher standards were not attainable in more densely populated areas. The absurdity of this numbers game eluded them, too. At a standard of 10 acres per 1000 population, Manhattan, at its peak population of 2,331,500 in 1910, would have required 23,315 acres of park, more than the island's entire 14,870 acres. Even at ASPO's lower standard of 10 acres per 3000, half of Manhattan would have to have been set aside for parkland.

The standards may have been helpful in determining how many facilities were necessary, what they should consist of, and where they should be located. Their principal importance, however, was in their role as information for park administrators to use in competing with other agencies that advocated other, often more pressing demands for scarce budget allocations. The statistics thus became a tool in balancing the public's desire for recreation facilities with politically achievable levels of funding. Of course, they also became a justification for maintaining and enhancing growing park department bureaucracies.

While park advocates and urban planners were trying to arrive at equitable standards, city officials were grappling

TABLE 3.2

NATIONAL PARK SERVICE ACREAGE STANDARDS (1938)

City population	Park acreage per number of people
More than 10,000	10 acres/1000
5000–8000	10 acres/750
2500–5000	10 acres/600
1000–2500	10 acres/500
Less than 1000	10 acres/400

SOURCE: National Park Service, *Recreational Use of Land in the United States*, National Resources Board, Washington DC, 1938.

Bronx, 1990. Playground equipment installed by Robert Moses at Orchard Beach. *(Alexander Garvin)*

with the problems of acquiring land for recreation, financing the purchase and construction of facilities, and ensuring they were adequately maintained. They knew, as New York City Parks Commissioner Robert Moses explained, that there is "no such thing as a fixed percentage of park area to population. . . . Sensible, practical people know that [it] depends upon the actual problems of the city in question."[16]

Moses acquired and developed more city parks and playgrounds than any municipal official in any American city at any time. While he was parks commissioner, between 1934 and 1960, he added 20,673 acres to the park system, including 17 miles of public beach, 218 tennis courts, 3 zoos, and 658 playgrounds. The location of these depended on the availability of funds, not on population within the service radius of the facility. For example, when federal relief workers were available, Moses employed them to create golf courses, swimming pools, and playgrounds in existing parks. While building bridges, tunnels, and highways, Moses usually condemned more land than absolutely necessary and transformed the excess into parkland. As anyone who has visited the Brooklyn Heights Esplanade, the parkland under Robert F. Kennedy (Triborough) Bridge, Manhattan's Henry Hudson Parkway, or any of these "nonpark" projects will tell you, what Moses created in the wake of his other public works makes major park projects in other cities appear insignificant.[17]

Design standardization was among the fundamental tools that enabled Moses to produce this vast inventory of facilities. He installed the same benches, swings, slides, seesaws, sandboxes, comfort stations, and sycamore trees over and over again. He relied on the same cheap, sturdy materials (asphalt, concrete, brick, slate, and wood) in every situation. Parks projects built simultaneously in other cities used similar materials and similar designs. The bureaucracies in those cities also had to standardize their product. The only difference was in Moses's phenomenal ability to get things done.

Midcourse Correction

By the 1960s, parks and playgrounds had become permanent fixtures of the cityscape. Their institutionalization in both the fabric and culture of urban areas rendered it easy to under-

Manhattan, 2012. Paley Park was one of the earliest examples of providing vest-pocket parks, rather than larger recreational facilities. Designed by Zion & Breen for a prominent location in midtown Manhattan, it became the model for the small neighborhood parks throughout New York City. Maintenance is paid for from an endowment. *(Alexander Garvin)*

value the administrative bureaucracies that had continued to produce and maintain standardized parks and playgrounds for half a century. People no longer believed, as had Jacob Riis, Jane Addams, and the reformers of a previous century, that parks would eliminate crime, delinquency, and contagious diseases. A new wave of reformers demanded more effective public open space. They had had enough of the "swing, slide, sandbox stereotype." Now that parks and playgrounds were abundant, critics could label them "dreary 'people-proofed'" yards that fail "to offer any sort of relief valve for the overwhelming sense of frustration in the youth of a neighborhood."[18]

The redefinition of our image of park and playground began in scattered projects around the country, but took its clearest shape in New York City. In 1966, the newly elected mayor, John Lindsay, and his newly appointed parks commissioner, Thomas P. F. Hoving, embarked on a program of creating what they called "vest-pocket parks." As Hoving explained, it was time to get rid of "the black-topped, link-fenced asphalt prison, that standard architecture that has made the W.P.A. style the longest art style of the 20th Century."[19] The idea was to work with community groups to select vacant lots and underutilized properties that were easily accessible to toddlers, teenagers, and the elderly. After sites had been chosen, the most talented designers would be hired to create exciting new play environments.

Robert Moses knew all the problems Lindsay would encounter. Except for design, the mayor's plan was not very different from the one Moses had implemented in the 1930s. Moses issued a nine-page analysis, calling the Lindsay initiative a grab bag of good ideas (for which there was no money) and many poor ones (that had been tried and discarded). Given predictable levels of funding and staffing, Moses felt vandalism could not be controlled, nor could ordinary maintenance be paid for, and by the 1970s it had become clear that he was right. New York City had neither the money nor the personnel to keep its capital plant in a decent state of repair.[20]

Hoving's ideas, like those of Moses before him, were not new. Park enthusiasts and community activists across the country had been calling for new parks policies, experimenting with abandoned lots in slum areas, and scheduling rock concerts and other events to attract young people who otherwise preferred the livelier activities of city streets. Nor was the design of the vest-pocket park original. Designers Paul Friedberg, Zion & Breen, Lawrence Halprin, and others were already at work on more inventive forms of play equipment and playground design.

Hoving ignored Moses. He hired creative designers and succeeded in scattering vest-pocket parks in poverty areas. Within months, many were in poor condition. The Parks Department could not keep up with ordinary repairs or provide necessary personnel for supervision. Benches were missing slats, paving blocks were gone, play equipment was broken, and graffiti disfigured the avant-garde designs.

The Lindsay administration, wanting both the new vest-pocket parks and the older traditional parks to attract people no longer using the city's public open space, used them as settings for celebrations, festivals, and performances, which Hoving called "happenings." Holiday picnics, dances, outdoor movies, kite flying, band concerts, and many of the other happenings had been going on in parks for years. What was different now was that these events took place almost every day, attracted throngs of young adults who had never before thought of parks as a setting for popular culture, and resulted in a level of use and, by extension, deterioration (crumbling pavement, ravaged lawns, eroded soil, exposed roots, etc.) for which no provision had been made. Neither funds nor personnel were sufficient to deal with the wear and tear. More important, no plans existed for regular, ongoing park rehabilitation.

Bronx, 1969. Vest-pocket park on Bryant Avenue soon after it opened. *(Alexander Garvin)*

Bronx, 1971. Vest-pocket park on Bryant Avenue in disrepair after less than two years in operation. *(Alexander Garvin)*

Hoving never had to face the sad results of his program, but in a strange way its terrible failure led to an innovative structure for long-term, successful management of parks. Within a year of his appointment as parks commissioner he became director of the Metropolitan Museum of Art. By the time Mayor Lindsay left office in 1973, the devastation was so serious that people were calling for a program of playground restoration and 10-year master plans specifying a regular cycle of renovation for the city's major parks. In the wake of the clamor for restoration, civic leaders established the Central Park Conservancy, which raised money and developed plans for the park's rehabilitation. Simultaneously, the recently elected Koch administration dramatically increased capital spending for park restoration.

In 1979, Elizabeth Barlow Rogers became the first Central Park administrator. Under her inspired leadership the effects of years of abuse were methodically eliminated, and adequate attention and funding were provided for continued improvements to the park. A similar effort initiated by Tupper Thomas took place in Brooklyn's Prospect Park. Many of Moses's playgrounds are being rebuilt throughout the city. But in New York and elsewhere, much more remains to be done.[21]

The reform effort of the 1960s may have had the flaw of ignoring maintenance and assuming unlimited budget allocations. However, it made several successful changes in our approach to public open space. It established once and for all that public open space is for daily use by large masses of people and that budget and personnel allocations for maintenance and restoration must be sufficient to accommodate such use. It also introduced the idea that recreation facilities are not unchanging artifacts, but instead have to be adapted to the needs of contemporary users. Finally, it buried forever the "swing, slide, and sandbox stereotype" and replaced it with the idea that play equipment has to grab a child's imagination and encourage adventure and exploration.

Community Gardens

During the later twentieth century, deteriorating, lower-income neighborhoods experienced an unexpected increase in open space whose use threatened to destroy whole neighborhoods. Large numbers of unprofitable slum properties were abandoned by their owners, and neither they nor their mortgagees believed these buildings were worth even the unpaid real estate taxes. Local governments repossessed these sites for failure to pay taxes. As the inventory of abandoned property owned by local governments grew, they were forced to demolish particularly hazardous buildings. Residents became increasingly alarmed about the impact of the activities that took place on these ill-maintained, city-owned lots. Other more entrepreneurial actors saw them as opportunities for community rejuvenation and took possession, sometimes illegally, sometimes pursuant to a month-to-month lease, sometimes as permanent managers. What emerged were "community gardens" that might include handsome floral displays or expensive, difficult-to-obtain herbs and spices, or even vegetable patches that supplied a bounteous harvest of zucchini, tomatoes, and other fresh produce.

Because there were so many opportunities to convert abandoned properties into community gardens, the situation offered an unexpected alternative to conventional public parks. In Philadelphia, 21,400 residential structures were demolished between 1970 and 1990. As a result, 9500 of the 23,000 city-owned properties were vacant land.[22] New York's inventory of city-owned apartments in occupied buildings had grown to 46,420 by 1980.[23] Some buildings were sold, often to the tenants. More often, cities demolished buildings that had become fire and safety hazards.

Haphazard but significant efforts transformed a few of these neighborhood hazards into locally managed and maintained open space. Every time an area of crime and blight

Manhattan, 2001. Community Garden at Avenue B and 6th Street. *(Alexander Garvin)*

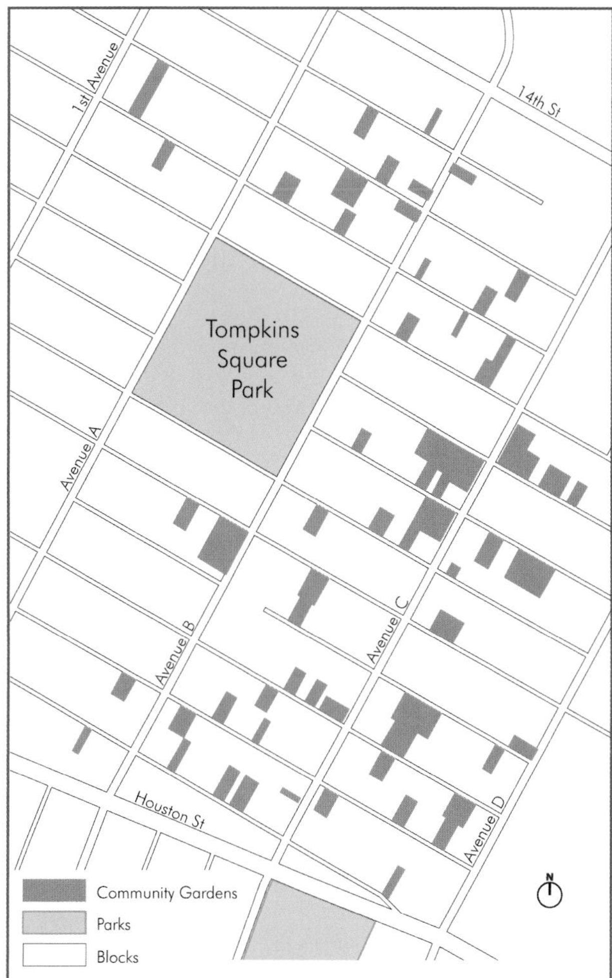

Manhattan, 2001. Map of Community Gardens on Lower East Side. *(Alexander Garvin and Joshua Price)*

was replaced by a meticulously cared-for garden, surrounding properties reaped benefits. To the surprise of residents and public officials, community gardens began to generate second growth in a surprising number of low-income areas.

During the 1980s, Manhattan's Lower East Side was typical of the low-income areas plagued by abandoned property. Its future was considered bleak. The emergence of community gardens on formerly garbage-strewn empty lots triggered a change in the perception of the area. For the first time there was evidence of care and concern for the neighborhood. This coincided with declining crime rates, a booming economy, and increased demand for apartments throughout New York City. As a result, newcomers were not as afraid of moving to low-rent, fixer-upper tenement apartments. Nonprofit and even some profit-motivated developers began to renovate or build new "affordable" co-op row houses and apartment buildings. Storefront retailers opened bistros, bars, and cafés in places that once housed drug dens. Within a decade, the areas affected had become desirable residential locations that were no longer thought of as being on the Lower East Side, and by 2012 the area was thought of as a gentrified neighborhood.

Community gardens have played an important part in reclaiming dilapidated neighborhoods, but they raise important and sometimes difficult questions for local governments. To work, they must be fenced and locked when not supervised. Because supervision is not always available, many community gardens are locked and, thus, not available to the general public much of the time. A few, like the fairly large one on Avenue B and 6th Street, post the times they are open and schedule yoga, story reading for children, live music performances, and other activities throughout the week. These conditions nonetheless provoke questions about equity, administration, and use: How does a city prevent these gardens from becoming the private preserve of those who have keys? To whom are their self-selected, volunteer supervisors accountable? During what periods of time should they be open to the public, and how can the public be assured of access? Under what circumstances should such interim uses be replaced by permanent facilities, such as affordable housing? These are only some of the more important issues that cities will have to resolve if they are to have successful community garden programs.

Reclaiming the Public Realm

Many people are under the misapprehension that there are few opportunities for adding parkland to already built-up

Philadelphia, 2012. By adding chairs, tables, and umbrellas, what was once a parking area for the 30th Street Railroad Station has become a place for people to stop and chat. *(Alexander Garvin)*

Manhattan, 2011. A portion of Broadway that was once occupied by cars is now actively used by people as their living room. *(Alexander Garvin)*

cities. Yet every American city includes a myriad of parking lots, empty buildings, unoccupied rail yards, empty factories, and warehouses. Few local governments are willing to spend the money to buy them. In fact, property acquisition is unnecessary because we already own more than enough land that could be reused for recreational purposes. It comprises everything from traffic islands and underutilized roadways to leftover public property and obsolete structures. They are as much a part of our public realm as actively used streets and highways, railroads, parks, and squares. At relatively low cost, they can become major public assets.

Toward the end of the twentieth century, cities began transforming this property into miniparks. Between 1996 and 2006, Chicago invested in more than 70 miles of planters, landscaped street medians, and trees, making its streets safer, more attractive places to drive, bicycle, and walk.[24] The same year that Chicago began its effort to improve the environment, New York City inaugurated a similar effort, which it called the GreenStreets Program. GreenStreets, however, went beyond landscaping existing streets and sidewalks. It identified nearly 3000 tiny patches of land, primarily traffic islands and small city-owned lots that were inappropriate for development. Most were paved with asphalt or concrete, creating harsh heat islands that increased the release of gases and particulates into the air. Over the next dozen years the parks department reclaimed 2300 leftover spaces into landscaped islands that average 2200 square feet in size. These miniparks usually include trees, shrubs, and groundcover that require a low level of maintenance and have a high tolerance for periods of drought, soil compaction, and pollution.

In both cities, the added greenery absorbs rainwater and carbon dioxide. It filters the air and traps dust and particulate matter, reduces noise, lowers the ambient temperature during summer months, and provides an attractive habitat for birds and small animals that control insect infestation. The miniparks have become important places for passive recreation, while simultaneously ameliorating the environment. By adding tables, chairs, umbrellas, and tubs of flowers and greenery, they become an outdoor living room. Then citizens can encounter one another, move around, sit, chat, play, or just admire the passing scene.

Mayor Michael Bloomberg's appointment of Janette Sadik-Khan as commissioner of Department of Transportation in 2007 accelerated New York City's program of minipark development. Commissioner Sadik-Khan's efforts were aimed at achieving the mayor's objective of reducing the city's carbon footprint 30 percent by 2030. Her approach combined reducing emissions from motor vehicle, by increasing the use of bicycles, with transforming territory not needed for vehicular traffic into usable, publicly accessible plazas—in the process, substantially increasing the city's miniparks. By 2013, this had resulted in the replacement of more than 300 miles of traffic lanes with bicycle paths and the conversion of 26 acres of roadbed into 50 new plazas.[25]

Parks as a Strategic Public Investment

Writing in 1861, Olmsted predicted, "The town will have enclosed the Central Park. . . . No longer an open suburb, our ground will have around it a continuous high wall of brick, stone, and marble." To evaluate its effectiveness, he suggested, "Let us consider, therefore, what will at that time be satisfactory, for it is then that the design will be judged."[26]

A continuous wall of brick, stone, and marble residences

Manhattan, 1887. The vacant lots on Central Park West, south of 72nd Street, indicate that even three decades after Central Park was completed, the huge crowds that used the park came from great distances. *(Collection of the New York Historical Society)*

has been in place around Central Park for over a century, and the residents of New York City have answered in exactly the same manner from the start. They use the park just as Olmsted and Vaux envisaged, and the aforementioned intensity of that utilization underscores how much they have embraced the park as a critical part of their lives in the city.

It is not surprising that the reasons for this phenomenal success are fundamentally linked with the ingredients of success. The market, location, and design of Central Park are an extraordinary match. Hundreds of thousands of people live and work within a few minutes' walk of the park. Olmsted and Vaux created a facility that was carefully designed to accommodate them. Moreover, the park has been able to flourish because it has so often been managed by individuals of entrepreneurial capacity, including Olmsted (superintendent of Central Park from 1857 to 1862 and superintendent or landscape architect from 1866 to 1878), Moses (commissioner of parks from 1934 to 1960), Elizabeth Barlow Rogers (administrator of Central Park from 1978 to 1995), Douglas Blonsky (administrator of Central Park since 1998), and the talented civil servants whom they attracted. Central Park's problems have occurred during periods in which either adequate financing was not provided or political support was temporarily strong enough to permit intrusive additional facilities to be erected or to allow park-damaging activities to take place. The Central Park Conservancy has helped to insulate the park from the problems created by momentary political fashion and financial stringency. By 2011, the Conservancy had an endowment of $140 million, managed over 700 volunteers, and had invested more than $480 million of privately raised funding into Central Park.[27]

Many vest-pocket parks and most adventure playgrounds, on the other hand, were built to meet the requirements of a population that was in place when they were conceived but in many cases has moved on. The result is a mismatch of market, location, time, and design. This is perfectly appropriate as long as society is willing to discard facilities when market changes render them obsolete. Most cities, however, are unwilling to tie up land and make capital expenditures for periods of such short duration. Even if they are prepared to invest in disposable parks, the financial and entrepreneurial requirements are beyond their capacity. It

Manhattan, 1898. Nineteenth-century strollers at the Central Park Mall. *(From* Greater New York Illustrated, *Rand McNally & Co., Skokie, Illinois, 1898)*

Manhattan, 2004. Twenty-first-century strollers at the Central Park Mall. *(Alexander Garvin)*

Manhattan, 1898. Sheep's Meadow, Central Park. *(From* Greater New York Illustrated, *Rand McNally & Co., Skokie, Illinois, 1898)*

Manhattan, 2005. The Sheep's Meadow in Central Park as it is used at the beginning of the twenty-first century. *(Alexander Garvin and Ryan Salvatore)*

Manhattan, 1894. Nineteenth-century boating in Central Park. *(Courtesy of the Library of Congress, J. S. Johnson Collection)*

Manhattan, 2009. Twenty-first century boating in Central Park. *(Alexander Garvin)*

Paris, 2006. Place des Vosges. *(Alexander Garvin)*

is unlikely that these facilities will ever be able to compete with other government functions for the necessary operating funds or with other activities for sufficient operating, maintenance, and supervisory personnel with the necessary entrepreneurial skills.

The best argument for additional spending on parks is that, as was the case with Central Park, the money will stimulate widespread and sustained private investment. Simple retention of open space in its natural state, however, is not enough. It must be acquired, designed, relandscaped, managed, and maintained for public use. Only when local governments establish programs and institutions that make this land available for active public use and deploy the land in a manner that reshapes surrounding settlement patterns will we begin to exploit its potential as a tool for actively contributing to the American city.

Paris and London

Strategic capital spending on landscaped public open space is responsible for some of the world's most admired urban open spaces: the squares, parks, and boulevards of London and Paris. The money for these facilities initially came from the aristocracy and the crown, and later from government. It was spent for the specific purpose of attracting a market (the growing populations of these cities) and generating further real estate development.

Among the first such investments were those made by the kings of France. Starting with Henry IV, each new king opened an unimproved section of Paris for development by laying out public squares. These *places royales* were geometrically regular in shape (i.e., circle, square, rectangle, and triangle), and each was accented at the center with a sculpture of the monarch who was responsible for its creation. Builders were required to maintain a uniform façade design behind which they could build as they pleased. Henry IV, starting in 1604, used the Place des Vosges to promote development of the Marais; Louis XIV created the Place Vendôme in 1677 and the Place des Victoires in 1684 to spur development of the second arrondissement; Louis XV fashioned the Place de la Concorde to encourage development north and west of the Tuileries.[28]

A similar approach was taken by the English aristocracy, whose income was in large measure earned by leasing land on their large estates. Many estates required development to proceed according to predetermined plans that included landscaped squares that had been initially set aside for the exclusive use of the occupants of surrounding buildings. This altered the character of the remaining property sufficiently to stimulate additional interest in its development.[29]

Once a plan had been decided upon, the estates offered long-term ground leases (usually 99 years) on the lots around the squares. The lessees were profit-motivated developers, who built town houses for sale or rent, and families, who built for their own use. At the end of the lease the estate either

London 2008. Cavendish Square. *(Alexander Garvin)*

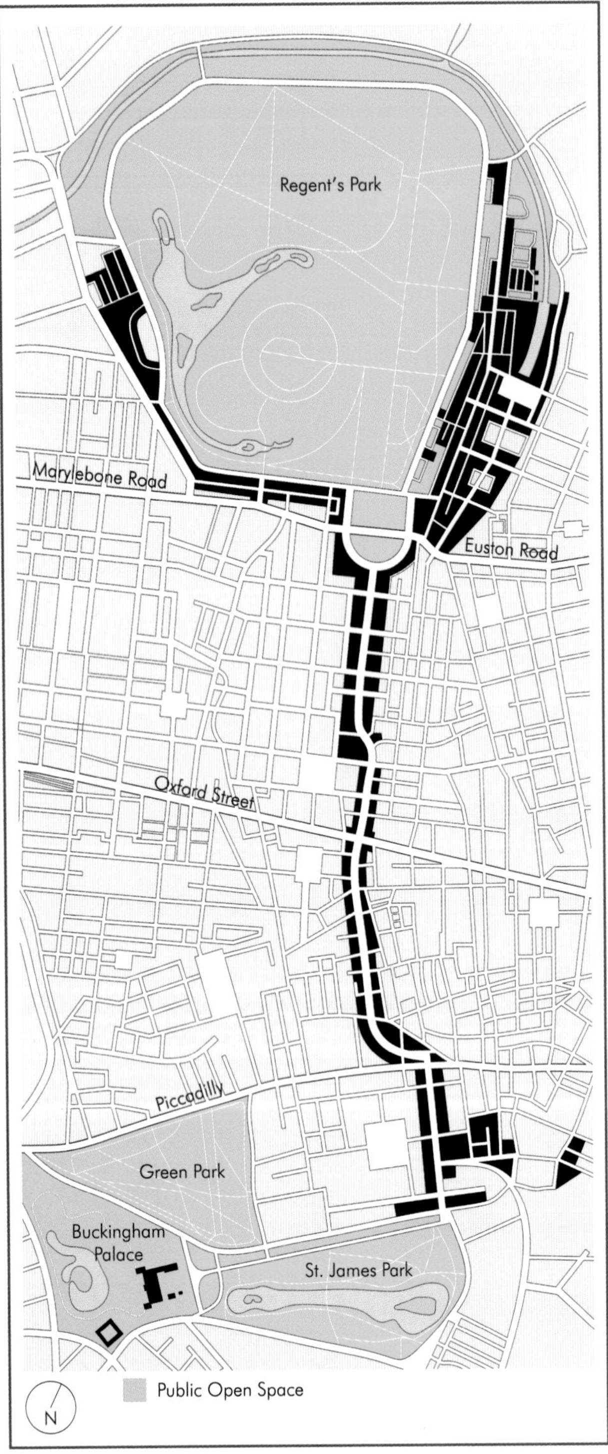

London. Sir John Nash reconceived St. James's and Regent's parks as central features in the redevelopment of crown property and connected them with an entirely new artery, Regent Street. *(Alexander Garvin and Ryan Salvatore)*

renegotiated the deal or took possession of both the land and the buildings that had been erected on it. Then it re-leased the property. In both cases, it usually obtained substantially higher income. This additional revenue reflected the additional attractiveness of a location that by then included both well-established, landscaped squares and the substantial buildings that surrounded them.

Covent Garden, laid out in 1630 by architect Inigo Jones for the Earl of Bedford, was the first of these London squares. It was followed by Leicester Square in 1635, St. James Square in 1684, and Grosvenor Square in 1695. By the eighteenth century, the landscaped square had become the accepted device for marketing estate property, a device that during the nineteenth century was primarily responsible for the development of Bloomsbury, Belgravia, and Islington. But it was an approach that was possible only at locations that included substantial amounts of undeveloped land near already built-up areas with growing populations ready to settle nearby.

Once vast territories had been developed, an entirely different set of problems emerged. By the start of the nineteenth century, London and Paris had become large, congested cities. The central sections of both cities had to accommodate rapidly increasing populations and expanding economies. Once again, public open space was used as a mechanism to accommodate growth by providing a framework around which to reconstruct existing city districts. Although the process of reconstruction was different in each city, the circumstances were the same. The cities' most valuable locations had been built up in a manner that no longer satisfied its needs. Individual parcels might be rebuilt, but growth was so rapid and market pressure so intense that entire districts needed to be redesigned.

In London, the process of reconstruction was initiated in 1811 when the prince regent, anticipating the expiration of the lease on his 543-acre Marylebone Estate, engaged architect-developer Sir John Nash to propose a scheme for its redevelopment. Nash's plan consisted of three parts: refashioning the district around the existing royal garden, known as St. James Park, creating a new district on the Marylebone Estate encircling a relandscaped Regent's Park, and connecting the two park districts with a broad new artery to be called Regent Street. Thereafter, the 2-mile stretch between the two

London, 2007. Regent's Park was conceived as an amenity that would increase the value of the rest of the Prince Regent's Marylebone Estate and induce residential development surrounding the park. *(Alexander Garvin)*

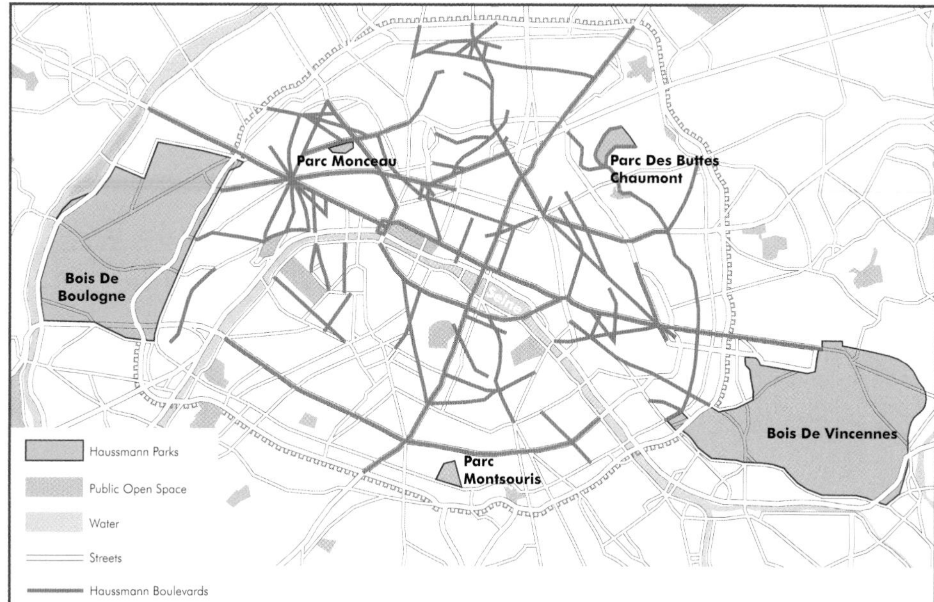

Paris. Haussmann created an entire network of tree-lined boulevards, green squares, and public parks. *(Alexander Garvin and Joshua Price)*

parks could be rebuilt by individual property owners to meet market requirements.[30]

The land around St. James and Regent's parks was bound to be prime territory for development. Without the new Regent Street, however, the scheme would have been far less successful. Prior to its creation, central London had three major east-west arteries (Oxford Street, Picadilly, and the Strand). Regent Street not only connected the parks but also provided, for the first time, a major north-south thoroughfare that linked these east-west arteries, thereby dramatically increasing accessibility to the entire district.

Nash's design concept for the land around the parks was intended to increase its already extraordinary marketability by creating a setting that gave the occupants of the surrounding buildings the illusion of nobility and great wealth. Opposite the park, he designed monumental residential structures, known among the English as *terraces.* From within the park the collection of these terraces appears as sumptuous palaces rather than middle-class residences, which is what they in fact are. From terrace windows, the park appears as any given resident's own landscaped estate. By creating these twin illusions, the design transformed a large amount of crown property into fashionable sites that could command high rents and attract further development on less attractive adjacent land.

Successful as the squares and parks of London were, they did not provide the most influential evidence that strategic investment in public parks could spur the improvement of surrounding areas. That evidence was provided by the parks that Baron Georges-Eugène Haussmann created in Paris a generation later. Haussmann was a public administrator who had worked in various parts of France before being appointed prefect of the Seine (Paris) in 1853 by Emperor Napoleon III. In the 17 years during which he occupied that position, Haussmann transformed Paris from a congested jumble of buildings into a modern metropolis with large parks and broad landscaped boulevards (see Chapter 19).

Like Nash, Haussmann worked for the crown. Like Nash, he refashioned two large royal forest preserves into public parks (the Bois de Boulogne and the Bois de Vincennes), connected them to the rest of the city with new avenues, and rebuilt much of the city in between. However, the scale of Haussmann's work dwarfed anything Nash dared dream about.

In 1850 Paris possessed only 47 acres of public park (1 acre per 22,300 inhabitants). The open-space system Haussmann and Adolphe Alphand, an engineer and landscape architect, established included two large regional parks (the Bois de Boulogne and the Bois de Vincennes), three district parks (the Parc Monceau, Les Buttes Chaumont, and the Parc de Montsouris), about 24 small landscaped public squares (ranging in size from a quarter of an acre to more than 6 acres), 90 miles of tree-lined boulevards, and the landscaped quays that lined the Seine River. When Haussmann left office, the park system included more than 100 times as much land as when he was appointed. In spite of a growing population, the parks acreage per capita was cut to 1 acre per 390 inhabitants.[31]

Each of the facilities Haussmann created was intended to stimulate development in the surrounding area. Avenue Foch (originally known as the Avenue de l'Impératrice in honor of Napoleon III's wife) provides an excellent illustration of Haussmann's approach. Ostensibly, the avenue was created to connect the Place de l'Etoile with the Bois de Boulogne. Its

Paris, 2008. Bois de Vincennes. *(Alexander Garvin)*

heroic width (460 feet from property line to property line) is far more than this objective required. As a result, the avenue has developed into far more than a heavily traveled traffic artery. It is really two linear extensions of the Bois de Boulogne that run alongside a central roadway. These linear parks are in turn flanked by boundary roads that service the buildings fronting on this extraordinary boulevard. Good vehicular access, however, is only one reason that Avenue Foch became the location of choice for some of the most elaborate mansions in Paris. The major reason is that the landscaped islands flanking the monumental central roadway are wide enough for residents to have the illusion of living opposite a park.[32]

Importing European Models

Some of the conditions that made the boulevards, parks, and squares of London and Paris so successful were similar to those in nineteenth-century American cities. Burgeoning populations provided a rapidly growing market. Facilities could be located on open territory very close to that market. However, neither the European conception of public open space nor the mechanisms for its establishment and maintenance were applicable in the United States. In Europe, parks were thought of as an integral component of the urban environment; in America, they were meant to stand in contrast to the city and provide a refuge from its noise, dirt, and confusion. In Europe, open space was specifically designed to meet the demands of surrounding building occupants; in America, parks were intended for widespread and active public use.

Even if the techniques used to create squares and parks in London and Paris were not transferable to the United States because the European strategies were based on completely different notions of property ownership and government responsibility, parks in all places were leveraged to sponsor increased development in the private market. Property development in substantial sections of London is based on the interrelationship between fee owners and their lessees, both of whom were financially dependent on the quality of commonly held open space. In America, most property owners avoid responsibility for property they do not control.

Haussmann's approach to project development also was inextricably linked to the development of surrounding land. The Prefecture of the Seine, however, did not maintain continuing ownership of the property adjacent to its parks. It prepared district development plans, acquired the necessary land, and, once the project was under way, it sold the then more valuable edges to developers. In the case of the Bois de Boulogne, Haussmann spent 14.3 million francs developing the park and sold excess land for 10 million francs. The investment in the Bois paid off handsomely because land that had previously been worth considerably less could command a new, higher price based on the existence of the park. To put it another way, property values increased as a result of proximity to a park, and property owners ultimately paid for that benefit.

The district development procedures that Haussmann employed were no more applicable in the United States than those of London's great estates. In nineteenth-century America, government acquisition of land for the purpose of selling it at a profit might have been quite effective. But state constitutions and the prevailing view of the role of government made it impossible to condemn privately owned property for resale at a profit. The biggest difference, though, was that no American city administration had the power of the great estates of London or of Napoleon III's prefectures. In the United States, park agencies were accountable to local legislatures, mayoral administrations, and voters who intentionally restricted their powers. Political considerations forced these agencies to make expenditures whose short-term benefits had to be immediately obvious to the electorate. In Europe, from the very beginning, the institutions that were responsible for park development conceived of parks as long-term investments that would produce continually increasing benefits for future generations.

One device, however, was transferable to the United States: a development plan with plenty of open space set aside for use by area residents. Such plans could reduce uncertainty about the future of the area, reassure potential investors, simplify the subdivision process, and allow lots to be put on the market in response to changes in market conditions.

Rather than develop such plans, however, most American cities simply mimicked European precedents. Most city plans included an initial public open space from which their designers expected the city to expand. New Orleans's Place d'Armes (today Jackson Square), the Boston Common, and the New Haven Green are some of the more famous examples. Until the latter part of the nineteenth century, however, little attention was paid to the role that public open space could play in shaping overall city development. The lovely public parks of New Orleans, Boston, and New Haven came centuries later, when they did indeed affect urbanization.[33]

Philadelphia and Savannah were different. They provided public open space in a more systematic fashion. The 1683 plan for Philadelphia, by William Penn and Thomas Holme, provided an open-space pattern (four 8-acre squares and a 10-acre central square) that its designers thought would determine the future character of the city. These five squares did shape development patterns within their immediate surroundings. However, the meager 42 acres they provided were insufficient to have any serious impact on the city as a whole.[34]

Only one major American city, Savannah, began with a plan that had the critical mass of open space necessary to affect the character of city life.[35] James Oglethorpe's 1733 design is based on a system of wards, each consisting of eight blocks and each of which is centered on a public square. Unfortunately, after the first 26 wards were completed, the design was discarded.[36]

The results were similar in cities where private land development included common open space. In New York City, a developer created Hudson Square between Hudson and Varick Streets in lower Manhattan. As the locus of development moved uptown, however, Hudson Square was replaced with a railroad freight depot. Fortunately, the city's other developer-created square, Gramercy Park, survived.

In Boston, privately developed squares fared no better. Charles Bullfinch's Tontine Crescent, designed around a crescent-shaped open space, was demolished in 1858 to make way for some stone warehouses. Pemberton Square, another privately developed open space, was eliminated during the 1960s to make way for a new office building in an urban renewal area. The destruction and disfigurement of the districts around these squares was possible because there was no entity responsible for any continuing relationship between the buildings and the open spaces they surrounded. Nor was any entity responsible for creating additional public open space in the immediate vicinity. Unlike the great estates of London, the initial developers retained neither ownership of the land nor any interest in what happened after they were through, and property owners were at liberty to replace buildings at will.

Using Parks to Promote Urbanization

American park advocates used the examples of Paris and London to buttress their argument that public open space could be an effective device for shaping the surrounding city. However, their principal argument had little to do with the quality of life in those cities. It was financial. They argued that as land values surrounding parks increased, owners either would develop their properties in a manner that justified that increased value or they would sell to somebody who would. Initially, real estate tax collections would increase as a result of the increase in land assessment, and later, as a result of development. In short, parks offered a productive return on the public investment that they required.

Park advocates during the nineteenth century were more sensitive to this thinking than we are today. In 1883, Horace William Shaler Cleveland, the initial designer of the Minneapolis park system, argued for its creation, saying,

> *In the ten years succeeding the commencement of work on Central Park in New York the increased valuation of taxable property in the wards immediately surrounding it was no less than $54,000,000, affording a surplus, after paying interest on all the city bonds issued for the purchase and construction of the park, of $3,000,000—a sum sufficient, if used as a sinking fund, to pay the entire principal and interest of the cost of the park in less time than was required for its construction.*[37]

His point was that at first the increased taxes could be used to pay for park acquisition and development. Later they could pay for other municipal activities. For this reason, park advocates argued that not only did parks not cost a penny, they generated additional taxes with which the city could pay for more services.

Others argued that investing in parks would relieve congestion. As newcomers in increasing numbers squeezed into already crowded areas, established city residents sought more comfortable surroundings. Property owners in outlying undeveloped areas seeking to profit from their land investments favored park development as a way of attracting this market. Businesses trying to find downtown locations for commerce, warehousing, and manufacturing favored creation of new parks just beyond developed portions of the city as a way of releasing built-up land for their reuse. Political machines saw park development as a source of jobs for unemployed voters and patronage for party members.

Cities everywhere wanted to outperform New York City. Park advocates reasoned that if their parks were more impressive than the new Central Park, their city would surpass New York in prestige and become America's premier city.

Brooklyn, 2005. View entering Prospect Park through Endale Arch. *(Alexander Garvin)*

Together, these nineteenth-century park enthusiasts mobilized a powerful coalition of social reformers, businesspeople, political bosses, and ordinary laborers who succeeded in creating thousands of acres of public open space. They understood what we have forgotten: that parks are more than patches of green or accumulations of recreation equipment and that they must be created with an eye to their impact on the city as a whole. The parks that this coalition supported were created during a time of phenomenal population growth and coincided with the establishment of mass transit systems that made accessible vast new territories. When a city chose to create a large park in the midst of all this activity, nearby property owners no longer faced the possibility of incompatible construction on what was now parkland. Consequently, the value of that property increased. The intense development that followed the establishment of Prospect Park in Brooklyn, Forest Park in St. Louis, and other, similar large parks was the natural market reaction to this increase in property values. It did not accompany the establishment of Griffith Park in Los Angeles or similar parks because building conditions in the surrounding area were not favorable and because other locations were in greater demand.

Prospect Park, Brooklyn

The independent City of Brooklyn was one of the first municipalities to use a new park to spark real estate activity in undeveloped territory. In response to its request for authority to create a major new park, the New York state legislature in 1859 authorized creation of a park commission similar to the one it had authorized for Central Park in neighboring New York City. The commissioners for the new Brooklyn park hired Olmsted and Vaux, who in 1866 completed the design of a 526-acre park designed to satisfy the needs of a city that in 1870 would have a population of 420,000, making it the third largest city in the United States, exceeded only by New York and Philadelphia.

Olmsted and Vaux tried to improve upon their earlier design across the East River by creating an even lovelier landscape and further reducing the possibility of intrusions

Brooklyn, 2008. Children playing on the longmeadow in Prospect Park. *(Alexander Garvin)*

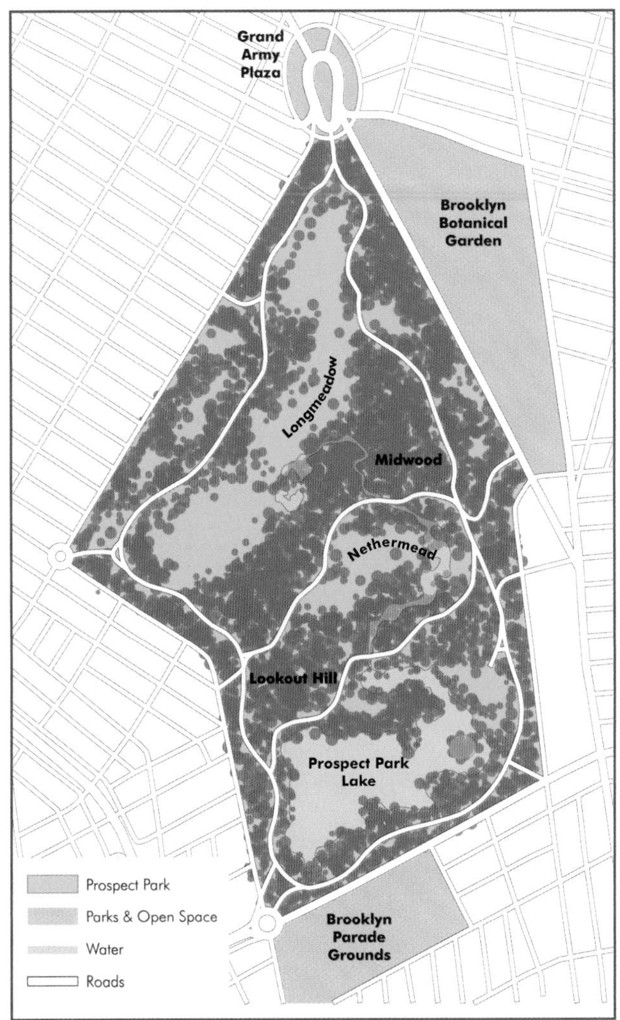

Brooklyn, 2006. Plan of Prospect Park. *(Alexander Garvin and Joshua Price)*

from the surrounding city. Their proposal included sweeping meadows, wooded hills, and an artificial lake. As in Central Park, the circumferential vehicular carriageway, bridle path, and pedestrian path are separate from one another. The landscape is artificially built up even more dramatically along the periphery to eliminate views of the surrounding city. Originally, conifers planted along the periphery of the park excluded any glimpse of the surrounding city, even during the winter, when the leaves had dropped. Eventually, the conifers died and began to be replaced only in the late 1980s.[38]

Approaching Prospect Park from Grand Army Plaza, one walks toward a landscaped berm, passes through an underpass that tunnels beneath it, and, like Dorothy opening the door of her black-and-white house to wander into Technicolor Oz, enters into a gorgeous landscape devoid of any trace of the city, its smells, its noise, or its vehicles. Visitors who take this route experience a magical "feeling of relief" from "the cramped, confined" city and gain what Olmsted called "a sense of enlarged freedom."[39]

Paris, 1998. Avenue Foch. *(Alexander Garvin)*

The only irreparable damage to the design is the result of the park's attractiveness. By the 1920s, its surroundings had become so popular that developers tore down some of the houses facing the park and built the handful of apartment towers that are now visible from the Long Meadow. They are the only intrusions that shatter Olmsted and Vaux's carefully designed separation of city and country.

Olmsted and Vaux also proposed three parkways, inspired by Haussmann's Avenue Foch: one leading to Prospect Park from Fort Hamilton at the Narrows (Fort Hamilton Parkway, which was not built as envisioned by Olmsted and Vaux), the second from the Atlantic Ocean at Coney Island (Ocean Parkway), and the third from Queens County (East-

Brooklyn, 2005. Ocean Parkway. *(Alexander Garvin)*

ern Parkway). Like Haussmann's boulevard, these parkways were initially conceived with a central roadway for through traffic, two flanking, landscaped, linear islands, and two service roads with sidewalks providing access to the buildings fronting on the boulevard, but their scale was much narrower and far less grandiose. In Paris, the flanking landscaped islands were wide enough to be real parks; in Brooklyn, they were paved walkways lined with benches and rows of trees. Nevertheless, Ocean and Eastern Parkways were sufficiently broad and alluring to attract quality buildings. They also provided an excellent setting for convivial chatter while promenading, sitting on benches, and walking the dog.

Olmsted predicted that Prospect Park and the parkways leading to it would spur development because

> *. . . advance in value will be found to be largely dependent on the advantages of having near a residence, a place where . . . driving, riding, and walking can be conveniently pursued in association with pleasant people, and without the liability of encountering the unpleasant sights and sounds . . . in the common streets.*[40]

He was right. Developers responded to his park and parkways by building single-family homes in Flatbush (south of Prospect Park), row houses in Park Slope (north and west of the Park), and apartment houses along Ocean and Eastern Parkways (south and east of the Park). In fact, by the 1890s Park Slope had supplanted Brooklyn Heights as the location of choice for many of the city's wealthiest residents. Market reaction was intense because these facilities had little competition. There were no other landscaped parks or boulevards of their size and convenience, or with anything like their facilities, anywhere in Brooklyn.[41]

Given that the neighborhoods around Prospect Park and Ocean and Eastern parkways were among the most affluent in Brooklyn, their residents were among the first to leave the city for the suburbs during the 1950s and 1960s. Though they were quickly replaced by new residents, the stock began to deteriorate, precipitating a period of significant neighborhood decline. When that trend was reversed during the 1970s and 1980s, though, public open space once again influenced the market reaction. As might be expected, market activity and price increases began with buildings that lined Prospect Park and Ocean and Eastern Parkways.

Forest Park, St. Louis

Forest Park in St. Louis sparked the same initial market reaction that was experienced around Central and Prospect parks. Unlike them, however, it was not as carefully designed for the needs of a growing population and failed to anticipate future uses.[42]

When Forest Park officially opened in 1876, developers invested in surrounding properties, which they correctly thought would benefit as much from the presence of the park as from the extension of mass transit running to the central business district. Within 20 years, the Forest Park area included seven streetcar lines, carrying more than 2.5 million passengers annually. Developers reacted to the new park and transit lines by subdividing their properties and selling lots to wealthy citizens (see Chapter 14).

The design of Forest Park by Maximillian Kern, who served as the park's superintendent and landscape gardener, as well as chief engineers Julius Pitzman and Henry Flad, was inspired by Olmsted. It included similar curvilinear pedestrian paths, winding carriageways, artificial lakes, broad meadows, and wooded rambles. Unfortunately, they copied these features without understanding how carefully Olmsted had tailored his design to fit the specific needs of the people who would use his parks. The designers of Forest Park did not provide separate and adequate paths for pedestrian, vehicular, and horseback traffic. The result is that even today it is difficult for people to walk around the park unless they use a perimeter pathway.

From the beginning, Forest Park had to be remodeled to meet consumer demand. Large areas of the park were used for the Louisiana Purchase Exposition of 1904, well known to moviegoers from the film *Meet Me in St. Louis,* and some of the structures left behind by the fair are still in use. Numerous other intrusions that followed reflected the determination of the St. Louis Department of Parks to add "the concept of social utility" to "the element of natural beauty." Had Forest Park, like Central and Prospect parks, been designed for mass recreation, it would not have required alteration to perform its function of "raising of men and women rather than grass and trees."[43]

Unlike Brooklyn, the population of St. Louis began to decline during the 1930s, dropping more than 50 percent over the next 60 years. Meanwhile, the park succumbed to the neglect that inevitably accompanied the reduction in spending for park maintenance and operations. Despite the resulting decline in market demand, the neighborhoods surrounding Forest Park have remained the location of the most sumptuous and elegant mansions in the city. Without Forest Park, the area would have succumbed to the same forces of decay that affected so many other parts of St. Louis. Perhaps that is why an $86 million public-private campaign for the park's restoration at the turn of the twenty-first century has been so successful. The Flora Conservancy of Forest Park, established in 1999 as a not-for-profit organization to work in partnership with the St. Louis Department of Parks, Recreation and Forestry, has taken a leading role in renovating Forest Park and continues to work to enhance and maintain the park.[44]

Despite the resulting decline in market demand, the area

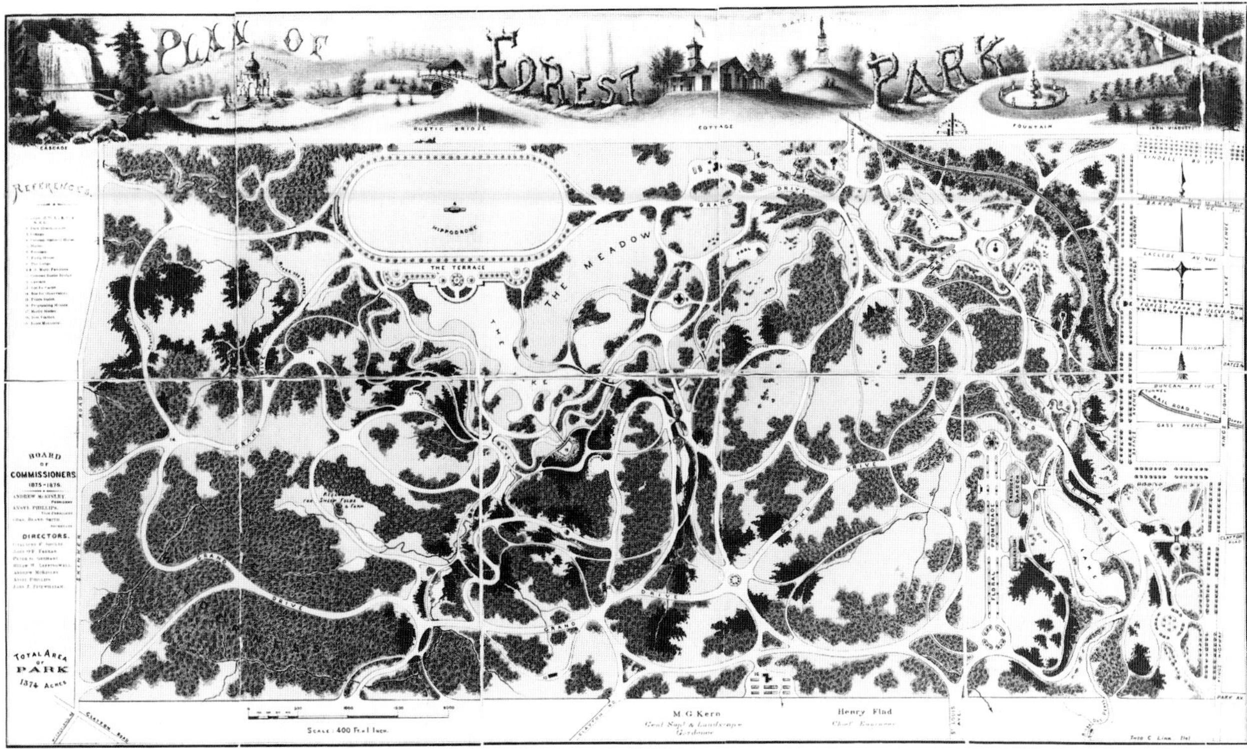

St. Louis, 1875. Plan of Forest Park. *(Courtesy of Missouri Historical Society, St. Louis)*

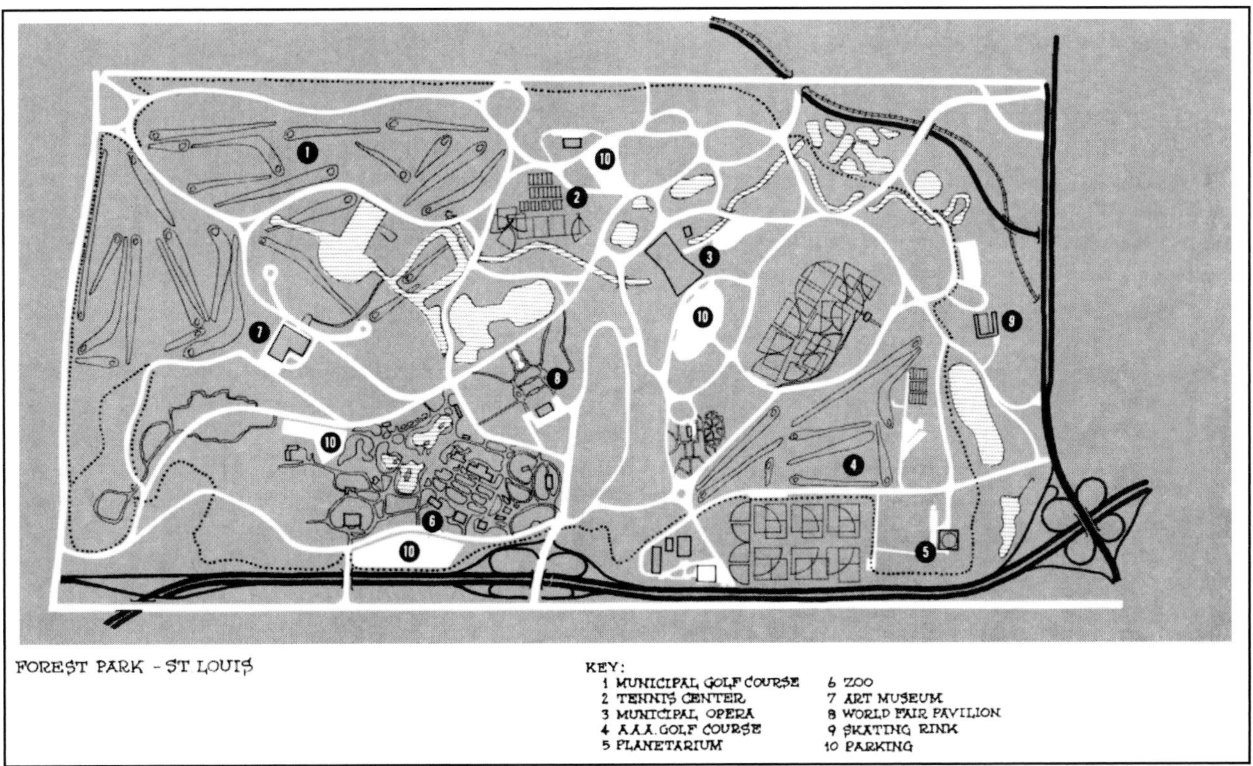

St. Louis, 1976. Map illustrating the changes to Forest Park intended to add the missing element of "social utility" to the original plan. *(From A. Heckscher,* Open Spaces, *1977; reprinted with permission from the Twentieth Century Fund, New York)*

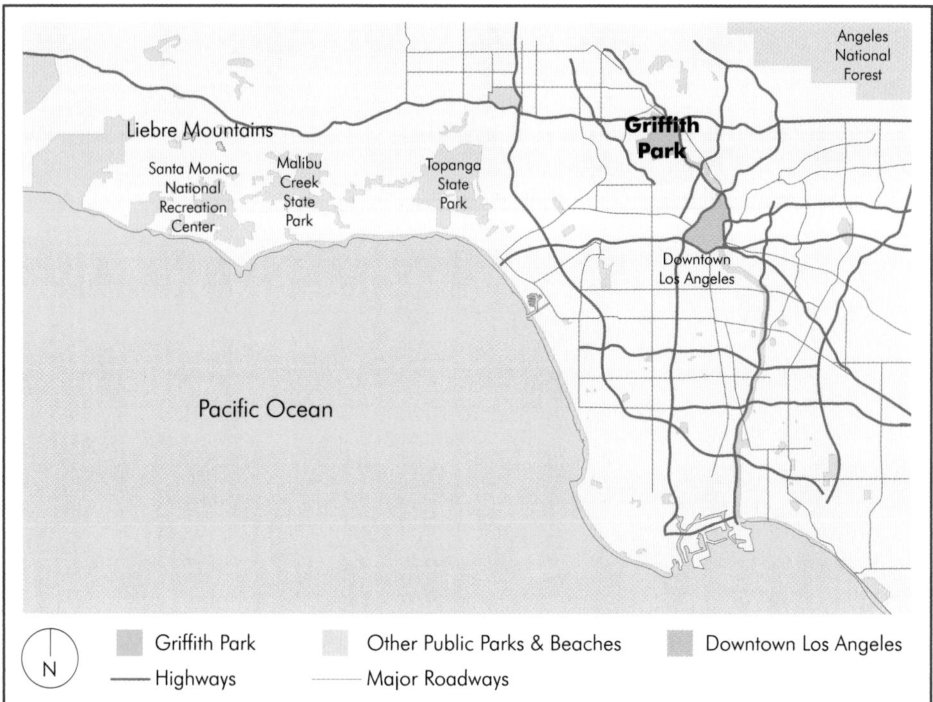

Los Angeles. 2013. Except for its ocean beaches, most of the city's parkland is located on steep slopes far from concentrations of population. *(Alexander Garvin and Ryan Salvatore)*

around Forest Park has not decayed to the extent of so many other parts of the City of St. Louis. The presence of Forest Park has stabilized the market values of the sumptuous homes nearby.

Griffith Park, Los Angeles

When Colonel Griffith J. Griffith donated a 3000-acre portion of Rancho Los Feliz to the City of Los Angeles in 1896, it instantly became one of the largest municipal parks in the country. Like Prospect and Forest parks, it was located in undeveloped territory a few miles from the center of a thriving metropolis, which had already reached 102,000. Like these parks, it was the only major facility in the city. But unlike them, the new Griffith Park failed to stimulate any significant market reaction because so few people lived nearby. Residential development in Los Angeles did explode after the creation of Griffith Park, but largely westward toward the Pacific Ocean, away from the park and not in any way related to the existence of the park.[45]

Griffith Park is in a mountainous area northwest of downtown Los Angeles. The steep topography on three sides limits the amount of construction that is possible. Furthermore, beginning in 1908, the City of Los Angeles enacted a series of land use regulations that restricted construction on the surrounding hillsides to one-family houses. Consequently, developers chose to concentrate their activity on the flat plain below the park and no adequate density of park-goers reside in close proximity to the park boundaries.

The greatest difference between Griffith Park and other large city parks is in design. Olmsted's parks and those of his followers are essentially artificially manufactured landscapes. From the beginning, however, most of Griffith Park was left in its natural state. Over the years, the city added acreage to the park as well as riding trails, vehicular roadways, an observatory, an open-air theater, five golf courses, a small picnic and playing area, two freeways, and considerable additional land. Nevertheless, Griffith Park remains a 4063-acre section of wilderness accessible by automobile and which is more suitable for hiking than intense utilization by masses of city dwellers.

The Chicago Lakeshore

In 1866, a group of civic reformers and businesspeople who owned property in the undeveloped Hyde Park Township, 6 miles south of downtown Chicago, began discussing the creation of a park system. Although the Illinois legislature approved their demand for a major park south of the city, it was rejected in a referendum. Finally, in 1869, separate parks commissions for North, West, and South Chicago were approved.[46]

The western facilities, Humboldt, Garfield, and Douglas parks, were established on open land. Designed by architect William Le Baron Jenney, they combined features of Haussmann's district parks with those Olmsted and Vaux had included in Central and Prospect parks (e.g., boathouses,

Chicago, 2005. Lincoln Park.
(Alexander Garvin)

formal tree-lined pedestrian malls, and circumferential carriageways).

The three most important parks, Lincoln Park in the North and Jackson and Washington parks in the south, anchored what was to become one of America's most extraordinary waterfront parks: 23 miles of grass, trees, lagoons, beaches, marinas, playgrounds, and landscaped roadway. When these parks were conceived, Lincoln Park was at the end of a horsecar line on the northern boundary of Chicago. The southern parks were not even in Chicago; they were in an undeveloped suburb that was developing around stations of the Illinois Central Railroad and which had a population of just over 1000.

The parks with the most natural advantages, those along the shore of Lake Michigan, should have attracted the bulk of the city's expansion. By the middle of the nineteenth century, large portions of the lakeshore were already occupied by the railroad or were being used as garbage dumps. Consequently, development activity was more intensely concentrated inland. Only by the middle of the twentieth century, when undesirable land uses had been removed, additional parkland had filled in many gaps, and Lake Shore Drive had united this parkland into a single, continuous facility, did an overwhelming amount of real estate activity finally shift toward the lake.

The South Park Commission acquired about 1000 acres consisting of a sandy marsh fronting on Lake Michigan and a desolate stretch of flat prairie farther inland. It hired Olmsted and Vaux to transform this land into parks, and the design that the firm submitted consisted of four parts: Jackson Park on the lakefront, Washington Park in the interior, a broad, linear Midway Plaisance connecting them, and three landscaped boulevards, similar to what the firm had recommended in Brooklyn.

As Olmsted and Vaux explained, "The first obvious defect of the site is that of flatness," for which they compensated by creating a series of artificial lakes and lagoons, but without resorting to artificial hills, depressions, or "trivial objects of interest." Much of the land was swampy, and the water table frequently was too close to the surface for large trees. This problem was solved by draining the swamps to create the lakes and lagoons, by using the scooped-out sand and mud to build up areas that were to be planted with large trees, and by cutting a drainage channel through to Lake Michigan. Their plan called for a public beach as a barrier to wind and wave damage from Lake Michigan.[47]

The South Park Commission chose not to create the elaborate lagoon that had been proposed for Jackson Park. Nor did it extend a waterway from Washington Park down the Midway to the lake at Jackson Park and out to Lake Michigan. The design, as carried out under the supervision of Horace William Shaler Cleveland, was far less ambitious.[48]

As usual, Olmsted's predictions about the activities that would take place in the parks and about the market reaction in the areas surrounding them proved to be correct. He had designed Jackson and Washington parks as urban facilities meant for active use by large populations rather than as "a distant suburban excursion ground." By 1889, when Hyde Park was annexed by the City of Chicago, the neighborhood's population had grown to 85,000.[49]

The reason the south parks were actively used had as much to do with their character as with the sizable and growing populations that surrounded them. Olmsted and Vaux had, for the first time, specifically designed important sections of a large park (Washington Park) as "an arena for athletic sports, such as baseball, football, cricket, and running games," which they thought would come into fashion.[50]

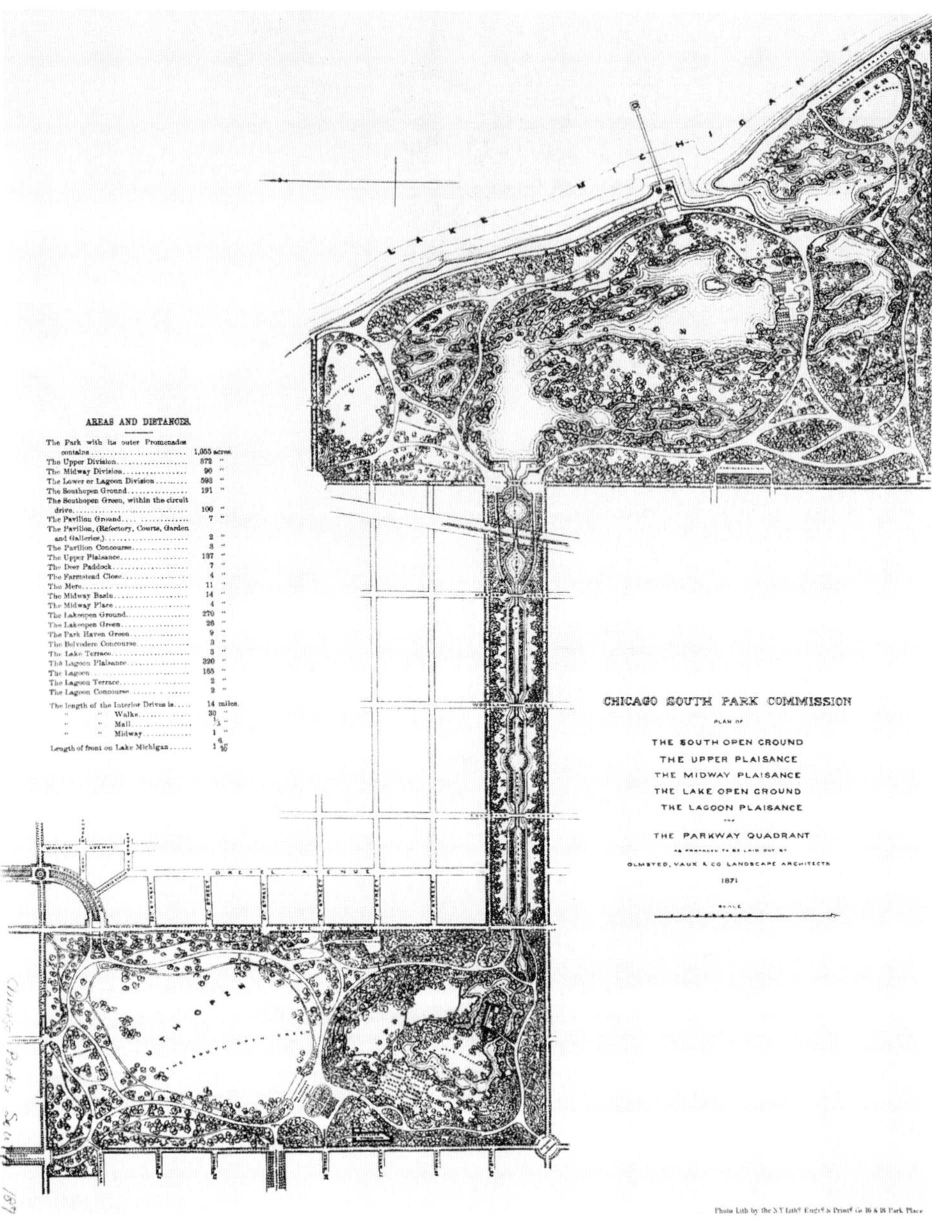

Chicago, 1871. Olmsted and Vaux plan of the South Park System including Jackson Park, Washington Park, and the Midway. *(Courtesy of Chicago Historical Society)*

It was not till Olmsted worked with Daniel Burnham on the Chicago Fair of 1893 that he returned to the design of these parks (see Chapters 5 and 19). Land had to be dredged, earth moved, and trees planted to create the elaborate lagoons that were included in the original proposal and which were now a major feature of the plan for the fair. Once the fair had closed, Jackson Park was relandscaped by the Olmsted firm in a manner that, except for the site of the Museum of Science and Industry and some sections along Stony Island Avenue, was very similar to the 1871 design.

Burnham, however, suggested far more radical action for the city's park system: connecting all of Chicago's waterfront parks to create a single facility. In 1894, he convinced a group of the city's most important businesspeople to advocate a landscaped South Shore Drive that would link Jackson Park to the Loop business district. Burnham's design envisaged the removal of some obsolete rail lines and the creation of a limited-access parkway. Between Lake Michigan and the remaining rail lines, he proposed a lagoon, a roadway, and a linear park. At the downtown end, he planned a boat basin and a major new park. The scheme was not implemented.

Thirteen years later, Burnham resurrected the drive, park, and boat basin for the *Plan of Chicago* (see Chapter 19). This time he proposed a broad South Shore Drive, a larger and more elaborate downtown park, and an additional North Shore Drive that included Lincoln Park and extended to the city's current northern boundary. In 1911, the city

Chicago, 2008. Crown Fountain, Millennium Park. *(Alexander Garvin)*

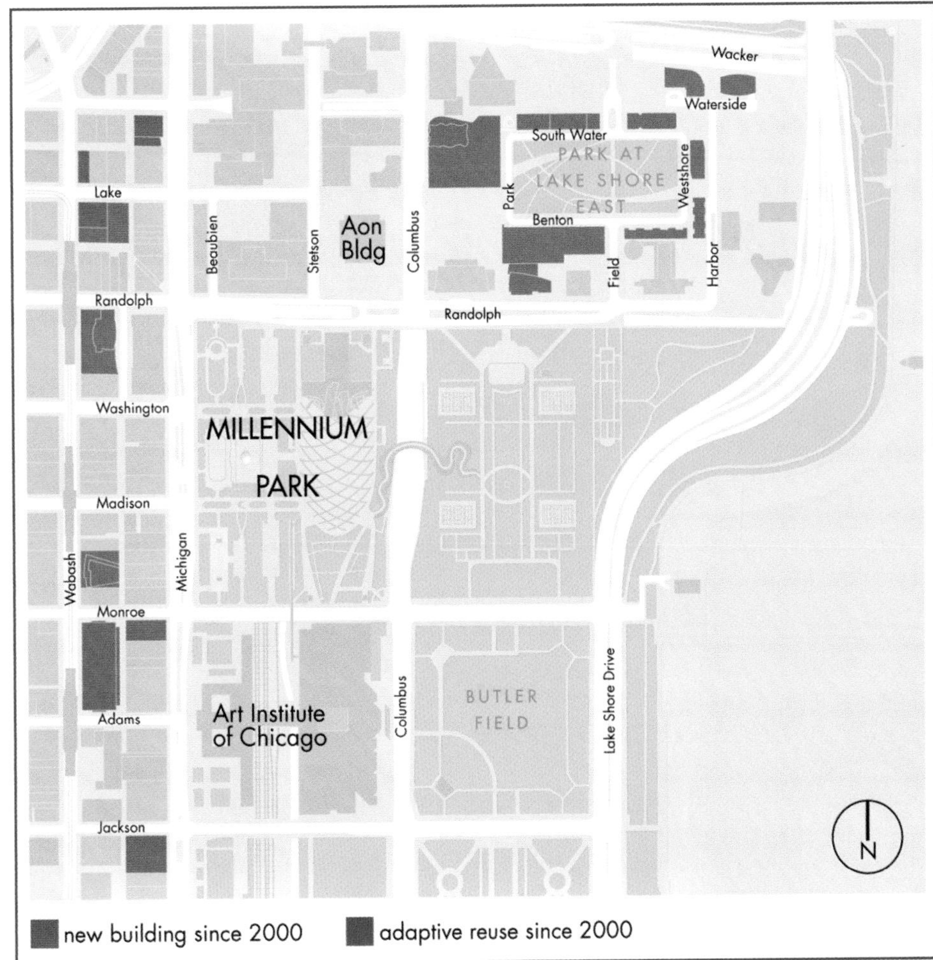

Chicago, 2012. Public investment in Millennium Park generated widespread private investment in nearby real estate. *(Chicago Cartographics)*

Chicago, 2008. Lincoln Park.
(Alexander Garvin)

council formally approved adoption of the *Plan of Chicago.* The downtown component, Grant Park, became the landfill centerpiece of this extraordinary continually evolving linear park, and over the past century the city has spent billions of dollars on landfill, landscaping, roads, beaches, and recreation facilities along the shore of Lake Michigan. Its most recent and spectacular expenditures have been the nearly $500 million in private contributions and city funds spent to create the 24.5-acre Millennium Park. The park, which opened in 2004, includes a dramatic concert pavilion designed by architect Frank Gehry, *Cloud Gate,* a popular sculpture by Anish Kapoor, and the Crown Fountain, by Jaume Penza. During the years since completion, the area around Millennium Park has been radically transformed with the construction of 3600 new rental and condominium apartments, the conversion of 1200 others, and the introduction of 1100 new hotel rooms. This private market reaction is estimated to have cost $2.45 billion and generated 70,000 jobs. Five million tourists (12 percent of the city's annual visitors) come to the park every year, generating $1.4 billion of direct spending and $78 million in tax revenue.[51]

The southern portion of Burnham's plan, especially around Jackson Park, attracted a few developers, who built apartment towers that capitalized on the parks and the lovely views of the lake. Along Lake Shore Drive to the north, however, the impact was spectacular. There, developers created a gold coast of expensive apartment buildings, rivaling New York's Fifth Avenue. No park improvement anywhere has had greater impact.

The neighborhoods around Chicago's inland parks prospered until the middle of the twentieth century, at which time they succumbed to the awesome social and economic forces that swept through much of the city, leaving in their wake a desolate scene of dilapidated buildings, vacant lots, and widespread abandonment. At the same time, market activity shifted back toward Lake Michigan, where, during the half century following publication of the *Plan of Chicago*, the city had spent so much money moving garbage dumps and railroad-related uses in order to create the linear park the plan had recommended.

New York's Hudson River Park and High Line

Real estate development along Chicago's lakeshore was a premeditated response to years of imaginative planning. But along the Manhattan's lower Hudson riverfront, substantial real estate activity has been an accidental response to failed planning and the community-based initiatives that often developed in response. The story begins with the Miller Highway, better known as the West Side Highway, begun by the Manhattan borough president in the 1920s. Half a century later it was falling apart. In fact, in 1973 one section collapsed, leading to its closure and creating a hardship for drivers who depended on this important commercial artery.

The city and the state of New York intended the replacement highway to accommodate regional traffic going from Brooklyn to New Jersey, as well as intracity traffic (especially trucks) then using New York's congested streets. It was included in the Interstate Highway System in 1970 as a result of a political compromise that included the administrations of Mayor Lindsay, Governor Rockefeller, and President Nixon.

What emerged was no ordinary highway project. In order to maximize the flow of federal subsidies and to obtain local approval of the highway, it was transformed into a 226-acre redevelopment scheme entitled: Westway. The crumbling, obsolete, elevated West Side Highway and the 27 rotting piers along the Hudson River were to be demolished and replaced with 178 acres of landfill. When Westway was formally unveiled in 1974, only a small portion was devoted to a new six-lane roadway and it was relegated to a tunnel running under the landfill. Of the land to be created along the Hudson, 82 acres were to be for parks, 53 acres for manufacturing, and 35 acres for new housing.[52]

The project received federal approval in 1976. At that time, its estimated cost was $1.2 billion and not one penny was to come from the city of New York. Instead, the project actually generated cash for the city. Ninety percent of the project's cost (acquisition, demolition of the abandoned piers, landfill, parks, new streets, etc.) was to come from the federal Highway Trust Fund. The rest would be paid for by New York State. Westway would assume responsibility for the city's Hudson River piers and pay for demolishing them. It would also pay for building a new municipal incinerator and bus garage. Better yet, the city would receive cash for the property that it had to contribute to the project. In 1981, President Reagan even arrived with a giant-size reproduction of the $85 million check the federal government intended to issue to pay for land acquisition.[53]

Almost from the beginning, city officials had to decide whether to exchange the billions of dollars that the High-

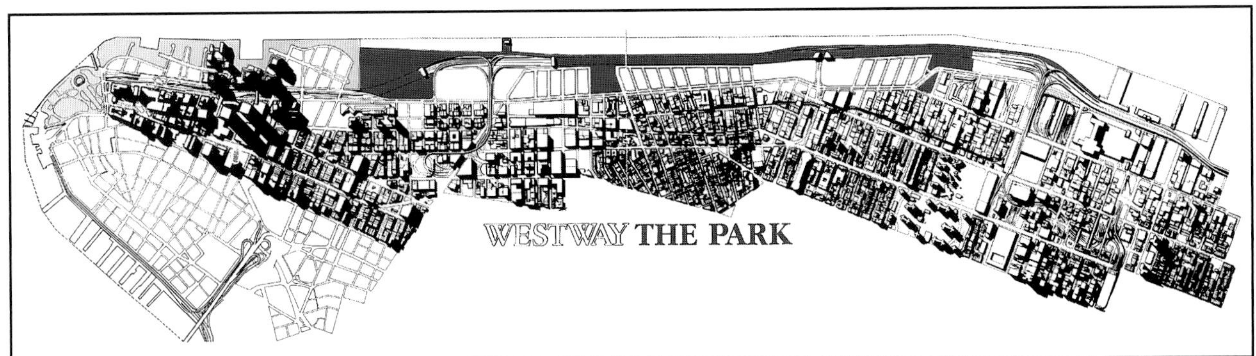

Manhattan, 1985. Plan of the defeated proposal for Westway—an underground highway and landfill project extending the West Side of Manhattan. *(Courtesy of New York City Department of Parks and Recreation)*

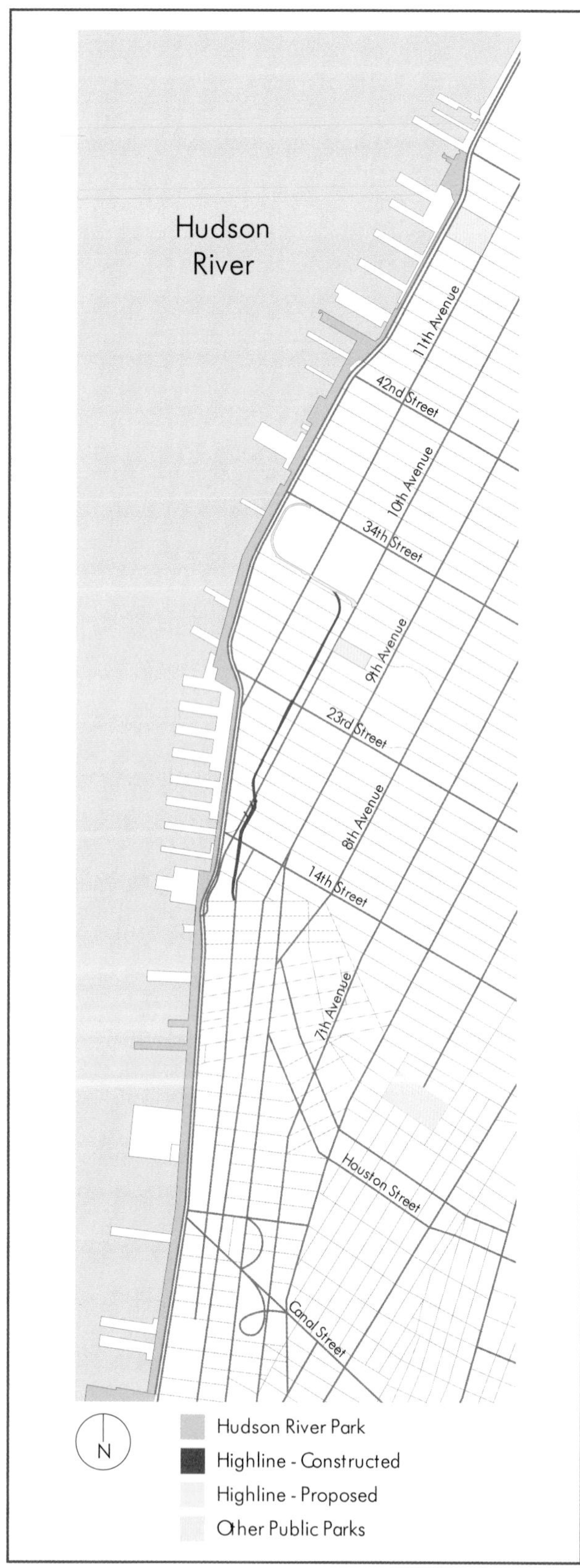

Manhattan, 2012. Map of Hudson River Park and High Line. *(Alexander Garvin and Ryan Salvatore)*

way Trust Fund would contribute to Westway for substitute transportation projects. Congress had authorized such trade-ins starting in 1973. Over the next decade, $6.3 billion had been paid out for trade-ins in 20 states, and, given its scale, Westway was a prime target for this type of exchange. Complicating its funding, Congress in 1982 eliminated the inflation escalator that previously had been applied to the estimated cost of projects from the point of trade-in. It also established a formula for the annual distribution of trade-in money. Under the new framework there was no way that New York City could get anything like the annual expenditures that the development of Westway would require. Furthermore, nobody was sure that Congress, which had to vote on appropriations for a trade-in but not for Westway, would agree to spend in excess of $2.4 billion (the amount to which the project had ballooned by 1981) in New York City, especially because the proposed roadway was so small a part of the project.

Community organizations representing residents and interest groups located along the proposed route, as well as civic groups (many concerned with environmental protection) opposed Westway. The most important weapon in their arsenal was litigation claiming failure to comply with city, state, and federal statutes. In 1982, a federal judge invalidated the environmental impact analysis that was a requirement before interstate highway funds could be released (see Chapter 17). Although the state and city dutifully commissioned the additional studies that were required, they were concerned that time would run out before they could exchange Westway funding for other transportation projects.

When faced with a congressionally imposed deadline for exercising the trade-in option, Mayor Koch and Governor Cuomo decided not to gamble on further environmental reviews. Their concerns were well founded because New York City, unlike Boston, was not represented in Congress by the Speaker of the House, who could fight for the huge expenditures needed to pay for its big-dollar highway project, the Big Dig, which ultimately cost more than $24 billion. So, in 1985, they traded Westway for $1.7 billion ($690 million of which was earmarked for a replacement highway, $690 for various transit projects, and $325 million for direct assistance to the Metropolitan Transit Authority).[54]

For the next two decades, as a result of often intense and contentious dialogues among several city and state administrations, interest groups, and community representatives, an overwhelmingly supported consensus emerged. Consequently, the New York state legislature created the Hudson River Park Trust in 1998 to design, develop, and administer a 5-mile-long, 550-acre park parallel to route 9W, an at-grade boulevard that replaced the collapsed West Side Highway. Its 13-member board (5 members appointed by the governor, 5 by the mayor, and 3 by the Manhattan borough president) works alongside The Friends of Hudson River Park, a non-

Manhattan, 2009. Hudson River Park. *(Alexander Garvin)*

profit citizen organization which represents the millions of people who use the park during the year.

The park offers the widest range of recreational opportunities including virtually every sport except football. Soccer, baseball, basketball, golf, swimming, and ice hockey are available at the 28-acre Chelsea Piers Sports & Entertainment Complex. In addition to specific sports, the park attracts tens of thousands of residents from all over the city to the 5-mile trail set aside exclusively for jogging, cycling, and Rollerblading. The other particularly popular activity is sunbathing on the many open piers and stretches of grass.

As of 2012, more than $350 million had been spent on the park's development, financed largely from the state and city capital funds. Operations, however, are funded from rents from commercial tenants, parking fees, concession revenues, grants, and donations. In 2012, the 34 privately leased piers adjacent to the park generated $7.7 million in lease payments and $6.8 million in parking fees. Chelsea Piers, for example, paid $3.75 million in rent during 2008.

Community-based activism similar to that which produced the Hudson River Park continued into the twenty-first century and helped to spur the High Line, a 1.5-mile-long elevated and landscaped promenade half a block east of the Hudson River Park. In addition, the High Line Park, like the Hudson River Park, replaced an obsolete transportation corridor—in this case, an abandoned elevated railroad rather than a collapsed elevated highway. Unlike the Hudson River Park, however, the city government had no specific plans for reusing the High Line; the plans originated within the community. The park began as an idea proposed by two local activists: Robert Hammond and Joshua David. The city initially opposed the community-sponsored idea for a High Line Park rather than the community opposing city plans, as had been the case with the Westway. In each case, however, the community triumphed in its conflict with the city. The city paid for the bulk of the cost of developing the park. Equally important, the emerging private market investment in adjacent property and the real estate taxes those new developments pay to the city more than covers the cost of their development.[55]

The original High Line, which opened in 1934, was built to replace an at-grade freight railroad that interrupted vehicular traffic whenever trains crossed city streets. Thus, elevating the tracks not only improved the efficiency and cost of operating manufacturing and warehousing in Manhattan, it also eliminated traffic accidents. By 1960, the interstate highway system was carrying so high a proportion of freight traffic that the southernmost section of the High Line was taken out of service. Twenty years later, trains stopped using this elevated railroad altogether. From that moment on, the structure began to deteriorate, and by the end of the century it had become a rusting hulk overrun with weeds. Neighboring property owners complained that it was a dangerous blight on surrounding neighborhoods and began legal action to have it demolished. In 1999, Hammond and David, who were not acquainted with one another, decided to save the High Line. They had met at a community meeting, joined forces, and founded the nonprofit organization Friends of the High Line to fight for its transformation into an elevated linear park. Three years later, after hundreds of meetings, they obtained City Council approval for their idea. In 2005, the city

Manhattan, 2012. The High Line. *(Alexander Garvin)*

took possession of the railroad. Four years later the first section of the park opened. By 2012, the park extended 1.5 miles from Gansevoort to West 30th Street and the city had gained control of the entire elevated rail extending to 34th Street.

The High Line varies in width from as little as 40 feet to as much as 100 feet. Most sections are 50 feet wide, sometimes with as little as 8 feet of sidewalk between planting beds. The skilled design team, led by James Corner, founder and director of the landscape firm Field Operations (the landscape architects for the project) and including Diller Scofidio + Renfro (architects) and Piet Oudolf (horticulture), has made this fairly narrow promenade augmented with occasional benches look much broader. It appears far more open because every 200 feet below the promenade is a 60-foot perpendicular opening for another east–west street, each providing expansive views of the Hudson River and the surrounding neighborhoods. At 17th Street, where the High Line cuts diagonally across Tenth Avenue, making use of the 100-foot width of the avenue, the designers created an amphitheater that provides seating with a view of the traffic below and the skyline beyond.

Although the High Line and the Hudson River Park are only half a block from one another, their primary users could not be more different. The thousands of mostly middle-income adults (including many tourists) that flock to the High Line every day come for a stroll and a chance to see the surrounding city from the air. The even greater number of people using the Hudson River Park, come for every form of active reaction, not just a stroll with spectacular views of the Hudson River and New Jersey. For the cyclists, joggers, and Rollerbladers, the trip from their neighborhood is as much a part of the park experience as the time they spend along the river.

Both parks have generated widespread investment in the area. Even before the High Line went into construction, property values adjacent to the Hudson River Park had begun to rise. Between 2003 and 2005 sales prices increased 80 percent for all of existing condominium apartments that were sold (285 apartments) along the first completed section of the Hudson River Park that is south of Gansevoort Street, where the High Line begins. During that same period the sales price of 657 condos along the uncompleted portion of the park increased in value only 45 percent.[56] While there are as yet no studies of the impact of the High Line, enough new buildings have been erected in the area since 2009 to justify the expenditures on the park.

Using Park Systems to Change Entire Cities

Like any part of a city, a park is subject to powerful local conditions and major national economic and social trends. Some parks also play an important role in determining how these forces will affect the surrounding city. Conceived and designed as part of a single unified system, they can direct market activity toward certain areas and away from others, shape the character of market activity in those areas, retard or stimulate shifts in population, and even alter the pattern of daily life.

Frederick Law Olmsted was the first American to think about parks in such urban-planning terms. The facilities he designed for Brooklyn, Chicago, and Buffalo were conceived as parts of broader schemes for municipal improvement. Despite creation of a series of remarkable parks, he was nevertheless unable to persuade civic leaders in these cities to let him implement all his ideas. Finally, in the last quarter of the nineteenth century, the City of Boston hired him to create a city park system that was conceived as a unified whole, and the city executed it.

Like the prototypes that Olmsted developed for the large urban park, the playground, and the landscaped boulevard, the park system he devised for Boston became the prototype for similar systems in Kansas City, Cleveland, Cincinnati, Seattle, and many other American cities. These park systems were spearheaded by crusading individuals who rarely understood the tremendous scope of what Olmsted actually had in mind or the importance of hiring designers of similar genius. The most serious difficulty, however, was that their passionate commitment to acquiring parkland usually blinded them to the need for continuing public support, steadily increasing streams of income, and careful nurturing by a dedicated public agency staffed with talented professionals. Only in Minneapolis did all the factors come together to produce a park system that demonstrated, as Olmsted's work in Boston would not, the powerful *continuing* role that a park system could play in shaping the evolution of the surrounding city.[57]

Boston's Emerald Necklace

In 1875, the Massachusetts legislature and the voters of Boston approved the creation of a Board of Park Commissioners, similar to the ones that were responsible for parks in other cities. Luckily for Boston, the board's desire for new public parks coincided with Olmsted's move from New York to Boston. In 1872, he ended his partnership with Calvert Vaux. Six years later, the commissioners of Central Park dismissed him from his role as superintendent/landscape architect, a position he had occupied in one form or another since the mid-1860s.[58]

In part because of the increasing number of commissions in the Boston area and in part because of a growing friendship and professional relationship with architect Henry Hobson Richardson, Olmsted spent an increasing amount of time there. His transition to Boston began during 1878, when he spent the first of four summers in the area. In 1881, he established an improvised office in Brookline. Two years later he

Boston 1866. Back Bay from Parker Hill in Roxbury prior to its transformation into a public park. *(Courtesy of the Boston Public Library, Print Department)*

acquired the farmhouse that would become his home and the headquarters of his firm.[59]

Boston's program of park development began as soon as the voters had endorsed it. Mayor Samuel Cobb appointed three prominent businessmen to constitute a board of commissioners of the Department of Parks. Its first project was a new park for the Fens, one of Boston's natural salt marshes. The Fens accommodated the outflow of the Muddy River, Stony Brook, on its way to the Charles River, and the saltwater backflow from the Charles during high tides. Because it also was used for storm drainage and all manner of waste, the area was rapidly becoming a breeding ground of epidemic diseases.

In 1878, the board held a design competition for the new park and asked Olmsted to evaluate the 23 submissions. His appraisal was that some of the designs failed to consider the flooding and others thought of the proposed park only as a flood-control problem. The board was so disenchanted with the submissions and impressed with his analyses that it engaged Olmsted to become its professional adviser and landscape architect.[60]

By the time his work was finished in 1895, Olmsted had created a 2000-acre Emerald Necklace, extending for 6 miles from the Common and Public Garden through a variety of

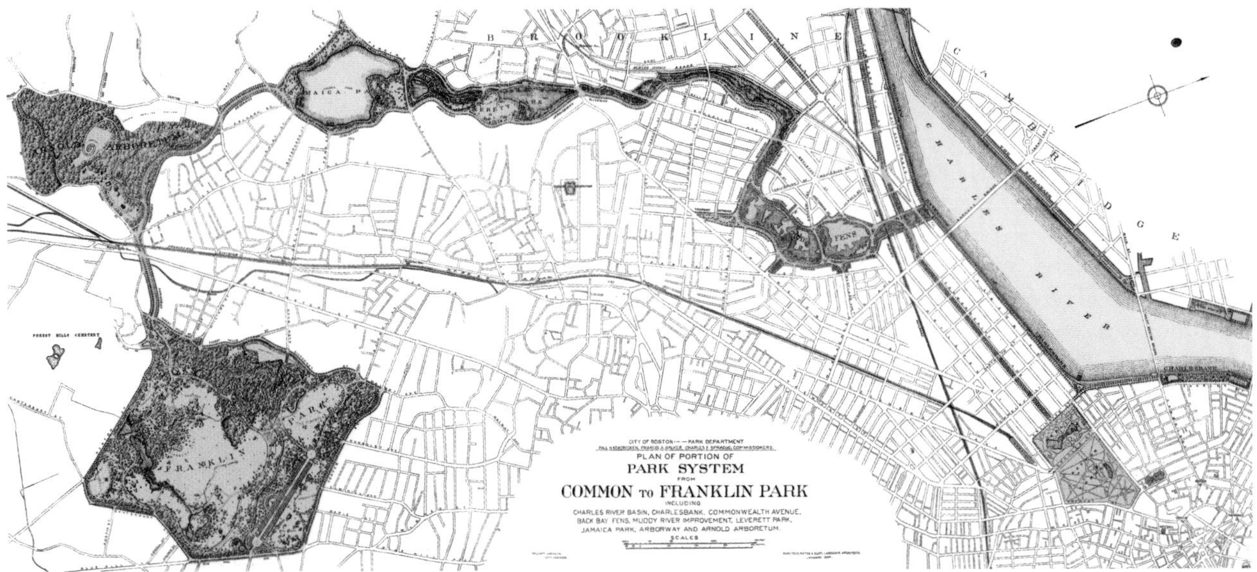

Boston, 1894. Plan of the Emerald Necklace. *(Frederick Law Olmsted National Historic Site, Courtesy of the National Park Service)*

Boston, 1989. Olmsted's design transformed the Back Bay–Fens from a brackish, sewage-filled breeding ground of pestilent diseases into an ingenious natural drainage scheme that used this winding waterway as a storage basin for storm water and drained its marshy banks as an integral part of a 6-mile-long park system. *(Alexander Garvin)*

Boston, 2005. Muddy River became the centerpiece of an "attractive suburban residence district," which generated an increase in "market and taxable values securing the city a rapid return for its outlay." *(Alexander Garvin)*

neighborhoods to the outer limits of Roxbury and South Boston. Commonwealth Avenue and the Charles River Embankment, the first links in the Emerald Necklace, were not Olmsted projects. Both had long been accepted parts of the park agenda dating back to the infill of the Back Bay. He included them because he wanted to be sure that they linked the new park for the Fens with Boston Common and Public Garden. In addition, he proposed linking the inland end of the Fens with five new public parks (Muddy River, Leverett Park, Jamaica Pond, Arnold Arboretum, and Franklin Park), which themselves were linked by new parkways. The resulting scheme was intended to eliminate pestilent swamps and other nuisances that retarded development, provide a convenient means of communication among adjoining districts, stimulate construction in surrounding communities, and integrate each part of the system into a single comprehensive design.

The philosophy behind this design was not different from that of Central Park. But when Olmsted applied it to the design of the Emerald Necklace, he had the benefit of two decades of practical experience with public parks. The first principle was derived from the work of Capability Brown: base the design on the "capabilities" of the site. As always, Olmsted was careful to see that park activities took place where the topography was appropriate. The second principle was market-driven. A park design that met the needs of one neighborhood might not be as effective in serving the needs of another. Since the park that was topographically and functionally most suited to an individual's needs might not be the one nearest to his or her residence, Olmsted contended that it was wrong to think of each park individually, "as if it were to be of little value except to the people of the districts adjoining it."[61] He felt the entire system must be easily accessible to all classes, whether they came by foot, on horseback, or in a vehicle, and must be easy to navigate in order to get to any type of facility.

These principles are clearly exhibited in the Fens. Only half the Fens could be used for active recreation. The rest was needed for Olmsted's ingenious natural drainage scheme, which used the winding waterway and marshy banks as a storage basin for storm water and introduced interceptor sewers to handle the flow of the Muddy River and Stony Brook. To accommodate high tides and tolerate sea spray, Olmsted planned for occasional flooding of up to 20 acres. He also restricted tree planting to the borders of the new roads that crossed the marsh or lined its edges.

The original raison d'être for Olmsted's design no longer exists. A dam, completed in 1910, eliminated the flow of saltwater. Later, construction fill eliminated some of the marshland. In the 1960s, the Charlesgate highway interchange and network of overpasses into Storrow Drive disfigured the northern end of the park. Nevertheless, enough of Olmsted's original concept remains for anyone to marvel at its ingenuity. Moreover, had the park not been created, it is doubtful that substantial further upland construction could have continued without major expenditures on an extensive sewer system. Even then, developers would have shunned the areas immediately surrounding the increasingly fetid swampland.

The park Olmsted created at Muddy River is an equally extraordinary design. He proposed this joint effort to the Brookline and Boston park boards in 1880. At that time

Boston, 2009. Jamaica Pond. *(Alexander Garvin)*

Brookline, whose name originates from the narrow, winding tidal *brook* that marked the boundary *line* between the two cities, had a population of about 7000. Muddy River, like the Fens, was fast becoming a brackish, sewage-filled breeding ground for mosquitoes. Olmsted's plan, revised several times prior to its completion in 1892, proposed transforming the area into an "attractive suburban residence district, agreeably connected" to a new park that would induce "advance of market and taxable values securing the city a rapid return for its outlay."[62]

To create this "attractive suburban residence district," Olmsted supervised substantial earthmoving and the planting of significant quantities of new shrubs, trees, and flowers, in the process converting a drainage disaster into a useful and attractive public park. The park itself is an artificially constructed landscape with a redesigned river channel, an independent sewer system, and a pedestrian promenade that extends along its entire length. At certain strategic points, he erected bridges that crossed over the park to connect the adjacent communities; at others, he narrowed the riverbed to accommodate broad meadows that could be used as sports fields. Along the edge of the park, which sometimes rises to a level 20 feet above the waterway, he built a parkway.

The best location from which to see Olmsted's radical transformation of the landscape is near the Longwood station of the contiguous Massachusetts Bay Transportation Authority (MBTA), originally used by Boston & Albany Railroad. Neither the station nor the tracks are visible from within the park, because a landscaped berm keeps them out of view. Overhead, a bridge connects unseen, adjacent residential neighborhoods. The pedestrian strolling through the park is aware only of the landscaped banks of the meandering Muddy River.

Jamaica Park includes 60 acres of parkland surrounding Boston's largest freshwater lake. This 70-acre lake had been a popular site for ice-skating and boating when it was acquired by the park board and remained so after Olmsted completed his work. Other than filling in some sections to provide for an attractive circumferential promenade, he did little more than add a few trees to the beautiful pines and beech trees that already grew there. But it was just enough to transform the shore into an attractive setting for strolling and jogging, a place to escape from the noise and confusion of the city.

From Jamaica Park, Olmsted extended a tree-lined "Arborway" with landscaped park islands similar to his parkways in Brooklyn and Chicago. The Arborway leads to the Arnold Arboretum, which already existed when Olmsted began his work on the Boston Park System. The arboretum was created to display the widest possible variety of trees from all over the world on the site of a 210-acre farm that was willed to Harvard University by Benjamin Bussey in 1842. However, because the donor's heirs retained a life tenancy, it did not become available until 1873. Five years earlier, James Arnold, a respected authority on trees, died and left $100,000 that his trustees agreed to donate to Harvard for the purpose of transforming the Bussey farm into Arnold Arboretum. Olmsted became involved with the project in 1874 before he moved to Boston, when the arboretum's director, Charles Sprague Sargent, wrote to him suggesting that it be donated to the city of Boston for use as a public park. In response, Olmsted emphasized his conviction that parks should be active spaces for people to recreate, writing, "A park and an arboretum seem to me to be so far unlike in purpose that I do not feel sure that I could combine them satisfactorily."[63]

Boston, 2005. Arborway. *(Alexander Garvin)*

Boston, 2010. Franklin Park provides city dwellers with broad meadows for active recreation and rugged forested areas for rambling. *(Alexander Garvin)*

In 1882, Harvard agreed to sell the property to the city of Boston, which acquired several critical additional acres and leased it all back to the University for 1000 years, in exchange for which Harvard would manage and maintain the facility. In spite of his doubts, Olmsted agreed to design a combination park-arboretum. The plan is based on roadways that wind their way up the hilly terrain. Specimen trees, grouped by family and genus, were planted along these roads so that any visitor can see the many varieties native to the northern temperate zone.

From the Arnold Arboretum, Olmsted extended another "arborway" leading to Franklin Park in West Roxbury. When the park board acquired this 500-acre site, the area contained little more than a few farms. Olmsted was delighted with the terrain, which he thought of as a perfect site for a country park that would be "within easy reach of the people of the city" and could counteract "a certain oppression of town life."[64] The money to develop the park came from a $2.5 million bond issue approved by the city council in 1886 and from a bequest from Benjamin Franklin. Franklin had willed a sum of money to his two favorite cities, Boston and Philadelphia. Upon his death, this money was to be invested for 100 years. The proceeds were to be applied "to some public work" when the investment matured in 1891–1892.[65]

Olmsted's design for Franklin Park is in many ways similar to his work for Central and Prospect parks. Each section of the park, described in writing along the margins of the General Plan of 1885, is intended for a different activity. There are broad meadows for active recreation (baseball, football, tennis, and, later, golf) contrasted with more rugged, forested areas for rambling. The circumferential road is like those in other Olmsted parks. There is a formal tree-lined promenade called the Greeting, similar to the Central Park Mall that was later converted into a zoo. The only real difference from Olmsted's other large parks is that he did not need to alter substantially the topography or install a major drainage system. Thus, with a minimum of money, he created a large country park with "breadth, distance, depth, intricacy, atmospheric perspective and mystery" at the end of a 6-mile pleasure route that extended all the way from downtown Boston.[66]

There is no doubt that Boston would have continued growing with or without its park system. The Emerald Necklace created the framework for that growth. By replacing breeding grounds for pestilent epidemics with lovely parks,

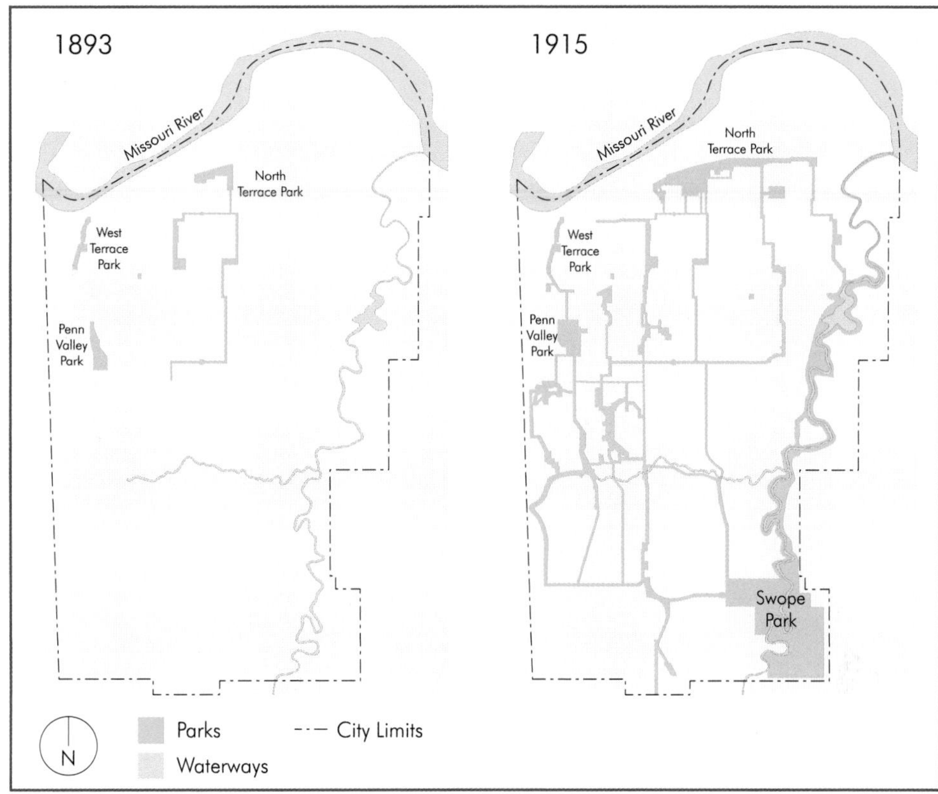

Kansas City, 1893 and 1915. George E. Kessler's proposal for a park system and plan showing newer and larger parkways and parks that were added to Kessler's original 1893 scheme. *(Alexander Garvin)*

Olmsted established surrounding territory, rather than more distant suburbs, as prime locations for initial development. Property owners were eager to build close to the new park system. Once the areas around the Emerald Necklace had been developed, however, its influence waned.

Boston's population has declined, dropping from 801,000 in 1950 to 618,000 in 2010.[67] During that time there has been little pressure for second growth outside the downtown business district. The Emerald Necklace begins just beyond this area of market activity. Thus, there was little or no second growth for it to influence. Nor could it restrain the region's continuing suburbanization.

The Emerald Necklace demonstrates more persuasively than Olmsted's work in Brooklyn, Buffalo, or Chicago, the appropriateness of developing a single comprehensive park system to meet the needs of an entire city. It has provided an element of stability during a period of significant social and economic change. Despite neglect and deterioration, the power of its design and the effectiveness of the engineering and landscaping have ensured that it will always be a heavily used park system. Its greatest importance, though, lies in its influence on the design of park systems for other cities. Too often, these park systems are parodies of Olmsted—connected swatches of green that appear significant on a map but, because of their location, topography, and design, cannot function as a park system for a large city with a heterogeneous population.

Kansas City

The initial section of the Kansas City park system demonstrates how easy it was to embrace Olmsted's rhetoric and imagery without understanding his philosophy or design practices. It was designed by George Kessler, a landscape architect who began his professional career as superintendent of parks for a small railroad outside Kansas City, then in 1890 as parks "engineer" for the newly established Kansas City Board of Park Commissioners.[68]

The park system Kessler proposed in a report published in 1893 consisted of landscaped boulevards connecting three new public parks: West Terrace Park, North Terrace Park, and Penn Valley Park. Civic leaders assumed that these proposed parks and parkways, like Boston's Emerald Necklace, were organized into a comprehensive system with a variety of

Kansas City, 1994. Linwood Boulevard nearly a century after it became a "parkway." *(Alexander Garvin)*

Kansas City, 1997. Ward Parkway, unlike Kessler's earlier "parkways," was broad enough to attract and retain some of the city's wealthiest residents. *(Alexander Garvin)*

facilities that were easily accessible to all its citizens, and they took it for granted that the new system would spur real estate development. Instead, they got an opportunistic assemblage of inexpensive land that had only a passing impact on surrounding neighborhoods.

West Terrace Park is a bluff, rising 200 feet above the stockyards, that in 1893 was occupied by unsightly billboards and weather-beaten shanties. Although the site provided splendid views of the Kansas and Missouri rivers below, it was unlikely to be developed because it was too steep and too near the odors of the stockyards. Most of the relatively flat land that might have become usable parkland was eliminated from the proposal in 1899 because it was equally usable for tax-paying real estate development and too many people objected to paying for such costly land.

The views of the Missouri River are even more spectacular from the much larger North Terrace Park. This site was even less likely to be developed because it was both steeper and farther from downtown Kansas City. Since most of its 200 acres are wild cliffs cut by high ravines, it is a wonderful place for hiking, but little else. Once again, topography prevented Kessler from accommodating the variety of active and passive recreation facilities that Olmsted designed for the Emerald Necklace.

Penn Valley Park was created out of a ravine in the southwestern part of the city. Transforming such ungrateful terrain into a public park required all of Kessler's ingenuity. He formed an artificial lake by erecting a 30-foot-high earthen dam and supplied proper drainage by building a system of cement gutters and underground conduits. The landscape that emerged, like that of the other two parks, provides lovely views. But unlike them, it also provides a setting for some of the activities that are typical of Olmsted's designs.

All three parks may have preserved dramatic topographic features and provided stunning views, but their major attraction to Kansas City's Board of Park Commissioners was that their land acquisition was relatively cheap ($2.3 million) and removed relatively little developable property from the tax rolls. Although the parks seemed easily financeable, these parks unfortunately lacked the brilliance of an Olmsted design that would have provided the level pedestrian paths, broad meadows, gentle wooded knolls, and ample playing fields that made his parks elsewhere so successful.

In 1896, Kansas City was unexpectedly supplied with land that could support the wide range of activities. Thomas Swope donated 1334 acres of land 4 miles southwest of the city limits for a public park. Within what would be named Swope Park, Kessler and his successors were able to create a landscape with some of the features of Franklin Park in Boston. During the century since Swope Park was added to the system, the city has annexed so much land that the park is now well within the city boundaries, and it has become a major recreation facility for a metropolitan region with more than 2 million people.

In addition to parks, Kessler's system proposed a "comprehensive, well-planned and thoroughly maintained" system of landscaped boulevards, which he thought would counter "the tendency to . . . build residences in the suburbs." As he explained,

> *The best and most expensive residences will go up along boulevards, but these avenues will exercise a decided effect upon the character of residences to a considerable distance on each side. They will, in fact, create compactly and well built-up residence sections.*[69]

With the exception of the elaborately landscaped Paseo, which was planned to extend from North Terrace Park south to 18th Street, the boulevards proposed in the 1893 report were not boulevards at all. Armour, Linwood, and Independence Boulevards were tree-lined, 100-foot-wide rights-of-way that in other cities would be called avenues. Kansas City was unwilling to authorize anything wider because of the high cost of acquisition and the probable opposition to the required condemnation.

Initially these tree-lined rights-of-way did attract "the best and most expensive residences." That success was also one reason for their demise. Kessler was able to use them as arguments for the acquisition of much wider rights-of-way and for the creation of the broader, better-landscaped parkways that were extended farther to the south (e.g., Ward Parkway, Meyer Boulevard, and South Paseo). He was able to obtain these broader rights-of-way because neighborhood improvement associations and local subdivision developers actively supported them. These realtors and the people who lived in their subdivisions argued that such multilane arteries provided easy access for automobile traffic going downtown. They also understood their importance in increasing property values, an objective that was so important that a few of them actually donated some of the land through which the new parkways were planned. These newer parkways became the focus of the handsome new residential communities created by the developers who had lobbied for more generous public open space. They also supplanted Kessler's older parkway neighborhoods, hastening their demise.[70]

The second set of parkways was so attractive that it soon supplanted the older, narrower "boulevards" as locations for the city's best and most expensive residences. The neighborhoods surrounding these later parks and parkways have retained their allure and remained residential communities of choice for the city's middle class. On the other hand, Kessler's first so-called boulevards are lined with vacant and abandoned buildings, and their once handsome trees have succumbed to disease.

The Kansas City park system provides a lesson in the difference between rhetoric and reality. In 1893, Kessler promised that his boulevards would "exercise a decided effect upon the character of the residences to a considerable distance on each side." In reality, they were no different from similar avenues throughout Kansas City, and they had no lasting impact. By the 1980s many of the earliest sections of the Kansas City park system were lined with vacant and abandoned buildings. One reason for the abandonment of these older neighborhoods was that Kansas City, like many American cities, experienced the outward movement of residents who could afford suburban homes in areas with plenty of open space.

Had the initial sections of Kessler's park system been better located and designed, this trend might have been retarded. The first three parks appeared to be lovely landscapes, enhanced by trellises and pergolas. In fact, they were steep cliffs with relatively little flat land that was usable by large crowds.

The system Kessler proposed in 1893, unlike Boston's Emerald Necklace, included few facilities that were designed to serve the city's future population. Kessler also failed to foresee how quickly the countryside would be engulfed by new residences. The later sections of the park system that were added as a result of private donations, community activism, and lobbying by real estate developers illustrate the importance of Olmsted's vision of a park system, not as a budget extra but as a framework around which the city could grow and develop. They, rather than Kessler's early efforts, demonstrate that a park system can become a convenient means of communication among adjoining districts, a variety of places for active and passive recreation, and development incubator for surrounding communities.

Atlanta's BeltLine Emerald Necklace

Atlanta's BeltLine began as an idea proposed by Ryan Gravel, a Georgia Tech graduate student. He observed that the city was ringed by railroads, some of which were out of use in 1999. Gravel proposed stringing them together into a transit line encircling the city. For the next five years he advocated the transit line at numerous public meetings. City Council President Cathy Woolard and Mayor Shirley Franklin were among its proponents. Another supporter, James Langford, then Georgia State Director of the Trust for Public Land (TPL), thought underutilized property along the BeltLine presented opportunities for adding much-need parkland.[71]

In July 2004, he persuaded TPL to hire Alex Garvin & Associates to examine those opportunities. Members of the firm spent weeks searching for properties that could be reused as parks. They flew in a helicopter in order to understand and appreciate fully the corridor's relationship with both the surrounding communities and the downtown. They also hiked, climbed, and even crawled through kudzu-covered ridges along the entire 20-mile BeltLine right-of-way. The firm came to the conclusion that by itself the transit ring was so expensive that it could not garner enough public support. The same was true for the huge sums required to purchase substantial amounts of new parkland, despite the fact that Atlanta had less park acreage than 85 percent of similar cities across America. Instead, four months later in *The Beltline Emerald Necklace: Atlanta's New Public Realm,* Alex Garvin & Associates proposed a public realm framework, which combined parks and transportation into a single entity.

This 141-page plan, illustrated with maps, photographs, charts, and cost estimates, presented a 2549-acre Emerald Necklace, inspired by the park system Olmsted designed for Boston. The plan tied together 13 park jewels with 46 neighborhoods and proposed three new MARTA subway stations. It was a public realm framework that provided access to every major destination in Atlanta and offered extraordinary strolling, jogging, Rollerblading, and cycling opportunities to every resident of the city. Four park jewels were proposed on 334 acres of publicly owned land not then being used for recreational purposes. In addition, the plan suggested adding 115 largely unused acres to four existing parks (including the city's premier park, Piedmont Park). The plan also

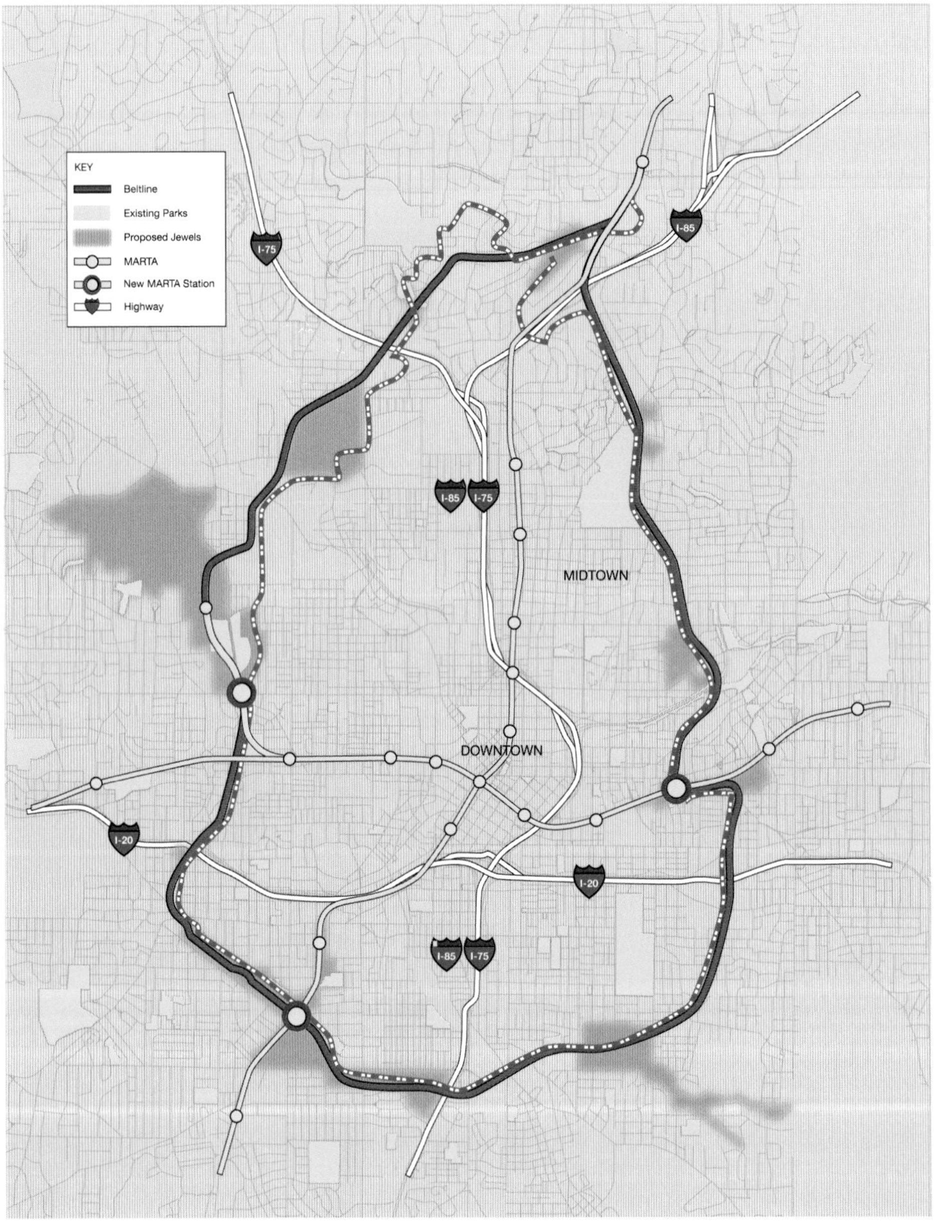

Atlanta, 2004. Beltway Emerald Necklace. *(Alex Garvin & Associates)*

recommended acquiring 778 acres of open space, including the *Bellwood Quarry,* which would become the city's largest public park and would function as park centers for five large mixed-use jewels.

The BeltLine took Atlanta by storm. During 2005 the BeltLine Partnership, a quasi-governmental development entity, was formed to provide community support, and the City Council approved a Tax Allocation District (TAD) to pay for the plan's implementation. The TAD will finance the BeltLine's acquisition and construction by allocating for those purposes any taxes over and above the amounts being paid by properties in the district in 2005. In 2006, the city and county voted to purchase the Bellwood Quarry (which had initially been acquired by TPL in anticipation of its public

Atlanta, 2004. Bellwood Quarry. *(Alexander Garvin)*

approval) and established Atlanta BeltLine Inc., a development corporation charged with implementing the project. Since then, this entity has purchased property for part of the "jewels" called *Boulevard Crossing* and *North Avenue Park* in the plan. Successive sections of the BeltLine trail have been developed and opened for public use. Despite setbacks, such as defeat of a statewide bond issue in 2012 that would have provided funding for the first leg of the transit loop, the BeltLine is creating a new public realm framework that is already having a positive impact on residents' quality of life and will continue to do so for generations to come.

Minneapolis: America's Outstanding Park System

The best-located, best-financed, best-designed, and best-maintained public open space in America is the Minneapolis

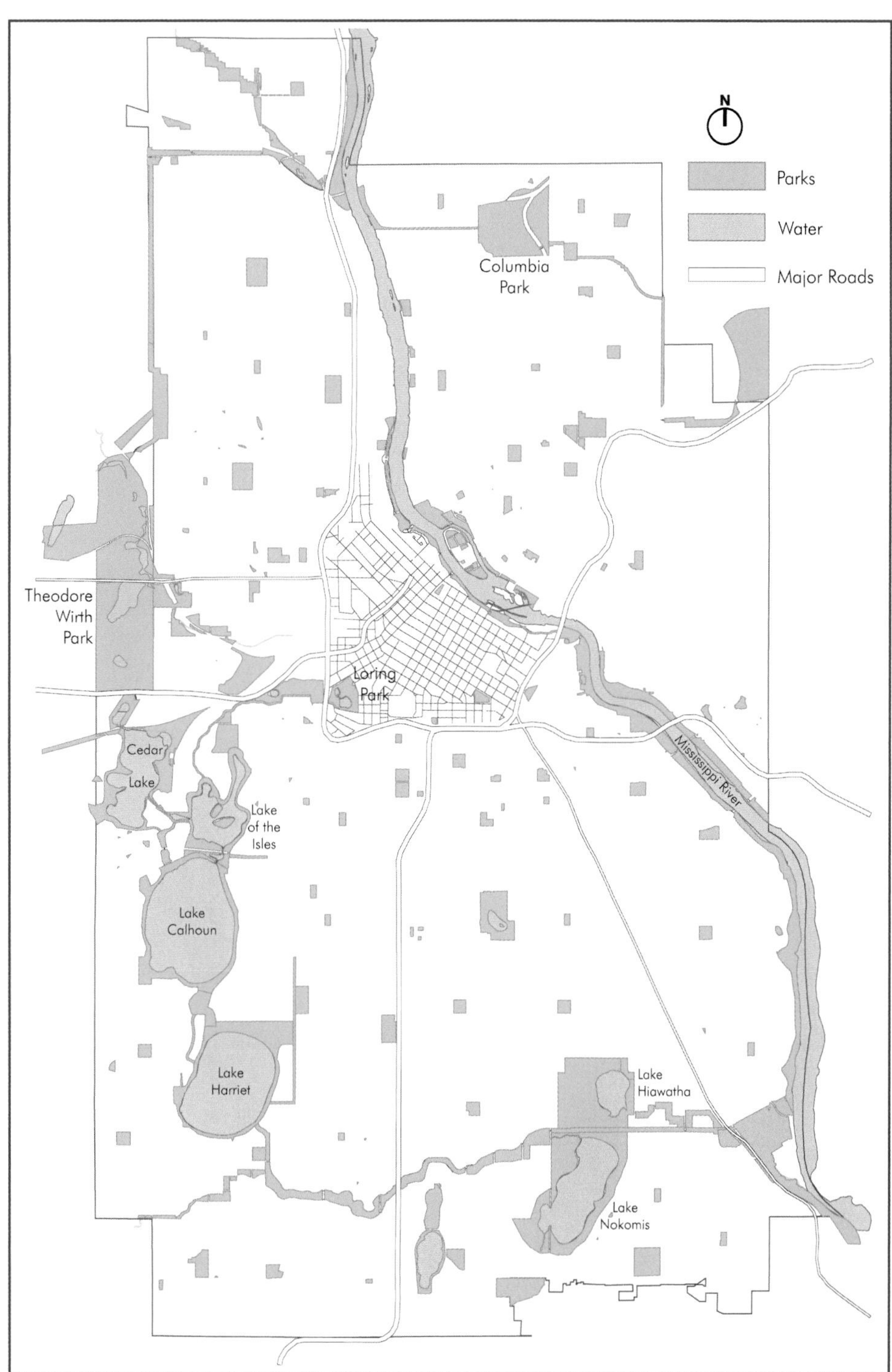

Minneapolis, 2013. Plan of America's finest park system. *(Alexander Garvin and Joshua Price)*

Minneapolis, 1908. Hydraulic dredge at work on Lake of the Isles. *(From T. Wirth,* Minneapolis Park System, *Minneapolis Board of Park Commissioners, 1945)*

park system. Its 6732 acres of land and water are organized around 22 major still-water lakes, ranging in size from 2-acre lagoons to 425-acre Lake Calhoun. They are part of an integrated system of 182 park properties that also includes 60 miles of generously landscaped parkways, nearly 50 miles of designated walking trails and bicycle paths, 6 public golf courses, 61 supervised playgrounds, 47 recreation centers, 21 supervised beaches and swimming pools, rose gardens, historic structures, athletic fields, fountains, picnic grounds, bird sanctuaries, waterfalls, and even a wildflower garden. This parkland extends throughout the city, and the greater portion of it is connected by a continuous system of jogging and walking trails, bicycle paths, and landscaped vehicular parkways. With a 2010 population of 383,000, Minneapolis has 57 people for every acre of public open space. More important, this parkland is easily accessible from anywhere in the city, provides the widest variety of facilities, and affords the user an opportunity to travel from facility to facility without leaving the park system. As a result, every section of the city is supplied with beautifully landscaped, well-maintained public open space that also has increased the desirability of adjacent land sufficiently that these areas have never lost their attractiveness.[72]

The city's comprehensive park system would probably not surprise the members of the Minneapolis Board of Trade, who in 1883 unanimously adopted a resolution calling for the establishment of an independent park commission. As they explained,

> *The rapid growth of our city . . . warns us that the time has come when, if ever, steps should be taken to secure the necessary land for such a grand system of Parks and Boulevards as the natural situation offers and will give to Minneapolis, not only the finest and most beautiful system of Public Parks and Boulevards of any city in America, but which, when secured and located as they can now be at a comparatively small expense, will, in the near future, add many millions to the real estate value of our city.*[73]

The Board of Trade was fought by the city council, which opposed any independent entity with the power to issue bonds, levy taxes, condemn property, and develop parks without the approval of property owners, taxpayers, or, of course, the elected members of the city council. Their objection was overruled by the Minnesota legislature, which authorized a referendum in which voters overwhelmingly approved the idea.

Soon after its establishment in 1883, the Board of Park Commissioners engaged H. W. S. Cleveland, a prominent landscape architect who had previously been hired by Chicago's South Park Commission, to revise and execute Olmsted's designs. Cleveland urged the Minneapolis Commission to be generous in its appropriations:

> *Do not be appalled at the thought of appropriating lands which seem now too costly, simply because they are far out of proportion to your present wants. . . . Look forward for*

Minneapolis, 2009. Lake Harriet. *(Alexander Garvin)*

Minneapolis, 2009. Bicycle riding along Cedar Lark Trail Park has become particularly popular. *(Alexander Garvin)*

a century, to the time when the city has a population of a million, and think what will be their want. They will have wealth enough to purchase all that money can buy, but all their wealth cannot purchase lost opportunity, or restore natural features . . . which would then possess priceless value, and which you can preserve . . . from the destruction which certainly awaits them.[74]

Cleveland may have been wrong in predicting that Minneapolis would have a population of a million. But he could not have been more correct about the prohibitive cost and difficulty of acquiring land for parks after the city had matured.

Like Olmsted, Cleveland urged the creation of an extended system of parks and boulevards "rather than a series of detached open areas." His arguments, though, were different. Having learned from the Chicago Fire of 1871 that "it is hopeless to try to contend with fire when it sweeps from block to block," he explained that when there are broad landscaped boulevards, 200- or 300-feet wide, "the attack is reduced to a skirmish with cinders, and the firemen have an opportunity to hold their ground against it."[75] Like Olmsted, he proposed to exploit the "capabilities" of the site, in this case, by displaying such unique features as the gorge of the Mississippi River, Minnehaha Falls, and the region's many beautiful lakes. Like Olmsted, he proposed to dredge and fill swampy marshes, not merely in order to eradicate the breeding grounds of pestilent epidemics, but also to make the extraordinary lakes of Minneapolis the central feature of the new park system.

H. W. S. Cleveland provided the seeds of the Minneapolis park system. However, the system is largely the work of Theodore Wirth, superintendent of parks from 1906 to 1935. He is the man who extended the system beyond the early parkways, Minnehaha Falls, and the initial lake parks. He dredged these and other lakes, graded their banks, eliminated flooding, and installed permanent paving. He supervised the addition of thousands of acres of new parkland, miles of connecting parkways, and dozens of smaller neighborhood facilities. He planted thousands of acres with grass, shrubs, trees, and flowers, built the park drives, and installed the recreation facilities that we see today.

The parks Cleveland and Wirth created were directly in the path of the city's growth. Without them, the character of that expansion might well have been different. The city's lakes included large, unattractive swampy sections that frequently flooded surrounding areas and were a favorite dumping ground for refuse. Once they had been dredged and landscaped, they became attractive public parks and thus a magnet for development.

The system Wirth built demonstrates the wisdom of Olmsted's view that each component in a park system should be individual in its character, reflecting the potential of the topography and the utility of its individual service function. Thus, decades after they became part of the park system the lake-centered parks continue to offer diverse activities for a population with many tastes: Lake Harriet is a family facility with picnic grounds, playing fields, a band shell, and

Minneapolis, 2009. Minnehaha Parkway. *(Alexander Garvin)*

other group recreation facilities; Lake Calhoun is a setting for fast-paced iceboating and sailboarding; and Lake of the Isles provides a setting for roller skating, cross-country skiing, and just strolling. The jogging trails, bicycle paths, and parkways connecting these and other facilities make them accessible to anyone coming from any part of the system.

The impact of the Minneapolis parks can be seen in the increasing cost of land acquisition. The first 30 acres of Loring Park were acquired in 1883 for $4904 per acre. An addition to this park, made 19 years later, cost $48,096 per acre. Even today, more than a century after the first park acquisitions, the neighborhoods that surround the parks contain some of the city's most valuable residential property.

The level of maintenance in the Minneapolis park system is perhaps its most impressive achievement. Paths and trails are repaved when they wear down. The grass is cut regularly and replanted when necessary. Signs are repainted or replaced when they are no longer legible. Boathouses, refreshment stands, picnic pavilions, benches, and playground equipment are all in good repair.

The excellent condition of the Minneapolis park system may be attributed to the benign habits of a relatively homogeneous population. I believe it also has a great deal to do with the unique and innovative system of administration. From its inception, the Minneapolis park system has been separated from the rest of city government. The Minneapolis Park and Recreation Board, as it has been known since 1967, is elected for staggered four-year terms. It consists of nine members, six elected by district and three at large. The board, not the city government, owns the land, enacts the ordinances governing the system, operates the recreation programs, polices the parks, and establishes the budget. It can issue bonds to pay for acquisition and development and has the power to levy taxes within limits set by the city's Board of Estimate and Taxation (on which it holds a seat, as it does on the city Planning Commission). There are only two significant governmental checks on its activity: the mayor and the city budget. Since 1975, the mayor has been able to veto its actions, provided the board cannot muster a two-thirds vote to override. The more important control over its activities is budgetary. Although the board has the ability to levy taxes, its tax revenues are insufficient to pay all expenses, making it dependent on further city appropriations.

The popularly elected Minneapolis Park and Recreation Board is probably more powerful than any other U.S. government entity administering city parks. Because it is accountable to the voters, unlike virtually every other city park agency, it is also more attentive to the wishes of the population it serves. This unique form of governance and financing has enabled the residents of Minneapolis to have the best-located, best-financed, best-designed, best-maintained park system in America.

Ingredients of Success

Few people still believe, as did nineteenth-century reformers, that parks can influence the character and development of an entire city. Nevertheless, parks can and should be used in that way. Chicago's lakeshore parks, New York's Central Park, and the Minneapolis park system continue to have a major impact on the surrounding city. The explanation for their influence lies in their manipulation of the same ingredients responsible for all successful planning: market, location, design, financing, time, and entrepreneurship.

Market

During the nineteenth century, when increasing amounts of territory were being developed in response to growing populations, establishing public parks seemed essential. Once a new park was established, development engulfed it and moved farther out toward the suburbs. Satisfying such growing demand for public open space is simple: establish public parks at a pace that keeps up with an ever-increasing population.

Public investment in city parks, however, will fail to generate private market activity in surrounding areas when the new facility fails to attract people who will spill over into those areas. The hikers in Kansas City's North Terrace Park could not possibly constitute a sufficient market for adjacent property owners.

As cities matured, however, some discovered that parkland could help to retain an existing market and even to attract activity that had not heretofore existed. Olmsted's parks, for example, supplied space for so many types of recreation that they were easily adapted by ever-changing populations, continued to retain their attractiveness, and became stabilizing forces for surrounding neighborhoods.

Sometimes parks can eliminate an impediment to private investment in an area. The High Line was specifically designed to attract pedestrians to an area that had up to that time repelled most nearby residents as well as visitors from other areas of the city. Once the park was completed, property owners tried to profit from this improvement to the area by erecting new apartment buildings and opening stores and restaurants. That symbiotic relationship between a public park and adjacent private property spurred new development that in turn attracted tourists and has even resulted in hotel construction. The substantial private market reaction, however, was only possible because there already was considerable market demand for art galleries, restaurants, and upscale residences in the Chelsea neighborhood surrounding the High Line.

Manhattan, 2010. The Hudson River Park provides ample room for walking, jogging, cycling, and driving—all at the same time. *(Alexander Garvin)*

Location

If market was not an issue in planning nineteenth-century parks, neither was location. There was plenty of relatively inexpensive, undeveloped land just outside already settled areas. Furthermore, facilities like Forest Park in St. Louis and Prospect Park in Brooklyn could be created at sites that were only a few minutes away by streetcar. Land availability, however, was not enough. Griffith Park in Los Angeles could not have real impact because much of the city's rapidly growing population was moving away from its location. Many people were settling west of downtown in areas made easily accessible by streetcar service (see Chapter 6). Others wanted to be near the Pacific Ocean. Griffith Park was in the wrong location.

Some parks are surrounded by properties that are inherently unattractive for development. The property at the bottom of the West Terrace Park in Kansas City, for example, could never stimulate much private market activity there because the park had been created above what was then the slaughterhouse district of the city.

While proximity, access, and terrain were important ingredients in the initial success of parks, the inherent characteristics of the location and of neighboring land uses may have even greater importance. Chicago's lakeshore parks, for example, provide wonderful views and summer breezes. Their beaches and marinas, and the opportunity for water sports provided in the parks, enhanced the already attractive location. Property owners, of course, responded by investing along these parks.

Design

In considering the design of Prospect Park, Frederick Law Olmsted thought that the critical considerations were (1) convenience of shape, (2) amplitude of dimensions, (3) topographical conditions, and (4) the surrounding circumstances.[76] These all played a role in the failure of the initial sections of the Kansas City park system to generate any continuing market reaction. Their location, shape, and topography made it difficult for them to affect more than a limited amount of surrounding territory. The importance of ample dimensions is even more evident when one compares the Hudson River Park with San Francisco's Embarcadero. The bike and jogging lanes of the Hudson River Park are protected from vehicular traffic by ample landscaping and trees. The Embarcadero's bike lanes are included within a roadway

San Francisco, 2006. So much of the Embarcadero is devoted to vehicular traffic that many cyclists leave the bike lane to the automobiles and jostle for space with pedestrians on a sidewalk that is not wide enough to comfortably accommodate both. *(Alexander Garvin)*

that is often jammed with vehicular traffic. Thus, it is common to see pedestrians dodging the bicycles that prefer to use the safer sidewalks.

Familiarity with Olmsted's parks and park systems leads me to add two other elements that were critical to their design: a variety of places that can accommodate a large range of recreational activities and a mutually reinforcing arrangement of those places. The significance of this complementary relationship becomes immediately evident in examining Forest Park in St. Louis. Had it been designed for a range of recreational activities, the parks department would not have made intrusions that added the element of "social utility." However, when it did so, it was too late to arrange them in any complementary sequence.

These same factors are among the reasons for the success of the Minneapolis park system. Its designers dredged lakes and filled swamps in order to prevent flooding and provide facilities for all manner of water-based recreation; created parkways that were broad enough to be actively used for jogging, cycling, strolling, and active sports (not just vehicular transportation); and landscaped the edges of parks and parkways in a manner that enhanced the illusion of living in the country.

Because they seem so familiar, most observers pay little attention to the extraordinarily effective design of Olmsted's parks. He knew how to create places that thousands of people loved to go to. That same ability to attract throngs of people is evident in the much more glamorous work on Millennium Park and the High Line. Regardless of whether the design is familiar or trendsetting, what is unarguable is that it is a necessary ingredient in great parks.

Financing

A surprising amount of parkland has been acquired through donation. Swope Park in Kansas City and Griffith Park in Los Angeles are notable examples. In most cases, however, governments need money to pay for acquiring property, transforming that land into usable public facilities, and then managing and maintaining those facilities after they have been created. They usually get the money for acquisition and development by selling bonds and then repay the bonds in the form of regular debt-service payments to the holders of those bonds.

Nineteenth-century park advocates understood this process very well. Whenever they proposed a new facility, they described it as an investment that could then be made at relatively low cost and eventually would produce revenues far in excess of initial cost. Park proponents argued that it was better to pay less for the land now rather than considerably more later, when an increasing population would insist on establishing additional facilities. They were usually right. Land costs, even in areas that to that point had not yet been developed, continually increased. Their second argument was that initial costs would be covered many times over by the increased taxes from surrounding properties whose value had increased as a result of the new park. The tax increment financing that is funding the Atlanta BeltLine is a contemporary example of this approach.

Minneapolis is among the very few park systems that are adequately financed. For over a century, the Minneapolis Park Board has received a dedicated stream of tax revenues that it can pledge as payment to bondholders who lend money for capital projects. As a result, it has few problems funding development or rehabilitation.

Some parks are financed from rental income and fees. The Hudson River Park, which gets most of its income from pier rental and parking fees, is just one example. Boston's Post Office Park is built over a parking garage, the income from which not only covers debt service on the bonds that financed the park's development but also pays the park's operating costs.

Other city park systems have to compete for capital funds. This money is usually obtained when a local government issues general revenue bonds to pay for everything from roadways to firehouses. Since parks are rarely priorities, few cities spend large sums on major park acquisition or restoration projects. New York City provides a particularly stark example. Between 1945 and 1955, while Robert Moses was still parks commissioner, "parks expenditures accounted for approximately 1.5 percent of each annual City operating budget." By fiscal year 2012, the share had dropped to 0.4 percent.[77]

Time

Some of the most wondrous parks are those in which visitors make their way through a structured sequence of experiences. Olmsted choreographed this sort of movement through Prospect Park. So have the designers of the High Line and the Hudson River Park.

Parks that function successfully over a 24-hour period, 7 days a week, are rare. Even Olmsted admitted that parks are unlikely to be safe at night. However, whether working with prairie land like Washington Park in Chicago or a narrow, topographically varied facility like Muddy River in Boston, he was insistent that parks include a wide variety of facilities that could attract people for different forms of recreation at different times.

The most common and serious error in planning parks, however, is thinking of them as development projects that terminate when the facility has been completed. As Olmsted explained not long after Central Park was opened to the public:

> *The people who are to visit the park this year or next are but a small fraction of those who must be expected to visit hereafter. If the park had to be laid out and especially if [it] had to be planted with reference only to the use of the next few*

years, a very different general plan, a very different way of planting and a very different way of managing trees would be proper.[78]

The life of every park is just beginning when its development is over. Minneapolis is one of very few cities that understood this simple idea and provided an administrative structure, a guaranteed stream of income, a regularly updated public mandate, and legal powers to ensure that its parks continued to be well maintained in good times and bad.

Entrepreneurship

Park developers with the imagination and drive of Robert Moses are as rare as park designers with the genius of Frederick Law Olmsted. Wherever they surface, cities will have wonderful parks. Good park systems, however, cannot be built on the hope that such individuals will be available. Moreover, they cannot succeed without the determined effort of dedicated public servants who operate within an environment that fosters public entrepreneurship.

During the nineteenth century, Brooklyn, Chicago, Boston, and many other cities were able to create wonderful parks because park development was a major part of the public agenda and undeveloped land was so readily available. Those cities also operated with administrative structures that were more open to public entrepreneurship. They had park commissions made up of citizen leaders who actively pursued their mandate and were deeply involved in directing operating agencies. Minneapolis is the only city with a major park system that is still operated by an independent park commission. The results are visible throughout the system.

Today, municipal bureaucracies staffed by tenured civil servants operate most city park systems. Bold initiatives are unlikely to emerge from this sort of institutional setting. Agency staff is just not in a position to oppose elected public officials who are eager to reallocate park funds to other pressing demands for municipal assistance, usually actively supported by their constituents. The inevitable result: declining budgets and deteriorating public parks.

Harnessing local initiatives is the most effective way to overcome inadequate budget allocations, poor maintenance, and inattentive management. A good example is the Joseph C. Sauer Playground, on Manhattan's Lower East Side. Built by Robert Moses with Public Works Administration (PWA) assistance in 1934, this playground was so badly deteriorated by the 1980s that the 12th Street Block Association had to organize its cleanup, lobby for its redesign, and fight to have its reconstruction added to the city budget.[79] When the enlarged and redesigned Sauer Playground reopened in 1993, the block association took responsibility for supervising its management and maintenance. It supplied the entrepreneurial role that had been missing for decades and demonstrated that small public facilities could be successful when there was active public participation in their operation. Similar community-based management is what has kept community gardens attractive places for neighborhood residents.

The role of entrepreneurship has become an increasingly important component in park development, restoration, and management. Sometimes it is provided by elected public officials such as Mayor Daley's efforts on behalf of Millennium Park or Mayor Franklin's support for the Atlanta BeltLine. Sometimes it comes from individuals who never expected to be major players. Without Robert Hammond and Joshua David, for example, there would be no High Line. Much of the entrepreneurial activity is coming from citizen groups that create conservancies that replace the government agencies that used to be responsible for the day-to-day stewardship of these parks. As of 2012, the Shelby Farms Park Conservancy operates its 4500-acre Memphis facility; the Piedmont Park Conservancy operates its 189-acre Atlanta facility; the Bryant Park Corporation operates its 6.5-acre New York facility (see Chapter 8). As with the Sauer Playground and most community gardens, these major community-based nonprofit institutions are exercising the leadership that great parks require.

Parks as Planning Strategy

Despite advocacy by the growing movement for environmental sustainability, public investment in public open space is no longer as fashionable as it was a century ago, when Minneapolis began creating its extraordinary park system. Nevertheless, public investment in parks was then and still is an effective means of stimulating a desirable private market reaction. As with all public investment intended to stoke the private market, the trick is to spend public money in ways that will reduce for private financiers the risk of investing in surrounding property or to devote resources to initiatives that will attract activity that will spill over into the areas surrounding parkland.

Olmsted's work in the Fens and Muddy River in Boston is a good example of park investment that reduces the risk of investing in surrounding areas. As long as these areas remained pestilent breeding grounds of disease, investment in surrounding areas remained unattractive. After these polluted waterways became public parks, surrounding property became attractive for development. The same thing happened in Minneapolis after it terminated the danger of flooding by dredging its lakes, landscaping their banks, and transforming them into public parks.

Parks also stimulate private investment by attracting customers to an area. By spending hundreds of millions of dollars over several decades, Chicago created a wonderful string of lakefront recreation facilities that brought tens of thousands of people to the waterfront. Developers were quick

Manhattan, 1994. By lobbying for the renovation of the Joseph Sauer Playground and then supervising its management and maintenance, the 12th Street Block Association has demonstrated the crucial entrepreneurial role that can be played by dedicated, community-based organizations. *(Alexander Garvin)*

to perceive the spending power of this growing market and made fortunes supplying it with new apartments. The city repeated this effective strategy when it invested nearly half a billion dollars in Millennium Park, and its allocation of resources has been rewarded with new spending, construction, and taxes far in excess of anything its supporters had dreamed of.

As Atlanta's BeltLine and New York's High Line so clearly demonstrate, there are plenty of inner-city properties that are no longer in demand but can be recycled as new city parks. There also still is open land available in rapidly suburbanizing areas. Public investment should be directed to these opportunities.

Existing city parks and playgrounds are particularly good candidates for new public investment. In many cases, like the Olmsted parks, they are priceless historical artifacts whose continuing benefits were paid for long ago. If they were put up for sale, cities could net tremendous amounts of money. Instead of protecting these valuable assets, we allow them to deteriorate. They should be treated with the same respect we

Seattle, 2007. Freeway Park built over a highway right-of-way. *(Alexander Garvin)*

accord the artworks in our museums. Once restored, they too can attract a substantial and profitable market.

Parks in inner cities can also trigger second growth. Such parks are relatively easy to create from underutilized property that has been left behind by previous users. Among the possibilities are rail yards and rights-of-way; waterfront areas that are no longer needed for shipping, warehousing, or manufacturing; streets, highways, and interchanges whose traffic can be rechanneled to other arteries; and even residential areas where local governments already own large blocks of vacant property repossessed for failure to pay taxes. Both the Hudson River and the High Line parks clearly demonstrate the enormous potential in intelligently adapting obsoleted property to new uses. Seattle created additional public open space in 1976 by building 5-acre Freeway Park over Interstate Highway 5. The park was so successful in tying together the sections of the city that had been separated by the highway that it was expanded in 1984 and again in 1989 with the completion of the Washington State Convention and Trade Center (see Chapter 5). This combination of commercial and recreational facilities was substantial enough to attract lots of people and, thus, to generate a significant market reaction from neighboring property owners, who recognize an opportunity to profit from the spillover.

There are also attractive sites for new parks outside center cities. Developers who convert countryside into sites for single-family houses and condominiums usually leave new residents with little more than their own yards for recreation. Larger subdivisions may add a swimming pool, clubhouse, tennis courts, or even golf links. These facilities are usually available exclusively to residents or club members. Only rarely will a subdivision include a genuinely *public* park. If developing areas were to include a continuous and varied public park system that could be adapted to the changing needs of future generations, it too would become a stabilizing force for surrounding communities.

Suburbs continue to miss these opportunities because they may not have park agencies with the necessary political mandate, legal authority, or financial wherewithal. Imitating the manner in which Minneapolis has devoted real estate tax receipts or the way that Atlanta has profited from tax increment financing can reverse the situation. All that is necessary is enabling legislation at the state level that authorizes local governments to map park districts, establish elected park boards, allocate a fixed proportion of sales and property taxes for park purposes, and authorize the park board to use that money for maintenance and debt service on bonds.

The revenue base for these elected park boards is central not only to their ability to develop new park facilities but, perhaps more important, to successful urban planning overall. A set-aside of property- or sales-tax revenues would allow park boards to make long-term plans and use the money to make debt-service payments on bonds issued for property acquisition and development. However, if the only source of revenue is a citywide tax, park boards will conceive and develop facilities that meet the insular requirements of single-function interest groups. Instead, they should be making park investments that will stimulate complementary private market activity in surrounding neighborhoods.

Notes

1. For a full discussion, see Alexander Garvin, *Public Parks: The Key to Livable Communities*, W. W. Norton & Company, New York, 2011, p. 224.
2. Ralph Waldo Emerson, "Nature" (1836), in *Selected Essays*, Penguin Books, New York, 1982, p. 43.
3. "Capability" Brown received his nickname because he was forever expounding on the "capabilities" of the site.
4. Toward the end of the eighteenth century, Capability Brown's soft, graceful landscapes were supplanted in popularity by more rugged, romantic visions of sublime nature. These newer gardens by Uvedale Price, Richard Payne Knight, and Humphrey Repton still tried to display the natural beauties of the site, but did so while minimizing their apparent interference with nature. They wanted gardens that appeared wild and undisturbed. Both esthetics were eminently picturesque and both continued well into the nineteenth century, when American cities began establishing public parks.
5. Frederick Law Olmsted, *Walks and Talks of an American Farmer in England*, University of Michigan Press, Ann Arbor, MI, 1967, pp. 52–53.
6. In 1851, Savannah set aside 10 acres for Forsythe Park. The park was later doubled in size, but its final 20 acres are tiny in comparison with the 843 acres of Central Park. In 1812, Philadelphia acquired for a municipal waterworks the first 5 acres of what was to become Fairmount Park. In 1828, to protect the purity of the city's water supply, the site was enlarged to 28 acres. However, it wasn't until 1855 that the Pennsylvania legislature authorized acquisition of any substantial amounts of land for purely recreational use.
7. Ambrose Kingsland, "Message to the Common Council," April 5, 1851.
8. John P. Rea, *Minneapolis Tribune*, May 23, 1880.
9. Jane Addams, *The Spirit of Youth and the City Streets*, Macmillan Co., New York, 1909, p. 103.
10. Jacob Riis, *The Peril and the Preservation of the Home*, Jacobs, New York, 1903.
11. Frederick Law Olmsted, "Report on the Charles River Embankment 'Charlesbank,'" *Twelfth Annual Report of the Board of Commissioners of the Department of Parks for the Year 1886*, City Document 24-1887.
12. Cynthia Zaitzevsky, *Frederick Law Olmsted and the Boston Park System*, The Belknap Press of Harvard University Press, Cambridge, MA, 1982, pp. 96–100.
13. Ibid.
14. Galen Cranz, *The Politics of Park Design*, MIT Press, Cambridge, MA, 1982, pp. 80–83.
15. Philadelphia City Planning Commission, *Comprehensive Plan for Swimming Pools*, January 1968.
16. Robert Moses, *Six Years of Park Progress*, City of New York Department of Parks, New York, 1940, p. 10.
17. Robert Moses, *26 Years of Progress 1934–1960*, City of New York Department of Parks, New York, 1960, p. 52.
18. Clare Beckhardt, "Proposed Redesign of the West 46th Street Playground," The Parks Council, New York, 1972, p. 1.
19. Thomas P. F. Hoving, "Think Big About Small Parks," *New York Times Magazine*, May 16, 1966, p. 12.
20. *New York Times*, May 11, 1966, p. 51.
21. Elizabeth Barlow Rogers, *Rebuilding Central Park: A Management and Restoration Plan*, MIT Press, Cambridge, MA, 1987.

22. Peter Harnik, "Philadelphia Green," in Alexander Garvin and Gayle Berens, *Urban Parks and Open Space,* ULI-The Urban Land Institute, Washington, DC, 1997, pp. 158–167.
23. New York City Citizens Housing and Planning Council, "Giuliani Confronts In Rem Dilemma," *The Urban Prospect,* New York City, January/February 1995.
24. Keith Schneider, "To Revitalize a City, Try Spreading Some Mulch," *New York Times,* May 17, 2006.
25. http://www.nyc.gov/html/dot/html and interview with Janette Sadik-Khan, March 27, 2012.
26. Frederick Law Olmsted, letter to Charles Brace, December 8, 1860.
27. Peter Harnik, *Inside City Parks,* Urban Land Institute, Washington, DC, 2000, p. 11; and "Annual Report 2011," Central Park Conservancy, www.centralparknyc.org.
28. Historical and statistical material on the squares of Paris is derived from Michael Dennis, *Court and Garden,* MIT Press, Cambridge, MA, 1986, pp. 43–51, 79–90, 128–136; and Pierre Lavedan, *Les Villes Françaises,* Editions Vincent, Freal & Cie, Paris, 1960, pp. 126–142.
29. Historical and statistical material on the great estates of London is derived from John Summerson, *Georgian London,* Barrie & Jenkins Ltd., London, 1988, pp. 73–86, 147–161, 181–187; Donald J. Olsen, *Town Planning in London—The Eighteenth and Nineteenth Centuries,* Yale University Press, New Haven, CT, 1982, pp. 27–96; and Steen Eiler Rasmussen, *London: The Unique City,* MIT Press, Cambridge, MA, 1967, pp. 165–201.
30. Historical and statistical material on the development of Regent Street and Regent's Park is derived from Summerson, op. cit., pp. 162–180; and Rasmussen, op. cit., pp. 271–291.
31. Historical and statistical material on Haussmann and his work on the parks, squares, and boulevards of Paris is derived from David H. Pinkney, *Napoleon III and the Rebuilding of Paris,* Princeton University Press, Princeton, NJ, 1958, pp. 75–104; George F. Chadwick, *The Park and the Town: Public Landscapes in the 19th and 20th Centuries,* Prager, New York, 1966, pp. 152–162; and Antoine Grumbach, "The Promenades of Paris," *Oppositions #8,* Spring 1977, MIT Press, Cambridge, MA, pp. 50–67.
32. Avenue Foch was known as Avenue du Bois prior to being renamed in honor of Marshal Foch.
33. Carl Feiss, "Early American Public Squares," in Paul Zucker, *Town and Square from the Agora to the Village Green,* Columbia University Press, New York, 1959, pp. 237–255.
34. John W. Reps, *The Making of Urban America,* Princeton University Press, Princeton, NJ, 1965, pp. 157–174.
35. See Chapter 14.
36. Reps, op. cit., pp. 185–202.
37. H. W. S. Cleveland, "Suggestion for a System of Parks and Parkways for the City of Minneapolis," read at a meeting of the park commissioners, June 2, 1883, reprinted by Theodore Wirth, *Minneapolis Park System 1883–1944,* Minneapolis Board of Park Commissioners, 1945, pp. 28–34.
38. I am indebted to Tupper Thomas, administrator of Prospect Park, for a multitude of insights on the design of the park and for information on the park's conifers.
39. Frederick Law Olmsted and Calvert Vaux, "Preliminary Report to the Commissioners for Laying Out a Park in Brooklyn, New York," (1866), reprinted by Albert Fein (editor), *Landscape into Cityscape: Frederick Law Olmsted's Plans for Greater New York,* Cornell University Press, Ithaca, NY, 1968, p. 98.
40. Frederick Law Olmsted and Calvert Vaux, "Report of the Landscape Architects and Superintendents to the President of the Board of Commissioners of Prospect Park, Brooklyn" (1868), reprinted by Fein, op. cit., p. 157.
41. Marine Park (1822 acres) is larger than Prospect Park (526 acres). However, it became a park long afterward (1924), involved little landscaping, and is difficult to reach by mass transit. *Source:* Office of the Borough President of Brooklyn.
42. Historical and statistical material on Forest Park is derived from Caroline Loughlin and Catherine Anderson, *Forest Park,* Junior League of St. Louis and University of Missouri Press, Columbia, MO, 1986; and August Heckscher, *Open Spaces: The Life of American Cities,* Harper & Row, New York, 1977, pp. 173–177.
43. *Report of the St. Louis Department of Parks* (1915), quoted by Heckscher, op. cit., p. 177.
44. "St. Louis Surpasses Goal in Raising Funds for Park," *Planning* 67(3):27 (March 2001); and http://stlouis-mo.gov/government/departments/parks/floraconservancy/.
45. Historical and statistical material on Griffith Park is derived from Robert M. Fogelson, *The Fragmented Metropolis: Los Angeles 1850–1930,* Harvard University Press, Cambridge, MA, 1967; and Heckscher, op. cit.
46. Historical and statistical material on Chicago's parks is derived from Olmsted, Vaux & Co., "Report Accompanying Plan for Laying Out the South Park" (1871), partially reproduced in S. B. Sutton, *Civilizing American Cities,* MIT Press, Cambridge, MA, 1971, pp. 156–196; Daniel Bluestone, *Constructing Chicago,* Yale University Press, New Haven, CT, 1991, pp. 7–61; Victoria Post Ranney, *Olmsted in Chicago,* R. R. Donnelley & Sons, Chicago, 1972; Harold M. Mayer and Richard C. Wade, *Chicago: Growth of a Metropolis,* University of Chicago Press, Chicago, 1969; and Jean F. Block, *Hyde Park Houses,* University of Chicago Press, Chicago, 1978.
47. Olmsted, Vaux & Co., "Report Accompanying Plan for Laying Out the South Park," in Sutton, op. cit., p. 161.
48. H. W. S. Cleveland had worked for Olmsted, Vaux & Co., on the design of Prospect Park. He left to establish his own firm and moved to Chicago in 1869.
49. Kenneth Jackson, "Foreword" to Block, op. cit., p. vii.
50. Olmsted, Vaux & Co., "Report Accompanying Plan for Laying Out the South Park," in Sutton, op. cit., p. 168.
51. Historical and statistical material on Millennium Park is derived from Timothy J. Gilfoyle, *Millennium Park: Creating a Chicago Landmark,* University of Chicago Press, Chicago, 2005; and Ryan Mikulenka (team leader), *Millennium Ark Quadrangle Net Value Report,* Texas A&M University and DePaul University, Chicago, 2011.
52. Historical and statistical material on Westway is derived from New York City Planning Commission, *Land Use and the West Side Highway,* New York, 1974; U.S. Department of Transportation Federal Highway Administration and New York State Department of Transportation, *Draft Environmental Impact Statement and Section 4(f) Statement for West Side Highway,* Washington, DC, 1974; Regina Herzlinger, "Costs, Benefits, and the West Side Highway," *The Public Interest,* no. 55, (Spring 1979) National Affairs Inc., New York, pp. 77–98; Sam Roberts, "Battle of the Westway: Bitter 10-year Saga of a Vision on Hold," *New York Times,* June 4, 1984, p. B1; Roberts, "For Stalled Westway, a Time of Decision," *New York Times,* June 5, 1984, p. B1; and Roberts, "Bass: Why Is Hudson So Important," *New York Times,* June 26, 1984, p. C1.
53. In its final version, Westway included 16 acres for institutional and commercial purposes, 24 acres for industry, 36 acres for the roadway, 57 acres for residential development, and 93 acres for a park. In every version, the proposed industrial uses were a fantasy. Manhattan had been hemorrhaging industrial firms for decades. The allocation of land for industrial purposes was there largely to attract support from labor unions.
54. Historical and statistical material on the Hudson River Park is derived from www.hudsonriverpark.org and Real Estate Board of New York and the Regional Plan Association, "The Impact of Hudson River Park on Property Values," unpublished report, 2007.
55. Historical and statistical material on the High Line Park is derived from Joshua David and Robert Hammond, *HIGH LINE: The Inside Story of New York City's Park in the Sky,* Farrar, Straus and Giroux, New York, 2011; and http://www.thehighline.org.
56. Real Estate Board of New York and the Regional Plan Association, "The Impact of Hudson River Park on Property Values," unpublished report, 2007.
57. In the late 1960s, Boulder, Colorado, initiated a similar effort. During its first two decades of operation, this program has added more than 12,000 acres to the park system. While it is too early to evaluate its success, it appears as though Boulder will provide a second example of the effectiveness of Olmsted's ideas.
58. Historical and statistical material on the Boston park system is derived from Cynthia Zaitzevsky, op. cit.; Frederick Law Olmsted,

Seventh Annual Report of the Commissioners of the Department of Parks for the City of Boston for the Year 1881, reproduced in Sutton, op. cit., pp. 221–227; Frederick Law Olmsted, *Suggestions for the Improvement of the Muddy River, Sixth Annual Report of the Board of Commissioners of the Department of Parks for the City of Boston for the Year 1880,* reproduced in Sutton, op. cit., pp. 228–233; and Frederick Law Olmsted, *Notes on the Plan of Franklin Park and Related Matters* (1886), reproduced in Sutton, op. cit., pp. 233–262.

59. Laura Wood Roper, *Flo: A Biography of Frederick Law Olmsted,* Johns Hopkins University Press, Baltimore, 1973, pp. 324–368, 383–392.
60. Hermann Grundel won the competition and was awarded $500. (Cynthia Zaitzevsky, op. cit., p. 54.)
61. Frederick Law Olmsted, Seventh Annual Report of the Commissioners of the Department of Parks for the City of Boston for the Year 1881, reproduced in Sutton, op. cit., p. 221.
62. Frederick Law Olmsted, Suggestions for the Improvement of the Muddy River, Sixth Annual Report of the Board of Commissioners of the Department of Parks for the City of Boston for the Year 1880, reproduced in Sutton, op. cit., p. 231.
63. Frederick Law Olmsted, letter to Charles Sprague Sargent, July 8, 1874, quoted in Zaitzevsky, op. cit., p. 60.
64. Frederick Law Olmsted, *Notes on the Plan of Franklin Park and Related Matters* (1886), reproduced in Sutton, op. cit., p. 248.
65. City of Boston Department of Parks, Twenty-first Annual Report of the Board of Commissioners for the Year Ending January 31, 1896, p. 50.
66. Frederick Law Olmsted, *Notes on the Plan of Franklin Park and Related Matters* (1886), reproduced in Sutton, op. cit., p. 249.
67. U.S. Department of Commerce, Bureau of the Census, *Statistical Abstract of the United States* (1978) and (1991), Washington, DC.
68. I would like to thank Marc and Mel Solomon for helping me to understand the difference between the earlier and later sections of the Kansas City parks system. Historical and statistical material on Kansas City and its park system is derived from William H. Wilson, *The City Beautiful Movement,* Johns Hopkins University Press, Baltimore, 1989, pp. 99–125, 208–212; William S. Worley, *J. C. Nichols and the Shaping of Kansas City,* University of Missouri Press, Columbia, MO, 1990; and Carla C. Sabala (editor), *Kansas City . . . Today,* Urban Land Institute, Washington, DC, 1974.
69. *Report of the Board of Park and Boulevard Commissioners of Kansas City, Mo.,* Board of Park and Boulevard Commissioners, Kansas City, Resolution of October 12, 1893, pp. 14–15.
70. William S. Worley, op. cit., pp. 78–85, 100–107.
71. Historical and statistical material on the Atlanta BeltLine is derived from Alex Garvin & Associates, *The Beltline Emerald Necklace: Atlanta's New Public Realm,* Trust for Public Land, San Francisco, 2004, pp. 141; and http://www.beltline.org.
72. Historical and statistical material on the Minneapolis park system is derived from Theodore Wirth, op. cit.; Sheila M. Speltz, "The Minneapolis Park & Recreation System," 1987, unpublished; League of Women Voters of Minneapolis, *Minneapolis—A Guide to Local Government,* Minneapolis, October 1977; and http://www.minneapolisparks.org.
73. Theodore Wirth, op. cit., p. 19.
74. H. W. S. Cleveland, "Suggestions for a System of Parks and Parkways for the City of Minneapolis," June 2, 1883, quoted in Wirth, op. cit., p. 29.
75. H. W. S. Cleveland, "Suggestions for a System of Parks and Parkways for the City of Minneapolis," June 2, 1883, quoted in Wirth, op. cit., p. 29.
76. Frederick Law Olmsted and Calvert Vaux, "Preliminary Report to the Commissioners for Laying Out a Park in Brooklyn, New York" (1866), in Fein, op. cit., p. 96.
77. http://www.nyc.gov.
78. David Schuyler and Jane Turner Censer, *The Papers of Frederick Law Olmsted,* vol. 6, Johns Hopkins University Press, Baltimore, 1992, p. 539.
79. Patricia Leigh Brown, "Reclaiming a Park for Play," *New York Times,* "The City" section, September, 12, 1993, p. 1.

4

Retail Shopping

Chicago, 2012. North Michigan Avenue. *(Alexander Garvin)*

People often measure the health of a community by the condition of its commercial areas. They believe that vacant and boarded-up stores indicate a withering economy, while busy shopping streets signify a prosperous municipality. Politicians and local officials are particularly sensitive to this indicator because it means changes in the tax base and, therefore, in the ability to pay for government services and jobs. Their most frequent response is to offer land and financing for new shopping facilities because, like sports stadiums, cultural facilities, and convention centers, they attract throngs of people who spend large sums of money. Furthermore, these customers come at times when other activity is less significant, bringing 24-hour vitality to streets, sidewalks, and parking facilities that would otherwise be empty and unsafe.

All shopping destinations try to attract customers who would otherwise spend their money elsewhere. They want to be where the action is. But the similarity ends there. Retail strips and public markets are as old as retailing itself. Nineteenth-century shopping arcades and their air-conditioned twentieth-century descendants evolved from the desire to establish a customer-friendly environment free of the noise, dust, and confusion of heavily trafficked public streets. Department stores, discount centers, and the more recent onset of e-retailing reflect fundamental changes in merchandising. Shopping centers are the result of a population that is completely dependent on motor vehicles. Festival marketplaces and entertainment centers are responses to changes in lifestyle and the demands that grow out of those changes. And all of them compete with one another for customers, just as cities and suburbs do. One of the most effective ways of enhancing downtown business districts is by taking public actions that increase the magnetic attraction of their retail outlets. For public action to be effective, however, planners must have a realistic understanding of how these very different forms of retailing work.

Most consumers choose among stores that compete within the same trade area. In those cases, public assistance to one facility over another produces neither new spending nor new jobs. It simply moves spending from one location to another, favoring one group of businesses over another. The justification for such favoritism can only be that it generates other public benefits. Retail activity may be more appropriate in one location because the infrastructure in that part of town may be underutilized. It may generate customers for nearby facilities that otherwise would not be able to survive. On the other hand, the noise and traffic it attracts may not be welcome in the surrounding area. All too frequently, new shopping facilities divert customers from other commercial districts, causing vacancies and property deterioration.

If we are to improve our cities and suburbs we must favor projects that recognize and respect surrounding communities and avoid public investment in retail facilities whose success is detrimental to other parts of town. We also need to be sure that the induced private market reaction justifies any public action that is necessary to restore or create new retail facilities. Then we can get beyond the identification of a venture's projected sales, jobs, and tax proceeds to concentrate instead on the role retail shopping may play in improving our communities.

Retail Strips

Roadside merchants were probably responsible for the earliest retail establishments. Later, local artisans and prosperous merchants used the ground floor of their homes to display and sell goods. The lively retail streets that resulted continue to fascinate city residents and professional planners alike. This is more than nostalgia. The eyes and ears that abound in active retail districts, as Jane Jacobs pointed out in *Death and Life of Great American Cities,* make for safe streets and safe cities. As a result, ground-floor retailing is a common prescription for ailing neighborhoods.

Retail uses cannot occupy every city street. Each retail establishment has a minimum amount of sales below which it cannot operate. In 2006, for example, a typical shoe store occupied between 2200 and 3200 square feet of floor area and grossed between $376,000 and $1,142,000 in annual sales.[1] For the volume of sales to reach these levels, a shoe store needs a critical mass of customers.

Los Angeles, 1996. Automobile-oriented retail strips along Ventura Boulevard in Encino. *(Alexander Garvin)*

Manhattan, 2011. The tall buildings along 86th Street provide a customer base that, when combined with people coming by subway and bus, supports intense retail activity. *(Alexander Garvin)*

Customers do not travel long distances to buy shoes. If the store is on a city street, it will depend on a great deal of walk-in business. This can occur only where nearby buildings are quite large and the surrounding district quite dense. Consequently, the shoe store is likely to be located in a high-rise office district, in a high-rise residential district, or on a busy neighborhood street lined with other stores. If it is located in the suburbs, where customers come by car, it is likely to be in a large shopping center or on a heavily trafficked arterial; in either case it will need plenty of parking spaces.

One way to increase customers for walk-in retailers is to stage street fairs. Every summer Saturday or Sunday, some New York City street is host to merchants who rotate their attractions. Vehicular traffic is obstructed for one day, but customers spill over from outdoor counters set up along the center of the roadway to the retailers that line the sidewalk. Delray Beach, Florida, has taken this idea one step further. On Thursday nights all sorts of entertainers perform along the length of Atlantic Avenue, attracting throngs of tourists and retirees from nearby communities who enjoy live entertainment and invariably also visit the retailers, whose stores and restaurants profit from the added business. This same concept of selectively heightening the retail environment either by intensifying its for-sale offerings or augmenting them with alternative entertainment can also be seen at the regular "sidewalk sales" hosted by suburban retailing centers.

The character of any retail street is a reflection of both pedestrian and vehicular traffic patterns, the customers it serves and the income of those customers, and the location and character of its competitors. There are enough customers on the streets of downtown Philadelphia, Chicago, or New York to support virtually continuous street-front retailing. But, even if the residents of a street in a suburban one-family-per-house subdivision wanted retail stores on their street (and most do not), there are not enough nearby customers to support such stores. The residents have no alternative but to do their shopping at places easily accessible by car. For this reason, twenty-first-century America is replete with busy arterials lined with gas stations, drugstores, supermarkets, shopping centers, big-box retailers, and fast-food restaurants. Making sense of this retail environment requires understanding how all its occupants operate.

Delray Beach, Florida, 2005
(Alexander Garvin)

Public Markets

Many public markets sprang up spontaneously where people traded goods. They bought and sold farm products, handicrafts, and a wide range of natural and manufactured products. Other markets, like the Agora in Athens or Trajan's Market in Rome, were government-created. Although new forms of retailing long ago supplanted the public market as America's main supplier of goods, all sorts of markets continue to flourish in the United States: farmer's markets, food markets, crafts markets, flea markets, antique markets, and so on. There are open-air markets, shed-roof structures with stalls for vendors, and market halls with a full range of utilities, refrigeration, and storage. Even whole market districts made up of independent businesses may still be found serving local needs.[2]

Prior to the twentieth century, public markets were a major component of American retailing. As late as 1918, the U.S. Census Bureau noted that there were 237 municipal markets in 227 cities with populations greater than 30,000 inhabitants. At that time New Orleans contained 23 neighborhood markets, Baltimore had 11, New York City had 9, and Pittsburgh 6. Some of the better known, like Cleveland's West Side Market, Philadelphia's Reading Terminal Market, and Los Angeles's Grand Central Market, are still in active use.

After World War I, however, the use of public markets declined precipitously. Widespread use of refrigeration allowed produce to be shipped much greater distances and to be kept fresh for longer periods of time. As a far greater proportion of the nation's produce came to be shipped in refrigerator trains or trucks, public markets lost business to truck-supplied retail stores, especially chain stores that offered a greater variety of produce at cheaper prices. Consequently, the importance of these public markets greatly declined.

Seattle, 2007. Pike Place Market. *(Alexander Garvin)*

Pike Place Market in downtown Seattle is probably the most prominent of America's public markets. Established as a street market in 1907 in response to citizen outcry against high prices, farmers began selling their produce in the first of several market buildings in only three months' time. Within 20 years more than 600 farmers had permits to sell their goods in Pike Place Market's 500 stall spaces. They served 25,000 people per weekday and 50,000 on weekends.[3]

Like so many other markets, Pike Place Market began a steady decline in the 1940s. In 1963, the mayor and city council proposed replacing its deteriorating market stalls with office buildings, apartments, a hockey arena, and a 4000-car garage. Opponents, organized by architect Victor Steinbrueck, collected enough signatures to place a "Save the Market" referendum on the ballot in 1971. It won 76,369 to 53,264. Pike Place Market is now a national landmark. In 2010, this 22-acre complex was visited by more than 9 million people, who patronized stalls operated by more than 100 farmers, 190 craftspeople, 240 street performers and musicians, and 300 food merchants, restaurants, and owner-operated businesses.[4]

In the closing decades of the twentieth century, the public market no longer seemed an obsolete form of retailing. Many

Charleston, 2011. City Market has served the city since the early nineteenth century. *(Alexander Garvin)*

Charleston, 2011. Local farmers, who cannot afford downtown rents, opened a green market that attracts residents from all over the city. *(Alexander Garvin)*

consumers placed a greater value on high-quality merchandise. Others, who sought direct contact with merchants and the intimacy of small stores, flocked to public markets in an effort to fight against the impersonal character of large, national chain stores. Immigrants, especially from the Far East, began growing a much larger variety of fruits and vegetables. A larger number of artists and artisans sought outlets for their wares. The combination has spurred a revival of public markets. Old markets like Charleston's City Market, which had been in operation downtown since 1839, experienced a boom in retailing to tourists. As a result, its green market merchants moved to a new open-air facility in another part of the city. In fact, the popularity of green markets is pervasive: the number of farmer's markets in the United States doubled between 2001 and 2011.[5]

Department Stores

Nineteenth-century merchants devised the department store as a way of supplying large numbers of customers with the widest possible range of goods at low prices. Its emergence revolutionized consumption and made possible the mass distribution of a quantity and variety of goods that previously had been unthinkable. Before the advent of department stores, consumers bargained individually with merchants. There was no uniformity in the quality or price of what was sold. Department stores, on the other hand, offered merchandise at prices that were fixed and clearly marked. Customers could have their purchases delivered. If a product were faulty, the store could exchange it or offer a credit. These innovations were possible because of the way department stores were managed and financed.[6]

Most department stores began as small dry-goods outlets. Lord & Taylor, for example, opened in 1825 on Catherine Street in Lower Manhattan and grew by accretion till it moved to Grand Street in 1853, to Broadway at 20th Street in 1872, and finally to Fifth Avenue at 39th Street in 1914. Its first suburban store opened in Manhasset, Long Island, in 1941. Seventy years later, Lord & Taylor operates more than 50 stores throughout the nation.[7]

The word *department* explains how a single store was able to offer so large a quantity and so wide a variety of goods. While department stores were owned and administered as individual companies, each department handled its own goods, operating as a semi-independent unit with its own budget, marketing strategy, and personnel.

Department stores are and always have been large businesses that purchase in bulk and, for that reason, obtain lower prices from their suppliers. The key to their success is fast turnover of the merchandise. They sell goods at prices not much higher than what they pay suppliers and thus low enough to sell in great quantities. They pay suppliers after sales have been completed. As a result, these cash transactions allow them to reimburse suppliers quickly enough to avoid inventory financing.

Mid-twentieth-century department stores followed their customers to the suburbs. In the beginning, the stores themselves were the major actors. In 1940, when the May Company opened a new store on the so-called miracle mile of Wilshire Boulevard in Los Angeles, its aim was to capture the area's emerging market. Seven years later, it opened another branch at the Broadway-Crenshaw Shopping Center a few miles away. Although the May Company was still following its customers into suburban Los Angeles, its role had shifted from property owner and developer to tenant. By the twenty-first century, department stores rarely owned and developed property. They no longer were at the cutting edge of retail innovation or urbanization. National chain stores and shopping center developers had become the major players. Nevertheless, the role of department stores is central to understanding urban and suburban retailing, because these stores continue to play a major role as anchors in regional malls.

Shopping Centers

Shopping centers are more than purveyors of everyday consumer goods. They are outlets for surplus leisure time, for singles looking for fun, for young parents seeking entertainment for their children, for customers seeking a "safe" taste of excitement, and for out-of-town visitors eager for a good time.

According to the International Council of Shopping Centers, as of 2012 there were 111,348 shopping centers in the United States. Their 7.4 billion square feet of leasable area generated more than $2.33 trillion in sales, supported 12.2 million jobs, and represented over 15 percent of U.S. gross domestic product.[8]

Country Club Plaza, Kansas City

By the 1920s it had become clear that many customers preferred shopping at a single location for all their needs. In many cities, huge numbers of families were traveling by automobile to shopping destinations, but none of the extant forms of retailing had been designed to accommodate this massive onslaught of cars. Then Jesse Clyde Nichols, a Kansas City real estate developer, found a way to combine modern retail merchandising with automobile travel, in the process inadvertently creating America's first planned shopping center.

Nichols began assembling and developing the 5000 acres that were to become Kansas City's Country Club District in 1906. At that time the area consisted of undeveloped land at the southern extremity of the city. Nichols thought he could improve lot sales by providing a retail center within which

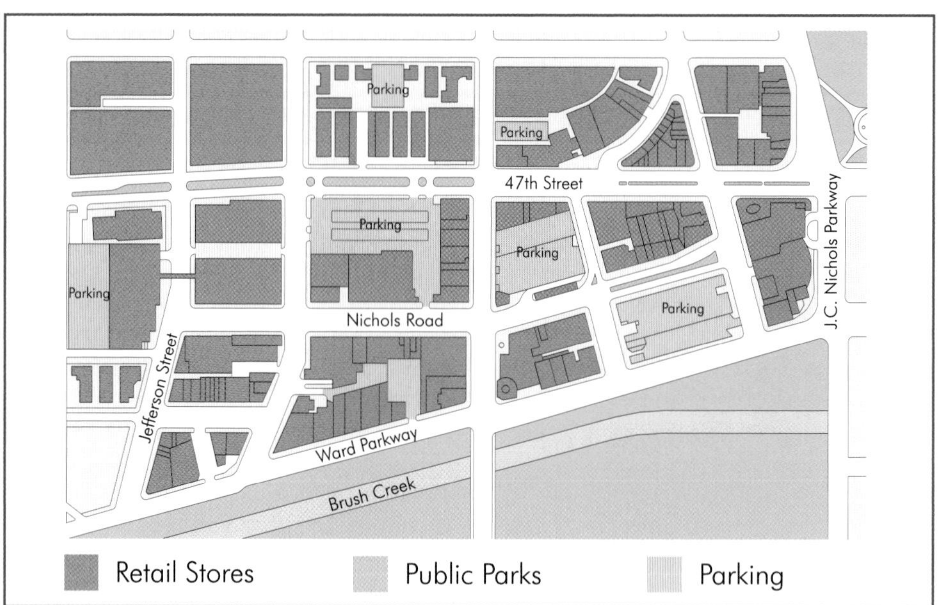

Kansas City. Country Club Plaza. *(Alexander Garvin and Ryan Salvatore)*

residents could do their shopping. In 1920 Nichols began to conceive a commercial center that would serve both the residents of his growing new community and the increasingly suburban population of Kansas City. Today, when residential development extends southward for miles, the idea appears obvious. At the time, establishing a major retail center in the middle of nowhere was thought to be preposterous. Nichols nonetheless recognized the potential impact that the automobile would have on the physical space of retailing, and since most homeowners traveled by car, he integrated off-street parking lots, off-street freight-handling areas, and garages into the retail district that he called Country Club Plaza.[9]

Cities had long included groups of stores within a single building in single ownership that were unified by a consistent architectural image. Riverside, Illinois (1871), Roland Park, Maryland (1896), and Lake Forest, Illinois (1916–1917) are among the early examples of such retail blocks in suburban communities. Country Club Plaza was fundamentally different, not because it eventually included a greater variety of commercial activity and occupied a much larger territory but because the incorporation of parking and freight handling into its design and Nichols's innovative merchandising techniques distinguished the plaza (and every true shopping center) from the earlier retail blocks.

As Country Club Plaza evolved, Nichols came to understand that it needed more than convenient shopping. By preplanning the location of major stores and carefully selecting tenants, Nichols was able to manipulate the flow of customers from their point of arrival (parking) to their ultimate destination (retail anchors). The anchors were the magnets that attracted the customers. Shoppers traveled to them to do their significant shopping. Consequently, retail anchors were in a position to bargain for lower rents. Nichols surmised that he could make up that lost revenue only if he could rent the remaining space at higher prices. For smaller retailers to afford high rents, however, they had to have high sales revenue. Nichols selected their locations so that nearby anchors would bring them the most customers. For example, he rented space near the movie theater to establishments that served food and beverages.

Retailers are entrepreneurs, and J. C. Nichols was a retail entrepreneur *par excellence.* At first the retailers to whom he was trying to rent stores were wary of leasing them "in the middle of nowhere." Although they were eager to capture the larger market that would eventually occupy the Coun-

Kansas City, 1981. Country Club Plaza—America's first planned shopping center. *(Alexander Garvin)*

Kansas City, 2011. Economic activity in Country Club Plaza has spilled over, generating substantial nearby construction. *(Alexander Garvin)*

try Club District, they worried about the period before there were enough homeowners to support the new stores. Nichols's solution was the *percentage lease.* A percentage lease is an agreement in which the rent that tenants agree to pay is equal to a stipulated percentage of their gross volume of sales. It allowed the Nichols Company and its tenants to gamble on and profit from the future success of the area.

Another feature Nichols pioneered was common merchandising. He established regular opening times, promoted special events (e.g., seasonal festivals, Christmas decorations, and outdoor musical performances), and advertised on behalf of the entire shopping center. Customers who would not otherwise have come to a particular shop were attracted to the center by these merchandising strategies. The resulting increased level of sales allowed the small stores to pay higher rents.

The least appreciated of Nichols's innovations was his insistence on an easily identifiable image for the plaza. He engaged Edward Buehler Delk, a young architect from Philadelphia, who submitted his proposals in 1922. The concept and image of the center, however, were Nichols's. He called his shopping center the "Plaza" because he wanted to reproduce the character of the colorful Spanish plaza marketplaces he had so admired on his trips to Europe. For this reason, Delk designed buildings with ornamental ironwork, tile, balconies, fountains, courtyards, and towers inspired by the architecture of Seville, Spain. The result was a 978,000-square-foot retail center with more than 100 stores, off-street parking for 4300 cars, plus 700 on-street parking spaces.

In order to facilitate automobile circulation and encourage pedestrian exploration and spending, Nichols minimized block size and maximized the number of streets and intersections. Only 54 percent of the plaza's 40 acres are in commercial use. The rest is used for circulation. This allows complementary groups of stores to attract customers who zigzag on streets and sidewalks from one block to another.

The area around Country Club Plaza grew and grew. By the beginning of the twenty-first century, it was surrounded by office buildings, hotels, and more than 5000 apartments. In fact, the plaza is not just a shopping center. It is a first-generation regional subcenter, much like Hollywood in Los Angeles or Oakland in Pittsburgh, all of which grew up long before the Interstate Highway System made possible the growth of a second generation of regional subcenters such as Post Oak Galleria in Houston, Tysons Corner, Virginia, and the Buckhead section of Atlanta (discussed later in this chapter). Creating and managing shopping centers has evolved considerably from Nichols's pioneering efforts.

A Shopping Center Hierarchy

The simplest and cheapest form of shopping center is a strip of stores that open onto at-grade parking lots visible from major roadways. These strip centers usually offer convenience goods (food, drugs, and sundries) and personal services that meet the daily needs of nearby neighborhood customers.

Supermarkets, drugstores, and small variety stores are typically the anchors of neighborhood shopping centers. They attract customers whose main reason for being there is proximity. Nobody wants to travel very far for a loaf of bread or a bottle of aspirin. For this reason, neighborhood centers usually serve a trade area of 3000 to 40,000 people (the greater the purchasing power, the smaller the number of people required to support the center; see Table 4.1).

Large discount stores, variety stores, and junior department stores initially were the typical anchors of community shopping centers. By the late 1980s, big-box stores specializing in books, toys, sporting goods, building supplies, hardware, and office supplies began to supplant them as anchors for community shopping centers that served a population of 40,000 to 150,000 people.

Regional shopping centers are usually organized around department stores and, like Country Club Plaza, include a large number of stores. Such centers require a supporting

Table 4.1
CHARACTERISTICS OF SHOPPING CENTERS

Type of center	Leading tenant (basis for classification)	Typical GLA, square feet	General range in GLA, square feet	Usual minimum area, acres	Minimum population support required
Neighborhood	Supermarket	50,000	30,000–100,000	3–10	3000–40,000
Community	Junior department store; large variety, discount, or department store	150,000	100,000–450,000	10–30	40,000–150,000
Regional	One or two full-line department stores	450,000	300,000–900,000	10–60	150,000 or more
Superregional	Three or more full-line department stores	900,000	500,000–2 million	15–100 or more	300,000 or more

source: Michael Beyard and W. Paul O'Mara, *Shopping Center Development Handbook*, Urban Land Institute, Washington, DC, 1999, p. 8.

population of more than 150,000 people. Their customers take advantage of the greater choice of products and more diverse range of retailers, and therefore will travel from greater distances to these centers. Toward the end of the twentieth century, however, department stores no longer provided a broad enough range of shopping opportunities. In response to this changing customer demand, a new generation of anchors has been emerging, including cineplex movie theaters, elaborate food courts, specialized entertainment attractions, and big-box and off-price retailers.

How Shopping Centers Work

Streets and sidewalks may have played an important part in Country Club Plaza's success as a shopping center, but the early developers of this form of retailing believed the key element was the off-street parking. Hugh Prather, the developer of the Highland Park Shopping Village (1929–1932) outside Dallas, Texas, literally made parking the center of his project. Customers could drive off one of two major arterials directly into a rectangular parking field onto which the stores faced. The Broadway-Crenshaw Center, which opened in Los Angeles in 1947 and was rebuilt in 1987, took a different approach.[10] Although its 550,000 square feet of stores appeared to front Crenshaw Boulevard, they really opened on the parking lot behind that had enough room for 2500 cars. A more common approach, however, was to completely surround the shopping center with cars.

In 1946, when the last segment of Hudson's Department Store in downtown Detroit was completed, it was the tallest and the second largest department store in the world (2.1 million square feet of floor space), exceeded only by Macy's in New York. Its owner, the J. L. Hudson Company, realized that its customers were moving to the suburbs and decided to follow them. Northland Center (1950–1954) was the first of several shopping centers it developed in suburban Detroit. The design, by Victor Gruen Associates, beautifully diagrams Nichols's ideas on the arrangement of the components of a shopping center.

Northland was a 147-acre site, surrounded on all sides by major arterials. Most of the site was occupied by a parking lot for 9500 cars, in the center of which Hudson's erected a three-story department store. Gruen arranged 143 small stores in a pinwheel shape that forced customers coming from their parked cars to pass these stores on their way to their department store destination. This scheme worked perfectly because of the relationship between the roads, the parking, the small stores, and the customer's ultimate destination: Hudson's. By the twenty-first century, Northland, which was enclosed as an air-conditioned mall in the 1970s, had been surrounded by thriving suburban communities built on vacant land.

I. M. Pei's design for the Roosevelt Field Shopping Center on Long Island also was based on a single initial department store (Macy's) with the small stores arranged in rows leading to it. When Roosevelt Field opened in 1956, the de-

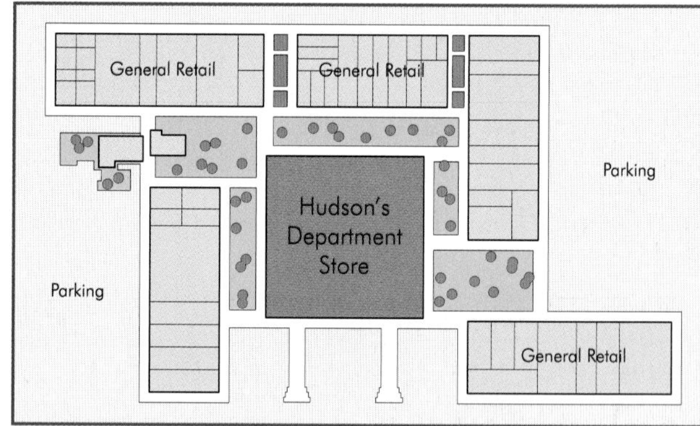

Detroit. Northland shopping center customers whose destination was Hudson's department store had to pass smaller stores to get there. *(Alexander Garvin and Ryan Salvatore)*

Detroit, 1967. Northland shopping center was the first of dozens of shopping centers that were created in the effort to attract the growing market of customers that was moving into suburban Detroit. *(Alexander Garvin)*

veloper (Webb & Knapp) had no trouble renting stores next to Macy's. Rents decreased relative to distance from the department store. The stores at the far end could not be rented until William Zeckendorf (CEO of Webb & Knapp) entered into an agreement in which Gimbel's agreed to occupy that end for the sum of $1. With the addition of Gimbel's, Roosevelt Field demonstrated the effectiveness of an old formula: anchor stores at each end of a pedestrian mall lined with smaller retailers. Like Northland, it was later remade as an air-conditioned shopping mall.

A range of other anchor-based formulas evolved. Among the more successful of these in suburban Chicago were Park Forest Plaza (1950), Old Orchard Shopping Center (1956), and Oakbrook Center (1962), designed by the architectural firm of Loebl, Schlossman, and Bennett for Philip Klutznick's Urban Investment & Development Company. These shopping centers were successful when they opened because they adhered to the sequential marketing principle that Nichols had pioneered at Country Club Plaza. As their principal designer, Richard Bennett, explained: a shopping center is "a frame for the picture of the love affair between a customer and piece of merchandise."[11]

Skokie, c. 1960. Old Orchard Shopping Center. *(Courtesy of Urban Investment and Development Company)*

Although Klutznick dropped Nichols's thematic decoration, his shopping centers were quite special because of lavish landscape designs by Lawrence Halprin. Many other developers built sparsely landscaped walkways lined with stripped-down retail structures. They had trouble competing with the growing number of air-conditioned shopping malls. Some, like Northland, Northgate, and Roosevelt Field, eventually enclosed and air-conditioned their pedestrian walkways. Others, like Old Orchard and Oakbrook, were expanded, renovated, and repositioned so they could compete with newer shopping facilities.

Another approach was to create projects with a distinctive architectural image, such as the one that Nichols devised for Country Club Plaza. Borgata (1981), an 87,000-square-foot specialty center in Scottsdale, Arizona, appeared in the garb of a walled Tuscan town. Irvine Spectrum Center (1995–1998), a 480,000-square-foot complex in Orange County, California, created a pseudo-Moorish image. By satisfying consumer nostalgia for exotic retail environments, they cre-

Scottsdale, 2010. The Borgata is a shopping center in the form of a Scottsdale walled Tuscan town. *(Alexander Garvin)*

ated the same sort of identifiable image as the ersatz version of Seville, Spain, which Nichols had pioneered in the 1920s. Once customers had been attracted to that image, it was simply a matter of adhering to the rest of Nichols's formula.

All these shopping centers were designed to be independent of anything in their surroundings except the roads on which their customers arrived. Because customers drive away when they are finished shopping, the impact of most shopping centers is largely one of generating traffic. In some cases, shopping centers attract enough traffic to induce neighboring property owners to erect other automobile-oriented retail establishments. The resulting suburban landscape is made up of popular islands of activity located in parking fields connected by traffic arteries.

Irvine, 2010. Shopping at Irvine Spectrum takes place in a Moorish-inspired setting. *(Alexander Garvin)*

Shopping Arcades

It may be difficult for contemporary readers to imagine, but at the start of the nineteenth century, sizable sections of the world's largest cities were without sewers, sidewalks, or paving. Horses, delivery carts, carriages, and other traffic made streets an inconvenient and sometimes impossible place for shopping. Removing retail activity from city streets was a good way to provide customers with an environment unimpeded by competition from other activity or by the vagaries of climate. Off the streets, in a safe, clean, attractive environment, customers would be free to examine goods and decide what to purchase.

Like Trajan's Market in Rome and the bazaars of Isfahan, Persia, the galleries, passages, and arcades erected during the first decades of the nineteenth century were an attempt to provide the consumer with precisely this sort of refuge. Seen from above, these early shopping arcades appear to be long, narrow buildings connecting existing streets. On the inside, they are essentially bright, skylit, interior walks flanked with stores.[12]

The contemporary air-conditioned shopping mall is a response to the same phenomenon. Our streets may be paved, sewered, and lined with sidewalks, but they are obstructed by trucks, buses, and automobiles filling the air with noise and pollutants. Often there is nowhere to park and no way to stroll along the street to compare goods and prices. As a result, developers provide structured parking that leads directly to shopping malls where goods can be purchased in an environment free from the noise, foul air, and obstructions of the city street.

The Galleria, Milan

The world's most famous shopping arcade, Milan's Galleria Vittorio Emanuele II, is an excellent example of how to use a shopping facility as the centerpiece of a municipal improve-

London, 2008. Burlington Arcade, completed in 1819—one of the earliest retail shopping arcades. *(Alexander Garvin)*

ment scheme. The ostensible reason for this project was to commemorate the 4500 people who died in the struggle for Italian independence when the French and Sardinians led by King Victor Emmanuel triumphed over Austria at the battle of Magenta in 1859.[13] Presumably, the project was intended to create a major public open space in the center of the city and a dramatic setting for the Duomo (Cathedral of Milan). In fact, the project provided a network of new streets wide enough to accommodate heavy traffic, a connection between the new Cathedral Square and the piazza in front of a similarly important institution, La Scala Opera, and the hub of an emerging downtown business district.

A royal decree in 1860 authorized a lottery whose profits were supposed to pay for the project. Simultaneously, the city held a design competition that attracted 220 submissions, none of which was accepted. Instead, an 11-member commission was appointed to find a better scheme. The commission held another competition, which was won by Giuseppe Mengoni, an architect from Bologna.[14]

Two problems remained: public opposition and financing. People protested the destruction of 6 acres in the middle of old Milan. One of the most vocal opponents was the newspaper *Pungolo,* whose offices were in a building scheduled for demolition. In those days, preservationists usually failed to stop public improvement projects. Financing was the more serious obstacle because the lottery had raised only 1 million of the 15 million lira required for demolition and construction. The government of Milan sought assistance from the private sector to finance the project.

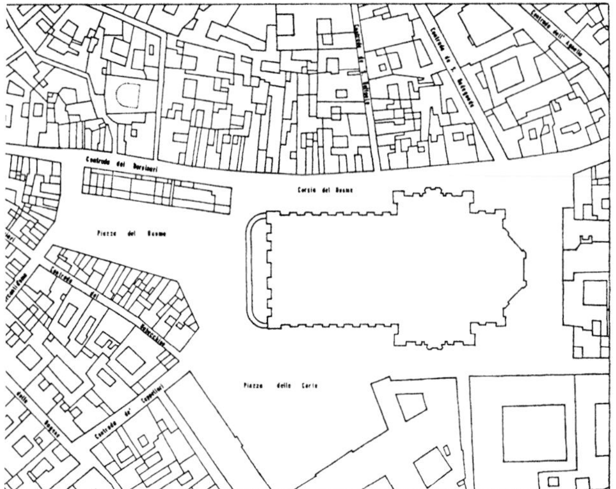

Milan, 1850. Plan of the area around the cathedral prior to redevelopment. *(From Johann Friedrich Geist,* Arcades, *MIT Press, Cambridge, MA, 1982)*

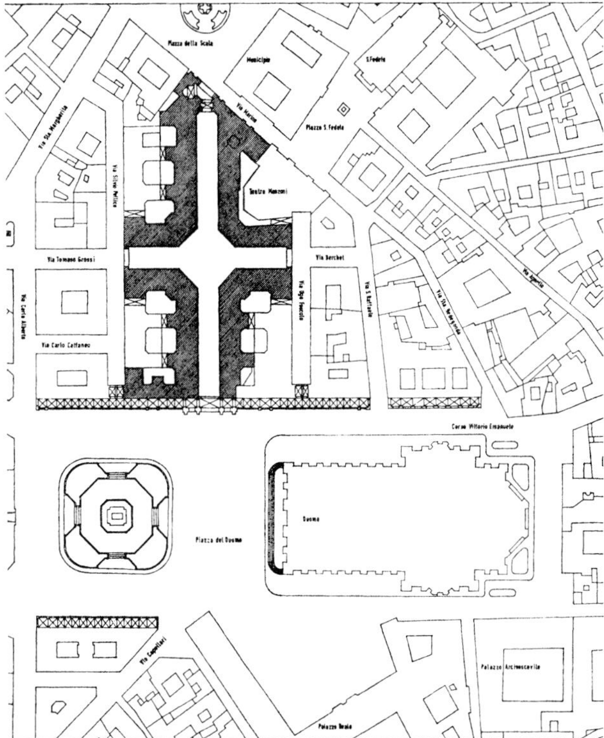

Milan, 1900. Plan of the area around the cathedral after redevelopment. *(From Johann Friedrich Geist,* Arcades, *MIT Press, Cambridge, MA, 1982)*

Milan, 2012. Pedestrians protected from the weather walk through the Galleria Vittorio Emanuele, sometimes making purchases or stopping for coffee, on their way to and from nearby sections of the city. *(Alexander Garvin)*

In 1864, the city entered into an agreement with an English firm, the City of Milan Improvement Company, Ltd., to finance and build what would become the city's central tourist attraction. In exchange, the developers were promised a 5 percent return on their 16 million lira investment.

The Galleria Vittorio Emanuele opened to the public in 1867. It is cruciform in shape and has a four-story interior façade concealing what is really a seven-story building containing 1260 rooms. The two arms, 645 and 345 feet long, respectively, are covered with a glass vault that becomes a glass dome at the crossing. The ground level and mezzanine are filled with shops, cafés, and restaurants. The next level consists of offices and studios, and the top four floors include residential apartments.

The galleria is no narrow pedestrian passage accommodating retail trade along a connection between streets. It is an entire district, built to a scale that even Caesar would have found impressive. Nor is the galleria intended to serve the needs of a particular class of customers or of a specific residential neighborhood. It is intended to serve the complex needs of the vast numbers of people who crowd their way through downtown Milan. More than a century after it was completed, it serves thousands of tourists who mingle with local residents as they stroll, shop, eat, and hustle through one of the world's most successful public spaces.

The Galleria Vittorio Emanuele was extremely important because of its extraordinary influence on later development. Newspapers treated its opening as a national event. Architectural journals printed elaborate analyses. Turin, Genoa, and Naples followed suit, building arcades on the pattern and scale set in Milan. GUM in Moscow, the Queen Victoria Building in Sydney, Australia, and the Old Arcade in Cleveland are some examples of its influence outside Italy.[15]

Building codes enacted at the beginning of the twentieth century required arcades to prevent smoke accumulation and meet modern safety standards. These code requirements made shopping arcades prohibitively expensive. While architects and planners continued to dream of vast, day-lit interior shopping spaces, they just could not get government officials

Cleveland, 2012. Old Arcade, completed in 1890, is still functioning as an attractive retail emporium set apart from the noise, fumes, and traffic of local streets. *(Alexander Garvin)*

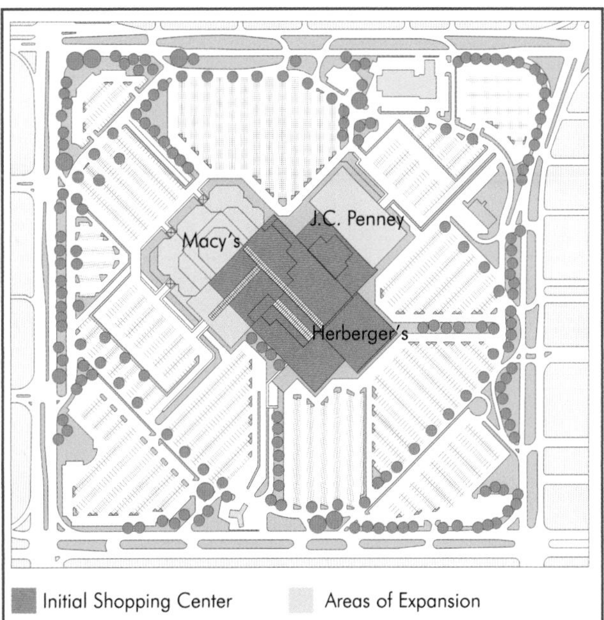

Edina, 1956–1991. This plan of Southdale shows both the original scheme and the 1991 expansion. *(Alexander Garvin and Ryan Salvatore)*

to permit their construction without prohibitively expensive means of egress, ventilation, or materials.

Southdale and the Revival of the Galleria

In 1956, the arcades of the nineteenth century were resurrected in Edina, Minnesota, a suburban community outside Minneapolis. Victor Gruen, then America's premier shopping center architect, understood that "with extremely cold winters and very hot summers . . . outdoor public pedestrian spaces would be attractive to shoppers during only a few days of the year."[16]

No suburban shopping center developer wanted to lose customers for even one day. Gruen proposed to solve the problem through climate control. His department store client (Dayton's) agreed. The result was Southdale, America's first air-conditioned shopping mall, a 679,000-square-foot complex that initially included two department stores, 64 shops, and parking for 5200 cars, all grouped around a two-story, skylit, climate-controlled "garden court."[17]

Southdale opened a new era in marketing. It demonstrated that climate-controlled shopping arcades, adapted to the requirements of modern merchandising, were more profitable than open-air shopping centers. From then on, many developers switched to building air-conditioned shopping malls.

Southdale was so successful that in 1972 the Dayton Hudson Corporation expanded the center, adding a third department store and 43 specialty stores. In 1978 it sold Southdale to the Equitable Life Assurance Society, which decided 10 years later to expand again. When the remodeling was completed in 1991, Dayton's had a new 370,000-square-foot store. Its old location was converted into 50 additional shops and a food court was built above them. Following yet another renovation in 2011, Southdale today boasts all the typical hallmarks of American malls: 4 anchor stores, 120 other retailers, a 16-screen cinema, and a food court populated by some of the nation's most fashionable large-scale eateries.

When Gruen designed Southdale, no retailer had had any experience with a climate-controlled garden court. By the time the 1990–1991 renovation took place, the design of air-conditioned malls had become routine. Experience had demonstrated that the 105-foot width at Southdale was excessive. The new extension was only 58 feet across. Additional recessed downlighting and indirect cove lighting augmented the new daylight provided by a steeply ridged skylight that replaced the old sloping skylight. The original pair of scissor escalators that had blocked customer flow and their views of some of the storefronts were replaced with two pairs of escalators that minimized these interruptions. The clutter of food stalls, kiosks, tables, benches, displays, and artwork that disrupted customer traffic trying to cross the garden court was replaced by a line of pushcarts that cleverly reduced the apparent width of the court and channeled consumers as they zigzagged between pushcarts and store windows.

It is more than half a century since Southdale opened. Because it has been skillfully managed, expanded, and remodeled, Southdale remains a major shopping destination in suburban Minneapolis despite its location a few miles

Edina, 1956. Southdale quickly became a destination for residents of nearby Minneapolis suburbs. *(Courtesy of the Victor Gruen Collection, American Heritage Center, University of Wyoming)*

away from the country's largest air-conditioned shopping center, Mall of America. Once Southdale had demonstrated the effectiveness of a climate-controlled shopping environment, air-conditioned malls proliferated, and not just in the suburbs. In Houston, Atlanta, and other cities, developers erected climate-controlled retail centers that grew into entirely new regional downtowns. Victor Gruen brought his design to downtown Rochester, where it eventually failed. Cities like Santa Monica and Portland were more successful in using an air-conditioned mall to attract the customers that they would otherwise have lost to the suburbs.

Suburban Malls

The construction of the Interstate Highway System made it relatively easy for developers to purchase large, vacant, highway-accessible sites with plenty of room for parking. Many simply followed Gruen's prototype: department stores at both ends connected by a two-story, shop-lined, skylit pedestrian mall. If there were more than two department stores, the mall might be bent, thereby creating additional sites for an anchor store at the bend. For example, the scheme devised between 1961 and 1965 for the 1.6-million-square-foot NorthPark Center in Dallas took the form of an L. Shoppers walked past 160 smaller stores on their way to a department store at the end of each arm of the L. In 2005 it was expanded by adding two more arms, thereby completing a rectangular, skylit shopping environment, plus 110 more small stores, a 15-plex movie theater, an international food court plaza, two more department stores, and a 1.4-acre central landscaped garden.[18]

Retailing is an activity in which the players are in continuous competition to satisfy changing consumer tastes. One approach is to expand and update, as Southdale and countless shopping centers have done. Another is to offer something that is not available at other malls. The 4.2-million-square-foot

Edina, 2005. The clutter that initially disrupted consumer traffic at Southdale was replaced by a line of pushcarts that channeled customers as they zigzagged between pushcarts and store windows. *(Alexander Garvin)*

Edina, 2005. The escalators at Southdale were repositioned to encourage window-shopping and a line of pushcarts was located to channel customers as they zigzagged between the pushcarts and store windows. *(Alexander Garvin)*

Dallas, 2009. NorthPark. *(Alexander Garvin)*

Mall of America adopted this strategy when it entered into competition with Southdale and every other shopping center in metropolitan Minneapolis. When it opened in 1992, its tenant roster of 4 department stores and more than 520 other stores made it the largest shopping mall in the United States.

Bloomington, 2009. Entertainment opportunities make the Mall of America unique among air-conditioned shopping malls. *(Alexander Garvin)*

The pedestrian area takes the shape of a rectangular corridor lined with stores, repeated on four levels. Each corner of the rectangle is occupied by a department store. All four floors open onto a huge central skylit space that differentiates the Mall of America from its competition. Nickelodeon Universe (a family amusement zone including bumper cars, a roller coaster, an aquarium, and a LEGO store) occupies this central space, helping the mall to attract 40 million visitors annually, more than any other attraction in the United States. Since it opened, more than half a billion people have visited the Mall of America, which helps to explain why the average amount spent there per visit is 50 percent higher than the national average.

Regional Subcenter Malls

Because air-conditioned malls tend to be larger and draw more customers than open-air shopping centers, they have played an even greater role in attracting other retailers and businesses to locations made more accessible by nearby highway construction. In the process, they have transformed many undeveloped fringe areas into active regional centers. A typical example is Tysons Corner Center, in Fairfax County, Virginia. In 1967, when Tysons Corner Center opened, it was an 1800-foot-long mall on one level anchored by three department stores. The site, then at the rural urban fringe of Washington, D.C., had been selected because it was enclosed by two major arterials and the I-495 beltway. A third of a century later, Tysons Corner Center had grown to include 5 department stores and 208 restaurants and retail stores, altogether occupying 2.2 million square feet and including a movie theater plus parking for 10,300 cars. Tysons Galleria, across Route 123, includes another 800,000-square-foot air-conditioned mall, anchored by three more department stores, a 400-room hotel, and four office towers. These two are just a few of the retail, office, and residential projects that have clustered together at Tysons Corner, which, with its 33 million square feet of commercial buildings, has become a super-regional center.[19]

Air-conditioned malls have also spurred second growth in already urbanizing areas. The Galleria at Post Oak was initially developed by Gerald D. Hines on a 35-acre site at the intersection of Westheimer and Post Oak Road, just off the Interstate 610 beltway that loops around Houston. When it opened in 1970, the project included a three-level, air-conditioned, skylit 420,000-square-foot retail mall anchored by Neiman Marcus and centered on a year-round ice-skating rink. In addition to 116 stores and restaurants, the Galleria included a hotel, two office buildings, and parking for 7000 cars. Six years later the Galleria at Post Oak was expanded by adding 220,000 square feet to the mall area, 355,000 square feet for Lord & Taylor and Marshall Field department stores, another hotel, two more office buildings, and parking for another 3000 cars. Galleria III opened in 1986, and the entire complex was again expanded in 1994. Following its latest renovation in 2003, this huge complex comprises 375 retail stores within 2.4 million square feet of space. It is the number one shopping and tourist destination in Houston, attracting approximately 24 million visitors per year. While Galleria III was expanding, Post Oak was transformed from the location of a pioneering mixed-use commercial complex into an urban center with high-rise office buildings, apartment towers, and stores for virtually every national chain. In all respects, it has become Houston's second downtown (see Chapter 6).[20]

Post Oak in Houston is but one example of the air-conditioned malls that have become the core of entire new commercial districts. The Buckhead section of Atlanta is another dramatic example of this phenomenon. The area already was the city's most elite residential suburb when Lenox Square (a shopping center anchored by two department stores) opened in 1959. Lenox Square was typical of so many shopping centers: two department stores joined by a landscaped plaza and flanked by more than 50 other stores. Nine years later another mall, Phipps Plaza, opened down the street. Both have been expanded and remodeled several times. As of 2012, Lenox Square contained 1.6 million square feet of air-conditioned retail space, anchored by three department stores; Phipps Plaza contained 825,000 air-conditioned square feet, anchored by three other department stores. During the decades in which these two malls kept expanding, two interstate highways and two branches of the Metropolitan Atlanta Rapid Transit Authority (MARTA) were extended into Buckhead. The market response to these investments has been substantial. As of 2012, Buckhead has become a high-density business district with tens of millions of square feet of commercial space (see Chapter 6).

Urban Malls

At first, air-conditioned shopping malls were suburban phenomena. Rochester's Midtown Plaza, completed in 1962, and New Haven's Chapel Square Mall, completed in 1965 (see Chapters 2 and 7), were urban exceptions. Unlike Southdale, Midtown Plaza was embedded into a block of existing buildings in downtown Rochester. The design, however, had the same flaws. The court was too vast; escalators and planting boxes obstructed zigzag shopping and blocked the view of merchandise displayed in store windows. From the begin-

Houston, 2007. Post Oak Boulevard 37 years after the Galleria opened. *(Alexander Garvin)*

Atlanta, 2007. Lenox Square (on the left) has helped to transform Buckhead into one of Atlanta's major business districts. *(Alexander Garvin)*

ning, Midtown Plaza failed to capture potential retail sales and thus failed to revive downtown Rochester. As in New Haven, the design flaws were serious, but the real problem was declining market demand. Rochester's 1950 population of 332,000 plummeted to 211,000 in 2010, while the Rochester Metropolitan Area increased from 488,000 in 1950 to 1,054,000 sixty years later. Downtown Rochester's potential customers, like those of downtown New Haven, did their shopping in the suburbs.

It was not until developers had perfected the design and merchandising techniques of the suburban shopping mall that public officials found ways of successfully applying them in urban settings. Some city governments sought to retain or attract shoppers by conceiving projects that they believed would contribute to a healthy downtown. Others thought of shopping malls as agents of area redevelopment. Even then, projects could be commercially successful (attractive to the consumer and profitable for the developer) but detrimental to the business district because they pulled customers off the streets and away from existing downtown shopping areas.

Santa Monica Place is a good example of one city's attempt to use a shopping mall to maintain its economic health. Downtown retail activity in Santa Monica during the early 1960s had been steadily declining. The first effort to reverse this decline came in 1965, when the city transformed three blocks of Third Street into a pedestrian shopping precinct with over 100 retail stores. While the new open-air Santa Monica Pedestrian Mall was itself initially successful, the rest of the business district continued to decline. The city council thought an air-conditioned shopping mall could complete the job that pedestrianization had started and capture the retail business that would otherwise go to the large shopping malls then being built throughout metropolitan Los Angeles. It selected a three-block site at the south end of the Santa Monica Pedestrian Mall, just off the Santa Monica Freeway. The idea was to replace the area's parking lots and scattered buildings (containing more than 35 businesses and 14 residences) with department stores, office buildings, and a hotel. In 1972, the city council designated it a redevelopment area and sought proposals from developers. The project, called Santa Monica Place, was developed by the Rouse Company and designed by architect Frank Gehry. It consists of 2 department stores, a three-story, skylit arcade with 163 shops, and 2 six-story garages with 2034 parking spaces.[21]

Just as in Milan a century earlier, there was immediate opposition from site tenants, including United Western

Rochester, 2006. Four years after Southdale opened, the Victor Gruen design for Midtown Plaza continued making the same mistakes. The court was too vast and escalators and other obstructions prevented customers from seeing merchandise and from zigzag shopping. *(Alexander Garvin)*

Santa Monica, 1979. The Third Street Pedestrian Mall prior to the opening of Santa Monica Place. *(Alexander Garvin)*

Newspapers, publisher of the *Santa Monica Evening Outlook*, the city's newspaper. Once again, financing, rather than opposition, proved to be the critical problem. The developers were unable to obtain financing for anything except retail shopping facilities. For this reason, in 1974, the hotel and office buildings were deleted and the site was reduced to two city blocks. Even in spite of this downsizing and the sizable market on the west side of Los Angeles, financial institutions were not willing to provide sufficient mortgage financing to cover the cost of development. A $14.5 million municipal bond issue covering the cost of acquiring the site and building the parking structures provided the necessary additional funds to kickstart development.[22] Thus, what had been a government-sponsored redevelopment project became a government-sponsored shopping mall.

When Santa Monica Place opened in 1980 it became one of the country's most successful retail malls. By 1989 it was generating $130 million in sales ($227 per square foot). It has been a commercial success because it is conveniently located, provides the necessary parking, and offers both variety and quantity of goods. Local customers drive through the streets of Santa Monica. The rest come by freeway from all over the western part of metropolitan Los Angeles. No wonder that during its first year of operation this shopping arcade attracted as many as 125,000 people per week.

Santa Monica Place certainly attracted shoppers who otherwise would not have been in downtown Santa Monica, thereby creating jobs and tax revenue. At least initially, these shoppers drove into the garage and went directly from their cars to the shopping arcade. They had little reason to go anywhere else before driving home. Their purchases had no impact whatsoever on the rest of the business district. Worse yet, Santa Monica Place also drew customers away from the rest of the city, especially from the Santa Monica Pedestrian Mall, which went into decline shortly after Santa Monica Place opened. Many stores were forced to shift their marketing strategy to cater to the lower end of the economic spectrum. Others went out of business. Vacant and boarded-up shop windows became so prevalent that in 1988 the city embarked on a program to revive its pedestrian mall, redesigning the paving and street furniture, building parking structures on both sides of each of the promenade's three blocks, providing zoning bonuses for entertainment-oriented uses, and renaming it the Third Street Promenade (see Chapter 17). These initiatives attracted additional customers to downtown Santa Monica and transformed the Third Street Promenade into one of the most popular attractions in that part of metropolitan Los Angeles. In an ironic twist, it was so successful that it began drawing customers away from Santa Monica Place.

In 1999, Macerich, one of the largest owners of regional and community shopping centers in the United States, purchased Santa Monica Place. Macerich hired the Jerde Partnership and ROMA Design to prepare a scheme that eliminated the skylight in favor of an open-air central space

Santa Monica, 1991. The supergraphics on the south façade of Santa Monica Place are aimed at attracting customers who come by car, shop, and can drive away without setting foot anywhere else in the city. *(Alexander Garvin)*

Santa Monica, 1985. Vacant and underutilized stores on Third Street five years after Santa Monica Place opened, attracting away its customers. *(Alexander Garvin)*

Santa Monica, 2011. Once Santa Monica Place was transformed into an open-air mall and connected to the Third Street Promenade, their synergistic relationship benefited all Santa Monica. *(Alexander Garvin)*

that exposed the complex to Santa Monica's benign climate, removed parking and other elements blocking the view from Third Street, and broke through a direct connection to the Promenade. When the repositioned Santa Monica Place reopened in 2010 it had been repositioned as a high-end part of the Promenade with such tenants as Nordstrom, Burberry, Tiffany, and Vuitton.[23]

If the strategy of building a shopping mall in the midst of the central business district initially failed to revitalize the downtown retail environment of Santa Monica, it succeeded in Portland, Oregon. The site was smaller, comprising only three 200- by 200-foot downtown blocks. Prior to redevelopment, it had been occupied by two- and three-story buildings that had to be purchased and assembled by the Portland Development Commission. As in Santa Monica, the Rouse Company was selected to be the developer in response to a request for proposals (RFP) issued in 1981. The city issued $32 million in bonds to cover the cost of land assemblage, tenant relocation, and construction of a 630-car public garage. The remaining $115 million was financed privately.

When the project was completed in 1990, it consisted of a 60,000-square-foot building for Saks Fifth Avenue, an underground food concourse, a 71-store glass retail pavilion (called Pioneer Place), a 16-story office building, and the public garage. The four-level retail pavilion deviated from the standard air-conditioned mall by locating the anchor department store off-site. Customers who went from the third level of the mall to the second floor of Saks crossed over Yamhill Street via a second-story glass bridge.[24]

Pioneer Place has been a busy addition to downtown Portland for more than a decade. It played an important role in linking the city's retail core, just to the west, with the specialty retailing district that was emerging around the Yamhill Historic District and Saturday Market to the east. More important, the national retailers (Williams-Sonoma, Sharper Image, Banana Republic, etc.) to whom Rouse rented were locating in Portland for the first time. They, along with the first Saks in the northwest United States, acted as a magnet to draw shoppers who ordinarily would have patronized suburban shopping malls. The project was so successful that it was extended across the street, where a second network of shops was built around a skylit rotunda. Pioneer Place did not single-handedly intensify downtown Portland's role as the region's retail hub or increase the number of people coming to the central business district. It was part of what has been a continuing and comprehensive effort to make downtown Portland the centerpiece of a thriving metropolitan area.

Despite the planning success of Pioneer Place and perhaps because of initial failures in Rochester, New Haven, and Santa Monica, many public officials recognize that the pri-

Portland, Oregon, 2007. Pioneer Place brought Saks Fifth Avenue and other national chains downtown. *(Alexander Garvin)*

Manhattan, 2008. The multistory shopping mall in the Time Warner Center benefits from customers from its office building, hotel, and luxury condos, as well as from the surrounding high-density district. *(Alexander Garvin)*

vate retail industry has a much better grasp on how to create climate-controlled shopping malls than any government-run development authority. Consequently, most new urban malls are entirely private-sector products. Experienced real estate companies such as Urban Retail Properties, the developer of suburban shopping centers like Old Orchard, believed that an urban mall would be more likely to succeed if it had its own customer base. Consequently, the 350 stores that constitute the seven-level retail mall at Water Tower Place in Chicago have the benefit of the population that occupies the components of the 74-story, 758,000-square-foot mixed-use complex comprising 350 luxury condo units, a Ritz-Carlton Hotel, and prime office space. Of course, the development also attracts shoppers from the entire Chicago metropolitan area who come to shop on North Michigan Avenue.

The Related Companies took the same approach to the development of the Shops at Columbus Center, the retail component of the Time Warner Center in Manhattan. Its 40 upscale retailers occupy a relatively small, curved, three-story part of the 750-foot-high, 2.8-million-square-foot center. But they benefit from the patronage occupants of the Time Warner World Headquarters, a 250-suite Mandarin Ori-

Las Vegas, 2010. Daniel Libeskind's Crystals provides City Center with an eye-catching icon that competes for the attention of the throngs of customers on the Strip. *(Alexander Garvin)*

ental Hotel, 225 luxury apartments, CNN's live broadcast production studios, the concert halls of Jazz at Lincoln Center, and numerous restaurants. The most inventive and profitable retail component of the TimeWarner Center is its transformation of usually unrentable basement space into a 60,000-square-foot Whole Foods Market, that draws office workers from Midtown and residents of the West Side of Manhattan.

A location in a high-density, mixed-use complex may not be enough to guarantee customers. City Center is a 16,797,000-square-foot, mixed-use complex developed by MGM Resorts International and Dubai World, covering 76 acres on the Las Vegas Strip. The obvious problem for any retail mall within such a complex is competition from everything else on the Strip. How do you keep the occupants of City Center's 4000-room hotel and casino (Aria), two 400-room boutique hotels, two residences, and condo-hotel from doing their shopping anywhere else on the Strip? How do you attract the crowds coming and going from all the other attractions on the Strip? Part of the answer is by offering fine restaurants and the biggest names in luxury retailing (Gucci, Hermes, Cartier, Versace, etc.). The rest is the work of Daniel Libeskind, the designer of Crystals, the amazing urban shopping mall at the complex. The developers recognized the value of design as a market differentiator, and exploited it to sustain demand for the retail environment and enliven the entire complex. Consequently, anybody going by on the Strip is immediately struck by this shopping mall's astonishing crystalline forms that stand out within the pervasive glitz of everything else in Las Vegas.

Reviving Urban Retail Districts

As early as the 1960s, shopping center development had become routine. Once the anchor stores had signed on, financing was quickly forthcoming. There was no need for fancy design. Occasionally, developers like Klutznick included a landscape setting that they felt improved sales. Most developers, however, removed any design features that might distract customers from shopping. By the late 1960s, people began to yearn for buildings of greater character. Recasting old buildings as an ersatz marketplace became a productive strategy for center cities that wished to successfully compete with increasingly efficient, standardized suburban shopping centers. The thirst for more colorful retail environments that also transformed historical structures began to be satisfied with the opening of Ghirardelli Square in San Francisco in 1968, the evolution of Old Market in Omaha between 1967 and 1970, and the repositioning of Quincy Market in Boston in 1976.

Rather than rely on anchor stores, Ghirardelli Square, Old Market, and Quincy Market rented to merchants who needed less space. The resulting agglomeration of vendors offering food and drink and impulse and specialty items provided a novel shopping experience that was not available in safe, clean, well-managed suburban shopping centers. They capitalized on nostalgia for a bygone era but sanitized the experience, eliminating the odors, the filth, and the dangers that initially had induced shoppers to flee from city shopping districts to the safety of air-conditioned malls.

Omaha, 2007. In 1967, when Ghirardelli Square opened, the owner of the six blocks of Old Market began restoring the lofts and warehouses of this early-twentieth-century business district. By the twenty-first century, it had become a major retail attraction for residents of the entire metropolitan area. *(Alexander Garvin)*

San Francisco, 2006. The factory and warehouse structures of the Ghirardelli Chocolate Company were converted into a 54,000-square-foot urban marketplace that successfully combined nostalgia for old San Francisco with the freshness of a new retail facility. *(Alexander Garvin)*

Ghirardelli Square, San Francisco

After World War II, San Francisco's waterfront, like those in most American port cities, was in decline. As shipping, warehousing, and manufacturing moved away, the northern section of San Francisco's waterfront, known as Fisherman's Wharf, began to be turned back to the seamen for whom it had been named. Even though vacant and underutilized upland warehouses and factories were available at attractive prices, there were not many users.

William Roth was the first developer to successfully adapt these empty buildings for use as restaurants, shops, and tourist-oriented retailing. In 1962, he purchased a 2.5-acre block of early-twentieth-century buildings from the Ghirardelli Chocolate Company, which had decided to transfer production to modern facilities in San Leandro (see Chapter 2).

Roth wanted to transform the various factory and warehouse structures into an attractive shopping facility for tourists. Roth's plan required cleaning building exteriors, gutting their interiors, creating a central plaza with a view of the port, adding new structures where appropriate, and slipping a 300-car garage underneath. What emerged in 1967 was Ghirardelli Square—a 54,000-square-foot urban marketplace that successfully combined nostalgia for old San Francisco with the freshness of a new retail facility.

This combination became an instant favorite with tourists and made Ghirardelli Square the prototype for similar projects in San Francisco and around the country. A few blocks away, the old Del Monte fruit cannery became a dining, shopping, and entertainment center. Seafood restaurants that had long been located opposite the piers began to multiply. Souvenir shops opened everywhere. Within a decade, Fisherman's Wharf had been transformed from a working waterfront into one of San Francisco's major tourist attractions, which it remains to this day.[25]

Quincy Market, Boston

Country Club Plaza established the effectiveness of a powerful (nostalgic) architectural image, structured parking, and sequential merchandising. Ghirardelli Square demonstrated the attractiveness of reusing ostensibly obsolete buildings in formerly congested mercantile districts. However, it was not until 1976, when the three granite market/warehouse structures behind Boston's historic Faneuil Hall were reopened as an urban marketplace, that this prototypical marketing strategy was perfected and came to be copied in practically every sizable American city.

The three buildings known as Quincy Market originally opened for business in 1826.[26] Until then, the area had been underwater. Like much of Boston, it had been reclaimed in an attempt to satisfy the city's voracious appetite for new land. The buildings, designed by Alexander Parris, continued to house the city's produce and meat markets until the site became too congested and the structures too outmoded to continue as an efficient food distribution center.[27]

In 1961, Quincy Market became part of an ambitious scheme by the Boston Redevelopment Authority (BRA) for the redevelopment of the downtown waterfront. Rather than demolish the market, the BRA designated its three structures for renovation. Nine years later, after a series of feasibility studies and a $2 million HUD grant for historic preservation, the BRA finally issued a request for proposals from interested developers. The winning development team, Benjamin Thompson & Associates (architect) and Van Arkle-Moss (developer), was unable to put together the necessary financing. In 1974, the city designated James Rouse (working with

Boston, 2010. Quincy Market. *(Alexander Garvin)*

Benjamin Thompson and Associates as architect) as the new developer.

Rouse was a successful, Baltimore-based suburban shopping center developer, better known nationally as the man behind the new town of Columbia, Maryland (see Chapter 16). When he became involved with Benjamin Thompson and his wife and business partner, Jane Thompson, Rouse had been looking for a site that could demonstrate that urban shopping centers could be as profitable as their suburban counterparts. He realized that Quincy Market was just that. It was located on the edge of the increasingly popular waterfront district, just behind the new City Hall and Government Center (see Chapter 5), next to the financial district, and not far from the city's department stores. There were more than enough daytime workers in the area to support a major new shopping facility, and another 20,000 people lived within walking distance of the site. The most significant factor, however, was the secondary market in the surrounding suburbs. The question he and the Thompsons set about answering was how to attract these affluent consumers to Quincy Market.

They rejected the notion of organizing the project's 6.5 acres of land and 370,000 square feet of interior space around two or three anchor department stores. There was no way to squeeze them into structures that were nearly 550 feet long and 50 feet wide. Moreover, Rouse did not want to compete with Filene's or any of Boston's other existing department stores. Instead, he proposed to base the project on small businesses. The scheme contained 160 small stores occupying 219,000 square feet of retail space plus 143,000 square feet of small office suites. Each occupant would lease a small area, do more business per square foot than more conventional tenants, and thus be able to pay more rent per square foot. Rouse called this new retail form a *festival marketplace.*

At first, financial institutions were reluctant to lend money for so innovative a venture. Except for Ghirardelli Square, there had been little experience with adapting older structures for urban retailing. Moreover, banks and insurance companies could not base their projections on the credit rating of the smaller, less-well-known vendors to whom Rouse proposed to rent. Their customary procedure was to require leases from department stores and major retail chains. Nevertheless, Rouse obtained a $21 million mortgage from Teachers Insurance and Annuity Association and raised $9 million in equity capital. The remaining $10 million came from city, state, and federal programs. In addition, the city of Boston abated property taxes until the project opened and leased the site to Rouse for 99 years at $1 per year. In exchange, the city received 20 to 25 percent of gross income in lieu of taxes.

The central structure at Quincy Market is reserved for food outlets. Adjoining structures are filled with shops offering fashionable clothing, accessories, jewelry, and gifts. Both outdoor and indoor areas are flooded with pushcarts. As Rouse explains,

> *We hired a bright young woman who went out all over New England identifying artists and craftsmen and small entrepreneurs with narrow specialties. She worked on 900*

prospects for those 43 pushcarts, evaluating and recruiting them. We designed the carts and provided boxes and baskets to hang on them.[28]

The pushcarts and food outlets, like the department stores in a suburban shopping mall, became major attractions, drawing office workers, suburban customers from all over the region, and passing tourists, who stop along the way to make additional purchases. Indeed, Quincy Market has been particularly successful because it also tapped Boston's substantial tourist market. Of the 12 million customers who came to Quincy Market in 1981, 60 percent were tourists.

Quincy Market's success exceeded even the most optimistic sales projections. By 1981, annual sales averaged an astonishing $377 per square foot per year for the food vendors and $345 per square foot per year for the other merchants. Rents ranged from $30 to $45 per square foot per year for the north and south buildings and $50 to over $100 per square foot per year in the central structure. Even three years later, median sales per square foot in most regional shopping centers had reached only one-third that of Quincy Market. For the past four decades Quincy Market has remained one of Boston's major generators of tourism and the jobs and taxes that come with it.[29]

The Revival of the Urban Marketplace

Most older cities have several districts that were once filled with people. Virtually all of these districts have buildings of character. For these buildings to be preserved, they need occupants. Ghirardelli Square and Quincy Market demonstrated that tourist-oriented retailers could be those occupants. At first, city governments saw the festival marketplace as a cure-all for decaying mercantile areas. Eventually it became simply another device among many to bring new life to troubled city districts.

Trolley Square in Salt Lake City was one of the earliest and more successful of these. In 1969, a local developer, Wallace Wright, Jr., acquired a 13-acre site that had been the storage and repair facility for the city's electric trolleys. The trolley barns had been built in 1908 but had been out of use since 1945. Wright transformed them into an air-conditioned retail complex he called Trolley Square. When it opened in 1972, this festival marketplace included about 80 shops and restaurants. Unlike Ghirardelli Square, however, over time it lost customers to competing retail outlets. By 2012, over 40 percent of Trolley Square was empty.

Bedford Stuyvesant, New York City's largest African-American neighborhood, became a symbol for the problems and hopes of urban America during the 1960s. While successfully campaigning for the U.S. Senate in 1966, Robert Kennedy initiated a major effort to revitalize the area. Together with Senator Jacob Javits, he organized the Bedford Stuyvesant Restoration Corporation. They successfully fought for extra federal assistance, attracting tens of millions of dollars to the neighborhood. Money poured in for health, education, job training, day care, housing, and every other program Washington had to offer.

Once it got going, the Restoration Corporation sought a project that would be the physical embodiment of the effort to revitalize Bedford Stuyvesant. In 1970, it settled on a vacant dairy located on Fulton Street, the neighborhood's primary retail street. The idea was to transform this empty building and the rest of the block into an urban marketplace that would be a catalyst for the revitalization of the surrounding area. Appropriately, the project was named Restoration Plaza, and it opened in stages between 1975 and 1980, by which time the entire block had been redeveloped. The final design, by architecture firm Arthur Cotton Moore/Associates, includes 115,000 square feet of retail space, 170,000 square feet of offices, an outdoor skating rink, a community center, and underground parking for 150 cars.[30]

Like Ghirardelli Square, the core of the project is a courtyard that leads to all the commercial tenants. The similarity ends there. Restoration Plaza was not geared to impulse buying or tourists. It was completely oriented to a local market. The prime retail tenant, at one end of the block, was one of Bedford Stuyvesant's few supermarkets. The other end is anchored by chain clothing stores.

Although Restoration Plaza began as an attractive shopping center, there is no evidence that it had any beneficial impact on Fulton Street or any other section of Bedford Stuyvesant. It has not attracted one dollar that was not already being spent by neighborhood residents. In fact, it may have drawn customers away from the marginal, locally owned shops that line Fulton Street.

The people who conceived of Restoration Plaza duplicated the courtyard and underground parking of Ghirardelli Square. They reproduced the nostalgic image of the older buildings. However, they failed to understand that design by itself is never an adequate ingredient with which an urban marketplace can generate new economic activity—it must attract a new market. Without the attractions that could bring customers from outside Bedford Stuyvesant, Restoration Plaza could never generate additional economic activity. As a result, during the 1980s the project succeeded only in moving customers from one part of the neighborhood to another and from one group of businesses to another. By the twenty-first century, like Trolley Square, it had lost customers to other sites in Bedford Stuyvesant that were better able to supply the demands of local residents.

Off-Price Retailers

Consumers have an insatiable appetite for a greater and greater variety of goods and services at lower and lower prices. Retail history is an ongoing saga of how that appetite

Collierville, Tennessee, 2007. Walmart. *(Alexander Garvin)*

is satisfied. Businesses appear, grow, wane, and die because of innovations in production, distribution, and marketing. Whether it's nineteenth-century mail-order merchandising or twenty-first-century Internet purchasing, retail business is forever changing.

Several new forms of retailing emerged at the end of the twentieth century that have changed the landscape, competing for customers with both urban retail districts and suburban shopping centers. Some manufacturers started offering goods directly to the consumer at lower prices that reflected the elimination of intermediaries. Developers soon found it profitable to combine these merchants into *outlet malls*. Large-scale, mass-merchandising retailers independently occupying *big boxes* on large sites offered the same goods available from traditional retailers, but at low enough prices to attract away a significant portion of their sales. A different set of developers combined them into *power centers*. Anyone trying to change our cities and suburbs must study these new forms of retailing, because they are rendering obsolete a large part of the previously built landscape, including some first-generation shopping centers.

Big Boxes

Stand-alone stores have always been a significant market force. They include single-purpose retailers selling appliances, furniture, hardware, and other merchandise, as well as variety discount stores. Traditional stand-alone stores were joined at the end of the twentieth century by large national chains occupying boxlike structures. They sell products without being involved with property owners or developers. They also offer a greater variety and wider range of goods in their category of merchandise. Examples include Home Depot, Toys "R" Us, and OfficeMax. They are destinations to which shoppers travel irrespective of any other purchases they may wish to make.[31]

The most important players among big-box retailers are warehouse clubs. Price Club launched the first of these in 1976. It was quickly joined by Sam's Club (owned by Walmart), Costco (which merged with Price Club in 1993), and BJ's. As of 1994, there were 600 warehouse clubs in the United States. Two years later the number had climbed to 1100.

The conventional wisdom was that big-box retailers depended on plenty of space for truck loading and customer parking. This meant suburban sites on major arterials. It did not take long for the industry to augment these sites with convenient city locations. They were not about to allow conventional retailers to monopolize the hundreds of billions of dollars spent annually by city residents. But by 2000 retailers like Staples and Bed Bath & Beyond occupied urban sites that had neither parking nor off-street truck-loading bays, and by 2012 big boxes had become the giants of U.S. retailing.

Five Largest U.S. Retailers	Total U.S. Sales[32]
Walmart	$309.7 billion
Kroger	$ 90.4 billion
Walgreen	$ 72.2 billion
Home Depot	$ 70.4 billion
Target	$ 69.9 billion

Power Centers

Once big-box retailers began to proliferate, the obvious next step was to combine them into a single complex, which came to be known as a *power center*. Power centers first appeared in the 1980s as a form of super shopping center anchored by

Atlanta, 2009. A power center anchored by a large Target store. *(Alexander Garvin)*

Park City, Utah, 2007. Tanger Outlet Center. *(Alexander Garvin)*

major national chain retailers. Much like early shopping centers, they usually were open-air facilities, but with less than 10 to 15 percent of their retail floor area devoted to small shops. Although some strip power centers are smaller, most range in size from 250,000 to more than 800,000 square feet and sit on more than 10 acres. They tend to draw customers from a trade area resembling that of a typical regional shopping center, roughly 200,000 people or more.

Outlet Centers

An outlet center is an aggregation of factory outlet stores, each owned separately by a different manufacturer, but (as with any shopping center) managed collectively. In 1936, Anderson-Little, a New England men's clothing manufacturer, opened the first factory-direct stores. Vanity Fair opened the first multitenant manufacturers' outlet center in 1974 in Reading, Pennsylvania. Since then, outlet centers without anchor tenants have become more and more popular. As of 2012, there were in the United States 337 outlet centers totaling 75 million square feet of retail space.

Initially, outlet centers were established in remote locations. They were used by manufacturers to dispose of excess production, goods that could not be sold to conventional retailers, or irregular or slightly damaged merchandise. As outlet centers began to proliferate, they started to include standard merchandise. They also began to occupy locations that were closer to large metropolitan areas.

Tempe, 1997. Arizona Mills is a 115-acre value mall that combines outlet stores, discount retailers, a 24-screen cinema, restaurants, and a food court, with more conventional national retailers. *(Alexander Garvin)*

Value Malls

The value mall concept was pioneered by the Mills Corporation in 1985 when it opened Potomac Mills, a 1.7-million-square-foot mall that included 16 anchor stores, 24 national specialty stores, and a 15-screen cinema. As of 2001, the Mills Corporation operated 11 value malls with 17 million square feet of store area. The Simon Property Group bought the Mills Corp. in 2007. At that time, the Mills Corporation operated 38 malls totaling approximately 45 million square feet of retail space. Value malls typically draw from wider markets than do conventional shopping centers.

In addition to their broader tenant mix and enclosed layout, value malls are unlike traditional manufacturers' retail outlet centers in that they include significant entertainment, category-killer, and discount components. Unlike manufacturers' outlets that prefer to be near tourist destinations primarily in rural areas, value malls are usually located within or at the fringe of major metropolitan areas. Moreover, apparel manufacturers and retailers searching for growth opportunities have been more willing to locate their outlet stores closer to their traditional stores. Value malls offer them that opportunity.

Prepackaged Urbanity

Although Country Club Plaza was created, owned, managed, and merchandised by a single company, in one respect

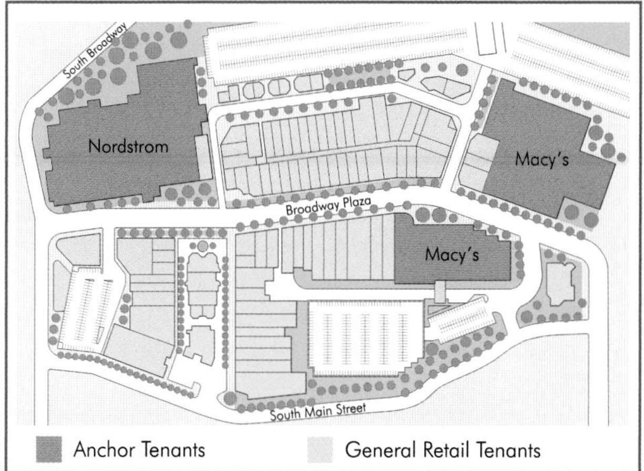

Walnut Creek, California. Broadway Plaza continues to be a retail center whose stores are located along a traditional urban shopping street. *(Alexander Garvin and Ryan Salvatore)*

Beverly Hills, 2013. Two Rodeo Drive combines luxury stores in a setting that is meant to recall their original settings in London, Paris, and Rome. *(Alexander Garvin)*

it is like a traditional city retail district: the plaza is laid out with streets and sidewalks connected to the rest of Kansas City. This approach was adopted by Westwood Village in Los Angeles and Suburban Square in Ardmore, Pennsylvania (outside Philadelphia), as they began to take shape in 1927 and then evolved over the next decade. With the exception of Broadway Plaza in Walnut Creek (outside San Francisco), very few post–World War II shopping complexes chose this route.

Broadway Plaza first opened in 1951 as a complex of 38 stores anchored by Sears and J. C. Penney. Unlike Westwood Village and Suburban Square, however, it was laid out along a single street. As the city and the region grew, Broadway Plaza became an increasingly important retail destination. Consequently, it was expanded and remodeled in 1956 and again in 1967. Its latest transformation occurred after it was sold in 1985. As of 2012, Broadway Plaza had evolved into a 660,000-square-foot complex of more than 90 shops, services, and restaurants, anchored by Nordstrom and Macy's.

The increasing nostalgia for the urban shopping ambience has led to a new group of urban shopping centers, like University Village in Seattle. Between 1985 and 1997, Sloan Capital Companies and Tom Croonquis, its director of development, transformed an obsolete shopping center into University Village, a 400,000-square-foot combination of more than 80 upscale stores that open onto streets with sidewalks and short-term parallel parking, as well as onto pedestrian-only shopping areas. The Reston Town Center (see Chapter 16) is a higher-density version built that way from the ground up.

The simplest way to prepackage an "urban" retail ambience is to provide convenient parking and build what appears to be a complex retail street but is really a shopping center managed and maintained by a single owner. A good example of this is Two Rodeo Drive, the 126,000-square-foot luxury shopping complex built atop an underground 600-car garage that opened in Beverly Hills in 1990. Its customers leave

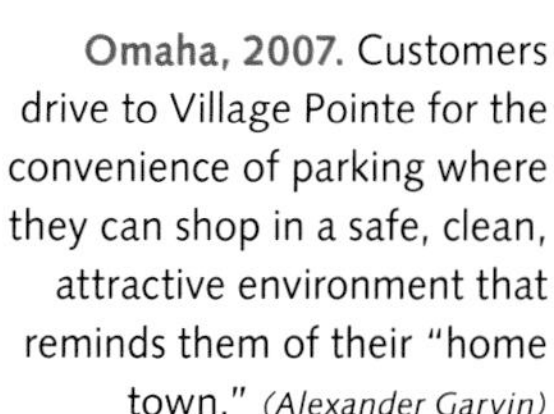

Omaha, 2007. Customers drive to Village Pointe for the convenience of parking where they can shop in a safe, clean, attractive environment that reminds them of their "home town." *(Alexander Garvin)*

Seattle, 2007. An entire entertainment center packaged into a single structure. *(Alexander Garvin)*

their cars to meander through a fantasyland of the most expensive retailers from London, Paris, and New York perched on a condensed version of Rome's Spanish Steps. Three years later, the even more ambitious City Walk at Universal City opened. Customers stroll between an 18-screen cinema and the Universal Studios amusement park. Along the way they can sample the best of Los Angeles without ever having to set foot in the city itself. Its designer, the Jerde Partnership, even added fountains in which children can romp. Visitors know it is safe because it is the location of a substation for the county sheriff. Shopping centers around the country, like Village Pointe in Omaha, are creating ersatz streets as a way of attracting customers seeking prepackaged urbanity—without the filth, noise, or dangers that many believe are common to cities.

Entertainment Centers

What better way to attract customers seeking a fun way to spend leisure time in an urban setting than to combine dining, retail, and entertainment within a secure, pedestrian-oriented environment. Shopping complexes of this sort are currently referred to as *entertainment centers.* City Walk at Universal City is a prime example, as is the Irvine Spectrum Center. They function in a manner similar to conventional shopping malls. Each of the constituent retail-restaurant-entertainment components attracts customers independently. When grouped into a single destination, the combination draws visitors from a greater distance and a wider variety of overlapping market areas than any of its occupants would individually.

These so-called entertainment centers generally include a few brand-name attractions, such as Hard Rock Café or Planet Hollywood; national retailers, such as Nike Town, Disney Stores, and Warner Bros. Studio Stores; large bookstores and music vendors, such as Barnes & Noble and HMV; a megascreen cinema complex; plus bars, restaurants, and nightclubs. In some places, these venues are assembled into a single building; in others, they spill over onto nearby streets. The largest of these is the Power & Light District in Kansas City, Missouri. This mixed-use complex covering nine blocks is adjacent to the Sprint Arena, the Kansas City Convention Center, and the Kauffman Center for the Performing Arts.

Denver, 2009. Entertainment centers like Denver Pavilions combine multiscreen cinemas and restaurants with major national retailers, while bringing customers who spill over onto 16th Street, generating sales in nearby stores. *(Alexander Garvin)*

Kansas City, Missouri, 2011. Power & Light District. *(Alexander Garvin)*

It includes more than 50 restaurants, bars, shops, and entertainment destinations, including two movie theaters, and Kansas City Live! (a covered outdoor plaza to be used for a wide variety of events).

Such prepackaged, decent, safe, and sanitary destinations guarantee that customers will be able to enjoy the frisson of urbanity without the dangers of setting foot in a real city. Calling them entertainment centers, however, is really a misnomer. There are marvelous true entertainment centers in the United States. Las Vegas, Disney World, the Vieux Carré, and Times Square are among the most popular. Like the growing number of prepackaged entertainment centers, they include much that is artificial and tacky. Unlike those standardized products, they pulsate with a vitality that cannot be mass-produced.

Mixed-Use Centers

Another growing form of retailing is the mixed-use center. It is not a new phenomenon. Country Club Plaza is a mixed-use center. So is Westwood Village. Old Orchard included an office building. Post Oak Galleria included hotels and, like Tysons Corner and Lenox Square, spearheaded the evolution of truly mixed-use regional centers.

As suburban areas mature, however, shopping center owners are realizing that the very locational advantages that produce retail success make their properties equally attractive as sites for office and residential buildings. Conversely, the occupants and visitors of residential and office buildings constitute an additional market for retail establishments. As a result, there are an increasing number of retail-based mixed-use developments.

Mizner Park, in Boca Raton, began as an effort by the city's Community Redevelopment Agency to resuscitate a failed 420,000-square-foot shopping center. The agency declared the area blighted and issued a request for a redevelopment plan. The winner, Crocker & Company, and its architects, Cooper Carry & Associates, created a mixed-use complex. Mizner Park opened in 1991. Customers approach by car from Federal Highway. There is a sign for a multiplex cinema and visible parking opportunities, but little to indicate the special character of Mizner Park. After parking, customers enter into a new, two-block-long space with landscaped center islands bounded on either side by one-way streets, on-street parking, and sidewalks. The ground floors of the arcaded buildings on either side of this quasi-boulevard are lined with shops and restaurants. On one side, the upper stories contain 106,000 square feet of offices; on the other side, 136 apartments occupy the upper floors. Mizner Park may not include much to surprise residents of Chicago, New York, and other cities accustomed to dense, complex, mixed-use districts, but it represents an effort to maximize the locational advantages of a shopping complex by adding housing, a hotel, and offices, while offering a prepackaged taste of urbanity to high-end shoppers eager for something more than the usual shopping mall. They are dwarfed by projects like Easton Town Center, outside Columbus, Ohio.

If Santa Monica Place and Quincy Market offer examples

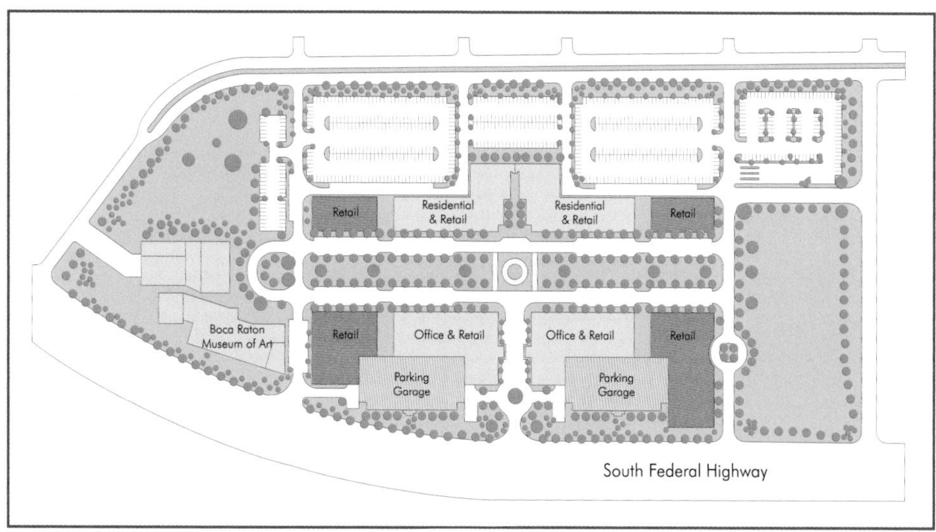

Boca Raton, Florida. At Mizner Park, retail stores occupy the ground floors of low-rise, suburban office and apartment buildings to create a clean, safe, carefully packaged "urban" experience. *(Alexander Garvin and Ryan Salvatore)*

of government-sponsored retail development, Easton provides one for a project driven by a large local business and the private market. Easton is the centerpiece of an ambitious mixed-use, retail-centered project. It began in 1992, when Les Wexner, president and CEO of The Limited Inc. asked the Georgetown Company, a New York–based development firm, to enter into a joint venture for the development of a 1300-acre district adjoining the national headquarters of The Limited. The site, on the Interstate 270 beltway encircling Columbus, Ohio, 10 minutes' drive from downtown, provided a unique opportunity to create a regional center for the northeast quadrant of the city.

As of 2012, Easton consists of over 9 million square feet of office buildings, 1.8 million square feet of retailing, hotels, power centers, recreational facilities, housing, a forest preserve, and, at the center of it all, the 90-acre Easton Town Center. Each section of Easton has been developed based on the financial and design requirements of that particular land use. The corporate office buildings on the oval, for example, open onto a roadway that encloses a very large landscaped and forested area. Visitors can park on the oval in front of an office building; employees park behind the building. One of the residential sections, Easton Commons, is a 330-unit complex of rental row houses whose design is inspired by the crescent-shaped apartment buildings of Bath, England. Both the office oval and the housing are quite different from their competition in the Columbus metropolitan area. But whatever the particular land use, it is part of a complex that is integrated into the arterial grid, frequently terminating the axis of one of Easton's streets.

At the heart of everything is the Town Center, a 1.5-million-square-foot retail complex. Its site design is the outcome of collaboration between the Georgetown Company and planner-architect Jaquelin Robertson, and the buildings themselves were designed by Development Design Group. Its 77 acres are organized around a 10-block-long complex of squares, streets, and sidewalks. The northern end encompasses a fashion district that includes Nordstrom and Lazarus department stores as well as the usual national retailers. At the center is an air-conditioned retail atrium inspired by nineteenth-century European train sheds. Its main feature is a 30-screen cinema. The southern end is a retail village in which Crate & Barrel, Barnes & Noble, 13 restaurants, and a variety of popular stores all open onto streets with sidewalks and on-street parking. The pièce de résistance, however, is a public square with a fountain that attracts both children and adults from early morning until the stores close.

Easton is the product of a remarkable team of dedicated professionals that is unlikely to be duplicated. However, the principles that they have used in organizing and designing Easton have become a model for the increasing number of mixed-use retail-based projects that are appearing around the country. CityPlace in West Palm Beach, Florida, is just one example. It was developed by The Palladium Company, a subsidiary of the Related Companies, in response to an RFP issued by the city following Nancy Graham's election as mayor in 1991. Its design, by Elkus/Manfredi Architects, retained existing streets and converted a Methodist church into the Harriet Himmel Gilman Theater. The 600,000-square-foot project, which opened in 2000, includes a Macy's, an IMAX and a multiplex movie theater, and more than 100 shops and restaurants, as well as 570 residences. In 2007, the 18-story CityPlace Tower was added to the site, Nevertheless, it has had difficulty attracting enough customers to justify development. CityPlace had offices above its retail stores and was conceived to be adjacent to the Dryfoos School of the Arts, the

Easton, 2001. Children from all over the Columbus metropolitan area come to play in this fountain in the Easton town center. *(Alexander Garvin)*

West Palm Beach, 2005. CityPlace. *(Alexander Garvin)*

Palm Beach County Convention Center, and the Kravis Center for the Performing Arts. Thus, unlike Mizner Park, which is basically stand-alone complexes, CityPlace is completely integrated into the life of West Palm Beach. But it has failed to be as profitable as Easton, a much larger and more varied mixed-use powerhouse that continues to grow and to attract customers from all over the Columbus metropolitan area.

Ingredients of Success

The health of a shopping district is a function of the market it can potentially serve and the strength of the competition. If an area is in trouble, it is because one or more of the following are true:

- The market that the district serves has declined in size, income, or both.
- The district no longer serves the tastes of the market.
- Alternative methods of merchandising have rendered the area functionally obsolete.
- Newer, more attractive commercial areas have captured its market.
- Adjacent conditions are having a negative frictional impact, chasing away the district's market.

Improving the health of a retail district by opening a shopping mall or festival marketplace requires that the actors understand which of these conditions is causing the problem and then manipulate the ingredients of project success to alter those conditions. As Pioneer Place demonstrates, construction of a retail facility that brings additional customers downtown can be a significant element in building a healthy business district. There is no guarantee, however, that those customers will spill over for the benefit of the surrounding area. Since retail marketing is in constant flux, as experience in Santa Monica has shown, opening a new facility is only the start. To be effective, every form of retailing must be continually adjusted to local market conditions.

Market

Population trends are central to the effectiveness of retail projects. If an area is experiencing population or income growth, building a new shopping facility (e.g., Pioneer Place) can harness the increasing spending power. It directs the additional spending to a part of town that will profit from that spending without having any serious negative impact on existing retail districts. If population and income are stable or declining, however, a new facility (e.g., Restoration Plaza) will simply shift spending from one part of town to another.

Sometimes spending is a matter of consumer taste. During the 1960s and 1970s, Boston and San Francisco had declining populations. Nevertheless, Ghirardelli Square and Quincy Market successfully attracted customers without damaging other parts of the city because many of these customers were attracted from outside the city.

As long as there are more convenient or attractive alternatives, no new retail facility is likely to capture any significant share of the market. Chapel Square Mall lost customers to suburban shopping centers that were built to attract the growing population of metropolitan New Haven. Like Harborplace in Baltimore (see Chapter 7) or Pioneer Place, new shopping facilities should offer something unique that cannot be obtained elsewhere. Then they will attract people who would not otherwise be in the city, thereby generating sales, jobs, and tax revenues that would not otherwise be available.

One big change in the market has been the growth in purchasing on the Internet. In 2011, international e-commerce alone represented $680 billion, up 19 percent for the year. Best Buy, for example, reduced its average store size from 50,000 square feet to less than half that size.[33] The obvious result is a decrease in retail occupancy and shrinkage in the size of the typical retail establishment, especially department stores and big boxes. That, in turn, is reflected in vacancies on traditional retail streets and in shopping complexes.

Location

Many urban shopping centers exploit the inherent characteristics of their location. Irvine Spectrum is able to offer a wide variety of outdoor amusements because most of the time Irvine's weather is benign. Harborplace and Ghirardelli Square benefit from the adjacent waterfront.

Other retail facilities are more dependent on proximity to their customers. The cafés and restaurants of the Galleria Vittorio Emmanuele, for example, could not survive if their only customers were the occupants of the offices and apartments on the upper floors. They need the additional spending of those who have come to visit La Scala or the Duomo, plus millions of tourists who visit Milan. Irvine Spectrum is located at the intersection of two freeways, which makes it a short drive for tens of thousands of Orange County residents.

Proximity to a new shopping facility does not necessarily bring commercial revitalization. CityPlace, which serves the Palm Beach metropolitan area, draws customers who might otherwise continue driving for another mile and a half to shop downtown. Consequently, it has not had any beneficial impact on the central business district. Clematis Street, once the pride of West Palm Beach, has had to struggle to survive competition from CityPlace. When Santa Monica Place opened, it drew so many customers away from the Santa Monica Pedestrian Mall that for many years the street was known for its vacant, deteriorating, and transient retail outlets. For those reasons, government should consider assistance only to projects that bring new customers, rarely pilfer customers from other retail facilities, and generate additional activity in the surrounding neighborhood.

Design

When an area is losing its market to other, better-functioning facilities, the best way to fight back is to augment existing retail structures with similar up-to-date alternatives. Santa Monica Place and Pioneer Place are examples of projects subsidized by cities that were not willing to lose customers to suburban malls.

Building a modern, air-conditioned shopping mall, however, is no guarantee of increasing retail sales. As the failures of Rochester's Midtown Plaza and New Haven's Chapel Square Mall demonstrate, the organization of a shopping center's component parts can determine its success. Moreover, in New Haven insufficient attention had been given to the movement of customers as they made their way from parking to their department store destinations and back again. They could have benefited from closer scrutiny of Country Club Plaza's disposition of retail outlets as a way of ensuring that customers on their way to their destinations had to pass tempting merchandise.

Character is as important to the design of shopping facilities as the organization of its constituent parts. Making open-air shopping centers look like Main Street may provide customers with amenities and a place to relax, linger, and socialize. From a commercial point of view, however, the aim is to keep customers in the shopping center for longer periods of time. The theatrical image of Country Club Plaza increases its competitiveness, as does the festive atmosphere and historic architecture of Ghirardelli Square and Quincy Market. Old Market in Omaha may be a very different interpretation of the festival marketplace, but it, too, capitalizes on a distinctive atmosphere, created from scratch by their designers.

Financing

The prototypes of the modern American shopping complex (Country Club Plaza, Southdale, and Ghirardelli Square) were entirely conventionally financed. So are virtually all shopping centers. On the other hand, many downtown malls and marketplaces rely on government assistance.

Banks and insurance companies are understandably wary of investing in areas with vacant stores and deteriorating retail facilities, so developers find it difficult to finance retail revitalization schemes. Government should reduce this risk by participating in project planning and assisting in property acquisition. The Power & Light District, Santa Monica Place, Quincy Market, and many other commercial revitalization projects would have been impossible without government assistance in planning, assembling sites, and holding them until financing is in place.

Financial institutions rarely provide enough mortgage financing to cover project costs. In privately financed projects the remainder is equity capital. Most retail-revitalization projects are not able to generate enough revenues to cover both debt service and return on equity. Government can fill the gap. After all, if these projects were able to raise the necessary financing on their own there would be no need for public assistance.

Government can provide money without appearing to subsidize the developer. One way is to assemble and hold the site until it is ready for development (thereby avoiding costly compensation of tenacious holdouts), eliminating real estate taxes during the holding period, and covering the costs of carrying the property until permanent financing can be arranged. Another way is to follow the example of Portland and finance required parking structures. Conventional shopping centers cover huge amounts of territory with inexpensive asphalt parking fields. In urban areas, paying for a public garage eliminates the competitive advantage of suburban sites whose open parking fields cost a fraction of any parking structure. In suburban areas, subsidizing structured parking can be justified because it conserves land for other uses. At Quincy Market, government assistance covered the cost of historic preservation. But, whatever the form of public assistance, it should be restricted to the minimum needed to guarantee financial feasibility.

Time

It may take years to put together a retail-revitalization scheme. Consequently, government has a crucial role to play. By using its power of condemnation, it can shorten the time needed to assemble a site. By purchasing the site and holding it until a project is ready to go, it can eliminate the cost of carrying the property (i.e., interim interest, real estate taxes, and operating expenses). Both were crucial to the feasibility of Quincy Market and Santa Monica Place.

However, the single most important time period for the financial success of any shopping facility is the time customers spend there. Country Club Plaza may have been different from the Galleria in Milan in terms of location, appearance, and tenancy, but it manipulated the time its customers spent in the same way. The cafés and restaurants of the Galleria captured visitors on their way to and from specific destinations (La Scala Opera and the Duomo), just as the small shops of Country Club Plaza captured them between the time they parked their cars and the time they entered a department store.

If the time that customers spend in a retail facility is central to its financial success, the time they do *not* spend there is crucial to the health of the surrounding city. Many business districts are dead at night and on weekends because they lack retail, entertainment, and leisure facilities that attract large crowds. As a result, a tremendous investment in infrastructure lies fallow except for one-third of the day five days a week. Moreover, the city is forced to pay for duplicating that infrastructure in other retail districts that themselves are underutilized during the workday and might be enhanced by addressing both markets. Consequently, retail-revitalization schemes that minimize downtime are very cost-effective. They also produce a level of vitality that cannot be duplicated by dividing land uses and activities among separate districts.

Market conditions will always continue to shift long after a project is finished. That is why the owners of Lenox Square and Phipps Plaza have continually renovated and updated these centers. Projects (e.g., Chapel Square Mall and Midtown Plaza) that ignore changing times inevitably lose customers to retailers who adjust to those changing conditions.

Entrepreneurship

J. C. Nichols understood, as no developer had before him, that the automobile made mass marketing possible. For it to take place, however, there had to be a destination (department stores) that attracted huge numbers of customers, a place of arrival (parking), and a path that forced these customers past goods they might wish to purchase on the way to their ultimate destinations (anchor stores). By sheer force of entrepreneurship, he overcame the skepticism of Kansas City's retail merchants, attracting them to an undeveloped part of Kansas City. More important, he persuaded these hesitant retail merchants to locate where they would have the greatest effect on the project as a whole.

When businesses are uncertain about an area or a project, government participation can help overcome their hesitation, coordinate the various players, and take the necessary risks. It can do this by analyzing the market, planning the project, acquiring the property, and, sometimes, relocating tenants and preparing sites for development. That leaves the ultimate developer with the no-less-difficult job of finding retail tenants, putting the financing together, building the project, and operating the facility.

When government entrepreneurship is inadequate, tremendous time is lost until a developer is able to put a deal together. In fact, packaging the project may prove impossible. It took the Boston Redevelopment Authority (BRA) nine years to put together an ostensibly feasible scheme for Quincy Market. When the developer selected was unable to bring the project to fruition, it took another four years for the BRA to restructure the deal and bring in James Rouse, who had the know-how to hit a home run.

Retailing as a Planning Strategy

Planners trying to revive downtowns, make them more competitive with other city or suburban locations, or just keep up with market changes have to be careful not to take actions which the market has already passed by. Such market changes were particularly vivid at the start of the twenty-first century. During the prior decade, about 200 million square feet of big boxes, power centers, regional malls, and other retail structures were being erected every year. For 2010 and 2011 combined, less than 11 million square feet of multitenant retail space was completed.[34]

Nevertheless, public officials are increasingly eager to prescribe retailing as a way of giving 24-hour life to districts in need of revival. They want to duplicate the success of Quincy Market or "downtown" Buckhead without ensuring that all six ingredients of success can be met, what the probable benefits of their actions are likely to be, and whether they are worth the money. Too often, they discover the impact of proposed retail projects after they open. In Santa Monica, the city government spent a decade getting an up-to-date air-conditioned shopping arcade. When it opened, the new mall attracted customers away from Third Street. As a result, the city spent another decade seeking to replace the customers Third Street had lost.

The unintended consequences of inadequately conceived retail-revitalization projects arise because public officials focus their attention on specific projects rather than on the community as a whole. Not only is this poor urban planning, it is unfair to businesses that do not get public assistance. There will always be cases when a project's citywide benefits outweigh the damage to local businesses. As long as the project involves a street closing, a rezoning, or some other government action that does not involve subsidization, there is no reason for more than public consideration of the impact of those actions. When government money also closes the gap between available private financing and project cost, the public should get a commensurate return on its investment. That investment ought to be approved by the appropriate legislative body as a way of ensuring that any negative local consequences are outweighed by benefits to the community as a whole. More extensive public action is unnecessary, because the private sector has little difficulty supplying the retail opportunities the public demands.

Notes

1. Anita Kramer (project director), *Dollars & Cents of Shopping Centers: 2006,* Urban Land Institute, Washington, DC, 2006, pp. 18–24.
2. Historical and statistical information on public markets is derived from Theodore Morrow Spitzer and Hilary Brown, *Public Markets and Community Revitalization,* Urban Land Institute, Washington, DC, 1991; and Valerie Jablow and Bill Horne, "Farmers' Markets," *Smithsonian,* Washington, DC, June 1999, pp. 121–130.
3. See www.pikeplacemarket.org.
4. See www.pikeplacemarket.org.
5. Michelle Kayal, "Number of farmers markets in the U.S. jumps 17 percent," *The Herald News,* August 22, 2011, http://heraldnews.suntimes.com/lifestyles/7217897–423/number-of-farmers-markets-in-us-jumps-17-percent.html.
6. Historical and statistical information on department stores is derived from John William Ferry, *A History of the Department Store,* The MacMillan Company, New York, 1960; Michael B. Miller, *The Bon Marché: Bourgeois Culture and the Department Store 1869–1920,* Princeton University Press, Princeton, NJ, 1981; and Richard Longstreth, *City Center to Regional Mall: Architecture, the Automobile, and Retailing in Los Angeles, 1920–1950,* MIT Press, Cambridge, MA, 1997.
7. See www.lordandtaylor.com.
8. Historical and statistical information on shopping centers is derived from Michael D. Beyard and W. Paul O'Mara, *Shopping Center Development Handbook,* Urban Land Institute, Washington, DC, 1999; www.icsc.org; Victor Gruen and Larry Smith, *Shopping Towns USA,* Reinhold Publishing Company, New York, 1960; Louis G. Redstone, *New Dimensions in Shopping Centers and Stores,* McGraw-Hill Book Company, New York, 1973; Geoffrey Baker and Bruno Funaro, *Shopping Centers Design and Operation,* Reinhold Publishing Company, New York, 1951; and International Council of Shopping Centers, "Country Fact Sheet: United States," July 20, 2012.
9. Historical and statistical information on Country Club Plaza is derived from William S. Worley, *J. C. Nichols and the Shaping of Kansas City,* University of Missouri Press, Columbia, MO, 1990; and Carla C. Sobala (editor), *Kansas City Today,* Urban Land Institute, Washington, DC, 1974, pp. 10–28.
10. The only original building that remained was the Broadway Department Store.
11. Philip M. Klutznick, *Angles of Vision,* Ivan R. Dee, Chicago, 1991, p. 190.
12. Johann Friedrich Geist, *Arcades—The History of a Building Type,* MIT Press, Cambridge, MA, 1982.
13. Victor Emmanuel, originally king of Sardinia, assumed the title of the first "King of Italy" in 1861. He is considered, together with Garibaldi, Mazzini, and Cavour, to be responsible for the creation of an independent, unified Italy.
14. Geist, op. cit., pp. 74–75, 371–401.
15. Carroll L. V. Meeks, *Italian Architecture 1750–1914,* Yale University Press, New Haven, CT, 1966, pp. 290–297.
16. Victor Gruen, *The Heart of Our Cities,* Simon & Schuster, New York, 1964, p. 194.
17. See Dean Schwanke, with Terry Jill Lassar and Michael Beyard, *Remaking the Shopping Center,* Urban Land Institute, Washington, DC, 1994, pp. 123–129.
18. Carla S. Crane, *Dallas/Ft.Worth Metroplex Area . . . Today,* Urban Land Institute, Washington, DC, 1979; and http://northparkcenter.com/.
19. Historical and statistical information on Tysons Corner is derived from Schwanke, Lassar, and Beyard, op. cit., p. 116.
20. Historical and statistical information on Post Oak is derived from Carla C. Sobala, *Houston Today,* Urban Land Institute, Washington, DC, 1974; Crane, Levitt, Finlay, Gregerson, *Houston, Texas, Metropolitan Area . . . Today,* Urban Land Institute, Washington, DC 1982; Stephen Fox, *Houston Architectural Guide,* American Institute of Architects, Houston, 1990; International Council of Shopping Centers; and Simon Property Group.
21. Historical and statistical information on Santa Monica Place is derived from Barbara Goldstein, "A Place in Santa Monica," *Progressive Architecture,* July 1981, pp. 84–88; *Santa Monica Evening Outlook,* March 6, 1972, p. 15, August 31, 1978, pp. 9–10, and October 15, 1980, p. 10; and Alexander Garvin, *The Planning Game: Lessons From Great Cities,* W. W. Norton & Company, New York, 2013, pp. 58–63.
22. The block between Ocean Avenue and Second Street was eliminated from the redevelopment area.
23. Garvin, op. cit., pp. 59–63.
24. Historical and statistical information on Pioneer Place is derived from Urban Land Institute, *Pioneer Place, Portland, Oregon,* Project Reference File, vol. 25, no. 4, Urban Land Institute, Washington, DC, January–March 1995.

25. In 1986, Ghirardelli Square was remodeled and expanded to include 91 retail establishments in 155,226 square feet. See W. Anderson Barnes, "Ghirardelli Square Keeping a First," *Urban Land,* Urban Land Institute, Washington, DC, 1986, pp. 6–10.
26. The buildings were named after Mayor Josiah Quincy, who had been responsible for this landfill and construction project.
27. Historical and statistical information on Quincy Market is derived from Walter Muir Whitehall, *Boston: A Topographical History,* Belknap Press, Harvard University Press, Cambridge, MA, 1963, pp. 95–98; Mildred F. Schmertz, "Faneuil Hall Marketplace," *Architectural Record,* December 1977, pp. 118–127; James Thomas Black, Libby Howland, and Stuart L. Rogel, Downtown Retail Development: Conditions for Success and Project Profiles Urban Land Institute, Washington, DC, 1983, pp. 50–52; and Garvin, op. cit., pp. 47–53.
28. Schmertz, op. cit.
29. Urban Land Institute, *Dollars and Cents of Shopping Centers: 1987,* Urban Land Institute, Washington, DC, 1987, p. 10.
30. Carla S. Crane (editor), *New York Metropolitan Area . . . Today,* Urban Land Institute, Washington, DC, 1980, p. 76.
31. Historical and statistical information on big-box retailers, power centers, outlet malls, and value retailing is derived from W. Paul O'Mara, Michael Beyard, and Dougal Casey, *Developing Power Centers,* Urban Land Institute,Washington, DC, 1996.
32. International Council of Shopping Centers, "Country Fact Sheet: United States," May 24, 2012.
33. Jeff Green and Jason Baker, "The Evolution of Retail," *Shopping Center Business,* April 2012, pp. 44–45.
34. Ibid.

5

Palaces for the People

Cleveland, 1999. Jacobs Field. *(Alexander Garvin)*

Over a century has passed since the trustees of the Boston Public Library explained that they wanted to build "'a palace for the people' and, as such . . . a monumental building worthy of the city."[1]

Such palaces for the people, whether libraries, stadiums, museums, city halls, courts, or other public facilities, are more than fashionable civic embellishments, municipal status symbols, or even promotional edifices. By attracting people who spend millions of dollars, they also become agents of economic development.

Monumental public structures have considerable appeal. Politicians get votes for awarding construction contracts and distributing construction jobs. When the project is completed, there are additional operating contracts, permanent jobs, increased retail sales, and new taxes with which to pay for government programs. If the project is a stadium, there is the additional status to be gained by bringing in or retaining a major league team. The question any municipality must answer is: Will it bring enough additional revenue, jobs, and status to justify the capital investment?

Any public facility can be claimed to be successful in and of itself as long as revenues exceed expenses. For a public facility to have beneficial impact on the rest of the city, however, it must attract and sustain a critical mass of customers and yet be located and designed so that these customers *cannot* be fully accommodated within the facility. Only by this seemingly contradictory set of attributes will a public facility generate beneficial interaction with the rest of the city, because only then will its customers have a reason to set foot or spend money anywhere else. In addition, its periods of operation must complement those of surrounding areas, thereby providing customers when these areas would otherwise be empty.

Too often, palaces for the people are conceived of as single-purpose facilities whose impact on surrounding areas (measured in additional traffic, pollution, noise, garbage, etc.) generates only opposition. Instead, they should be planned and financed in conjunction with improvements to neighboring properties, thereby avoiding a good deal of political conflict. More important, facilities so planned would then spark further market-generated improvements in their immediate vicinity and become engines for continuing prosperity.

World's Columbian Exposition of 1893

One new municipal palace is good; more are better. The World's Columbian Exposition that opened in Chicago in

Chicago, 1893. The monumental exhibition structures grouped around the Court of Honor at the World's Columbian Exposition became the model for well-planned civic and cultural centers. *(Courtesy of Chicago Historical Society)*

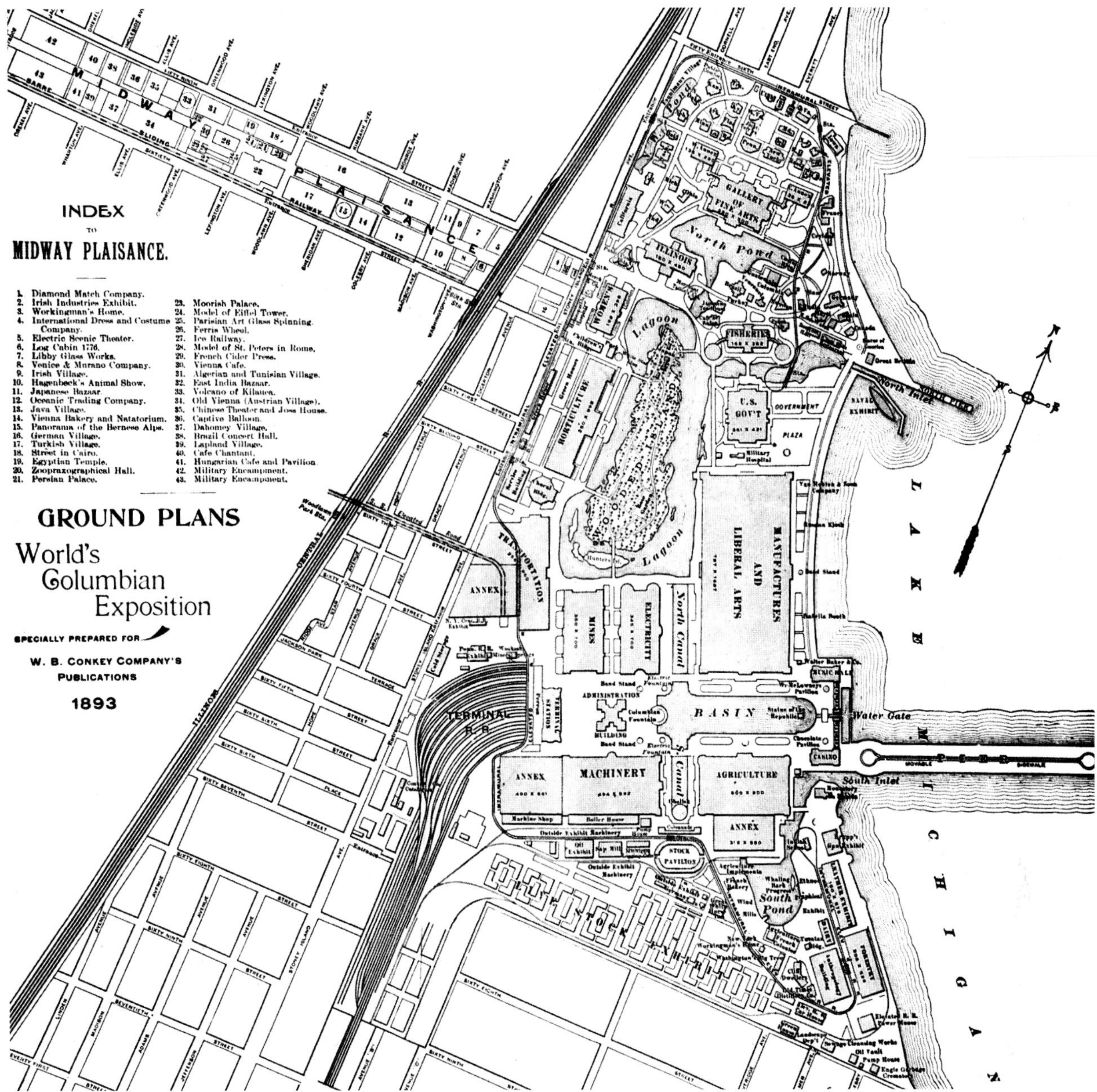

Chicago, 1893. Plan of the World's Columbian Exposition. *(Courtesy of Chicago Historical Society)*

1893 demonstrated this concept as nothing had before. Visitors saw that the attractions had been combined into a coherent, powerful whole and concluded that this was the key to successful urban planning.[2]

Such planning became conventional wisdom because the people who went to the Chicago Fair propagated it. Its 27.5 million admissions (paid and unpaid) equaled 43 percent of the population of the United States.[3] Obviously, 43 percent of the U.S. population did not go to the fair. But anytime even 10 percent of the population of the country sees something, the impact is phenomenal.

Preparation for the fair began in 1889, when Chicago established a corporation to lobby for, plan, and develop an exposition to celebrate the four-hundredth anniversary of the discovery of America by Christopher Columbus. The corporation promptly issued $5 million in bonds. The following year, after considerable controversy, Congress designated Chicago as the site for a World's Columbian Exposition "of arts, industries, manufactures, and the products of the soil, mine, and sea."

The Exposition Corporation selected the Olmsted firm to be its landscape architect and Burnham and Root as coordinating architects. As a site for the exposition, they recommended Jackson Park, which Olmsted had designed more than two decades earlier. It was a particularly attractive location because the South Side Rapid Transit Company agreed to extend the elevated railroad from its 63rd Street terminus to the park (see Chapter 3).[4]

Olmsted and Burnham proposed gutting the existing landscape, building the exposition, then replacing it with an improved park when the fair was over. They conceived a three-part exposition: a formal group of major exhibition structures organized around a Court of Honor containing a large water basin (350 feet wide and 1100 feet long) decorated with sculpture and fountains; picturesque pavilions grouped around the park's irregular ponds and lagoons; and an amusement area strung out along the seven-eighths of a mile of Midway Plaisance leading west into Washington Park. The buildings themselves were designed by the nation's most prominent architects, who transformed this exhibition of the world's progress in science, art, industry, and agriculture into a major architecture show.

The artists and architects who designed the Chicago Fair created what Henry Adams called "the first expression of American thought as a unity."[5] Ironically, it did not look like the uniquely American steel and glass skyscrapers then going up in Chicago. The architects of the fair combined styles then fashionable at the École des Beaux Arts in Paris. The result was an eclectic confection: Roman in civic presence, Baroque in axial organization, and Renaissance in surface decoration. The architects adopted a common design vocabulary and used similar materials. Most structures were built to a common cornice height, included built-in outdoor incandescent lamps to provide lighting for pedestrians, and were covered with the white plaster-cement cladding for which the exposition was nicknamed the "White City."[6]

When the exposition opened (a year late), it was both a popular and a financial success. The 21.5 million paid admissions and numerous concessions produced almost $33 million in gross revenues and a $2.25 million return for the exposition's investors. Tens of millions more dollars were spent by visitors to Chicago hotels, restaurants, and stores and by real estate developers who rushed to build on undeveloped sites made attractive by their proximity to the fair and easily accessible by the elevated railroad. The most important result of the exposition, though, was not its popular renown or its profitability or its contribution to tourism and real estate development; it was its extraordinary impact on American architecture and urban planning.

At the conclusion of the fair, all the buildings that occupied the grounds were removed except for Burnham's Fine Arts Building, which was reconstructed in the 1930s to house the Museum of Science and Industry. The site was transformed into the verdant, pastoral landscape Olmsted had initially planned for the site. Even when the buildings were gone, though, memories of the exposition convinced generations of architects and public officials that careful, coordinated planning resulted in a more convenient and efficient environment (see Chapter 19). It also launched a vision of the "City Beautiful" as the appropriate appearance for municipal improvements, and it demonstrated that public buildings could have even greater impact when collected into a coherent district and designed to function as an ensemble.[7] This idea of grouped public buildings quickly became part of the progressive agenda of municipal reform. Today it lives on in every city as the civic center, the cultural center, the sports center, and the convention center.

Civic Centers

Civic reformers who wanted to implement the City Beautiful strategy in the years immediately following the Chicago Fair faced a major obstacle. There was as yet no way for government to direct the location and design of privately owned real estate. However, government could ensure the thoughtful organization of municipal offices, courthouses, and facilities for legislators. Thus, the early manifestations of the City Beautiful were usually proposals for new civic centers.

The idea of clustering government buildings was not new. It went back to the Campidoglio in Rome, and before. Gathering such facilities together into a civic ensemble provided citizens with a physical and symbolic representation of local government. In America, the reason for a civic center was more a matter of efficiency and economy than of symbolism. Expanding municipal, county, and state governments needed to replace scattered, inadequate, outworn facilities. The people who went to grouped facilities to obtain government services and the government employees who served them could be counted on to patronize nearby businesses. Tourists might be attracted by the opportunity to see government in action and thus make a small contribution to the local economy. It was beyond imagination, however, that 120 years after the Chicago Fair, government (federal, state, and local) would employ more than 17 million people and annually spend more than $6 trillion. In the process of such substantial expansion, government took possession of a significant portion of every city's land area and office space and became a major component of its economy.[8]

The McMillan Plan for Washington, D.C.

The first American city to implement a major plan for the clustering of government buildings was Washington, D.C. By 1900, the federal government already owned more than enough structures to create several civic centers. Unfortunately, they had been built with little regard to their relationship to one another, to the city as a whole, or to generating

Washington, D.C., 1901–1902. Model showing the Capitol and the irregularly shaped open spaces that at the time included a railroad station and numerous other intrusive structures. *(Courtesy of National Commission of Fine Arts)*

further private development. For more than a century, private as well as public development in the national capital had been loosely based on a plan originally conceived in 1791 by French engineer Pierre-Charles L'Enfant. His design, based on the château gardens of the French landscape architect André Le Nôtre and the emerging modern city of Paris, was an amalgam of diagonal boulevards and a rectilinear grid. The diagonal boulevards allowed carriages and horseback riders to reach their destinations by cutting through whole neighborhoods. The grid permitted the simple survey, sale, and reconveyance of property for the construction of homes and public buildings.[9]

L'Enfant's plan specified the sites of major public buildings. The Capitol was to be placed at the top of Jenkin's Hill, where it could dominate the city. It was connected by Pennsylvania Avenue (a broad thoroughfare intended to be lined with major public structures) to the "Presidential Palace," located on high ground at the north end of a vast public open space leading south to the Potomac. A second "Grand Avenue," lined with sloping gardens that extended from the houses on either side, was to run east-west, connecting Capitol Hill with the Potomac.

During the course of the city's growth, a variety of intrusions significantly altered L'Enfant's initial conception. A canal was extended through the site of Grand Avenue to connect the eastern and western branches of the Potomac River, and the Smithsonian Institution obstructed the path of the proposed boulevard, which itself had become part of an

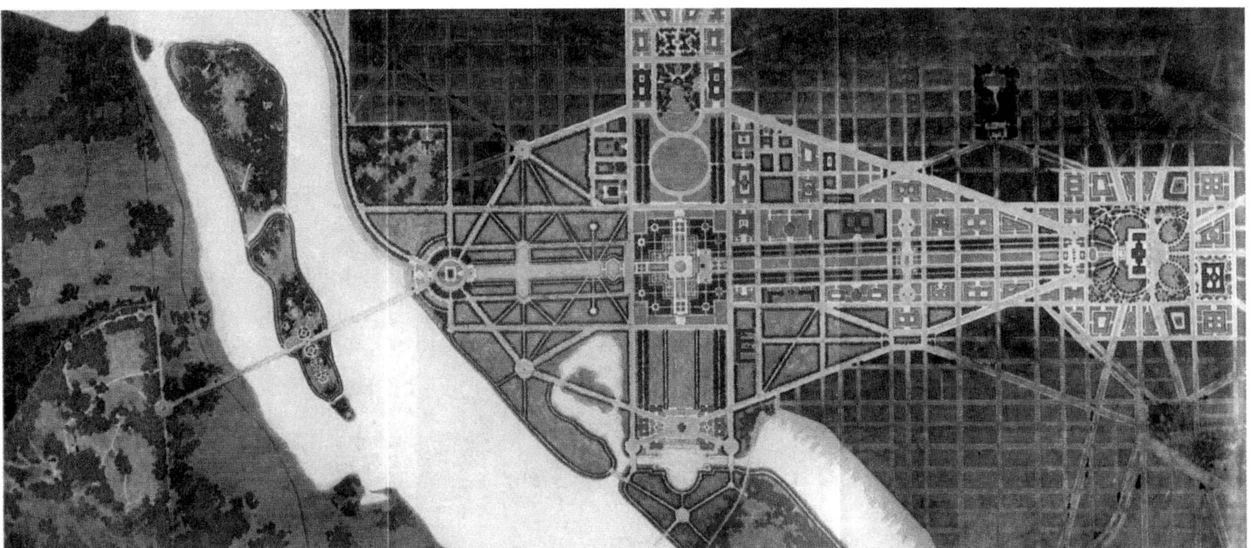

Washington, D.C., 1901–1902. The McMillan Plan for the redevelopment of the Mall created a setting for America's largest concentration of majestic public buildings. *(Courtesy of Senate Park Commission)*

Washington, D.C., 1901–1902. Bird's-eye view of the McMillan Plan illustrating the monumental design framework for future development in the national capital. *(Courtesy of Senate Park Commission)*

irregular, asymmetrical public garden. When the Treasury Building was erected, it blocked the view between the Capitol and the White House. The Washington Monument, which required foundation conditions that could support an immense obelisk, was built several hundred feet off the crossing of the north-south axis of the White House and the east-west axis of the Capitol. The most serious intrusion was the Baltimore & Ohio Railroad, which, in 1872, built its tracks and station right in the middle of this picturesque park.

Proposals for a suitable centennial celebration of the founding of the national capital culminated in 1901 with a Senate resolution directing the Senate Committee on the District of Columbia to report on the development and improvement of the city's park system. The committee's chairman, Senator McMillan, appointed Daniel Burnham, Frederick Law Olmsted Jr., Charles Follen McKim, Charles Moore (McMillan's secretary), and Augustus St. Gaudens to prepare the report. With the exception of McMillan and Moore, they had worked together on the Chicago Fair and saw their current project as a logical continuation of these earlier efforts.[10]

During the summer of 1901, Burnham, McKim, Olmsted Jr., and Moore traveled together to Europe to examine what they conceived to be the world's finest gardens, palaces, and civic complexes: Versailles, Vaux-le-Vicomte, and the Champs Élysées (which had so influenced L'Enfant); Hadrian's Villa at Tivoli, the Villa Medici, and the grand piazza of St. Peter's; Vienna, Budapest, and Paris. These European royal monuments continue to provide the inspiration for the layout and buildings of the capital of the American republic.[11]

At every stop, the team discussed the sites, their design philosophies, and specific proposals for Washington. The designers' common approach is marvelously captured in an event that occurred in Venice. One night McKim disappeared. When Olmsted wondered how they could find him, Burnham replied, "That's easy. . . . We will go to the Piazza di San Marco and find him on the axis."[12] Indeed, the plan they devised is essentially a series of imperial axes intended to enhance the symbolic significance of Washington's major monuments.

Washington could not continue to function at the speed and scale of a leisurely pedestrian. It could not continue to close down for the summer. Nor would it remain what John F. Kennedy later referred to as the perfect combination of southern efficiency and northern charm. Washington had international responsibilities. So Burnham, McKim, Olmsted Jr., and the others proposed a monumental scheme that would transform L'Enfant's design of the capital for a provincial republic into the permanent administrative center of an international empire. It included conversion of L'Enfant's Grand Avenue into a spectacular mall; creation of a second, similarly symbolic axis perpendicular to it; relocation of the railroad tracks and terminal; establishment of three new groups of public structures; and development of an integrated park system.

As was the case at the Chicago Fair, Burnham was responsible for coordinating their work. Together with McMillan and Moore, he handled the promotional aspects of obtaining the support of the administration of Theodore Roosevelt, the Congress, and a variety of civic organizations. Olmsted's firm again prepared the landscape plan; McKim developed and refined the commission's design proposals; St. Gaudens planned the sculptural decoration. In January, 1902, the commission presented its report in the form of a written document, drafted by Charles Moore, and an exhibition of scale models, plans, and perspective drawings, prepared under McKim's direction in New York.[13]

L'Enfant's Grand Avenue was to be doubled in width to 800 feet and lined on both sides with four rows of trees, a service road, and a procession of majestic public buildings. The axis of the proposed Mall was shifted slightly to accommodate the awkward placement of the Washington Monument. It was extended farther westward along a reflecting pool, longer than the one at the Chicago Fair, through a new park (on reclaimed land) to a proposed Lincoln Memorial and from there diagonally across the Potomac on a low Memorial Bridge to the National Cemetery at Arlington.

The imperial character of the new Mall was matched by the enhancement of L'Enfant's north-south vista from the White House to the Potomac. This open space was broadened and extended across reclaimed land to a memorial building group, or pantheon, on what is today the site of the Jefferson Memorial. Since there was no way to compensate for the off-axis placement of the Washington Monument, McKim designed an elaborate terraced formal garden with a reflecting pool marking the crossing of L'Enfant's two axes.

These organizing axes would have been impossible without Burnham and McMillan. As the architect of the proposed new Union Railroad Station, Burnham persuaded Alexander Cassatt, president of the Pennsylvania Railroad, to move its railroad tracks and station north of the Capitol, provided Congress would pay for a tunnel under Capitol Hill. McMillan obtained Congressional approval for this scheme just prior to his death in August 1902. The following year, Congress appropriated the necessary funds. While Burnham's Union Station and the diagonal boulevard connecting it with Capitol Hill were not formally part of the commission proposal, they were an integral part of the plan and were intended to provide a majestic gateway to the proposed monumental government center.

Three new groups of public buildings were to be established. The first was to be created by replacing the old houses on Lafayette Square with large-scale government offices, the second by locating additional public edifices in the triangle between Pennsylvania Avenue and the Mall, and the third by turning Capitol Hill into a vast square defined by huge new structures for the Supreme Court and other government entities.

The most ambitious element of the plan was the park system in which Olmsted Jr., like his father, proposed to exploit topographic features, provide neighborhood recreation facilities, and tie together Washington's disparate neighborhoods. He achieved this with substantial new park areas, riverside drives, and scenic park boulevards encircling the city.

Few of the individual projects, when completed, followed exactly the details presented in the drawings. Lafayette

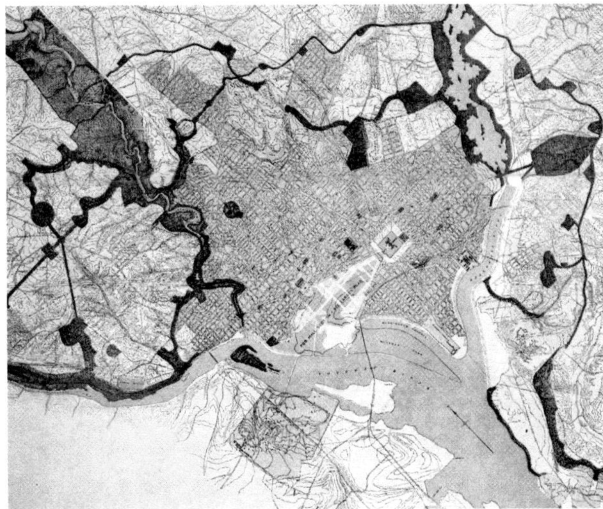

Washington, D.C., 1901–1902. Park system proposed by the McMillan Plan for Washington, D.C. (proposed, executed, and additions). *(Courtesy of Senate Park Commission)*

Square was not demolished. The Smithsonian Institution was not moved to a Beaux Arts palace.[14] The proposed pantheon of monuments became a single Jefferson Memorial. However, the monumental city that we know today would have been impossible without the conceptual framework provided by the McMillan Plan.

With or without the plan, Washington's massive bureaucratic engine guaranteed a critical mass of expenditures that had to spill over into the surrounding city, as well as an avalanche of tourists who would have even more impact on the local economy. The monumental design framework of carefully considered groups of public structures substantially increased the city's symbolic meaning, attractiveness, and usefulness to millions of visitors who might otherwise have had much less to see, would have come for shorter visits, and would have spent far fewer dollars.

Virtually every American city prepared plans for a civic center in an attempt to emulate either the Chicago Fair or the McMillan Plan. The authors of these plans throughout the rest of the nation seem to have failed to understand that there were fundamental forces that drove development in Chicago and Washington that were either not available or not transferable elsewhere. No complex of government buildings could attract even a fraction of the 27.5 million who visited the World's Columbian Exposition during the short time it existed. The designers of these new civic centers also misunderstood the McMillan Plan. Government is Washington's very raison d'être. In most cities, government can be only a small segment of the local economy. A municipality might need a city hall, a courthouse, a police station, a central post office. But in most instances these facilities can cover only a fraction of the city's land surface, occupy only a minor amount of building space, and employ only a small portion of its population. Nevertheless, architects and planners sought to use these buildings to provide a monumental framework for urban development, and politicians were quick to distribute the resulting patronage.

Ironically, Daniel Burnham was the man responsible for promoting this faulty planning strategy. Upon completion of the McMillan Plan, he became one of three commissioners appointed to advise Cleveland on questions of urban planning and went on to propose similar civic centers for San Francisco and Chicago (see Chapter 19).

The Group Plan, Cleveland

For nearly a decade after the Chicago Fair, Cleveland's leadership tried to create a group of public structures inspired by the fair. They believed that this ensemble would attract tourists to "visit the city and enjoy the wonderful picture of municipal enterprise and beauty."[15] In 1902, with the help of Mayor Tom Loftin Johnson, they persuaded the Ohio legislature to allow the governor to appoint an advisory commission that, as expected, proposed a new civic center.

The governor appointed three architects who had been recommended by Mayor Johnson and other Cleveland civic leaders: Daniel Burnham, Arnold Brunner, and John Carrère. They repeated the procedures that had been so successful in preparing the McMillan Plan. Once again Burnham (during visits from his Chicago office) coordinated the planning taking place at Brunner's and Carrère's New York offices as well as the promotion and lobbying efforts in Cleveland. In this instance, he also wrote the text.[16]

The Group Plan they released in 1903 was a portfolio-sized brochure filled with handsome, detailed plans and perspective drawings of an entirely new civic center. It showed a monumental group of public buildings organized along a 400-foot-wide mall extending one-third of a mile from the corner of Public Square to a new Union Railway Terminal (for which Burnham had already been designated architect) on Lake Erie. Along the lakeshore, the plan proposed a new park promenade. The drawings also depicted new buildings that extended for blocks in every direction. Presumably, these were to be privately financed and built. Like the structures at the Chicago Fair, they were designed in the appropriate Beaux Arts style and to a common cornice height.

The civic center proposed in the plan was intended to replace the slum that lay between Public Square and Lake Erie.

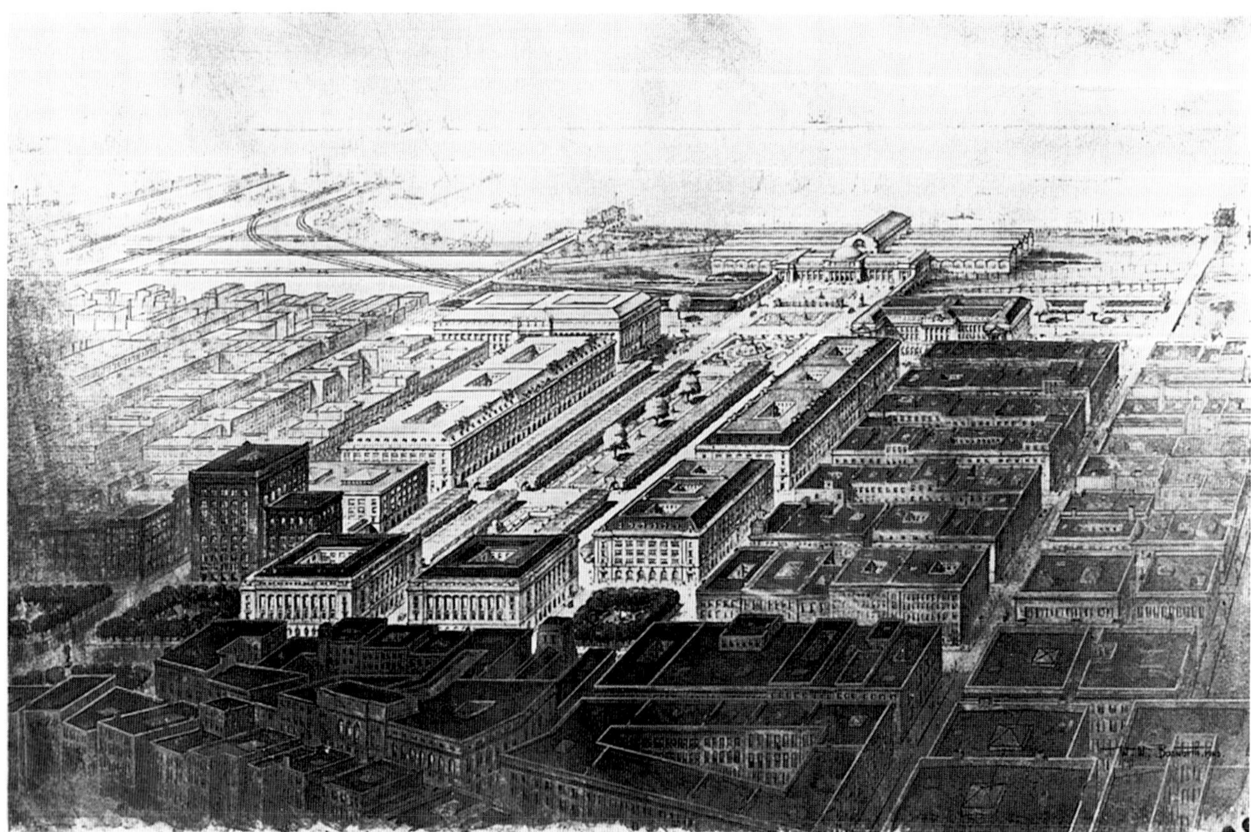

Cleveland, 1903. Group Plan for the redevelopment of downtown Cleveland. *(From Burnham, Carrère, and Brunner,* Report on the Group Plan of Public Buildings of Cleveland, Ohio, *Board of Supervision for Public Buildings and Grounds, Cleveland, 1903; Courtesy of Avery Library, Columbia University, New York)*

However, the plan's brief text was vague in explaining how all this would happen. It dismissed acquisition and relocation problems, saying, "The present population of the district . . . will have to be moved elsewhere."

In 1903, when the Group Plan was released, only three buildings were in the planning stage. Over the ensuing decades, other civic structures were erected for occupants who needed space and had the money to pay for it, but the plan's most important features were ignored. Burnham's terminal was never built because the railroads serving Cleveland could not agree on its location, design, or funding. The park promenade along Lake Erie remained a dream because the railroads were unwilling to pay the vast sums that would be required to reroute their facilities.[17]

The authors of the Group Plan had boldly announced that "the new public buildings" would spur property owners to "develop this territory and extend the business center of the city toward the Lake." They could not have been more wrong. The civic structures that were erected produced little or no market reaction. Only the unbuilt railroad station would have generated enough travelers going through the Mall on their way to work, to business appointments, and to hotels and other visitor services to fill this vast public space.

Some privately financed property redevelopment might have occurred in response to this new market. But even with the additional traffic, the civic center probably would have failed to attract the city's business center because it was located at the wrong end of town. The path of city growth was to the east, out Euclid Avenue toward the wealthier suburbs. Any chance of moving the business district to the blocks surrounding the new civic center disappeared when the city's railroad station was established at Terminal Tower, completed in 1930.

During the century following publication of the Group Plan, not only did it fail to generate any market reaction in the surrounding area, it also failed to contribute to the vitality of downtown Cleveland. After 5:00 p.m. and on weekends, when all the government buildings are closed, the area is deserted. Even during the day, it appears empty. It is 560 feet wide, too wide for a casual walk to one of the other buildings. Instead of a "wonderful picture of municipal enterprise and beauty," the Group Plan proved to be a large, rectangular, hollow core surrounded by monuments for a government bureaucracy.

Municipal leaders in Cleveland had mistakenly thought that civic centers, in and of themselves, provided the critical mass needed to generate both increased commercial activity and further real estate development. This was a convenient delusion that helped civic leaders in Cleveland and elsewhere to obtain political support for bond issues to pay for proposed civic centers. They argued that the bonds would be paid off with the increased tax revenues from private development in surrounding areas. This strategy failed in Cleveland and other cities when the location and design of these centers precluded any significant spillover spending and therefore also precluded sufficient market activity to justify further development.

The situation began to change during the last decade of the twentieth century. Burnham & Root's Society for Savings Building was renovated in conjunction with construction of the Key Center that included a 57-story, 1.25-million-square-foot office tower, a 424-room Marriott hotel, a multilevel underground garage, and renovation of the adjacent portion of the mall—all of which brought a great many more people to the site. In 2010, the city established a 15-member commission to develop a plan that would result in a new underground development with a 235,000-square-foot Medical Mart, 230,000 square feet of exhibition facilities, 90,000 square feet of meeting rooms, and a grand ballroom underneath a section of the mall. The mall itself will be relandscaped and will now slope up 17 feet to allow pedestrian access and natural light to enter from the street. All these new facilities should bring thousands of people downtown, some of whom will spill over into the rest of downtown Cleveland.[18]

Memorial Plaza, St. Louis

The clustering of government buildings for St. Louis was initially proposed by the Civic League in 1907 and finally completed in 1960. A 1923 bond issue paid for the first 10 blocks; further bond issues in 1933 and 1944 paid for the rest. The resulting procession of public structures extends for more than half a mile along Market Street between City Hall and the railroad station. Commonly referred to as "Central Parkway" or "Memorial Plaza," it includes a Soldier's Memorial, an auditorium, courts, government offices, and the post office. The only special attraction along its entire length is Carl Milles's "Meeting of the Waters," a delightful fountain with 14 bronze figures representing the meeting of the Mississippi and Missouri rivers.[19]

Memorial Plaza was as unsuccessful in stimulating a market reaction in St. Louis as the *Group Plan* was in Cleveland. This time it was not a matter of location. Central Parkway linked the business district with the expensive residential areas around Forest Park (see Chapter 14). Nor was it the absence of the railroad station, which was already at the site. In the beginning the failure could have been ascribed to an incomplete project with an inadequate number of public buildings and too few government employees. When the plan was completed it became clear that it failed to generate a private market reaction because of faulty design. The buildings extended over such a long distance that they could not generate the critical mass of activity needed to spark adjacent private development. Even today, when Union Station has been converted into a successful retail-tourist center, there

St. Louis, 2010. The public buildings of Memorial Plaza are strung out over too great a distance to produce a critical mass of customers. *(Alexander Garvin)*

are not enough people to animate this half-mile-long public open space.

The Los Angeles Civic Center

In Los Angeles, as in Cleveland and St. Louis, the civic center took decades to complete. The first proposal for a civic center appeared in 1909 in a city plan prepared for the Municipal Arts Commission by Charles Mulford Robinson. He proposed to locate a combined civic and cultural center at the northern end of the business district, below Bunker Hill, which was then a crazy quilt of frame dwellings, rooming houses, and commercial buildings. The mayor's Civic Center Committee, directed by William Mulholland, chief engineer for the city's Public Service Department, made a second proposal in 1918. It also suggested building on the northern edge of downtown, but eliminated the libraries, museums, and other cultural facilities of the earlier plan and proposed a separate cultural center for the middle of the city. Neither plan was adopted.[20]

In 1923, a voter referendum approved creation of a civic center, ratified a $7.5 million bond issue, and authorized construction of a new city hall. The City Planning Commission hired the firm of Cook and Hall to prepare a plan for the new administrative center. The Allied Architects Association made another proposal. In neither case were there sufficient occupants to fill the proposed structures nor money to pay for them. So construction proceeded on a piecemeal basis, starting with the Hall of Justice in 1925 and City Hall in 1926–1928.

Over the next four decades government buildings were erected along a plaza consisting of three landscaped open spaces interrupted by streets and connected by a series of stairs and ramps. The northern end of this axis, a half mile from City Hall, was completed during the 1960s. It includes

Los Angeles, 2013. Civic Center after reconstruction as Grand Park. *(Courtesy of Chicago Historical Society)*

the Music Center and the Water and Power Building, which, like the other parcels, are separated from one another by traffic arteries and parking facilities. The Los Angeles Civic Center has had little influence on the development of the city. Initially there were not enough government buildings to make an impact. Once there were, the sloping site and distances between structures ensured isolation rather than a critical mass of mutually reinforcing activity. Construction of the Hollywood Freeway during the 1950s cut off any potential private market investment on the other side of the freeway. Government acquisition and clearance of then-seedy Bunker Hill in the 1960s eliminated any possible influence downtown. With the addition of the Disney Concert Hall, the government-assisted redevelopment of Bunker Hill and the creation of the Music Center at the top of the hill are now largely completed.

The last major development parcels on Grand Avenue, now planned for a mix of housing and retail uses, were stalled by the Great Recession. As part of that project, a $56 million redesign of the Civic Center's central space opened in 2012. This space, renamed Grand Park, provides a dramatic fountain; ramps providing access from the park to the four streets that cross the park to those unable to use stairs; a Starbucks café; areas of lawn, flowers, plants, and trees; and bright pink movable chairs and tables. But it does not provide park users other than tourists and office workers from the buildings that had been there for decades, nor attractions that could attract them from elsewhere in the city. It remains to be seen whether programming events and activities that will take place within the park will overcome the problems that were unsolved before the remodeling: the absence of a market for the park, the sloping topography, and the heavily trafficked streets that cross the park. Like the civic centers of Cleveland and St. Louis, the Los Angeles Civic Center remains a hollow core attracting little activity during the day and nothing at night and generating virtually no private market investment in the surrounding city.

Government Center, Boston

Most American civic centers are theatrical settings for the day-to-day activities of government. Boston's is one of the few exceptions. Its government center really did influence city growth and development. Part of its success was a matter of timing. It was created in the 1960s, when government and its customers were becoming a rapidly growing market. It also exploited a critical location, where the business district converges with the waterfront, Beacon Hill, and the West and North Ends, and it was designed to encourage anybody who traveled to the new government center to make use of these surrounding districts.[21]

Boston's City Planning Board first proposed clearing Scollay Square, then a notorious red-light district, for the purpose of building a civic center in 1917. The idea was revived during the Great Depression and again in the 1950s. Finally, in 1961, Mayor John Collins and Edward Logue, his development administrator, obtained federal urban renewal funding for a new government center consolidating federal, state, and municipal facilities at a reinvented Scollay Square, now referred to as Government Center.

New Deal and post–World War II legislation had expanded government activity to the point that federal, state, and city offices were scattered in rented space and converted annexes throughout the city. Consolidation offered the prospect of greater efficiency and reduced cost. More important, with the help of House Speaker John McCormick and Presi-

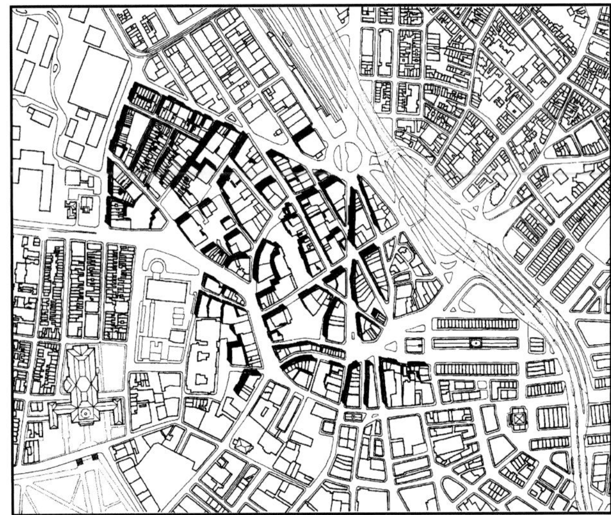

Boston, 1961. Map of the streets and buildings replaced by Government Center. *(Courtesy of Pei Cobb Freed & Partners)*

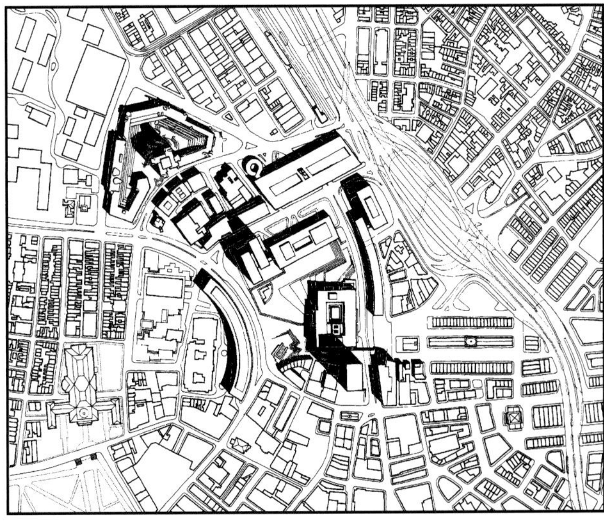

Boston, 1961. Redevelopment plan for Government Center. *(Courtesy of Pei Cobb Freed & Partners)*

Boston, 2005. Government Center. *(Alexander Garvin)*

dent John F. Kennedy (both from Boston), Mayor Collins was able to attract federal and state buildings that otherwise might have been located elsewhere.

The master plan, prepared by architect I. M. Pei, proposed transforming 26 decaying city blocks into a modern townscape composed of 15 large buildings. Its centerpiece was a spacious new plaza and city hall, designed by Kallman, McKinnel, and Knowles, the unanimous choice as winner of a major international design competition.

Pei's design made Government Center an integral part of the cityscape, not an obviously separate district. Unlike earlier City Beautiful civic centers, buildings were not uniform in height, color, material, and scale. Nor did Pei specify a consistent style. Most important, the plan included more than just government buildings. It preserved a few existing private commercial structures, introduced some new ones, and shuffled them together with the new government buildings. Instead of depending on axial symmetry, the design unified this disparate collection of structures through the use of pedestrian walkways, arcades, and open spaces, all leading to City Hall Plaza.

Government Center brought 25,000 workers where there had formerly been 6000, and it located them on 60 acres in the middle of the business district. Downtown Boston is so tightly concentrated that Government Center is an easy walk from almost anywhere. Thus, Government Center could benefit from proximity to the financial, general office, and shopping districts, and they in turn could profit from the customers that the new center would provide. The interaction of this critical mass of customers and activity was exactly what was needed to spark developer interest in additional construction on sites adjacent to Government Center.

Nevertheless, for more than half a century Government Center has suffered from two of the same problems that afflicted earlier civic centers: land use segregation and oversized open space. Because Government Center consists almost entirely of commercial and institutional offices, the plaza remains empty except on those very few occasions when it is the setting for major public events. Like other civic centers, it closes down for the night. During the day, City Hall Plaza may provide a monumental setting for government. But, like the vast central spaces of other civic centers, it is in scale with surrounding public buildings—not with the people who use it.

Federal Center, Chicago

Chicago's 4.6-acre Federal Center was built in three stages between 1959 and 1974. It fills one entire block and a portion of a second, on either side of Dearborn Street. More than 14,000 people work in its two high-rise office slabs, one of which includes 12 floors of federal courtrooms. Thousands more do business in court and in federal offices or visit the one-story-high post office. Mies van der Rohe, the designer of this redevelopment project, arranged its buildings in a way that purposefully channels the substantial pedestrian activity they generate.[22]

Unlike the vast monumental spaces of the Los Angeles

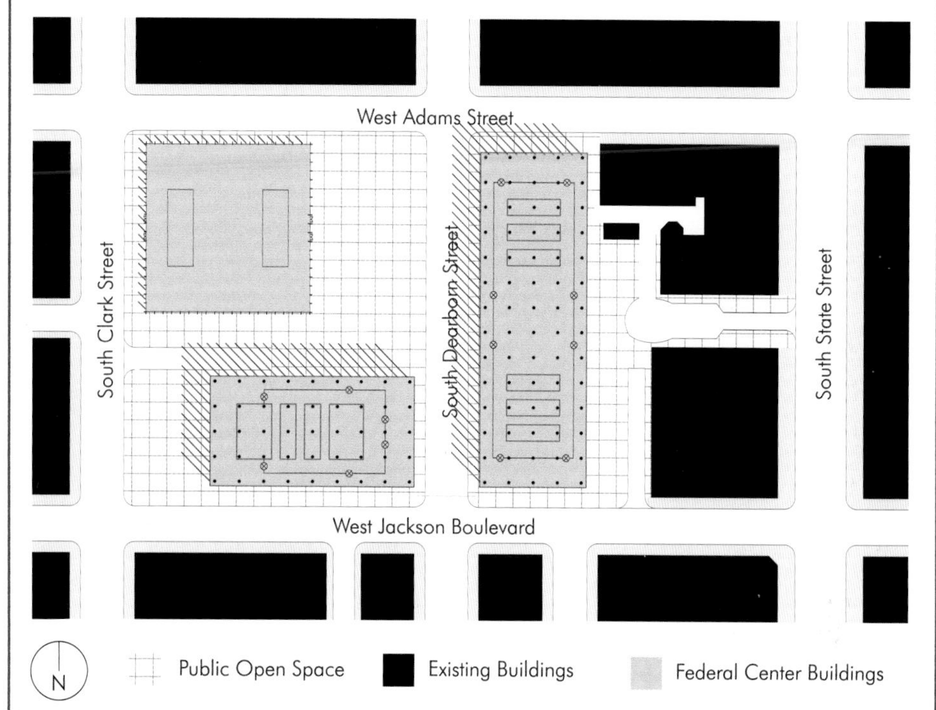

Chicago, 2012. The Federal Center creates an interlocking series of inviting public spaces by using the older office buildings across the street and the buildings in the center as bounding walls. *(Alexander Garvin and Ryan Salvatore)*

and Cleveland civic centers, Chicago's Federal Center creates a pedestrian-friendly environment for both building users and passersby. Adams Street and Jackson Boulevard, which bound the project on the north and south, are only 66 feet wide. To provide more light and air to the sidewalk along these streets, the office buildings are set back slightly from the property line. Conversely, both buildings extend to the property line along Dearborn Street, whose right-of-way is 80 feet across, wide enough to provide plenty of natural light. The design extends the pedestrian realm farther, under 26-foot-high building arcades, thereby encouraging passersby to walk through the Federal Center on their way to other destinations.

A large public plaza is created on the west side of Dearborn Street by recessing the post office 122 feet. People walk-

Chicago, 2012. The red Calder sculpture draws pedestrian attention to the Federal Center. *(Alexander Garvin)*

ing by the plaza can stop to look at Alexander Calder's large, freestanding, red-painted steel sculpture entitled *Flamingo* or peer at what is going on behind the glass façade of the post office. A second public plaza at the corner of Jackson Boulevard and Clark Street is connected to the post office plaza by a broad walkway. Both plazas are part of a much larger volume of space defined, not by property lines, but by the walls of the older buildings across the street and the new glass office building on the east side of Dearborn Street. Thus, not only is the Federal Center's pedestrian activity integrated with the rest of the city, so are its public spaces. Indeed, this is a civic center conceived and designed to be so much a part of the daily life of Chicago that people walk through the Federal Center without noticing that it is anything special. Its 14,000 workers contribute as much to the economic and social life of the surrounding district as the people who come to do business there.

Cultural Centers

The cultural center is a recent phenomenon. Prior to the mid-twentieth century, theaters were built where profit-motivated owners thought they could maximize box office receipts; so were privately owned music halls. Opera houses, museums, and other cultural institutions occupied sites their donors already owned or found convenient to purchase. Occasionally, city governments provided sites for these buildings, often in an existing public park. Cultural facilities simply were not thought of in terms of their impact on the growth and development of the city. Today they must be. According to the National Assembly of State Arts Agencies, the $166 billion nonprofit arts industry in the United States supports 3 million jobs annually.[23]

When the Chicago Fair popularized the notion of clustering public buildings, no individual cultural facility had a reason or a mechanism for relocating into a single arts district. With the exception of theaters, which tended to be clustered as a method of profiting from the resultant aggregate market, the impetus had to originate elsewhere. Occasional clustering of cultural facilities occurred in civic centers that could not otherwise obtain tenants for the sumptuous facilities that had been envisioned. This happened in San Francisco in 1932, when the Opera and Veteran's Auditorium were added to the scattered government buildings in its civic center, and in Washington, D.C., during the New Deal, when the National Archives and National Gallery began filling the gaps in the McMillan Plan. However, it was not until the advent of Lincoln Center that consolidating disparate arts institutions into a single cultural center became popular.[24]

Lincoln Center, New York City

When Robert Moses, chairman of the Committee on Slum Clearance, proposed a renewal plan for Lincoln Square in 1955, New York City was not in decline. More than 9 million square feet of new office space had been built since the end of World War II.[25] The assessed value of taxable real estate had climbed 31 percent, from $13.8 billion in 1945 to $18.1 billion in 1955.[26] The 1950 census had reported a population increase of 436,000. Nobody doubted a continued rosy future. Instead, people worried about specific neighborhoods. Moses proposed creating a cultural center that would completely change one such threatened neighborhood.

The city's white middle class, especially on Manhattan's West Side, was confronted with a massive influx of African-American and Puerto Rican newcomers.[27] In an effort to accommodate them, once-fashionable brownstone row houses had been converted into rooming houses or single-room occupancies (SROs), with as many as eight households (and often many more) sharing the same kitchen and bathroom. Apartments had been subdivided and rented to tenants whose only way of paying was by packing in more people than these accommodations were designed for. Overcrowding only increased the wear and tear on the buildings. New and old populations, now jammed into the neighborhood, were soon at each other's throats. The most insightful depiction of the scene is the Leonard Bernstein–Stephen Sondheim–Arthur Laurents 1957 Broadway musical, *West Side Story*, which took for its setting the very spot that Moses proposed for redevelopment and that later became the location for the film version.[28]

The Lincoln Square Urban Renewal Project was intended to be the centerpiece of Moses's effort to "save" the West Side from destruction by the newcomers. When he proposed it in 1955, he already had 11 renewal projects under way, including 2 that were intended to reverse deterioration of the West Side: West Park at the northern end of the area and Columbus Circle in the south. The West Park Renewal Project (better known as Park West Village) was scheduled to clear 3700 apartments, occupied primarily by poor African Americans and Puerto Ricans, and replace them with 2700 apartments for the middle class. Columbus Circle was to remove two blocks of shabby commercial buildings and tenements accommodating 300 households for the purpose of building a convention center, an office tower, and 600 middle-income apartments.[29]

With the extremities of the neighborhood theoretically protected from further intrusion by redevelopment projects at Park West Village and Columbus Circle, Moses needed something to spark the revival from within. He knew that the Metropolitan Opera was looking for a new home and that the New York Philharmonic had been told to vacate Carnegie Hall when its lease expired in 1959. He understood that they could provide the economic and psychological impetus needed to revitalize the West Side, so he persuaded them to relocate to the area and financed their move with federal renewal funds.[30]

The project Moses first presented to the public was a $160 million complex that covered 18 blocks and included new

Manhattan, 1957. Site of Lincoln Square Uran Renewal Project prior to redevelopment. *(Courtesy of the Citizens Housing and Planning Council, New York)*

buildings for the Metropolitan Opera, the New York Philharmonic, and Fordham University, as well as 3800 apartments, and miscellaneous community facilities. Over the next two years he proposed adding a 10-story fashion center, headquarters for a national engineering society, a shopping center, a skyscraper hotel, and a legitimate theater complex with five new halls. None of these additions survived the proposal stage.

When Moses finally submitted a redevelopment plan for public review in 1957, four northern blocks along Broadway had been removed from the project scope. Nevertheless, at least 678 businesses and 5268 households (4600 in apartments and the remainder in rooming houses) had to be relocated.[31] In their place he proposed a truly powerful stimulant: the Metropolitan Opera, Philharmonic (now Avery Fisher) Hall, the New York State Theater (now David H. Koch Theater and home of the New York City Ballet), the Vivian Beaumont Repertory Theater, the Juilliard School, the Library and Museum of the Performing Arts, Fordham University Law School, headquarters for the American Red Cross, an 800-car garage, new public schools, a fire station, a public bandshell, and 3800 middle-income apartments to be known as Lincoln Towers. If the activity from these facilities could not revitalize the West Side, nothing could.

Opposition was vociferous. Area residents formed the Lincoln Square Citizens Committee, which held protest rallies in front of City Hall. They knew they would have to pay for Lincoln Center, not in dollars or taxes but in uprooted lives. As one opponent put it: "We are planning to take away what they [the residents] have with the reason being that they

Manhattan, 1955. Rendering of the proposed Lincoln Center for the Performing Arts in a form that recalls St. Peter's Square in Rome. *(Courtesy of the Citizens Housing and Planning Council, New York)*

are living under bad conditions."[32] Opponents seized on the arrogant dismissal of any concern for the lives of people living on the site that was, to them, behind the proposed mandatory relocation.

Advocates for the relocatees were not the only critics. There was opposition to the very concept of a cultural supermarket. As Jane Jacobs complained,

> *[Lincoln Center] is planned on the idiotic assumption that the natural neighbor of a hall is another hall. Nonsense. The natural neighbors of halls are restaurants, bars, florist shops, studios, music shops, all sorts of interesting places. Look for instance at what has been generated by Carnegie Hall on West 57th Street, or by some off-Broadway theaters.*[33]

She and other critics felt that the components should be separated and used to revive many declining neighborhoods. Moses knew this to be impractical. He had had difficulty persuading the Philharmonic to move a few blocks north of its previous home at Carnegie Hall. Getting the Metropolitan Opera or the Juilliard School to move to one of his renewal projects in Brooklyn or the Bronx was out of the question.

With all this controversy, the City Planning Commission and Board of Estimate hearings on the Lincoln Square Urban Renewal Project were tumultuous. It was time to bend. Moses amended the plan again. This time he added 420 tax-exempt cooperative apartments. Finally, more than two years after it had been proposed, Lincoln Center was approved.

Designing Lincoln Center was as difficult as obtaining project approval. The constituent institutions had selected some of the country's best-known architects: Philip Johnson, Wallace K. Harrison, Max Abramovitz, Pietro Belluschi, Eero Saarinen, and Skidmore, Owings & Merrill. Like their predecessors at the World's Columbian Exposition, the architects of Lincoln Center chose to emulate European models. One version was inspired by St. Peter's Square in Rome; a later one, by Piazza San Marco in Venice. What finally emerged was a $175 million group of neo–Beaux Arts containers, built to the same cornice height, out of the same travertine and glass, and differentiated primarily by their structural supports.

The rest of this $280 million project is so disparate in scale, material, height, and color that most New Yorkers have long forgotten that the 45-acre Lincoln Square Urban Renewal Project is more than an internationally known performance emporium. In fact, the only connection between the Fordham University Law School, the American Red Cross Building, Public School No. 199, and the 3800 apartments of Lincoln Towers is that they were built on land acquired and cleared for the same urban renewal project.

Ironically, Lincoln Center demonstrated the validity of Jane Jacobs's observation that "the natural neighbors of halls are restaurants, bars, florist shops, studios, music shops, all sorts of interesting places." It also proved her criticism wrong. The planners of Lincoln Center were after precisely the frictional effect she had described. They understood that concentrating so many cultural institutions in one place would not preclude "all sorts of interesting places" but rather result in greater activity than the individual components could generate separately. But even they underestimated the impact of this critical mass of 12,000 ticket holders and 6800 musicians, dancers, actors, costumers, stagehands, ushers, porters, and other people who put together the performances.

Manhattan, 2005. The 12,000 ticket holders and 6800 musicians, dancers, actors, costume makers, stagehands, ushers, porters, and other people who put together the performances at Lincoln Center have transformed the area into one of the city's liveliest neighborhoods. *(Alexander Garvin)*

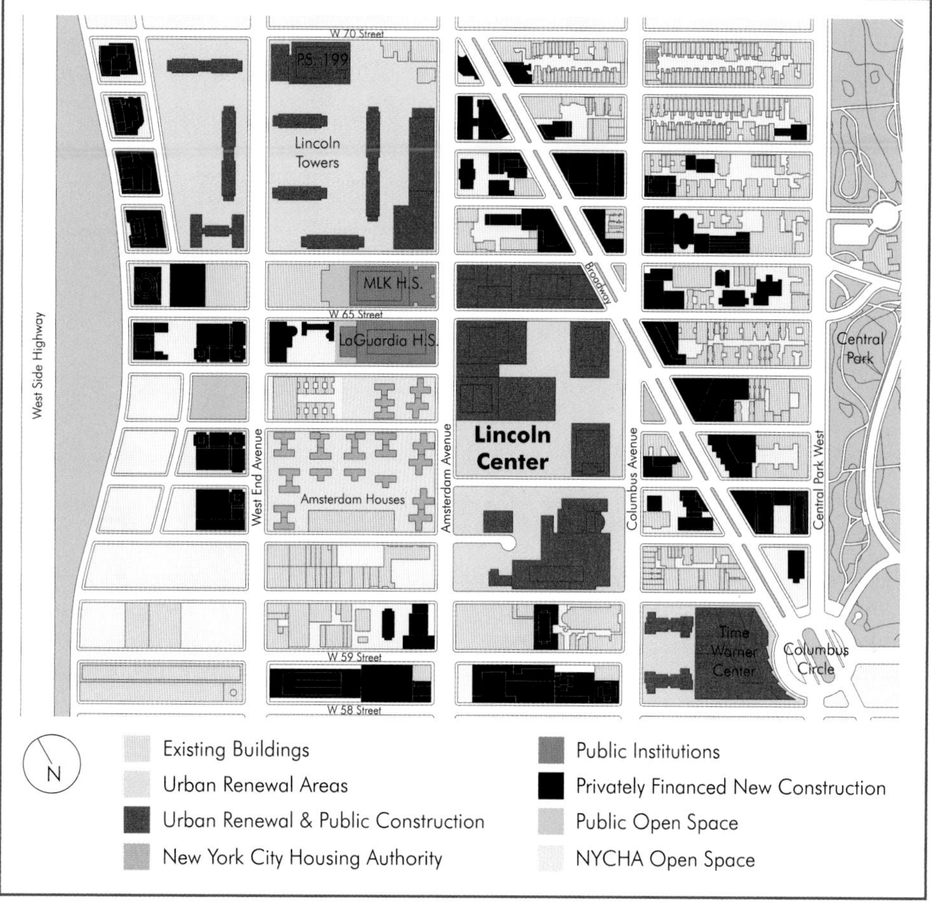

Manhattan, 2012. Lincoln Center area. Buildings in red were created as part of the Lincoln Square and Columbus Circle Redevelopment Projects. Buildings in dark green are schools that were part of the Lincoln Square Redevelopment Project. Public housing is light blue. Everything in black was erected by independent private developers. *(Alexander Garvin and Ryan Salvatore)*

Many people, like Jacobs, continue to believe that redistributing the various Lincoln Center institutions to other neighborhoods would have provided enough of a critical mass to revive those slum districts. We do not know whether the huge sums of money raised to pay for their relocation to Lincoln Center would have been donated to individual component institutions if they had been scattered around the city. But we can see the massive changes they brought to the Upper West Side.

Within a few years, Lincoln Center had forever banished the world of *West Side Story* to the musical stage. Today, Broadway and Columbus Avenue, opposite Lincoln Center, are lined with fashionable boutiques, restaurants, and new, privately financed apartment towers. The magnitude of the change was reflected by the $650 million increase in the assessed value (1950 to 2012) of just the four blocks between Lincoln Center and Central Park.[34] The blocks to its west were occupied by public institutions or the 1084 public housing units in the Amsterdam Houses and, thus, were not available for the private market activity that exploded along Columbus Avenue and the rest of the West Side.

Today Lincoln Center is part of a very different city from the city it helped to change. When the project was conceived, cultural institutions were not considered to be significant components of the city's economy. At that time, manufacturing constituted 30 percent of New York City's labor force. By 2010, when Lincoln Center was reporting annual revenues of more than $300 million, manufacturing employment in the city had fallen to 4 percent of the labor force, tourism had become one of its leading industries, and the arts had become an even more potent force in American society. Clearly, Lincoln Center had become a major player in a multi-billion-dollar sector of the city's economy.

Lincoln Center, like most Broadway hits, went on the road. Los Angeles, Washington, Dallas, and other cities produced their own versions. These might involve a concert hall, a repertory theater, or some other players. Frequently, they were little more than pallid groups of official buildings. The location, the staging, the participants, the design, or something else kept them from being more than just arts facilities. Most cities did not have the critical mass of cultural institutions to produce another *West Side Story*.

The Los Angeles Music Center

Los Angeles opened its Music Center four years after Lincoln Center debuted. It was built at the top of the hill, at the far end of the Civic Center. At first glance, the Music Cen-

Los Angeles, 2005. People come to the Music Center by car, park, attend a performance, and never set foot in downtown Los Angeles. *(Alexander Garvin)*

ter resembles its New York antecedent. As at Lincoln Center, travertine and glass containers flank a public plaza, this time containing both a fountain and an impressive sculpture by Jacques Lipchitz. Once again, the buildings are differentiated by the design of their exterior colonnades. On one side is the 3250-seat Dorothy Chandler Pavilion; on the other, the Ahmanson Theater and the smaller Mark Taper Forum, together providing another 2850 seats. However, when the center was conceived, there were not enough players to put on a complete show. Consequently, separated from these facilities by Hope Street, stands, not an opera house, but the offices of the Water and Power Company.

The location on a hill at the northern end of downtown prevents the Music Center from having a real impact on its surroundings farther down the hill. Topography, traffic, and distance prevent much interaction with the Civic Center to the southeast. The Harbor and Hollywood Freeways cut off interaction across these busy traffic arteries. The substantial hike up Bunker Hill through the vacant lots and parking fields of the still-incomplete urban renewal area prevents interchange with the business district to the southwest. Even if the business district were not so far away, there is nothing to interact with because intervening blocks have remained vacant since they were cleared for redevelopment during the 1960s. Even construction of 2515 seats in the two performance halls at the Disney concert facility and additional parking has failed to overcome topographical and land use obstacles. As a result, over more than half a century, the Music Center has been a much appreciated island of culture with no effect on the rest of downtown Los Angeles.

The failure to repeat the remarkable impact of New York's arts emporium is not just a matter of location and topography. It is also a matter of conception and design. Angelenos go everywhere by car. Therefore, the Music Center has to be easily accessible by automobile. By building a 2000-car garage beneath the plaza of the Music Center and across the street from the Disney Concert Hall, automobiles are able to drive off the adjacent Hollywood and Harbor freeways directly to their parking destination. Once safely parked, these motorists have no reason to go anywhere else.

Lincoln Center had been planned to throw off sparks that would catch fire along Broadway, Columbus Avenue, and throughout Manhattan's West Side. One day, the Los Angeles Music Center also may throw off sparks. We will find out only when there is something around it that can catch a bit of the fire.

Kennedy Center, Washington, D.C.

Los Angeles is not the only city to demonstrate that performance halls do not always have "natural neighbors" or induce "all sorts of interesting places." Kennedy Center in Washington, D.C., completed in 1971, proves the same point.

When President Eisenhower appointed the District of Columbia Auditorium Commission in 1955, the general presumption was that the capital's cultural center would become one of the (as yet unbuilt) monuments proposed in the McMillan Plan. The site most frequently mentioned was located on the Mall, opposite the National Gallery, but it had already been set aside for the Air and Space Museum. So, in 1957, the commission recommended a lovely 28-acre site in Foggy Bottom, overlooking the Potomac. The site had the advantage of requiring the condemnation of only a few structures, the demapping of some streets, and the rerouting of a small section of Rock Creek Parkway. However, it was a location on the edge of the city, where its impact would be minimal.[35]

The design, by architect Edward Durrell Stone, has been described by critics as the largest box of Kleenex in the world.

Washington, D.C., 1994. The Kennedy Center is too far from surrounding neighborhoods to have much impact. *(Alexander Garvin)*

It is essentially a rectangular solid, 630 feet long, 300 feet wide, and 100 feet high, that contains the opera house, concert hall, two theaters (one large, the other more intimate), the American Film Institute's projection hall, a performing arts library, restaurants, and reception rooms. The base provides parking for 1400 cars, from which the Kennedy Center's 6100 ticket holders can proceed by elevator to the single 600-foot-long, air-conditioned lobby. They need never even step outside on their way upstairs. Consequently, this $70 million monument, which makes a major contribution to the cultural life of the national capital, makes no impact on anything except possibly the street level of the neighboring Watergate residential complex.

Dallas Arts District

Cultural centers also can shape areas that lie in the path of development. This did not happen in Los Angeles or Washington because their cultural centers were located and designed to be self-contained. Like them, Dallas is creating a cultural center located along a highway on the edge of the business district. But rather than build an expensive arts emporium, Dallas created a 68-acre, mixed-use district to be developed by different landowners for a variety of purposes.[36]

In 1982, after a decade of discussion, the city hired Sasaki Associates to prepare a master plan for a 17-block area at the northeastern edge of the central business district. They proposed a design scheme, a zoning ordinance, a financing strategy, and a management plan for an arts district that left existing streets in place. New buildings for the Dallas Museum of Art and the Dallas Symphony are intermingled with existing institutions such as the cathedral of Santuario de Guadalupe, the Dallas Bar Association, and the Dallas Arts Magnet High School. Other sites have been set aside for later private development as office buildings and retail centers. The design scheme establishes Flora Street as a linear axis along which to group new buildings. At one end is the Dallas Museum of Art, in the middle the Morton Meyerson Symphony Center, and at the other end a shed used by the Dallas Theater Center. Flora Street is a sidewalk-lined city street that is intersected by other sidewalk-lined city streets rather than a vast City Beautiful mall.

The Planned Development District Ordinance, approved in 1983, ensures that the district will remain pedestrian in scale and character by specifying building height, setbacks, ground-floor uses, parking, and loading requirements. Land uses on Flora Street, for example, are restricted to shops, restaurants, plazas, and fountains. Street walls cannot exceed 50 feet. The effect of the ordinance can be seen in the pavilions, terraces, fountains, and sculpture collection that make up the Flora Street front of the 50-story Trammell Crow Center (formerly the LTV Center).

Dallas issued more than $100 million in general obligation bonds to pay for land acquisition, infrastructure im-

Dallas, 2009. The performance palaces at the Dallas Art Center are located along city streets that link directly downtown, but separated from the rest of the city by the Woodall Rogers Freeway. *(Alexander Garvin)*

provements, a 1600-car garage, and new buildings for the Dallas Museum of Art and the Dallas Symphony. Among the more important additions to the District are the Nasher Sculpture Center with its 1.4-acre sculpture garden, designed by Peter Walker, and 55,000-square-foot building, designed by Renzo Piano, added in 2003; the 2300-seat Winspear Opera House designed by Norman Foster and the 600-seat experimental Wyly Theater, designed by Rem Koolhaas, both added in added in 2009; and 10-acre Sammons Park, added in 2010, that ties them all together into a lovely, high-design package.

An Arts District Management Association is responsible for planning, implementation, and operations. Civic groups have set up two nonprofit foundations to raise money from the public and provide support for the arts. In addition, the city council has appointed an arts district coordinator to be a liaison between city hall, private developers, and arts interests.

The goal of the Dallas Arts District, like that of Lincoln Center, is to attract customers to a location from which they will spill over into the surrounding city and stimulate a desirable market reaction. But the design, unlike Lincoln Center, rejects a separate precinct for arts-related buildings. So far, this framework has failed to intermix sites used for conventional commercial purposes with sites for arts activities. Projected office buildings have yet to appear because Flora Street is too far from the center of business activity and too many other potential construction sites are available there. Restaurants and cafés have not sprung up because demand is satisfied inside the Dallas Museum of Art and the Trammel Crow Center. Entertainment and other arts-related retailing has continued to flourish in the West End, Knox-Henderson, and other sections of the city. Consequently, Flora Street combines prestigious arts institutions with parking lots and other temporary uses. It is not a combination that can produce the active street life the Arts District's planners had hoped for. By 2013, however, the new 5.2-acre Klyde Warren Park will have covered a three-block-long stretch of the Woodall Rodgers Freeway, between the Arts District and the increasingly popular residential area north of the freeway. This connection may begin to increase the interaction between the district and the city's growing population.

Aronoff Center, Cincinnati

Cesar Pelli's design for the Aronoff Center achieved objectives similar to those of the Dallas Arts District by incorporating all its constituent institutions in one building, much like the Kennedy Center. Unlike Washington's arts emporium, however, the Aronoff Center is located on a city block right in the midst of downtown Cincinnati. The $82 million complex, which opened in 1995, consists of three performance halls, a rehearsal facility, and an art gallery. Its largest theater, Procter & Gamble Hall, seats 2719. The 437-seat Jarson-Kaplan Theater provides a setting for medium-sized performances and meetings. The Fifth Third Bank Theater can be configured for a wide variety of functions, from 150-seat intimate studio theater productions to conferences, meetings, and parties for up to 280 people. The same versatility is true of 3500-square-foot Weston Art Gallery, which is used for dinners and receptions,

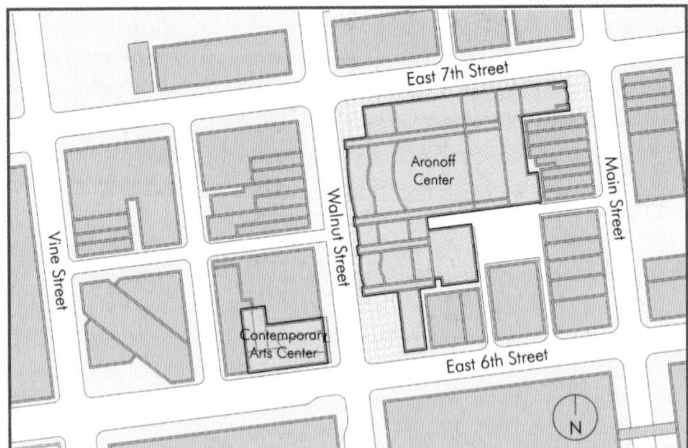

Cincinnati. The Aronoff Center. *(Alexander Garvin and Ryan Salvatore)*

Cincinnati, 2010. The Contemporary Art Center and the Aronoff Center. *(Alexander Garvin)*

not just exhibitions. In 2003, a new Contemporary Art Center, designed by Zaha Hadid, opened across the street. As with the Aronoff Center, it provides no parking and depends on the surrounding streets for visitor services.

Like Lincoln Center, Cincinnati's performing arts complex does not supply everything visitors seek on-site. Unlike Lincoln Center, though, most patrons come by car, park in garages that are used by office workers on weekdays, and walk to their destination on Walnut Street. The effect of this pedestrian traffic became visible within months. A by-the-slice pizzeria was transformed into a full-service Italian restaurant. Several new restaurants opened up. Suburban theatergoers come downtown for dinner and a show; office workers remain downtown in the evening for the same reason.

The Aronoff and Contemporary Art Centers represent a radical departure from most cultural centers. They are not opulent marble monuments set on pedestals and apart from the rest of the city. These simple buildings open directly onto the street, just like the bars and restaurants that their patrons enjoy on their way to and from a show. Perhaps that is why these institutions have been so quickly and inconspicuously integrated into the daily life of Cincinnati.

Sports Centers

At the start of the nineteenth century, athletics were an informal activity engaged in by those with enough leisure time. Organized sports such as baseball, football, and basketball were just being invented. As professional athletics grew in popularity, civic leaders sought sports palaces in order to enhance municipal self-esteem and project an image of being "in the major leagues."

For many citizens, stadiums and arenas bring an improvement in the quality of life. They contribute customers and jobs that spill over into surrounding areas, stimulating property development by businesses interested in capturing this new market. For others (especially those who live and work in the area), the increase in traffic, noise, and pollution is unjustifiable.

The first spectator sports facilities were usually financed, built, owned, and operated by professional teams or franchise owners. They were expensive in the past. The Miami Dolphins' open-air Joe Robbie Stadium cost $90 million in 1987; Riverfront Stadium in Cincinnati cost $44 million in 1970; the Houston Astrodome cost $45 million in 1964. Today, however, large stadiums and arenas are usually far too expensive to be privately financed. The MetLife Stadium in the New Jersey Meadowlands, which opened in 2010, cost $1.6 billion; the Dallas Cowboys Stadium, which opened in 2009, cost $1.3 billion. The money to pay debt service on the bonds that finance these stadiums comes from ticket receipts, parking fees, food and beverage sales, and the very high prices paid for club seats and suites. Because these facilities are so

MetLife Stadium, Meadowlands, New Jersey, 2012. Aerial view. *(Alexander Garvin)*

costly, franchise owners are usually able to avoid the entrepreneurial challenges of developing these structures, and because the competition for major league teams is so fierce, most city governments are only too happy to offer them publicly developed and financed facilities. Indeed, according to a recent analysis, since 1995 about four-fifths of the cost of NFL stadiums in the eight smallest media markets have been paid by publicly funded subsidies.[37]

Most sports palaces, like cultural centers, are *not* financially self-sufficient. Typical stadium revenues (rentals, concessions, parking, advertising, etc.) may cover operating costs, but even when they are able to earn substantial revenues from their programmed events, sports centers often fail to cover debt service on the bonds that financed their development. Proponents justify subsidizing stadiums and arenas because they bring money to the local economy—many times the money brought by a concert hall or a theater. America's largest cultural center, Lincoln Center, accommodates 12,000 ticket holders. Major league baseball parks seat 40,000 to 50,000; football stadiums, 65,000 to 100,000. According to a 2007 study, visitors to ordinary games spend about $140. Those attending a Super Bowl stay for several days and spend thousands of dollars on transportation, lodging, meals, and game tickets.[38]

No wonder every city wants a major league team. The cost-effectiveness of these sports palaces, however, depends on whether the business revenues, taxes, and jobs that they bring with them more than compensate for the government subsidies that pay for them. Even when the tax revenues are insufficient to pay the debt service on the government-issued bonds that financed the facility, the subsidy may be justified in terms of jobs created and business revenue generated.

But, for a city to reap the entire potential of this lucrative market, its sports palaces must be conveniently located, accommodate thousands of vehicles, and encourage spectators

Atlanta, 2004. Most of the time, Turner Field is surrounded by empty parking lots too far from the city's business district to have any effect on game days. *(Alexander Garvin)*

to support downtown merchants before and after the games. If, like the 53,000-seat Atlanta–Fulton County Stadium or the 50,000-seat Turner Field that replaced it, a facility is placed along a highway far from most commercial activities, the city will be unable to profit from the millions of fans who come there for a game.

Arenas and stadiums cover vast territories, especially when a sea of automobiles surrounds them. Atlanta–Fulton County Stadium required 19.4 acres to accommodate 6500 cars. When Turner Field was built, parking was increased to 8500 cars. Consequently, it is critical to design each facility and its accessory parking in a manner that avoids smothering its surroundings. Buffalo deftly avoided the problem by placing Coca-Cola Field within the downtown street system. Cleveland did the same with Gund Arena and Progressive Field. The idea that a stadium should be integrated into the downtown fabric is not very radical; Roman arenas were sited that way.[39] The only difference is that modern sports palaces have to provide for spectators *and* their cars. All the cars, however, need not be accommodated on-site. Coca-Cola Field provides parking for only 2000 cars. The garage for Progressive Field and Gund Arena has only 3000 spaces. Everybody else parks in the surrounding business district or uses mass transit. They walk through downtown to and from a game, often stopping on the way for a meal or a beer or to purchase something.

The Los Angeles Coliseum

Los Angeles was one of the first cities to conceive of a sports arena as a device for municipal improvement. Civic leaders were anxious to do something about Exposition Park, a fairground and racetrack in operation since 1872. They felt it was having an undesirable impact on both area residents and adjacent University of Southern California (USC) students. In 1898, they successfully persuaded the state, county, and city jointly to purchase the 90-acre site.[40]

There was no consensus on the site's reuse. The university wanted a sports arena. However, public officials were doubtful that Los Angeles (whose population in 1900 was just 102,000) could generate enough ticket sales to cover debt service on the necessary bonds. For this reason, development of Exposition Park began in 1910 with the Los Angeles County Museum of History, Science, and Art (today the Museum of Natural History).

In 1920, when Los Angeles had grown to 577,000 people, the city finally decided that there was a sufficient market for a stadium, but even in 1923, when the Los Angeles Coliseum first opened, sports events could not attract anywhere near the 76,000 spectators that it could accommodate. Its primary

Los Angeles, 1992. The Coliseum is set back too far from surrounding neighborhoods to affect more than vehicular traffic before and after big games. *(Alexander Garvin)*

Buffalo, 2009. Coca-Cola field brings thousands of fans downtown, generating sales and income taxes nights and weekends when most workers are no longer downtown. *(Alexander Garvin)*

user was the USC football team, which at that time attracted fewer than 14,000 spectators.

Only in 1946, when the Rams moved from Cleveland, did the stadium finally begin to attract big-league crowds. At one time or another, the Coliseum has been the home stadium for major league football (the Rams, the Chargers, the Raiders) and baseball (the Dodgers). The Coliseum was enlarged for the 1932 Olympics and remodeled again for the 1984 Olympics. Today, with 93,000 seats, it can accommodate anything from professional football to a papal mass and, therefore, is one of the few Olympic stadiums that did not become a white elephant after the games had ended.

Exposition Park also includes the Los Angeles Sports Arena, the California Museum of Science and Industry, the Los Angeles County Natural History Museum, the California Aerospace Museum, the California Science Center Imax Theater, the California Museum of Afro-American History, the Multicultural Center, and the Exposition Park Rose Garden. This extraordinary assemblage, a few blocks from the Harbor Freeway, is easily accessible from anywhere in the Los Angeles metropolitan area. It should be a powerful force generating all sorts of economic and social activity. Instead, it is an island of separate public structures, 2 miles south of downtown Los Angeles. It is too far to permit any interaction with the business district and too self-contained to have much impact on USC or the surrounding deteriorated, low-density neighborhoods.

Coca-Cola (formerly Pilot) Field, Buffalo

Buffalo, like so many Rust Belt cities, had been in decline for decades when Mayor James Griffin announced a major effort to build a new downtown stadium. In 1980 he appointed a private-sector committee, established a multiyear development timetable, and initiated the market, environmental impact, parking, and traffic studies that public assistance programs required.[41]

Bringing baseball back to Buffalo was more than an attempt at a psychological shot in the arm. It was an economic development project that was intended to recapture recreational spending that had been lost to cities as far away as Toronto, to provide service employment in a city that was in a 10-year period of double-digit unemployment, and to generate additional retail sales and tourist spending. The mayor had the support of the chamber of commerce, local labor unions, sports fans, and the media. The missing ingredient was a sports team. A baseball team was supplied in 1985 when the Rich family (owner of the Rich Product Corporation, a frozen-foods conglomerate that produces Coffee Rich creamer) purchased a Wichita minor league baseball franchise for $1 million, renamed it the Bisons, and moved it to Buffalo.

Coca-Cola Field and its two garages opened in 1988. It was designed by the HOK Sports Facilities Group and cost $56 million. The money was patched together from a variety of sources: a New York State Urban Development Corporation (UDC) loan, city bonds, federal Urban Development Action Grant (UDAG)-backed revenue bonds, and contributions from the Bisons and the city, county, and state governments. Because its backers hoped to attract a major league team in the future, the design allows for future expansion.

Thus far, Buffalo has lost out to other competitors for major league franchises. Nevertheless, the city continues to reap real benefits from Coca-Cola Field. The Bisons have a payroll of over 900 (mostly part-time) workers from April through September. In recent years they have attracted an annual paid attendance of about 550,000 fans. During the first year that Coca-Cola Field was in operation, these fans generated $21.7 million in direct and indirect spending in downtown Buffalo and $1.65 million in additional sales and income taxes.

Gateway Sports and Entertainment Complex, Cleveland

By combining several sports under one roof, Cleveland's civic leaders hoped to attract fans throughout the year and thus

Cleveland, 2012. Quicken Loans Arena and Progressive Field. *(Alexander Garvin)*

generate enough traffic to support additional downtown business. The idea of a 100 percent publicly funded, domed, multipurpose stadium for professional baseball and football, however, was rejected by a countywide referendum in 1984. Despite this setback, Cleveland's civic leadership continued to advocate the project, spawned in large part by the politically desirable opportunity to return to downtown Cleveland professional teams that had decamped to suburban Richfield 20 years prior. To that end, the city created a public-private partnership to assemble a 28-acre downtown site for the Gateway Sports and Entertainment Center.[42]

Six years after the original referendum, the Gateway Center was redefined as a baseball park that would retain the Cleveland Indians franchise, adjusted to include an arena that would bring the Cleveland Cavaliers basketball franchise back downtown, and coupled with an adjacent development. In 1990, Cuyahoga County voters approved a 15-year tax on alcohol and cigarette sales. The tax revenue thus produced was allocated to cover debt service on $169 million worth of bonds and became the core funding for a public-private financial package that paid the $450 million cost of the Gateway Center.

Gateway Center, which opened in 1994, includes Progressive (formerly Jacobs) Field, home of the Indians; the Quicken Loans (formerly Gund) Arena, home of the Cavaliers basketball team; and 3200 parking spaces in two garages. It adjoins and is connected to Tower City Center, a 6.5-million-square-foot complex of offices, hotels, and retail stores, as well as Horseshoe Casino, which opened in May 2012. Visitors to Gateway and Tower City Center come by car,

Cleveland, 2012. Fourth Street profits from customers generated by sports facilities a block away. *(Alexander Garvin)*

by bus, or on one of the Regional Transit Authority trains that stop at the station below the Tower City. They can stay at one of the 708 guest rooms in Tower City's two hotels, see a movie in its 11-screen cineplex, dine at one of its 10 full-service restaurants or at the food court, gamble at a casino, and attend an event at the ballpark or the arena.

Annual attendance at the 43,000-seat Progressive Field set a Major League attendance record between 1995 and 2001, selling out 455 straight games. Attendance, like winning, is never certain. In 2010 attendance was just under 1.4 million; in 2011 it was just above 1.8 million.[43] Attendance at the 20,500-seat Quicken Loans Arena has set attendance records over 150 times since its opening in 1994, but attendance is subject as much to the success of the teams playing in the arenas as to the cities in which they are built.[44] Specific numbers aside, the steady, nonseasonal 12-month flow of spectators to the Gateway venues and Tower City Center has supported significant commercial redevelopment in the surrounding area. Before Gateway opened there were 6 food and beverage establishments in the area; five years later there were 36. To preserve blighted historic buildings and encourage their reuse, the city adopted legislation severely restricting building demolition for surface parking lots provided tax incentives and bond financing, as well as assisting building renovation, through the use of its powers of condemnation. As a result, a residential neighborhood called Historic Gateway has emerged. As of 2012, the area included more than 1000 occupied apartments; there will be 500 more, plus at least 300 hotel rooms when six of its office buildings have been converted into loft-style residences.

The most dramatic changes in the area occurred on East Fourth Street, one block from the arena. Through the efforts of developer Ari Maron, this back alley once characterized by abandoned properties has become a lively, pedestrian-only concentration of renovated lofts and busy ground-floor retail stores, bars, and restaurants. The revitalization of Fourth Street began at the beginning of the twenty-first century with the residential conversion of an empty multistory loft building and the transformation of another into the House of Blues. By 2012, this short, dense block had become an entertainment venue, with customers eating and drinking at outdoor tables or strolling to and from live comedy clubs or just window-shopping.

Gateway has demonstrated that downtown sports facilities need not be surrounded by a sea of parking. If they are properly located and designed, sports centers will generate a steady, year-round flow of potential consumers. It is only a matter of time before these potential customers attract entrepreneurs who open new businesses. What better evidence is there that well-planned sports palaces can be powerful catalysts for downtown development?

Coors Field, Denver

Of all the recently built sports palaces, Coors Field has had the most impact on its surrounding city, albeit by accelerating investment that was already under way. This 50,000-seat baseball stadium opened in 1995, at the same time that Denver's economy was surging into a major boom. The customers it attracted increased the already substantial interest in the surrounding district, known as Lower Downtown (LoDo). LoDo had been the first section of Denver to develop. In 1863, a major fire destroyed most of the wooden structures that had been built in the area. The city responded by enacting an ordinance requiring brick or stone construction. The restaurants, brothels, and gambling halls that were built during the latter part of the nineteenth century turned LoDo into a busy but tawdry part of the Old West. Unfortunately, by the 1970s, many of these buildings had lost their occupants. Thus, LoDo had the same underutilized appearance as older loft districts in other cities.

LoDo's revival began in the late 1960s with the renovation of a block of Victorian structures along Larimer Street and its designation as a historic district. This revival, however, was stalled in the mid-1980s by a combination of downtown overbuilding and a crash in the energy industry. Civic leaders

Denver, 2011. Coors Field is located on city streets in the LoDo section of the city. *(Alexander Garvin)*

Denver, 2011. After Coors Field opened, property owners in LoDo began renovating historic structures; opening brewpubs, bars, and restaurants; and erecting new buildings. *(Alexander Garvin)*

decided that a baseball stadium was just what was needed to trigger LoDo's revival. How right they were!

In 1990, one year before Denver was awarded its baseball franchise, the six-county Denver area approved a 0.1 percent sales tax to fund a baseball stadium. This tax generated $168 million. The rest of the project's $215 million cost came from the Denver Rockies baseball team and from the Coors Brewing Company, which paid $15 million for the naming rights.

The 76-acre site selected for Coors Field was right on the edge of LoDo, two blocks from Union Station and four blocks from the 16th Street Mall. The design, by the HOK Sports Facilities Group, offered something few other stadiums could supply: spectacular views of the nearby Rocky Mountains. It also differed from most sports palaces by not providing on-site parking. Fans coming by car park in nearby lots otherwise used by daytime workers and walk to the stadium. On their way to or from a game they can go shopping or stop for a meal or a drink.

The year Coors Field opened, sales taxes from LoDo increased 86 percent over the previous year. This was because more than 70 percent of this tax revenue came from the sale of food and beverages to consumers who were attracted to the stadium area. In a poll taken of the 4 million fans annually attending games, 37 percent of respondents reported spending between $41 and $75 per game in the downtown area, much of that in the immediate vicinity of the stadium. Within three years, 25 new restaurants and six additional microbreweries had opened in LoDo. More than 800 housing units had been created in either new or rehabilitated buildings, and another 250 were under construction. As of 2012, 39 run-down structures in LoDo had been converted into occupied residential buildings with loft-style apartments.[45]

Convention Centers

Cities invest in convention centers because these very large exhibit halls are perceived as municipal moneymaking machines. In fact, convention centers make the other palaces for the people seem like small potatoes. The "meeting" industry in the United States directly supports 1.7 million jobs and indirectly supports an additional 4.6 million jobs. The industry accounts for approximately $263 billion in annual spending, generating more than $14 billion in federal and $11 billion in state and municipal tax receipts. Moreover, the 1.8 million meetings in the United States and the 200 million people who attend them annually create a demand for 250 million overnight hotel stays.[46] Like a stadium, a convention center, to be financially successful, needs to attract enough of a market to pay operating costs and debt service. Convention centers usually collect sufficient fees to meet operating costs, but often cannot cover the debt service. In those cases, government officials try to justify subsidizing debt service out of public funds by estimating the additional tax revenues that otherwise would not be collected. They also enumerate the jobs that have been created, but usually without any dependable estimate of the proportion of those jobs that will actually go to city residents.

A modern convention center requires a single-level exhibition hall of several acres. Creating such a vast space often requires removing huge chunks of downtown land and, with it, huge numbers of people. The result is curiously paradoxical. Conventions bring customers and economic activity to the city, whereas convention centers often remove those same customers from the city streets. Thus, one must be careful to prevent the edges of any convention center from reducing pedestrian activity and thereby having a blighting impact on surrounding areas.

Conventions may be an economic bonanza. But this bonanza can be tapped only when, like the Washington State Convention and Trade Center in Seattle, convention facilities are strategically located, appropriately designed, and purposefully integrated into the physical structure and economic life of the surrounding city. Convention facilities, like those along the Detroit waterfront, that are separated from the surrounding city and are designed to satisfy visitors' every desire without those visitors setting foot in the rest of the city will never have much beneficial impact. They occupy such large sites that (unless they are built over untenanted sites such as highways, streets, or railroads) they also remove customers and activity from their periphery, in the process blighting contiguous businesses.

Renaissance Center, Detroit

Detroit's vast visitor complex is located at the southern end of the city along the Detroit River. Initially it included Veteran's Memorial Hall (opened in 1950), the 2900-seat Henry and Edsel Ford Auditorium (opened in 1955), and the 2.5-million-square-foot Cobo Exhibition Hall and Convention Arena (opened in 1960). The pièce de résistance opened in 1977. It is the Renaissance Center, a $350 million complex designed by architect John Portman. It includes a 73-story hotel (at the time of its completion, the world's tallest), four 39-story towers containing 2.2 million square feet of office space, and a 14-acre, 4-story podium containing additional retail, convention, and parking facilities. During the 1980s the complex was amplified by two additional 21-story office towers, the Joe Louis Arena, and the renovated Cobo Center.[47]

Millions of tourists make use of these facilities. Businesspeople profit from the revenues they generate. Middle-class workers benefit from thousands of jobs that are located there. The city government raises tremendous amounts of tax revenue. The only loser has been downtown Detroit.

When Henry Ford II announced the Renaissance Center in 1971, Detroit was in trouble. Its population had declined to 1,511,000, a drop of 339,000 since 1950.[48] He rallied the business community, persuading 50 of the city's major firms to invest $1 million each. In addition to this $50 million in equity, Ford obtained $200 million in mortgage financing ($180 million from a consortium of insurance companies and $20 million from Ford Motor Credit Company).

Detroit, 2008. Traffic on Jefferson Avenue separates the Renaissance Center from the rest of the business district. *(Alexander Garvin)*

The strategy was to allow Detroit's business district to profit from a lucrative tourist and convention business. Renaissance Center's backers selected a site along the riverfront, where the new development could augment Cobo Hall and the other facilities that were already there. More important, the project would also clear 33 acres of blighted property and hook onto Jefferson Avenue, which was being transformed into a 12-lane traffic artery that connected directly into the Interstate Highway System.

When Renaissance Center opened in 1977, it was unable to obtain projected rents or levels of occupancy. Planners had overestimated the convention and tourist market that could be attracted to the city. Nor were there enough tenants for office space at high rent levels or enough customers to support the vast network of retail outlets. Consequently, revenues would not cover project debt service. The mortgagees chose not to foreclose. They became equity partners of a financially restructured venture in which they also had an active management role.

Meanwhile, the Detroit business district was dying a slow death. Hudson's, the nation's tallest and second-largest department store, closed its 2.1-million-square-foot store on Woodward Avenue in 1983 and demolished it in 1998. The 18-story Hilton Hotel on Washington Boulevard was shut down. Vacant office space and retail frontage became the norm. By 2010, the city's population would drop to 714,000,

Detroit, 2008. The creation of the Renaissance Center did not prevent the deterioration and abandonment of nearby buildings. *(Alexander Garvin)*

less than half of what it had been when Renaissance Center had first been proposed.[49]

Renaissance Center had exacerbated an already bad situation. The project's office space was designed to attract the city's major firms. Consequently, their move from existing downtown buildings drew customers away from the already declining business district, leaving the area pockmarked with vacant structures and parking lots where there had once been buildings. Still worse, the project was separated from downtown Detroit by the traffic on Jefferson Avenue. Downtown Detroit is just too far away for the 16,000 office occupants of Renaissance Center to use downtown stores and restaurants.

Convention visitors had no reason to leave Renaissance Center. They came from the airport along convenient modern highways. Once safely inside, they were unlikely to risk a visit downtown. Cobo Hall and every other place they might need to go were happily isolated between the Detroit River and Jefferson Avenue.

In an attempt to remedy the lackluster market response to Renaissance Center, the project's administrators hired real estate consultants, who recommended eliminating design and marketing flaws. Retail facilities became more appropriate to the project's tenantry, and the hotel began to operate more successfully. In order to remedy the center's isolation from downtown Detroit, the city tried to tie the complex to the rest of town by building an elevated "peoplemover" transit line circling the central business district and connecting Cobo Hall and Renaissance Center with the rest of downtown Detroit. But the damage could not be undone. From the time Renaissance Center opened, the only place in Detroit that developers have been willing to erect new buildings is directly opposite it. Meanwhile, downtown firms have continued to move away, affecting an even greater need for occupants and activity in downtown Detroit.

In 1995, General Motors purchased Renaissance Center, retrofitted the office buildings as its national headquarters, and, in 2000, relocated to the complex. Ford Motor Company businesses moved from Renaissance Center to a new headquarters in Dearborn. Some occupants moved to the suburbs. Others moved into vacant downtown office space. This game of musical chairs did not revive downtown Detroit, whose future remains very much in doubt. Even if Detroit's decline is reversed, it is still unlikely that the occupants of Renaissance Center will cross Jefferson Avenue more frequently than their predecessors or that their presence will make much difference to nearby retailers.

Jacob Javits Convention and Exhibition Center, New York City

New York City's civic leaders wanted a convention center that would not result in the loss of site tenants who brought customers and traffic to the surrounding area. They also wanted to avoid losing time in litigation with relocatees and paying the substantial costs involved in their relocation. Initially, they sought to achieve these goals by planning a 40-acre convention center on a platform in the Hudson River between 43rd and 47th streets. It was to be connected to midtown Manhattan by ramps and bridges extending across the West Side Highway to 11th Avenue. The project was unveiled in 1972 and abandoned during the city's ensuing fiscal crisis. When Mayor Edward Koch revived the project in 1978, it was reconceived for a 21-acre site in Manhattan, east of the highway, between 33rd and 39th Streets. The new site was occupied by an underutilized railroad yard and thus also avoided displacing site tenants.[50]

The Jacob Javits Convention Center, which opened in 1984, was built to host trade shows, large conventions, heavy

Manhattan, 2004. The Jacob Javits Convention and Exhibition Center is far from the theaters, hotels, restaurants, and retail stores that cater to New York City's tourists. *(Alexander Garvin)*

industrial exhibits, banquets, and large special events. Its 814,000 square feet of exhibit space and 100 meeting rooms can accommodate 85,000 visitors at one time. The complex is serviced by 50 covered loading docks, three drive-in ramps, and a 4-acre truck-marshaling area used for loading and unloading activities.

Compared to the 2,670,000 square feet of Chicago's McCormick Place, America's largest convention center, the Javits Center is small. Nevertheless, it was relatively successful when it opened, hosting approximately 60 conventions and trade shows annually through the 1990s. During the second decade of the twenty-first century, Javits was hosting 80 major trade shows and 70 major events annually.

Like Detroit's Renaissance Center, the Jacob Javits Convention Center has brought convention business without generating the hoped-for benefits for the surrounding area. It is more than half a mile from the nearest hotels, which are opposite Pennsylvania Station or in the Theater District. Visitors have to travel by taxi, limousine, or bus to get to the Javits Center, and access to New York's subway is inconvenient and distant. When visitors arrive, they find little or nothing of interest in the surrounding blocks. To the south are rail yards, to the north a bus depot, to the west the Hudson River. Until 2003, the blocks east of the Javits Center were zoned for manufacturing and contain parking lots, warehouses, and low-rise loft buildings. Thus there was nowhere else for conventioneers to go. Consequently, no businesses were profiting from the thousands of people who come to see a show or attend a meeting at the convention center. The rezoning has led to the construction of new apartment buildings. More important, the extension of the subway to 34th Street and 11th Avenue will greatly reduce travel times from the city's many hotels and from the theater district.

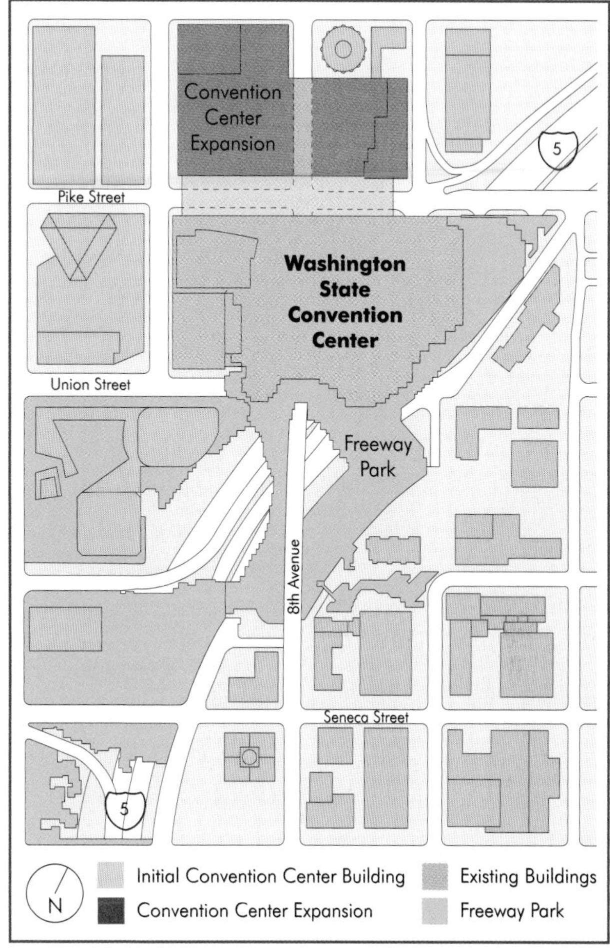

Seattle. The Washington State Convention and Trade Center straddles an interstate highway. *(Alexander Garvin and Ryan Salvatore)*

Washington State Convention and Trade Center, Seattle

The enormous size of convention centers, like those of Detroit and New York, often condemns them to fringe locations, where they become barriers to street traffic and deaden rather than generate activity. The Washington State Convention and Trade Center demonstrates how intelligent planning can transform such behemoths into a force for municipal improvement. Without moving a single customer from the site, this convention center was built right in downtown Seattle where it could affect its surroundings and bring together sections of the city once separated by an interstate highway.[51]

Downtown Seattle is built on a steep, hilly site overlooking Elliott Bay. It is cut off from the eastern sections of the city by an interstate highway that was completed in 1965. Ever since, civic leaders have sought to reconnect both sides of town by bridging the highway. Their first effort, completed in 1976 and expanded in 1984, was the delightful Freeway Park. To that, in 1988, they added nearly 1 million square feet of exhibit halls, meeting rooms, retailing, parking, and support facilities of the Convention and Trade Center. In 2001 a further expansion added 100,000 square feet of exhibition space, a 990-car parking garage, loading docks, a 450-room hotel, an office building, and a new home for Seattle's Museum of Science and Industry. These additions are connected to the original structure by a 90-foot-wide bridge over Pike Street, topped with a glass canopy.

Seattle, 2007. Highway ramps and city streets run right through the Washington State Convention and Trade Center. *(Alexander Garvin)*

Seattle, 2007. Visitors to the Washington State Convention and Trade Center stay at downtown hotels, patronize nearby restaurants, and use local streets and sidewalks to go to and from its events. *(Alexander Garvin)*

The Washington State Convention and Trade Center was developed by a public corporation created by the state legislature in 1982 and designed by a team of architects led by the Seattle architectural firm TRA. LMN Architects designed the 2001 expansion. The total development cost for the original section of the center was $157 million, of which $121 million came from state general obligation bonds, $1 million from the City of Seattle, and most of the rest from private funds. Its operations are paid primarily by a tax on hotel rooms in Seattle and surrounding King County. The state of Washington covered most of the cost of the $147 million expansion, which has been financed with hotel-motel tax and convention center income.

Seattle's convention center bridges the highway and encompasses within itself three city streets, ramps that access the highway, and a variety of landscaped sitting areas and pedestrian paths. It supplies the hotels, restaurants, and retail stores on surrounding blocks with customers, and they in turn supply conventioneers with everything that is missing from the center itself. As a result, instead of deadening activity along its edges, the convention center fills the vacuum created by the highway by attracting pedestrian and vehicular traffic and uses them to tie together different sections of the city.

From an economic perspective, both the Detroit and the Seattle combination of convention center, hotels, restaurants, and retail stores is desirable. In both cases, vast facilities draw a substantial convention and visitor market downtown, bringing with them hundreds of millions of dollars in expenditures that would not otherwise be made in those cities. From an urban planning perspective, however, Seattle's facility has become a major force for municipal vitality, whereas Detroit's draws away the city's lifeblood.

Pennsylvania Convention Center, Philadelphia

Philadelphia provides one of the best examples of a convention facility that is changing the character of its surrounding central business district. The story begins in the mid-1950s, long before the Pennsylvania Convention Center was even conceived of. Edmund Bacon, executive director of the City Planning Commission, proposed balancing Penn Center, the new transit-retail-office hub west of City Hall, with a similar project on the east (see Chapters 6, 12, and 19).

The new transit hub that Bacon envisioned was opened by the Southern Pennsylvania Transit Authority (SEPTA) in 1984. SEPTA had taken over the Reading Railroad's commuter rail service after the Reading had declared bankruptcy for the fourth and final time in 1972. So, when the $330 million City Center Commuter Connection opened and operations ceased at the old Reading Terminal, it was assumed that the 1893 railroad station and 344,000-square-foot headhouse

Philadelphia, 2011. The Pennsylvania Convention Center. *(Alexander Garvin)*

would be torn down and replaced with a revenue-generating building.

Opposition to demolishing the station came from architecture buffs, preservationists, and thousands of Philadelphians who patronized the market located at street level, underneath the station's single-span, arched train shed. While this opposition was gaining strength, the Pennsylvania Convention Center Authority, created by the state legislature in 1987, decided to build just north of the now unused train shed. Civic leaders had been trying to replace the undersized, obsolete convention center located in West Philadelphia since the 1950s. In 1990, the Convention Center Authority purchased the market and agreed to renovate the train shed and make it a part of the design.

The 1.3-million-square-foot Convention Center complex opened in 1993. The first two floors of the nine-story brick Reading Terminal's headhouse act as a grand entrance to the complex. They lead directly to the Grand Hall and Ballroom that occupy the terminal's soaring train shed. The Marriott Convention Center Hotel occupies 210 rooms on the upper floors of the headhouse as well as a separate 1200-room building across the street. To the east, spanning three blocks, is the four-story Gallery at Market East shopping complex. Underneath lies the rail and subway complex that brings more than 20,000 commuters every day. The new building, just north of the train shed, hosts all sorts of regional and national exhibitions and meetings, including the annual Philadelphia Flower Show.

Construction of the Pennsylvania Convention Center has led to a major increase in tourism. Within three years the facility was hosting 32 major conventions, making Philadelphia the seventh busiest convention city in the United States, behind Las Vegas, Chicago, Orlando, San Diego, New York, and San Francisco. The number of downtown hotel rooms doubled. Between 1992 and 2000, the number of fine dining restaurants increased from 65 to 175. In 2000 alone, the Convention Center generated more than 581,000 hotel room nights, 7300 jobs, and $2.6 billion in total overnight visitor spending in the five-county Philadelphia area.[52]

An additional $786 million wing that extended the Convention Center to Broad Street was opened in 2011. It increased exhibit space by 550,000 square feet to more than 1 million square feet, nearly tripled the size of the ballroom, and added three exhibition halls and 29 meeting rooms. During that year "Center City's existing inventory of 11,159 hotel rooms generated 45% of all regional hotel-room revenue in 2011, up from 38% in 2008."[53]

One reason the complex has been so successful is its convenient location right in the middle of downtown Philadelphia; the other is that its large footprint, like that of Seattle's Convention Center, is penetrated by transportation corridors. To the east are Chinatown and historic Independence

Hall with its Liberty Bell. To the south and west are city hall, the arts district, and more retail and dining opportunities. Most of the city's downtown hotel rooms are within a 20-minute walk of the Convention Center. Its success is not just a matter of location. The center is completely integrated into the fabric of downtown Philadelphia because several streets pass right through the project, keeping it from walling off nearby sections of the business district, and because it is linked to subway and commuter rail lines that tie it directly to the entire metropolitan region.

Ingredients of Success

Cities continue to invest in ever more monumental palaces for people to visit because they correctly conceive of them as magnets attracting a lucrative market that can generate jobs and desperately needed taxes. As these facilities grow larger and more expensive, their increasing appetite for government subsidies makes it important to understand the requirements for their success. After all, why should a city subsidize something that does not produce sufficient benefits?

To be successful, palaces for the people must attract a market that would otherwise not exist, and they must be located and designed in a manner that forces that market to interact with the rest of the city. The financing, development, and management issues are somewhat different from profit-motivated private ventures because so many of these facilities may not be immediately self-supporting. That condition, however, is all the more reason to demand that public officials and the development agencies established to build and manage these facilities consider and plan for the private market reactions that they generate.

Market

In creating palaces for the people, civic leaders must determine the size and character of the market that will be attracted. How much will that market spend and on what? During what periods of time will it be there? What sorts of support services and activities will it require? Most important, how can the rest of the city profit from the people attracted to that particular facility? If these questions are not answered, cities will not be able to plan intelligently for the spillover of that market and consequently will not be able to take public actions that will generate desirable private-sector reactions.

For planning purposes, the most important characteristic of any market is its source. Attracting conventioneers from out of town clearly results in a net addition to the local economy. The people coming to all but a few sports events come from the same market area. Nevertheless, a new stadium may add a great deal to the economy of the surrounding city when it attracts sports fans who would otherwise be attending games in other parts of the region. Most civic centers, on the other hand, merely move an existing market from one part of the same city to another.

Only when a market is sufficiently large will it produce any reaction from surrounding property owners. A single auditorium could never have generated the changes that Lincoln Center brought to the West Side of Manhattan; just as a single sports arena could not have brought enough weekend customers to enliven downtown Cleveland all year long. Every instance requires a critical mass without which little will change.

The most important consideration in accommodating any of these markets is *not* to supply everything within the planned facility. A concertgoer who can obtain a meal, a snack, a drink, or a souvenir right there has no reason to go anywhere else. Similarly, if a convention center includes enough hotel rooms on-site to satisfy every conventioneer, there will be little or no market spillover to surrounding properties. That is why, from the very beginning, the Renaissance Center was guaranteed to flop, and the Seattle and Philadelphia convention centers were bound to generate activity in surrounding areas.

Location

Palaces for the people need to be easily accessible to enormous numbers of users, to the people who work there, and to the vehicles that deliver the goods and services that they provide. That is why cultural centers, stadiums, and convention centers are located along major highways.

Highway access alone is not enough. Vehicles need to be able to enter and exit quickly and easily during peak periods. Nearby city streets and parking lots all contribute to making Cleveland's sports facilities and the Seattle Convention Center easily accessible. Peak-period traffic at the Kennedy Center and Turner Field, on the other hand, is largely dependent on highway access ramps. As a result, when everybody leaves, serious congestion ensues.

Sufficient parking and loading space is essential, though it need not be supplied entirely on-site. Coca-Cola Field achieves this by developing a reciprocal relationship with existing facilities in downtown Buffalo; Lincoln Center, by depending on surrounding blocks of the Upper West Side. The sea of parking around Turner Field and the Los Angeles Coliseum serves only to further isolate them from the surrounding city.

For planning purposes, the most important characteristic of any location is its periphery. Not only are the Los Angeles Civic and Music Centers located at the far end of the central business district, too far away to have much effect on downtown life, they are cut off from their surroundings on two sides by freeways. The Renaissance Center is similarly cut off from downtown Detroit by Jefferson Avenue.

Thus, the key to successfully locating palaces for the people lies in artfully combining convenient highway access with a site in the midst of an existing downtown business district and then encouraging a mutually reinforcing relationship with that district.

Design

Designing successful public facilities is largely a matter of dimensions and arrangement of components. The importance of dimensions cannot be overemphasized. The distances across the plaza in Cleveland's Group Plan made interaction among the buildings difficult. The length of Memorial Plaza in St. Louis precluded the development of a critical mass of customers affecting its surroundings. These civic centers create impressive public open spaces. Unlike the plazas of the Federal Center in Chicago, the ability of these spaces to come alive requires vast crowds that are unlikely to visit simply to attend a trial, renew a driver's license, or obtain a passport.

The size of the facility itself is also critical. Convention centers, stadiums, and their parking facilities cover acres of land, in the process removing the customers who previously patronized the facilities and institutions that remain along their periphery. In many cases, sports fans or conventioneers cannot replace the customers that these businesses lose. Seattle accommodates the large footprint required by its convention center by placing it over a highway that itself had removed the customers who previously had been on the site. The Washington State Convention and Trade Center also illustrates the importance of relationship among traffic arteries, vehicular storage, exhibition halls, meeting rooms, and support services for visitors. The design specifically incorporates an interstate highway and its access ramps, city streets and their access ramps, and pedestrian walkways in a manner that prevents convention center traffic from interfering with other vehicular and pedestrian circulation, even during those times that downtown traffic is at its most intense.

Palaces for the people can be sited in three ways. They can be freestanding like the Los Angeles Coliseum or Denver's Coors Field; they can be clustered together like the performance halls at Lincoln Center or the civic buildings in Cleveland's Group Plan; or they can be completely integrated into the city's street system like the Federal Center in Chicago. Each of these strategies can generate or fail to generate further improvements to the surrounding city, depending on the arrangement of their contents. Coors Field depends on parking scattered throughout LoDo. Fans on their way to and from a game inevitably are attracted by the bars, restaurants, and souvenir stands that line their route. At Turner Field there is nothing between the cars and the stadium except more cars. That is why businesses in LoDo have thrived, whereas nothing much has happened to the properties near Turner Field. The 800 parking spaces under the plaza at Lincoln Center are not enough to service the 18,000 people who attend performances and work there, so those who do not come by public transportation park in the surrounding neighborhood. Like the fans going to Coors Field, the restaurants and stores that line their route attract them. In Cleveland, on the other hand, ample parking near each of the buildings in its civic center means there is little reason to walk any distance to get to one's car or patronize the small number of businesses in the surrounding district. Chicago's Federal Center, like any facility that is an integral part of a street grid, automatically funnels customers to surrounding buildings.

Financing

Neither civic nor cultural centers make any pretense of being self-financing. Federal buildings are paid for by congressional appropriation. General obligation bonds, whose debt service is paid from tax revenues, usually finance state and municipal buildings. Museums, concert halls, theaters, opera houses, and other cultural institutions tend to be paid for through charitable donations.

Financing, however, is a major issue in the case of sports and convention centers. In most instances, revenue bonds, naming rights, advertising revenues, and exclusive marketing rights for specific products finance them. As of 2008, either Coca-Cola or Pepsi had paid to be the exclusive provider of soft drinks available at nearly every stadium and arena in the country. The price of naming rights has been escalating. In 1995, the Coors Brewing Company paid $15 million for the naming rights to the stadium used by the Denver Rockies baseball team; in 2008, Progressive Insurance paid $57.6 million for the naming rights to the stadium used by the Cleveland Indians baseball team. In Los Angeles, Farmers Insurance committed to pay $700 million for 30 years for a future NFL stadium. Nevertheless, naming rights can expire. Thus, in 2005 Cleveland received a new infusion of money when Quicken Loans purchased naming rights for what had previously been the Gund Arena.[54]

Because these facilities are occupied by revenue-generating businesses, their revenues are often set aside to cover operating costs and debt service on the bonds. Stadium developers are becoming more and more effective at generating these revenues. One reason so many arenas and stadiums have been rebuilt recently is that their operators can earn greater revenues from selling skyboxes and club seats for an entire season than they can from selling general admission tickets. A large portion of the financing for Joe Robbie Stadium in Miami, for example, came from the sale of executive suite leases.

Too often, when a facility is completed, it cannot generate sufficient revenues to cover debt service. As a result, municipal governments use general revenues and dedicated taxes to make up the difference. Hotel and car rental taxes are popular with local legislators as a device to cover debt service on convention centers. Presumably, their constituents pay such costs only when they are out of town. Better still, they can say that the businesses that profit from the additional customers are simply sharing some of the revenues generated by the new facility. Taxes on alcohol and tobacco (often called *sin taxes*) are also popular. Cleveland helped to pay for Progressive Field with a 4.5-cent-per-pack tax on cigarettes and a $3 per gallon tax on alcohol. No municipality starts out by promising to subsidize these profit-making businesses. They are lured into it by the promise of additional employment and taxes. Indeed, the indirect expenditures by conventioneers on food, hotel rooms, transportation, and retail purchases may generate the taxes needed to subsidize these convention facilities. Only when that proves to be the case are government subsidies justified.[55]

Time

Civic, cultural, sports, and convention centers are all dependent on the patronage of large numbers of people. The way they move through these facilities is critical. No city will benefit if people simply park their cars, go about their business, and return home. The goal is to get visitors to spend some time in town on their way to and from their destinations. The Federal Center in Chicago and the cultural complex in Cincinnati accomplish this by mixing commercial office buildings with sites designated for government or cultural activity. Coca-Cola Field and the Seattle Convention Center force their customers into the city by failing to supply all their needs within the facility itself. Whichever strategy is adopted, the result is to prolong the visit and force visitors to interact with the rest of town during their stay.

The most important benefit derived from palaces for the people is the activity generated during those times that the city would otherwise be without it. Lincoln Center brings people to Manhattan's West Side at night and on weekends. The Pennsylvania Convention Center brings additional customers to a section of that city's business district at night and on weekends, when it would otherwise be devoid of much activity. However, the opposite can also be true. Football stadiums, for example, only bring customers eight or ten times a year. If they cannot be used for other purposes, they create a vacuum more than 350 days a year. Baseball stadiums, on the other hand, operate throughout a long season. Civic centers have a deadening impact on weekends, when most government offices are closed. To be successful, these public palaces need to bring people at times that the surrounding district would otherwise be devoid of activity.

Most stadiums and convention centers bring a mass of customers that is large enough to generate a market reaction if those customers spill over into the rest of town. But civic centers can achieve this only when they include enough individual buildings to aggregate a similar critical mass of customers. In most cases, aggregating a critical mass of buildings takes many years. It was three decades before Cleveland's civic center included enough government buildings to develop this critical mass, and it took half a century in St. Louis. In the interim, neither city could benefit from a critical mass of customers. By the time each civic center had been completed, it had become obvious that faulty design would preclude the necessary critical mass from ever developing. Chicago's Federal Center, on the other hand, is specifically designed so that there is no need to wait decades before its effects are felt. From the day it opened, 14,000 government employees and the thousands of others who have business with them have been generating customers who patronize nearby establishments.

Entrepreneurship

Planning prescriptions are never implemented spontaneously. They require a private or public entrepreneur. Sometimes, business leaders supply the entrepreneurship. Buffalo's Coca-Cola Field would not have been built without the backing of the Rich Product Corporation, nor would Detroit's Renaissance Center without Henry Ford II and the 50 major corporations that supported him.

The difficulty with depending on corporate entrepreneurship is that it usually confuses financial self-sufficiency with urbanistic success. Coca-Cola Field may generate enough revenues to cover operating costs and debt service. Its contribution to the city, however, comes from attracting a market that would not otherwise be downtown and accommodating that market in a manner that forces consumers to spill over into the business district, where they may purchase something. The Renaissance Center may have been a financial failure, but its disastrous effect on downtown Detroit has nothing to do with that financial failure. That came about because its very existence skimmed the cream off the downtown office market, while its location and design ensured that its convention business had little or no effect on the rest of the city.

Some of the most successful public facilities are the product of public, rather than private, entrepreneurship. Without the political acumen of Senator McMillan or the persuasiveness of Daniel Burnham, the Mall in Washington, D.C., would be a very different place.

Just as private developers often confuse financial feasibility with desirability, public entrepreneurs often confuse public support with good planning. The creators of the St. Louis and Los Angeles civic centers were able to obtain support and funding for their plans without understanding that the monumental spaces they created produced hollow cores rather than vibrant contributors to the economy and character of the surrounding business district.

Robert Moses never had the luxury of relying on easy public acceptance of Lincoln Center. From the day he first announced this project, he had to fight with citizen organizations, community groups, and skeptical politicians. They forced him to adjust and readjust his proposal until it took into account some of the interests they represented. The same thing happened in Buffalo after Mayor Griffin proposed building Coca-Cola Field. He had to adjust the project to the demands for union workers, minority employment and contracting, cheap replacement parking, and sensitive relocation practices.

The process of public review forced Moses and Griffin to take into consideration conflicting demands and to broaden their view of the public interest. As a result, Lincoln Center and Coca-Cola Field were altered until they each became feasible both financially and politically. More important, they also were altered to accommodate some of the interests of surrounding communities.

Palaces for the People as a Planning Strategy

Civic, cultural, sports, and convention centers are magnets that attract customers from the surrounding metropolitan area and beyond. They are also very expensive—increasingly expensive as they grow in size and complexity. Too often, the revenues they generate are not enough to cover operating costs, maintenance, and debt service on the bonds issued to pay for them.

Whenever projects become a drain on city or state governments, critics legitimately question the wisdom of investing in such facilities. Why should the public subsidize privately owned, profit-making sports teams that pay enormous salaries when thousands of children are being left behind in inadequately funded schools? Why should the city maintain uncompetitively high taxes to cover the losses of a convention center? The answers will be negative as long as the desirability of these projects continues to be measured primarily in terms of their own financial feasibility. But if other businesses and residents could obtain tax-exempt financing on terms that were similar to those of the proposed new facility, there would surely be voices offering quite different answers.

Many large public palaces have negative consequences. As a result, they trigger opposition from area residents, workers, businesses, and property owners, who object to traffic congestion, displacement, and other problems caused or exacerbated by the new or expanded facility. More important, these negative spillover effects frequently dominate the public dialogue.

Yet the year Coors Field opened, sales tax collections in LoDo increased 86 percent, and another 8 percent the following year. This was because of the increase in retail establishments. By 1998, six new microbreweries and 25 new restaurants had opened. During that same period, the number of housing units in LoDo increased from 270 to 1074.[56] Philadelphia's Convention Center was followed by the creation of 4000 new hotel rooms. The new bars, restaurants, and hotel rooms could not be part of the debates over the creation of these facilities because they were not as yet in place. Besides, skeptical opponents could point to the relative absence of sports-fan-related businesses around such facilities as Turner Field in Atlanta, or the lack of hotel construction around the Javits Convention Center. Perhaps because Buffalo's Coca-Cola Field occupies less territory than a big league stadium, because it is easier to fill 19,500 seats than to fill a full-size 50,000-seat facility, or because it brings spending on weekday night and weekend game days, when downtown Buffalo would otherwise be deserted, the $56 million spent for Coca-Cola Field seems well worth the money—particularly when compared with the $1.6 billion spent on the 80,000-seat MetLife Stadium that generates neither activity nor spending outside the huge parking fields in the New Jersey Meadowlands on the approximately 20 days a year that it hosts a home game for the Jets or the Giants, unless it can attract additional events that seat 80,000 people.

Conventional planning procedures will never ensure that public facilities will enrich the cities that surround them, and neither will the parochial efforts of arts organizations, sports enthusiasts, the tourist industry, or local chambers of commerce. The best way to ensure that these large public projects generate widespread and sustained spillover benefits is to require an approval process that includes an analysis of the effects on the surrounding area.

Notes

1. Thirty-seventh Annual Report of the Trustees of the Public Library, City Document #13, Boston, 1888, p. 6.
2. Historical and statistical information on the Chicago Fair is derived from James Gilbert, *Perfect Cities, Chicago's Utopias of 1893,* University of Chicago Press, Chicago, 1991; R. Reid Badger, *The Great American Fair,* Nelson Hall, Chicago, 1979; Stanley Appelbaum, *The Chicago World's Fair of 1893,* Dover, New York, 1980; William H. Wilson, *The City Beautiful Movement,* Johns Hopkins University Press, Baltimore, 1989, pp. 53–74; Thomas S. Hines, *Burnham of Chicago,* Oxford University Press, New York, 1974, pp. 73–138; Leland M. Roth, *McKim, Mead & White, Architects,* Harper & Row, New York, 1983, pp. 174–179; and Laura Wood Roper, *FLO—A Biography of Frederick Law Olmsted,* Johns Hopkins University Press, Baltimore, 1973, pp. 425–433, 444–450.
3. The 1890 Census reported a total population of 62,947,714. See U.S. Department of Commerce, Bureau of the Census, *Statistical Abstract of the United States 1989,* Washington, DC.
4. Henry Codman, John Wellborn Root, Charles Follen McKim, and Augustus St. Gaudens also played major roles in determining the shape of the fair. Codman, one of Olmsted's partners, accompanied

him on his first trip for the Exposition Corporation and continued to share management responsibilities until his sudden death at the age of 29, several months before the fair opened. Olmsted, Codman, Burnham, and Root met regularly for over a year, during which Root acted as a sort of graphic stenographer recording their ideas. Root died equally suddenly in early 1891, two years before the fair opened. Burnham never got over the loss. Perhaps this led him, acting as the exposition's director of works, to consider McKim his right-hand man during the later stages of the fair's development. St. Gaudens was responsible for the coordination of the fair's opulent sculptural decoration and thus made a major contribution to its distinctive appearance.

5. Henry Adams, *The Education of Henry Adams,* Boston, 1918, p. 343.
6. The major exception was the Transportation Building designed by Adler and Sullivan.
7. Charles Mulford Robinson first used the phrase "City Beautiful" in 1899 in an article he wrote for the *Atlantic Monthly.* This is one reason that William H. Wilson argues that the City Beautiful movement did not originate with the Chicago Fair (op. cit., pp. 64–65, 70–71). However, the Chicago Fair surely popularized many of the concepts and esthetic devices that later became known as the City Beautiful.
8. Deirdre A. Gaquin and Mark S. Littman (editors), *1998 County and City Extra—Annual Metro, City and County Data Book,* Bernan Press, Lanham, MD, 1998.
9. Historical and statistical information on the L'Enfant Plan is derived from John W. Reps, *Monumental Washington,* Princeton University Press, Princeton, NJ, 1967; Frederick Gutheim (consultant), *Worthy of the Nation,* National Capitol Planning Commission, Washington, DC, 1977; Pamela Scott, "This Vast Empire," in Richard Longstreth (editor), *The Mall in Washington 1791–1991,* National Gallery, Washington, DC, 1991, pp. 37–60; and Christopher Tunnard, *The Modern American City,* Van Nostrand, Princeton, NJ, 1968, pp. 53–54.
10. Frederick Law Olmsted Jr. was in his early twenties when his father's firm began work on the Chicago Fair. Not only was he familiar with its initial planning and design, he spent one summer working as an aide-de-camp to the superintendent of construction, and he was later to say that it was one of the three most stimulating experiences of his professional life (Roper, op. cit., p. 431).
11. Historical and statistical information on the McMillan Plan is derived from Reps, op. cit., pp. 71–157; Gutheim (consultant), op. cit., pp. 111–136; Hines, op. cit., pp. 139–157; Roth, op. cit., pp. 251–259; and David C. Streatfield, "The Olmsteds and the Landscape of the Mall," in Longstreth, op. cit., pp. 117–141.
12. Charles Moore, *The Life and Times of Charles McKim,* Houghton Mifflin Co., Boston, 1929, p. 194 (quoted in Reps, op. cit., p. 97).
13. Roth, op. cit., p. 254.
14. This Gothic "obstruction" (never referred to in Moore's text) was carefully eliminated from every drawing and model of the proposed mall.
15. Herbert B. Briggs, "Municipal Improvement, Cleveland," *The Inland Architect and News Record,* 34, August 1899, pp. 4–5 (quoted in Hines, op. cit., note 5, p. 160).
16. Historical and statistical information on the Group Plan is derived from Daniel Burnham, John Carrère, and Arnold Brunner, *Report of the Group Plan of the Public Buildings of the City of Cleveland, Ohio,* Board of Supervision for Public Buildings and Ground, Cleveland, 1903; Holly M. Rarick, *Progressive Vision: The Planning of Downtown Cleveland 1903–1930,* The Cleveland Museum of Art, Cleveland, 1986; and Hines, op. cit., pp. 158–173.
17. The federal building and post office were completed in 1911; the Cuyahoga County courthouse in 1913; city hall in 1916; the Cleveland public library in 1925; the public auditorium, music hall, and convention center in 1927; and the Board of Education Building in 1930.
18. http://www.clevelandmedicalmart.com.
19. Historical and statistical information on Memorial Plaza is derived from The Civic League of St. Louis, *A City Plan for St. Louis,* St. Louis, 1907; Harland Bartholomew, *Comprehensive City Plan—St. Louis, Missouri,* St. Louis, 1947; and George McCue, *The Building Art in St. Louis,* St. Louis Chapter of the American Institute of Architects, Knight, St. Louis, 1981.
20. Historical and statistical information on the Los Angeles Civic Center is derived from Robert M. Fogelson, *The Fragmented Metropolis—Los Angeles 1850–1930,* Harvard University Press, Cambridge, MA, 1967, pp. 262–271; and Paul Gleye, *The Architecture of Los Angeles,* Rosebud Books, Knapp, Los Angeles, 1981, pp. 102–104.
21. Historical and statistical information on the Boston Government Center is derived from Walter Muir Whitehill, *Boston: A Topographical History,* Harvard University Press, Cambridge, 1968, pp. 200–217; Rachelle I. Levitt (editor), *Cities Reborn,* Urban Land Institute, Washington, DC, 1987, pp. 9–53; and Donlyn Lyndon, *The City Observed: Boston,* Vintage Books, New York, 1987, pp. 32–42.
22. Peter Carter, *Mies van der Rohe at Work,* Phaidon Press Ltd., London, 1999, pp. 69, 132–135.
23. "Facts and Figures on the Creative Economy," National Assembly of State Arts Agencies, Washington, DC, http://www.nasaa-arts.org, retrieved August 6, 2012.
24. Although the Brooklyn Academy of Music, erected in 1908, includes four performance halls, it does not include the variety of arts institutions that characterize a true cultural center.
25. Real Estate Board of New York, Inc., *Office Building Construction Manhattan 1947–1967,* New York, 1968, p. 1.
26. *The City of New York Official Directory* (1946 and 1956), The City Record, New York, 1946 and 1956, p. 8.
27. The southern part of the West Side, known as San Juan Hill, was the last of a series of nineteenth-century African-American neighborhoods to be established prior to Harlem. It had been named after the well-known battle in the Spanish-American War in parody of the racial conflicts that took place on the slopes leading to 60th Street.
28. Historical and statistical information on Lincoln Center is derived from New York City Department of City Planning, *Transcript of Public Hearing before the Planning Commission,* New York, September 11, 1957; Edgar Young, *Lincoln Center: The Building of an Institution,* New York University Press, New York, 1980; *Community Development Program Progress Report, 1968,* New York City Housing & Development Administration, New York, 1968; Hart, Krivatsy & Stubee, *Lincoln Square Community Action Planning Program,* Lincoln Square Community Council and New York City Department of City Planning, New York, 1970; Robert Moses, *Public Works: A Dangerous Trade,* McGraw-Hill, New York, 1970, pp. 519–533; and Alexander Garvin, *The Planning Game: Lessons From Great Cities,* W. W. Norton & Company, New York, 2013, pp. 158–161.
29. Officially reported statistics on these projects can be found in *Community Development Program Progress Report, 1968.* However, like all official reports, it undercounts the number of relocatees because it included only those tenants on the site at the time of title vesting. A large number of residents vacate between the time a redevelopment project is announced and final condemnation approval by the court.
30. The owner of Carnegie Hall intended to tear it down and build a more profitable office tower. His plans were upset when violinist Isaac Stern organized a movement to save Carnegie Hall. By then the Philharmonic had made its commitment to Lincoln Center.
31. *New York Times,* July 29, 1958, p. 50.
32. New York City Department of City Planning, *Transcripts of Public Hearing before the Planning Commission,* New York City, September 11, 1957, p. 24, quoting Harris L. Present, chairman of the New York City Council on Housing Relocation Practices.
33. Jane Jacobs, speech given at the New School, April 20, 1958.
34. The Assessed Value of Tax Blocks 1115–1118 (62nd to 66th Streets, Columbus Avenue to Central Park West) was $25.6 million in 1950. By 2012 it was $675.2 million. *Source:* Bureau of Real Property Assessment, Department of Finance, New York City.
35. Historical and statistical information on Kennedy Center is derived from Brendan Gill, *John F. Kennedy Center for the Performing Arts,* Harry N. Abrams, New York, 1981, pp. 23–32.
36. Historical and statistical information on the Dallas Arts District is derived from The Arts District Associations, "Dallas Arts District-

Fact Sheet," Dallas, May 1985; "Dallas Arts District—Project Summary," Dallas, April 1986; and http://www.thedallasartsdistrict.org.

37. David C. Petersen, *Convention Centers, Stadiums, and Arenas,* Urban Land Institute, Washington, DC, 1989, pp. 45–48, 98–100, 111–113, 117–119; and Jack Nicas, "St. Louis's Dome Dilemma," *Wall Street Journal,* February 28, 2012, p. A3.
38. See Petersen, *op. cit.*, p 5; also Anthony C. Krautmann and David J. Berri, "Can We Find It at the Concessions? Understanding Price Elasticity in Professional Sports," *Journal of Sports Economics* 2007, vol. 8: pp. 183–184.
39. The Arena of Nimes seated 21,000; the Coliseum in Rome, more than 50,000.
40. Historical and statistical information on the Los Angeles Coliseum is derived from David Gebhard and Robert Winter, *Architecture in Los Angeles,* Gibbs M. Smith, Layton, UT, 1985, p. 257; and Petersen, op. cit., pp. 101–104.
41. Historical and statistical information on Pilot Field is derived from Charles F. Rosenow (president, Buffalo Development Companies), *Presentation to the 1991 ICMA Sports & Events Management Conference,* April 10, 1991.
42. Historical and statistical information on Gateway Sports and Entertainment Center is derived from David M. Hirzel, "Cleveland Gateway Revisited," *Urban Land,* Urban Land Institute, Washington, DC, April 1996, pp. 42–43; *Project Reference File* (volume 24, number 1), *Tower City Center,* Urban Land Institute, Washington, DC, 1994; the Cleveland City Planning Commission; and interviews with Hunter Morrison III, director of city planning for Cleveland.
43. "Franchise Attendance, Stadiums and More," Major League Baseball, http://www.baseball-reference.com/teams/CLE/attend.shtml.
44. "Quicken Loans Arena," Encyclopedia of Cleveland History, http://ech.case.edu/index.html, retrieved August 6, 2012; and Linda S. Henrichsen, Cleveland City Planning Commission.
45. Downtown Denver Partnership; and http://www.hometodenver.com/Lodo_Denver.
46. "The Economic Significance of Meetings to the U.S. Economy," a report prepared by PwC US, February 2011, http://www.pcma.org/Resources/Research/Study-The-Economic-Significance-of-Meetings-to-the-US-Economy.htm.
47. Historical and statistical information on Renaissance Center is derived from Louis G. Redstone, *The New Downtowns,* McGraw-Hill, New York, 1976, pp. 130–137; Carla Crane (editor), *Detroit . . . Today,* Urban Land Institute, Washington, DC, 1977, pp. 49–59; Meyer and McElroy (editors), *Detroit Architecture,* Wayne State University Press, Detroit, 1980; and Stephen A. Horn, "Detroit's Renaissance Center—Redevelopment Rescues City Symbol," *Urban Land,* Urban Land Institute, Washington, DC, July 1987, pp. 6–11.
48. U.S. Department of Commerce, Bureau of the Census, *Statistical Abstract of the United States 1978,* Washington, DC, 1978, p. 24.
49. Bureau of the Census.
50. Historical and statistical information on the Jacob Javits Convention Center is derived from Heywood Sanders, *Space Available: The Realities of Convention Centers and Economic Development Strategy,* The Brookings Institution, Washington, DC, 2005; and http://www.javitscenter.com.
51. Historical and statistical information on the Washington State Convention and Trade Center is derived from the Urban Land Institute, "Washington State Convention and Trade Center," *Project Reference File,* vol. 19, no. 10, the Urban Land Institute, Washington, DC, April–June 1989; "Only Connect," *Architectural Record,* February 1989, pp. 112–117; and Petersen, op. cit., pp. 77–79.
52. Historical and statistical information on the Pennsylvania Convention Center is derived from Center City District & Central Philadelphia Development Corporation, *State of Center City, Philadelphia, 2001,* Center City District & Central Philadelphia Development Corporation, Philadelphia, 2001; Center City District & Central Philadelphia Development Corporation, *State of Center City Philadelphia, 2010,* Philadelphia: Center City District & Central Philadelphia Development Corporation, 2010; and "Philadelphia's new convention digs are topic at tourism talk," *Philadelphia Inquirer,* April 26, 2011.
53. Center City District & Central Philadelphia Development Corporation, *State of Center City Philadelphia 2012,* Philadelphia: Center City District & Central Philadelphia Development Corporation, 2012.
54. "Beverage Battle: Venue Pouring Rights for '08," *Sports Business Daily,* April 16, 2008, http://www.sportsbusinessdaily.com/Daily/Issues/2008/04/Issue-143/The-Back-Of-The-Book/Beverage-Battle-Venue-Pouring-Rights-For-08.aspx#, retrieved August 6, 2012.
55. Commonwealth of Pennsylvania, Governor's Sports and Exposition Facilities Task Force Report, August 1996; and Mark S. Rosentraub, Major League Losers: The Real Cost of Sports and Who's Paying For It, Basic Books, New York, 1997.
56. Downtown Denver Partnership.

6

Retrofitting the City for a Modern Commercial Economy

Philadelphia, 2010. Market Street. *(Alexander Garvin)*

Cities continually try to retrofit their business districts to better satisfy the needs of their changing economies. This allows them to compete successfully with other nearby communities as well as those around the world. They succeed wherever their actions make it easier to obtain financing, goods, services, and a talented labor force and, by extension, where they make it easier to supply goods and services wherever they are in demand. That is why cities invest in street, highway, railroad, and mass transit networks. These transportation investments do more than increase the number of people who can travel to or from a city or circulate within it. Along with parks and public buildings, they create a public realm that attracts and keeps people downtown.

Intelligent public realm investments always have a transformative effect, particularly when they improve the quality of life for the people coming downtown. The Boulevards of Paris, created while Baron Georges-Eugène Haussmann was prefect of the Seine (effectively the chief executive officer of the Paris metropolitan region), are a good example. Under the patronage of Emperor Napoleon III, Haussmann transformed the dirty, foul-smelling, congested clutter of medieval Paris into the city of light and air that we know today. As explained by Daniel Burnham, the man who proposed a similar street widening initiative for Chicago 50 years later, those physical improvements transformed the city so that it was "fit to sustain the army of merchants and manufacturers which makes Paris today the center of commerce as wide as civilization itself."[1]

At Burnham's recommendation, in 1920, Chicago completed a bridge over the Chicago River connecting the downtown section of Michigan Avenue with a widened North Michigan Avenue. This new artery opened up a district just beyond the downtown Loop to additional customers. Developers responded by erecting structures that eventually attracted Chicago's upscale retailing and offices away from State Street in the Loop to North Michigan Avenue.

Transportation investments can draw customers away from or into downtown. During the 1950s and 1960s, highway construction made shopping trips to suburban Minneapolis easy. At that time about 9000 people a day did their shopping downtown. By 1977, ten years after the city had reinvented its major downtown shopping street, Nicollet Avenue, as a pedestrian-friendly transitway, the number of daily downtown shoppers had risen to 40,000.

This chapter presents the different ways in which the transportation and public realm improvements made by cities can both contribute to and detract from the competitiveness of business districts. It opens with an account of how Paris was transformed and ultimately became an inspiration for some American cities. It subsequently analyzes the effect of more contemporary investments in street extensions, highways, bridges, and tunnels in a variety of American cities. It then examines the impact of railroad and transit investments. A fourth section concentrates on the improvements to the pedestrian environment. Finally, putting everything together, it demonstrates how coordinated investments in rail, transit, streets, and sidewalks made between the 1950s and the present has led to the reinvention of downtown Philadelphia.

Retrofitting Paris

By 1850, Paris had become a city of 1,050,000 that was producing nearly one-quarter of France's exports. Its 350,000

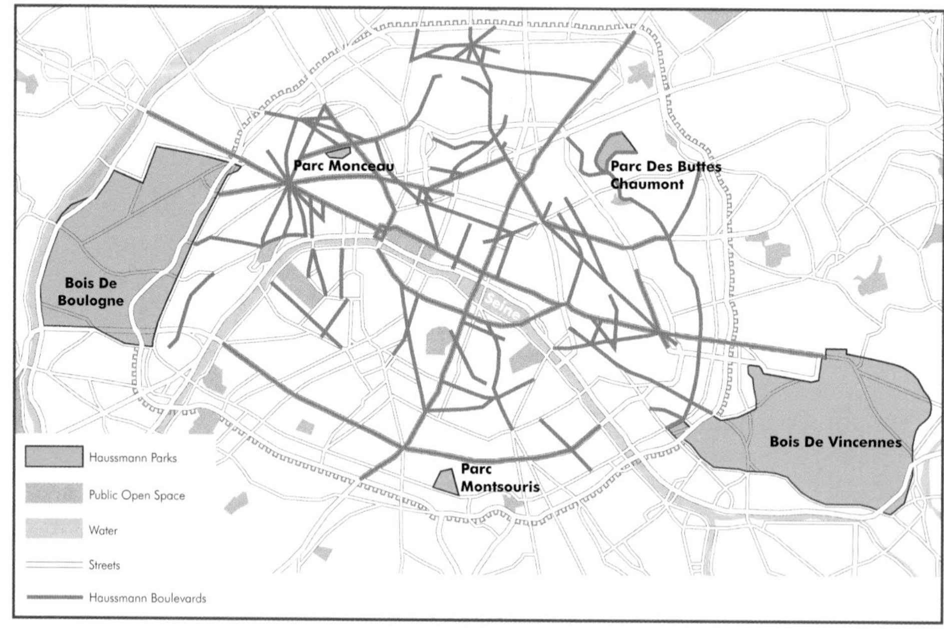

Paris. Map (parks and boulevards). *(Alexander Garvin and Joshua Price)*

workers produced clothing, furniture, and jewelry and built hundreds of buildings for its businesses and residents. Work on retrofitting Paris to accommodate its burgeoning economy started long before Haussmann came to office in 1853. Existing streets were paved and more than 150 new streets were constructed. New canals, drainage systems, bridges, and gas-lit street lamps were an integral part of the process. These investments increased the population that could inhabit the city and made transportation of people and goods much easier. The most important action to make the city more competitive during the first half of the nineteenth century, however, was building six railroad stations that could service the city's residents, workforce, and its business's customers.[2]

The most dramatic changes to Paris, however, came after 1848, when Louis-Napoléon Bonaparte, nephew of the great Napoléon, returned from exile and was elected president of the new Second Republic. Three years later he and his associates overthrew that republic and replaced it with the Second Empire; Bonaparte proclaimed himself emperor Napoleon III, and in 1853 he placed Haussmann in charge of retrofitting the city. The changes initiated and managed by Haussmann permitted the city to double in population by the time the Second Empire was replaced by the Third Republic in 1870.

Haussmann accomplished this transformation by the strategic installation of a water-distribution system, a sewer system, a street system, a park system, and a vast array of community facilities and monuments. These public improvements refashioned Paris into a city of uniform façades and axial roadways. At the intersections of these arteries Haussmann created public squares that provided monumental settings for railroad terminals, churches, schools, theaters, government offices, the Opera, the Louvre, and other major public structures.[3]

Paris buildings rarely exceeded five or six stories because developers would not build higher than their customers would willingly climb. Louis XVI codified this practice in 1784 in an ordinance that limited the maximum height of a cornice to 17.5 meters. From that point, buildings could rise another 4.9 meters, but the additional rise had to be set back (i.e., recessed) at a 45-degree angle from the cornice line to the topmost roofline.[4] Even after the invention of the elevator in 1853, many Paris developers continued to build lower buildings than the seven or eight stories permitted by the regulation. Elevators were too expensive. Besides, before Haussmann began his work, only 1 of every 5 buildings was supplied with running water, fewer than 150 pumped it above ground level, and none could depend on a steady supply of potable water, making convenient access to the ground-level a de facto requirement. Moreover, where pipes were not buried deep enough, winter freezes interrupted the flow. Even in good weather many conduits were inadequate to meet peak period demand, with the result that water flow often declined to a mere trickle. Worst of all, much of the water was polluted because it came from points along the Seine River that were downstream from sewers that emptied directly into the river.

Haussmann devised, financed, and supervised construction of a system that provided Paris with a dependable daily supply of 80 million gallons of unpolluted water at an even temperature during all seasons. The water came from springs and rivers east of Paris along a system of tunnels, bridges, siphons, aqueducts, and reservoirs. It was distributed through a system of underground mains that delivered it to buildings at a constant pressure, which allowed it to climb naturally to 230 feet above sea level. As a result, soft, cool, fresh water could be provided to the top floor of any building in Paris without the expense of pumping.

Haussmann's record of roadway construction is as impressive as the lasting impact made by his improvements to the water system. He constructed 90 miles of broad boulevards and avenues, of which 57 miles were entirely new streets and 33 miles were existing arteries that were widened to as much as 100 feet. They provided easy access to and from new railroad stations, markets, public institutions, and parks, thereby also enabling the distribution of goods and services needed by a city with a population of 2 million. To acquire the necessary rights-of-way, Haussmann had to condemn lots that extended beyond the projected roadways. In the process he destroyed 12,000 structures and created large, attractive building sites that were easy to sell to developers. It did not take long for opportunistic Parisians to line Haussmann's new arteries with six-story buildings, supplied with running water and serviced by Haussmann's new sewer system.[5]

Paris, 2010. The Boulevard des Italiens is more than an artery on which trucks, buses, and cars move large quantities of goods and people. It is a living room where people meet one another, move around, shop, do business, play, sit in a café, or just wander. *(Alexander Garvin)*

The movement of goods and people that these new avenues made possible was essential to the city's booming economy. Their impact on the quality of life for the labor force and population is as important. Those 90 miles of tree-lined boulevards are not just traffic arteries. People go there to shop or sit in sidewalk cafés where they can meet a friend or linger over the newspaper and a coffee. The investment in the boulevards was intended to shape the experience of daily life for the residents of Paris and to provide a framework for the changes that would take place for decades thereafter. Combined with the government's investments in parks and squares, these street improvements made Paris *Paris*. As the inspirational example set by Haussmann began to migrate across the Atlantic to the United States, American public officials initially concentrated on the transportation strategy without fully understanding that such public investments in public property actually constitute a *public realm approach* to planning.[6]

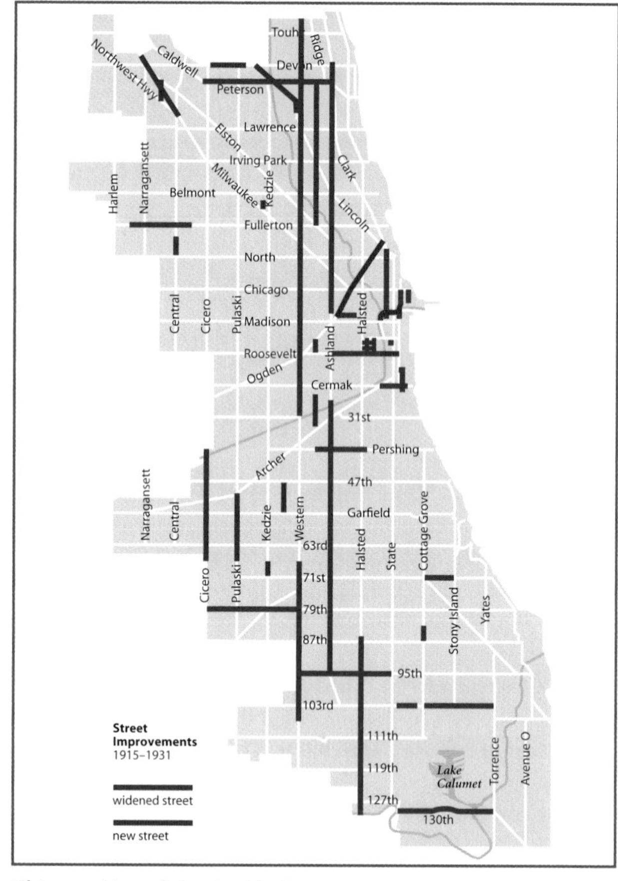

Chicago. Map of street widenings. *(Chicago Cartographics)*

Arterial Improvements

Without safe, convenient, and efficient traffic arteries, people have a difficult time getting to work, shop, play, or home. In 1800 there were no paved streets in the United States. That began to change around 1820 with the invention of macadam. Even when city streets had been paved, though, they often were too narrow or too poorly networked to carry the traffic that was developing in many growing cities. Street widening was an obvious solution. Chicago, for example, widened 120 miles of streets primarily during the 1920s.[7]

Extending arteries into yet-to-be densely developed territory was another approach. That is why the city also built bridges across the Chicago River connecting its downtown Loop with the rest of the city. New York's East River bridges and tunnels were erected as much to open territory for development as to connect the growing labor force in Brooklyn and Queens with the Manhattan Central Business District and to make the shipment of goods easier, faster, and cheaper. In some cities, the new arteries were combined with development of new parkland.

The landscaped roadways that Frederick Law Olmsted and Calvert Vaux designed for Buffalo, Chicago, and Brooklyn are among the best. As with the Avenue Foch in Paris, these broad boulevards were intended as much to provide

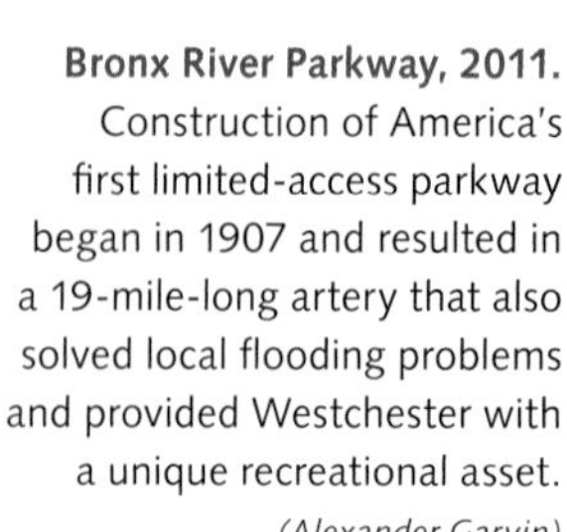

Bronx River Parkway, 2011. Construction of America's first limited-access parkway began in 1907 and resulted in a 19-mile-long artery that also solved local flooding problems and provided Westchester with a unique recreational asset. *(Alexander Garvin)*

Florida, 2007. I-95 interchange. *(Alexander Garvin)*

a public realm framework for the development properties outside the already developed sections of the city as to initiate the park experience as soon as people left their homes or workplaces. These parkways turned driving into a pleasurable recreational activity and simultaneously insulated nearby residents from the noise and pollution generated by commercial traffic. These street extensions, however, were interrupted at frequent intervals when they intersected other arteries, slowing down traffic and reducing capacity.[8]

The nation's first limited-access parkway, the Bronx River Parkway (begun in 1907, extended in stages, and completed in 1923), served as the model for parkways intended to relieve congestion on major arterials. They diverted automobiles to more pleasant, landscaped, limited-access roadways designed exclusively for their use. The parkways built in Westchester, Long Island, and Connecticut during the 1920s and '30s opened New York City's accessible labor pool to new houses being built still farther into the suburbs. They, in turn, were augmented by more than 47,000 miles of highways created and financed by the National Interstate and Defense Highways Act of 1956. This road system connected America's metropolitan regions with one another along uniform arteries; the legislation mandated roadways that included at least two 12-foot-wide lanes in each direction, each of which was to be flanked by a 4-foot-wide shoulder and a 10-foot-wide breakdown lane. These specifications were designed to move traffic everywhere at 50 miles per hour. Outside urban areas, the highways all had banked curves that made them safe for trucks, buses, and cars traveling at 70 miles per hour. As explained in Chapter 7, however, they caused serious disruption and environmental problems when they penetrated into major cities.[9]

When arterial extensions increase accessibility to a business district market or its labor force, they can provide a city with a substantial advantage over its competitors. That certainly was the case in Chicago and New York. In Boston, New Haven, and many other cities, however, they opened vast territory in the suburbs that attracted developers, who erected housing and commercial projects that then successfully attracted customers and workers who might otherwise have remained downtown. In Houston and Atlanta, those extensions increased the city's competitiveness by making possible the creation of an additional business district within city limits. Downtown Minneapolis and Denver could well have experienced a loss of customers to competing subcenters, but they made intelligent investments in the public realm that increased their regional dominance.

The Impact of Initial Street Grids

Unlike Paris, most cities in the United States had rectilinear *plats* made up of broad streets.[10] Since each colony had been established by a different entity, plats differed in street width, block and lot size, and organization. By far the most common design was the rectangular grid, either oriented to the points of the compass or parallel and perpendicular to rivers, cliffs, and other major geographic features. Such plats were easy to survey, easy to divide into rectangular lots that could be subdivided or recombined, easy to describe when conveying title, and easy to add onto if the town outgrew its borders.

The rectangular grid that was adopted in much of the nation was a direct response to the Northwest Ordinances of 1785 and 1787. It provided the basis for surveying and conveying virtually all land outside the original 13 colonies. This north-south, east-west grid, 6 miles square, was extended across the landscape, without reference to hills, mountains, cliffs, lakes, rivers, or any topographical features. The 6-mile squares thus created were called *townships,* sometimes—but frequently not—synonymous with political designations. Each geometrical township was in turn divided into 36 sections, each 1 mile square and each totaling 640 acres. These squares were then further divided into quarters, and again into smaller segments, resulting in a mostly rectangular landscape. Anybody who has flown across the country will instantly recognize the image.

U.S. Grid. Aerial view. *(Alexander Garvin)*

The Land Ordinances provided templates for the platting of urban areas, but many cities adopted plats with different dimensions, and sometimes the block and street dimensions acted as constraints rather than as accelerants to their growth. Philadelphia's blocks were too large, so streets were added to the grid to increase both the amount of frontage and the capacity of traffic the streets could handle. Chicago and Dallas widened existing arteries. A few cities considered introducing Parisian-style diagonal avenues, but they were considered inappropriate in Chicago, Seattle, Brooklyn, and other cities where they had been proposed because they involved too much dislocation, not to mention money.

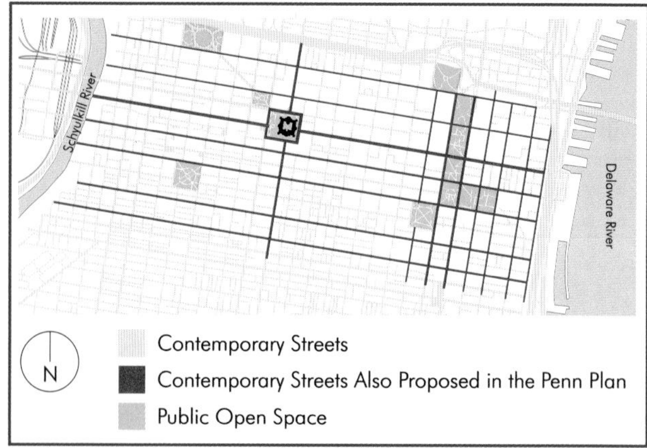

Philadelphia, 2012. The street, block, and lot system that William Penn initiated had to be retrofitted almost from the start to meet the needs of the city's growth and development. *(Alexander Garvin and Ryan Salvatore)*

Supplementing the Grid: Philadelphia and Dallas

The initial plat of Philadelphia did not have enough streets to accommodate traffic in the city. As devised in 1682 by William Penn and Captain Thomas Holme, the plan extended about two miles from the Delaware River to the Schuylkill River. Everything was centered on the crossing of 100-foot-wide Market Street, which runs east-west, river-to-river, and 100-foot-wide Broad Street, which runs north-south and where city leaders eventually erected City Hall. The rest of the grid was made up of 50-foot-wide streets that created blocks of variable dimensions, ranging from 350 to 600 feet on a side. The row houses that were erected along the streets left too much inaccessible property in the middle of the block. As a result, some streets were relocated and others augmented or replaced. Many blocks were subdivided with further streets and alleys that improved traffic flow and increased the number of row house sites. The continuing opening of new streets kept pace with downtown expansion until the mid-twentieth century, when Edmund Bacon, the city's planning director, conceived the redevelopment programs discussed at the conclusion of this chapter. Between 1907 and 1911, George Kessler, the landscape architect who prepared a plan for the initial park system of Kansas City (see Chapter 3), worked for the Dallas Board of Commissioners on proposals to connect or widen 14 streets. All of Kessler's interventions were intended to improve traffic flow in this growing city. Only four of the street extensions were realized because the dislocation and condemnation involved was either too disruptive or too expensive.[11]

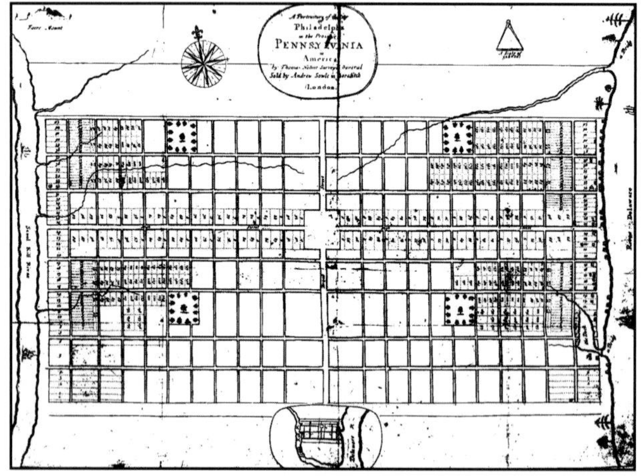

Philadelphia, 1682. William Penn's proposed street grid established a framework for the evolution of downtown Philadelphia. *(Courtesy of Free Library of Philadelphia)*

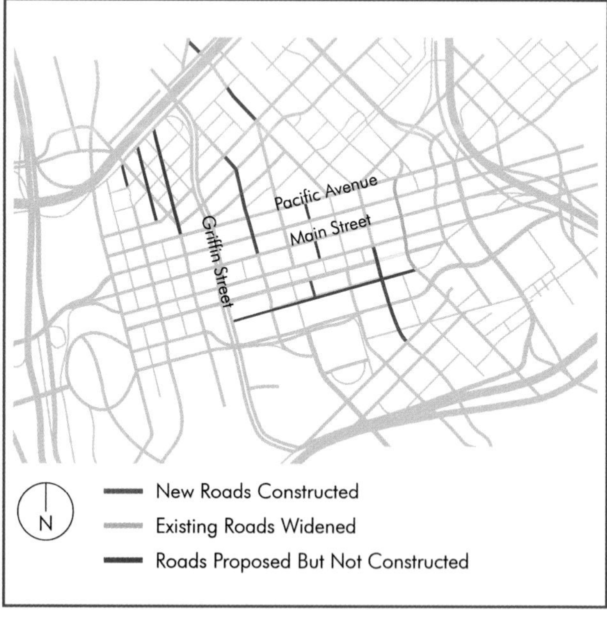

Dallas, 1911. Map of added streets. *(Alexander Garvin and Ryan Salvatore)*

Tying Streets Together: Chicago's Michigan Avenue

Chicago had plenty of streets, but many of them were constrained by their narrowness and the limited number of bridges across the Chicago River. Ever since 1884, about half the traffic moving north and south across the Chicago River had squeezed into the 24 feet of roadway and two 7-foot-wide sidewalks on the Rush Street Bridge. The congestion was sufficiently bad to generate a number of plans, culminating in 1909 with the proposals made by Daniel Burnham and Edward Bennett in the *Plan of Chicago.* They proposed both street widening and new bridges and major diagonal boulevards.[12] In 1913 the city council decided to build an additional bridge connecting both sides of Michigan Avenue and widen substantially the rights-of-way at either end of the new bridge. The section north of the river, then known as Pine Street, was nearly a mile long and between 64 and 75 feet wide. Michigan Avenue's new right-of-way was 150 feet wide. It was not entirely straight because the city council had insisted on preserving the landmark Water Tower & Pumping Station, which had survived the Fire of 1871 that had leveled so much of the city. Widening the street cost $14.9 million. It required the complete demolition of 34 buildings and the partial demolition of 33 others. The bridge and new avenue were opened to traffic in 1920. In the ensuing decade, 31 major buildings were constructed or remodeled along this portion of North Michigan Avenue, and property values increased sixfold. During the same period, more valuable properties on State Street, south of the bridge, had only doubled in value. The new bridge and widened avenue established North Michigan Avenue as the location of the future and set in motion the forces that, by the end of the century, would shift a substantial portion of downtown retailing from State Street to North Michigan Avenue, transforming it into the "Miracle Mile" of retailing in Chicago and one of the nation's most vibrant urban destinations.[13]

Chicago, 2008. Property owners responded to the extension of Michigan Avenue by building office towers. *(Alexander Garvin)*

Chicago, 1909. Rendering of North Michigan Avenue as proposed in the *Plan of Chicago. (From Burnham and Bennett,* Plan of Chicago, *1909)*

Adding Diagonals: Philadelphia's Benjamin Franklin Parkway

Almost none of the vast network of Paris-inspired diagonal boulevards proposed in Burnham's *Plan of Chicago* were implemented because they involved too much expensive condemnation and relocation and they were not really needed in a city with a rectilinear grid of wide streets that were rarely more than 350 feet apart. The diagonal arteries in Virgil Bogue's 1911 *Plan of Seattle* were rejected for many of the same reasons (see Chapter 19). Only one major diagonal boulevard was ever cut through the rectilinear street grid of a major American city: Benjamin Franklin Parkway in Philadelphia.

Once Fairmount Park (see Chapter 3) had been established in the northwest corner of town, Philadelphians wanted a convenient route to get there from the center of the city. In 1892, Philadelphia's director of public works persuaded

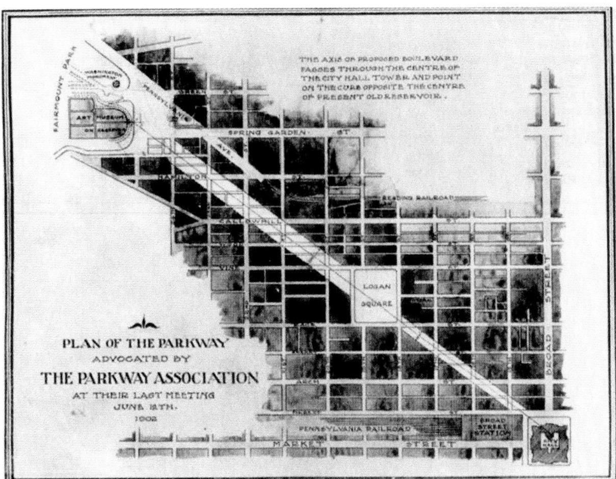

Philadelphia, 1902. Creating Benjamin Franklin Parkway required acquisition and demolition of dozens of private properties. *(Courtesy of University of Pennsylvania Fine Arts Library)*

the city to map a grand diagonal boulevard going from City Hall through Logan Circle (originally conceived as one of the squares in William Penn's plan) to Fairmount Park. The Fairmount Park Art Association commissioned a new plan, prepared in 1907 by a panel of architects that included Paul Cret, Clarence Zantzinger, and Horace Trumbauer. The final scheme for what became the Benjamin Franklin Parkway was designed by the French architect Jacques Grébier, approved in 1917 and completed in time for the opening of the first phase of the Philadelphia Museum of Art in 1928.[14]

The city spent $35 million to pay for acquisition and demolition of some 1300 properties, construction of the new roadway, and landscaping around it. Like Olmsted's parkways, the Paris's Avenue Foch inspired Franklin Parkway. However, unlike Olmsted and more like Haussmann, Grébier provided ample landscaped park islands on either side of the roadway, making at least an effort at the creation of a public realm and not solely a boulevard. The areas to the south and east of Franklin Parkway included commercial buildings, manufacturing lofts, warehouses, and tenements. The northern and western sections were not yet intensely developed. Nevertheless, the city's expanding residential market, which preferred the suburbs, largely ignored both sides of the parkway.

Eventually, the sparsely developed area to the north of the parkway attracted a few apartment buildings. Without active market demand, however, there was little reason for developers to assemble and redevelop property in the triangle to the south. After World War II, when urban planners became convinced that private redevelopment was unlikely, the city declared the area's sheds, warehouses, and tenements to be "inefficient and uneconomical," designated it a renewal area, and cleared the worst section for an apartment complex.

Benjamin Franklin Parkway had little effect on surrounding areas because its grass and trees were not enough to attract people who were on their way to other destinations. They had reasons to go to City Hall, the Free Library, the County Courthouse, the Franklin Institute, churches, offices, and apartment buildings at the eastern end of Franklin Parkway, right in the hub of downtown Philadelphia. However, people going to these places had no reason to walk more than half a mile to the Art Museum. As long as the bulk of city life remained downtown, the grand boulevard could never be a busy pedestrian precinct. As a result, Benjamin Franklin Parkway became a monumental axis: green, filled with motor vehicles, but devoid of people. Its symbolic purpose, connecting a palace of the arts with a palace of politics, may have made geometric sense but had little effect on the daily lives of downtown workers or neighboring apartment dwellers and, therefore, unlike the Avenue Foch or any of the boulevards of Paris, could not stimulate second growth.

Extending Highways, Bridges, and Tunnels into the Suburbs: New York City

New York's arterial extensions, however, altered an entire region. By 1920, New York had become the world's largest city, with a population of 5.6 million. Its metropolitan population of 7 million also made it the world's largest metropolitan area. Its future growth, however, was dependent on access to this growing market and, more important, an ever more extensive and talented labor force. That future was guaranteed by regional investment that extended arterials into New Jersey, Connecticut, and upstate New York. These extensions were more than simple city streets or even grand ones like Benjamin Franklin Parkway. They originated as streets, and, as technology and demand evolved, these modest street projects evolved into tunnels and bridges, landscaped, limited-access parkways, and, later, broad interstate truck and bus routes.

The region began the process in 1921 by creating the na-

New York City, 2007. Bridges and highways connect New York City with an entire metropolitan region. *(Alexander Garvin)*

tion's first semi-independent, bi-state agency with the ability to acquire property and make transportation-related investments without legislative approval: the Port Authority of New York and New Jersey. The Port Authority is governed by a board that has an equal number of directors from each state. By tradition, its chair is appointed by the governor of New Jersey and its executive director by the governor of New York. Port Authority projects are financed by bonds whose debt service is funded from user fees. Its activities include management of freight movement in and out of the harbor, construction and management of bridges and tunnels connecting New York City with New Jersey, and the operation of the region's three airports.[15]

The region's many parkways were created by a series of separately created agencies. The 19-mile-long Bronx River Parkway is the work of the Bronx River Park Commission. Six other parkways are the product of the Westchester County Park Commission, established in 1923. The Long Island parkways are the work of the Long Island State Park Commission, whose initial, unpaid chairman, Robert Moses, served from 1924 to 1963. Seven Brooklyn and Queens parkways were built by the New York City Parks Department while Moses was commissioner of parks, his only paid job between 1934 and 1960. Many of them are connected to the two tunnels and six bridges, built by the Triborough Bridge and Tunnel Authority while Moses was chairman of that agency from 1933 to 1968.[16]

This combination of arteries, bridges, and tunnels connected the Manhattan business district with its labor force in the other boroughs, New Jersey, Westchester, Connecticut, and Long Island. Together, they allowed the city to grow to almost 8 million people in 1960 and its metropolitan area to reach a population of almost 19 million. That system was further expanded by highways built by state highway departments and, before adoption of the New York City Charter that took effect in 1961, by the city's borough presidents. As a result, New York has remained one of the most important global economies, with a gross domestic product of $1.28 trillion; even though it is a city, it has a GDP larger than all but 12 *countries* in the world.[17]

Adding Regional Highway Loops: Route 128 Boston and I-610 Houston

Widespread and increasing private automobile travel was interfering with commercial traffic throughout the Boston metropolitan area, just as it was in New York. Because the problem was entirely contained within the Massachusetts communities encircling the city, there was no need for new metropolitan agencies. Moreover, prior to World War II, Massachusetts political leaders were more comfortable making minor changes to a series of already existing roads in different communities that could become a semicircular

Boston. Map of Route 128. *(Alexander Garvin and Ryan Salvatore)*

artery, which they named Route 128.[18] In 1950, Massachusetts invested in repaving and transforming Route 128 into a four-lane highway with on- and off-ramps that could accommodate heavy traffic from trucks and buses in addition to automobiles. Eighteen months later, during the first three days of operation, more than 76,000 vehicles traveled the new highway. In 1955 it was included in the federal Interstate Highway System and in 1974 made a part of I-95. Two years after it opened, there were more than 200 companies that were located around the highway, including Polaroid, Sylvania, and Canada Dry. Twelve years later, another 500 firms had moved there.

Route 128 forever altered the Boston metropolitan area. Boston lost one-quarter of its population between 1950 and 2010. During that time, the metropolitan area's population essentially doubled, from 2.3 million to 4.5 million, while employment in the metropolitan area increased by a factor of 2.5 times what it had been in 1950. Downtown employment, however, stayed virtually the same, at just over 300,000 jobs, a number that was substantially affected by the presence of more than 40 educational institutions with 175,000 enrolled students and the faculty, staff, and associated employees to support them. We cannot know whether Boston's downtown business district would have been doing better or worse without this arterial extension, but it is clear that growth in the region was concentrated outside of the center of the city because Route 128 was so accommodating.

Boston dropped from being the tenth largest city in the United States in 1950 to the twenty-first largest in 2010, while Houston, with a 1950 population of less than 600,000, went

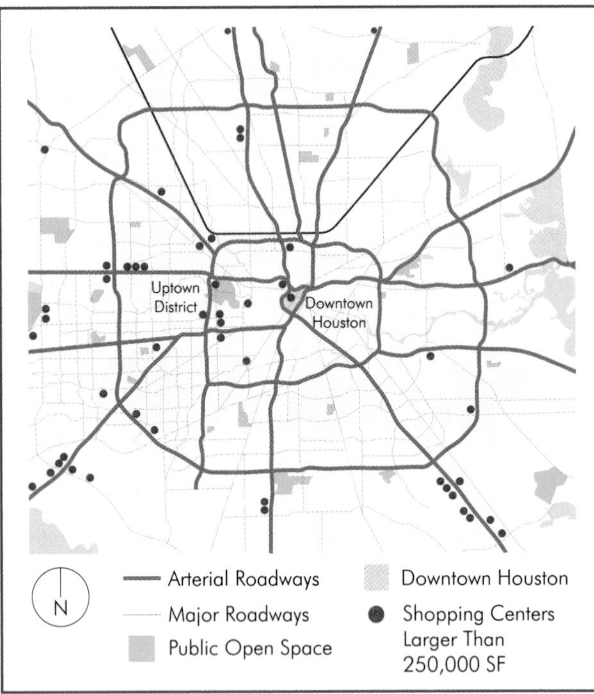

Houston. Map of Route I-610. *(Alexander Garvin and Ryan Salvatore)*

from fourteenth largest to fourth largest, with a 2010 population of nearly 2.1 million. The arterial extensions created for Houston consisted of three highway rings intersected by a series of diagonal arterials, like the spokes of a wheel. Their impact on this growing metropolis was very different from that of Route 128 on Boston.

Unlike Boston, Houston had regularly enlarged its city limits by annexation, going from 160 square miles in 1950 to 599 square miles in 2012. Even so, its population growth has been enormous and so too has its economic expansion. During that 60-year period, the number of jobs within its expanding city limits more than quadrupled, to over 1 million in 2010. Fueled in large part by demand from the growing energy industry, nearly 20 million square feet of office space has been built since the early 1980s, representing almost 60 percent growth in office space alone.

In 2012, there were nine large shopping centers located outside downtown Houston, usually at highway intersections. Five of these centers are located on I-610, clustered along highways in the affluent, western part of the city. One center, known as the Galleria (see Chapter 4), was not even within the city limits in 1950. Since its construction in 1970, the Galleria has been expanded several times. It has spawned 31 hotels and more than 23 million square feet of office space. Today, that area is called *Uptown*. By 2010, it was generating in excess of $4.37 billion in retail sales annually and had become the seventeenth largest business district in the country. (Downtown Houston is the nation's twelfth largest business district.)

The automobile, and its relatives the bus and truck, are fundamental to the culture and economy of American cities. By no means, however, has the expansion of road networks been the only, or even the primary, driver of urban expansion since the nineteenth century. Indeed, rail transport of people and goods pushed urban boundaries well into less developed areas, with the automobile often filling in access to spaces in between.

Railroad Stations

Many planners prefer railroads because 12-foot-wide corridors occupied by trains can deliver so many more people downtown than the same corridor when it is used to move private automobiles or even buses. A 12-foot-wide highway lane can carry up to 2200 cars per lane per hour. Assuming each vehicle is occupied by 2 passengers, equal to 4400 people per lane per hour; for 4-person vehicles, 8800 people per lane per hour. Alternatively, it can accommodate 300 buses that carry up to 14,000 people per lane per hour. The same width of railroad right-of-way can deliver as many as 30,000 people per hour.[19]

There are, however, drawbacks and benefits to rail service, just as there are for travel by motor vehicle. It is not absolutely conclusive that a few train lines can provide downtown access to a larger portion of the labor pool or customer base than motor vehicles, which can pick up passengers from vast territories and deliver them to a broad range of locations. Their impact on a downtown business district, however, is dramatically different. People coming by train do not need miles of extra roadway (underutilized except during rush hour) to get to and from parking, which requires 300 square feet to store one automobile. Nor do people coming by train pollute the air with exhaust fumes.

Like motor vehicles, trains can also have a negative impact. The noise and vibration of a train passing a building is not something that most businesses or residents appreciate. Unless great sums of money are spent providing underground access and mechanical tunnel ventilation, train tracks can be obstacles to through traffic and create barriers, cutting up a city into separate districts that are difficult to connect without expensive bridges or underpasses—and even despite such connections, urban areas are often considered "the wrong side of the tracks" because of the social, cultural, and economic divisions that railroad rights-of-way often impose.

The benefits of rail travel become clear at destinations. Trains deliver large numbers of people to a specific location where property owners will be eager to build facilities that can profit from a convergence of commuters at the station. The concentration of travelers at the point of arrival provides the basis for the very high densities that support specialized services and institutions that could not survive in low-density suburban locations. As explained later in this chapter, the impact on a business district can be extended by public transit to its relatively compact service area.

Whatever the advantages of railroads, America's love affair with the automobile has continued unabated ever since mass production made cars affordable to the overwhelming majority of the country. In 2010, there were 242 million cars registered in the United States (essentially one car for every American over the age of 18). As a result, there are now 125,000 miles of railroads, down from a peak of 430,000 miles in 1930. As a result, train service to many cities has terminated. When a railroad station goes out of service, the negative impact on the district may be as significant as the positive impact of bringing huge numbers of people downtown. However, the impact on the business district does not have to be negative if the station or the territory it formerly occupied can be adapted and reused for desirable purposes that can be of benefit to its surroundings.[20]

New York City, 1911. Park Avenue and Grand Central Terminal. *(Courtesy of the Avery Architectural and Free Library, Columbia University)*

New York City's Grand Central Terminal

Railroad stations, both in use and repurposed, can transform entire cities. New York City's central business district, for example, shifted from Lower Manhattan to Midtown, during a period of less than half a century after the current Grand Central Terminal opened in 1913. On the other hand, the trains arriving on elevated tracks into Philadelphia's Broad Street Station in the center of that city deterred developers from building along West Market Street until the station was demolished in the 1950s.

Chicago's Dearborn Station, unlike the Broad Street Station, was not demolished, but when it went out of service, demand for nearby loft space evaporated. Building owners consequently abandoned their property. After Union Station in St. Louis went out of service, it was converted into a major regional shopping mall. Each of these four cases demonstrates a different planning situation arising from proximity to a railroad station.

The New York Central Railroad had been planning the replacement of its 1871 terminal structure on 42nd Street in Midtown Manhattan when a serious accident forced it to accelerate the project. On the morning of January 8, 1902, a local train crashed into an express train in the tunnel under Park Avenue, filling it with smoke, dust, and debris. Fifteen people were killed and 36 injured. Not long afterward, the state legislature passed a law requiring conversion of train service from steam locomotives to electric power. Electrification eliminated smoke, fumes, and any danger of passenger asphyxiation, thereby allowing trains to arrive and be stored underground on more than one level.[21]

The site was expanded from 23 to nearly 48 acres, an area three times that of the World Trade Center. The expansion required demolishing more than 200 buildings, removing 1.6 million cubic yards of rock and 1.2 million cubic yards of earth, and reconfiguring below-grade rail yards between 42nd and 50th streets to accommodate two levels of tracks and platforms. Park Avenue, Vanderbilt Avenue, and nine east-west streets had to be erected over the tracks, and all of this had to take place while the trains continued to operate.[22]

The development rights that were created over the tracks provided sites for the Waldorf Astoria, Biltmore, and Commodore hotels, the Yale Club, and numerous office and apartment buildings; a total of over 21 million square feet of space was created, an amount nearly twice the area of the World Trade Center when it was destroyed by terrorists in 2001.

New York City, 1929. Development along Park Avenue sixteen years after the new Grand Central Terminal opened. *(From* Grand Central, the World's Greatest Railway Terminal, *by William D. Middleton)*

When the new terminal was opened in 1913, the railroad carried 44,000 passengers daily. By 1929, it carried 145,600.[23] The Grand Central redevelopment project triggered a sustained and widespread private market reaction. The developers who purchased and built over the rail yards created an urban area that has grown to be the largest business district in the world by any measurement. They attracted other developers, who built office buildings, department stores, and hotels on Madison, Fifth, and Lexington avenues. Park Avenue north and south of the yards became sites for new apartment buildings that were, in turn, replaced by office buildings. The investment in Grand Central, together with Pennsylvania Station (1909), the East and Hudson River tunnels that connected the stations with the rest of the country, and the Port Authority Bus Terminal at 42nd Street (1950), helped Midtown to replace Lower Manhattan as the prime business district of the New York metropolitan region.

Broad Street Station, Philadelphia

In 1881, when Philadelphia's central railroad station opened opposite City Hall, it provided access for suburban commuters and travelers from around the country. They could continue traveling along one of the city's numerous trolley lines to virtually anywhere downtown. As demand grew, the station eventually accommodated 12 and finally 16 tracks, covered by a single shed, which, when it was completed in 1892, had the largest single span of any station roof in the world.

Unlike New York, where property owners raced to build office buildings that could be filled with as many office workers as arrived at the station, in Philadelphia, passengers rushed to get away from the station and the elevated structure that supported the tracks coming into the train shed, commonly known as the "Chinese Wall." They hated noise and soot from the 530 trains per day that arrived on the overhead tracks.

Philadelphia, 1923. Broad Street Station. *(Courtesy of Free Public Library of Philadelphia)*

As in New York, the situation changed in 1923 as a result of a damaging fire, this time in the train shed of the Broad Street Station. It ended daily service for several weeks. As a result, the Pennsylvania Railroad decided to reduce the danger of fire and relieve congestion by building the 30th Street Station across the Schuylkill River and creating an underground Suburban Station, one block north of the Broad Street Station. Unlike the downtown station, which was a terminus, 30th Street Station was a stop on a through railway station that did not require intercity trains to turn around to exit the station. By itself, the new 30th Street Station could handle all the traffic coming in and out of Philadelphia. Commuters could avoid the Broad Street Station and come in to Suburban Station, designed specifically to serve them. Because the downtown terminal had become superfluous, in 1925, the Pennsylvania Railroad entered into an agreement with the city to demolish both the Broad Street Station and the Chinese Wall.[24] Nevertheless, the Chinese Wall remained in service for another three decades, repelling all development from West Market Street until, as explained at the end of this chapter, it was finally demolished, making possible the complete reconstruction of the western half of downtown Philadelphia.

Dearborn Station, Chicago

The closing of Dearborn Station in Chicago led to the demise of a thriving portion of the business district, popularly known as Printers Row. Its lofts and warehouses had been erected in response to the opening of the station in 1885. These buildings had been designed by major Chicago architects and exhibited what were then the latest construction techniques. The area got its name from the many printing companies that had been attracted to the location because it was convenient to ship and receive paper products, printed materials, and other goods by rail. As trucks replaced railroads as the prevalent method of shipment, and one-story factories with horizontal production lines replaced multistory lofts as preferred sites for production, the buildings north of Dearborn Station lost their appeal. Companies began moving away after World War II. Their departure was accelerated by the demolition of 12 buildings to make way for Congress Parkway, which provided access to the Eisenhower Expressway. By the 1970s, Printers Row consisted of dilapidated hulks with broken windows and crumbling façades. Prime downtown locations of this sort do not remain unused for long. As explained in Chapter 13, the abandoned lofts were eventually purchased and renovated, transforming the neighborhood that appealed initially to artists and designers and eventually to swinging singles and empty nesters who enjoyed the hip character of the area. A group of dedicated business leaders organized to transform the rail yards south of the station into the planned "new-town-in-town" discussed in Chapter 15.[25]

Chicago, 2012. Dearborn Station. *(Alexander Garvin)*

Union Station, St. Louis

Like 30th Street Station, Union Station in St. Louis was one of the nation's premier rail facilities. When it opened in 1894 it was the world's largest and busiest railroad terminal. The headhouse contained a hotel, a restaurant, passenger waiting rooms, and railroad ticketing offices that led to an 11.5-acre train shed with the largest roof span in the world. Located literally in the center of the United States, the station provided service to 22 rail companies, more than were handled in any terminal in the world. The heyday of St. Louis rail fever was nonetheless short-lived: while rail travel to Philadelphia continues to this day, automobile and air travel diverted so many of the St. Louis rail passengers that Amtrak ceased operations in downtown St. Louis in 1978.

Over the next seven years it was transformed into a shopping mall with a variety of restaurants, a 539-room hotel, a museum, and an entertainment complex. Although it is a popular stop on the city's Metrolink light-rail system, Union Station does not attract the business typical of major suburban shopping malls. It is located at the far end of the half-mile-long civic center (see Chapter 5) and is but one of the city's scattered collection of tourist destinations that includes the Gateway Arch (Jefferson National Expansion Memorial), Busch Stadium, Laclede's Landing, America's Convention Complex, and Edward Jones Dome, all of which are just far enough apart to keep people in their cars after going there. Thus, Union Station brings nothing like the 100,000 people a day that used to enter the business district.[26]

Mass Transit

The most commonly advocated method for extending the customer base and labor pool beyond a downtown business district is to provide mass transit service. *Mass transit* may refer to an underground or elevated heavy rail system; a trolley, streetcar, or light-rail network; or buses that provide local or express service on city streets, in a dedicated bus lane or within a specially built bus rapid transit corridor. Each of them involves a very different level of service and different capital and operating costs; all of them can and often do affect business district competitiveness.

The number of passengers required for transit service to be financially self-sufficient usually exceeds the number currently using the system. Therein lies a chicken-or-egg conundrum: should a city wait until there are enough customers to justify investing in public transportation, or should it make a major investment and wait until there are enough customers to justify transit service? One answer was supplied by Henry Huntington, who was a prime mover in the creation of the mass transit system that shaped the development of Los Angeles prior to widespread automobile ownership. Referring to electric railways, he explained in 1904:

St. Louis, 2010. Union Station. *(Alexander Garvin)*

> *It would never do for an electric line to wait until the demand for it came. It must anticipate the growth of communities and be there when the builders arrive—or they may very likely never arrive at all, but go to some other section already provided with arteries of traffic.*[27]

Subway and elevated mass transit require major capital expenditures and high-density corridors to supply enough customers to make service cost-effective. New York and Chicago are unique among American cities whose economies are dependent on this form of transit. Investments that they made a century ago have paid off as manifest in residential development along subway corridors that is sufficient to provide their business districts with access to a huge labor pool and consumer base. The downtown business districts of both cities have continued to grow ever since.

Half a century ago, Washington, D.C., San Francisco, and Atlanta made similar investments. METRO has provided a significant stimulus to downtown Washington. BART has allowed downtown San Francisco to continue as the region's major business district during a period when it has faced increasing competition from Silicon Valley and communities in the East Bay that can tap the same market. Atlanta's MARTA System, however, has had a unique, very different effect.

Most supporters of mass transit fail to understand the importance of bus service. One reason buses are so popular is that they can provide service to a much broader area than railroads, at a lower capital cost, and without as much dislocation. Nonetheless, even buses using existing traffic arteries are unlikely to be financially sustainable by suburban populations. Hence, buses can be particularly cost-effective in high-density business districts where customers can be concentrated within a very short walk of the bus. The free bus that services 16th Street in downtown Denver provides a vivid example of such cost-effectiveness.

Los Angeles Streetcars

It is more than four decades since architectural historian Reyner Banham explained that Los Angeles was the product

Los Angeles, 1925. Street railway map. *(City and County of Los Angeles)*

of its streetcar system. Nevertheless, the myth that its form and growth are the products of the automobile has continued to flourish.[28] Streetcar service in the Los Angeles metropolitan area began in 1874; within 13 years, the network included 43 separate franchises. In 1898 Henry Huntington's syndicate purchased the Los Angeles Railway Company, and in 1901 Huntington created a second company, the Pacific Electric Railway, to sponsor real estate development around the rail hubs in an effort to profit from the increase in property values on land he already owned that would accompany introduction of transit service. As Venice, Glendale, Watts, and Hollywood all grew up around mass transit, so did Huntington's fortune and that of everybody around him. By 1925, the Los Angeles Railway Company operated on 700 miles of transit service and the Pacific Electric Railway was the largest electric interurban transit system in the world. This system, which at its peak included 1164 miles of track, connected more than 50 communities.

In 1910, the population of Los Angeles was 310,000; one decade later, it had nearly doubled to 577,000. Looking back from the twenty-first century, it is difficult to imagine that in 1914 there were only 42,720 registered automobiles in Los Angeles (only one car for every eight people), especially when the dramatic population growth during that decade is factored into the analysis. Most of these people and the businesses that provided their employment had filled in the territory that had been platted into streets and blocks extending 4 to 5 miles from the downtown. They walked a block or two to the major arteries from which they took streetcars to their destinations. Thus, long before Los Angeles became the freeway capital of the universe, its extensive mass transit system had already determined the settlement pattern of the metropolitan region, as well as the location and character of it business districts. But it also contained the seeds of its own destruction.

As more and more real estate was developed, more and more intersections sprang up where intersections and grade crossings could delay a person's commute, disrupt service, and cause traffic accidents. Increasingly popular private cars provided even more convenient service if those intersections could be avoided by traveling instead on limited-access freeways. The desire for a freeway system began to emerge in the 1920s and was given shape in 1930 by the Olmsted Brothers and Bartholomew and Associates' report: *Parks, Playgrounds and Beaches for the Los Angeles Region*. When the freeway system began to emerge in the second half of the twentieth century, it served the residences of existing populations and places of employment that had previously been served by public transportation, reinforcing the settlement patterns that had initially been established half a century earlier rather than creating altogether new ones.[29]

The electric interurban transit system had created a multifocal metropolitan region long before Los Angeles's freeways appeared on the landscape. Despite competition from Hollywood, Beverly Hills, Santa Monica, and the other subcenters that matured around the transit system, the region's most intense economic activity continued to take place downtown—as it does today. Freeway construction merely reinforced existing land use patterns and increased the amount of economic activity that could take place within the region. That is not what is happening in Atlanta.

Atlanta's Subway

If the construction of the freeway system reinforced historical development patterns and maintained the hegemony of the downtown in Los Angeles, Atlanta offers an example of a process with opposite results. The subway system created by the Metropolitan Atlanta Rapid Transit Authority (MARTA) in the 1970s and 1980s initially delivered service to areas that were roughly parallel, similar to an existing highway "X" composed of interstate highways I-75/I-85 and I-20. While MARTA did not pass through many of the city's low-density residential neighborhoods, it did connect Hartsfield International Airport with downtown Atlanta. More important, MARTA also connected the airport and downtown Atlanta with Midtown and Buckhead, the city's other major employment centers.

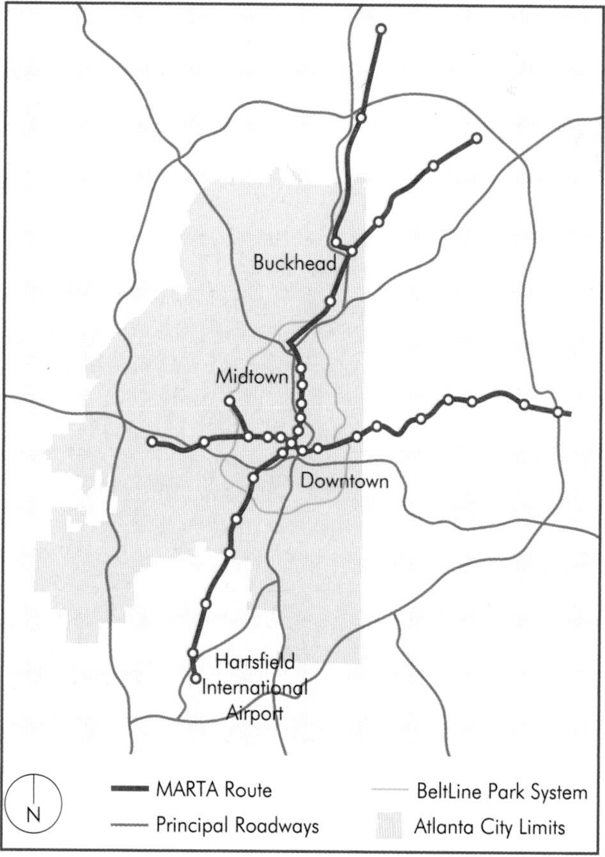

Atlanta, 2007. MARTA. *(Alexander Garvin and Ryan Salvatore)*

Atlanta, 2007. Buckhead, aerial view. *(Alexander Garvin)*

This combination of highway and transit service spurred real estate development in both Midtown and Buckhead, making them major competitors with downtown Atlanta. Buckhead is served by two MARTA lines and interstate highway 400, which branches off I-85 before it passes east of Buckhead. This transportation network, combined with two shopping centers (see Chapter 4), has transformed Buckhead into a major regional subcenter. As of 2012, Buckhead included more than 23 million square feet of office space, 8.3 million square feet spread among 1400 retailers and 300 restaurants, 5275 hotel rooms in 25 hotel properties, and 23,000 elevator-serviced apartments. If Buckhead were not within the city limits of Atlanta, it would be the ninth largest city in Georgia.[30]

Atlanta reversed the pattern set in Los Angeles by locating MARTA along existing highway corridors. While three or four more decades must pass before Atlanta's subway system is as old as the Los Angeles freeway system, MARTA has already contributed significantly to the city's transformation into several major subcenters. Midtown has become a major competitor with downtown Atlanta for business offices and, because of post-MARTA construction of the High Museum and the Woodruff Arts Center (home of the Atlanta Symphony), it is now the city's cultural center. Moreover, at the rate that Buckhead is growing, it may surpass both Downtown and Midtown to become the city's premier center of economic activity.

Denver's Sixteenth Street Bus

During the early 1950s, Denver was beginning a period of more than 30 years of expansion. Developer William Zeckendorf Sr. opened a new $8 million department store, a modern 884-room convention hotel, and a 457,000-square-foot office tower. Between 1950 and 1964, the amount of office space in Denver doubled. However, despite a rapidly growing and increasingly prosperous population, retail sales were weak and transit ridership was down 35 percent.[31]

In 1955, business leaders who wanted to improve the stagnant commercial environment formed Downtown Denver, Inc. They brought in a study group from the Urban Land Institute, which recommended a new freeway network, one-way downtown traffic, parking structures, improved transit, and federally subsidized redevelopment. In 1961, a Master Plan Committee was formed to propose major redevelopment activity. Finally, in 1967, after years of controversy, voters approved a 27-block downtown urban renewal program. Despite these efforts, downtown Denver lacked focus that would knit together all of the development activities.

The missing ingredient slowly took shape during the 1970s. It began as a proposed pedestrian precinct with a free bus running for nine blocks along 16th Street. By 1979, when the 16th Street Mall received Urban Mass Transportation funding, the mall had been transformed into a 13-block transitway connecting two suburban rapid transit stations.

The 16th Street Mall, designed by I. M. Pei and Partners and Hanna/Olin Ltd. Landscape Architects, was completed in 1982. Commuters who arrive on regional and express buses emerge from one of the two transit terminals at either end of the mall. Those who come by car use one of the 44,000 parking spaces within walking distance of the mall. They can take a free shuttle bus that leaves every 70 seconds. The buses travel on 10-foot rights-of-way set aside exclusively for their use and stop at every intersection until they

Denver, 2006. Free bus, 16th Street. *(Alexander Garvin)*

arrive at one of the two transit terminals and reverse direction. The 16th Street Mall demonstrates the key relationship between location and time because the time/distance that can be covered between a parking place and an office, store, restaurant, or entertainment venue is dramatically increased by frequent free bus service. The project cost $76 million to develop: $29.5 for the mall, $5.1 million for a fleet of 19 shuttle buses, and $41.5 million for the two bus transfer stations and renovation of the Regional Transportation District offices. Customers using the shuttle bus can stop to patronize Writer Square (a shopping center with 3 restaurants and 25 upscale shops), Denver Pavilions (a 350,000-square-foot entertainment center with a multiplex movie theater, a nightclub, 5 restaurants, 5 specialty food retailers, and 20 stores), and Larimer Square (a district of renovated Old West Victorian buildings with 35 shops and 12 restaurants). Every major downtown office building, hotel, and store is within three blocks of this pedestrian/transit spine, as are Colorado University, the convention center, the state capitol, and the Denver Center for the Performing Arts. The 16th Street Mall successfully created a focus for the 90,000 people per day who work downtown and provides an attractive destination for people in the entire Denver metropolitan area. It was so successful that it generated even more market demand for transit service.

The impact of Denver's twentieth-century investments in mass transit cannot be overestimated. In 1950, when downtown businesses were searching for a way to stimulate the city's stagnant economy, 167,000 worked in Denver. Sixty years later that number had more than doubled. During the same period, the city's population increased by only 44 percent, reaching 600,000 in 2010.

In 2002, free bus service was extended two blocks to Union Station and to the area being developed over the station's rail yard and beyond. More important, a new light-rail system named FasTracks was superimposed on 16th Street. FasTracks is a $6.5 billion rail and bus rapid transit network covering the entire metropolitan region. The initiative will be paid for with federal subsidies and sales tax revenues approved by referendum in 2004. When the system is completed (projected for 2019), it will consist of 57 new transit stations, 18 miles of bus rapid transit, and six rail lines that extend in all directions and cover almost 120 miles.

It is difficult to separate the impact of all the things that Denver has done to make its downtown business district so successful, much less to identify the specific impact of the free bus or FasTracks. Certainly Coors Field, the Convention Center, the Performing Arts Center, and the Civic Center have all played a part. When FasTracks is complete, it will have added immeasurably to the impact by extending transit service to the entire metropolitan area. But virtually everything comes together within three or four blocks of the free bus traveling on Sixteenth Street, demonstrating how the city was able to rebound from its early inconsistent and poorly

Denver, 2011. FasTracks light-rail system station. *(Alexander Garvin)*

Cleveland, 2012. Euclid Avenue bus shelter. *(Alexander Garvin)*

coordinated efforts to rally around a unified strategy to revitalize and sustain its downtown.

Bus Rapid Transit in Cleveland

There is no better demonstration than in Denver that a transit investment combined with a dense concentration of activities and people will result in a successful business district. It is in Cleveland, however, that one can see the specific impact of a transit investment: bus rapid transit (BRT) on Euclid Avenue. In the nineteenth and early twentieth centuries, Euclid Avenue provided a grand home for the rich and powerful, who lived in sumptuous mansions on this boulevard leading out of the city. By the 1950s, however, many residents had moved to Shaker Heights (see Chapter 16) and other suburban communities, leaving the formerly luxurious residential street lined with abandoned buildings and vacant lots. Thus, during the last quarter of the twentieth

Cleveland, 2012. Euclid Avenue. Once bus service along the HealthLine began, developers started purchasing and converting lofts to residences. *(Alexander Garvin)*

century, Euclid Avenue was easily characterized as down and out.[32]

In 2008, the city opened the Euclid Corridor Transportation Project. This $200 million investment in the 6.8-mile roadway begins at Public Square, in the downtown heart of Cleveland, and is home to the seven historic theaters that make up Playhouse Square, Cleveland State University, the Cleveland Clinic, Case Western Reserve University, and University Hospitals, and connects the downtown to the neighboring city of East Cleveland. The Federal Transit Administration provided $82 million from its New Starts program toward the capital cost of HealthLine, the BRT that runs the length of Euclid Avenue. The street itself has been redesigned by Sasaki Associates to include bus shelters, bike lanes, planters, lighting, and public art.

Within the first four years after the HealthLine began operating, property owners and developers were erecting new buildings and investing in the derelict buildings that lined Euclid Avenue. The new buildings are still largely the result of institutional expansion. But the office building and loft conversion to residences is already altering perceptions of the corridor. Many new tenants are students, faculty, or health care workers; some of the new occupants are younger people who like the proximity to the student population; others are empty nesters who like the proximity to downtown attractions. Not unexpectedly, a transit line that is well designed and smartly positioned in the market has become the perfect complement to the Gateway Sports and Entertainment Center, the additions to the Group Plan, and the repositioning of East 4th Street: all of these efforts are improving the competitive position of a downtown business district that only 40 years ago was in doldrums.

Reclaiming the Public Realm

The movement of goods and people along the boulevards of nineteenth-century Paris was an important aspect of retrofitting Paris for a modern commercial economy. The other important aspect was the extraordinary public realm framework within which that movement took place. People who lived in cramped quarters (often without heat, hot water, and bathrooms) and worked in ill-lighted, poorly ventilated sweatshops made active use of the broad, tree-lined boulevards that were essential to efficient operation of the city's economy. It was along these boulevards that they could go for a stroll, or meet a friend in a sidewalk café, or enjoy a glass of wine. Much of America's effort to retrofit cities to operate within an increasingly global economy, like that of Paris, was devoted to the role of street, highway, rail, and mass transit systems. Until the second half of the twentieth century, it paid little attention to the nonvehicular aspects of retrofitting the city.

Civic leaders began to get concerned with the health of the central business district with the emergence of competition from suburban shopping centers. Transportation improvements were no longer enough to guarantee a greater variety and lower cost of goods and services at downtown locations. As the competitive edge of downtown business districts began to disappear, their merchants began to lose sales to shopping centers that were closer to the homes of a rapidly expanding suburban population and their businesses began vacating downtown offices for suburban locations that provided an easily accessible, safe, clean, attractive environment. One obvious way to compete was to retrofit downtown so it would be as safe, clean, and attractive for the labor force its businesses required and the customers its retailers served.

At first, government's response was to make downtown streets more customer-friendly by reducing the presence of automobiles. Public officials adopted this prescription as a relatively inexpensive way of helping merchants who were desperately trying to stanch the hemorrhaging of their businesses. They believed that downtown workers would be more likely to remain in the business district if they could do their shopping on newly pedestrianized streets and that suburbanites would be less likely to patronize outlying shopping centers if they were offered a vehicle-free downtown alternative.

The new outdoor pedestrian environment took shape as fully pedestrian malls, semi-malls, transitways, and skywalks. Fully pedestrianized shopping streets, or *pedestrian malls,* eliminated all ordinary vehicular traffic, permitting only emergency access and service vehicles. *Semi-malls* were really avenues with widened sidewalks and narrowed roadways. Their aim was to create promenades, which, like the Champs-Élysées in Paris, offer an especially attractive setting for the shopper. *Transitways* combined the improved shopping environment with public transportation. In this case, a narrow portion of the old streetbed was dedicated to public transportation. *Skywalks* removed pedestrians from the street and created an entirely new, climate-controlled, vehicle-free system of walkways connecting shops, hotels, restaurants, office buildings, apartment houses, and garages, all one or more levels above the city's vehicular traffic.

America's first pedestrianized shopping mall opened in 1959 in Kalamazoo, Michigan. Like most later pedestrian malls, it was intended to revive what everybody thought was a decaying downtown. Over its first 10 years of operation, retail sales increased annually by 10 percent. Some stores even reported 150 percent increases in gross sales during the first five years. Pedestrian activity throughout downtown Kalamazoo increased 30 percent. This caused sufficient wear and tear to require a $300,000 renovation, completed in 1970. In 1974, a fourth block was added, and a year later, Kalamazoo Center, a multiuse convention facility including a hotel, restaurants, shops, and a theater. Eventually, these improvements were insufficient to maintain a lively shopping district. In 1998, Burdick Street was reopened to limited vehicular traffic. Nevertheless, as America's first pedestrian shopping mall, it played an important role in demonstrating that a vehicle-free shopping street could spark but not sustain downtown revival.

Cities across the country thought that pedestrianizing a few blocks was enough to reverse downtown decline. In Miami, Minneapolis, Houston, and other cities, it was effective because it was combined with other efforts to revitalize the business district. But in Chicago, Providence, Louisville, and many other cities, it failed to restore spending. As these projects reveal, a great public realm requires more than just exterior decoration. Brick paving, trees, benches, and new street furniture may demonstrate public concern for a declining area. Like Mercurochrome, brick paving is red-colored and easy to apply, but it is ineffective for curing serious ailments.

Lincoln Road, Miami Beach

Lincoln Road in Miami Beach had been a fashionable retail street from its establishment during the second decade of the twentieth century until the 1950s. This "Fifth Avenue of the South" had been known for department stores such as Saks and Bonwit Teller as well as the stately palm trees that lined its broad sidewalks until it began to be supplanted as a prime tourist location by a section of Collins Avenue a couple of miles north. Threatened merchants and property owners commissioned the then-popular architect Morris Lapidus to reconceive the street as a refurbished, vehicle-free mall. Their slogan was "a car never bought anything," which made sense for this once-glamorous vacation venue.[33]

When the newly pedestrianized Lincoln Road opened in 1960, its sidewalks were enhanced by tropical landscaping, fountains, unusual concrete canopies, and dramatic lighting. At first, this was enough to attract some of the customers that had been enticed to other locations. But by the late 1970s, the street was once again in decline.

Miami Beach, 1993. By the start of the twenty-first century, Lincoln Road was no longer popular and included empty storefronts and third-rate retail outlets. *(Alexander Garvin)*

Miami Beach, 2009. Once Lincoln Road had been renovated, visitors from all over the metropolitan area flocked to spend time at its stores, restaurants, and nightspots. *(Alexander Garvin)*

In 2001, a heterogeneous combination of merchants, pioneering artists, designers, entrepreneurs, and civic leaders persuaded the Miami Beach City Commission to spend $16 million on physical improvements that would transform Lincoln Road from a run-down 1960s pedestrianized shopping street into a tourist destination. Once rehabilitation had been completed, its restaurants and nightclubs attracted visitors from all over the metropolitan area. In 2011, it was further enhanced by the addition of New World Center, a concert hall designed by architect Frank Gehry, attached to a 2.5-acre park designed by the Dutch architectural firm West 8 for regular outdoor presentations that bring even more customers to Lincoln Road.

Nicollet Avenue Mall, Minneapolis

When business and political leaders in Minneapolis formed the Downtown Council in 1955, the city was suffering from the same problems that faced Kalamazoo: declining retail sales, declining downtown office occupancy, and a declining tax base. The strategy for dealing with these problems emerged from a collaborative effort between the business community and the city planning department. It began in 1957 with the Downtown Council's Nicollet Avenue Survey Committee and continued in 1959 with the Minneapolis Planning Department's publication of its *Central Area Plan*.[34]

Both survey and plan called for a totally reorganized business district in which the mass transportation system would be restructured to provide more direct service to retail areas, traffic congestion would be reduced, adequate parking would be readily available, and downtown pedestrian circulation would be convenient and comfortable. To achieve a separation between pedestrian and vehicular traffic, they called for Nicollet Avenue to become a transitway and for a system of skyways that would bridge over downtown streets in order to connect buildings at the second-floor level.

Southdale, America's first air-conditioned shopping mall, which had just opened in nearby Edina, revolutionized American retailing (see Chapter 4). For the first time, consumers could shop in a totally climate-controlled, naturally lit shopping mall that included every sort of retail store. Minneapolis is the coldest of the country's 45 largest cities. Its merchants knew that retail sales declined during bad weather, a problem that Southdale had solved. If the merchants were to withstand this competition, they had to offer their customers a similar climate-controlled shopping environment. The skyway system appeared to be just what they needed.[35]

The first two skyways opened in 1962 as part of Northstar Center, an office-hotel-parking complex developed by Baker Properties. Other skyways followed in 1964 and 1969. However, it was not until 1973, when the IDS (Investigators' Diversified Services) Center opened, that a real pedestrian system was created.[36]

The IDS Center, designed by architects Philip Johnson and John Burgee, included a 51-story office tower, a 19-story hotel, and a 525-car parking garage, all enclosing an 8-story, skylit atrium with cafés, restaurants, and shops, called the Crystal Court. IDS provided what had been missing from the skyway system: a central attraction, a critical mass of customers, and connections to anchor stores. Every part of the skyway system now led to the multilevel Crystal Court and from there directly into Minneapolis's two major department stores. For the first time, the skyways, like any suburban shopping mall, included sufficient consumer attractions to induce heavy pedestrian traffic. During a typical March day in 1980, for example, more than 20,000 pedestrians crossed the skyways that connected the IDS Center with the department stores.

While there were no accepted principles for such privately built facilities when the first three skyways were built in 1962, it did not take long for common characteristics to emerge. A minimum street clearance of 17 feet has been established to permit all trucks and buses to drive underneath. All sidewalls are made of clear glass so that pedestrians can identify the streets they cross, determine their destinations, and relate to the city around them. Where possible, bus stops are located adjacent to the parking garages that connect to the skyway system. Consequently, rents in buildings connected to the skyway system are 10 percent higher than in those without skyway connections.

Minneapolis, 2005. The concentration of employers in the buildings along Nicollet Mall provides enough customers to keep both the street and the skyways full of customers. *(Alexander Garvin)*

The second ingredient of pedestrianized Minneapolis was the 10-block-long Nicollet Mall, which opened in 1967. In landscape architect Lawrence Halprin's design, private cars and trucks were banished, sidewalks were widened and lined with trees, flowers, benches, litter baskets, kiosks, sculptures, fountains, and bus shelters. A sinuously undulating two-lane roadway was set aside exclusively for buses and taxis.

Pedestrianizing Nicollet Avenue cost $3,874,000, one-fourth of which was paid by the federal government with a $513,000 Urban Mass Transportation Grant and a $484,000 Urban Beautification Grant. The rest came from an 18-block special assessment district in which adjacent property owners paid in proportion to the benefits they were to receive. The district continues to be assessed annually for about 90 percent of the cost of maintenance. The city pays the remaining 10 percent.

When Nicollet Mall opened in 1967, there were only 9000 daily shoppers downtown. Ten years later, there were over 40,000. Retail sales had increased by 14 percent. Millions of square feet of new office space were added to the business district. By 1982, the skyway system connected 35 city blocks and more than 81,000 jobs. Ten years after that, it connected 44 city blocks that included 94 percent of downtown retail space, 85 percent of its rentable office space, 70 percent of its hotel rooms, and 48 percent of its parking stalls. By 2011, the 8-mile-long skyway system included 83 enclosed bridges that connected 73 blocks.[37]

Because of this success, the streetscape was no longer adequately handling pedestrian and bus traffic along the mall. In 1987, the city formed the Nicollet Mall Implementation Board, which included public officials, property owners, tenants, and downtown business leaders. In 2009, it was folded

Minneapolis, 2012. Interior retailing has sprung up along the skyway system. *(Alexander Garvin)*

into the newly established Downtown Improvement District, that operates much like business improvement districts around the country (see Chapter 8). Minneapolis decided to spend $22 million to refurbish the Nicollet Avenue Mall. The project was completed in 1991. Changes to Nicollet Avenue keep happening, but its role as the prime retail/office center of the Upper Midwest continues. Although the city's 2010 population of 383,000 was 27 percent lower than it was in 1950, the improvements to its public realm have helped to keep city employment at 221,000—virtually the same as it was in 1960, just before Minneapolis decided to invest in the Nicollet Avenue Mall and skyway system.

State Street, Chicago

Pedestrianization may have succeeded in Minneapolis and Miami Beach, but it didn't work on State Street in Chicago. State Street had been Chicago's premier shopping street since the Fire of 1871. Crowds of people came downtown to see a show, to shop at Marshall Field's and all the other department stores, or just to enjoy the sights. After World War II, though, increasing numbers of suburban alternatives became available. In 1957, Old Orchard Shopping Center opened in the northern suburbs, followed by Oakbrook Center in the western suburbs in 1962 (see Chapter 4). A growing number of better stores on North Michigan Avenue began to attract away downtown customers. When Water Tower Place opened in 1976, the shift away from State Street to Michigan Avenue became palpable.[38]

An obvious response was to copy what had been done (at that time, apparently with success) in so many other cities: pedestrianize State Street. Consequently, civic leaders decided to spend $17 million (much of it federal transportation subsidies) transforming it into a gently undulating two-lane transitway. The Department of Public Works widened and repaved the sidewalks and installed contemporary streetlights, modern subway entrances, and bus shelters. Private automobiles and taxis were banished, forcing department store and hotel customers off State Street and onto side streets. Many of the Loop's bus lines were rerouted to State Street, simultaneously concentrating noisy, polluting diesel buses during the morning and evening rush hours, when the street was filled with potential retail customers. Unfortunately, conditions deteriorated. In Chicago's rigorous climate, the new sidewalk pavements often cracked. Repairs were poorly executed, emphasizing the street's run-down, low-rent appearance.

There is no way to be sure that making State Street a transitway hastened its decline as a shopping destination, just as there is no way to be sure that the construction of large garages at strategic locations along State Street might have been a more effective way of enhancing its competitive position as a convenient downtown destination. But decline it did. In 1981, two years after the transitway opened, Goldblatt Brothers Department Store closed, followed by Sears in 1982, Montgomery Ward in 1984, and Wieboldt's in 1987. The Greater State Street Council, made up largely of Loop merchants, commissioned a study that resulted in a 1987 report titled *Greater State Street.* It blamed the decline on pedestrianization and maintained that the absence of auto traffic and the uninviting wide sidewalks limited access and isolated State Street from the rest of downtown.

Despite its difficulties, State Street had real assets. It was located in the midst of one of America's most successful business districts, occupied by 500,000 office workers. It still included Marshall Field's (1 million square feet) and Carson Pirie Scott (0.5 million square feet), two of the country's flagship department stores. Clearly, State Street was still in a position to compete for business from Chicago's 3 million customers and the metropolitan area's 8 million.

The turnaround began with a series of unrelated actions. In 1988, the city selected a design for its new central library, to be located at the southern end of the Loop on State Street. The 10-story Harold Washington Library, designed by Hammond, Beeby, and Babka, would become the largest municipal library building in America. It contains a circulating library of more than 2 million volumes, an auditorium, a video theater, a restaurant, and a bookstore. When the library opened

Chicago, 1979. Transforming State Street into a transitway removed enough customers to accelerate the street's loss of customers to North Michigan Avenue. *(Alexander Garvin)*

Chicago, 2012. Remotorizing State Street triggered its rebirth as a major shopping destination. *(Alexander Garvin)*

in 1993, it established a major attraction to anchor the southern end of the district, bringing 6000 people per day to State Street. That same year, one block north of the library, DePaul University completed its conversion of the long-vacant Goldblatt's Department Store into a 650,000-square-foot complex of university classrooms, offices, and a Music Mart with a dozen music-oriented stores and three restaurants. The DePaul Music Mart Center attracts more than 10,000 students, employees, and customers every day. Three blocks north of the library, Palmer House Hilton completed a $30 million renovation. Even more important, in 1994 Marshall Field's invested $50 million in the renovation of its flagship store.

In 1996, the city began work on a nine-block, $24.5 million reconstruction of State Street. It included reintroducing cars and taxis to the street, reducing the size of the sidewalks from 40 to 52 feet to their historic width of 22 to 26 feet, replacing the streetlights with facsimiles of the 30-foot-high cast-iron light fixtures that had been installed on State Street in 1926, and replacing the tree grates, air vents, and Transit Authority stair/escalator enclosures with new furnishings that were more in keeping with the historic character of the area.

There is no way of determining whether remotorization hastened State Street's revival. Nevertheless, that revival is under way. At the northern end of the street, a theater district is taking shape, anchored by the Chicago Theater, restored initially in 1987 and acquired and further restored by Disney Theatrical Productions in 1996, and the Oriental Theater, restored by the Livent Company in 1999. Also in 1999, the Reliance Building (1891–1995), designed by D. H. Burnham & Co., was renovated and converted into a hotel. Burnham's nineteenth-century design inspired the construction in 2000 of a new residential hall for the School of the Art Institute of Chicago. Sears had moved back to State Street after an 18-year absence.

As of 2012, State Street was still changing. It had become an appealing location for such big box retailers and off-price chains as Old Navy, T.J.Maxx, Urban Outfitters, and Nordstrom Rack (see Chapter 4). Louis Sullivan's famous Carson Pirie Scott department store was being remodeled for Target. The once-vacant block 37, between Washington and Randolph streets, had been built up and was slowly filling up with tenants. The opening of the Michigan Avenue Bridge had attracted upscale retailing away from State Street, and the rehabilitation of older department stores for university use had changed the character of its clientele north of the Harold Washington Library. There surely would be further changes.

The character of the public realm, so important to the evolution of Lincoln Road and Nicollet Avenue, affected State Street only during the 17 years it had been a transitway—and that impact was entirely negative. This experience provides an excellent illustration of the relationship between market forces (such as the character of the customer base, the strength of competing retail areas, and changes in marketing and consumer taste) and public investment. It also underscores the importance of understanding and profiting from those market forces.

Uptown, Houston

Market forces affecting the Uptown section of Houston, unlike those that for 17 years prevented street improvements to State Street in Chicago from generating additional customers, transformed an area six miles from downtown Houston that in 1950 was not yet part of the city and in 1964 was selected by the Gerald D. Hines Interests for the 230-acre suburban multiuse City Post Oak complex (see Chapter 4), into the seventeenth largest business district in America. Among the most important of those forces were the region's rapidly expanding population, creation of interstate highway loop I-610, and construction of millions of square feet of office and retail space. Thus, major improvements to the public realm were likely to attract even more economic activity.[39]

The money to pay for investments in the public realm came from a Tax Increment Reinvestment Zone (TIRZ) created in 1987 by the Texas legislature. As with all tax increment financing (TIF), for the life of the TIF any real estate taxes in excess of the amount collected prior to establishment of the TIF (presumably the result of increased property val-

Houston, 1979. Construction of the Galeria began the transformation of Post Oak Boulevard into a major shopping destination. *(Alexander Garvin)*

Houston, 2007. Post Oak Boulevard became the center of one of the nation's major regional subcenters. *(Alexander Garvin)*

ues that reflect the impact of public improvements made to the district) are transfered to the TIF. That stream of income goes to pay debt service on bonds issued to pay for public improvements.

Uptown Houston used tax increment proceeds to pay for a $235 million program that added specialty street lamps, signage, stainless steel gateway arches, stainless steel halos over major streets and intersections, bus shelters, sidewalk improvements, and landscaping to Post Oak Boulevard and nearby streets. As a result, it is now much easier to find your way and drive around the district. Once people have parked their cars, it is a much pleasanter place to be. In fact, not everybody drives from destination to destination—they often walk. Together, these investments have contributed to making this business district the home of more than 23 million square feet of office space, 5 million square feet of retail space, 31 hotels with more than 7000 rooms, and 166,000 residents in 2011. Every day, 200,000 people work, dine, and shop there.

Putting It All Together to Reinvent Downtown Philadelphia

Up to now, this chapter has been devoted to the use of streets and highways, railroads and mass transit, and other investments in the public realm as devices for retrofitting business districts to the needs of their evolving economies. As Walter Moody, the Chicago Plan Commission's first managing

Philadelphia, 1950. Aerial view of the approach to the old Broad Street Station (popularly known as the "Chinese Wall") that became the site of Penn Center. *(Alexander Garvin)*

Philadelphia, 1951. Dock Street Market. *(Alexander Garvin)*

director, explained, these individual, haphazard public improvements cannot be called planning. He called "a street widening unrelated to any other improvement or purpose" or a civic center created without reference to everything else in the city "unplanning."[40]

Among the public investments presented so far, only Denver put them together into a self-reinforcing combination. That combination, however, was not premeditated. It evolved over half a century in response to changing regional issues. During the second half of the twentieth century in Philadelpha, however, public officials, led by Edmund Bacon, executive director of the City Planning Commission (1949–1970), conceived and implemented a series of public investments that reinvented the city's downtown business district.[41]

Bacon understood that downtown Philadelphia would not remain one of the nation's leading cities unless it eliminated major impediments to private development, improved railroad and subway access and circulation, and introduced

Philadelphia, 1945. The elevated railroad, popularly known as "the Chinese Wall," and the Dock Street Market inhibited investment in downtown Philadelphia. *(Alexander Garvin, Joshua Price, and Ryan Salvatore)*

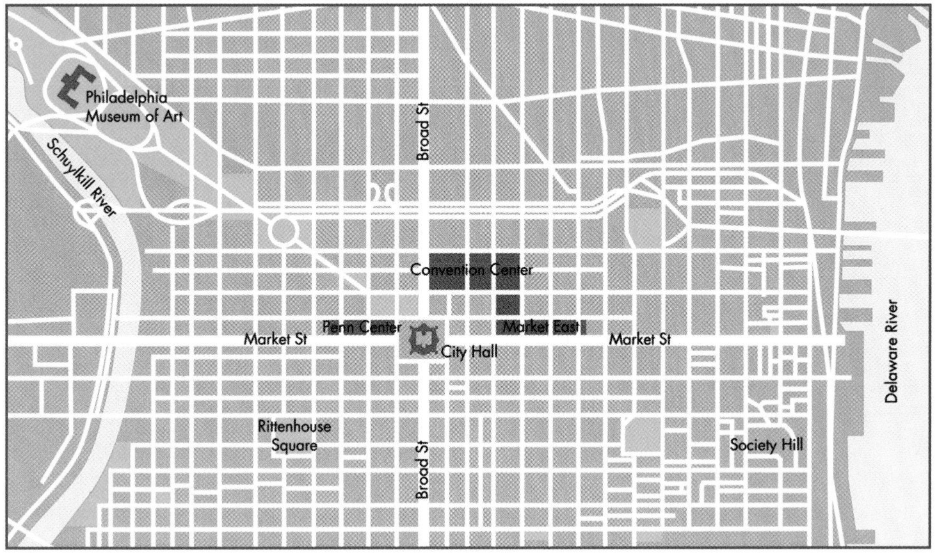

Philadelphia, 2010. The combination of Penn Center, rail-subway restructuring, Market East, a new convention center, and the reestablishment of Society Hill as an attractive residential neighborhood was intended to attract substantial investment to downtown Philadelphia. *(Alexander Garvin, Joshua Price, and Ryan Salvatore)*

Philadelphia, 2011. Penn Center. *(Alexander Garvin, Joshua Price, and Ryan Salvatore)*

magnets that would attract private investment. Those impediments were the Chinese Wall leading to Broad Street Station (discussed earlier in this chapter) on the western side of the business district, and the Dock Street Market on the eastern.

Bacon wanted to replace Broad Street Station and the Chinese Wall with a sunken pedestrian concourse, one level below the street, lined with retail stores, connecting the suburban commuter railroad station, City Hall, several high-rise office buildings, the east-west subway line under Market Street, the north-south subway line under Broad Street, and a new regional bus terminal. The Pennsylvania Railroad still owned and operated the railroad terminal, and had promised to tear it down decades earlier.

In early February 1952, after working with architect Vincent Kling on a variety of schemes for the site, Bacon convinced James Symes, the Pennsylvania Railroad's executive vice president, to announce that it was about to start demolition of the Broad Street Station. Then, after two years of tough negotiations that involved the mayor, the business community, and civic leaders, the railroad proposed a scheme, quite different from what Bacon had initially proposed. It was, however, something that everybody could accept.

Philadelphia, 1998. Market East Railroad Station. *(Alexander Garvin)*

When the project, eventually named Penn Center, was completed, Bacon, like so many others, was deeply disappointed. The pedestrian areas between the office towers were uninviting. Real street-level and below-grade open-air retailing never materialized. Although Penn Center was not exactly what Bacon had conceived, it included many of his ideas. The underground pedestrian precinct did connect Suburban Station, the subway, City Hall, and Penn Center. It represented the sort of intelligent, premeditated combination of public investments that triggered a wave of new office and hotel construction west of City Hall that has continued into the twenty-first century.

Along Market Street, east of City Hall, much the same happened—only it took Bacon several decades to bring it to fruition. Removal of the Dock Street Market, however, hap-

Philadelphia, 2010. The Gallery is a shopping destination for many of the city's low- and moderate-income shoppers. *(Alexander Garvin)*

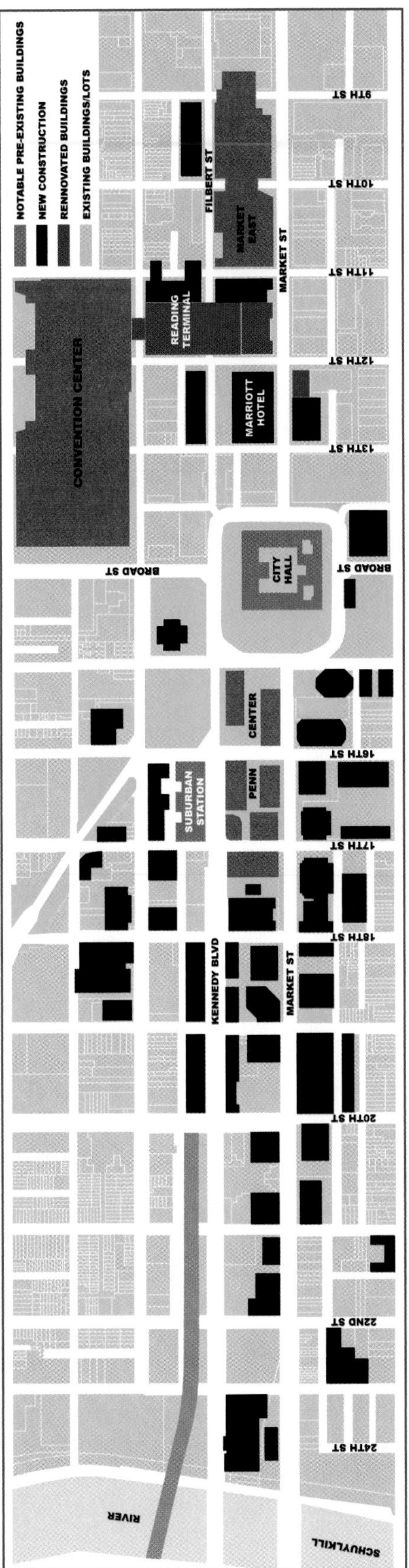

Philadelphia, 2013. Brilliantly conceived public investment in downtown Philadelphia triggered a widespread and sustained private market reaction. *(Alexander Garvin, Joshua Price, and Ryan Salvatore)*

Philadelphia, 2011. East Market Street. *(Alexander Garvin)*

pened quite quickly because it had become truly outmoded. Not only was its downtown location inconvenient to truck, rail, and air freight, it had become a traffic bottleneck. "Even the loading docks were not of a height adapted to trucks which brought the produce in." Philadelphia needed a functional produce market in a convenient location easily accessible to trucking.[42]

In 1954, Bacon, along with the Greater Philadelphia Movement, an organization of public-spirited business leaders, proposed moving the market. The site they selected was a 388-acre garbage dump in South Philadelphia, easily accessible by highway and railroad to the rest of the country. The Philadelphia Redevelopment Authority acquired it for $7 million. After spending $10 million on streets and utilities, it transferred the site to the newly formed Food Distribution Corporation, which spent another $100 million to create the country's first municipally sponsored industrial park for processing, packaging, distribution, and wholesaling of food products. The site of the old market provided the first development in Society Hill, the urban renewal project discussed at length in Chapter 12.

The magnet Bacon had in mind for East Market Street was a mixed-use complex of shopping facilities, hotel, and convention facilities connected to suburban commuters, shoppers, and visitors who traveled the Reading Terminal. Bacon's initial idea was to tear down the Reading Terminal, connect it underground to Suburban Station, west of City Hall, and create a retail and office complex covering a rebuilt transit center.

For two decades, the idea was thought to be impractical. What finally emerged was a variation on what Bacon originally suggested. It began to take shape after Bacon retired from government in 1970. That, however, did not stop him from doing all he could to push it along. The results, along with the huge convention center that was added later, are discussed at length in Chapter 5.[43]

The combination of public investments that Bacon set into motion around 1950 reinvented downtown Philadelphia. At that time, downtown was an obsolete collection of structures providing employment for people who went home and left the city empty after 6 p.m. and on weekends. By the second decade of the twenty-first century, downtown had been virtually rebuilt and was alive with tourists (more than 29 million a year), downtown residents, and suburbanites who came downtown for entertainment and restaurants. Together, they supported nearly 700 restaurants, take-out establishments, and other food-related businesses, 1400 retail establishments, and, particularly significant, 12,250 apartments (the third largest downtown residential population in the nation).[44]

Philadelphia, along with Detroit, Cleveland, and St. Louis, were among the top 10 cities in America in 1950. Philadelphia still is among the top 10 because of the public improvements that were made at Edmund Bacon's behest. Is there any better demonstration that coordinated investments in the public realm made with intelligence and foresight can make a real difference?

Ingredients of Success

The negative impact of Broad Street Station on downtown Philadelphia indicates that making transportation investments may not bring improvements to the business district, just as failure of Chicago's State Street Mall demonstrates that just banishing automobiles from a city's major shopping street is unlikely to spur downtown revival. To be successful, investments in the public realm must be located and designed in a manner that will attract people (customers, workers, and residents) who would otherwise not be there. Among the reasons that pedestrianization was successful in Denver is that it was augmented with public transit that extended accessibility beyond the territory that could be covered on foot in five minutes. Moreover, its safe, clean, well-managed environment kept people downtown, where they could make purchases on their way to their ultimate destination.

Market

Pedestrianization cannot attract a market when none exists. As Jane Jacobs so eloquently explains, without "tremendous numbers of people . . . there would be no downtown to amount to anything."[45] To bring such "tremendous numbers" downtown, business districts must contain more than stores. They need a dense concentration of office workers, conventioneers, tourists, and residents. Otherwise, there will not be enough customers to support the desired level of retail activity. They also need to provide those customers with easy access to their desired downtown destinations. That is exactly why the combination of investments made in downtown Philadelphia were so effective.

Major investment in the public realm should be considered only where there is already a large concentration of potential retail customers or when they can be combined with new facilities that will attract a critical mass of consumers. Sometimes, concentration is brought about by property owners. In Minneapolis, construction of the IDS Center at the intersection of Nicollet Mall and the city's premier department store triggered an increase in demand. The office workers and tourists brought by IDS created a critical mass that generated additional hotel, department store, and shopping arcade developments.

At Grand Central Terminal in New York, the combination of railroad and subway service attracted a critical mass of customers and triggered private development of office buildings and hotels on nearby blocks. That high a concentration of people does not happen often. But it is that very same combination that allowed Penn Center to provide a boost to West Market Street in Philadelphia.

Every business district faces competition. In many instances the competition is suburban. Sometimes matching the competition involves only minor changes to the business district. Denver was able to attract customers simply by providing a free bus, supplying additional parking, and pedestrianizing a few downtown blocks. Sometimes the only way to compete is to make radical physical changes to the business district. Minneapolis had to create a skyway system with a climate-controlled retail environment that rivaled Southdale and, later, air-conditioned shopping malls.

Location

When it comes to the public realm, proximity can either spur or repel investment. If public realm investments enhance mobility and increase people's desire to be there, the result is an increase in demand and, with it, additional development. That is precisely what occurred on Post Oak Boulevard in Houston.

Investments in the public realm by themselves, however, are futile if the proposed pedestrian district is not easily accessible to its market. Cleveland increased accessibility by building strategically located bus rapid transit along Euclid Avenue and encouraging private development expansion by the educational and health-related institutions along its route. Denver built accessibility into the 16th Street Mall by combining it with suburban bus terminals at each end of its transitway.

Locational benefits can shift. Closing Dearborn Station led to building abandonment on Printers Row in Chicago. The opposite happened when Broad Street Station closed in Philadelphia. The elimination of the noise and soot from the trains that came in on the Chinese Wall created a market for sites that had lain fallow for a century.

Design

Every successful public realm must be easily accessible and inviting. This is not a matter of charm or glitter. The Kalamazoo Mall was not a great work of architecture. It attracted customers because it was easy to reach and compact enough to allow walking from one end to the other in a few minutes. In time, its components became outmoded. Nothing was done to reposition and remarket Burdick Street, so it became easy to remotorize. There is nothing particularly noteworthy about Denver's 16th Street, although it is the work of a famous architect, I. M. Pei. It is successful because the market it serves is so greatly enlarged by the free bus and light-rail service. On the other hand, the very distinctive, shiny street furniture along Houston's Post Oak Boulevard attracts people who would not otherwise be there.

Financing

It is difficult to raise money for public investments unless it comes from government subsidies or from payments by businesses that will ultimately profit from increased business transactions and retail sales or property owners from increased rents. Government subsidies are always scarce. So, to the degree that it makes financial sense, investment in the public realm should come from its beneficiaries. The Pacific Electric Railway in Los Angeles invested in streetcars as a way of increasing the prices for which it could sell land that it owned to which it extended rail service. The Triborough Bridge and Tunnel Authority in New York paid for its bridges and tunnels with the tolls paid by the cars, trucks, and buses that used them. Uptown Houston pays for its street improvement with increment financing.

Too often, however, public investments rely on government subsidies. Then, when the benefits prove to be illusory, there is no money to fix the situation. That is why Dearborn Station in Chicago remained empty for so many years.

In general, city governments and businesses begin to think about improving the pedestrian environment only when an existing shopping area begins to decline. By that point, some firms will be in financial difficulty and unable to invest in improvements. If they were not in trouble, there would be little reason to invest in better management of physical improvements. Other firms will doubt that the increase in business will justify the expense. Thus, business groups or government agencies will often avoid making a decision until federal funds are available.

Among the common sources of public subsidy are the Intermodal Surface Transportation Efficiency Act of 1991 (ISTEA) and the Transportation Equity Act for the 21st Century (TEA-21), enacted in 1998. The money for these programs comes from the gasoline tax established under the Federal Aid Highway Act of 1956, which established the Interstate Highway System.

Financing problems do not end when construction does. Somebody has to pay for maintenance. In Minneapolis, 90 percent comes from the assessment district. Too often, irresponsible public officials choose to redistribute money from projects that do not need 100 percent of their user charges to other services that do not get enough money to be self-sustaining. Inevitably, there comes a time when one or the other, or maybe both, will not be self-sustaining. Then the surrounding district suffers.

Time

Well-conceived, privately owned, privately managed shopping centers manipulate the flow of customers from their point of arrival to their destinations. For the most part, cities are unable to do this because they neither own the properties that abut public streets nor determine who will lease them. It is, therefore, particularly difficult to manipulate the time that consumers spend between arrival and destination. Denver has placed pushcarts, kiosks, and even upright pianos along Sixteenth Street to encourage people to move from one destination to another. But most public realm projects, unlike Sixteenth Street, are considered in terms of what happens during the time that people are passing through.

Because the Minneapolis skyway system was conceived of as a way for people to move around despite bitter cold, heat waves, and rain, sleet, and snow, it generates downtown activity during every season of the year. Even more important, the Minneapolis leadership understands that tastes change and so does competition. Therefore, after two decades it redesigned and updated the Nicollet Avenue Mall so it would not lose customers to newer suburban alternatives. It is doing the same for 2013.

Entrepreneurship

Changes to the public realm don't just appear; neither does the public realm run itself. Most public realm investments are the product of a close working relationship between downtown businesses and local government. In some cases, it is the business community that initiates the process; in others, government takes the lead. Both have to participate because the public realm is government owned, while the money to pay for improvements usually has to come from business. Government, led by Edmund Bacon, pressured the Pennsylvania Railroad to tear down the Chinese Wall and Broad Street Station in Philadelphia. Ideas for reconstruction came from Bacon and his colleagues. But they neither owned the property nor had the money to pay for bringing their ideas to life. The money to reinvest to create Penn Center came from the railroad, and from developers and property owners for the reconstruction of the rest of West Market Street. The station would have been demolished one way or another. Without Edmund Bacon's determined public entrepreneurship, however, it would have taken longer—perhaps too late to attract enough investor capital away from suburban alternatives to reinvent downtown Philadelphia.

Retrofitting Downtown as a Planning Strategy

America continues to spend billions planning highways, installing traffic systems, and building garages but devotes too little attention to pedestrian movement after the vehicles have arrived. We need to do more to make our cities friendlier to the automobile's occupants. This requires more than redecorating a few downtown arteries. It means creating entire networks that encourage interaction among shopping facilities, convention centers, hotels, office buildings, and all the other components of a healthy business district. The combination of FasTracks with the free bus on Denver's Sixteenth Street Mall, wayfinding signage and traffic enhancements with landscaping in Uptown Houston, and Philadelphia's synergistic combination of trains, transit, and pedestrian enhancements with shopping, convention, and hotel facilities demonstrate that retrofitting business districts for a modern economy is as possible in twenty-first-century America as it was in nineteenth-century Paris.

Notes

1. Daniel H. Burnham and Edward H. Bennett, *Plan of Chicago* (first published in Chicago in 1909), Da Capo Press, New York, 1970, p. 17.
2. For a more detailed account of the retrofitting of Paris during the nineteenth century, see Alexander Garvin, *The Planning Game: Lessons from Great Cities*, W. W. Norton & Company, New York, 2013, pp. 65–96.
3. Historical and statistical material on Paris is derived from Colin Jones, *Paris: The Biography of a City*, Viking, New York, 2002; Pierre Pinon, *Atlas du Paris Haussmannien*, Éditions Parigramme, Paris, 2002; David H. Pinkney, *Napoleon III and the Rebuilding of Paris*, Princeton University Press, Princeton, NJ, 1958; Anthony Sutcliffe, *The Autumn of Central Paris*, Edward Arnold, London, 1970; Henri Malet, *Le Baron Haussmann et la Renovation de Paris*, Les Editions Municipales, Paris, 1973; Sigfried Giedion, *Space, Time and Architecture*, Harvard University Press, Cambridge, MA, 1956, pp. 641–679; and François Loyer, *Paris Nineteenth Century: Architecture and Urbanism*, Abbeville Press Publishers, New York, 1988.
4. Loyer, op. cit., pp. 129, 234, 407–408.
5. In his comprehensive atlas of Haussmann's Paris, Pierre Pinon provides five reasons for creating these broad avenues: (1) to traverse the city from north to south and east to west; (2) to connect major destinations, such as railroad stations, bridges, and important intersections; (3) to bypass central Paris; (4) to create monumental axes to major buildings, such as the Arc de Triomphe or the Opéra; and (5) to open up the eight arrondissements annexed in 1860 to real estate development. Other writers emphasize that these broad arteries provided canons with an unobstructed path of fire and therefore could play a major role in suppressing insurrections.
6. For a discussion of the *public realm approach* to planning, see Garvin, *The Planning Game*, op. cit., pp. 11–14.
7. Garvin, *The Planning Game*, op. cit., p. 105.
8. Alexander Garvin, *Public Parks: The Key to Livable Communities*, W. W. Norton & Company, New York, 2011, pp.137–142.
9. Historical and statistical material on American highways is derived from Dan McNichol, *The Roads That Built America*, Sterling Publishing Company, New York, 2006; J. R. Meyer, J. F. Kain, and M. Wohl, *The Urban Transportation* Problem, Harvard University Press, Cambridge, MA, 1965; George M. Smerk, *Readings in Urban Transportation*, Indiana University Press, Bloomington, IN, 1968; and Mark H. Rose, *Interstate Express Highway Politics 1939–1989*, University of Tennessee Press, Knoxville, TN: 1979.
10. A *plat* can be defined as a plan for the actual or proposed territorial organization of a city or any of its parts, including the arrangement and dimensions of public spaces, streets, blocks, and building lots.
11. George Kessler, *A City Plan for Dallas*, Board of Commissioners of the Park Board, Dallas, 1911.
12. John W. Stamper, *Chicago's North Michigan Avenue*, University of Chicago Press, Chicago, IL, 1991, pp. 1–27, 197–201.
13. A discussion of State Street can be found in Chapter 19.
14. Historical and statistical material on Benjamin Franklin Parkway and its surroundings is derived from David B. Brownlee, *Building the City Beautiful: Benjamin Franklin Parkway and the Philadelphia Museum of Art*, Philadelphia Museum of Art, Philadelphia, 1989; and John F. Bauman, *Public Housing, Race, and Renewal: Urban Planning in Philadelphia 1920–1974*, Temple University Press, Philadelphia, 1987.
15. Historical and statistical material on the Port Authority of New York and New Jersey is derived from Julius Henry Cohen, *Developing the Port of New York: An Address*, National Rivers and Harbors Congress, Washington, DC, December 6, 1922; Jameson W. Doig, *Empire on the Hudson: Entrepreneurial Vision and Political Power at the Port of New York Authority*, Columbia University Press, New York, 2001; and Garvin, *The Planning Game*, op. cit., pp. 128–132.
16. Historical and statistical material on the arterial system of New York and New Jersey is derived from Michael R. Fein, *Paving the Way: New York Road Building and the American State, 1880–1956*, University Press of Kansas, Lawrence, KS, 2008; and Garvin. *The Planning Game*, op. cit., pp. 133–164.
17. Robert Moses was certainly a major supporter of the city's highway system, but he neither conceived, developed, nor operated that system. See Garvin, *The Planning Game*, op. cit., pp. 146–156.
18. McNichol, op. cit., pp. 148–151.
19. Ketterlson & Associates, Inc., KFH Groups, Inc., Parsons Brinckerhoff Quade and Douglass Inc., and Dr. Katherine Hunter-Zaworski, *Transit Capacity and Quality of Service Manual, 2nd Edition*, Transportation Research Board, Washington DC, 2003; and American Transportation Association, *2012 Public Transportation Fact Book*, Washington DC, 2012.
20. U.S. Department of Transportation, Office of Highway Policy Information, Highway Statistics Series, "Licensed Drivers, Vehicle Registrations, and Resident Population," http://www.fhwa.dot.gov/policyinformation/statistics/2010/dvlc.cfm, retrieved August 15, 2012. On rail track mileage, U.S. Department of Commerce, Bureau of the Census, *Historical Statistics of the United States: Colonial Times to 1970* (Part 2), 1975.
21. Christopher Gray, "How a Deadly Train Accident Created an Elite Street," *New York Times*, Section 11, p. 9, January 14, 2001.
22. John Belle and Maxinne R. Leighton, *Grand Central: Gateway to a Million Lives*, W.W. Norton & Company, New York, 2000, pp. 63–75.
23. Ibid.
24. Edmund Bacon, "Philadelphia in the Year 2009," reproduced in Scott Gabriel Knowles (editor), *Imagining Philadelphia: Edmund Bacon and the Future of Philadelphia*, University of Pennsylvania Press, Philadelphia, 2009, p. 9.
25. Historical and statistical information on Dearborn Station is derived from Lois Wille, *At Home in the Loop*, Southern Illinois University Press, Carbondale, 1997, pp. 113–126.
26. Historical and statistical information on Union Station is derived from Norbury L. Wayman, *St. Louis Union Station and Its Railroads*, Evelyn E. Newman Group, St. Louis, MO, 1987; and Carolyn Hewes Toft with Lynn Josse, *St Louis: Landmarks and Historic Districts*, Landmarks Association of St. Louis, St. Louis, MO, 2002, pp. 67–70.
27. Henry E. Huntington originally quoted in the *Los Angeles Examiner*, December 12, 1904; see William B. Fredericks, *Henry E. Huntington and the Creation of Southern California*, Ohio State University Press, Columbus, OH, 1992, p. 7.
28. Reyner Banham, *Los Angeles: The Architecture of Four Ecologies*, University of California Press, Berkeley, CA, 2009, pp.13–18, 64–65.
29. Los Angeles Department of City Planning, *A Parkway Plan for the City of Los Angeles and the Metropolitan Area*, Los Angeles, CA, 1941.
30. Historical and statistical information on Buckhead is derived from Susan Kessler Barnard, *Buckhead: A Place for All Time*, R. Bemis Publishing, Ltd., Marietta, GA, 1996; Buckhead Coalition Inc., *Buckhead Guide Book*, Buckhead, GA, 1997; and Buckhead, Inc., www.buckhead.net; Matt Hennie, "Buckhead: Experience in Growth," *GeorgiaTrend*, July 2008, http://www.georgiatrend.com/July-2008/Buckhead-Experience-In-Growth/; U.S. Census Bureau; and interview with Jim Durrett, Executive Director, Buckhead Community Improvement District, August 8, 2012.
31. Statistical and historical information on downtown Denver is de-

rived from Melvin D. Moore (editor), *Downtown Denver: A Guide to Central City Development,* Technical Bulletin #54, Urban Land Institute, Washington, DC, 1965; Leo Adde, *Nine Cities: The Anatomy of Downtown Renewal,* Urban Land Institute, Washington, DC, 1969, pp. 165–193; Donna McEncroe, *Off the Mall Step by Step,* The Denver Partnership, Inc., Denver, CO, 1987; Denver Metro Convention and Visitors Bureau; Harvey M. Rubenstein, *Pedestrian Malls, Streetscapes, and Urban Spaces,* John Wiley & Sons, New York, 1992, pp. 186–188; and "Denver's 16th Street Mall," *Urban Land,* Urban Land Institute, Washington, DC, May 1985.

32. Statistical and historical information on Euclid Avenue is derived from Jan Cigliano, "Euclid Avenue," pp. 93–128, in Jan Cigiano and Sarah Bradford Landau (editors), *The Grand Avenue 1850–1910,* Pomegranate Artbooks, San Francisco, 1994; and Vernon Mays, "Health on Wheels," *Landscape Architecture Magazine,* June 2011, pp.76–83.
33. Statistical and historical information on Lincoln Road is derived from Nicholas N. Patricios, *Building Marvelous Miami,* University of Florida Press, Gainesville, FL, 1994; Allan T. Shulman, Randall C. Robinson Jr., and James Donnelly, *Miami Architecture,* University of Florida Press, Gainesville, FL, 2010; and Roberto Brambilla and Gianni Longo, *For Pedestrians Only: Planning, Design, and Management of Traffic-Free Zones,* Whitney Library of Design, New York, 1977.
34. Statistical and historical information on downtown Minneapolis is derived from Lawrence W. Irwin and Jeffrey B. Groy, *The Minneapolis Skyway System: What It Is and Why It Works,* City Planning Department, Minneapolis, MN, 1982; Brambilla and Longo, op. cit., pp. 132–135; David Gebhard and Tom Martinson, *A Guide to the Architecture of Minnesota,* University of Minneapolis Press, Minneapolis, MN, 1977, pp. 24–38; Jennifer Waters, "The Minneapolis Story," *Urban Land,* vol. 52, no. 4, Urban Land Institute, Washington, DC, April 1993, pp. 34–40; and "The New Nicollet Mall," *Urban Land,* vol. 52, no. 4, Urban Land Institute, Washington, DC, April 1993, pp. 70–71.
35. One heating degree day is accumulated for each degree that the mean daily temperature drops below 65°F. Minneapolis records 8007 heating degree days per year; San Francisco, 3161. See Lawrence O. Houstoun Jr., "Weather Report," in *Planning,* American Planning Association, Chicago, December 1990, pp. 19–21.
36. Baker Properties later became Investors' Diversified Services Properties (IDS) and then Oxford Properties.
37. Irwin and Groy, op. cit., www.ids-center.com/pages/architecture.html; Leif Pettersen, "Take the Skyway," February 17, 2011, http://www.vita.mn/story.php?id=116353724.
38. Statistical and historical information on State Street is derived from Adrian Smith and Phil Enquist, "State Street, Reviving the Heartbeat of the Loop," *Urban Land,* vol. 55, no. 2, Urban Land Institute, Washington, DC, February 1996, pp. 14–19; "That Great Street Hopes for a Comeback," *Planning,* vol. 63, no. 1, American Planning Association, Chicago, January 1997, pp. 12–15; Greater State Street Council, *Greater State Street,* Chicago, October 1987; Blair Kamin: "Stately Street Retro Renovation Puts a Once-Great Shopping Mecca on the Road to Economic and Aesthetic Recovery," *Chicago Tribune,* November 15, 1996; Robert Sharoff, "Putting 'That Great' Back in State Street," *New York Times,* May 14, 2000; and Bruce Japsen: "State Street Revival Marches to Steady Beat," *Chicago Tribune,* October 20, 2000.
39. Statistical and historical information on the Uptown Section of Houston is derived from Carla Sabala, *Houston Today,* ULI-Urban Land Institute, Washington DC, 1974; ULI Local Arrangements Committee, *Houston Metropolitan Area . . . Today 1982,* ULI-Urban Land Institute, Washington DC, 1982; and www.Uptown-Houston.com/.
40. Walter Moody, *What of the City,* A. C. McClurg & Co., Chicago, 1919,p. 18.
41. For a more complete description of the creation of Penn Center, see Garvin, *The Planning Game,* op. cit., pp. 175–181.
42. For a more complete description of the relocation of the Dock Street Market, see Garvin, *The Planning Game,* op. cit., pp. 182–183.
43. For a more complete description of the creation of Market East, see Garvin, *The Planning Game,*op. cit., pp. 188–191.
44. Central City District and Central Philadelphia Development Corporation, *State of Center City Philadelphia 2010,* Philadelphia, 2010.
45. Jane Jacobs, *The Death and Life of Great American Cities,* Random House, New York, 1961, p. 4.

7

The Life and Death of the City of Tomorrow

Atlanta, 2004 *(Alexander Garvin)*

After World War II the American city experienced cataclysmic change. Broad superhighways were thrust into its heart. Spacious pedestrian plazas with shiny glass towers replaced familiar neighborhoods. The impetus for this radical transformation was the idea that cities were terminally ill. The disease seemed obvious wherever there were deteriorated or vacant buildings, decreasing populations, and declining employment, retail sales, and office occupancy. Many experts thought replacing what they observed to be a functionally obsolete physical plant could reverse the decline. They prescribed clearing and rebuilding decaying cities, section by section, until every corner had been transformed and an efficient, modern metropolis had been created and tied together by superhighways that would connect surrounding metropolitan areas and the rest of the nation.[1]

This redevelopment prescription was made available to the nation when Congress enacted the Housing Act of 1949, which offered to reimburse two-thirds of the cost of redevelopment for any city that used the remedy. Over the next quarter of a century, the federal government paid $12.7 billion to nearly 1000 cities that took advantage of the offer. Some cities had a vision of the modern metropolis that they wished to become. Others just wanted to clear a slum or move its occupants. Many tried redevelopment simply because money was available. In each case, the Housing Act of 1949 financed the elimination of large sections of the city, paved the way for new highways and garages, and subsidized the development of massive superblocks.[2]

Access by motor vehicle to the rebuilt sections of a city's business district was financed largely by the National Interstate and Defense Highways Act of 1956, which paid for 90 percent of the cost of an Interstate Highway System that by 2010 extended throughout the United States for 47,182 miles. In some cities, like Atlanta, New Haven, and Hartford, the highway skirted the edge of the central business district. In others, like Houston, Los Angeles, and Portland, Oregon, the interstate highways often consist of ring roads, providing easy access to anywhere downtown.

In most cities, retrofitting downtown for trucks, buses, and private cars, as well as redeveloping the business district, was a painful process. From the beginning, there was opposition from relocatees, preservationists, and people who opposed public intervention into the private market. Nevertheless, across the country, city governments established redevelopment agencies to rebuild entire districts. With every new project, that opposition increased. By 1973, when the Nixon administration abruptly terminated the program, few people wanted their cities rebuilt: too many of the program's 2532 urban renewal projects had been shameful failures.

The all-encompassing diagnosis that cities were dying and that the cause was an obsolete physical plant was flawed. In many cities, such as Dallas, Phoenix, and San Diego, the problems that required attention were caused by growth rather than decline. Those cities that had shrinking populations, such as Cleveland, Cincinnati, and St. Louis, were not dying, nor was their population decline caused by outmoded structures or inadequate highway access.

In Pittsburgh, Baltimore, and other cities where redevelopment projects did not satisfy the entire demand for new space, they stimulated further private investment and triggered genuine urban renewal on surrounding private property. In many instances, though, redevelopment projects caused hardship for the residents, businesses, and workers who were displaced, subsidized the creation of arid districts dominated by mediocre high-rise buildings, and retarded further private market activity. This history should deter civic leaders from initiating further government-assisted redevelopment except where the decline is truly caused by physical and functional obsolescence, where individual entrepreneurs cannot overcome that obsolescence without government intervention, where the cost and time of obtaining approvals and assembling a project are beyond the capacity of individual developers, or where the project is located and designed in a manner that ensures that it will generate additional market activity in surrounding areas.

Le Corbusier's Vision

The physical paradigm for America's redevelopment program was provided by the Swiss-born French architect, painter, and writer Charles-Edouard Jeanneret, better known as Le Corbusier. As early as 1924, Le Corbusier had written that "the city of today is a dying thing," called for "a frontal attack on the most diseased quarters," and demanded their replacement by new districts, "vertical to the sky, open to light and air, clear and radiant and sparkling."[3]

Le Corbusier summarized his vision in *The City of Tomorrow, and Its Planning,* first published in 1924. He proposed the creation of new cities that would consist of three separate districts: a business center of office towers, a residential area of elevator apartment houses, and a manufacturing-warehousing district. The residential and office areas were essentially a

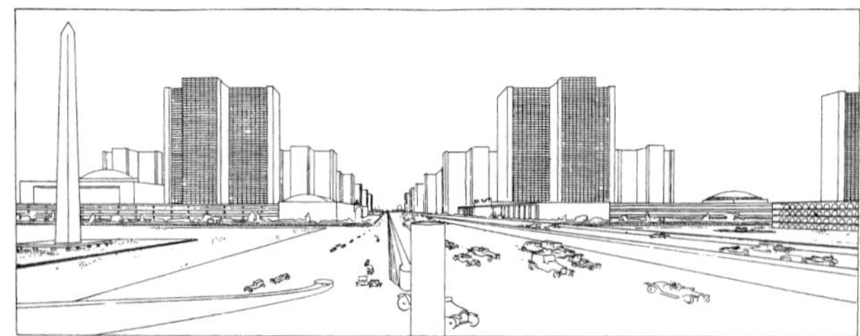

Paris, 1922. The highways and towers of Le Corbusier's City of Tomorrow inspired the urban renewal program in America. *(Courtesy of Artists Rights Society [ARS]/SPADEM, 1995, Paris)*

Buenos Aires, 1929. Le Corbusier's towers-in-the-park vision of a new living environment for Buenos Aires. *(Courtesy of Artists Rights Society [ARS]/SPADEM, 1995, Paris)*

continuous green park divided by elevated highways into superblocks. Citizens were to speed along the highways in their automobiles until they reached the appropriate branch road leading to an "auto-port." There they would park their cars and enter a green pedestrian precinct containing all the amenities of modern life.

Le Corbusier described the proposed skyscraper business district thus:

> *. . . a great open space 2,400 yards by 1,500 yards, giving an area of 3,600,000 square yards, and occupied by garden parks. . . . In these parks, at the foot of and round the skyscrapers, would be the restaurants and cafes, the luxury shops, housed in buildings with receding terraces; here too would be the theaters, halls and so on; and here the parking places or garage shelters.*[4]

Each 1200-square-foot residential superblock was supposed to include all the necessities of a healthy family life:

> *. . . communal services (catering and household supplies), nursery, kindergarten, open-air playground in the park, primary school . . . complete stadium, large swimming pool and sand beach + tennis courts + infants' playground + covered play areas underneath buildings + immense ribbons of sunbathing beaches on roof-gardens. . . .*[5]

At ground level these superblocks flowed together to form one immense park, only 15 percent of which was covered with buildings.

Along with its revolutionary combination of public open space, superhighways, and skyscrapers, the City of Tomorrow provided a brand-new way of modernizing cities. In sixteenth- and seventeenth-century Rome, the popes acquired privately owned land to provide necessary public thoroughfares. In nineteenth-century Paris, Napoleon III and Baron Georges-Eugène Haussmann did so to provide an infrastructure of public works that would spur private reconstruction of appropriate parts of the city (see Chapter 6). Le Corbusier proposed the next step: condemnation of whole districts to allow the efficient reconstruction of an entire city. In physical terms, this evolution to full-scale redevelopment may have been relatively obvious. In economic and political terms, it meant rejecting capitalist market economics and assigning to government the functions of the real estate developer.

Urban redevelopment required condemnation of thousands of privately owned properties. Le Corbusier did not specify whether they would be acquired by negotiation or expropriation. He was equally unclear about which entity would implement the plan. However, he was clear, if naive, in explaining how to pay for it. Eventually, the money would be earned by "the initiator of this change" (national government?) from the "fourfold or tenfold increase" in land value that would result from more efficient, high-density use of land. How "the initiator" would raise the gargantuan amounts of cash needed to pay for initial land acquisition or the even more massive costs of reconstruction was never explained. Nor did Le Corbusier give any consideration to the problem of relocating millions of residents and thousands of businesses.[6]

Although Le Corbusier's ideas may have been well known to intellectuals and to the architects who would be hired to design any redevelopment project, most Americans had never heard of them. Nevertheless, millions of Americans were familiar with the brave new world of highways and skyscrapers that he advocated. It was this image of the new utopia, rarely drawn by Le Corbusier himself, which had been hammered into the public consciousness by countless magazine articles, exhibitions, and Hollywood movies. Its most widely known version is the one that was presented to millions of visitors at the New York World's Fair of 1939. There, at the General Motors exhibit, they could see a model of "Futurama," architect Norman Bel Geddes's depiction of America in 1960:

> *. . . traffic mov[ing] at designated speeds of fifty, seventy five, and a hundred miles an hour along highway surfaces . . . skyscrapers spaced far apart, the base of each one occupying a full city block. On the roofs of some were landing places for airplanes and autogyro [as helicopters were then called]. Parks occupied a third of the total city area. It was a utopia of abundant sunshine, fresh air, and recreational opportunity.*[7]

Pre–World War II office buildings were not designed to meet the rapidly growing needs of their occupants. Tenants wanted up-to-date electrical systems, air-conditioning, and

Plan Voisin, Paris, 1925 *(Courtesy of Artists Rights Society [ARS]/SPADEM, 1995, Paris)*

large office floors that could accommodate increasingly sophisticated business machines. Workers expected a pleasant working environment that was easily accessible by automobile. Most older office districts could not provide these conditions. The City of Tomorrow, with its highways, parking garages, broad plazas, and office towers, seemed to be what was needed.

Adapting the City of Tomorrow to Postwar America

Whether in its pristine version by Le Corbusier, the commercial version presented by General Motors at the 1939 World's Fair, or the more generalized schemes depicted in the mass media, the City of Tomorrow had to be adapted to the needs of America's ever practical population. This meant providing a rationale for a national urban redevelopment program, reducing its size to fit financial and political reality, and then creating financial inducements for local governments and private developers to build it.

Local government was needed for its power of eminent domain. There is no way to redevelop blighted city districts without condemnation of privately owned real estate, because some property owners will hold out for prices that make redevelopment financially unfeasible and others will refuse to sell at any price. Moreover, government ownership of redevelopment properties eliminates both the risk that the project may not be approved and the interim costs of paying taxes, insurance, interest on the purchase price of the property, and operating expenses while redevelopment plans are being developed, debated, and approved.

The Fifth Amendment to the Constitution clearly states that government can take private property "for public use" by condemnation (eminent domain) only if it complies with "due process of law" and provides "just compensation." The Fourteenth Amendment further states that in such instances, all property owners must be afforded "equal protection of the laws."

The taking issue comes into play whenever a property owner objects to the legitimacy of a government agency's compliance with due process, payment in compensation for taking the property, or equitable treatment, or when the property's reuse is not by a public agency. That is the reason the Housing Act of 1937 only allows property reuse by local government housing authorities.

In the 1954 case of *Berman v. Parker*, the Supreme Court deferred to the local legislature in determining what constitutes "public use," by unanimously upholding the legitimacy of taking private property from one private owner in order to transfer it to another private owner under Title I of the Housing Act of 1949.[8]

The use of eminent domain for redevelopment in which the end use is not "public" continues to be subject to controversy because local officials sometimes are not careful enough in determining public *benefit* or do so without sufficient regard to public acceptance of that *benefit*. For example, the New London Development Corporation (NLDC) established by the City of New London, Connecticut, to foster economic development, decided it needed to recoup more than 1500 jobs that were lost in 1996 when the federal government closed its Naval Undersea Warfare Center. Two years later, after acquiring the center and other privately owned properties in the area by eminent domain, the NLDC transferred the site to Pfizer Inc., a huge pharmaceutical company, which promised to build a $300 million research facility on the site. The NLDC believed that the site's redevelopment would create additional jobs, increase tax revenues, and revitalize the area.

Susette Kelo and several other property owners objected to the taking of their property, claiming that reuse by Pfizer Inc. was not "public use." The case became the centerpiece of a growing property rights backlash movement that objected to overreaching government agencies. Finally, in 2005, in a five-to-four decision, the Supreme Court held that the NLDC's action was indeed constitutional.[9]

Though Pfizer did occupy the facility for a short while, only four years after the Kelo decision the company closed the research facility and relocated its 1400 jobs elsewhere (presumably to nearby facilities in Groton, Connecticut). The NLDC had "cleared dozens of homes to make room for a hotel, stores and condominiums that were never built."[10] Given the end result, one has to wonder if local officials had taken sufficient care to see that public action would result in public benefit. That concern has led many states to tighten the taking requirements in state laws allowing the use of eminent domain.

Most government agencies have little experience with property development and even less experience in erecting office buildings, retail stores, or apartment houses. The real estate industry does both every day. On the other hand, government has always assembled property for public buildings and, starting in the 1930s, had engaged in slum clearance for Public Works Administration (PWA) and public housing projects (see Chapter 10). Thus, Title I of the Housing Act of 1949 assigned to local government the central role in planning and property acquisition and to private developers the responsibility for project design and construction. This combination had already been tried in Pittsburgh and provided the basis for a national program.

The Golden Triangle, Pittsburgh

By World War II, Pittsburgh's Triangle, the point at which the Allegheny and Monongahela rivers join to form the Ohio, had become a jumble of rail yards and lofts. Without easy highway access and parking facilities, the entire business district was becoming intolerably congested. Surrounding rivers periodically overflowed. On St. Patrick's Day 1936, they rose 24 feet, with property damage estimated at at $200 million. Flooding was one thing; everyday smoke, smog, and grime another. Eager to maintain employment in steel mills and coalfields, the city had refused to implement effective smoke-control regulations. As a result, day often differed little from night, and streetlights remained in continuous use.[11]

Richard King Mellon, one of the city's most powerful businessmen, decided that conditions had to change. In 1943, he helped to form the Allegheny Conference on Community Development (ACCD), a group of elite business and civic leaders determined to clean up the environment and transform Pittsburgh into an efficient, modern corporate center. The ACCD developed a comprehensive improvement program that was embodied in a "Pittsburgh Package" of state legislation. Eight of its proposed ten bills were passed, including county smoke control, a public parking authority that

Pittsburgh, 1945. Fifth Avenue at 11 a.m. Air pollution so darkened the skies that street lamps had to be kept lighted 24 hours a day. *(Courtesy of Allegheny Conference on Community Development)*

Pittsburgh, 1900 *(Courtesy of Allegheny Conference on Community Development)*

Gateway Center, Pittsburgh, 2008 *(Alexander Garvin)*

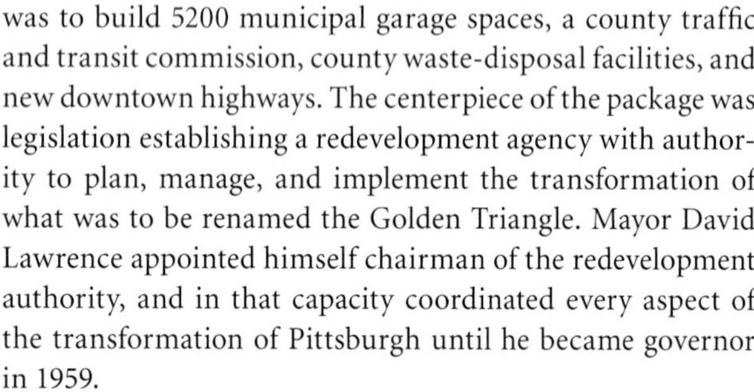

was to build 5200 municipal garage spaces, a county traffic and transit commission, county waste-disposal facilities, and new downtown highways. The centerpiece of the package was legislation establishing a redevelopment agency with authority to plan, manage, and implement the transformation of what was to be renamed the Golden Triangle. Mayor David Lawrence appointed himself chairman of the redevelopment authority, and in that capacity coordinated every aspect of the transformation of Pittsburgh until he became governor in 1959.

The redevelopment plan included restoration of Fort Pitt (the 1759 fort from which modern Pittsburgh had grown), the new 36-acre Point Park, new highways and bridges connecting the city with the surrounding suburbs, Gateway Center (a 23-acre office complex financed by the Equitable Life Assurance Society), several independent office buildings, a 750-car underground garage, a hotel, and a 27-story apartment building.

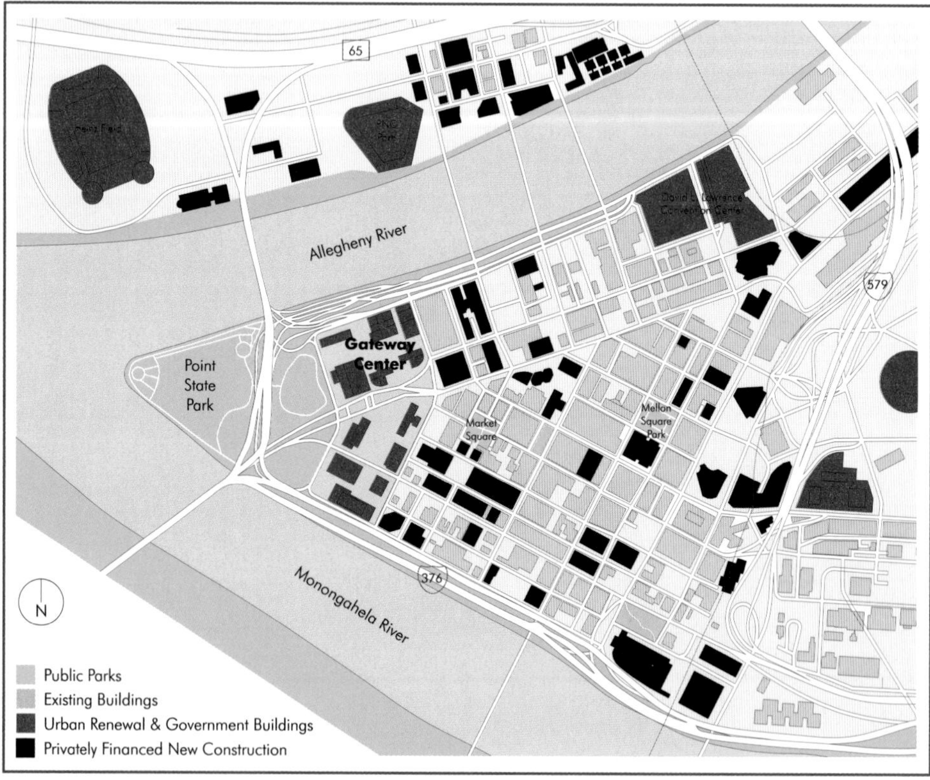

Pittsburgh, 2012. Gateway Center helped to trigger new office construction elsewhere in the business district. *(Alexander Garvin and Ryan Salvatore)*

During the decade in which this program was getting started, the city's population grew a mere 7000 to 677,000 in 1950. As redevelopment progressed, Pittsburgh's population began a steady decline to 306,000 in 2010, a drop of 55 percent. During the 60 years in which the city's population decreased by 371,000, most American cities were experiencing a significant migration of residents toward the suburbs, resulting in a population rebalancing between the two. However, during the same period, the population of the metropolitan area also decreased to 2,356,000, but only by 49,000. Thus, unlike the situation in so many other cities, Pittsburgh's population decline was not matched by soaring suburban growth.

Manufacturing employment in Pittsburgh, once an industrial powerhouse, had declined to a mere 7900 jobs by 2010. As a result of downtown redevelopment, Pittsburgh was able to retain its role as one of America's major business centers. That year, the city ranked eighth for cities with the most Fortune 500 companies, most of which occupied the new buildings that had been erected downtown. *The Economist* called Pittsburgh the most livable city in the United States in 2011, and *Forbes* did the same in 2010. The Pittsburgh package of flood control, improved air quality, highways, parks, parking, and modern buildings allowed the city to reinvent itself.

A National Program

Redevelopment of the Golden Triangle began as a strictly local enterprise. If it was to be used on a national scale, there had to be a convincing public policy rationale. In postwar America, that proved to be very easy. Cities across the nation were experiencing increasing physical deterioration, loss of population, decreasing employment, declining retail sales, decreasing office occupancy, and a rapidly disappearing tax base. The flight to the suburbs was accelerating. All sorts of people were clamoring for action: businesspeople who worked downtown, property owners who saw values plummeting, the poor who could not afford to move to the suburbs, and politicians who depended on all these people for votes.[12]

Faced with the demand for action to save America's "decaying" cities, Congress decided on a program based on the formula that was working for Pittsburgh. Its purpose was to assist localities to "clear blighted areas" and to develop "well-planned" communities. That meant assembling large tracts of land, which would take private developers many years, during which they would have to pay real estate taxes, insurance, and other project expenses—often making redevelopment financially infeasible. Moreover, market conditions could have changes when all the property had been assembled. Assembling all the property could be difficult and quite expensive because individual owners might hold out for outrageous prices. Since the project required assembling all the property, the developers would have no choice but to pay what was required.

Because the ultimate cost of assembling a site and paying all the costs of preparing it for development is often beyond what is affordable by whatever will be built on the site, government may have to take a loss. Of course, the money has already been expended. Thus, when it is time to develop the property, those losses (taken over many years) are no longer apparent to city residents, who are then far more interested in getting the project built.

Title I of the Housing Act of 1949 provided a subsidy of two-thirds of the loss on the sale of property in redevelopment projects.[13] Once the Housing Act of 1949 had been enacted, two-thirds of that loss was absorbed by the federal government. So it appeared to be a very good deal to city residents.

Creating a New Working Environment

City governments, then as now, had neither the money nor the entrepreneurial skills to provide this brave new world. They hoped to obtain both by participating in the federal urban renewal program. This required a redevelopment agency with the ability to plan, acquire, and prepare sites for development and to sell those sites at artificially low prices to developers who would agree to execute their redevelopment schemes; federal approval of their redevelopment plan; and enough money to cover a one-third share of project expenses.[14]

In Philadelphia, Boston, and New Haven, the impetus for redevelopment came from mayors who had been elected promising to redevelop what most people thought of as obsolete business districts. In Baltimore, San Diego, and Cincinnati, the business community led the campaign for redevelopment. It saw urban renewal as the best way of winning a market that would otherwise go to the suburbs. In Hartford, Cleveland, and San Francisco, government officials saw redevelopment as a device for obtaining federal grants. Whether politicians, businesspeople, or bureaucrats took the lead, Title I provided the subsidies to pay for replacing congested downtown districts with local visions of a new working environment. The only exceptions were New York City and Chicago, both of which had huge, healthy business districts and continued to rely on the private market to continue providing up-to-date commercial space.

With or without federal subsidies, government-managed redevelopment made sense where assembling a site for a mutually supportive combination of uses was prohibitively time-consuming or expensive. Baltimore adopted this approach for the redevelopment of Charles Center and the Inner Harbor. San Diego did the same in redeveloping its entire downtown. Whether in Baltimore or San Diego, these projects grew out of the need to satisfy local market demands rather than to replicate Le Corbusier's vision. New Haven, however, badly misjudged market demand. Thus, when it adopted the redevelopment strategy that Le Corbusier had set forth, it was unable to halt the continuing decline of its central business district.

New Haven, Connecticut

By the late 1930s, public officials in New Haven began to be concerned about potential competition from cities that were developing better highway access to national markets and from retailers opening stores in its growing ring of residential suburbs. A 1941 report predicted that without highway access, New Haven's importance might be reduced to that of a "a small local center."[15] Its proposed solution was construction of highway connectors that would allow trucks, buses, and cars to reach regional arteries leading to suburban and national markets. At that time, highways were the responsibility of state and federal governments rather than city officials. Besides, New Haven did not have the resources to pay for major highway construction.

In 1953, the State of Connecticut embarked on a program to build the Connecticut Turnpike, which became I-95, a major link in the Interstate Highway System. That November, Democrat Richard C. Lee was elected mayor, and when he took office in 1954 he began a campaign to extend a highway spur along Oak Street from the turnpike into downtown New Haven. Despite objections from state highway officials and from some downtown residents and businesses, Lee succeeded. One year later, acquisition of the necessary property began.

Lee thought of the connector as an integral part of redevelopment of downtown New Haven. That redevelopment program was an outgrowth of the 1953 mayoral election, in which a central feature of Lee's winning campaign had been the promise to implement redevelopment plans that had been discussed for years but had never moved beyond the planning stage. Two years earlier, Lee had lost the election to the city's incumbent Republican mayor. During the campaign he had visited the "slums" that occupied the site that would become the Oak Street Connector. As he later explained:

> *I came out of one of those homes on Oak Street, and I sat on the curb and I was just sick as a puppy. Why the smell of this building; it had no electricity, it had no gas, it had kerosene lamps, light had never seen those corridors in generations. The smells. . . . It was just awful and I got sick. And there, there I really began . . . right there was when I began to tie in all these ideas we'd been practicing in city planning for years in terms of the human benefits that a program like this could reap for the city.*[16]

Following his election Lee appointed Edward Logue as development administrator. Logue, a recent graduate of the Yale Law School, had been active in Lee's campaign. While the mayor rallied public support for the redevelopment program, Logue packaged the projects. Together, they bulldozed Washington into paying for them and were so successful that New Haven received more Title I subsidies per capita than any other city in the country (see Chapter 12).[17]

New Haven, c. 1955. Oak Street prior to redevelopment. *(Courtesy of the Yale University Library)*

The Oak Street Urban Renewal Project was the first component of what would become "a comprehensive development program" intended to free New Haven "of slums and create an environment that [would] stimulate economic growth and promote the welfare of its citizens."[18] The renewal project, which was conceived as an integral part of the redevelopment of the Oak Street corridor, received federal approval in 1956. Properties in the section of Oak Street closest to downtown New Haven were assembled as sites for buildings to be erected by corporate developers of offices and retail stores or providers of health services. Two sites were for Yale University buildings. The rest provided for new apartment buildings containing 812 dwelling units. Development of these properties required the relocation of more than 522 families, 289 single individuals, 284 rooming house residents, and countless small businesses. Little attention

New Haven, 2012. Oak Street Connector leading out of New Haven. *(Alexander Garvin)*

was paid to them because slum clearance was not yet generally perceived as causing difficult to justify problems for the relocatees. Then, only private development that occurred in the general vicinity happened several decades later, when a one-story drugstore opened on York Street and a small hotel opened on George Street.

The results certainly reflected the strategies prescribed in *The City of Tomorrow.* However, the finished project had little to do with the problems of downtown New Haven, which already had begun to lose businesses to the suburbs. One department store had closed in 1952, another had moved to suburban Hamden, and the rest would go out of business.

It was the next component of the city's comprehensive redevelopment program, the Church Street Redevelopment Project, however, that was conceived as the downtown cornerstone of the city's renewal. The project's major objective was to maintain New Haven's role as one of Connecticut's premier retail centers. A detailed market survey had concluded that New Haven's population would remain at its 1950 level, 164,000, for the next 40 years. The surrounding suburban market was projected to grow from 156,000 to 746,000. The authors of the accompanying report had reasoned that if the city became more accessible to these suburbanites, it would remain a major retail center. However, access provided by the Oak Street Connector alone was not enough. Easy downtown circulation, parking, and modern buildings were thought to be as important.[19]

The project called for a three-block section of Church

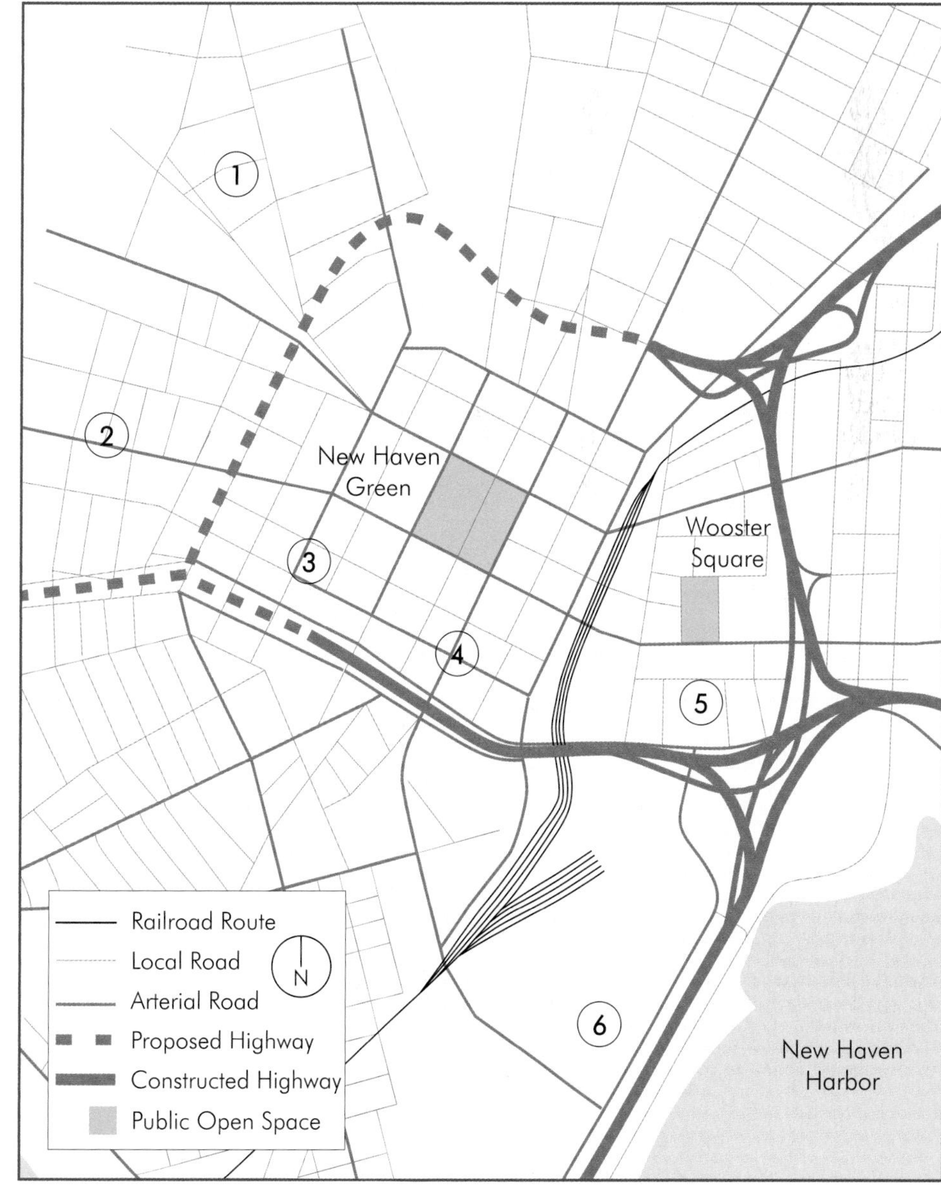

New Haven. The city was awarded more federal urban renewal dollars per capita than any other city, as demonstrated by the near complete coverage of its six urban redevelopment areas within the city: (1) Dixwell Urban Renewal Area, (2) Dwight Urban Renewal Area, (3) Oak Street Urban Renewal Area, (4) Church Street Urban Renewal Area, (5) Wooster Square Urban Renewal Area, and (6) Long Wharf Urban Renewal Area. *(Alexander Garvin and Ryan Salvatore)*

New Haven, 1954. Church Street prior to redevelopment. *(Courtesy of New Haven Redevelopment Agency)*

Street to become a destination for retail customers driving downtown on the new Oak Street Connector. New Haven's existing streets and blocks were to remain in place. George Street was to be bridged by the proposed six-story garage that connected directly into two department stores. Crown Street was to be bridged by a climate-controlled walkway connecting the garage and department stores with an air-conditioned shopping mall, office building, and hotel (see Chapter 2). Once in town, customers would park in a wonderful new garage, do their shopping, and go to a show at the Shubert Theater or at one of the movie palaces that were within a block or two of the site.

The well-known Broadway producer Roger Stevens was designated as the project's developer. Stevens had been chairman of the Finance Committee of the Democratic National Committee during the 1952 presidential campaign. More important, he was a successful real estate investor who had recently put together the syndicate that purchased the Empire State Building. His expertise, however, was in financing and operating existing revenue-producing real estate, not in developing complex projects.

Stevens was involved in many ventures and could devote only limited time to Church Street. Consequently, Lee and Logue took the lead in developing the project. They worked diligently to overcome political opposition, bureaucratic inertia, and lawsuits. They even persuaded one of the city's surviving department stores, Malley's, to give up its prime downtown location and participate in the renewal project. Despite these efforts, when Church Street was cleared and ready to go in 1961, the only component of their plan that was under way was the 1280-car municipal garage. Lack of progress became a campaign issue, and Lee came within 4000 votes of losing the election.

Construction costs and interest rates had increased during the six years since the project had been announced. Logue had left New Haven to direct the Boston Redevelopment Authority. Mayor Lee had no choice but to take an even more active role in packaging Church Street. In 1962, Lee persuaded Macy's, which had discussed participating in the project for five years, to open a department store on the site. The financial deal that Stevens had envisioned was no longer feasible, and he wanted out. In 1964, Lee convinced the Fusco-Amatruda Construction Company to take over the development from Stevens. Meanwhile, retail shopping in New Haven continued to decline, going from 2961 stores in 1933 to 1714 stores in 1963.[20]

Finally, more than a decade after the project had been announced, the redevelopment of Church Street was complete. But it did not become the commercial center its planners had envisioned. In 1982, Malley's Department Store went out of business. The building remained vacant until it was demolished in 1997. The shopping mall was a failure until the late 1980s, when a new city administration provided substantial subsidies and brought in the Rouse Company to renovate and remarket the project. Even then, though, the mall could never really be successful because its customer base was consistently relocating to the suburbs, where it could shop more safely and conveniently. In 1993, the problems became even more acute when Macy's closed its doors. The Macy's building remained a vacant hulk for 14 years, providing a haunting reminder of the hopes and dreams of the urban renewal era. By the time it was torn down, only one new privately developed building had been erected on the other side of Church Street.

The rebuilding of Oak and Church streets was part of a comprehensive approach to downtown redevelopment, which was as ineffective as its individual components. The Oak Street Connector was to be extended to provide access from the suburbs west of downtown, and a highway loop was planned around the center city. Both were stopped from happening because of community opposition to displacement. There never were enough downtown commuters to justify the half-mile-long parking garage that was planned for State Street.

In 1973 and 1987, after redevelopment efforts had ended,

New Haven, 1993. Empty department stores on Church Street three decades after redevelopment. *(Alexander Garvin)*

two privately financed office buildings were erected on the block that contained City Hall. Then, in 1990, the city began building additional office space around its existing property. This public investment was certainly not stimulated by the prior private development.

In the late 1980s, the city began one more effort at redevelopment, this time in the Ninth Square, east of the failing Church Street Redevelopment Project. Finally, in the early 1990s, it persuaded the State of Connecticut and Yale University to help finance the construction of a garage with 628 parking spaces, 50,000 square feet of ground-floor retail space, and 335 privately built apartments (193 of which were heavily subsidized). This investment did little more than provide decent housing for 335 households. Their purchasing power could support only about one-quarter of the retail space, assuming there were not already vacant stores in the surrounding area. The housing has remained in good condition, and there are now some restaurants that fill ground-floor space. But by the start of the twenty-first century, the Oak Street, Church Street, and the Ninth Square redevelopment efforts had generated little if any private market activity.

From the start, the redevelopment strategy for downtown New Haven had been a mistake. The market survey on which the project was based had incorrectly predicted that the city's

New Haven, 2013. Gateway Community College replaced the two failed department stores that were built on Church Street during the 1960s. *(Alexander Garvin)*

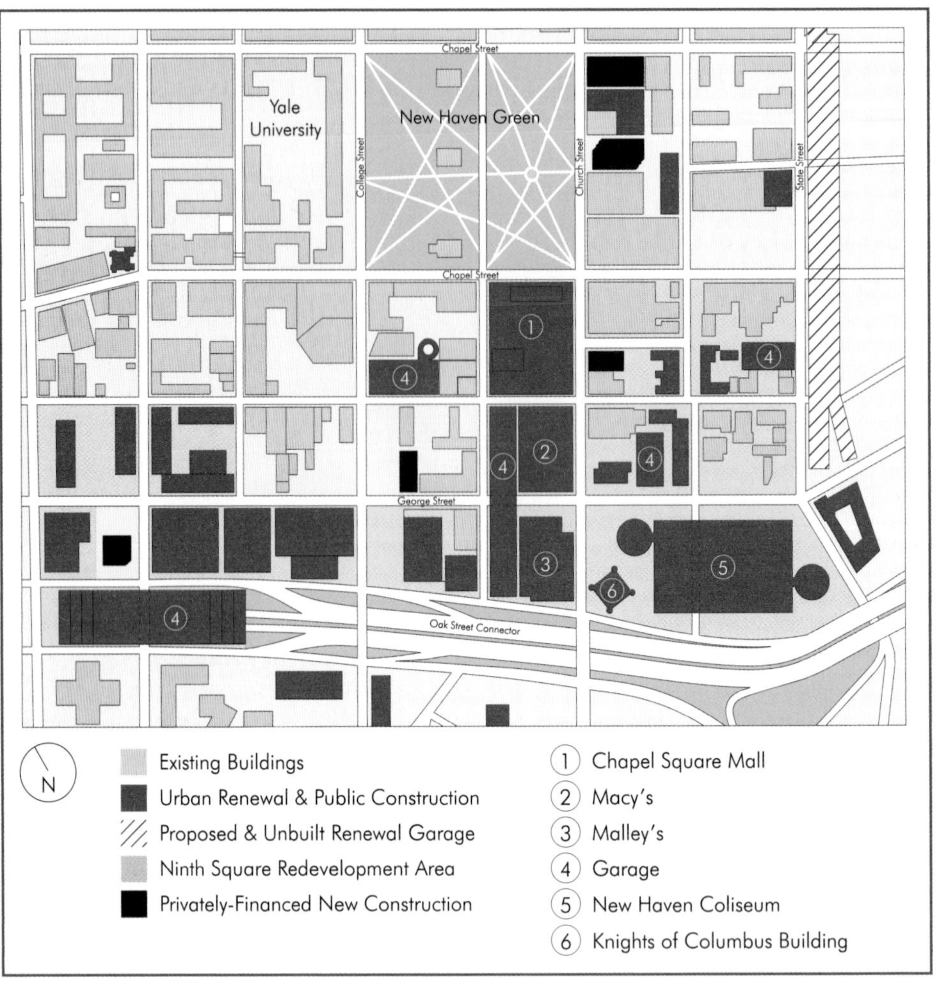

New Haven, 2012. Massive public investment produced very little private market reaction. *(Alexander Garvin and Ryan Salvatore)*

population would remain stable for 30 years. By 1980, it had shrunk 23 percent, to 126,000. Although this shrinkage meant fewer and relatively poorer customers, the suburban market, which in 1980 had grown to 635,000, should have been more than enough to fill the gap.[21] Instead, this suburban population had chosen to shop elsewhere. Between 1960 and 1973, seven major shopping complexes, containing 3,342,000 square feet of floor area, opened in surrounding suburbs. They attracted the market New Haven had lost. Moreover, other suburban shopping venues continued to open during the following decades.

The situation began to change around 2005. Immigrants, particularly from South America, were partially responsible for the city's total population increasing to 130,000 by 2010. Bars, restaurants, and entertainment venues began to attract large numbers of young people on nights and weekends. A 500-unit, privately financed apartment tower with a supermarket on the ground floor opened on lower Chapel Street in 2011. In 2012, Gateway Community College opened on the site that had once been occupied by Macy's and Malley's, bringing more than 7000 students downtown. New Haven is again growing—but not as a result of its redevelopment program.

Just as the city replaced the department stores that it thought incorrectly would attract shoppers downtown, in 2013 it began replacing the Oak Street Connector, which it also thought incorrectly would carry their cars into the business district. After spending millions of federal dollars on the City of Tomorrow, the city's leadership had concluded that Le Corbusier's prescription did not work. It remains to be seen whether an old-fashioned, at-grade boulevard will do better.

There is no way to know whether New Haven would be better or worse off today had the redevelopment not taken place. We do know that the highway extensions and most of the new buildings are still in use. We also know, though, that the supposedly comprehensive effort at redevelopment generated virtually no private market reaction. Thus, Walter Moody would have identified the New Haven version of the City of Tomorrow as an example of "unplanning."

Constitution Plaza, Hartford

Hartford, Connecticut, unlike New Haven, did not adopt a comprehensive redevelopment strategy. It decided to transplant its first component of the City of Tomorrow on the edge of its downtown. Presumably, that would be sufficient to generate enough of a private market reaction to allow the state capitol and center of the insurance industry to triumph over its competitors.[22]

When Hartford established a Redevelopment Agency in 1950, the organization's first task was to study possible sites for a federally assisted Title I project. Within a year, business leaders, government officials, and their consultants had agreed on the 11-acre Front–Market Street district, a "slum" that was only a block east of the city's shopping district. In comparison to other proposed sites, the relocation load (108 "marginal" businesses employing 1037 workers, 187 families, and 31 individuals) was relatively low. More important, its location between a proposed highway and the business district seemed ideal for modern office buildings.

In 1952, the federal government granted Hartford Title I assistance with which to plan Constitution Plaza, a complex of glass office buildings, retail stores, and a hotel, all to be built on the Front–Market Street site. The first hurdle came in a challenge to the constitutionality of taking property from one owner for sale to another (see discussion earlier in the chapter). This issue was overcome in 1954, when the state Supreme Court ruled in favor of the Redevelopment Agency. In 1956, voters approved an $800,000 bond issue to cover the local one-third share of the project's cost. Only in 1958, when relocation was completed, could demolition begin.

Four developers presented plans for the project. F. H. McGraw and Company, the winning bidder, spent two years trying to arrange financing for the development. When it failed, the Travelers Insurance Company came to the rescue, organizing a subsidiary, Constitution Plaza Inc., to finance, build, and operate the project. Travelers, one of Hartford's

Hartford, 1953. Front–Market Street prior to clearance for Constitution Plaza. *(From Louis Redstone,* The New Downtowns, *McGraw-Hill, New York, 1976; courtesy of McGraw-Hill)*

many insurance companies, needed better office space, understood the local market, and had a commitment to improving the city. Most important, it was in the business of financing real estate development. In 1963, the Phoenix Life Insurance Company opened its headquarters, connected to the project by a bridge. Eight years after completion of the redevelopment project, the Travelers opened additional offices connected by bridge to Phoenix Life Insurance Company and to Constitution Plaza.

In Hartford, a misguided redevelopment strategy and the flawed design that it ultimately produced proved to be a recipe that damaged rather than helped the central business district. Constitution Plaza turned out to be an independent enclave built on a platform that covers an 1875-car garage. It was conceived in conjunction with an elevated highway, structured parking, and retail facilities that were intended to serve its occupants and limit their need to ever enter the rest of the city.

Hartford's planners had erroneously thought that bringing the City of Tomorrow downtown would solve what they

Hartford, 2012. The overpass that became an empty pedestrian entry level for commercial structures at Constitution Plaza. *(Alexander Garvin)*

believed was the city's problem: an obsolete physical plant. But the marginal businesses and deteriorating residences that had occupied the site of Constitution Plaza *were* not the cause of Hartford's decreasing building occupancy or its declining retail sales. Demand for office space was declining because companies like the Connecticut General Life Insurance Company (2300 jobs) and the Fuller Brush Company (1500 jobs) were moving to the suburbs. Retail sales were decreasing because office workers also were moving to the suburbs and shopping closer to home. Clearance could not restore this market.

For five decades, Hartford had continued to deteriorate. The city's population shrunk from 177,000 in 1950 to 122,000 in 2000, while the surrounding metropolitan region grew from 407,000 to 855,000. City employment dropped from 82,000 to 44,000. The number of retail stores in the city declined from 2468 to 367. But the region nonetheless continued to grow, going from a population of 855,000 in 2000 to 1,212,000 in 2010.

The firms that relocated to Constitution Plaza did not fill the gap because they moved to an enclave that was designed to be separate from the rest of downtown Hartford. Workers and visitors could park in the project and go directly to their offices without having to set foot in the city. In the beginning, if they needed to purchase something, they could do so at one of Constitution Plaza's underutilized stores, all of which eventually failed because they had so few customers. Those who did visit the business district had to cross Market Street, which was widened as part of the project, only further distancing Constitution Plaza from the city.

A decade and a half after Constitution Plaza opened, three new office buildings were built just east and also connected by bridges to the plaza level of the redevelopment project. Some additional office space was erected downtown, but there was so little demand for the office space at Constitution Plaza that some of its vacant space could be rented at low enough prices to be appealing to government agencies. The hotel had so few customers that it closed in 1994.

In yet another effort to revitalize the downtown, in 2005 Hartford opened a grand new park along the Connecticut River, a 200,000-square-foot convention center and 409-room hotel, and the 140,000-square-foot Science Museum. Even though downtown employment increased by 34 percent, to 59,000, between 2000 and 2010—enough to make downtown office construction once again attractive—most of Constitution Plaza continued to deteriorate. The market may have strengthened in Hartford, but the project's design deficiencies and isolation from the rest of the city continue to handicap its ability to sustain any tenant base.

Despite its failure to revitalize Hartford's business district, Constitution Plaza proved to be a bonanza for its government sponsors. Prior to redevelopment, the 11 acres that were cleared for the project had been assessed at $2.3 million and paid $103,000 in annual real estate taxes. Upon completion, in 1964, the renewal area was reassessed at $23.4 million and paid $1.05 million in annual real estate taxes. Since the City of Hartford needed only $800,000 to cover the local share of Title I subsidies, it was able to recoup its initial investment in less than one year. The Constitution Plaza version of the City of Tomorrow may have had little impact on the health of downtown Hartford. But the federal government had provided a very profitable way to discover its ineffectiveness.

Baltimore

Baltimore adopted a very different redevelopment strategy. Instead of trying to transform the entire downtown into the City of Tomorrow like New Haven or transplanting it to a circumscribed site like Hartford, Baltimore's leaders decided to start with selective surgery. Here downtown redevelopment meant retaining old buildings, while at the same time replacing obsolete sections with structures for which there was substantial consumer demand.

Like Hartford and New Haven, Baltimore had been experiencing declining retail sales (a 10 percent drop in department store sales between 1952 and 1957) and declining downtown tax assessments (a similar 10 percent drop between 1952 and 1957). Although office occupancy had remained a steady 97 percent since 1942, the business community was sufficiently concerned that, in 1954, it formed the Committee for Downtown, Inc., and, a year later, the Greater Baltimore Committee, Inc. Both organizations were dedicated to fighting what they perceived as alarming deterioration in the central business district, where three-quarters of the office buildings had been erected prior to 1920. In fact, nothing new had been built since 1928.[23]

The two groups came together to form the Planning Council of the Greater Baltimore Committee. This private, nonprofit planning organization became the technical consultant for the redevelopment of downtown Baltimore and, under contract, for other civic groups and government bodies. Because the planning council was not a municipal agency, it was less likely to succumb to political pressure or government inertia. More important, by avoiding out-of-town consultants, it was forced to face local realities and get directly involved with project implementation.

The planning council hired David Wallace, an architect-planner who would later head the firm of Wallace-McHarg Associates. He directed a team of experts that included George Kostritsky (later a partner in the architectural firm of Rogers, Taliaferro, Kostritsky, and Lamb) and Dennis Durden, who would later spearhead the redevelopment of downtown Cincinnati. Their goal was to devise a project that would generate further private redevelopment elsewhere in the business district. That meant a project that was

Baltimore, 1964. The business district, with Charles Center, after clearance for redevelopment and after completion of the first new office building. *(Courtesy of Urban Land Institute [ULI])*

big enough to have real impact but not big enough to fully satisfy the demand. They wanted to avoid project success at the expense of other downtown development and did so by planning to satisfy only two-thirds of identified demand for additional office space.

The 33-acre site they selected for the project, which became known as Charles Center, was right in the middle of the city, where it could create a focus for and link together Baltimore's retail, government, and financial districts. Because the site dropped 66 feet from one end to the other, there was room to superimpose an entirely pedestrian level that bridged over Baltimore's busy streets and sidewalks without disturbing existing traffic and to simultaneously slip new garages underneath the new pedestrian level.

The planning council was determined to avoid cutting out the heart of downtown Baltimore and then waiting for a decade, as New Haven and Hartford did, while developers tried to package their projects. As a result, it began by retaining five older buildings whose assessed value represented 47 percent of the entire 33-acre site. These pre–World War II buildings would continue to attract activity both during and after redevelopment.[24]

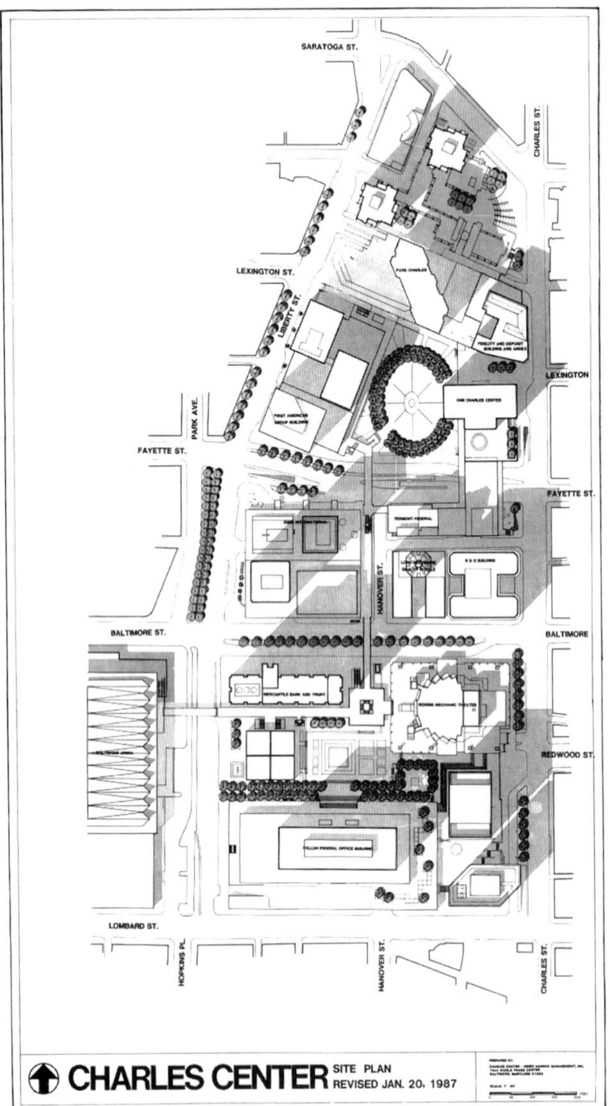

Baltimore, 1987. Redevelopment plan for Charles Center. *(Courtesy of City of Baltimore Development Corporation)*

Baltimore, 2011. By including a new legitimate theater, the plan attempted to generate visitors at different times of the day. *(Alexander Garvin)*

Instead of a single redevelopment site with one prime developer, Charles Center was subdivided into 16 parcels whose development could be timed to absorb market demand. The size of these sites made them easier to finance and, therefore, attractive to a broader range of developers. The decision to subdivide also required careful integration of the project's many components.

Although the Baltimore Urban Renewal and Housing Agency was nominally responsible for Charles Center, and designation as a Title I project involved supervision by federal agencies, the project was implemented by a special management office directed by J. Jefferson Miller, a recently retired businessman. The office was responsible for generating public support, supervising 148 separate acquisitions involving 216 different properties, overseeing the relocation of 350 businesses with 8789 jobs, managing the disposition of the 16 development sites, coordinating public works construction, and making sure everything happened in a timely manner. As Miller explained:

> *The entire project became a Chinese puzzle. For example, the Hamburgers retail store had to be completed so that Hamburgers could move [from its original site within the redevelopment area], before we could start work in the hotel area. The Vermont Federal Savings and Loan Association had to be moved to the Sun Building, which is located in the future theater area, so that they could do business while their former home was being demolished and their new home being erected on the same spot.*[25]

The developer of the first office building was selected by competition. Six organizations submitted proposals. The winner, Metropolitan Structures of Chicago, proposed a scheme by architect Mies van der Rohe. It was selected because the jury felt that the design of this lead site should set a standard for the rest of the project. Later projects, whether selected by design competition, by displaced site tenants, by government agencies, or by owner-builders, were carefully adjusted to fit the scheme. Only one of them, the Mechanic Theater designed by John Johansen, can be characterized as architecturally distinctive.

When Charles Center was completed over a decade later, it included 1.8 million square feet of office space, 430,000 square feet of retailing space, 800 hotel rooms, an 1800-seat legitimate theater, 367 apartments, and parking for 4000 cars. This new working environment was just what local businesses were looking for: convenient parking; easy access to downtown shopping, restaurants, and entertainment; and large office floors equipped to provide all the needs of modern business machines. Where there once had been 12,000 jobs, there were now 17,000. Real estate taxes quadrupled.

Baltimore, 2011. Old and new buildings lining the Charles Street edge of the Charles Center Urban Renewal Project. *(Alexander Garvin)*

More important, Charles Center was accomplished without causing undue hardship to the site tenants that were displaced. Although 44 establishments went out of business, "93 percent of the businesses in Charles Center were relocated without liquidating. Of those that moved, 80 percent were relocated within the city of Baltimore and only 3 percent left the metropolitan area."[26]

By demonstrating that there was a market for new office space but satisfying only a portion of that demand, Charles Center triggered even more investment outside the renewal area. Encouraged by the initial success of Charles Center, further government redevelopment projects focused on shifting development activity to the harbor. As a result, Charles Center is no longer the centerpiece of downtown Baltimore. Continuing changes in retailing as well as changes to the city itself have required alterations to Charles Center. Hamburger's department store was unable to compete successfully with national chains. Its building was demolished and replaced in 2001 with Johns Hopkins School of Professional Studies in Business and Education.

Unlike Hamburgers, the new building does not bridge over Fayette Street. In fact, most of the system of walkways and overpasses has been removed, without much impact on pedestrian or vehicular circulation, demonstrating the wisdom of having retained the streets and sidewalks that had been in place prior to redevelopment. It also demonstrated that separating pedestrian and vehicular traffic, however easy to do on a sloping site, had never been necessary. More important, because not enough office space was built within the redevelopment project, private developers erected additional office buildings east of Charles Center. Thus, Charles Center, like the Golden Triangle in Pittsburgh, demonstrated that when market demand is frustrated by an obsolete or inappropriate physical plant, intelligently conceived and managed reconstruction can revive a deteriorating business district. However, by concentrating later redevelopment around the city's waterfront, Baltimore shifted most downtown business activity to the Inner Harbor.

The redevelopment of Baltimore's declining industrial waterfront originated in the late 1950s. The Greater Baltimore Committee, Inc., decided to continue its urban renewal efforts, then successfully under way at Charles Center two blocks to the north. While, this time, the strategy called for complete reconstruction, except for the waterfront promenade and the Port Authority Headquarters, each element of its Inner Harbor Plan either could not get voter approval or could not obtain financing. Nevertheless, the city proceeded to acquire the necessary 95 acres.[27]

Inner Harbor languished for more than a decade until James Rouse (chairman of the Greater Baltimore Committee,

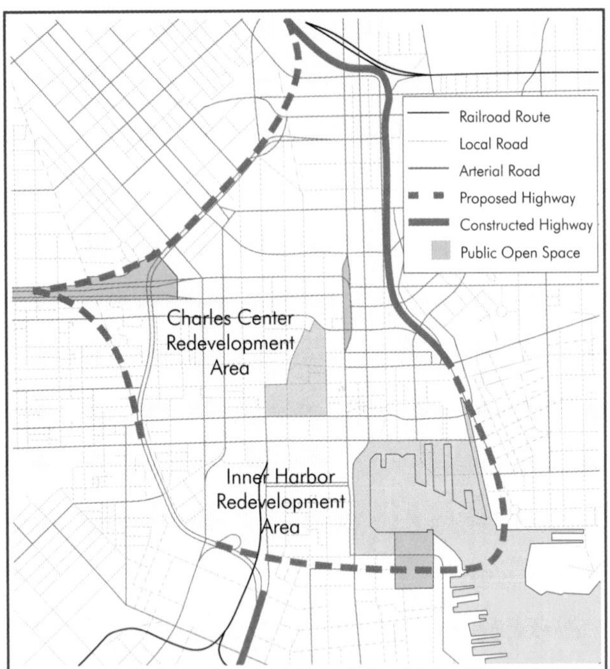

Baltimore. Like New Haven, Baltimore proposed a highway encircling downtown that failed to be approved. *(Alexander Garvin and Ryan Salvatore)*

Inc., in 1965, when it published its initial waterfront plans) proposed to provide Inner Harbor with an urban marketplace like the one he had just created at Quincy Market. At that time, the plan for Inner Harbor included a new convention center, 9600 parking spaces, a 500-room Hyatt Regency Hotel, a major aquarium, and a marina. In 1978, Rouse negotiated the long-term rental of a 3.2-acre site for Harborplace, his proposed urban marketplace. This improved version of Quincy Market was just what was needed to transform the Baltimore waterfront into the major downtown tourist center envisioned by the plan.[28]

The Washington-Baltimore suburbs were one of the fastest-growing markets in the nation. The boats in Baltimore harbor had always been a regional attraction. The new waterfront promenade, with its decorative paving, benches, and streetlights, along with the aquarium, old ships anchored in the harbor, and new parking facilities, made the waterfront attractive, accessible, safe, and convenient for that regional market. Adding an urban marketplace with its many commercial attractions would provide the excitement and gaiety needed to bring that enormous market to Inner Harbor. The adjacent convention center and hotel meant Harborplace would profit from another major market: out-of-town tourists. Moreover, because Inner Harbor had been carefully planned to fit together with Charles Center and the rest of downtown Baltimore, Harborplace would also profit from a third market: daytime office workers.

Harborplace is based almost entirely on its Boston predecessor. Rouse again called on Benjamin Thompson and Associates, but in Baltimore they transformed the opaque masonry of Quincy Market into two glass pavilions surrounded by covered porches and terraces right along harbor.[29] The larger of its two pavilions, like the central structure at Quincy Market, includes a food court offering a wide variety of ethnic, prepared, and fresh foods from restaurants, fast-food counters, market stalls, delicacy shops, and pushcarts. The smaller pavilion is for specialty stores and more formal restaurants. Together, the two pavilions provide 142,000 square feet of gross leasable area for 142 merchants, including 49 eating places, 20 food stores, 36 specialty shops, 2 florists, and 35 pushcarts and kiosk vendors. By night, the brightly lighted glass façades provide a sparkling enticement for outsiders to come in, join the fun, and (naturally) spend money. By day, the glass expanses also open up the view to the waterfront for the customers inside doing their shopping. The roll-up exterior doors allow direct contact with outside activities when the weather permits. Porches and terraces provide additional places from which to enjoy the harbor.

During 1980, its first year of operation, Harborplace attracted 18 million visitors. Annual sales per square foot exceeded those of Quincy Market, which had had four years to build its clientele. The city gained $3 million in new real estate taxes, 2500 new jobs (one-third held by previously unemployed Baltimore residents), and 6 million tourists from outside the greater Baltimore area. Rouse went on to create festival marketplaces in Miami, New York, Richmond, and other cities. But none of these other projects has transformed the city around them in the same dramatic way.

Baltimore, 2011. New buildings lining Pratt Street across the street from the Inner Harbor Redevelopment. *(Alexander Garvin)*

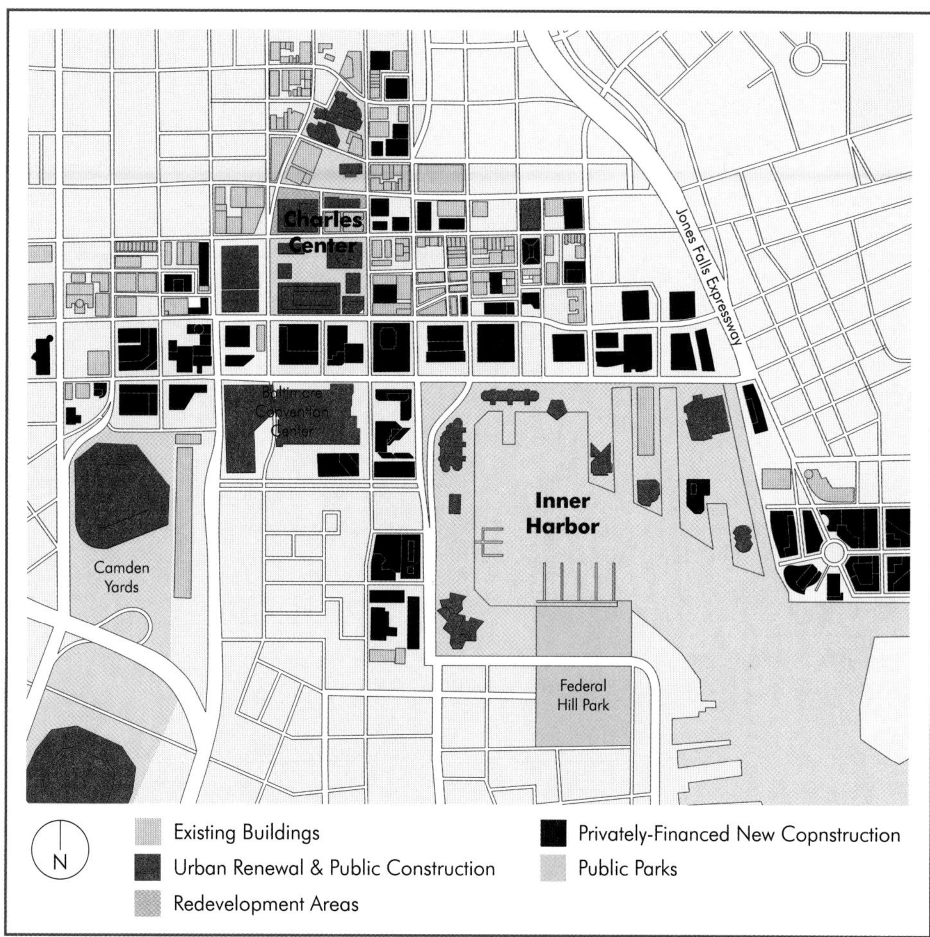

Baltimore. In Baltimore, unlike New Haven, public investment produced a substantial private market reaction. *(Alexander Garvin and Ryan Salvatore)*

One reason for the initial success of retail shopping at Inner Harbor was that a year earlier the 425,000-squarefoot Baltimore Convention Center had opened, bringing tens of thousands of tourists downtown. It was followed by the National Aquarium in 1981, an IMAX theater in 1987, the Oriole Park baseball stadium at Camden Yards in 1992, an 800,000-square-foot addition to the Convention Center in 1997, the M&T Bank football stadium in 1998, and an expansion of the aquarium in 2005. Together, these palaces for the people provide an avalanche of customers. Their activity spills over across Pratt Street, which has become the commercial heart of twenty-first-century Baltimore. This is one of the most obvious examples of public investment generating a widespread and sustained private market reaction.

San Diego

During the 1960s, downtown San Diego experienced the same difficulties that Hartford, New Haven, and Baltimore had in the 1950s: loss of business to the suburbs, decreasing retail sales, and general deterioration. The city's long-established red-light district, which attracted sailors from the busy harbor and carousing vagrants who enjoyed the balmy weather much of the year, extended farther and farther into the central business district. In an effort to reverse these trends, the city embarked on a series of public projects that generated billions of dollars of private investment. Unlike those cities, which were all shrinking in population, San Diego quadrupled in size between 1950 and 2010, reaching a population of 1,307,000 in 2010. And, unlike those cities, San Diego's redevelopment strategy was not based on the City of Tomorrow.[30]

The redevelopment effort began in 1969 as a 3-block plan to improve the area around Horton Plaza Park. Public pressure successfully forced expansion of the project to the 15-block area approved by the city council in 1972. Three years later, as a result of a nationally advertised competition, Ernest W. Hahn Inc., a California-based realtor responsible for dozens of shopping centers, was designated the developer.

It took Hahn and the city seven years to put the project together. Ultimately, the city provided $39 million for property acquisition, relocation, infrastructure improvements, and development of two theater facilities within the redevelopment area. The rest came from the private sector: $85 million in permanent financing, $15 million in equity

San Diego, 2007. Horton Plaza has become a nine-block shopping complex with 140 tenants, a seven-screen multiplex cinema, a legitimate theater, a nightclub, 19 eating establishments, and a constantly changing number of pushcart and kiosk vendors.

from Hahn, and $40 million from the department stores that would anchor the project.

When Horton Plaza finally emerged in 1985, it was a nine-block, 885,000-square-foot shopping complex with 140 tenants, including four department stores, a seven-plex cinema, a legitimate theater, a nightclub, 19 eating establishments, and a constantly changing number of pushcart and kiosk vendors. Two historic structures, the old Balboa Theater and the Spreckels Building, were incorporated into the building along with two reconstructed façades of Italianate buildings that had been on the site.

The design by the Jerde Partnership bears no resemblance to anything Le Corbusier envisioned. In fact, its only relationship to earlier redevelopment projects was a belief in the need to replace an obsolete capital plant. Horton Plaza consists of 10 clusters of shops, each characterized by a different color scheme and architectural image, each open to San Diego's benign climate, and each on a different level and tied together by a series of stairs, ramps, escalators, and elevators that cross the S-shaped central open space. The combination is a confusing and colorful array of merchandise displayed in pastel-tinted Mediterranean piazzas, neon-clad Victorian arcades, stucco-and-tile Indian pueblos, multicolored Gothic palazzi, and ornamented, postmodern emporia—a circus atmosphere that is amplified by mimes, jugglers, bands, aerobics teams, and the like. It is also one of the country's most effective examples of sequential merchandising. Ramps and walkways force customers to pass all sorts of spending opportunities on their way to any destination.

Horton Plaza was commercially successful because it supplied a cleverly designed, unique attraction with plentiful merchandise for a huge market: 60,000 to 100,000 downtown office workers, 310,000 in-town residents, 1.9 million fashion-oriented metropolitan shoppers, and millions of tourists and conventioneers. New hotels, office buildings, retail projects, and apartment houses continue to be built

San Diego, 2009. The city's redevelopment projects continue to stimulate widespread private market activity (shown in white on the model) throughout downtown San Diego. *(Centre City Development Corporation, San Diego)*

on surrounding blocks. Adjacent older buildings are being renovated. In the nearby historic Gaslamp Quarter, upscale restaurants and other, more conventional, tourist-oriented retailing have replaced porn shops and cheap hotels. Thus, Horton Plaza helped to make downtown San Diego a lively attraction virtually 24 hours a day, 365 days a year.

Like Baltimore, San Diego did not stop investing downtown with one urban renewal project. In 1989 it opened a 616,000-square-foot convention center. One block away, it erected 42,000-seat Petco Park, home of the San Diego Padres baseball team. The San Diego Centre City Development Corporation (CCCD) spearheaded numerous investments in parks and street improvements and assisted developers in creating thousands of downtown apartments. The CCDC was the principal force behind these public actions from its inception in 1975 until 2012, when, pursuant to a state law eliminating all government development corporations, it ceased to exist. At that time, this vibrant 1500-acre business district included 7000 businesses, nearly 75,000 workers, and 35,000 residents in six distinctive residential districts.

Since 1975, San Diego has made more than $1.5 billion in downtown investments that generated more than $13 billion private development and $50 million annually in sales and transient occupancy taxes. There are few more effective examples of downtown redevelopment in the United States.

Transfiguration and Death

There had been opposition to urban renewal from the beginning. Because of opposition from relocatees and their advocates, Congress amended the Housing Act in 1959 to require relocation plans for every renewal project. It kept increasing relocation benefits until the Uniform Relocation Act of 1970 finally pushed costs to a level that severely restricted anything but the most selective slum clearance.

Civic organizations and community groups often argued for indigenous solutions and questioned the effectiveness of replicating similar project designs for every situation. Congress reinforced this criticism when it amended the Housing Act in 1959 to guarantee the participation of site tenants in the planning process. The Act now required that each renewal project have a Citizen Project Advisory Committee. As a result, when Edward Logue left New Haven to take over Boston's redevelopment program he established project offices in every community slated for renewal. The staff in these offices worked with the designated Citizen Project Advisory Committee and with local businesses, settlement houses, and every other interest group in the area. The proposals that emerged had little to do with any vision of utopia. Even so, they usually sparked opposition and were often altered to meet neighborhood political reality prior to any formal approval.[31]

Finally, on January 5, 1973, President Richard Nixon unilaterally declared a moratorium on all new housing and renewal projects. He was able to do so because the projects had become too costly and the opponents too powerful. It would be convenient to hold Nixon accountable for the death of urban renewal. However, when he declared his moratorium, urban renewal itself had become a dying thing.

Until the Nixon moratorium, every city had to compete for its share of national urban renewal funding. As a result, the projects kept coming. Congress ended the competition for urban renewal grants and guaranteed each locality its fair share of federal funds when, one week after Nixon left office, it passed the Housing and Community Development Act of 1974 (see Chapter 12).

The new act created block grants that could be used for a variety of purposes, not exclusively redevelopment. Everybody assumed that new and less expensive urban renewal projects would be forthcoming. They were not. America had lost faith in government-subsidized redevelopment. Besides, property owners across the country were rebuilding city districts without renewal assistance. Local governments, noticing this unassisted private redevelopment, chose to spend their block grants on all possible other purposes. The effort to rebuild cities section by section was over. San Diego and a few other cities initiated redevelopment projects like Horton Plaza. The limited size of the sites reduced the required relocation and, along with it, the opposition. These redevelopment projects were, however, exceptional cases that involved only a few blocks rather than ambitious efforts to rebuild entire downtown districts.

Ingredients of Success

What critics of urban renewal fail to understand is that the problems faced by cities like Pittsburgh or New Haven, where population and employment opportunities were shrinking, were very different from those of cities like San Diego or Houston, which had growing populations and expanding economies. Critics who point to continuing deterioration are usually referring to those parts of town where redevelopment either did not take place or was inappropriate from the moment it was conceived. Moreover, the fate of cities like New Haven and Hartford could well have been worse had there been no redevelopment.

In some places, urban renewal worked, but its critics were so vocal that they drew attention away from its successes. Projects like Pittsburgh's Golden Triangle and Baltimore's Inner Harbor demonstrated that by removing portions of a city that were impediments to a prosperous economy, an entire city could be transformed. They also demonstrated that replacing whole cities section by section was not only politically and financially impossible, it was also unnecessary. Le Corbusier had been mistaken. The city itself was not a "dying thing."

During the later twentieth century, cities as diverse as Seattle, Atlanta, and New York experienced increases in population. Downtown employment was increasing in San Diego, Los Angeles, Miami, and many older cities. Although suburbanization of these cities had not abated, it was accompanied by substantial immigration (especially from Asia and Latin America) and in-migration of empty nesters and singles in their 20s and 30s. In cities like Detroit, St. Louis, and Hartford, however, urban shrinkage continued.

Market

Physical reconstruction is not a solution to urban shrinkage. It may make city districts safer or more convenient. But, as the redevelopment of Church Street in New Haven demonstrates, it cannot provide retail customers, especially when competitors supply rapidly growing suburbs with more convenient alternatives.

Where consumer demand is strong, redevelopment can provide cities with the means to retain their customers. The Golden Triangle, for example, allowed Pittsburgh to supply the business community with the physical environment it demanded and otherwise would have had to seek elsewhere. Because the project satisfied the needs of a only few large corporations, once reconstruction of the Triangle was under way, developers sought sites for additional new buildings that were needed by other firms and thereby triggered the Pittsburgh Renaissance. Similarly, Horton Plaza capitalized on its rapidly growing metropolitan area by providing a different kind of shopping environment than was available in the suburbs.

Location

State laws usually restrict redevelopment to blighted areas. Consequently, projects are automatically saddled with locations where the market is not vigorous. This can be overcome by selecting blighted sites in the midst of otherwise healthy districts. Charles Center, for example, could tap demand from contiguous downtown shopping, government, and financial districts. At the same time its occupants spilled over to increase market demand in those districts. Conversely, Hartford's government and shopping districts, although nearby, were just far enough away to prevent Constitution Plaza from having the same synergistic impact on the rest of downtown. A blighted location on the edge of a downtown business district may be just what is needed for successful residential redevelopment. That is why Horton Plaza could benefit from all the residential building going up in downtown San Diego.

In most cities, however, the locations chosen for redevelopment were already occupied by hundreds of people and businesses that had to be relocated. During the 1950s, relocation expenses were low enough that they did not affect project feasibility. When pressure from citizen groups and their elected politicians forced Congress to adopt the Uniform Relocation Act of 1970, such inner-city redevelopment became feasible only where, as at Charles Center, it did not involve complete clearance and retained economically healthy businesses or, as in downtown Pittsburgh, it removed land uses that had serious negative impact on surrounding properties.

Design

A complete City of Tomorrow remains a vision that can be experienced only in Le Corbusier's drawings. Since it was never brought to fruition, it is a mistake to place the entire blame for the failures of specific redevelopment projects on his image of the future. Too often, the designers of these projects copied his brave new world without fully understanding what he had in mind.

Their most serious error was to assume that a new city could be created one superblock at a time. The cultural and consumer activities needed by any successful city require a critical mass. That critical mass cannot be mathematically divided and apportioned to individual projects. Thus, when redevelopment consists of an independent project rather than a network of interconnected projects, neither the project nor the surrounding city can benefit from that redevelopment. Constitution Plaza, set apart from the rest of downtown Hartford, is such a project. Charles Center and Inner Harbor, on the other hand, were conceived as an integral part of downtown Baltimore. Together with the surrounding busy blocks, it attracts a sufficient concentration of people to support the widest array of land uses.

It is also a misunderstanding to equate spacious public open spaces and shiny glass towers with urban renewal. There is nothing inherently wrong with glass skyscrapers. Designed by fine architects, they can rival the greatest of architectural monuments. In the hands of those who designed most of the high-rise boxes of our redevelopment projects, they are as uninspired as the somewhat smaller boxes of their suburban competitors.

Replacing old buildings within redevelopment areas with new buildings, whether well designed or not, is not the key to economic or social revival. The redevelopment of Church Street replaced hundreds of "obsolete" structures with brand-new buildings. Nevertheless, New Haven continued to decline. On the other hand, Charles Center successfully revived downtown Baltimore precisely because it integrated old buildings with the new. Horton Plaza introduced a cacophony of styles that recalled civilizations Le Corbusier had thought of as long dead.

Large, open areas are no more inherently incompatible with an exciting urban environment than new buildings. But when, as at Constitution Plaza, they take the form of the vast emptiness of many redevelopment projects, they deaden rather than revitalize the city around them. The purported appeal of light, air, and dramatic vistas are not enough. Urban areas need open spaces that attract people and the activities that they bring with them. At Horton Plaza and Inner Harbor they were supplied by stadiums and a convention center that were just outside the redevelopment project, while at Constitution Plaza they were built at street level, just far enough away unconnected to the platform level where people entered the project's buildings.

The success of a redevelopment project depends more on the arrangement of its components than on generous open space or new buildings. The parking garage in the Church Street Redevelopment Project, for example, was located so that shoppers could go directly into a department store without setting foot anywhere else in the project or in downtown New Haven. As a result, New Haven did not profit much from the customers attracted to Church Street. At Charles Center, on the other hand, underground parking is scattered throughout the project. This results in mutually reinforcing interaction among the project's retail customers, theatergoers, residents, and office workers. Because people coming and going from other destinations in downtown Baltimore also use these garages, they increase the synergistic relationship between Charles Center and the rest of the city.

Financing

By transferring to government agencies the cost of carrying a project (interest on the money used for site acquisition, real estate tax payments prior to resale to the ultimate developer, professional fees, and site development costs), Title I greatly reduced the risk of failure and, in effect, provided financing when it was least likely to be forthcoming from the private sector.

Ordinarily, development sites are sold at a price that reflects the "highest and best use" of the property. In redevelopment projects, sites are sold at a lower price that reflects the reuse specified in the redevelopment plan in the interest of providing for the city something that the private market is not otherwise incented to provide. Inevitably, the actual costs of acquiring, carrying, and preparing the property for this reuse will exceed even that lower price. By subsidizing the difference between the gross project cost and a resale price that is justified by this planned reuse, government makes that reuse financially feasible.

Even carrying the cost of project development and subsidizing reuse price is no guarantee of success. Early projects like Constitution Plaza and Church Street had difficulty obtaining long-term financing because all the buildings had to be financed as a single project and thus required extremely large sums of money. Charles Center overcame this difficulty by subdividing the redevelopment area into smaller parcels that required smaller mortgages and lower equity investments.

Time

Most redevelopment projects were conceived as mechanisms to provide modern office space in a park-like setting rather than as places in which people spend their waking hours. At Constitution Plaza, visitors parked their cars under the project and climbed stairs to a vast platform, which they crossed on their way to an office building. Shopping center developers would have used the walk as an opportunity to attract them to stores, restaurants, and perhaps even movie theaters.

Early planners of the urban renewal program thought little about what would occur over a 24-hour period, 365 days a year. As a result, Constitution Plaza and many other downtown redevelopment projects are empty at night and on weekends. Consequently, they have a deadening effect on the surrounding city. Later projects considered their impact throughout an entire week. Charles Center, for example, included apartment buildings, a department store (that was eventually demolished), restaurants, and even a major theater because its planners hoped these facilities would keep downtown Baltimore alive after 6:00 p.m. and on weekends. Inner Harbor went further by providing an aquarium, restaurants, and a place to rent boats or take boat tours and provided much of the rationale for other public investments in the convention center and two stadiums.

The greatest inadequacy of the renewal program, however, was that it failed to consider the impact of the lengthy periods of time required for planning, project approval, relocation, clearance, construction, and marketing. Mayor Lee nearly lost an election because New Haven businesses had lost the customers that had once occupied the Church Street Renewal Area and because citizens objected to years of maintaining an unproductive hole in the ground.

Entrepreneurship

Just as financing has to be provided both for the site and for development, so does entrepreneurship. Most early redevelopment projects had neither the public entrepreneurs to ram through their plans nor private entrepreneurs with the ability to package financing, design, and construction. Lee and Logue in New Haven were particularly effective public entrepreneurs when it came to obtaining local political support and federal subsidies for their redevelopment schemes. But even they ran into trouble with Church Street because Roger Stevens, the private developer to whom they had entrusted project development, was not effective enough in putting together the package.

The planners of Charles Center, who had the benefit of seeing this problem in New Haven and other cities, understood that market demand and a suitable site for development, by themselves, did not automatically result in project completion. Without skilled public and private entrepreneurs, "the project itself could fail, and the market growth could express itself elsewhere within the metropolitan area."[32] Their solution to this problem was to offer important parcels for development by public competition. Each developer/architect team had to demonstrate its ability to design, finance, and build its proposal. Other sites were sold directly to users who had the necessary financial strength to develop them.

The City of Tomorrow as a Planning Strategy

Today it is clear that America's cities were not terminally ill. Some of them were losing a part of their population, employment base, and retail sales to surrounding suburbs. In cities with decreasing populations, downtown redevelopment did not reverse this decline. Nevertheless, where there was steady or increasing demand for downtown business locations, as in Baltimore or San Diego, redevelopment did simplify, accelerate, and reduce the cost of supplying sites that could satisfy the demand for new office buildings, retail facilities, and hotel and convention business.

Where functional obsolescence is a problem, redevelopment projects can help to reverse downtown decline. New highways can provide access for thousands of trucks that supply the myriad of goods and services needed in any modern city. Multilevel garages can provide parking for tens of thousands of office workers and downtown shoppers, without whom the modern city would have little or no economic base. The most important part, however, is determining when functional obsolescence is actually the problem and not simply a convenient scapegoat.

Whether such redevelopment requires federal subsidies is not at all clear. In cities with and without downtown renewal projects, private developers continued to assemble sites for large, new office buildings. Highways and garages, not government-created and subsidized redevelopment parcels, were the critical elements in retaining existing businesses and attracting the growing market for new office space.

From the start, urban redevelopment has been a tantalizing mirage. There never was any possibility of clearing America's great cities and rebuilding them in Le Corbusier's image, nor of squeezing his utopian vision into the scaled-down projects that met Title I requirements. Instead, the federal urban renewal program became a mechanism for meeting the needs of the politicians, municipal officials, downtown business interests, and developers who wished to alter the character of their cities.

Politicians saw redevelopment, often correctly, as a mechanism for retaining a competitive position within a growing but increasingly suburbanized metropolitan region. Certainly, that was what happened in Pittsburgh, Baltimore, and San Diego. Besides, it provided patronage jobs. Politicians supported the program as long as the voters remained in favor of redevelopment.

Municipal officials advocated redevelopment because it generated tremendous amounts of federal money, often at little or no cost to the city government. The city government, for example, was able to recoup its investment in Constitution Plaza in less than one year. New Haven was so adept at manipulating its noncash contributions that redevelopment probably resulted in a cash surplus (see Chapter 12).

Downtown business interests in cities like San Diego favored redevelopment because it subsidized improvements to their competitive position. The program allowed government to move businesses, land uses, and people that interfered with their market. It also provided these businesses with new locations and modern buildings that might not otherwise have been affordable.

Many developers who participated in the program found it very profitable. They did not have to waste their time assembling property because they could obtain sites that were already approved for development. The price they paid for these sites did not reflect their true cost. Prices were artificially set at a level that made development attractive. More important, developers did not have to spend a penny until the necessary government approvals were in place. Their competitors still had to struggle with conventional real estate development.

Eventually, opponents of redevelopment made the program extremely difficult and expensive to execute. Environmental protection and historic preservation statutes precluded indiscriminate use of the power of eminent domain. Relocation requirements greatly increased project costs. Most important, by making redevelopment one among many permitted uses of the Community Development Block Grant, Congress forced its proponents to compete with other demands for federal subsidies.

While redevelopment remains a viable prescription for some situations, further legislation is not needed. When and where cities wish to replace a functionally obsolete physical plant, the necessary enabling legislation is already in place. Projects like Baltimore's Inner Harbor and San Diego's Horton Plaza can serve as models for future development.

Only a handful of cities will have good reasons to choose redevelopment. The experience, both good and bad, with Title I will stand them in good stead. The rest should avoid it, because there are very few situations in which government-subsidized redevelopment is the proper mechanism for fixing the American city.

Notes

1. This chapter is dedicated to Vincent Scully, in whose class, more than 50 years ago, I first was presented with the comparison of Le Corbusier's City of Tomorrow and America's urban renewal program.
2. Title I of the Housing Act of 1949 subsidized 2532 projects in 992 cities. In addition, 429 cities (virtually all of which also received Title I money) participated in the NDP program enacted in 1968. These two programs distributed $12,680,880,000 in federal renewal funds. *Source:* U.S. Department of Housing and Urban Development, Program Completion Division.
3. Le Corbusier, *The City of Tomorrow, and Its Planning* (first published in Paris in 1924), MIT Press, Cambridge, MA, 1971, pp. 171, 280.
4. Ibid., p. 167.
5. Le Corbusier, *The Radiant City* (first published in 1933), The Orion Press, New York City, 1967, pp. 162–163.
6. Ibid., p. 71.
7. Mel Scott, *American City Planning Since 1890,* University of California Press, Berkeley and Los Angeles, 1969, p. 361.
8. *Berman v. Parker,* 348 U.S. 26 (1954).
9. Susette Kelo, et al., *Petitioners, v. City of New London, Connecticut, et al.,* 545 U.S. 469 (2005).
10. Patrick McGeehan, "Pfizer to Leave City That Won Land-Use Case," *New York Times,* November 13, 2009.
11. Historical and statistical information on the redevelopment of the Golden Triangle is derived from Roy Lubove, *Twentieth Century Pittsburgh,* John Wiley & Sons, New York, 1969, pp. 106–141; Robert C. Alberts, *The Shaping of the Point,* University of Pittsburgh Press, Pittsburgh, PA, 1980; Julie Percha, "Move over, Honolulu; Pittsburgh's No. 1 in U.S.," *Pittsburgh Post-Gazette,* February 22, 2011, http://www.post-gazette.com/pg/11053/1127102–53.stm; and Francesca Levy, "America's Most Livable Cities,“ Forbes.com, April 29, 2010, http://www.forbes.com/2010/04/29/cities-livable-pittsburgh-lifestyle-real-estate-top-ten-jobs-crime-income.html.
12. Jon C. Teaford, *The Rough Road to Renaissance,* Johns Hopkins University Press, Baltimore, 1990.
13. In urban renewal projects subsidized by the Housing Act of 1949, *net project cost* (the difference between *gross project cost* [acquisition, relocation, demolition, site preparation, and required infrastructure and community facilities] and the subsidized *resale price* needed to make the planned construction financially attractive to private developers) is divided between the federal and local governments. The federal government pays two-thirds. In many states, the local share is split 50–50 between state and city. This local contribution did not have to be made in cash. A city could get a *noncash credit* for the cost of infrastructure and community facilities. Since most urban renewal took place in areas where infrastructure and community facilities were obsolete and required replacement, the expenditures eventually would have been made anyway. So the noncash credits effectively reduce local cost still further.
14. Title I restricted eligibility to sites that were "predominantly" residential prior to or after redevelopment. Cities whose proposed downtown projects did not meet either test applied under a 10 percent exception provision. Congress expanded eligibility when it amended the Housing Act in 1954.
15. Maurice Rotival, "*Report to the City Plan Commission,*" Part III, section B3a, New Haven, 1941.
16. Mayor Richard C. Lee, quoted in Robert A. Dahl, *Who Governs?,* Yale University Press, New Haven, CT, 1961, p. 120
17. Historical and statistical information on the redevelopment of Church and Oak streets is derived from Alan Talbot, *The Mayor's Game,* Harper & Row, New York, 1967; Dahl, *op. cit.*; Raymond Wolfinger, *The Politics of Progress,* Prentice-Hall, Inc., Englewood Cliffs, NJ, 1974, pp. 298–356; L. Thomas Appleby, interview, November 21, 1988; Jeanne Lowe, *Cities in a Race with Time,* Random House, New York, 1967; *Redevelopment Oak Street Area,* New Haven, November 4, 1955, revised March 1959; *Church Street Redevelopment and Renewal Plan,* New Haven, September 3, 1957, revised through

May 27, 1964; *New Haven Redevelopment Agency Annual Report,* New Haven, 1967; *CRP New Haven Community Renewal Program,* New Haven, March 29, 1968; *Ninth Square Project Executive Summary,* New Haven, November 1, 1993; and numerous discussions with A. Tappan Wilder.

18. Dwight Street Redevelopment Plan, New Haven, 1958.
19. Homer Hoyt Associates, Market Survey of Stores in the Church Street Project, Washington, DC, 1958.
20. U.S. Census Bureau, 2007 Economic Census.
21. U.S. Department of Commerce, Bureau of the Census, *State and Metropolitan Area Data Book 1991,* U.S. Government Printing Office, Washington, DC, 1991, p. 192.
22. Historical and statistical information on the redevelopment of Constitution Plaza is derived from Joel R. Fredericks, *The Front-Market Urban Renewal Project: An Evaluation,* unpublished, 1975; and James Armstrong, *Constitution Plaza: A Development Analysis,* unpublished, 1978.
23. Historical and statistical information on the redevelopment of Charles Center is derived from Martin Millspaugh (editor), *Baltimore's Charles Center: A Case Study in Downtown Development,* Urban Land Institute, Washington, DC, 1964; Baltimore Department of Housing and Community Development, *Prospectus, Charles Center Development Area 16B,* Baltimore Department of Housing and Community Development, Baltimore, 1968; John Morris Dixon, "Charles Center," *Forum,* New York, May 1969, pp. 48–57; and Michael Halle, "Charles Center," unpublished, 1977.
24. The buildings that were retained were the Lord Baltimore Hotel, the Fidelity Building, the Baltimore Gas & Electric Building, the Baltimore & Ohio Railroad Building, and the Eglin Parking Garage.
25. J. Jefferson Miller, "Management, Land Acquisition, Relocation," in Millspaugh (editor), op. cit., p. 40.
26. Ibid., p. 43.
27. Historical and statistical information on Harborplace is derived from Greater Baltimore Committee, Inc., and the Committee for Downtown, Inc., *The Inner Harbor and City Hall Plaza,* Baltimore, 1965; Douglas M. Wrenn, *Urban Waterfront Development,* Urban Land Institute, Washington, DC, 1983, pp. 146–155; and Alexander Garvin, *The Planning Game: Lessons from Great Cities,* W. W. Norton & Co., New York, 2013, pp. 54–58.
28. The first proposals for the Inner Harbor included an east-west expressway cutting across the harbor, a civic center organized around an open-air mall connecting the harbor with city hall, a new headquarters for the port authority, an east-west "minirailway," a theater/museum "playground," new apartment towers, and a waterfront promenade. See Garvin, *The Planning Game,* pp. 20–21
29. A more detailed discussion of Quincy Market can be found in Chapter 4.
30. Historical and statistical information on Horton Plaza is derived from *Urban Land Institute Project Reference File,* vol. 16, no. 19, October–December 1986, Urban Land Institute,Washington, DC; Bernard Frieden and Lynne Sagalyn, *Downtown, Inc.: How America Rebuilds Cities,* MIT Press, Cambridge, MA, 1989, pp. 123–131, 145–153, 191–197; Centre City Development Corporation, *Downtown Today,* Winter/Spring 2010, Issue 6; and http://downtownsandiego.org/.
31. Langley Carleton Keyes Jr., *The Rehabilitation Planning Game,* MIT Press, Cambridge, MA, 1969.
32. Dennis Durden, "The Feasibility Study," in Millspaugh (editor), op. cit., p. 32.

8

Downtown Management

Denver, 2012 *(Alexander Garvin)*

By the 1980s, business and civic leaders began to believe that physical improvements by themselves often were not enough to make downtown districts competitive. Decorative paving, handsome street furniture, and effective lighting did not remove litter and graffiti or discourage pickpockets and shoplifters. Business districts needed good management if they were to compete with shopping malls that were owned, managed, and marketed as a single unit. The solution that evolved was the *business improvement district* (BID), a nonprofit organization formed by property owners and businesses who banded together to provide the services they could not afford on an individual basis but needed if they were to compete successfully with the suburbs.

In Philadelphia, New York, Denver, and dozens of other cities, BIDs have provided the additional security, sanitation, and promotional services needed to restore confidence in formerly declining business districts. The services they supplied (often in combination with the same brick paving, street furniture, and other physical improvements typical of demotorized streets) created attractive, pedestrian-friendly precincts that enticed customers, thereby demonstrating that entrepreneurial management plays as important a role as intelligent design or cost-effective redevelopment in making downtown more competitive.

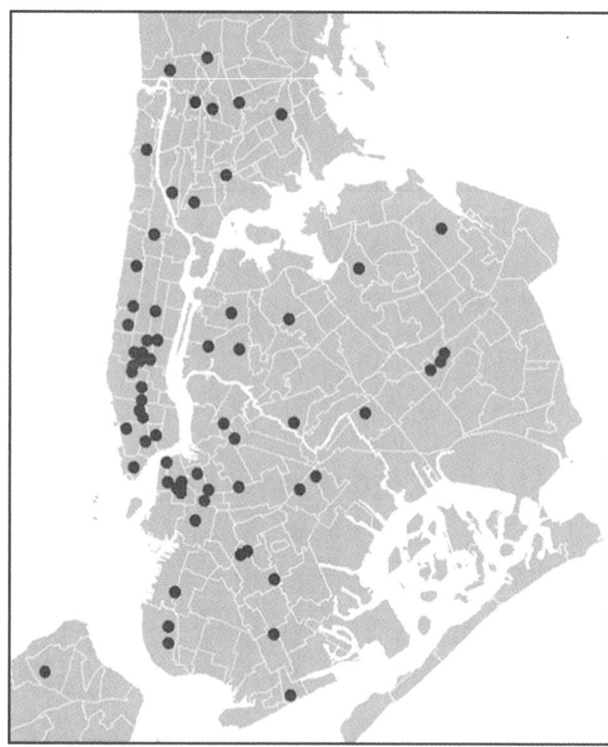

New York City, 2013. Map of the city's 67 business improvement districts. *(Courtesy of New York City Department of Small Business Services)*

Business Improvement Districts

America's better suburban shopping malls are clean, safe, and attractive. It is easy to shop and get around in them. The climate is controlled, the lighting is pleasant, and signage is easy to understand. There usually are fountains, artwork, flowers, benches and restrooms, a variety of places to eat, and interesting products to purchase. Mothers pushing strollers, senior citizens, and young people like to spend time there. Too few city destinations are as customer-friendly.

The public officials in the nineteenth and twentieth centuries who widened arteries, installed railroads and transit, and invested in other infrastructure improvements correctly expected the private sector to follow. But they did not anticipate the problems that accompanied the high-density development that followed. The developers of the 1950s and 1960s who demanded automobile-friendly districts with garages that were easily accessible by interstate highway mistakenly thought that these public investments guaranteed they would overcome suburban competition. The planners who created the automobile-free streets of the 1960s and 1970s erroneously thought that physical improvements were all that was needed to create a successful pedestrian precinct. Removing the noise and pollution caused by motor vehicles is not enough; neither is installing decorative paving, attractive street furniture, and beautiful planting. To some degree, all of them were right, but retaining existing businesses and attracting new ones requires more than an auspicious physical environment. It also requires a minimum level of public services (street sweeping, garbage collection, traffic control, security services, etc.).

During the second half of the twentieth century, downtown business interests were increasingly at a disadvantage in competing successfully with other recipients of government funds. Local agencies were required by state and national governments to make increasingly large mandated expenditures for activities that had never before been the province of local government. They were also under pressure from neighborhood residents to provide a higher level of local service.

The easy solution was to increase spending in neighborhoods that legitimately claimed they had been neglected and reduce allocations to commercial districts that were inhabited by fewer voters. The usual result was a business district with increasing amounts of uncollected litter, overflowing trash receptacles, broken lighting fixtures, and less police protection.

Business leaders began working together because they were losing customers to safe, clean, attractive, suburban shopping centers that were controlled by a single developer-owner-manager-marketer who routinely provided the services that had been curtailed in business districts: regular trash collection, street cleaning, and security services. Some merchants associations had engaged in common marketing activities (e.g., late-night-opening agreements, Christmas decorations, and newspaper advertising), but they did not have the right to provide an entire district with security guards, sidewalk sweepers, trash removal, and information officers, and they could not individually afford to do so. In 1970, Bloor West Village in Toronto, invented a device to deal with this situation: the business improvement district (BID).[1]

A BID is a membership organization established and operated pursuant to an agreement with the city government. Its activities are paid for through a surcharge on real estate taxes, collected by the city government and spent by a board of directors consisting of property owners, businesses, and site occupants from the district. It took a few years for BIDs to become common. By 2012, there were more than 1400 BIDs in the United States and Canada, 67 in New York City alone. More notable BIDs include Denver's 16th Street Management District, New York City's Bryant Park Corporation, Philadelphia's Center City District, and the Downtown Stamford Special Services District (DSSD). Denver's BID provides standard services to a 120-block area surrounding its pedestrian mall. The Bryant Park Restoration Corporation began as an effort to provide a safe environment to a crime-saturated 6-acre downtown park and evolved into a multi-million-dollar agency that redesigned the park and provides a wider variety of recreational and entertainment opportunities. Philadelphia's BID provides virtually all the services needed by the entire central business district. The DSSD in Stamford, Connecticut, has pioneered promotional and programming activities that bring people downtown nights and weekends.[2]

Denver's 16th Street Management District

Civic leaders in Denver understood from the very beginning that the 16th Street Pedestrian Mall would require management and maintenance. They began in 1978 with an amendment to the city charter creating a special mall benefit district to extend one block on either side of the mall. Downtown Denver Partnership Inc. (a nonprofit business organization created in 1955 to promote and market downtown Denver) provided management and maintenance services from the opening of the mall in 1982. In 1984, the mall was expanded to cover 865 properties within a 70-block area. When the benefit district's 10-year charter expired in 1992, it was replaced by a business improvement district covering 120 blocks. Its seven-member board of directors (representing a range of property-owner categories) is appointed by the mayor and approved by the city council.[3]

Management and maintenance services, which continued to be provided by Downtown Denver Partnership Inc., include daily sidewalk sweeping; trash, snow, and graffiti removal; paying for additional services from the Denver Police Department; operating a variety of promotional and marketing programs; and installing and maintaining street furniture, planting, and lighting. The money for the BID's activities comes from annual tax assessments on privately owned commercial property. Assessments decrease with their distance from the 16th Street Mall, with its free municipal bus service. In 2000, the budget for the BID was $2.3 million.

When the BID was created, Denver was just beginning to emerge from a recession and downturn in the energy business. So it is difficult to ascribe the downtown improvements entirely to the BID or the opening of Coors Field (see Chapter

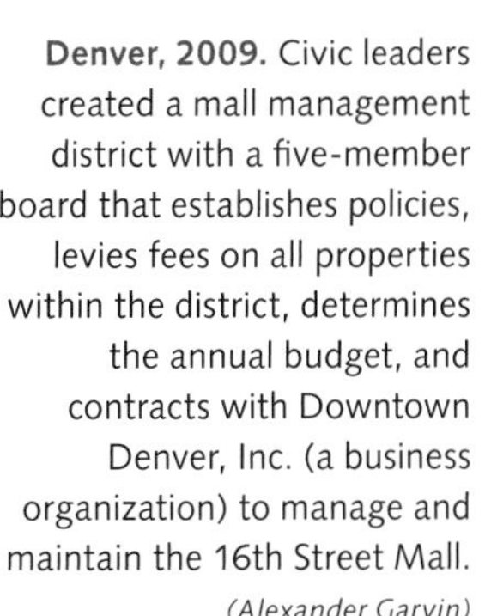

Denver, 2009. Civic leaders created a mall management district with a five-member board that establishes policies, levies fees on all properties within the district, determines the annual budget, and contracts with Downtown Denver, Inc. (a business organization) to manage and maintain the 16th Street Mall. *(Alexander Garvin)*

Denver, 2009. Virtually every attraction in downtown Denver is within walking distance of the 16th Street Mall. *(Alexander Garvin)*

5) or Denver Pavilions (see Chapter 6). However, the change has been dramatic. Downtown employment in 2000 is more than 25 percent greater than it was when the pedestrian mall opened. Between 1990 and 2000, 2433 new apartments were built downtown. Housing construction and residential renovations are now a continuous process. As a result in 2012, more than 10,600 households lived downtown. Downtown Denver used to close down at the end of the workday. But over the life of the BID, dozens of empty buildings and parking lots disappeared and have become shops, restaurants, office buildings, and residences. During 2012, downtown Denver was filled with more than 13 million tourists who spent more than $3 billion, as well as swinging singles in shorts and sandals, couples pushing baby strollers, business-suited office workers carrying briefcases, and teenagers on their way to Niketown, Hard Rock Café, or the multiplex movie theater. The annual tax collection from more than 1000 downtown retail sales and restaurants increased from $10.5 million in 1992 to $34 million in 2011.

New York City's Bryant Park BID

Bryant Park is a 6-acre island of green behind the New York Public Library on Fifth Avenue at 42nd Street. The original formal design had been updated in the 1930s, during Robert Moses's tenure as Commissioner of Parks. Material from the construction of the Sixth Avenue subway was used as fill to raise the level of the park 4 feet above grade. The low shrubs that were planted grew in and blocked views into the park. Consequently, neither passing pedestrians nor passing police patrols could see much of what was happening in the park.[4]

When Moses stepped down as commissioner in 1960, the parks department entered a period of continuing budget cuts. Its only possible response was to defer maintenance. Laurie Olin, the designer responsible for Bryant Park's later relandscaping, described the situation:

> *Trees overgrown—ground beaten bare—trash overflowing the waste cans, stuffed into the long-abandoned light boxes—lights broken off and missing—pavement not repaired—hedges allowed to grow up to hide the ugly lights, themselves neglected and ugly.*[5]

Bryant Park became a dangerous place. In 1973, the parks department began to close park entrances with wooden police barricades every night at 9 p.m. Three years later, the park's first murder occurred. There was a second murder in 1977. The police recorded 900 crimes in this 6-acre park during 1979. Finally, Andrew Heiskell, chairman of the New York Public Library, which bounds Bryant Park, hired Daniel Biederman, a recent graduate of Wharton, to spearhead the park's reclamation. Biederman asked William H. Whyte to analyze the situation. Whyte's report explained that the park was "dominated by dope dealers" and was therefore underused. His solution: remove iron fences and shrubbery, facilitate pedestrian circulation by providing additional entry points, and relandscape the entire park.[6]

Bryant Park's redesign and renovation is largely the work of Hanna/Olin (overall landscape design), Hardy, Holtzman, Pfeiffer (new structures), and Lynden B. Miller (plantings). The installation of a café, a restaurant, and several snack kiosks was as responsible as redesign and relandscaping for the project's success. The businesses in these new structures attracted people who would otherwise have had no reason to be in the park and transformed it into a safe haven visible to pedestrians, taxi drivers, and anybody passing by.

The project was financed through the creation of a business improvement district surrounding Bryant Park in 1984. At that time, the Department of Parks was spending $210,000 a year on management and maintenance. The city agreed to collect an additional annual real estate tax payment, equal to about 11 cents per square foot of floor area, from the owners of the 7.5 million square feet of prop-

Manhattan, 2009. Bryant Park is filled with people even on a Sunday afternoon when surrounding office buildings are closed. *(Alexander Garvin)*

erty surrounding Bryant Park (the payment amounted to 14 cents per square foot in 1999). That tax surcharge was passed through to the Bryant Park Corporation, which had been established to operate the BID. The money was used initially for the physical improvements that resulted in the park's successful reclamation. Thereafter, the tax surcharge was combined with other funds to pay for two shifts of security officers, two shifts of sanitation workers, regular planting, and all aspects of management and maintenance. By 2012, when the Bryant Park Corporation's budget was $7.9 million, only $1.1 million (14 percent) came from the BID. The rest was paid for with concession revenue, rent, sponsorships, and park usage fees. The Department of Parks had not spent any of its scarce resources on Bryant Park in almost two decades.

In the summer, people lie in the center lawn sunning themselves; in the winter, they ice-skate there on the temporary rink. Summertime strollers enjoy the shade of the sycamore trees that line the park; in November and December, they go shopping at the handicraft stalls that are temporarily installed under the trees. Adults play chess; children ride the carousel; office workers use the lending library or the Internet hot spot; and everybody can play Ping-Pong, arrange movable chairs so they can talk to one another or get some privacy, get a sandwich or a cup of coffee, or just snooze. On a typical winter day you can go skating; ride a carousel; watch a display of juggling; play chess, Ping-Pong, or petanque; dine in a restaurant or café, or in front of a kiosk; attend a concert; or just stroll until you find a chair in a place you feel like sitting.

No crimes were reported in Bryant Park during 2011. There were too many people in the park. The reason is not the permanent presence of a security guard. It is that it is impossible to conceal anything from the more than 2000 people in the park on a nice summer afternoon.

Manhattan, 2010. Christmastime ice-skating in Bryant Park. *(Alexander Garvin)*

Manhattan, 2011. The Bryant Park Corporation sets up small shops, which it rents to artisans and retailers for the Christmas season. *(Alexander Garvin)*

Center City District, Philadelphia

Rather than concentrate on a single shopping street like Denver's 16th Street, or a small park like Bryant Park, however important they may be, Philadelphia has directed its efforts to increasing the competitiveness of the entire downtown as a regional, national, and international business center. The strategy was to create a BID that would enhance center city as an arts and culture, shopping, dining, and entertainment district and reinforce the district as a prime residential location. This BID, Center City District (CCD), was established in 1990. Four years later it was reapproved by property owners and the City of Philadelphia through the year 2015. CCD's objective, like that of the BIDs in Denver and New York City, was a "*clean, safe, attractive and well-managed* public environment."[7]

As of 2012, the Center City District includes virtually all of downtown Philadelphia: 233 blocks, 9900 firms, a 265,000-person workforce, 40 million square feet of office space, 38 hotels, 3200 retail shops, 713 restaurants, and 273 outdoor cafés. It functions as a quasi-independent municipal authority overseen by a 23-member board of directors representing downtown property owners, businesses, institutions, and residents. Paul Levy, the energetic executive director who has managed the CCD from the start, put together an ambitious and very effective combination of programs that focus on streetscape enhancement, cleaning and maintenance, safety, crime prevention, marketing, communications, planning, and administration.

The CCD is financed through a 5 percent mandatory surcharge on real estate taxes. Its $19.4 million 2012 budget consisted of a wide array of public services plus debt service on a $21 million bond issue devoted to improving the pedestrian environment. More than 64 uniformed workers operate mechanical sidewalk sweepers at least three times a day, seven days a week. They also remove posters, stickers, stains, gum, and grime, and they remove or paint over graffiti on sidewalks, street furniture, and the ground floor of building façades. Forty-two uniformed community service representatives act as goodwill ambassadors, providing information and directions to visitors, workers, and residents. They also offer outreach services to the homeless population and work in partnership with social service agencies to help people in need obtain relevant services. The CCD provides enhanced police coverage through a police substation (open 15 hours a day, 7 days a week) that combines police officers from two precincts with supplementary uniformed on-street patrols. That effort requires public and private professionals to work with property owners, managers, and employers to decrease overall vulnerability to crime.

The record on capital improvements is no less impressive. Over the past 15 years, the CCD has invested $68.7 million in the public environment; another $50 million is currently under construction. Streetscape improvements include 1000 pedestrian-scale lampposts, 400 new trees, and improved store façades, lighting, and window displays. The CCD maintains an ever-changing array of banners on light poles and posters in transit shelters. However, the most effective change has been the installation of directional signs and maps that provide visitors with information on how to get to every important downtown destination. It even promoted the remotorization of the unsuccessfully pedestrianized Chestnut Street.

The cumulative effect of this 22-year operation has been

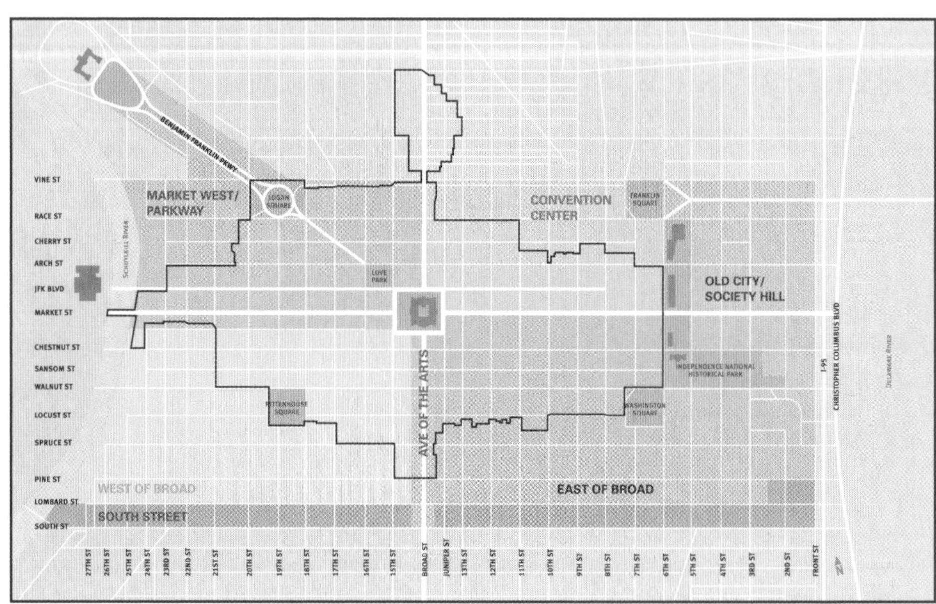

Philadelphia, 2013. Center City District. *(Central City District and Central Philadelphia Development Corporation)*

Philadelphia, 2001. Street sweepers keep sidewalks in the business improvement district free of litter. *(Alexander Garvin)*

remarkable. To be sure, its activities have been accompanied by the opening of a convention center (see Chapter 5) and major capital investments. But neither of these could be responsible for the decline in serious and quality-of-life crimes: decreasing from 20 per day in 1993 to 11.7 per day in 2011, a drop of 42 percent.

Philadelphia, 2010. The staff of the Center City District keeps downtown Philadelphia safe, clean, attractive, and convenient for everyday use. *(Alexander Garvin)*

When the BID was established, much of the Center City was empty after 5 p.m.; in 2012, it was alive with activity generated by thriving hospitality, convention, and tourist industries. The number of downtown hotel rooms has increased by 80 percent, serving more than 29 million tourists a year. The revival of in-town living is equally impressive: 12,946 new residential units have been built since 1997. The unmistakable conclusion of this extraordinary record has to be that under Levy's leadership the CCD has demonstrated that a properly operated BID will indeed produce a "*clean, safe, attractive and well-managed* public environment."

Stamford, Connecticut

Like downtown Philadelphia two years earlier, in 1992, when Stamford established its business improvement district, downtown Stamford was "perceived as dangerous and dirty, with few pedestrians after 5 pm, where retail store window-grates rolled down at dusk."[8] Like Downtown Denver, the DSSD took responsibility for daily street cleaning, trash and snow removal, and street security. Like Philadelphia's CCD, it arranged seasonal planting of gardens, street planters, small parks, and median dividers. In some ways, it has taken programming events even beyond the Bryant Park BID.

Every year since 1994, the DSSD has presented outdoor sculpture exhibits. In 2000, the artwork consisted of a multicolored Cow Parade; in 2012, it consisted of multicolored horses displayed throughout downtown. The DSSD organizes Restaurant Weeks, one of which, in February 2012, was responsible for 14,725 meals on Bedford Street alone. It sponsors Jazz Up July, which presents concerts to hundreds of spectators in Columbus Park and patrons of outdoor cafés across the street. On Thursday nights during the summer, it sponsors Alive@Five Stage, which brings about 7000 people

Stamford, Connecticut, 2012. The events that are programmed by the DCCD, like the Jazz Up performances, attract hundreds of customers downtown. *(Alexander Garvin)*

Stamford, Connecticut, 2012. People attending the various events sponsored by the DCCD usually stay on for lunch or dinner in one of the growing number of downtown restaurants. *(Alexander Garvin)*

downtown. In November, it sponsors the nation's second largest helium balloon extravaganza: the UBS Parade Spectacular. In 2011, these activities attracted approximately 360,000 people to DSSD events.[9] (See Table 8.1.)

This combination of active programming, stewardship of the public realm, and targeted retail assistance has paid off. During DSSD's first 20 years of operation, the amount of downtown office space has increased by 25 percent; retail occupancy has climbed from 78 percent to 91 percent; the number of downtown residences has tripled; and, as a result of increased property values, the city government now collects $15 million more in taxes from downtown commercial properties. Thus, in addition to increased employment, reduced crime, and an overall improvement in the city's quality of life, all of Stamford is receiving the benefit of increased tax revenues that pay for additional municipal services outside the business district.

Table 8.1

1992	Downtown Stamford	2012
2	major hotels	4
40	restaurants + clubs	85
7	movie screens	18
0	signature events	12
2,011	residential units	5,955
0	universities	1
6 million	office square footage	7.5 million
$328 billion	property assessment	$1.9 billion

Ingredients of Success

City residents and their local representatives always say, "There is no money." That is not correct. The Philadelphia City Budget for 2013 is $3.6 billion; New York City's is $68.5 billion. What *is* true is that competition for that money gets more intense every year. Consequently, local governments increasingly turn to BIDs to augment and even substitute for public-sector efforts.

The activities of these BIDs tend to be successful because they are conceived and administered by the businesses that pay for and benefit from them. The degree of that success, however, is a reflection of the same ingredients of success that characterize any planning initiative.

Market

There were huge available markets in the areas serviced by the BIDs discussed in this chapter. The reason the BIDs came into existence was to divert spending away from other areas and bring as much of the market as possible downtown. Denver began the process by making transit and other public investments that improved downtown access and circulation. The planning for the Bryant Park BID led the way in demonstrating how to greatly reduce downtown crime—something picked up in Denver, Philadelphia, and Stamford.

Philadelphia's mantra of "safe, clean, and attractive" has been adopted throughout the country. Its leaders will tell you that they would not have been as successful without stressing *attractive.* Tourists just won't come in significant numbers to unsightly places. Bryant Park and downtown Stamford, on the other hand, might place more emphasis on programming as a way of attracting customers who might otherwise go elsewhere.

Location

BIDs start with a huge locational advantage. They are right in the middle of their market area. But these BIDs were invariably created because customers found the site unattractive and sometimes were even repelled by conditions at the site. Thus, one of a BID's primary tasks is to transform a terrible site at a great location into a wonderful place at that very same location. That is the reason for the huge successes of Bryant Park Corporation and Stamford's DSSD.

Design

Consumer-oriented design is essential in any BID. Philadelphia's CCD has spent billions of dollars on decorative paving, improved lighting, way-finding signs, street trees, and decorative planting. This transformed a compact business district contained between the Delaware and Schuylkill rivers into a downtown that is as pedestrian-friendly as it gets.

Financing

Some people are amazed that during a period in which the United States has experienced a widespread tax revolt, people agreed to pay an additional tax. The explanation is easy. Downtown businesses wanted services that government was not supplying or was supplying in inadequate quantities. Their owners also understood that they were at a numerical disadvantage when it came to voting for those services. Consequently, they were prepared to tax themselves for the services they weren't getting from government.

The more interesting financial question is why that tax surcharge took on less importance over time. In 2000, the Bryant Park Restoration Corporation covered 31 percent of its annual budget from the real estate tax surcharge. The rest came from restaurant and concession rent, sponsorships, grants, and fees charged for commercial usage of the park. In 2010, the tax covered only 12 percent of the corporation's expenditures; it had been able to expand the other sources of revenue. Moreover, as a BID succeeds, the value of surrounding real estate increases. Thus, over the 20 years that the DSSD has existed, property assessments in downtown Stamford have increased by a factor of close to six, from $328 million to nearly $2 billion.

Time

Unlike many planning agencies, BIDs are deeply concerned with time. The Downtown Denver Partnership took an active role in getting the Denver Pavilions entertainment center located downtown (see Chapter 5), because this facility would bring in people at night and on weekends, when most office workers had gone home. They also bring entertainers to Sixteenth Street during happy hour for the same reason. Summertime movies on Monday nights in Bryant Park bring an average audience of 6000 people. On opening night of 1997, *The Wizard of Oz* attracted double that amount. Stamford's summer entertainment achieves similar results.

Entrepreneurship

Downtown Denver began and has continued its revival efforts with remarkable street and transit improvements. Many of its other activities emerged over time in support of those initial investments. In New York, Daniel Biederman was hired in 1979 to spearhead Bryant Park's reclamation at a time when BIDs were just emerging as a device with which to pay for security and sanitation along downtown shopping streets. In those days, very few public parks were managed by nonprofit entities. Biederman provided the inspiration to overcome skeptics, hired the designers and contractors whose specific proposals often conflicted with one another, and coordinated the many participants without whom the complex project would never have come to fruition. Paul Levy performed this role for Center City Philadelphia, as did Sandra (Sandy) Goldstein in Stamford. Without their energetic entrepreneurship, little would have happened.

Inspired entrepreneurship is not easy to come by. Bryant Park Corporation, Philadelphia's CCD, and Stamford's DSSD are lucky. Their admirable current condition is the result of the entrepreneurship of remarkable people who conceived, obtained approval, and implemented their activities.

Managing Downtown as a Planning Strategy

The strategies discussed in Chapters 6 and 7 (infrastructure and public realm investment, redevelopment and pedestrianization) all began with plans, often prepared by professional architects and planners. When it comes to BIDs, as well as most planning, President Dwight Eisenhower was right: "Plans are worthless, but planning is everything."[10] Plans that change nothing *are* worthless. The planning that went into the four BIDs discussed in this chapter, on the other hand, transformed those cities.

That planning was directed by three people who were not trained as planners. Biederman is a product of business school; Levy went into education; Goldstein was educated to be a schoolteacher. They conceived of the strategies that could change their cities; then they obtained the necessary approvals as well as the money to pay for them, and, finally, they took the actions needed to implement those activities.

All three concentrated on those activities that would improve the competitiveness of their respective districts. As a result, the businesses that paid for those activities reaped huge benefits, as did the people who worked in those businesses and the consumers who patronized them. Surrounding property values increased and, with them, the city's tax revenues, which paid for services elsewhere in town. No planner could hope for a more widespread and sustained private market impact.

Notes

1. Howard Kozloff and James Schoder, "Business Improvement Districts (BIDS): changing the Faces of Cities," *The Next American City 11* (Summer 2006), americancity.org/magazine/article/web-exclusive-kozloff/.
2. Statistical and historical information on BIDs is derived from Carol Jean Becker, Seth A. Grossman, and Brenda Dos Santos, *Business Improvement Districts: Census and National Survey,* International Downtown Association, and ULI-The Urban Land Institute, Washington DC, 2011.
3. Statistical and historical information on Denver's BID is derived from Lawrence O. Houstoun Jr., *Business Improvement Districts,* Urban Land Institute, Washington, DC, April 1997, pp. 150–163. Promotional materials distributed by Downtown Denver Inc., www.downtowndenver.com/bid_serv.htm; Renate Robey, *The Denver Post,* September 5, 1999; Cindy Christensen, Director of Economic Development, The Downtown Denver Partnership, April 2001; and Downtown Denver Partnership, *The State of Downtown Denver,* Denver, September 2012.
4. Statistical and historical information on Bryant Park is derived from Alexander Garvin and Gayle Berens, *Urban Parks and Open Space,* ULI—The Urban Land Institute, Washington, DC, 1997, pp. 23, 44–57; J. William Thompson, *The Rebirth of New York City's Bryant Park,* Spacemaker Press, Washington, DC, and Cambridge, MA, 1997; Bryant Park Restoration Corporation Cash Basis Operating Statements for the years ending June 30, 1999 and 2000; and interviews with Daniel Biederman, June 1, 1999 and January 2, 2001.
5. Laurie Olin, quoted in Thompson, op cit., p. 8.
6. For a discussion of Whyte's principles, see William H. Whyte, *The Social Life of Small Urban Spaces,* The Conservation Foundation, New York, 1980.
7. Statistical and historical information on Philadelphia's Center City District is derived from Houstoun, op. cit., pp. 133–149; *Center City Digest,* Center City District, Philadelphia, Spring 1996, Summer 2000, and January 2001; *State of Center City 1991–2001,* Center City District, Philadelphia, 2000; *Making It Center City Philadelphia,* Center City District, Philadelphia, undated; *Center City Streetscape,* Center City District, Philadelphia, undated; *Center City Design,* Center City District, Philadelphia, undated; Linda K. Harris, "Over a Decade, Downtown Finds New Life," *Philadelphia Inquirer,* March 20, 2001; and Center City District & Central Philadelphia Development Corporation, *State of Center City Philadelphia 2012,* Center City District & Central Philadelphia Development Corporation, Philadelphia, 2012.
8. Stamford Downtown Special Services District, *Stamford Downtown Twenty Years,* Stamford, 2012, p. 2.
9. Statistical and historical information on Stamford's Downtown Special Service District is derived from Stamford Downtown Special Services District, *Stamford Downtown Twenty Years,* Stamford, 2012; Stamford Downtown Special Services District, *2012 Work Plan,* Stamford, 2012; and interview with Sandy Goldstein, president of the Stamford Downtown Special Services District, July 11, 2012.
10. President Dwight D. Eisenhower, speech to the National Executive Reserve Conference (Washington, DC, November 14, 1957.

9

Increasing the Housing Supply

San Francisco Bay, 1997. Housing construction. *(Alexander Garvin)*

Housing doesn't just happen. Somebody has to build it, take care of it, and pay for it. Few people do all three. Fewer still have enough money to pay for their residence when they purchase it. They usually borrow money to purchase it and give back a mortgage to the lender, promising to pay interest on the loan and reimburse the lender what they have borrowed within a set period of time. If the borrower fails to do either, the mortgage entitles the lender to foreclose, taking back the property. The process of increasing the housing supply therefore requires a lender willing to provide money, a developer interested in building a house, and a future building occupant who is willing to borrow money. In the case of a residence within a multiunit apartment building, it also requires somebody who will take care of the building, especially those parts of the building that are not individual residences (hallways, stair, elevators, and its grounds). In a well-functioning market economy the demand for decent shelter is matched by enough residences supplied at a price that its customers are willing and able to pay. Sometimes a market failure occurs and government steps in to correct the situation. At the federal level, public intervention into the housing market dates back to the Great Depression, when banks were foreclosing mortgages at an alarming rate. By the twenty-first century, that intervention had become the primary influence on the mortgage market—sometimes sensible and sometimes not; sometimes with desirable results and sometimes with not.[1]

When major public action to increase the supply of housing began in 1933, one-half the home mortgages in the country were in default, annual housing production had dropped below 93,000 units, and mortgage lending had come to a virtual halt. If housing production was to be restored even to pre-Depression levels, mortgage financing had to be made easily available to both developers and consumers. This required a stable, orderly, easily accessible mortgage market.

Congress began the process of creating a stable supply of mortgage money by enacting legislation that insured bank deposits, thereby giving depositors the confidence they needed to keep their money in the bank. It went on to assure home buyers and builders that they could obtain this money from lending institutions by insuring mortgages that met standard lending practices. It also created a secondary market for federally insured mortgages, allowing financial institutions that needed cash to sell standard mortgages to those with enough surplus cash to buy them.

In addition to ensuring that financing would be available, Congress greatly expanded the market for additional housing by offering federally insured mortgages that had substantially reduced requirements for down payments. By extending the term of these mortgages, it also reduced the amount of the monthly debt-service payment on that mortgage. As a result, millions of individuals and families could afford to own a house.

Federal deposit insurance provided a reason for people to keep money in the bank and, thus, a stable supply of money that banks could lend. Banks were induced by greatly reduced risk to lend money for housing purposes because they were protected from losing more than one-tenth of the money. Businesspeople, who would not otherwise have entered the home-building industry, were induced to reenter the housing market because so many consumers could afford to purchase a house, paying for it with only 10 percent in cash and 90 percent from an FHA-insured bank mortgage. Moreover, because houses had preapproved federal mortgage insurance commitments, developers had an easier time borrowing money with which to build those houses and get into the home-building business. Thus, banks, developers, and homeowners alike were given access to the money they needed.

By creating an easily accessible mortgage market, Congress transformed millions of people who dreamed of homeownership into customers with the ability to pay for new houses, and thousands of struggling businesspeople became developers with an ability to supply them with the houses they desired. By 1941, national housing production had climbed to 619,000 units, more than six times the 1933 level.[2]

The real impact of this legislation, however, became evident after World War II. In response to burgeoning demand, housing production soared. By 1960 there were 53 million dwelling units in the United States, up 66 percent from 1940. Homeownership had increased from 44 to 62 percent.[3]

The reform in lending practices initially applied to one- to four-family houses, not to apartment buildings. Consequently, this extraordinary increase in the housing supply occurred largely outside dense urban areas. By 2010, single-family houses represented 67 percent of the nation's housing stock.[4] Had a similar approach been adopted for financing multifamily housing, millions of apartments would have been created and the shortage of housing in center cities would not have become so serious.

Providing a Stable Supply of Mortgage Money

The creation of a stable mortgage money market began with laws that were intended to increase depositor confidence in financial institutions. The Federal Home Loan Bank Act of 1932 established the Federal Home Loan Bank Board, with regulatory powers over savings and loan institutions similar to the powers that the Federal Reserve System had over commercial banks. The Home Owners Loan Act of 1933 established the Home Owners Loan Corporation (HOLC) to refinance home mortgages in default or foreclosure. The

Glass-Steagall Banking Act of 1933 created the Federal Deposit Insurance Corporation (FDIC), which eliminated the risk of depositing funds in participating banks. The Banking Act of 1934 did the same for thrift institutions through the creation of the Federal Savings and Loan Insurance Corporation (FSLIC).[5]

Without these actions, depositors would have withdrawn all their money, leaving most financial institutions without sufficient capital. Instead, depositors maintained savings accounts that provided the money that banks used to refinance home mortgages when they came due. In its first three years of operation, for example, the HOLC provided more than $3 billion to refinance over 1 million mortgages, impacting 10 percent of all nonfarm, owner-occupied residences in the United States.[6]

FHA-Insured Home Mortgages

The new banking laws helped financial institutions to attract deposits. However, these institutions needed to be coaxed into investing a major portion of that capital in housing. Congress provided the necessary inducement by enacting the National Housing Act of 1934, which created the Federal Housing Administration (FHA). Section 203 of this act created a mortgage insurance system that, for a small premium charge, provided participating lenders with insurance on 90 percent of the appraised value of one- to four-family houses.[7] When a bank foreclosed on a mortgage, it could transfer the mortgage to the FHA and in exchange obtain most of the money it had lent (90 percent of the appraised value of the building). By covering so large a part of the downside risk, Congress made home loans a safe investment.[8]

By reducing the down payment on a home mortgage to 10 percent, Congress dramatically increased the number of people who had the cash to make a down payment on a house. By requiring the mortgage to be fully self-amortizing, it eliminated the risk of facing a hostile mortgage market when the loan came due. By extending the amortization over a period of up to 35 years, it lowered monthly debt-service payments and increased the number of people who could afford a home mortgage. Thus, the most important effect of this legislation was that it converted the desire for homeownership into consumer demand.[9]

Construction lenders could depend on the eventual sale of a house that met FHA specifications because the purchaser could depend on an FHA mortgage. Consequently, banks decreased the amount of developer equity required for construction financing, thereby dramatically increasing the number of entrepreneurs who had the equity capital with which to enter the home-building industry. No housing program has been more successful in increasing housing supply. Between 1934 and 2012, the Federal Department of Housing and Urban Development (HUD) and FHA insured more than 34 million home mortgages.[10]

Initial Principal:	$100,000			
Annual Interest Rate	6%			
Term (Years)	25			
Payment Amount	$7,823			

Year	Payment	Interest Paid	Principal Paid	Principal Due After Payment
1	-$7,823	$6,000	$1,823	$98,177
2	-$7,823	$5,891	$1,932	$96,245
3	-$7,823	$5,775	$2,048	$94,197
4	-$7,823	$5,652	$2,171	$92,026
5	-$7,823	$5,522	$2,301	$89,725
6	-$7,823	$5,384	$2,439	$87,286
7	-$7,823	$5,237	$2,586	$84,701
8	-$7,823	$5,082	$2,741	$81,960
9	-$7,823	$4,918	$2,905	$79,055
10	-$7,823	$4,743	$3,079	$75,975
11	-$7,823	$4,559	$3,264	$72,711
12	-$7,823	$4,363	$3,460	$69,251
13	-$7,823	$4,155	$3,668	$65,584
14	-$7,823	$3,935	$3,888	$61,696
15	-$7,823	$3,702	$4,121	$57,575
16	-$7,823	$3,455	$4,368	$53,207
17	-$7,823	$3,192	$4,630	$48,577
18	-$7,823	$2,915	$4,908	$43,669
19	-$7,823	$2,620	$5,203	$38,466
20	-$7,823	$2,308	$5,515	$32,951
21	-$7,823	$1,977	$5,846	$27,106
22	-$7,823	$1,626	$6,196	$20,909
23	-$7,823	$1,255	$6,568	$14,341
24	-$7,823	$860	$6,962	$7,379
25	-$7,823	$443	$7,379	$0
TOTAL:	-$195,567	$95,566	$100,000	

25-Year Amortization Schedule. The annual debt-service payment on a 6 percent, self-amortizing, $100,000, 25-year mortgage is $7823. *(Alexander Garvin and Ryan Salvatore)*

Initial Principal:	$100,000			
Annual Interest Rate	6%			
Term (Years)	10			
Payment Amount	$13,587			

Year	Payment	Interest Paid	Principal Paid	Principal Due After Payment
1	-$13,587	$6,000	$7,587	$92,413
2	-$13,587	$5,545	$8,042	$84,371
3	-$13,587	$5,062	$8,525	$75,846
4	-$13,587	$4,551	$9,036	$66,810
5	-$13,587	$4,009	$9,578	$57,232
6	-$13,587	$3,434	$10,153	$47,079
7	-$13,587	$2,825	$10,762	$36,317
8	-$13,587	$2,179	$11,408	$24,909
9	-$13,587	$1,495	$12,092	$12,817
10	-$13,587	$769	$12,817	$0
TOTAL:	$135,869	$35,868	$100,000	

10-Year Amortization Schedule. The annual debt-service payment on a 6 percent, self-amortizing, $100,000, 10-year mortgage is $13,587. *(Alexander Garvin and Ryan Salvatore)*

VA-Guaranteed Home Mortgages

Congress adopted a similar approach for veterans' housing. It was eager to help GIs avoid the economic and social problems of post–World War II readjustment. In particular, it wished to supply millions of returning members of the armed forces with the credit necessary to obtain a home, business, or farm. This assistance took the form of the Serviceman's Readjustment Act of 1944, which established the Veterans Administration (VA) guaranteed loan.

The act authorized the VA to guarantee veterans' loans with no down payment or insurance premium requirement. This virtually assured lenders that they would not lose money. The federal government guaranteed 60 percent of the value of a VA mortgage loan. This guarantee was the equivalent of 60 percent borrower equity. In case of foreclosure, the bank needed to recoup only 40 percent of the amount of the loan from the sale of the foreclosed property. The federal government covered the rest.

Although VA home mortgage loans were essentially risk-free, lenders continued to require a nominal cash payment by the borrower (often less than 5 percent). Millions of veterans jumped at the chance of putting a few dollars down in order to own their home. As of 2006, more than 18 million dwelling units had received VA loans.[11]

Standardized Mortgages for a National Market

Prior to the Depression, mortgage instruments were quite different from what we know today. A first mortgage seldom covered more than 50 or 60 percent of any transaction. Since borrowers usually wished to reduce their equity investment, they frequently obtained additional second and third mortgage loans with higher interest rates that reflected the greater risk of these auxiliary loans. These were short-term loans, usually lasting less than three years. Unpaid principal was due in a single *balloon payment* when the loan came due. At that time, property owners would have to refinance their mortgage and pay the expenses involved with each new loan. They were at the mercy of new market conditions in which they might not be able to refinance the balloon mortgage and thus could lose the property through foreclosure. Mortgage

Chicago suburbs, 2000. Homes in Hoffman Estates have the benefit of FHA-insured home mortgages.
(Alexander Garvin)

instruments and loan requirements varied. One lender might require up-to-date electrical wiring; another might be satisfied with minimum code compliance. Even the language of mortgages varied. By requiring every mortgage to comply with its printed regulations, the FHA provided standardization that revolutionized the American housing industry.

All buildings with FHA-insured and VA-guaranteed mortgages had to comply with FHA *minimum property standards* regarding location, neighborhood conditions, subdivision design, structure, room size, quality of materials, mechanical equipment, even sewage disposal. Thus, both borrower and lender could depend on the quality of the product and were protected from subsequent, unexpected hazards.

Each FHA-insured mortgage was essentially the same, no matter which bank extended the loan, and no matter in which state it originated. It was a first mortgage that precluded additional mortgage liens, carried a fixed rate of interest, required the same debt-service payments every month, and was fully self-amortizing. Without this uniformity in lending practices, there could not have been mass generation of mortgage loans, mass production of new houses, or mass consumption of those houses.

Fannie Mae

The legislation that created the FHA also authorized private individuals to establish mortgage associations that could borrow money from the public for the purpose of purchasing and reselling FHA mortgages. It was thought that standard FHA-insured mortgages would be sold by financial institutions in areas with high demand for mortgages to institutions in other areas with surplus capital seeking safe, predictable sources of income. By selling these mortgages and making new ones with the money they received, lenders would be able to earn more than their assets would have ordinarily allowed.

A national mortgage market did not develop. Banks were comfortable underwriting mortgages in their own area but ignorant of market characteristics elsewhere. Furthermore, they were skeptical about tying down their funds for very long periods. Their needs might change. There might not be a buyer for the mortgages when they wanted to sell.

To correct this situation, in 1938 Congress established the Federal National Mortgage Association (FNMA, or Fannie Mae). Fannie Mae's job was to buy FHA mortgages from participating institutions in need of additional mortgage capital and in turn sell them to others with surplus capital. These lending institutions used the proceeds of the sale of these FHA-insured mortgages to Fannie Mae to fund new mortgages, constantly replenishing the pool of funds available for residential lending. In 1972, Fannie Mae extended its role to purchasing uninsured mortgages whose size was limited to prices that were affordable to low-, moderate-, and middle-income home buyers. As of mid-2012, Fannie Mae held a mortgage portfolio valued at more than than $675 billion.[12]

Most of the initial capital with which Fannie Mae purchased home mortgages came from the sale of its stock to the Reconstruction Finance Corporation and the U.S. Treasury. The balance was raised by requiring institutions whose FHA mortgages were bought by Fannie Mae to purchase a small amount of its stock. The income it paid on its securities came from the interest on the mortgages and the commitment fees it charged participating institutions. In 1968, Fannie Mae was rechartered as a private shareholder-owned company. Two years later, its stock began to be traded on the New York Stock Exchange.[13]

Freddie Mac

In 1970, Congress established the Federal Home Loan Mortgage Corporation (Freddie Mac), as a stockholder-owned corporation. Freddie Mac operates like any corporation obtaining capital from stock and bond markets, except it uses that capital exclusively to finance residential mortgages. Its function is to purchase mortgages from lenders (commercial banks, savings institutions, credit unions, etc.), package them as securities, and sell those securities (guaranteed by Freddie Mac) to investors. As had been the case with Fannie Mae, Congress established Freddie Mac to make available to residential lenders a continuous, reliable, low-cost source of mortgage capital. The difference is that the mortgages Freddie Mac purchases are not FHA-insured. At the end of fiscal year 2011, it had a total loan portfolio of more than $1.7 trillion.[14] During its first 30 years of operation, Freddie Mac provided financing for one-sixth of the nation's homeowners, and in the last decade alone it has helped finance mortgages for over 35 million American homeowners.[15]

Ginnie Mae

When Congress passed the Housing and Urban Development Act of 1968, it rechartered Fannie Mae as a private corporation that dealt with conventional mortgages and established the Government National Mortgage Association (Ginnie Mae) as an agency within HUD to guarantee FHA, VA, and Rural Housing Service (RHS) mortgage loans. The guarantee is backed by the full faith and credit of the U.S. government. As of 2012, Ginnie Mae had guaranteed more than 40 million home mortgages.[16]

Mortgage-Backed Securities

As the nationally chartered programs became an integral part of the financial markets, other financial institutions identified and pursued opportunities to enter the mortgage market. Beginning with the Real Estate Investment Trust Act of 1960, which allowed the creation of real estate investment trusts (REITs) that could sell securities and use the proceeds to invest in real estate, Congress enacted a series of laws to expand mortgage lending beyond conventional banking. Its most significant action was passage of the Financial Institutions Reform, Recovery and Enforcement Act of 1989 (FIRREA). As a result, privately owned financial institutions were encouraged to issue bonds collateralized by the cash flows from an underlying pool of mortgages. Their bonds were often divided into classes with different maturities, different priorities for the receipt of principal and sometimes of interest, and, thereby, different levels of risk. Equally important, the institutions originating mortgages became dependent on the purchasers of securities. As a result, the securities market, rather than the federal government, became the key player in the money supply and the biggest influence on the housing supply.

By the twenty-first century, mortgage-backed securities (MBSs) had become a major source of mortgage capital and the origin of the near collapse of the real estate market and the banking system in 2008. At first, banks bundled together mortgages that conformed to federal standards. Accordingly, these MBSs were insured by Fannie Mae, Ginnie Mae, and Freddie Mac. Some banks that originated the loans (usually federally insured) bundled them together as securities and then sold them. Other *originators* sold their mortgages to investment banking companies that bundled them together and sold the securities whose revenues came from monthly mortgage payments.

Since most MBSs were insured by Fannie Mae, Freddie Mac, or Ginnie Mae, the purchasers of the securities considered them to be "U.S. government backed" and thus absolutely safe investments (although only Ginnie Mae has an absolute federal guarantee). Fannie Mae and Freddie Mac insured a large number of mortgages that did not conform to the strict principles and tight standards that had once characterized FHA-insured mortgages. Moreover, there was no specific guarantee of a congressional bailout if the loans went sour.

In 2008, when enough mortgages failed to generate the expected payments from borrowers, there was no way to pay the holders of the securities that had financed those mortgages. That was the basis of the banking crisis discussed later in this chapter. The worst of that crisis is over, but even three years afterward, most of its problems were still unsolved. Of the nearly $14 trillion in total U.S. mortgage debt in 2011, more than 60 percent of outstanding mortgage debt came from mortgage-backed securities, all but $1.5 trillion of which was securitized or guaranteed by federal agencies.[17]

Multifamily Housing

A similar program for unsubsidized multifamily rental housing accompanied the New Deal program for one- to four-family houses: FHA 207. Like the 203 Program, it provided insurance on 90 percent of value.[18] That is where the similarity ended. The program was applicable only to new construction and excluded existing, multiple dwellings. Thus, unlike the owner of a house with an FHA-insured mortgage, the owner of an apartment building with an FHA-insured mortgage could not depend on finding a purchaser who could obtain similar financing. Naturally, developers were far more likely to risk equity capital in a safer market. As of 1940, fewer than 30,000 apartments had been built under the 207 Program.[19]

The 608 Program

In 1948, hoping to stimulate apartment house construction, Congress revived the little-used FHA 608 Program, which had originally been enacted during the war. It was successful in spurring new construction because its liberal underwriting standards attracted entrepreneurs who often did not need cash up front. During the six years it was in existence, this program financed 464,000 new dwelling units.[20]

The 608 Program provided 90 percent insurance on the estimated cost of development. Land values were established on the basis of an appraisal of current market value. Developers who had purchased land some years earlier at a substantially lower figure were able to withdraw in cash the difference between the appraised value of the property at the mortgage closing (a sum now higher than at the time of their purchase) and required equity investment plus the insured bank mortgage. Had this not been the case, they would have sold their land at a profit and never contemplated the risks of apartment house construction.

Cash advanced during construction was based on the estimated cost of the work. Consequently, those builders who were able to build at costs below those prevailing in the area (and below the estimates of FHA appraisers) made money during construction. If this had not been possible, competent builders would never have entered the program. They would have resented the penalty for being more skilled than their competition.

Some public officials were scandalized by such practices. More important, they were outraged by the scheming that was made possible by collusion among loan officers, appraisers, contractors, and developers who fraudulently overestimated project costs. Rather than blaming the crooks who had profited from scams, public officials questioned the validity of the whole program and, in 1954, allowed it to fade away.

Chicago, 2008. 1350–1360 North Michigan Avenue was the city's first apartment building with an FHA-608 mortgage. *(Alexander Garvin)*

In its stead, Congress revitalized the 207 Program, this time with cost certification and regulation of initial rents. Far fewer developers were willing to deal with the additional requirements, paperwork, and processing time. The new procedures increased the opportunities for discretionary action by government officials, a few of whom were willing to act only when helped along with an extra "fee" to cover their trouble. Thus, while the new procedures did not eliminate corruption, they did terminate the mass generation of FHA-insured market-rate mortgages for multifamily housing.

The Savings and Loan Crisis

For half a century, Congress regulated both thrift institutions (the savings banks and savings and loan associations that provided homeowners with permanent mortgages) and commercial banks (the larger institutions that made loans to developers who built houses for customers who financed their purchase with FHA-insured and VA-guaranteed mortgages). The resulting stable supply of mortgage money provided financing for a steadily increasing supply of decent, safe, and standard housing.

Since the federal government insured the deposits made to savings accounts and regulated the manner in which this money could be lent to homeowners, thrift institutions had to comply with federal regulations that determined the terms they had to offer both depositors and borrowers. Thrifts were not permitted to offer checking accounts (which were the purview of commercial banks and on which such banks were not permitted to pay interest). They were, however, able to offer a rate of interest on savings deposits that was slightly above the rate that commercial banks were permitted to offer. Because the government so carefully supervised their mortgage lending, they were allowed to maintain lower cash reserves (4 percent of deposits). Commercial banks, whose investments ranged far beyond home mortgages and were rarely government-guaranteed, were required to maintain more generous cash reserves (15 to 17 percent of deposits).

The strict guidelines for accepting deposits and making loans allowed thrift institutions little latitude and required virtually no expertise on the part of bank officials or regulators. Home mortgage lending became so routine that loans were, in effect, mass-produced. Cynical observers called it "the 'three-six-three' business: take in deposits at three percent, lend them out at six percent, and tee up at the golf course at 3:00 P.M."[21]

For four decades, interest rates remained stable and thrift institutions faced no difficulties. They could accept short-term deposits on which they paid market interest and then lend their depositors' money to homeowners for long periods of time at fixed rates of interest. As with so much else in the United States, this comfortable situation was profoundly affected by the period of inflation that began during the Vietnam War and was fueled by a succession of energy crises. Short-term interest rates, which had slowly fluctuated between 3 and 6 percent since the Great Depression, began to change, sometimes on a monthly basis, reaching an all-time high of 21.5 percent.

In a world of wildly fluctuating rates of interest, thrift institutions (whose rates of interest were regulated) found themselves facing intense competition for deposits from money market accounts, interest-paying checking accounts, and all sorts of new instruments that paid much higher rates of interest. The result was a massive outflow of money from savings accounts. A far more serious problem, however, was that thrift institutions were suddenly forced to pay depositors high rates of interest even though they were still earning far lower rates of interest on their long-term mortgages. Unless money market conditions changed, they would soon be out of business. Indeed, "in 1972 the nation's savings and loans had a combined net worth of $16.7 billion. By 1980 that figure had plummeted to a *negative* net worth of $175 billion."[22]

Thrift institutions demanded the right to compete on a level playing field. Their demands coincided with the passion for deregulation that hit the United States during the late 1970s and early 1980s. Congress responded by enacting the Depository Institutions Deregulation and Monetary Control Act of 1980 and the Garn–St. Germain Depository Institutions Act of 1982.

This legislation was intended to help thrifts attract deposits by increasing federal insurance from $40,000 per depositor to $100,000 per account. Reserve requirements for thrift institutions were lowered to 3 percent. In order to help thrifts increase the return on their investments, they were no longer restricted to mortgage lending (a large proportion of which had been federally insured). They were even permitted to become joint-venture partners with their borrowers. In order to attract more entrepreneurial management, they did not have to be operated as widely held membership organizations. For the first time, federally chartered thrift institutions could be purchased, owned, and operated by individuals.

The results of these banking reforms were catastrophic. Brokerage firms offered small investors participation in federally insured accounts, combined these deposits into $100,000 packages, and then sought the highest possible rates of interest. Thrift institutions that needed additional deposits had no choice but to pay top dollar for these accounts. Meanwhile, the Federal Reserve, in an attempt to combat inflation, forced interest rates to rise far above the average rate of interest paid by long-term mortgages.

Thrift institutions responded by making investments (e.g., junk bonds and joint-venture deals) that were supposed to produce higher returns than conventional government-regulated and -insured home mortgages. However, their employees were not accustomed to making risky investments. They were familiar with the routine world of FHA-insured loans, as were the federal regulators responsible for overseeing their activities. Both of these parties discovered the risks of such investments when the promised returns failed to materialize. By then it was too late.

In some cases long-established thrift institutions were acquired by crooks who exploited deregulation. Their scams, like many criminal exploits, make interesting but reprehensible stories. Some of these frauds even banded together to avoid regulatory supervision, trading bad loans among themselves so that when bank inspectors showed up, the books would look clean.[23]

Fortunately, most deposits were federally insured; consequently, relatively few depositors lost their hard-earned savings. However, when insurance premiums failed to cover losses, Congress had to make good on its guarantee and was forced to appropriate hundreds of billions of dollars. The bailout began in the late 1980s and extended for about a decade. Definitive data on the consequences of the savings and loan crises does not exist, but reputable analysts estimate that it cost taxpayers between $160 billion and $180 billion or more.

In reaction to the S&L crisis, thrift institutions have tightened lending policies and chosen to invest a substantial portion of their deposits in safe federal securities. As a result, during the early 1990s there was far less money available for home mortgages and thus a major decrease in home building. What money was available was no longer available on the same easy terms. In 1991, new housing production had decreased to a rate of 1,014,000 dwelling units per year. However, at the close of the century, when the crisis was a thing of the past, the supply of housing was again growing at more normal rates. As a result, in 1999 new housing was being produced at the rate of 1,665,000 dwelling units per year.[24]

The Subprime Mortgage Crisis

It did not take long for another threat to a stable money market to emerge—this time because the players involved in creating, insuring, rating, selling, and regulating mortgage-backed securities ignored the realities of real estate. They actually never dealt with the mortgages; they dealt with securities that came from streams of income paid by borrowers who had mortgaged their residences. Those streams of income were insured by Fannie Mae, Freddie Mac, Ginnie Mae, or private insurance companies. Because the mortgages were insured, the banks that took back the mortgages did not have to verify the creditworthiness of the borrowers or the value of the pledged collateral. They bundled the mortgages together into securities. They sold those bundles of mortgages as securities, thereby getting back the money they had lent, plus a fee for packaging the mortgages, all without any responsibility for how any of the mortgages would subsequently perform.[25]

The income streams from mortgages were rated by agencies that assigned their ratings without examining the creditworthiness of the borrowers or the value of the pledged collateral. The rating agencies had no responsibility for the performance of the securities, but they did receive a fee for their rating service. They relied on the rectitude of the institutions whose divisions issued the securities and the same insurance that the sellers of the securities had relied on. The securities were sold by companies and individuals who had no responsibility for their performance, but received a fee for selling them and were purchased by institutions and individuals who also relied on the rectitude of the rating agencies, the rectitude of the institutions whose divisions issued the securities, and the same insurance everybody else had relied on.

The only entity in this chain of financing that could lose money was the ultimate purchaser of the securities. Everybody else made a fee from selling their component of the chain, and some received bonuses for their high level of sales. The more securities they sold, the more money they made. Since they could not lose any money, they found more and more ways to justify issuing mortgages. Additional (often inexperienced) mortgage brokers were needed to deal with the flood of home buyers.

With so many people who were now able to obtain home mortgages, housing demand increased, and prices started to climb. The players in the chain began to believe that home prices had only one way to go: UP! In order to comply with market demand, they used the most recent sales prices to make the property appraisals on which to determine the size of the mortgage, sometimes inflating the size of the mortgage to 105 or 110 percent of the amount needed to purchase the house. Borrowers, sure that they could sell the property in a year or two for a nice profit, sought short-term or adjustable-rate loans that offered more competitive interest rates than the fixed-rate, self-amortizing mortgages made popular by the FHA programs. By 2007, homeownership in the United States had soared to 68.4 percent, and the average price of a house climbed from $145,000 in 1997 to $254,400 in 2007—a 75 percent increase. Average household income over that same period increased 43 percent.[26] Since incomes had not grown at the same rate as home prices, the inflated mortgage market could not be sustained.

When the crash finally came in 2008, it affected the entire world and upended the U.S. economy. As of 2012, for the first time in its 78-year existence, the FHA was projecting a loss of $16.3 billion; Fannie Mae and Freddie Mac had required $137 billion in federal funds to cover losses and been placed under conservatorship.[27] Some people blame the Great Recession that emerged on the complexity of new financial mechanisms (including derivatives) that had been invented in order to diffuse and minimize risk. Others claim it was the result of mismanagement (or worse) by the institutions involved in providing mortgage money, deregulation, or lax supervision by government agencies. There is truth to these observations. As Paul Volcker, chairman of the Federal Reserve from 1979 to 1987, has explained, there were many other, simpler reasons for the collapse of the mortgage market and the slowdown of housing production from more than 2 million dwelling units in 2005 to just over 600,000 dwelling units in 2011.[28] Among them he cites "unjustified faith in rational expectations, market efficiencies, and the techniques of modern finance . . . stoked in part by huge financial rewards that enabled the extremes of borrowing . . . and the pretenses and assurances of the credit-rating agencies . . . along with exceedingly large compensation for traders."[29] Put more frankly: Nobody had any "skin in the game."

Ingredients of Success

Anybody who experienced the S&L and subprime mortgages crises knows that there is no mystery to increasing the supply of housing. It requires a stable, orderly, easily accessible capital market such as was created by the banking reforms of the 1930s. Accessibility is as important as stability. As long as equity requirements and interest rates remained at the low levels established during the New Deal, millions of consumers maintained a steady level of demand for one-family houses. The same low-equity requirements and low interest rates allowed thousands of builders to supply that demand.

Regulation created the necessary stability. Consumers were able to depend on a standard product and standard lending procedures. More important, regulation also allowed financial institutions to use a relatively unskilled staff to establish millions of savings accounts and provide millions of home mortgages.

However, this money market had a distinctly suburban bias. With the exception of the six years during which the 608 Program was in operation, apartment house developers were unable to obtain easy access to FHA financing. Nor was there any workable program for the refinancing or rehabilitation of existing apartment buildings. Only during the 1970s and 1980s, when the money market was destabilized and obtaining financing became more difficult, did this extraordinary suburban success story slow down. It remains to be seen how long it will take for an easily accessible, stable mortgage market to be restored after the disastrous subprime mortgage crisis of the early twenty-first century.

Market

In a market economy like ours, government can encourage a level of demand that allows the real estate industry to supply that demand. The banking reforms of the 1930s generated that stable level of demand by making homeownership affordable to two-thirds of the population, by establishing physical and financial standards that met its requirements, and then by ensuring the availability of credit.

Lowering the required down payment on a house to 10 percent made purchase affordable. Requiring all FHA and VA mortgages to be fully self-amortizing over a long period made debt service affordable. As a result, millions of people who had previously been unable to own a house were able to purchase one. Equally important, when they were ready to move, they had a standard product that could easily be resold to another buyer who could count on financing the purchase on similar terms. But, as the subprime mortgage crisis demonstrated, creating an equityless mortgage market, based on continually increasing prices is a recipe for disaster.

Location

The beauty of the banking legislation of the 1930s was that it allowed market forces to supply housing at suburban locations that were easily accessible and inherently attractive. However, those market forces were precluded from operating in the central sections of our cities, with their preponderance of older apartment buildings. The bias against cities was not only a matter of inadequate FHA programs for existing multifamily housing. It was also the product of underwriting practices.

FHA-insured mortgages could not exceed 90 percent of "appraised value." If the appraised value was too low, the mortgage would be insufficient to justify a mortgage of the size the applicant needed. As a result, the project could not proceed. While the FHA had standardized the elements of required bank appraisals, the amount of the loan depended on the judgment of those approving it. That judgment involved an estimate of the property, the borrower, and the neighborhood. If the property failed to meet FHA standards, the mortgage insurance was denied.

Borrowers themselves might be deficient. This was not just a matter of net worth, income, or credit history. The FHA *Underwriting Manual* specifically stated that "if a neighborhood is to retain stability, it is necessary that properties shall continue to be occupied by the same social and racial classes" and recommended "suitable restrictive covenants." The *Underwriting Manual* also specified neighborhood criteria, which downgraded "older properties," "crowded neighborhoods," and "lower-class occupancy" common in urban areas. Simply put, the FHA (without the specific approval of Congress) used its underwriting practices to discriminate against cities and to finance further suburbanization.[30]

Design

FHA design standards brought certainty and predictability to the housing market. Consumers purchasing houses with FHA-insured mortgages could depend on structural soundness, decent construction materials, minimum room sizes, modern plumbing, and adequate electrical wiring. Developers who produced buildings that met these standards could depend on their customers obtaining long-term, low-interest mortgages. Banks could depend on selling mortgages on these standard products to Fannie Mae.

Developers of one-family houses were able to predict the FHA-insured mortgage that could be obtained. This allowed them to budget development expenditures. Consumers could predict the mortgage that would be available when they sold it. Similarly, banks could predict the physical characteristics of the product they were financing and minimize the time and effort devoted to underwriting. The result was mass production of standard houses, mass generation of standard mortgages, mass consumption of both, and a rapidly increasing housing stock. By the twenty-first century, local building and fire codes had supplanted the need for FHA minimum standards. But during the 1930s and 1940s, such federal standards were very much needed.

The minimum property standards that were so important to creating a market for one-family houses proved to be equally central to the bias against multiple dwellings. Section 2 of the Housing Act of 1934 authorized the FHA to insure loans up to $2000 for repairs and improvements. Given FHA procedures, existing buildings that were in need of restoration were bound to be given low appraisals, usually too low to cover the cost of acquisition and/or refinancing.

If a low appraisal was not enough to discourage the borrowers, minimum property standards would preclude any further desire for an FHA-insured mortgage. When a property was deficient in a few respects, the borrower (often at high cost) was required to remedy the inadequacy or forgo the loan. The cost of the work necessary to bring most urban multiple dwellings into conformance with minimum property standards (suitable for new construction) usually raised required rents beyond marketable levels. Moreover, the minimum property standards also eliminated major categories of housing. Many Philadelphia and Baltimore row houses, for example, could not meet FHA requirements. They were too narrow.

No serious FHA market-rate mortgage program for the rehabilitation of older buildings has ever emerged. None will emerge until there is recognition of the essential difference between building new housing and renovating structures that were built to the standards and tastes of another era. It may be cost-effective to require all new buildings to have smoke-free fire stairs designed for four hours' survival during any conflagration. However, it certainly is not cost-effective to relocate existing tenants, eliminate existing apartments, and reorganize the circulation patterns of most older multiple dwellings in order to accommodate four-hour-rated, smoke-free fire stairs.

Financing

Without deposit insurance, banks would not have had money to lend. Without FHA insurance, thrift institutions would not have invested nearly as much in home mortgages. Without Fannie Mae to buy FHA mortgages, there would have been no way to get additional mortgage money to banks that needed to satisfy additional demand. The mass origination of home mortgages, however, would not have been possible without standard FHA lending practices, underwriting techniques, and mortgage instruments.

These new lending practices did not apply only to federally insured mortgages. The fixed-rate, long-term, self-amortizing mortgage became the most common form of loan. In fact, lending institutions adopted the underwriting techniques standardized by federal programs for all their mortgages. Consequently, until the deregulation of the 1980s, thrift institutions were able to generate a steady stream of business that could be managed without hiring an expensive team of financial wizards.

However, the problems and cost of complying with minimum property standards and the bias against lending in "older neighborhoods" prevented the revolution in housing finance from extending to existing multiple dwellings. Balloon mortgages, second, third, and even fourth mortgages, short-term lending, and other pre–New Deal characteristics continued to be common to the financing of existing apartment buildings.

Far more serious problems arose in the twenty-first century when the players in the mortgage market concerned themselves with issues of financing and ignored the underlying real estate. To ensure compliance with FHA minimum property standards, loan officers routinely visited the site, studied engineering reports on the building, and checked the creditworthiness of borrowers. By the turn of the century, few players in the housing finance chain bothered with such details because a reputable bond-rating agency had at least purportedly certified the triple-A status on the income stream from a bundle of insured mortgages.

Time

One of the least appreciated results of the banking reforms of the 1930s was that they helped to insulate real estate from a changing money market. Millions of borrowers were able to avoid the hazards of trying to refinance balloon mortgages. FHA-insured mortgages were fully self-amortizing and extended over several decades. They also reduced the hazards of trying to sell a house during periods of tight money. If the FHA already had insured a mortgage on the house, it was likely to do so again. Similarly, thrift institutions no longer had to worry about liquidity. Whenever they needed cash, they could sell FHA-insured mortgages to Fannie Mae. Thus, lending institutions were insulated over many years from changing demands on their resources.

Developers, however, found that time mattered a great deal; they were not immune to changing market demand. When real estate prices plummeted in 2008, however, borrowers discovered that potential buyers were not willing to pay enough to satisfy their existing mortgage. Thus, they were frequently forced to abandon the property. The institutions with the mortgages could rely on insurance to make up the difference between the money they could obtain on resale and the amount of mortgage insurance. This, however, did not protect insurers from huge losses. Fannie Mae and Freddie Mac owned or guaranteed such a high proportion of the mortgage market when the mortgage crisis was at its height that the government placed them under federal conservatorship, dismissed their chief executive officers and boards of directors, and issued new Treasury senior preferred stock and common stock warrants to protect the financial soundness of the two agencies.

Entrepreneurship

In devising a program for one-family houses, public officials sought to create conditions that would attract small businesses into real estate. The simplicity of obtaining FHA 203 home mortgages greatly reduced the risk to home builders. They no longer worried about customers who could not obtain a mortgage. Nor did they need much equity up front because they could borrow against an FHA mortgage commitment that covered 90 percent of the purchase price (and presumably an even higher percentage of development costs). The only thing government officials had to do was verify that banks were complying with regulations. As a result, thousands of entrepreneurs who had relatively little cash were attracted to the home-building industry.

Public officials dealt with multifamily rental housing in

a very different fashion from owner-occupied one-family houses. They desired unlimited apartment-house production but were uncomfortable with windfall profits. As soon as it became clear that the 608 Program allowed clever developers to put up little or no cash and make lots of money, it was terminated. Replacement programs may have minimized developer risk and allowed paper equity contributions, but they also limited profit, specified the labor force, regulated tenantry, required time-consuming bureaucratic review, and thus were unable to generate the massive amounts of new urban multiple dwellings that resulted from the 608 Program.

During the S&L and subprime mortgage crises, an entirely new group of entrepreneurs emerged. These new players concentrated on financial instruments, rather than the housing that it paid for. Many of them played fast and loose with very much diluted government regulations and made huge amounts of money. As of 2012, there do not appear to be mechanisms in place to prevent a new group of entrepreneurs from doing exactly what their predecessors did to reap huge rewards.

Increasing Housing Supply as a Planning Strategy

The administration of President Franklin Roosevelt was responsible for the most fundamentally possible market intervention: creating conditions that ensured a smoothly functioning market economy. In the case of housing, Roosevelt's policies made possible a continuing high level of demand for decent conditions and the continuing availability of enough money to allow the real estate industry to supply that steady level of demand. This has resulted in a steady increase in the supply of housing. Consequently, in 2010 there were over 130 million housing units in the United States, more than four times the number in 1940.[31]

This ongoing increase in the housing stock makes possible the existence of houses and apartments in a very wide range of sizes, types, and prices. It also allows citizens to purchase a much wider range of other goods and services with the money remaining after they have paid for their residences. But, despite a well-functioning money market that allows resources to move from locations in which supply is adequate to lenders in areas in which there is growing demand for housing, in some areas of the country the supply of housing is inadequate and citizens cannot afford decent shelter. Reducing the cost of housing requires different forms of government intervention into the private market (see Chapter 10). Dealing with a local inadequacy in the supply of housing, however, requires action at the local level. It may require altering local regulations to make it easier to build, to build more than was previously permitted, or to build in areas where housing had not formerly been allowed. But without knowing the situation in a particular area, it is impossible to know what form of government action will be effective.

Notes

1. I am indebted to my friends Michael Piore and Alan Beller for many of the ideas in this chapter. See the discussion of New Deal legislation in Michael J. Piore and Charles F. Sabel, "Stabilizing the Economy," in *The Second Industrial Divide,* Basic Books, New York, 1984, pp. 73–104.
2. Kenneth T. Jackson, *Crabgrass Frontier,* Oxford University Press, New York, 1985, p. 205.
3. U.S. Department of Commerce, Bureau of the Census, *Statistical Abstract of the United States,* Washington, DC, 1978, pp. 789–792.
4. U.S. Department of Commerce, Bureau of the Census, 2010 Census, http://factfinder2.census.gov/faces/tableservices/jsf/pages/productview.xhtml?pid=ACS_10_1YR_DP04&prodType=table Retrieved 16 July 2012.
5. Milton Semer, Julian Zimmerman, Ashley Foard, and John M. Frantz, "The Evolution of Federal Legislative Policy in Housing: Housing Credits," in *Federal Housing Policy and Programs Past and Present,* J. Paul Mitchell (editor), Center for Urban Policy Research, Rutgers University Press, New Brunswick, NJ, 1985, pp. 69–106.
6. Kenneth T. Jackson, op. cit., p. 196.
7. Prior to 1938, FHA insured only 80 percent of a mortgage.
8. FHA premium charges cover the cost of operations and any losses from the sale of foreclosed mortgages. The FHA has been self-financing since 1938. By 1954 it had repaid with interest all initial advances made by the Treasury.
9. Most lenders issued mortgages with a 20- or 25-year term.
10. http://fanniemae.com/portal/about-us/loan-limits/index.html and http://portal.hud.gov/hudportal/HUD?src=/program_offices/housing/fhahistory.
11. Legislative History of the VA Home Loan Guaranty Program, U.S. Department of Veterans Affairs, August 23, 2006, http://www.benefits.va.gov/homeloans/mission.asp, retrieved July 16, 2012.
12. Federal National Mortgage Association (Fannie Mae "Monthly Summary, May, 2012," http://www.fanniemae.com/portal/about-us/investor-relations/monthly-summary.html, retrieved July 16, 2012.
13. Roger Starr, *Housing and the Money Market,* Basic Books, New York, 1975, pp. 167–181; and www.fanniemae.com/company/index.html.
14. Federal Home Loan Mortgage Corporation (Freddie Mac), 2011 Annual Report.
15. See www.freddiemac.com.
16. "About Ginnie Mae," http://www.ginniemae.gov/about/history.asp?subTitle=About, retrieved July, 2012.
17. www.federalreserve.gov/econresdata/releases/mortoutstand/current.htm.
18. As with single-family mortgages, FHA insured only 80 percent of a multifamily mortgage prior to 1938.
19. Subcommittee on Housing and Urban Affairs, Committee on Banking and Currency of the United States Senate, *Progress Report on Federal Housing Programs,* U.S. Government Printing Office, Washington, DC, 1967, p. 38.
20. Subcommittee on Housing and Urban Affairs, Committee on Banking and Currency of the United States Senate, *Progress Report on Federal Housing Programs,* U.S. Government Printing Office, Washington, DC, 1967, p. 39.
21. Paul Zane Pilzer with Robert Deitz, *Other People's Money: The Inside Story of the S&L Mess,* Simon & Schuster, New York, 1989, p. 63.
22. Stephen Pizzo, Mary Fricker, and Paul Muolo, *Inside Job: The Looting of America's Savings and Loans,* McGraw-Hill, New York, 1990, quoted by Michael M. Thomas in "The Greatest American Shambles," *The New York Review of Books,* January 31, 1991, p. 31.
23. Pilzer with Deitz, op. cit., pp. 80–122.

24. U.S. Bureau of the Census, www.nahb.com/facts/forecast/annual_starts.htm.
25. For a more detailed account of the subprime mortgage crisis, see Gillian Tett, *Fool's Gold*, Free Press, New York, 2009, and Michael Lewis, *The Big Short: Inside the Doomsday Machine,* W. W. Norton & Company, New York, 2010.
26. Homeownership rates from the U.S. Census Bureau, http://www.census.gov/hhes/www/housing/hvs/historic/index.html, retrieved July 16, 2012. Median house sales prices from the U.S. Census Bureau, www.census.gov/const/uspricemon.pdf, retrieved July 16, 2012. Median household income data for 1997 from the U.S. Census Bureau, http://www.census.gov/hhes/www/income/data/incpovhlth/1997/tablec.html, retrieved July 16, 2012; median household income data for 2007 from the U.S. Census Bureau, www.census.gov/prod/2009pubs/acsbr08-2.pdf, retrieved July 16, 2012.
27. Nick Timiraos, "Report: FHA to Exhaust Entire Capital Reserves," *The Wall Street Journal,* November 16, 2012, p. A4.
28. National Association of Homebuilders, http://www.nahb.org/generic.aspx?sectionID=819&genericContentID=554&channelID=311, retrieved July 16, 2012.
29. Paul A. Volcker, "Financial Reform: Unfinished Business," *The New York Review of Books,* November 24, 2011, pp. 74–76.
30. These quotations, reproduced in Kenneth Jackson's *Crabgrass Frontier* (op. cit.), are only part of the FHA's remarkable record of prejudice in lending described on pp. 207–218 of this indispensable account of the suburbanization of the United States.
31. U.S. Census Bureau, 2010 Census, http://factfinder2.census.gov/faces/tableservices/jsf/pages/productview.xhtml?pid=ACS_10_1YR_DP04&prodType=table, retrieved July 16, 2012. For 1940 statistic, see "Population, Housing Units, Area Measurements, and Density: 1790 to 1990," U.S. Census Bureau, August 23, 1993, http://www.census.gov/population/www/censusdata/hiscendata.html, retrieved July 16, 2012.

10

Reducing Housing Costs

Bronx, 2012 (Alexander Garvin)

There are two ways for government to close the gap between the cost of supplying decent shelter and a price that people consider affordable. One is to lower the cost of supplying housing. The other is to lower the cost to the consumer.[1]

The term *economic rent* refers to what is required to justify any real estate venture; it is the amount which will cover all the expenses of running the property. Those expenses consist of: (1) maintenance and operating costs, (2) real estate taxes, (3) debt service on mortgage loans, and (4) payments to equity investors. Some supply-side programs reduce the cost of providing housing by lowering the price of land or construction. Others reduce economic rent by lowering real estate taxes, debt service, or return on equity. Whether the reduction is passed through to the consumer depends on how the program is structured.

Although government issues detailed specifications for determining affordable rent within its programs, there is a much simpler and accurate approach to explaining affordability. This view defines *affordable rent* as the amount tenants are able and willing to pay for shelter, and it translates directly into consumer demand. Demand-subsidy programs either provide consumers additional money to purchase better shelter or provide supplemental payments to cover the gap between affordable and market rent. More money does not necessarily result in more or better housing. Like supply-subsidy programs, demand-subsidy programs result in improved housing only when they are structured to do so. The money must be spent on housing rather than on something else. It also must be sufficient to justify action by property owners. Otherwise, the additional demand that will result may produce higher prices rather than better accommodations.

Millions of families have benefited from these supply- and demand-side programs. Some are more economical than others, some more efficient, and some more responsive to consumer desires. The difficulty is that these programs do not necessarily improve our cities and sometimes damage them.

One of the most unfortunate aspects of subsidized housing programs has been their tendency to locate low- and moderate-income people in buildings that are visibly different from housing occupied by everybody else. As a result, subsidy recipients are identifiable as living in a "project," which also stigmatizes the surrounding community.

Most subsidized housing programs are conceived with little consideration of anything but shelter. The subsidies are used to produce affordable rents, not organic communities where residents can eat, shop, gossip, play, or be involved with their neighbors. For that reason, when the subsidies are translated into physical form, the result is a building that has nothing to offer the surrounding neighborhood.

Worse yet, some subsidy programs concentrate people with acute social problems where they may obtain decent shelter but not the services they desperately need. Residents of surrounding areas fear subsidy recipients will spill over into the neighborhood, bringing their problems with them. Consequently, they vigorously oppose subsidized housing in their community.

The housing-subsidy programs that have been of benefit to urban and suburban areas are those that allow subsidy recipients to enter the marketplace on terms that are similar to those of nonsubsidized citizens. They also provide a subsidy that is large enough to stimulate property owners to make further investments. A good example is the tax incentive that the federal government offers to homeowners. Since the enactment of the federal income tax, homeowners have not had to pay taxes on the income they use to pay mortgage interest and real estate taxes. In response, developers built millions of homes for buyers who would not otherwise be able to afford owning their residence. The Section 8 Moderate Rehabilitation Program, terminated in the 1990s, is another example. This program subsidized the difference between 25 percent (later, 30 percent) of income and economic rent for those recipients who lived in apartments that were upgraded to include decent plumbing fixtures, electrical wiring that could handle modern appliances, and other improvements that would not otherwise have been installed.

A particularly effective device for lowering the price of housing is the federal Low Income Housing Tax Credit (LIHTC), created in 1986 under the Tax Reform Act. It grants developers of low-rent housing the right to make deductions from federal income tax payments, thereby increasing the availability of affordable housing by requiring that any apartment receiving tax credits be rented to persons of low income. In many cities, developers use the subsidy for only 20 percent of the apartments. Since the program requires apartments occupied by low-income residents to be identical in design to the 80 percent rented on the open market, there is no way to know whether any low-income tenants reside in the building or (if so) where their apartments are located. Thus, any stigma connected with living in a "project" is completely eliminated.

In areas that are losing population, supply-side subsidies are unlikely to be needed except to upgrade older buildings. However, those residents who cannot afford the housing that is available will need them. Similarly, in cities with an increasing number of residents where housing production is not keeping up with the growth in population, demand-side subsidies will increase the price of whatever housing is available. Subsidies will be needed if every American is to have a decent home. The only question is, what sort of subsidy?[2]

Subsidizing Supply

Most federal housing assistance has been directed at lowering the cost of supplying decent shelter. Theoretically, if the lower cost of production is passed on in the form of lower prices, there will be additional demand. Increased demand will stimulate developers to increase the housing supply. Construction means jobs and therefore support for and from the housing industry.

Some supply-side programs require relatively little subsidy but create substantial controversy, especially when private property is taken through public condemnation. In other instances, the need for subsidies will be overcome by political attractiveness. The subsidies that go to housing built by nonprofit organizations, for example, are often politically more palatable than subsidies that go to housing built by the private sector. Much like the discussion of every program and project discussed in this book, the appropriateness of supply-side subsidies must always be measured for political as well as financial feasibility.

The changing character of the political environment means that few programs remain in place for long. The ways in which supply-side subsidies can be used, however, do not change. They directly reduce one of the components of economic rent (operating costs, real estate taxes, debt service, and return on equity) or lower them indirectly by reducing development costs.

Reducing Development Costs

Development costs can be grouped into a few categories: property acquisition, tenant relocation, demolition, site preparation, professional services (e.g., architecture, engineering, legal, accounting), actual brick-and-mortar construction, fees and taxes, marketing, and interim financing to cover these costs until the property has occupants and a permanent mortgage. Sometimes one must also add the cost of supplying the required infrastructure (streets, sewers, water mains, transit systems) and community facilities (schools, parks, hospitals, fire and police stations, libraries, etc.).

Reduction in development costs does not produce a proportionate reduction in economic rent. For example, a 10 percent reduction in development costs will produce a 10 percent reduction in debt and equity requirements. However, debt service and return on equity are only two of four components of economic rent. If they represent half the required economic rent, then a 10 percent reduction in development costs will reduce economic rent by only 5 percent. Despite this limitation, many supply-side housing programs concentrate on reducing the land and construction costs.

Land

The initial cost for any development is property acquisition. If plenty of sites are available, developers negotiate with property owners until they are offered land at a price that is low enough to justify a financially feasible project. When developers need to assemble a site by purchasing specific parcels from a variety of owners, the cost of land can become prohibitive. Property owners may get wind of interest in their land. They will naturally refuse to sell, holding out for the highest possible price—often threatening the feasibility of the project.

One of government's earliest attempts at reducing the cost of assembling a housing site involved condemnation of privately owned land by the federal government. Faced with "wide-spread unemployment and disorganization of industry," Congress in 1933 enacted the National Industrial Recovery Act. It authorized the Public Works Administration (PWA) to create jobs through a comprehensive program of public works, including "low cost housing and slum clearance." In most cases, land acquisition for these housing projects involved condemnation by the federal government.

Within three years, the Housing Division of the PWA had begun 50 projects in 35 cities, totaling 25,000 dwelling units. All 50 projects consisted of walk-ups, mostly one- and two-story buildings that covered less than half their carefully landscaped sites. Although some of these projects eventually succumbed to problems that afflict publicly owned and operated housing (discussed later in this chapter under "Public Housing"), most have provided decent homes for families of low income for nearly three-quarters of a century.[3]

College Court, in Louisville, designed by a team of local architects led by E. T. Hutchings, is a good example of the high quality of these PWA housing projects. A 5-acre city block was acquired for 126 one- and two-story buildings, grouped around beautifully landscaped open areas at the center of the block. White-painted wooden front porches and pitched roofs give the redbrick structures a handsome domestic appearance. The buildings themselves stretch out along the bounding streets, set back from the property line by green lawns. The consistent siting, common materials, and landscaped open space are the only indication that College Court is any different from its neighboring structures. Thus, it is difficult to identify "project tenants." In 1987, College Court was converted into a resident-owned, low-income condominium, completing its integration into the residential fabric of the inner neighborhoods of Louisville.

Louisville's second PWA project, on Algonquin Parkway, may not be as handsome, but the approach is similar. It consists of a group of one- and two-story buildings, in scale with the surrounding neighborhood, sited around landscaped open space.

The city's third project became one of the most significant projects never to be built. In 1935, a property owner whose land was condemned by the PWA objected and went to court, claiming that the federal government had no right to take his property. He invoked the Fifth Amendment to the Constitution, which guarantees that no person's prop-

Louisville, 2008. Consistent siting, common materials, and landscaped open space are the only indication that College Court Public Housing, completed in 1938, is any different from its neighboring structures. *(Alexander Garvin)*

erty shall be "taken for public use without just compensation." The implication is that taking private property for a *nonpublic* use is forbidden, even with compensation. In the case of *United States v. Certain Lands in the City of Louisville,* the property owner contended that taking one individual's property in order to provide housing for another was not a public use authorized by the Constitution.

The district court and, a few months later, the Sixth Circuit Court of Appeals ruled in his favor. They held that any taking had to be for the purpose of occupancy by a public agency (e.g., the U.S. Postal Service) performing a statutory or constitutional purpose (e.g., mail delivery). Since the Roosevelt administration was in the midst of a major battle with the Supreme Court and feared yet another New Deal program might be declared unconstitutional, it chose to withdraw the case rather than carry the appeal further.[4]

The case of *United States v. Certain Lands in the City of Louisville* is important because it terminated condemnation and housing construction by the federal government. At the same time, another court, in the case of *New York City Housing Authority v. Muller* (see Chapter 11), held that local authorities, pursuant to the New York State Constitution, could condemn property for the purpose of eliminating slums and blight and providing decent shelter for people of low income. This helped the Roosevelt administration switch to a program that looked to local authorities (with condemnation powers provided by state constitutions) to own, build, and operate housing for persons of low income. The necessary subsidy would come in the form of direct federal grants to the local housing authority. This same approach was adopted after World War II for the federal urban renewal program.

In addition to condemnation of privately owned land for housing construction by government agencies, there have been hundreds of instances of government condemnation for construction by private developers. In 1943, Missouri, Pennsylvania, and New York became the first states to enact legislation allowing eminent domain to be used for the purpose of condemnation of property for private housing development.

Under the terms of the Missouri Urban Redevelopment Law of 1943, a developer submits a detailed redevelopment plan, including evidence of the existence of blight, the presence of a market for the proposed housing, and the project's financial feasibility. The city council then issues a certificate transferring the power of eminent domain to the developer. If the developer fails to negotiate property acquisition on favorable terms, the court approves condemnation and determines the price paid to the former property owner.

In other states, the local redevelopment authority drafts a renewal plan. When the local legislative body approves the plan, the redevelopment authority goes to court to obtain possession of the required sites. The court decides on a fair level of compensation. Then the redevelopment authority sells the property for the amount of the condemnation award to a private developer who has agreed to execute the plan.

Quality Hill was the first housing project in Kansas City to be completed under the Missouri Urban Redevelopment

Kansas City, 2008. Acquisition costs for the Quality Hill Redevelopment Project, completed in 1954, were minimized because the city's powers of condemnation were used to prevent property owners from demanding astronomical prices. *(Alexander Garvin)*

Law. The site had originally been developed as a high-income residential district during the second half of the nineteenth century. As the population of Kansas City moved southward, this once-fashionable neighborhood began to deteriorate. By the time the bluffs overlooking the river were proposed for redevelopment, the area had been labeled a "slum." Given the housing shortage after World War II, slum clearance for the purpose of housing construction was easy to justify.

When Quality Hill was completed in 1954, it consisted of five 11-story redbrick apartment houses containing 510 apartments, off-street parking for 250 cars, two swimming pools, and two picnic areas. More than four decades later, Quality Hill is still in good condition. It provides decent, affordable shelter for many more families than had previously lived on the site. Whether Kansas City might have been better off if the area's residents had not been forced to relocate or if its Victorian houses had been restored rather than cleared is still debated. What cannot be debated is that condemnation proved to be an effective means of lowering the cost of housing development.

It is by definition impossible to reduce the cost of acquisition greater than would be possible through condemnation. However, reducing the cost to the developer is a matter of subsidy. When Congress enacted the Housing Act of 1949, which created the urban renewal program, it provided just such a subsidy. In urban renewal projects conceived and executed by local redevelopment authorities but subsidized by the act, a site was sold to a developer at a price that made redevelopment economically feasible. The difference between actual cost and the sales price was subsidized, two-thirds by the federal government and one-third by the locality.[5]

Corlear's Hook, one of the first projects to use federal urban renewal subsidies, illustrates how the program worked. Robert Moses had proposed the project soon after passage of the Housing Act of 1949. The 15-acre site, along the East River on Manhattan's Lower East Side, consisted of dilapidated tenements containing 878 apartments. Moses's development plan called for creation of 1668 cooperative apartments in four towers built by a union-organized non-profit housing corporation.

Affordable co-op apartments on the Lower East Side were feasible only if land costs were minimal, which at that time meant no more than $675 per apartment. Accordingly, the site was sold for $1,126,000 (1668 × $675 = $1,125,900). However, project costs (acquisition, clearance, and site work prior to resale for housing construction) were $6,411,000. Thus, the required subsidy was $5,285,000 ($6,411,000 – $1,126,000 = $5,285,000). The federal government provided $3,523,000, two-thirds of the required subsidy; New York City and the state of New York subsidized the rest.[6]

Construction

Attempts to reduce brick-and-mortar costs are often unsuccessful because they increase operating costs. The initial cost of an electric heating system, for example, may be less than

Manhattan, 2004. Corlear's Hook was one of the nation's earliest urban renewal projects subsidized with funds appropriated under Title I of the Housing Act of 1949. *(Alexander Garvin)*

a comparable oil or gas system. However, the lower initial cost may be more than compensated by the higher cost of electricity. Reducing room sizes, using poor materials, or installing cheap equipment will be more than recompensed by increased maintenance and replacement costs during the project's life. It also runs counter to the goal of increasing the supply of good housing.

Since World War II, there have been two important attempts to lower the cost of construction. Both concentrated on labor practices, standardization of components, and shortening development time (and thus interim financing and taxes). Only the first, an entirely private initiative, proved successful.

Between 1947 and 1951, William and Alfred Levitt built and sold 17,442 homes on former potato fields in Long Island. They named the project Levittown. It was the first of several similar Levittowns built across America. By applying factory production techniques to housing construction, the Levitts were able to sell their houses at prices 20 percent below the competition (see Chapter 16).

House production in Levittown was divided into 26 separate operations and subcontracted to 80 different firms that were supervised by the Levitt staff. Construction was scheduled in a manner similar to the assembly line that Henry Ford had used to reduce the cost of automobile production. But instead of components coming down the assembly line to the workers, each of the subcontractor firms brought its workers to the site at the point it was ready for such involvement. This reverse production-line approach was possible because there were only five house models and because every component was standard and prefabricated (lumber, windows, doors, roofing, etc.).[7]

The Levitts needed cheap and continuous supplies of materials and equipment. For this reason, they bought timberland and a lumber mill on the West Coast and precut the lumber. They also established a supply company that manufactured cement block, nails, and other needed construction components. They were able to purchase fixtures and appliances at lower prices by ordering in bulk from suppliers who gave handsome discounts to important clients.

The resulting economies in construction allowed the Levitts to sell houses with two bedrooms, a living room, a bathroom, and a fully equipped kitchen (including sinks, a stove with two ovens, a refrigerator-freezer, and a Bendix washing machine) on prelandscaped lots for about $7990. Nothing comparable was available in the area for less than $9000—and then without washing machines or landscaping.

In the late 1960s, George Romney, secretary of Housing and Urban Development, tried a similar approach. Romney, who had headed the American Motors Corporation, wanted to introduce mass-production techniques to government-assisted housing. He called this approach "Operation Breakthrough." Unfortunately, Operation Breakthrough had no way of overcoming traditional labor practices, failed to guarantee sufficient continuous production to justify capital investment in large new factories, and was used at sites and for designs that required development periods long enough to wipe out most of the savings in interim costs. It remains the only government effort to transfer housing production to the factory.

Soft Costs

Because the price of land and buildings is certain upon sale and the cost of construction is specified by contractors in

Levittown, 1947. Standardized design, prefabricated building components, and mass production allowed the Levitts to sell houses for $7990. *(Courtesy of the Levittown Public Library)*

HARD COSTS				
	ACQUISITION		$40,000,000	
	CONSTRUCTION			
	1. Gross Residential & Commercial Floor Area @ $400/sq.ft.	$176,000,000		
	2. Underground Garage @ $280/sq. ft.	$33,600,000		
	3. Elevator 5 Cabs @ $90,000/cab	$450,000		
	4. Landscaping	$50,000		
		SUBTOTAL	$210,100,000	
			TOTAL HARD COSTS	$250,100,000
SOFT COSTS				
	FEES & TAXES			
	1. Permanent Lender (1% of mortgage)	$2,255,104		
	2. Construction Lender (1% of mortgage)	$2,255,104		
	3. Mortgage Broker (negotiated)	$300,000		
	4. Lender's Legal Charges	$100,000		
	5. Borrower's Legal + Accounting Charges (negotiated)	$150,000		
	6. Closing Costs	$150,000		
	7. Mortgage Recording Tax (2.5% of mortgage)	$5,637,759		
	8. Engineer/Architect (negotiated)	$2,000,000		
	9. Studies, Surveys, Title Insurance (1% of hard costs)	$2,501,000		
	10. Misc. Permits & Fees (0.5% of hard costs)	$1,250,500		
	11. Overhead (3.5% of hard costs)	$8,753,500		
		SUBTOTAL	$25,352,967	
	DEVELOPMENT COSTS			
	1. Construction Interest (8.0%)	$18,040,830		
	(.5)(2 yrs.)(initial mortgage)(.08)			
	2. Real Estate Taxes (2 yr.)	$2,268,000		
	3. Contingency (5% of construction)	$10,505,000		
	4. Leasing, Marketing & Advertising (2% of hard cost)	$5,002,000		
		SUBTOTAL	$35,815,830	
			TOTAL SOFT COSTS	$61,168,796
TOTAL DEVELOPMENT COST				**$311,268,796**
Total Development Cost/built sq.ft.	$556			
Total Development Cost/DU	$859,859			
Total Development Cost/room	$226,542			

Soft and hard development costs In this example, soft costs represent about one-fifth of the total development cost. It is not unusual for soft costs to represent a much higher portion of the total development cost. Smart developers minimize the development time as a way of reducing interim interest and taxes and thereby substantially reducing soft costs.

a bid, these are referred to as *hard costs.* All other development costs are referred to as *soft costs* because they are dependent on the cost of land and construction, as well as the time it takes to complete the project. Soft costs include fees, mortgage interest, marketing expenses, taxes, and the cost of managing the development. Fees have to be paid to architects, engineers, lawyers, accountants, and other professionals for their work on the project; to government agencies for permits; to lenders for processing mortgage loans; and to a variety of companies that provide needed surveys, studies, and other services.

Developers rarely put up all the money needed to cover the total cost of a project. They borrow as much as they can and repay that money from sales proceeds or, in the case of a rental project, from a permanent mortgage issued when construction is completed and occupancy is under way. Consequently, one of the largest soft costs is the interim mortgage interest paid while a project is under development. Even though developments cannot generate revenues until they are completed, governments expect taxes to be paid. Those tax payments can be substantial, especially when construction takes a long time. Finally, there are the expenses incurred by the developer, who has to run an office, employ a staff, and pay sales or rental commissions.

Effective developers devote substantial time and energy to reducing soft costs. Cutting fees may be self-defeating if this action results in reducing the quality or timeliness of the services needed to bring a project to completion. Cutting marketing expenses may result in increasing the time it takes to sell or rent all the units, thereby increasing the cost of interim interest and taxes. Banks and governments do not readily reduce their charges. The only way to reduce these costs is to shorten the development period.

Reducing Economic Rent

Most government efforts to lower housing costs have been directed at reducing three of the four components of economic rent: real estate taxes, debt service, and return on equity. Buildings can be designed to reduce the fourth component (operating and maintenance costs). However, little can be squeezed from maintenance without seriously affecting either the longevity of the structure or the quality of life within it.

Real Estate Taxes

Almost every local government pays for some of its costs by collecting taxes on real estate. It assesses the value of each property, determines the amount of revenue it requires, and then establishes a tax rate that will produce the required revenue. The tax to be paid by a specific property is arrived at by multiplying the *assessed value* by the *tax rate.* For example, if the assessed value of a property is $100,000, and the tax rate is 9 percent, the annual tax payment will be $9000.

All owners expect their property to be assessed fairly. Few properties are. In part, this is because property values fluctuate. Accordingly, some localities reassess all properties annually. Others—for example, California—reassess only upon resale. Still others do so on an irregular basis. If properties are not all reassessed at the same time, then contiguous identical houses with identical market values could be paying different amounts of tax depending on when they were last assessed.

Some localities consciously choose to treat properties unequally as a policy matter. New York City, for example, wishes to encourage middle- and upper-class homeownership. To this end, one- and two-family houses are assessed at a fraction of their market price. Rental apartment buildings are assessed at much higher proportions of their market value. In slum areas, some apartment buildings are assessed for more than market value. As a result, particularly between 1965 and 1995, the city has taken possession of thousands of properties that their owners feel are not worth even the annual real estate tax payment, much less their total "assessed value."

Determining value is not easy. For a property recently sold, the market value has been established. However, most properties were last sold years ago. Since tax assessors do not know what income the property is producing or what somebody would be willing to pay for that income, they cannot base the assessment on the property's income. At best, they can be guided by recent sales of similar properties in comparable locations.

Local governments have an interest in continually increasing property assessments. In most states the aggregate amount a locality can borrow for capital construction is a percentage of the value of real property. Since there is always a desire for a new school, firehouse, or library, or an immediate need for road repaving, sewer replacement, or new water mains, there is persistent pressure for increases in real estate tax assessment.

If government is to lower the cost of housing, however, one obvious way is to cut real estate taxes. Reducing real estate taxes provides a subsidy without direct government expenditures and, therefore, often is the most politically expedient method for local government to reduce economic rent. Legislators can avoid the often unpopular position of voting for budget appropriations or specific projects, and the local governments can avoid the cost of staff to administer the development process. The owner does everything.

Tax programs usually subsidize only improvements. The justification is that, absent the improvement that is to be subsidized, there would be no additional tax to be paid. When the tax reduction expires, the municipality will collect additional revenues by taxing the improvements induced by initially lowering taxes.[8]

In 1920, New York became the first state to adopt enabling legislation that permitted *tax exemptions* for new housing development. The following year, New York City passed a law providing all housing that was either under construction or would be started between 1921 and 1927 with an exemption from any increase in real estate taxes because of the planned improvements. The exemption expired in 1932. By the time the program was terminated, it had provided $917 million in tax exemptions for the construction of 574,000 apartments.[9]

A similar program, called 421A after its section in New York City's Real Property Tax Law, has provided a 10-year exemption for new multiple dwellings started after 1971. The exemption begins at 100 percent and declines 20 percent every two years till the eleventh year, at which time the project pays full taxes. Between 1971 and 1992, 105,616 newly built apartments received 421A tax exemptions.[10]

Real estate tax reduction, usually referred to as *abatement,* is used much less frequently than tax exemption, and then usually in conjunction with other government programs. For example, real estate taxes in federally assisted public housing projects are abated. The nominal tax that is paid is determined by formula to be 10 percent of *shelter rent* (gross rent less utility costs).

Debt Service

Debt service can be reduced either by lowering monthly amortization payments or by reducing the rate of interest.

From the beginning, FHA programs reduced amortization by extending the term of the loan. Obviously, if a loan is repaid in equal monthly installments over a 25-year period rather than over 5 or 10 years, the payments for a longer term will be significantly lower. Of all government projects, the 50 housing projects started by the PWA between 1933 and 1935 had the benefit of the greatest reduction in annual amortization payments: a 60-year mortgage term.

Reducing interest rates did not become a common strategy until 1959, when the FHA 221(d)(3) Program was enacted. The money to pay for the project was raised by issuing bonds. At first, interest on the bonds was lowered only to the average cost of federal borrowing. In 1965 the maximum interest rate for 221(d)(3) projects was set at 3 percent.[11]

The federal government was able to offer lower interest rates because it could borrow money a few percentage points below what conventional borrowers had to pay. The lower interest rates reflected the greater security and tax deductibility of government bonds. Theoretically, this lower interest rate would be passed through to developers of moderate-income housing. However, the institutions that lent money for development had to pay market-rate interest to their depositors. In order to make up the difference, banks were allowed to sell their 221(d)(3) mortgages to Fannie Mae at a discounted price reflecting this below-market-rate interest.[12] The differential between the face amount of the mortgage and the discounted sale price was subsidized by congressional appropriation. Congress increased the subsidy for the FHA 235 and 236 programs enacted in 1968 to subsidize the difference between market interest and 1 percent.

Only 190,000 apartments were financed through the 221(d)(3) Program and 463,000 through the 236 Program. One reason for the relatively low number of apartments produced under these interest-subsidy programs was Congressional reluctance to pay for them. This was not just a matter of budget priorities or housing philosophy. It was also the result of unhappiness with their inefficiency compared with privately financed housing. Processing an FHA-subsidized mortgage for an apartment building, for example, took much longer than it took to obtain an FHA-insured mortgage on a one-family house.[13]

Another reason for the relatively low level of production was that the program was available only to nonprofit or limited-profit developers. Relatively few developers are willing to accept a return on equity of 6 percent or less, especially during periods when the prevailing rate of interest is much higher. They agree to do so because these programs often provide unusually attractive tax benefits and allow them to contribute "builder's profit" and "professional fees" in lieu of cash equity. In exchange, they are willing to accept the added costs and frustrations of government loan processing.[14]

In addition to criticizing the time required for FHA processing, Congress found it difficult to justify construction costs that were more than 20 percent above comparable, privately financed projects. Not surprisingly, when Richard Nixon unilaterally terminated these housing programs in 1973, Congress chose not to replace them.[15]

Return on Equity

Many people believe that return on equity is the obvious place to cut economic rent. However, if private, profit-motivated developers are to be attracted in sufficient numbers to generate major housing production, it is the last place to cut. Developers will simply flock to other businesses that produce a higher return. The goal is to keep the rate of return on equity high by lowering the amount of cash that the developer must put at risk.

At first, housing reformers concentrated on cutting the rate of return without understanding that this would reduce development activity. One of the earliest such "limited-profit" projects was the Tower and Home Apartments built between 1876 and 1878 in the Cobble Hill section of Brooklyn. This "model tenement," designed by William Field, provided direct access to apartments from an open gallery. Each apartment extended to the other side of the building, thereby providing through-ventilation.

The philosophy behind the Tower and Home Apartments is summed up in the catchy slogan coined by its developer, Alfred Treadway White: "philanthropy plus 5 percent." White wanted to demonstrate that capitalists could provide decent housing for working people and still make a 5 percent return.

Brooklyn, 2012. Tower Apartments, completed in 1876, kept rents low by limiting the owner's return to a nominal 5 percent. *(Alexander Garvin)*

The slogan may have been appealing, but not enough to generate much developer activity. Nevertheless, housing reformers remained convinced that reducing return on equity could produce substantial amounts of housing at reduced cost to its tenants. Rather than continue to demonstrate this by building individual model tenements, they decided to try legislation encouraging developers to build housing that limited the return on their investment.

In 1926, New York State enacted the Limited-Dividend Housing Corporations Law. This statute provided a 25-year exemption from any increase in real estate taxes for any housing project built by limited-dividend corporations. Over the next seven decades, only 22 limited-dividend projects, containing 10,300 apartments, were built—nothing like the production for which reformers had hoped.[16]

Another try at creating a major housing program by limiting the return on equity was made during the New Deal. The PWA experimented with this approach even before developing projects on its own. Eventually, the Roosevelt administration produced eight limited-profit projects in six states, containing 4100 dwelling units. All but one of the projects was provided with PWA mortgages covering about 85 percent of total development cost.[17]

In contrast, many more entrepreneurs can be enticed into producing low-cost housing by reducing the amount of cash equity required rather than by reducing the rate of return on that equity. For example, a project with a development cost of $1 million, a $750,000 mortgage whose annual debt service is 10 percent, and $250,000 in equity with an annual return on that equity of 20 percent requires $75,000 to cover debt service and $50,000 to cover return on equity, totaling $125,000 ([$750,000 × 0.1] + [$250,000 × 0.2] = $125,000). The same project with only $100,000 in equity requires merely $110,000 ([$900,000 × 0.1] + [$100,000 × 0.2] = $110,000). Furthermore, since the rate of interest paid on a mortgage is likely to be lower than the rate of return on equity, there is a reduction in the economic rent. A greater number of people will be able to afford this lower price. Consequently, the developer's risk will be reduced.

Another method of decreasing equity requirements without decreasing the rate of return is direct subsidization of the gap between development cost and the amount of the mortgage. The PWA, for example, provided a direct subsidy of 45 percent of the development costs of its own projects. The remaining 55 percent was covered by a 60-year mortgage at a low rate of interest.

Such capital write-downs have become increasingly popular with local government agencies. The Los Angeles Community Redevelopment Agency (CRA) used this approach in 1982 in redeveloping a dilapidated block on 11th Street, south of the downtown office district. The project, Vista Montoya Condominiums, provides 180 apartments for low- and moderate-income households. Twenty percent of the condominium purchasers were low-income families displaced by this and other nearby projects. When Vista Montoya was completed in 1984, they could not afford the monthly payments on the new condos. So the CRA provided a capital write-down in the form of a non-interest-bearing second mortgage covering up to $45,000 of the purchase price. These mortgages come due only when the buyer sells the unit, at which time the CRA will receive the full amount of the mortgage plus 50 percent of the appreciated value of the unit.[18]

Public Housing

The obvious way to reduce economic rent to the absolute minimum is to virtually eliminate real estate taxes, debt service, and return on equity—leaving little more than maintenance and operating costs. The Housing Act of 1937 did just that, calling it "public housing." The purpose of this legislation was as follows:

> . . . *to provide financial assistance to states and political subdivisions thereof for the elimination of unsafe and insanitary housing conditions, for the eradication of slums, for the provision of decent, safe, and sanitary dwellings for families of low income and for the reduction of unemployment and the stimulation of business.*

The act provided the subsidies to local housing authorities (created by state enabling statutes) that built, owned, and operated housing for "families of low income."

As government entities, local housing authorities need no financial return on equity—their return is better measured in the social utility generated by the creation of better housing for the citizenry. The money to pay for the project is produced by selling long-term bonds. Debt service on these bonds is covered by an "annual contributions contract" between the local housing authority and the federal government that makes the federal government responsible for paying amortization and interest payments on bonds issued by the local housing authority. In exchange, local governments are required to reduce real estate taxes to a nominal amount (10 percent of shelter rent). Thus, the rent paid by the tenants is reduced to slightly more than maintenance and operating costs.

In 1987, there were more than 1.45 million public housing units in the United States.[19] By 2009, that number had shrunk by 270,000 units following the widespread impact of HUD's HOPE VI program, which provides federal funding for demolition of existing housing stock (discussed in the following section).[20] For most people, the conception of public housing conjures up an image of millions of people living in high-rise dormitory stockades. The reason for this image is that in New York, Chicago, Boston, and many other cities, public housing took the form of depressing brick boxes sur-

Brooklyn, 1969. The Williamsburg Public Housing Project, completed in 1938, was designed to the most modern international standards of the period and included generous recreation facilities, a public school, day care, and 50 retail shops. *(Alexander Garvin)*

rounded by fenced-in patches of grass. There is nothing in the Housing Act that mandates, or even recommends, this design.

Local housing authorities also build projects that become local architectural assets. Williamsburg Houses, built in Brooklyn between 1935 and 1938, is an example of a design built in the latest architectural fashion of that era. Woodside Gardens, built in San Francisco in 1968, and 2440 Boston Post Road, built in the Bronx in 1972, won design awards. Similar outstanding architecture has been produced throughout the 50-year history of public housing. Unfortunately, as is the case even in private-sector development, design excellence is the exception rather than the rule.

A vivid image of public housing was provided with the dynamiting of Pruitt Igoe, a 2762-unit public housing project consisting of 33 eleven-story buildings, completed in 1954 in St. Louis. This project had so many problems that, in frustration, the St. Louis Housing Authority ordered its demolition only 18 years after it had been completed. Conditions similar to Pruitt Igoe have arisen in public housing projects in Boston, Newark, Chicago, and elsewhere. It is these notorious situations that remain in the public consciousness, not the

Bronx, 2006. Most public housing built after 1950 consisted of high-rise apartment buildings located within projects that looked quite different from adjacent neighborhoods. *(Alexander Garvin)*

Bronx, 2012. Public housing at 2440 Boston Post Road, completed in 1972, was intended to set a standard of architectural excellence for the surrounding neighborhood. *(Alexander Garvin)*

hundreds of well-managed projects that provide hundreds of thousands of good apartments for families who otherwise could not afford a decent home.

One explanation for the notorious failures is the poor quality of construction and design. Congress established room-cost limits for public housing but no controls on the cost of land or the quality of construction. In order to satisfy these cost limits, some projects (especially where land costs were high) were built as cheaply as possible. In other cases, local authorities, as a matter of social policy, chose to minimize housing quality. They found it difficult to justify providing poor tenants with "amenities" that working people could not afford.

It is wrong to blame the failure of public housing on inadequate design or quality of construction. Many projects that are now in trouble were built to optimum standards. In fact, when completed in 1950, Pruitt Igoe was hailed by contemporary architecture magazines as an example of excellence in housing design. The real explanation for the failure of specific public housing projects involves fiscal policies, tenant selection procedures, lack of tenant diversity, maintenance practices, and project management.

When local housing authorities run out of money, they often cut back on services and maintenance. They develop cash-flow problems because the federal government forbids the accumulation of reserves. Rental income in excess of gross expenses (plus a transfer to reserves of no more than 50 percent of rent) must be used to reduce federal contract contributions (i.e., debt service). As a result of such reductions, between 1945 and 1953 the federal government paid less than half the nominal amount of its annual contract contributions.[21]

Public housing is usually conceived without consideration of eventual replacement requirements. Not only do stoves and refrigerators require replacement, so do boilers, plumbing risers, windows, and the like. Without the budgeted funds to take care of these items, many local housing authorities choose to defer replacement and repairs. Such deferred maintenance is a recipe for deterioration, and those projects in which maintenance has been deferred inevitably become slums. Had local authorities been able to use rents to build up proper replacement reserves, many projects would not be in such poor physical condition.

Another problem was created in 1969, when Senator Edward Brooke of Massachusetts succeeded in amending the Public Housing Program to require that no tenant pay more than 25 percent of income for rent. Congress agreed to pay the difference between 25 percent of income (currently 30 percent) and rent, but it has never appropriated enough money to cover this commitment. Local housing authorities have to cover the gap. It also directed subsidies to the neediest of the needy, creating even more dysfunctional communities.

Faced with increasing fiscal problems, some housing authorities cut back on personnel. When local politicians force them to hire political cronies, they become employers of last resort. Without personnel competent to manage and maintain the buildings, conditions become even worse.

Not all problems are fiscal or administrative. Initially, public housing was a temporary haven for the "deserving" poor, en route to stable jobs and houses of their own. Since the end of World War II, however, public housing increasingly has become a haven of the dependent poor. These poor need more than annual contract contributions from the federal government. When they move into public housing, they bring a myriad of other problems with them. This is bad for them, bad for the project they move to, and bad for the surrounding city.

Housing authorities across the country have overcome the problems that forced the Boston Housing Authority into receivership and the St. Louis and Newark Housing Authorities to demolish once-sound apartments. This success is due to their relatively strict admission and occupancy standards, administration that has been relatively unaffected by politics, and budgets that are supplemented by city governments to cover the costs of repairs, replacement, renovation, and shortfalls in federal funding. Hundreds of thousands of families are happy to get apartments in public housing, and millions more wish they could. It is the housing of choice for tens of thousands of poor New Yorkers. Indeed, there is so much demand in New York that the waiting list for a New York City

Manhattan, 2012. Twenty percent of the apartments in many of the buildings in the northern part of Battery Park City receive federal Low Income Housing Tax Credits. The rest are rented at market prices. There is no way to determine which are rented to persons of low income. *(Alexander Garvin)*

- At least 40 percent of all the units will be occupied by persons/families earning 60 percent or less of the median income (the *60/40 test*).

The Low Income Housing Tax Credit is prorated according to the proportion of the development that is set aside for low-income tenants. The remaining apartments can be rented to anyone willing and able to pay market rents. Residents of the low-income units must be "evenly distributed" among the properties' overall units and thus be indistinguishable from units renting at market prices. Occupants of these units cannot be charged rents that are higher than 30 percent of their income.

There are two types of tax credits: (1) *9 percent credits,* granted to projects financed by a mortgage whose lender is subject to federal income tax on the interest it receives, and (2) *4 percent credits,* for projects whose mortgage is tax-exempt (usually from a state housing finance agency). Both work the same way. The recipient of a 9 percent credit receives benefits equal to 9 percent of development costs (excluding land), and the 4 percent credit receives 4 percent (also excluding land). The federal government believes this eliminates double-dipping into federal subsidies. Once they begin, the tax credits may be claimed for 10 years. A development that qualifies for the 9 percent credit and receives an $850,000 allocation from the state, for example, will provide its owner with $850,000 annually for 10 years, assuming that 100 percent of its units are for low-income tenants.

Boulder, 2000. All the units in The Woodlands are subsidized with federal Low Income Housing Tax Credits. *(Alexander Garvin)*

Subsidizing Demand

There is no way of ensuring "a decent home and a suitable living environment for every American family" unless every American family can afford it. Thanks to the federal programs that greatly increased housing supply, most Americans do not need financial assistance to be able to live in a decent home. Nevertheless, many low- and moderate-income people do!

There are many ways of helping the poor. The services strategy does so by supplying them with whatever society considers to be essential. Building publicly owned and operated housing for persons of low income is an example of this approach.

Another way is to help by increasing incomes. This allows low-income people to select the services they feel satisfy their needs. Giving poor people more money with which to purchase a decent home, however, does not guarantee that they will use it for that purpose. Poor families, especially those in acute need of improved clothing, food, and other necessities, may choose to spend on something other than housing. Even if they do spend this additional money on housing, there is no way of being sure that the added expenditure will result in a move to higher-quality housing (if it is available) or in an improvement in the quality of the housing they continue to occupy. The billions of dollars appropriated annually by Congress for welfare payments illustrate this point quite well.

Where there is insufficient decent housing, giving poor people more money with which to purchase it increases the price of available housing for everybody else. Prices should start to decline when the housing industry builds enough additional housing to meet the increased demand. Since it takes a long time for developers to perceive the increased demand, acquire property, hire architects to make the necessary plans, obtain the financing, and complete construction, the inflationary period can last many years.

If the subsidy is insufficient to cover the cost of supplying new housing, additional income will merely allow poor people to outbid those next up on the economic ladder, causing unintended hardship and opposition. Once again, the welfare program illustrates this quite well. Not one unit of housing has been built by developers seeking to satisfy the increased purchasing power of welfare recipients. Furthermore, such subsidies create considerable animosity among working families who do not get supplemental income to pay for better housing.

Without limiting occupancy to people who could not otherwise afford decent housing, no government program to lower the cost of supplying housing can guarantee that the subsidy will be passed through to the poor. For this reason, many housing programs (public housing, 221(d)(3), 236, Section 8, etc.) restrict occupancy to persons within defined income limits. Restricting occupancy also segregates the beneficiaries from the surrounding neighborhood. It is doubtful that Congress ever intended or even understood that income-specific housing programs increase segregation. Nevertheless, that has been the result in many of the nation's older public housing projects in which there has been little or no effort to integrate the subsidized residents with the market-paying residents.

Supplementing Rent

Demand-subsidy programs need not be inflationary or increase segregation. The Section 23 Leased Public Housing Program avoided not only segregation but also the stigma of conventional public housing projects. Section 23, enacted in 1965 and terminated in 1974, provided local housing authorities with annual contribution contracts covering the cost of renting apartments in existing buildings.

Theoretically, Section 23 guaranteed widespread tenant dispersal by limiting the number of apartments that could be leased to 10 percent of any building or project. In fact, there was little dispersal. In many areas, rents exceeded program limits. Even when rents were at or below program limits, landlords often refused to accept public housing tenants. Any housing authority could waive the 10 percent limit, and many did. Congress never appropriated enough money to disperse more than a few families. The program's most serious flaw was that it did nothing to reduce the inflationary impact on the local housing market.

The Rent Supplement Program, enacted at the same time as Section 23, tried to eliminate both segregation and inflation. It provided a 40-year subsidy covering the difference between "fair market rent" and 25 percent of tenant income.[34] Initial occupancy was restricted to families whose incomes were at or below public housing levels. Inflationary pressures were eliminated by tying the 40-year subsidy exclusively to newly built apartments. Thus, the increase in demand is matched by an equal increase in supply.

The Rent Supplement Program avoided economic segregation within a project by allowing the tenants receiving the subsidy to remain in their apartments even if their incomes rose. The subsidy simply decreased by 25 percent of any increase in income until rent equaled 25 percent of income. However, the Rent Supplement Program failed to prevent geographic and, by extension, racial segregation. Rent Supplement projects had to receive approval of any locality in which they were built. To nobody's surprise, during the three years of its operation, virtually no projects were proposed or approved in suburban areas.

In 1973, unhappy with the administrative complexity and expense of government housing programs, the high cost of government-financed development, and the inequitable geographic distribution of subsidy programs, the Nixon administration decided to experiment with direct payment of housing allowances. The Housing Allowance Experiment spent $160 million to provide 25,000 families in 12 metropolitan areas with housing vouchers for periods of 3 to 10 years.[35]

The affected families failed to generate sufficient demand, either to increase rents or to induce housing construction in any of the 12 participating areas. Most participants chose to remain where they were already living and not to spend the allowance on improved housing. They simply reduced the proportion of recipient family income going to rent. Recipients who did choose to move relocated to areas with populations that had higher incomes. One has to wonder why it was necessary to spend $160 million on an experiment that came up with such obvious conclusions.

The most recent demand-subsidy program to be terminated is Section 8 of the Housing and Community Development Act of 1974. The Section 8 Program applied only to persons of low income and subsidized the difference between "fair-market rent" and 25 percent (later 30 percent) of income. It came in four varieties: *existing housing* (which was essentially a 15-year housing allowance tied to the recipient in the form of a five-year contract, renewable for two additional five-year periods and payable for any apartment in acceptable condition that was within fair-market rent levels); *moderate rehab* (which tied the same 15-year subsidy to properties undergoing moderate renovation); *substantial rehab* (which provided the same subsidy for 20 to 40 years for gut rehabilitation of apartment houses); and *new construction* (which provided the same 20- to 40-year subsidy for construction of new multiple dwellings). To qualify, each locality had to decide on an annual mix of Section 8 projects based on a "Housing Assistance Plan" designed to meet the objectives of the 1974 act.

There is little difference between the Housing Allowance Experiment and the Section 8 Existing Housing Program except its nationwide character. On the other hand, Section 8 Moderate Rehabilitation eliminated major deficiencies in previous demand-subsidy programs. Because it was tied to existing buildings, it guaranteed significant improvement in the quality of housing supplied to all recipients. Because the amount of subsidy depended on tenant income, there was genuine economic integration. Because existing buildings (rather than large projects) received the subsidy, there was no stigma attached to residing in a Section 8–assisted property. Most important, there was no inflationary impact on a city's housing stock because no residents could take their subsidy and bid for housing elsewhere. The same benefits applied to Section 8 Substantial Rehab. In addition, since the subsidy was applied to structures that had previously been unoccupied, it also effectively increased the supply of available apartments by making previously vacant units habitable. However, it required several times as much subsidy per recipient family. The Section 8 New Construction Program was different from the other three programs only in that the subsidy per family was even greater.

During the 1980s, faced with huge budget deficits and very high cost per recipient, the Reagan administration terminated funding of all Section 8 programs that were tied to construction or rehabilitation. As of 1992, 1.5 million families had received Section 8 Existing Housing subsidies, 109,000 apartments had received Section 8 Moderate Rehabilitation subsidies, and 831,000 apartments had received Section 8 Substantial Rehabilitation and New Construction subsidies.[36] It is doubtful that many people in Congress, HUD, or local government perceived that Section 8 moderate rehab had achieved previously unfulfilled goals, without the usual ill effects, at a substantially lower cost per unit than any of the previous demand-subsidy programs.

Although HUD had terminated Section 8 production programs, it continued to issue Section 8 certificates and vouchers that accompanied recipients when they moved from one apartment to another. Since there was no longer certainty that Congress would continue providing these subsidies to ongoing occupants of Section 8 projects, HUD began negotiating refinancing agreements with project owners. Project mortgages were recalculated based on market rent rather than economic rent. HUD subsidized the difference. Since then, Congress annually reauthorizes whatever Section 8 subsidy is still required by the remaining qualified tenants.

Ingredients of Success

For most of the twentieth century we have been initiating housing assistance programs that promise once and for all to provide a decent home for every American family. As soon as developers and government officials learn how one program works, a new one takes its place. It takes several years for everybody to understand the new program. The resulting lag in production is unnecessary. After spending hundreds of billions of dollars on housing-subsidy programs, we have more than enough experience to know what will be successful.

Market

Most housing assistance programs affect the price paid by the occupant without having much impact on anybody else. Consequently, they are supported by program participants and ignored by the rest of the population until neighborhood opposition or budget appropriations become intolerable. Only those housing assistance programs (e.g., federal income tax deductions for mortgage interest payments) that improve market conditions throughout an area are likely to remain politically and financially feasible for any length of time and thus make much of an inroad in providing a decent home for every American family.

Occasionally, as in New York City's real estate tax-exemption programs, a locality will provide housing assistance that genuinely alters local market conditions. Budget constraints make widespread adoption of such local programs unlikely. Consequently, most programs that reduce housing costs are likely to originate with the federal government.

Congress has never funded housing-subsidy programs at a level high enough to produce substantial additions to any city's housing stock and, therefore, to lower prices. As a result, benefits were passed through to building occupants without affecting the housing market of the surrounding city and without gaining widespread public support. Nevertheless, the overall housing market has steadily improved. By 2010, 65 percent of the population of the United States lived in resident-owned housing.

Location

Consumers naturally favor residences with inherent advantages such as good views, plenty of light and fresh air, and beautiful landscaping. They prefer to be able to get to shopping, recreation, and employment facilities easily and quickly. Local policies that artificially lower residential real estate taxes, as well as the benefits of federal income tax deductions, only intensify these locational advantages. Consequently, buildings that receive such benefits tend to profit from continuing consumer interest and remain in good condition.

Subsidized housing projects, on the other hand, do not have to be built in attractive or convenient locations. More often than not, the limited subsidies that are available result in selection of sites whose low cost reflects their unattractiveness or inconvenience. Artificially low rents overcome most locational disadvantages. Consequently, most subsidized housing ends up on cheap land, far from "better" neighborhoods. Despite the availability of suburban land that is cheap enough to satisfy federal housing assistance requirements, very little subsidized housing has been located in the suburbs. It is almost as if the Department of Housing and Urban Development and its predecessor agencies had no idea there were housing problems or poor people outside center cities. Consequently, public assistance recipients tend to be concentrated in the least desirable center-city locations. The result is communities that are not only sequestered but also therefore easily stigmatized as "public projects."

The Low Income Housing Tax Credit, like other tax benefits, can be used in a way that does not distort market-based locational decisions. When it is applied to the minimum number of the apartments (i.e., either 20 percent of its units occupied by persons below 50 percent of median income or 40 percent of its units occupied by persons below 60 percent of median income), every apartment in the project has to be marketable. The program requires all apartments of each type to be the same. Because a minimum number must be occupied by persons of low income, and management cannot be sure which apartments will be occupied by low-income residents, it has no choice but to make sure that all apartments can be rented at market prices, regardless of whether the resident is a person of low income. In this way, LIHTC replaces bureaucratic review with consumer choice. It also sidesteps time-consuming political controversy, increases program efficiency, and eliminates the cost of project review.

Design

It is impossible to distinguish houses that receive income tax benefits from houses that do not because renters, rather than their owners, occupy them. The same is true of LIHTC apartment complexes, where 20 percent of the units are occupied by persons below 50 percent of median income *or* 40 percent of the units are occupied by persons below 60 percent of median income. Publicly assisted housing projects, on the other hand, are usually visibly different from the neighborhoods around them. Public housing projects are even more different. Consequently, they become lightning rods for discrimination and opposition.

These differences extend to the way homes and projects affect the landscape. Normally, urbanization is incremental: a few houses are replaced by an apartment building; several years later a movie theater is replaced by an office building; and so on. The resulting urban fabric is a complex mix of colors and materials, periods of construction, styles of architecture, land uses, and human activities. Publicly assisted projects, on the other hand, are usually large enough to vary significantly from the surrounding neighborhood.

If recipients of housing subsidies are to avoid the stigma of public assistance, they must be able to participate in the housing market without being identifiable. The only way for this to happen is for government to stop financing "projects" and operate within the context of incremental development by conventional developers whose buildings are financed by conventional lending institutions.

Financing

Using state and local bonds to create a pool of money for long-term mortgages at below-market interest rates is an efficient method of lowering debt service and, thus, the cost of housing. A similar result is obtained with 9 percent LIHTC projects financed with an ordinary bank mortgage. Most federal and state mortgage loan programs, on the other hand, employ underwriting procedures that are based on allowable costs, fixed fees, and limited returns on equity. In such cases, there is no reason for borrowers to reduce the size of their loan and, therefore, to further reduce economic rent. For this reason, some local agencies transfer their bond proceeds to local banks whose conventional underwriting procedures avoid this problem.

Nearly all nonprofit groups are unable to provide the substantial equity contributions most lenders require. The LIHTC deals with this problem by allowing them to sell the project's tax credits. In return for the right to use the tax credits over the 10-year period, the purchaser will make an upfront payment that usually provides the bulk of the developer's equity in an affordable housing development. Since tax credits taken over several years in the future are worth less than the whole amount would be if it could be used immediately, they are usually sold at a discount (typically 65 to 75 cents on the dollar).

Time

The simplest way to reduce housing cost is to reduce the time it takes to produce the housing. During the development period, every project incurs soft costs (expenses for interest on the already borrowed money, taxes, professional fees, insurance, rent, and salaries). If hard costs (land and construction) amount to $1 million and soft costs constitute an additional charge of 25 percent per year, then completing the project in six months instead of one year will save $125,000 ($1,000,000 × 0.25 × 0.5).Perhaps the most neglected aspect of time is the time it takes government to take the actions that are a necessary part of creating, maintaining, or paying for housing. The longer government takes providing a permit or an approval, the more the project will cost. Every additional day government takes to perform its role results in additional carrying costs for that same period of time. Thus, government has the ability to either reduce the cost of housing by reducing the time it consumes or to increase housing cost by not performing its role expeditiously.

In the case of subsidized housing, even when the additional amount of time and the resultant increased development costs are substantial, they may not be apparent, because, upon completion, the increased cost of development is rolled into the permanent mortgage, and its effect is thereby mitigated by other government subsidies. Since that increase is offset by project subsidies, there is no urgency for government personnel to approve the project. Thus, subsidies neutralize the effect of time on public and publicly assisted housing projects. The ultimate impact, however, is not neutralized. Since the amount of subsidy money available is finite, it has to be spread among fewer housing units, thereby reducing the number of people who benefit from the reduced price of housing and increasing the marginal cost of each unit to the taxpayer.

None of these additional costs apply to income tax or real estate tax programs that lower housing prices. Developers receiving tax subsidies have a financial incentive to reduce development time: the faster they complete the development, the lower their tax payment. Consequently, projects whose only subsidy comes from income tax or real estate tax benefits are invariably less expensive and take less time to complete. Since they use less subsidy money per unit, the total government allocation produces a greater number of apartments. Thus, a good way to obtain the maximum amount of low-rent housing is to shift all processing and financing to conventional lending institutions.

Entrepreneurship

Approximately 75 percent of American houses are constructed by home builders that construct fewer than 15 houses per year. The small size of their operations allows them to respond quickly to changes in demand, to maintain tight supervision over every aspect of their business, and to exploit unusual opportunities for minor economies. Most government-subsidy programs clash with this low-overhead approach to development. Small builders are not willing to comply with time-consuming requirements, prior reviews of proposed activity, and post-factum audits. Consequently, they avoid most government-subsidy programs. Perhaps that is why the largest home builders have so much market share; in 2011, the ten largest home-building companies were responsible for more than 20 percent of housing starts.[37]

The tax deduction for homeownership is a benefit that all home builders profit from but do not have to apply for. Their customers claim it on their income tax forms. This cost-effective approach completely eliminates the need for a government agency or for time-consuming operations that true entrepreneurs make every effort to avoid.

Developers of subsidized housing projects must withstand government scrutiny. Consequently, they need well-organized staffs and routinized business procedures. They employ a different labor force and very different construction practices from the more informal home builders. It is therefore relatively easy for them to comply with the provisions of the Davis-Bacon Law, HUD cost certification, Equal Opportunity regulations, and whatever else Congress or a state legislature decides to require. Compliance with these requirements is nonetheless time-consuming and increases the cost of development. In a competitive market, this situation would reduce profitability and deflect entrepreneurs to other businesses. It has no effect on the subsidized housing business because, with the exception of LIHTC, the government fixes both allowable costs and the percentage of return on equity. Thus, there is no incentive to bring costs below government-approved, cost-certified limits. Moreover, because the allowable return is limited to a fixed percentage of project cost, there is every reason to reach those cost limits.

When housing assistance programs do not limit cash equity requirements, costs, or return on equity, the developer has an incentive to use all his or her entrepreneurial skills to do what conventional developers do: try to minimize the development cost. That is one reason the FHA 608 Program (see Chapter 9) was so efficient. LIHTC has an additional efficiency incentive. During the first 15 years of occupancy, any noncompliance with program requirements (such as not having the approved number of apartments occupied by persons of low income) results in the recapture of all the tax credits. No investor will willingly risk such action by the Internal Revenue Service.

Reducing Housing Costs as a Planning Strategy

Even if there were a suitable residence for every American, some people would not be able to afford it. As this text demonstrates, there is a rich history of ways to provide them with the necessary subsidies. But if housing subsidies are to have a beneficial effect on our cities and suburbs, they must be embodied in programs that are economical, efficient, equitable, responsive to the desires of the recipients, able to generate desirable market activity, and not inclined to create problems for healthy neighborhoods. They must be economical because minimizing the subsidy per unit maximizes the number of people who can benefit. They must be efficient, or relatively few entrepreneurs will be interested in supplying subsidized shelter. Thus, it will take longer to produce the same number of dwelling units and fewer people will have access to affordable shelter. They must be equitable, or there will be legitimate opposition from those who do not receive subsidies and from those who believe their subsidy is inadequate. They must be responsive to the desires of the recipients, or recipient dissatisfaction will be directed at the housing they are forced to accept. They must also generate desirable market activity; otherwise, those who do not receive subsidized housing will not benefit from the program and therefore will oppose it.

The Low Income Housing Tax Credit Program is the most economical federal program providing housing for low-income residents yet enacted by Congress. It allocates subsidies by formula and leaves the distribution to state governments. Federal personnel become involved only if the Internal Revenue Service has to recapture tax benefits from a project that is not in compliance with program requirements. The minimal personnel and processing make it inherently efficient. Distribution by formula guarantees that it will be equitable. Those projects that are not occupied entirely by persons of low income have to be responsive to consumer demand; otherwise, market-rate apartments would be vacant. Moreover, the requirement that apartments occupied by low-income residents be identical to those occupied by persons with higher incomes eliminates the stigma of living in a "project." Program effectiveness on a local level, however, is entirely the product of the procedures used by state agencies distributing subsidies and of the manner in which local developers use them. Thus, the projects themselves might or might not meet the test of being economical, efficient, equitable, and responsive to consumer demand.

South Bronx Nehemiah

Ironically, one of the most effective local efforts to provide affordable (though not the most beautiful) housing, the South Bronx Nehemiah Project involved no federal money whatsoever. I. D. Robbins proposed what he called the Nehemiah Plan in a newspaper article (see Chapter 2). He first tried it building homes in the East New York section of Brooklyn. Its strategy was to cut every possible component of the development cost as well as every possible component of economic rent.[38]

The land price was set at a nominal $500 per dwelling unit. This was what the city charged most nonprofit developers. By developing one-family and two-family houses, Robbins reduced construction cost by approximately $30 per square foot. This lower cost was the result of different building code requirements that eliminated the need for elevators, fire stairs, specialty mechanisms and spaces to provide access for the disabled, and many other items. By repeating the same design and using prefabricated components and common construction techniques, he further lowered the hard costs. In addition, he reduced construction costs by purchasing in large quantities from original sources rather than through intermediaries, paying suppliers within nine days of delivery and making shrewd use of loyal contractors. The bids he received from excavation, foundation, electric, plumbing, and carpentry contractors were far below industry standards because he kept them steadily employed and paid their bills so quickly.

Speeding up construction further reduced costs. He did this by carefully scheduling the excavation contractor who used a backhoe to excavate all the basements on a single block front, followed by a foundation contractor

Bronx, 1994. The Nehemiah Plan produced affordable dwellings that were purchased by residents of the South Bronx. *(Alexander Garvin)*

who set footings and poured concrete for basement walls. Factory-manufactured steel frames were delivered to the site so that the carpenter could install them in two days. As Robbins explained,

> *Everyone is lined up and ready to go. If the land is ready to build on, we can put up ninety-three homes on four and a half acres and can move families in within five months. No, wait a second. Make that six months. We have to take into account foot dragging on the part of the housing bureaucracy.*[39]

Soft costs in government-assisted projects usually range from 20 to 40 percent of project cost. Robbins cut them to 6 percent by reducing fees, interim interest, taxes, and marketing and closing costs. Most projects include a fee of 10 percent for the builder-sponsor's overhead and profit. Rather than charge this routine $5000 to $6000 per unit, Robbins cut it to $1000 per unit. He was able to operate at this low figure because of low overhead. The staff consisted of Robbins, an assistant, two clerk-secretaries, a part-time building inspector who approved contractor vouchers, and a full-time expediter to obtain approvals from city agencies. Builders usually include a "general conditions" fee to cover security arrangements and other incidentals during construction. Robbins established a staging area where building materials and supplies were stored, hired security guards, and used an unoccupied apartment for his staff, thereby reducing costs to $500 to $600 per house. Most architects charge a fee that ranges from 5 to 15 percent of construction cost, depending on the size and type of building. Robbins paid $300 every time he repeated the same design. When a house is connected to the city's water and sewer systems, the builder pays a standard fee. Robbins persuaded the city to allow a single connection for all the houses on a block, thereby saving thousands of dollars.

Another reduction in soft costs came in interim interest, interim taxes, and marketing costs. The churches that sponsored South Bronx Nehemiah established a $3.5 million interest-free revolving fund that allowed Robbins to pay bills quickly and obtain lower bids from contractors. It also entirely eliminated interim interest on all the money borrowed during the development period. The city effectively eliminated real estate taxes, something that it did routinely for nonprofit housing. Marketing and selling was done through the churches. Standard application forms and closing documents curtailed legal and bank processing fees and other closing costs.

Bronx Nehemiah automatically lowered the cost to the homeowner by reducing economic rent. The homeowner took care of the maintenance costs that would usually pay project management and maintenance personnel. Real estate taxes were largely eliminated through the city's 421 Tax Exemption Program. Debt service was reduced because the State of New York provided the purchaser with a below-market interest rate $40,000 first mortgage and the City of New York provided an interest-free, $15,000 second mortgage that came due only when the house was sold.

The price paid by homeowners would have been even lower if the city government had agreed to Robbins's desire to round out sites. Neighborhood impact would have been far broader as well. Nevertheless, between 1993 and 1995 South Bronx Nehemiah provided 512 units of housing (224 one-family houses and 288 units in three-story condominium apartment buildings) at prices that were nearly half those of contemporaneous government-subsidized projects on nearby blocks.

Many of the parishioners who purchased Nehemiah

Bronx, 2012. Prices of Nehemiah Plan buildings were low enough that some residents of the nearby McKinley public housing project were able to purchase them. *(Alexander Garvin)*

homes moved from nearby public housing, in the process freeing up low-rent housing for which there was a waiting list of 200,000 families. This was a case of affordable housing that was very much in demand trickling down to a low-income household every time a Housing Authority tenant purchased a house. Had there been more Nehemiah projects, they might have triggered further improvements in nearby areas of the South Bronx, but government agencies thought the density was too low. They were uncomfortable working with procedures that other developers rejected. Consequently, they were not eager to continue providing cheap land or subsidized mortgage financing. Unfortunately, Nehemiah required this mortgage assistance. Houses could be built at low enough prices only when sites were built at low densities. I. D. Robbins died in 1996 at the age of 86, before he could persuade long-reluctant city officials to support another round of development.

Over the next 16 years, the New York City Partnership acquired infill sites in the neighborhood where it erected more than 70 additional units of affordable housing. It was followed by private developers who erected buildings on six sites and expanded the local grocery store. By 2012, this section of the South Bronx had become a vital neighborhood supplying hundreds of families with more-than-decent homes.

A Market-Based Approach to Reducing Housing Costs

South Bronx Nehemiah and LIHTC demonstrate that a comprehensive, market-based approach to housing development can reduce the required subsidy per dwelling and thereby generate a greater amount of affordable housing for the same amount of subsidy money. These two programs also minimized the role played by government agencies and the time and money they spend, thereby reducing the cost of distributing that subsidy. But, laudable as this may be, nothing in either strategy guarantees that benefits will accrue to anyone except program participants. Widespread and sustained market reaction requires that a critical mass of beneficiaries must be concentrated in the market area in which those beneficiaries are located. Only then can these strategies generate the spillover benefits that constitute serious planning.

Notes

1. The "rent" people refer to may mean different things and represent different sums of money. *Affordable rent* is the amount a tenant can pay and still retain enough to pay for food, clothing, medical care, transportation, and other necessities. *Market rent* is the amount that people will pay for similar facilities, similarly situated. *Contract rent* is the payment required by written agreement between the owner and tenant. *Regulated rent* is the amount established by those local governments that control everything, from the price the owner can charge to the level of service that must be provided. *Economic rent* is the amount required to justify any real estate venture, the amount that will cover all the expenses of running the property. Those expenses include (1) maintenance and operating costs, (2) real estate taxes, (3) debt service on mortgage loans, and (4) payments to equity investors.
2. The projects discussed in this chapter are intended solely to explain the techniques used in reducing housing costs and to illustrate some of the effects of those techniques. Evaluating the projects themselves would require far more space than is appropriate for this volume.
3. James Ford, *Slums and Housing,* Harvard University Press, Cambridge, MA, 1936, pp. 716–736.
4. *United States v. Certain Lands in the City of Louisville,* 9 F. Supp. 137, 141 (D.C.W.D. Ky. 1935) and 78 Fed. 2nd. 684 (C.C.A. 6, 1935). See Lawrence M. Friedman, *Government and Slum Housing,* Rand McNally, Chicago, 1968, pp. 102–103.
5. See Chapter 6, note 12.
6. New York City Housing and Development Administration, *Community Development Program Progress Report 1968,* New York City, New York, p. 215.
7. Historical and statistical material on Levitt and Sons and their new towns on Long Island and in Pennsylvania and New Jersey is derived from Eugene Rachlis and John E. Marqusee, *The Landlords,* Random House, New York, 1963, pp. 228–256; Herbert Gans, *The Levittowners,* Pantheon Books, New York, 1967; and John T. McQuiston, "If You're Thinking of Living in Levittown," *New York Times,* November 27, 1983.
8. Government reduction of real estate taxes can take two forms: (1) exemption for a specified period of time from any tax increase in the value of the property due to an improvement and (2) abatement for a specified period of time of some or all of the taxes to be paid. Sometimes, to ensure that the property owner passes this reduction on to the tenants, the tax reduction will be restricted to buildings with regulated rents or whose occupancy is exclusively for persons of low, moderate, or middle income.
9. Initially, the benefits were restricted to multiple dwellings of four or more stories and to increases in tax assessment up to $1000 per room and $5000 per apartment. Later, benefits were restricted to $15,000 per building. See Citizens Housing and Planning Council, *How Tax Exemption Broke the Housing Deadlock in New York City,* New York, 1960.
10. New York City Department of Housing and Development, Division of Financial Services. The figures are for fiscal years ending June 30. In 1986, the 421 Program was amended to exclude certain sections of the city (thought to be high-rent areas) unless the developer also provided "low-rent" apartments. In other geographic areas the exemption period was extended to 10 years.
11. Initially the 221(d)(3) Program was restricted to families "displaced from urban renewal areas or as a result of government action." In 1962 it was broadened to also cover all "low- and moderate-income families."
12. Later, Congress removed Fannie Mae from this role and established a new institution, Ginnie Mae (Government National Mortgage Association), specifically for the purpose of acting as a secondary market for such federally subsidized mortgages.
13. U.S. Department of Housing and Urban Development, *H.U.D. Statistical Yearbook,* Washington, DC, 1978.
14. U.S. Commission on Urban Problems, "Public Assisted and Subsidized Housing," in *Federal Housing Policy and Programs Past and Present,* J. Paul Mitchell (editor), Center for Urban Policy Research, Rutgers University Press, New Brunswick, NJ, 1985, pp. 319–336.
15. U.S. Department of Housing and Urban Development, "Interest Rate Subsidies: National Housing Policy Review," in *Federal Housing Policy and Programs Past and Present,* J. Paul Mitchell (editor), op. cit., pp. 337–364.
16. New York City Department of City Planning, *Public and Publicly Aided Housing 1927–1973,* New York, 1974.
17. With the exception of Knickerbocker Village, which was financed by

the Reconstruction Finance Corporation, they were low-rise complexes, designed and developed by individuals interested in housing reform. The best of these projects, Hillside Homes in the Bronx, was designed by Clarence Stein. It replaced 26 undeveloped acres with 1416 apartments primarily in four-story, walk-up structures organized around a series of landscaped recreation areas connected by a pedestrian spine. Community rooms, workshops, and a nursery were provided in the basements.

18. Urban Land Institute, *Project Reference File,* vol. 15, no. 4, January–March 1985.
19. U.S. Department of Commerce, Bureau of the Census, *Statistical Abstract of the United States,* Washington, DC, 1991, p. 732.
20. 2009 data from the U.S. Department of Housing & Urban Affairs, American Housing Survey, "Picture of Subsidized Households 2009," www.hud.gov, retrieved July 18, 2012.
21. Eugene J. Meehan, "The Evolution of Public Housing Policy," in *Federal Housing Policy and Programs Past and Present,* J. Paul Mitchell (editor), op. cit., pp. 287–318.
22. http://www.nyc.gov/html/nycha/html/about/factsheet.shtml.
23. Congress directly appropriated money to HOPE VI until 1999, when it passed Section 535 of the Quality Housing and Work Responsibility Act (Public Housing Reform Act) of 1998. This act amended Section 24 of the Housing Act of 1937 and authorized HOPE VI as a federal program; see www.hud.gov/pih/programs/ph/hope6/hope6.html.
24. U.S. Department of Housing & Urban Affairs, http://www.hud.gov/budgetsummary2010/fy10budget.pdf.
25. Historical and statistical material on Townhomes on Capitol Hill is derived from Daniela Deane, "A New Face for Public Housing," *Washington Post,* May 8, 1999, p. G01; Adam Katz-Stone, "Can Other Areas Repeat Success of Hill Renewal?" *Washington Business Journal,* February 19, 1999; www.archrecord.com/PROJECTS/DEC00/BTS/CAPITAL.ASP; and www.qualityplaces.marc.org/4a_studies.cfm?Case=6.
26. In 1999, when Townhomes on Capitol Hill opened, the median income in Washington, DC, was $34,000.
27. Historical and statistical material on Elm Haven is derived from Carrie Melago, "The Elm Haven Projects Have Been Reborn as a Sparkling Monument to Tenant's Patience," *New Haven Register,* February 25, 2001; Jamie Schuman, "Elm Haven Project to Receive a Total Facelift," *Yale Daily News,* February 18, 1999; and www.hud.gov/oig/ig811002.pdf.
28. Robert Carroll, John F. O'Hare, and Phillip L. Swagel, *Costs and Benefits of Housing Tax Subsidies,* June 2011 The Pew Charitable Trusts, http://www.pewtrusts.org/our_work_report_detail.aspx?id=85899361402; retrieved July 18 2012; and Charles Wallace, "Is It Time to Kill the Mortgage Interest Deduction?" *Daily Finance,* April 20, 2011, http://www.dailyfinance.com/2011/04/20/eliminate-mortgage-interest-tax-deduction/, retrieved July 16, 2012.
29. I have used 20-year and 40-year straight-line depreciation for simplicity and because any attempt to use current depreciation schedules would be obsolete as soon as Congress altered the tax laws.
30. California Housing Finance Agency, "Statistical Supplement to AnnualReport2010–2011,"http://www.calhfa.ca.gov/about/financials/reports/index.htm, retrieved July 18, 2012.
31. "45 Years of Housing Opportunities," Massachusetts Housing Finance Authority 2011 Annual Report, http://www.youblisher.com/p/272395-MassHousing-s-2011-Annual-Report/, retrieved May 29, 2012.
32. Statistical and historical information on LIHTC is derived from Susan Hobart and Robert Schwarz, "Housing Credits—A Leading Financial Tool," *Urban Land,* November 1995, pp. 37–42; "The Low Income Housing Tax Credit," *Urban Land,* November 1997, pp. 48–51, 78–80; and U.S. Department of Housing and Urban Development, www.huduser.org/portal/datasets/lihtc.html.
33. Congress originally set the formula of the allocation of low-income tax credits at $1.25 per capita. It was increased to $1.50 for 2001 and $1.75 for 2002.
34. *Fair-market rent* is a number determined by HUD to meet the objectives of whatever program is financing the project in question. In the case of the Rent Supplement Program, fair-market rent was computed to include up to 6 percent interest plus the 0.5 percent FHA fee.
35. See Bernard J. Frieden, "Housing Allowances: An Experiment That Worked," and Chester Hartman, "Housing Allowances: A Bad Idea Whose Time Has Come," in *Federal Housing Policy and Programs Past and Present,* J. Paul Mitchell (editor), op. cit., pp. 365–389.
36. Section 8 program statistics are derived by Find/SVP information services from HUD printouts for December 13, 1991. The 1.5 million recipients of Section 8 existing housing include Section 23 and Rent Supplement conversions, public housing demolition, and reallocations from other programs.
37. Barry A. Rappaport and Tamara A. Cole, *Housing Starts Statistics—A Profile of the Homebuilding Industry,* U.S. Census Bureau, Manufacturing and Construction Division, Washington, DC, 2000; Sarah Portlock, "Smaller Homebuilders Constructing New Ways to Find Work in Sour Economy," *The Star-Ledger* (New Jersey), April 21, 2012, http://www.nj.com/business/index.ssf/2012/04/smaller_homebuilders_construct.html; and John Gittelsohn, "Biggest U.S. Homebuilders Take Over Market as U.S. Home Sales Begin to Rebound," *Bloomberg News,* January 19, 2011, http://www.bloomberg.com/news/2011-01-19/biggest-u-s-homebuilders-to-gain-market-share-as-demand-returns.html.
38. Statistical and historical information on South Bronx Nehemiah is derived from interviews with I. D. Robbins during May 1994 and February 1995; and Jim Rooney, *Organizing the South Bronx,* State University of New York Press, New York, 1995.
39. I. D. Robbins, quoted in Rooney, op. cit., p. 104.

11

Housing Rehabilitation

New York City, 1974 *(Alexander Garvin)*

In Boston's South End, San Francisco's Haight-Ashbury, and Washington's Dupont Circle, rehabilitation is an ongoing activity. In too many other neighborhoods, buildings deteriorate because lending institutions are not prepared to finance improvements, occupants are not willing or able to pay for those improvements, and, thus, property owners cannot justify capital investment. The dilapidated housing that results may be more susceptible to fire, affect the health of its occupants, or generate other circumstances that result in remedial government spending. When property owners maintain and renovate their buildings, government can save money that it would otherwise spend. That alone may justify public intervention into the real estate market. Usually, both the occupants of deteriorating housing and residents who do not want to live next to dilapidated buildings demand government action. Until the second part of the twentieth century, the common government reaction was slum clearance. Housing rehabilitation is not only cheaper but also generates far less opposition.

We know how to stimulate the private sector to improve housing quality through rehabilitation. The programs that are needed will vary from city to city because the housing stock and the applicable laws vary. The techniques, however, will be the same: pursuing real estate tax policies that do not discourage property improvement, enacting rent regulations that allow property owners to recoup capital investments, and fostering an investment climate that encourages lending institutions to provide financing.

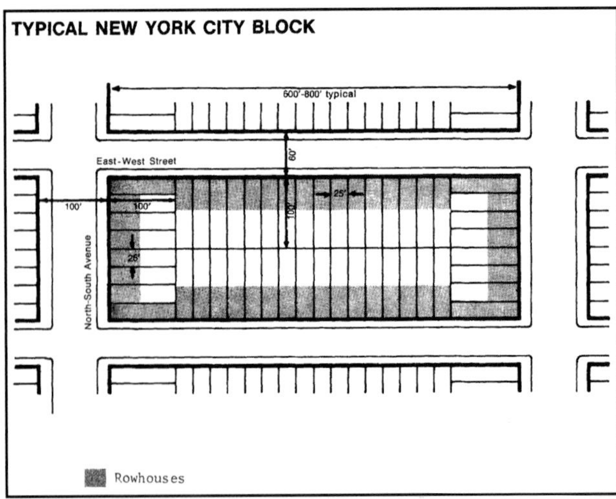

New York, 1811. Lot and block dimensions of the commissioners' plan. *(Courtesy of New York City Department of City Planning)*

The Housing Stock

Housing rehabilitation programs must be local in nature. They depend on the design and condition of specific buildings, the characteristics of their owners and occupants, the laws that govern what can be done with them, and the lending practices in the area. I have chosen to examine housing rehabilitation in New York City because any account of its rehabilitation programs will cover most situations facing other cities.

There is a more personal reason, though. As an architect, city planner, real estate developer, and native New Yorker, I have very specialized knowledge of the city's building stock and laws. I also have the unique perspective of a former government official charged with the responsibility of improving New York's housing stock. Between 1974 and 1978, I was Deputy Commissioner of Housing, in charge of all New York City's housing rehabilitation and neighborhood preservation efforts.

Block and Lot Patterns

America's first cities were founded on virgin territory. Colonists established plats that identified public thoroughfares, public open space, and the boundaries of individual lots for development. Thereafter, everything was at the option of the property owner.[1]

Since each colony was established by a different entity,

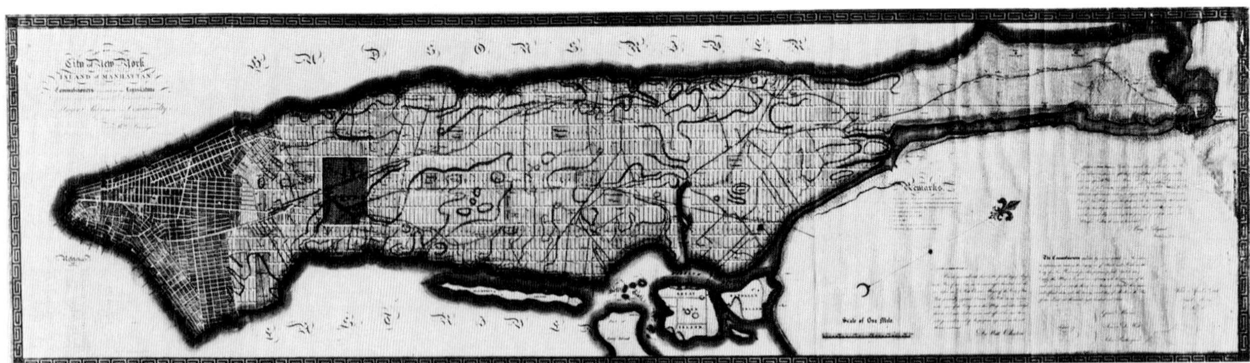

New York, 1811. Commissioners' plan showing how the grid was fitted to the topography of Manhattan island. *(Courtesy of the Museum of the City of New York, J. Clarence Davies Collection)*

Manhattan, 1882. Row houses in construction at Lenox Avenue and 133rd Street, illustrating what the commissioners had envisioned for Manhattan. *(Courtesy of Collection of the New York Historical Society)*

Manhattan, before 1885. Jacob Riis's photograph of the rear yard of an unidentified Lower East Side tenement. *(Courtesy of Museum of the City of New York, Jacob A. Riis Collection)*

plats differed in street width, block and lot size, and organization. By far the most popular design was the rectangular grid, either oriented to the points of the compass or parallel and perpendicular to rivers, cliffs, and other major geographic features. Such plats were easy to survey, easy to divide into rectangular lots that could be subdivided or recombined, easy to describe when conveying title, and easy to add onto if the town outgrew its borders.

There is no direct connection between a predominantly rectilinear system of property subdivision and slums. Nonetheless, since property owners have to fit their buildings into specific sites, block and lot dimensions play a major role in determining what is built. Sometimes the constraints imposed by these dimensions produce building configurations that are undesirable. In New York City, for example, because buildings are squeezed into long, narrow lots, it is difficult to provide a desirable level of natural light and ventilation to the rooms in the middle of these lots.

Manhattan's plat was designed by John Randall Jr., a professional surveyor, who chose to ignore the island's existing roads, streams, and sharp changes in elevation. Approved by the New York state legislature in 1811, his plat envisioned handsome, airy neighborhoods with blocks separated every 200 feet by 60-foot-wide east-west streets and divided at intervals of 350 to 800 feet by 100-foot-wide north-south avenues. The blocks were partitioned into 25- by 100-foot lots, on which it was believed developers would build row houses with spacious rear yards. Since the row houses were expected to be two rooms deep, the buildings would extend 40 to 60 feet back from the street curb. Thus, the rooms inside would receive light and air from the wide streets in front or the ample yards behind.

While much of New York was built on this pattern, property owners soon found it more profitable to do otherwise. They tried to fit in as many apartments as possible, covering as much of each lot as possible, and building as high as their customers were willing to climb. Four 25-foot lots were combined to fit five 20-foot buildings, three 25-foot lots to fit four buildings 18 feet 9 inches across, or worse. There was even

Manhattan, 1900. Riis's photograph of a model of a block of tenements on the Lower East Side. *(Courtesy of Museum of the City of New York, Jacob A. Riis Collection)*

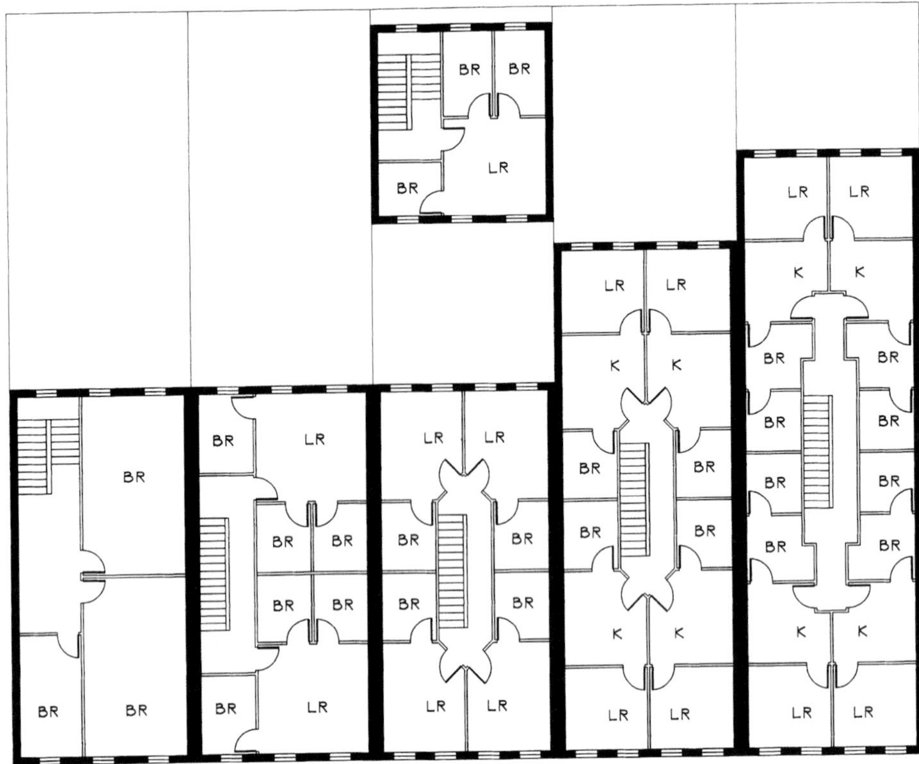

New York City, before 1867. Prior to regulation, property owners tried to maximize their rental revenue by jamming as many rooms as possible into tenement buildings (many without natural light and ventilation) and placing additional buildings at the rear of their lots. *(Alexander Garvin and C. Christopher Koon)*

a building that was only 8 feet wide. Instead of row houses facing other row houses at least 60 feet away, developers built tenements that routinely extended almost to the full 100-foot depth of the lot, leaving only a narrow alley between buildings five or six stories high and sometimes taller.[2]

The resulting apartments were similarly long and narrow. Some tenements had as many as sixteen rooms to the floor, only four of which (two in front and two at the rear) had windows. Living in such conditions was not just unpleasant; it was dangerous. Thousands died from tuberculosis, cholera, smallpox, and a variety of infectious diseases that spread easily in poorly lit and ill-ventilated apartments. Others perished in fires. Small children, the elderly, and others who were not able to move quickly, suffocated when flames sucked the oxygen from their rooms. The rest were unable to escape because there was no emergency means of egress. The prevalence of such unsafe and unsanitary living conditions explains in part the contemporaneous movement for public parks discussed in Chapter 3.

Reformers, horrified by these conditions, built "model tenements" that were not much better. At Gotham Court, a model tenement built in 1850 for the express purpose of rescuing the poor from such noxious conditions, Jacob Riis reported:

> *[T]en years after it was finished, a sanitary official counted 146 cases of sickness . . . and reported that of the 138 children born in it in less than three years 61 had died, mostly before they were one year old. . . . Seven years later the inspector of the district reported to the Board of Health that "nearly ten percent of the population is sent to the public hospitals each year."*[3]

While some developers adopted ideas from these "demonstration" projects, they continued to build dark and deadly dens because demand for housing was so intense that virtually anything could be rented.

Construction Regulations

Conditions grew so bad that, in 1867, the New York state legislature passed the Tenement House Act, which regulated multifamily housing construction. It required at least 3 square feet of transom window for each room and mandated that the window open onto another room that had a window with "a connection with the external air." At least one source of water was required in the house or yard for every 20 apartments. Most important, the law required installation of fire escapes in all nonfireproof buildings that did not already have a secondary means of egress. While it took years to enforce this last provision, block after city block eventually was populated with the pervasive character of fire escapes to meet this regulation.[4]

This statute, also known as the Old Tenement Law, was amended in 1879 to require every room to have a window that opened directly onto a street, yard, or air shaft. These

Manhattan, 1889. Jacob Riis's photograph of living conditions in a Bayard Street tenement, on the Lower East Side. *(Courtesy of Museum of the City of New York, Jacob A. Riis Collection)*

shafts, often barely 2 feet across and more than 60 feet high, were more effective as garbage chutes and sound amplifiers than as conduits for light and air.

In 1901, the state legislature decided that even Old Law tenements were unfit for human habitation. It enacted the New Tenement Law requiring new multifamily buildings to include one toilet per apartment. The air shaft was replaced with a courtyard of not less than 25 square feet (with a minimum dimension of 4 feet). The New Law also limited lot coverage to 70 percent and building height to one-and-one-third the width of the street it faced.[5]

Finally, in 1929, the new state Multiple Dwelling Law prohibited further construction of tenements. It required that every building with three or more apartments have a toilet and a bath in each dwelling unit and a sink with running water in every kitchen. It also mandated that every room have at least one window that opened onto a street or courtyard. Inner courts had to be at least 30 by 36 feet if completely enclosed or 20 by 30 feet if on a lot line.

As a result of these laws, New Yorkers speak in terms of Old Law tenements, New Law tenements, and post-1929 multiple dwellings. Anybody familiar with these terms can identify probable apartment layouts within. In other cities, building design may be different, but the relationship between the evolution of the built environment and the enactment of local construction regulations is the same. Thus, the Boston "triple-decker," the Chicago "six-flat," the Los Angeles "courtyard house," and the San Francisco "painted lady" refer to housing types shaped by the interaction between

Manhattan, 1915. Interior room in a pre-1879 tenement. Natural light and ventilation are provided by a window opening onto another room. *(Courtesy of Museum of the City of New York)*

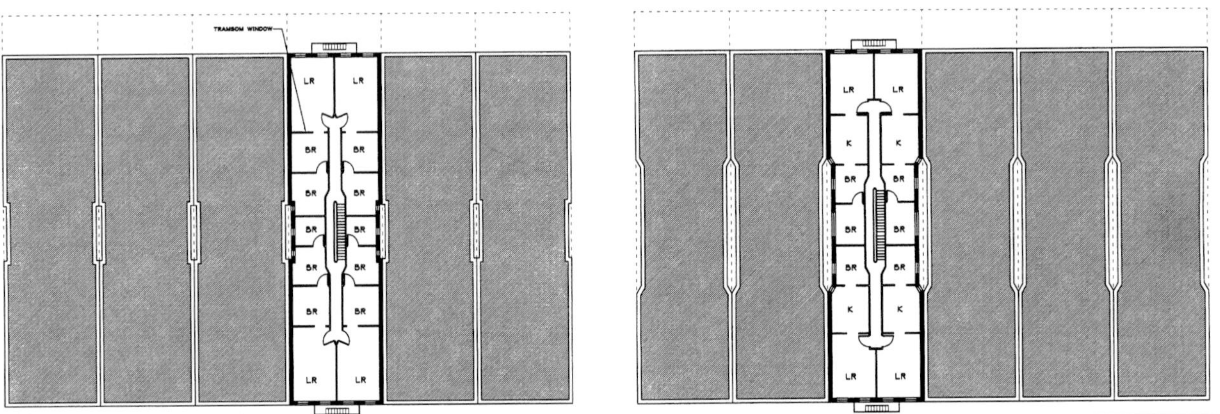

New York City, 1867, 1879. Floor plans of a typical tenement built under the provisions of the Tenement House Act of 1867 and after the 1879 amendments to the Tenement House Act. *(Alexander Garvin and C. Christopher Koon)*

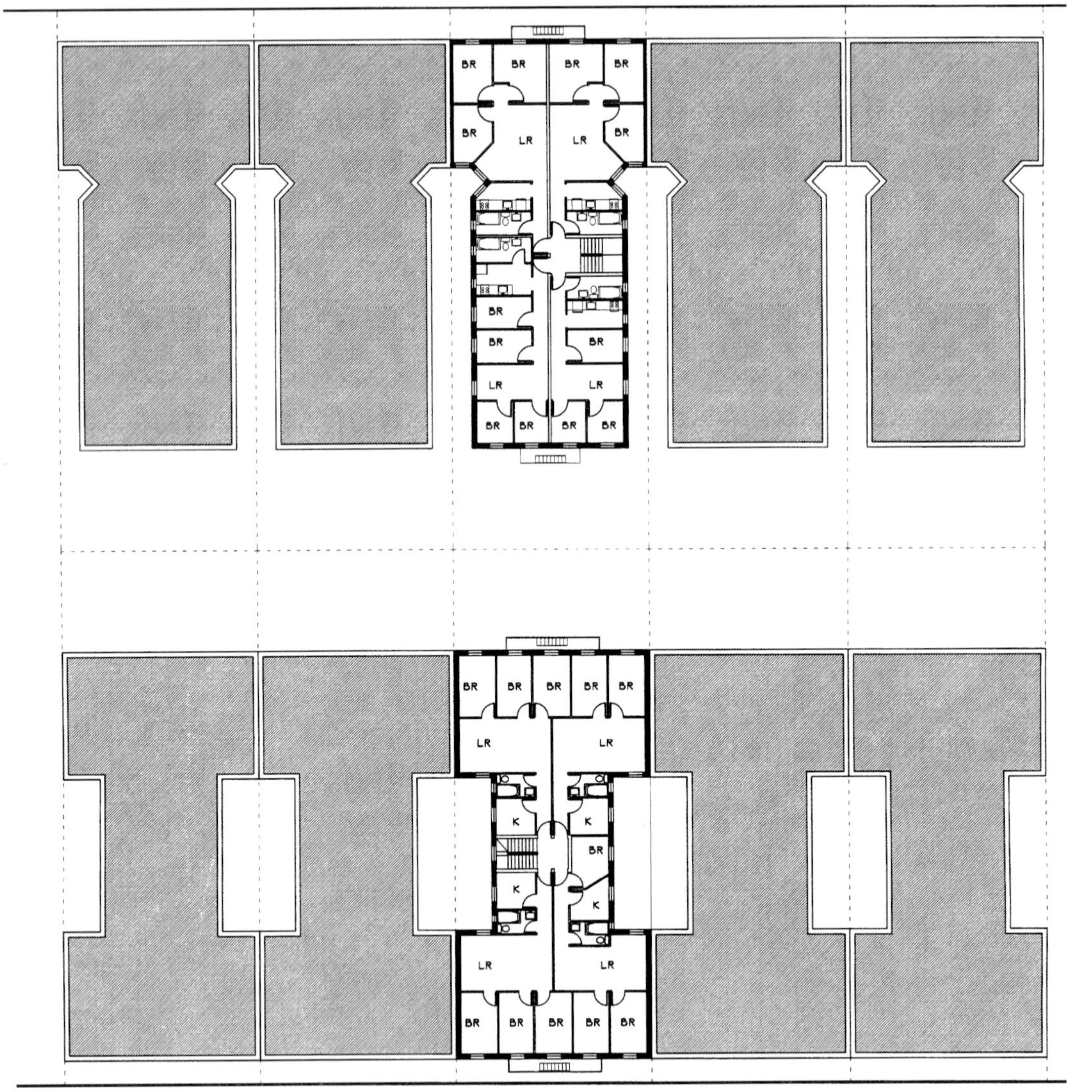

New York City, after 1901. Floor plans of typical tenements built under the provisions of the New Tenement Law of 1901. *(Alexander Garvin and C. Christopher Koon)*

home builders and the local block and lot dimensions and building regulations that govern their developments.

Renovating Old Buildings

Buildings everywhere are renovated all the time because equal or better housing cannot be produced at the same cost. Indeed, that is why people are ready to pay enough to justify the improvements and lenders are willing to provide the necessary financing. In districts where one-family homes predominate, it is relatively easy to determine whether these conditions exist. When multiple dwellings are involved, however, the parties concerned may not interpret conditions in the same way. Some tenants may desire improvements; others may not. Some may be willing and able to pay more rent; others may not. Whether tenants desire improvements or not, property owners will rehabilitate only if they receive a return that maximizes any additional investment of time and money. Lending institutions will provide mortgage financing only if they are sure the scope of work is sufficient for the property to remain in good condition for the duration of the loan, if building revenues will more than cover expenses, and if the property will not be adversely affected by conditions in the surrounding neighborhood. Thus, any housing rehabilitation program will have to balance often conflicting physical, social, and economic constraints.

Physical Constraints

Rehabilitation includes a continuum of possible work—starting with minor repairs and extending all the way to complete reconstruction. That continuum can be divided into three levels of rehabilitation: moderate, gut, and extensive.

Moderate rehabilitation includes new wiring, plumbing, mechanical equipment, bathroom fixtures, kitchen appliances, windows, bell-and-buzzer intercom systems, entry doors, roofing, and the repointing of existing masonry. In-

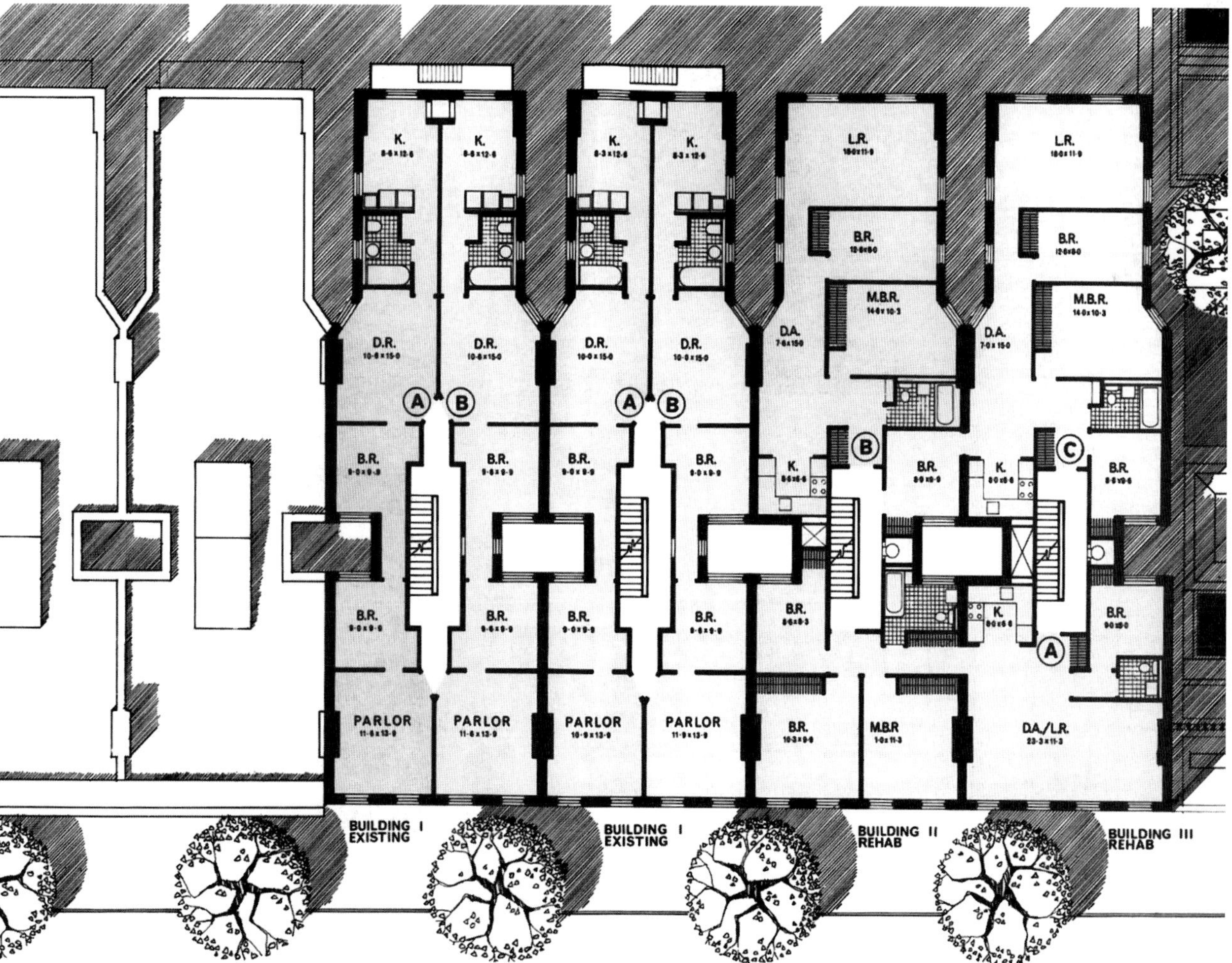

Bronx, 1967. Old Law tenements before and after gut rehabilitation, which improved apartment layouts but did nothing to increase the inadequate natural light and ventilation supplied by narrow air shafts. *(Courtesy of New York City Model Cities Program)*

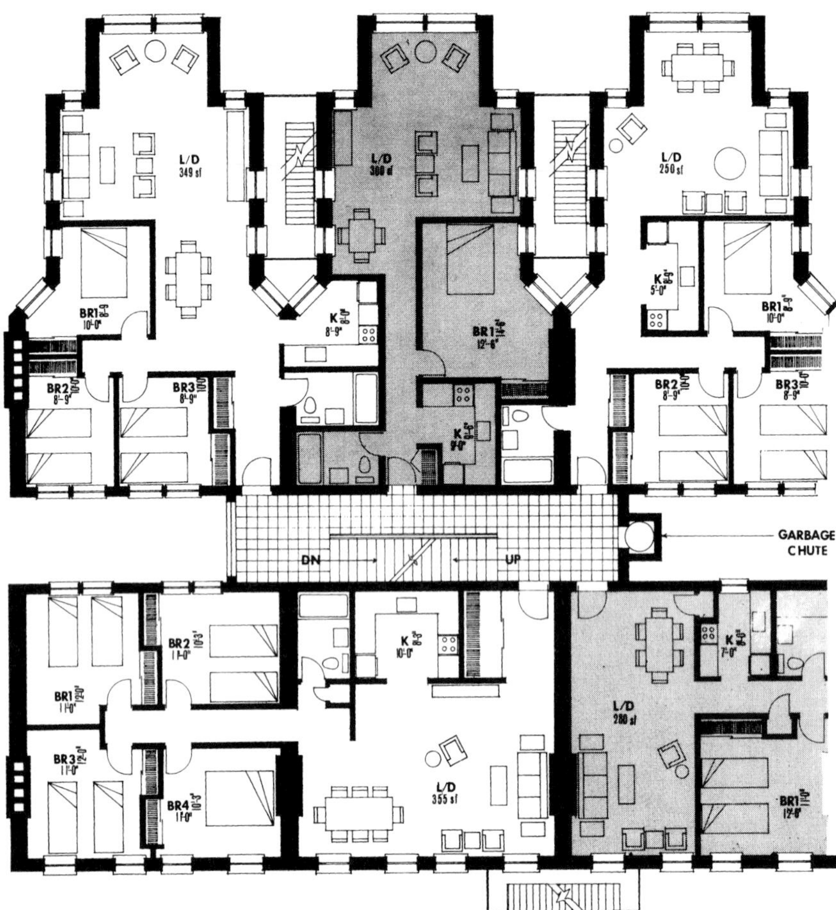

Manhattan, 1967. Old Law tenements after extensive rehabilitation, which improved circulation, apartment layouts, and ventilation. *(Courtesy of New York City Model Cities Programs)*

convenient though it may be, tenants can remain in place while work is in progress because apartment layouts remain the same. This not only eliminates moving expenses and rent paid for temporary quarters, since the tenant usually continues to pay the rent, it maintains the cash flow needed by the owner to pay real estate taxes and debt service on the mortgage. If the existing apartments have inadequate natural light and ventilation, poor layouts, and minuscule rooms, however, moderate rehabilitation will not be an adequate antidote.

Gut rehabilitation involves the same items as moderate rehabilitation but also includes new apartment layouts and therefore new walls, floors, ceilings, and partitions.

Extensive rehabilitation reconfigures whole structures, often removing large sections of buildings to permit improved light, ventilation, and apartment layouts. Neither gut nor extensive rehabilitation can take place with tenants in occupancy.

Because apartment layouts and room sizes in masonry row houses are acceptable, moderate rehabilitation usually is all that is needed. Other building types may have to be gutted if they are to provide decent apartments. Old Law tenements, for example, tend to contain apartments with inconvenient layouts. Small, narrow rooms with minimal light and ventilation are strung out one after another like railroad cars. Reconditioning these so-called railroad flats perpetuates designs that were already deemed unfit for human habitation in the era of Grover Cleveland. Moreover, retaining awkward apartment layouts may not be worth the cost of gut rehabilitation.

The best way to obtain good apartment layouts in Old Law tenements is to cut away a substantial portion of the structure, removing the rear portion of the building or carving out a broad courtyard in the middle. Sometimes such extensive rehabilitation can be accomplished only by combining buildings. This requires nearly as much work as building from scratch. Extreme reconstruction of this sort is justifiable only if it produces better or cheaper apartments than new construction.

Social Constraints

Because housing rehabilitation permits the physical characteristics of neighborhoods to remain intact, people often erroneously ignore the social problems that also come with it. All residents have to move from a building undergoing gut or extensive rehabilitation. Moreover, property improvement inevitably leads to higher rents. Thus, improvement of the housing stock inevitably results in changes to the social composition of the neighborhood in which it occurs.

Although living conditions will not be ideal while moderate rehabilitation is in progress, most tenants will accept the inconvenience. They usually desire building improvements enough to pay moderate rent increases. Inevitably, some tenants will not want the inconvenience or may not be willing to pay increased rent. The more serious problem is that some tenants may not be able to pay higher rents. In such cases, the scope of work can be reduced sufficiently to lower rents to an acceptable level or subsidies can be found to make the new rent affordable or to permit tenants to relocate. Without actions of this sort, physical improvement will be accompanied by tenant hardship.

Gut rehabilitation requires empty buildings and, like clearance and redevelopment, it therefore requires tenant relocation. Unless displaced tenants are paid enough for them to afford improved accommodations, they will be forced out, often to even more dilapidated housing. Even if relocation payments are sufficient to compensate for these and other hardships, most displaced tenants will make long-term moves and rarely return to their former residences.

Intelligently conceived rehabilitation projects often stimulate neighboring property owners to follow suit, thereby attracting new residents. Such gentrification is at once the cost and benefit of policies that improve the quality of the existing housing stock. The best way for government to cushion the impact of gentrification is to pursue policies that increase the supply of decent housing. That gives everybody a better chance to find affordable accommodations.

Financial Constraints

Moderate rehabilitation is much less expensive than complete restoration, and the difference is not only a matter of construction cost. Labor practices and wage rates for moderate rehabilitation are usually quite different from gut rehabilitation (especially for larger projects that mirror new construction practices). The labor force is predominantly nonunion and includes a high proportion of minority workers. Moderate rehab also requires significantly less time and therefore less in interest and tax payments during construction. Most important, tenants continue paying rent during renovation, helping to defray carrying costs during the development period.

Most structures requiring gut rehabilitation are in neighborhoods suffering from a multiplicity of problems. Banks are reluctant to risk their depositors' money in such areas. Consequently, gut rehabilitation projects usually involve additional expenditures for lighting, security, and other community improvements that reassure the permanent lender.

Given the significantly lower cost of moderate rehabilitation, one would expect this to be the predominant form of renovation. Instead, financial institutions concentrate on gut and extensive rehabilitation. They lack the expertise to balance the messy physical, social, and financial constraints involved in renovating tenanted buildings. The resulting difficulty in getting mortgage financing for renovation is one reason for the continuing deterioration of the existing housing stock.

Renovating Buildings in the Twenty-first Century

Buildings don't renovate themselves. Somebody must want to live in an older building; somebody else must want to fix up and operate the building for them. The money to do it does not just materialize—it comes from somewhere. In late 2008, when the subprime mortgage crisis was at its worst, two budding entrepreneurs decided it was the right time to raise the money with which to acquire and renovate an old building, Their experience reveals what's involved in housing rehabilitation.

Charles Dickens's description of the beginning of the French Revolution is applicable to the housing market after 2008, when these entrepreneurs decided to begin their real estate careers. "It was the best of times, it was the worst of times, it was the age of wisdom, it was the age of foolishness . . ."[6] It appeared to be a terrible moment to go into real estate because banks were unwilling to lend money, but it was an auspicious moment because building prices had fallen precipitously from their peaks just a few years earlier. Moreover, the foolishness of the previous decade still permeated the market (see Chapter 9). Most people—and, worse yet, most people in finance and real estate—still thought that building prices were only momentarily down and, with a little stimulus, prices could be restored and resume their "traditional" upward spiral. A few wise businesspeople thought this was a perfect time to invest and reap the benefit of the inevitable market adjustments that were to come. Among them were two friends, one who had practiced law and the other who had worked in the finance industry. They decided to find a building to renovate.

Their first requirement was a neighborhood that was easily accessible by subway, was already the home of a few hip retail stores and restaurants, and was attracting a new, younger population. Second, they sought a building whose occupants were paying substantially less than what other residents in the area were paying for newly renovated apartments. Finally, they wanted a property to which they could add value through rehabilitation and retenanting.

After considering a variety of places, they settled on the Crown Heights section of Brooklyn (see Chapter 13). They were convinced they had found the right neighborhood when they stopped at a local bar-restaurant that was jammed with people under 25. There they ran into a pal of theirs who had

Brooklyn, 2012. Rehabilitation under way on one row house in Crown Heights. *(Alexander Garvin)*

just graduated from Yale. "What are you doing here?" one of them asked. "I live across the street," she replied.

Their next problem was finding the right building. They found a four-story tenement which had once had a retail store on the ground floor. The residential portion upstairs was occupied by tenants who paid as little as $800 per month for a one-bedroom apartment, many of whom were months behind in paying their rent. Although at that time renovated apartments in the neighborhood were renting for around $1600 per month, the building was in such bad shape that nobody would pay that much. Given the low rents, the rent arrears, and the unwillingness of banks to provide financing because of the subprime mortgage crisis, nobody was interested in buying the property. With initial financing from the previous owner, the two friends purchased it in early 2011 at a price that they correctly believed would allow them to provide their investors with a comfortable return once the property had been renovated and retenanted.

They knew that higher rents were not possible without completely refinished stairs and hallways, new apartment layouts, new plumbing and wiring, and exposed brick walls that were so popular among younger adults. The new apartment layouts eliminated bedrooms without windows and provided completely new bathrooms and kitchens (see illustration of gut rehabilitation plan). To achieve this, the tenants had to be compensated to terminate their leases early, the building gutted, and everything rehabilitated. They were smart enough to do this in stages, though, so that they could preserve at least some rental revenue while vacated apartments were being renovated.

By the autumn of 2012, the building had been completely rehabilitated and the ground floor rented to a restaurant that was paying $3700 per month (contributing 25 percent of building operating expenses). Most of the new tenants were under the age of 25 and relatively new to their jobs and, thus, were unable to provide a long history of creditworthiness. They paid as much as $2700 per month, with financial guarantors (usually a family member) cosigning their leases. The total cost of the project was $1,000,000.

Once the apartments had been renovated and reoccupied, there was enough cash flow, after paying operating expenses and taxes, to obtain a $950,000 long-term bank mortgage, return most of the investors' equity, and provide them with 10 to 15 percent annual returns on their original investment (now largely paid back) for years to come. More important, there was enough left over to provide the developers with a handsome profit.

These entrepreneurs, who have since this initial project started new developments elsewhere in Brooklyn, have begun to build a portfolio of real estate investments that provides them with a steady income and the freedom to build an independent life. Never again will they be dependent on employment in a large firm over which they have no control. They have found a way to own a piece of America, to avoid a

Brooklyn, 2012. An apartment in a Crown Heights row house before and after gut renovation. *(Alexander Garvin)*

nine-to-five office job, to satisfy their intense desire to make something they could point to as their own, to leave their mark on the landscape, and to do something for the city they love. These rewards are why people renovate buildings.

Two Government Rehabilitation Experiments

Government can make it easy or difficult to renovate buildings; it can encourage rehabilitation by private developers or nonprofit organizations; it can even sponsor the renovation itself. When there is a huge pent-up demand for housing, government may get involved with rehabilitation because it is cheaper and faster than new construction. It may do so when a neighborhood is filled with deteriorating or abandoned buildings that are depressing the local real estate market. It may be responding to criticism of the poor quality of new construction, or the drab character of new buildings, or the destruction of our architectural heritage. New York City has a rich history of such projects. Their strengths and weaknesses can best be explained by retelling the stories of three experiments, all of which involved the Lower East Side tenements that reformers already considered unsatisfactory when they were originally built. They are First Houses at 112–138 East Third Street and 27–41 Avenue A (1933–1935), and Sweat Equity at 507–509, 517–519, and 533 East 11th Street (1976–1978).

First Houses

During the mayoral campaign of 1933, Fiorello La Guardia promised to eliminate slums and provide decent housing. Within weeks of his inauguration, he established the country's first municipal housing authority.[7] Two years later, the New York City Authority completed "Experiment No. 1," America's first housing project created, owned, and operated by a municipal housing authority. It was appropriately named First Houses.[8]

The Housing Authority wished to provide housing that low-income people could afford—apartments that rented for about $6 per room per month. It thought this could be done by replacing slums with new apartment buildings but soon discovered that sites could not be acquired at prices that allowed it to offer apartments at low enough rents. It decided to try rehabilitation. In December 1934, the authority persuaded Vincent Astor to sell some four- and five-story Old Law tenements. These were the sort of buildings with windowless rooms and long, dark corridors that Riis had so bitterly denounced. When the owner of two tenements separating the Astor properties refused to sell, the authority proceeded to condemnation. In March 1935, the New York State Supreme Court held in the case of *New York City Housing Authority v. Muller* that condemnation by a duly constituted local authority for the purpose of providing housing for persons of low income was permitted by the New York State Constitution.[9]

Experiment No. 1 demonstrated the utility of rehabilitating Old Law tenements provided that they were part of an overall redevelopment scheme that created functional apartment layouts. To provide adequate light and ventilation, the Housing Authority demolished every third building and cut away the rear third of all those remaining. It installed windows along the side of each building, 25 feet away from the neighboring structure. On Avenue A, the demolished buildings were filled in with one-story shops in order to maintain the commercial character of the street. On Third Street, the vacant lots provided access to the rehabilitated buildings and the rear yard. The new, enlarged rear yards were combined into a single play area, with trees, benches, and WPA-style artwork.

The original structures were reinforced, the roofs were replaced, and the façades were rebuilt using some of the original bricks. The interiors were redesigned to accommodate only two apartments per stair landing, thereby minimizing building circulation and maximizing cross ventilation. Modern wiring, plumbing, and heating systems were installed. Even the kitchens and bathrooms were completely rebuilt.

A monthly rent of $6.05 per room covered all the expenses of this 123-unit project. It was a bargain for the working-class tenants who moved in. Their average income was $23.20 per week.

Given the pervasive image of public housing as high-rise dormitory stockades segregated from surrounding communities, it is indeed ironic that the nation's first public housing was a rehabilitation project—a walk-up, in scale with the surrounding neighborhood and with stores on the ground floor.

Manhattan, 1934. Avenue A frontage of First Houses before rehabilitation. *(From* The Livable City, *courtesy of the Municipal Art Society)*

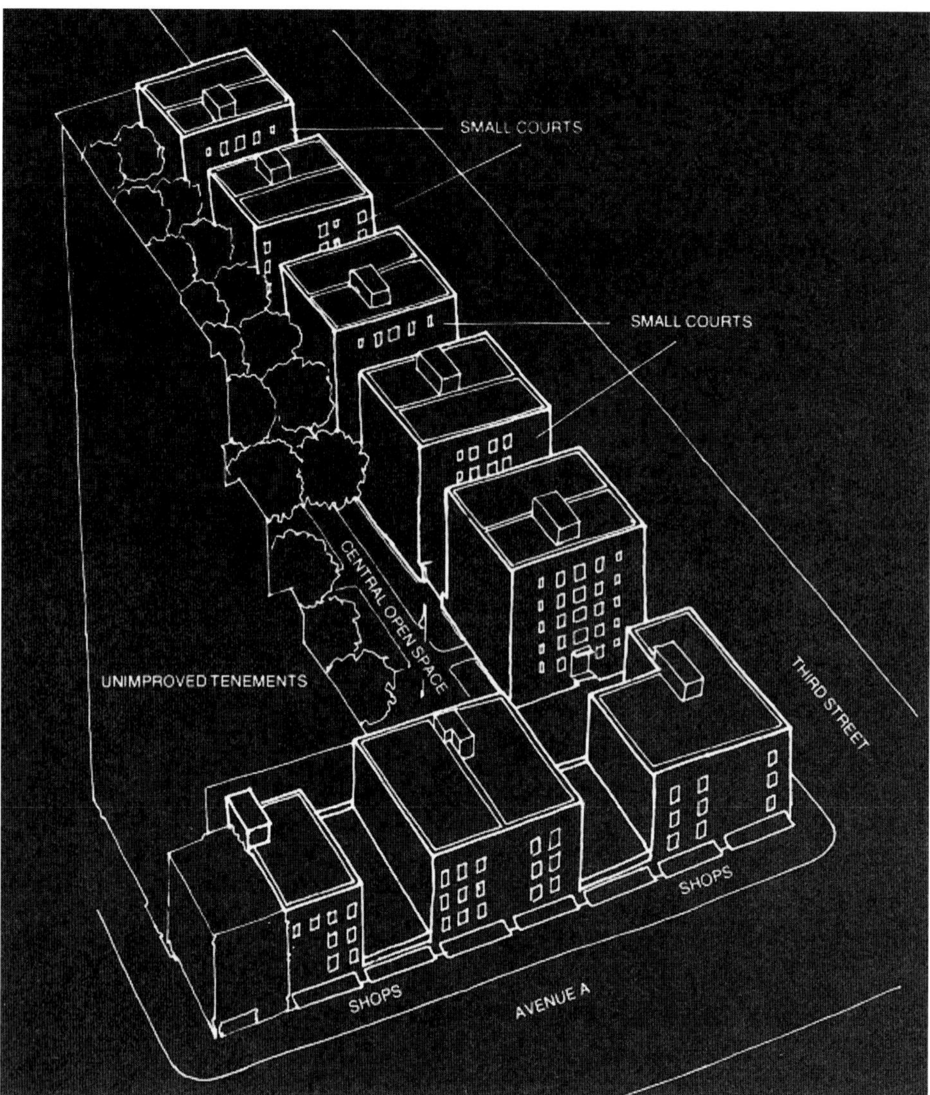

Manhattan, 1935. Isometric drawing of First Houses, showing the elimination of every third building. *(From* The Livable City, *courtesy of the Municipal Art Society)*

More than 60 years after completion, First Houses remains in excellent condition. If there were not a plaque on the buildings, passersby would never guess that it is public housing, much less a landmark in the history of government housing.

Sweat Equity on East 11th Street

In 1974, when I was deputy housing commissioner, a neighborhood group named Interfaith-Adopt-A-Building approached the city demanding to rehabilitate some vacant, city-owned Old Law tenements on East 11th Street, eight blocks north of First Houses. The buildings had been taken by the city because their owners had failed to pay real estate taxes. Not only were they a physical hazard, they attracted drug addicts and were subject to frequent and dangerous fires. It was easy to agree with neighborhood residents that everybody would be better off with the buildings repaired and occupied.

Interfaith-Adopt-A-Building persuaded the city to donate the buildings. Since it had neither money nor construction skills, Adopt-A-Building was in no position to rehabilitate these buildings. But the group did have enthusiasm, charismatic leadership, and superior technical assistance. It organized block-cleanup days, cleared the buildings of garbage and debris, protected them from squatters and addicts, and obtained assistance from the Urban Homesteading Assistance Board (U-HAB), the Consumer Farmer Foundation, and the federal Comprehensive Employment and Training Act (CETA).

U-HAB had been organized in 1974 by a group of former city employees who were unhappy with the government's lack of commitment to restoring older buildings and providing housing for the poor. It proposed to use government funds

Manhattan, 2005. Avenue A frontage of First Houses six decades after rehabilitation. *(Alexander Garvin)*

to finance housing rehabilitation and to train prospective low-income homeowners to restore their own homes. It also argued that required equity should be calculated in "sweat" rather than currency. As U-HAB explained it:

> *Over half of the ghetto youth have no employment prospects. Few incentives exist for positive, socially productive behavior. A decent home of one's own is an almost unreachable dream. . . . For many homesteaders the opportunity to own a decent home and to learn marketable job skills represents a turning point in their lives. The realization comes that there is now a new source upon which to depend and it is they, themselves.*[10]

In other words, since these young people had no cash of their own to provide the equity, which together with the cash from a lender would finance the acquisition and rehabilitation of the building, they would contribute their labor instead–hence: *sweat equity.* U-HAB expected far more than renovated apartments from sweat equity. The program was expected to provide training, employment, income, and a sense of community for young people who had no other chance of acquiring skills, jobs, or money.

Once established, U-HAB developed the know-how to prepare cost estimates; to hire architects, lawyers, appraisers, accountants, and construction supervisors; to draft grant applications; to process loan documents; and to negotiate with the municipal bureaucracy. However, it could not provide Adopt-A-Building or any other group with the cash required to initiate a rehabilitation project. The necessary seed money came from the Consumer Farmer Foundation.

CETA underwrote construction-training stipends for the self-helpers working on the buildings. This funding reduced substantially the amount of money required to cover the labor costs of any rehabilitation. A municipal rehabilitation loan covered 100 percent of the remaining cost. The extent of rehabilitation was reduced to ensure that development and operating costs were affordable for future occupants. In addition, the federal government provided a grant to install a solar-powered hot-water heating system and a rooftop electricity-producing windmill. This energy conservation and production equipment was intended to lower dramatically the monthly fuel expenditures and, by extension, further lower operating costs.[11]

Adopt-A-Building began with the rehabilitation of one Old Law tenement at 517–519 East 11th Street. Within two years the group had obtained the money to rehabilitate another three buildings and transform an adjacent vacant lot into a garden. The projected cost of rehabilitation ranged from $8000 to $13,700 per apartment.

Despite lower costs, free labor, tenant ownership, and neighborhood support, only one of the buildings—the one with the solar hot-water system and the electricity-producing windmill—was completed. It was also the only building without a CETA training component.

CETA labor stipends were for a fixed period, and stipends were terminated once the training period was over. Recipients

Manhattan, 1988. East 11th Street a decade after the sweat-equity rehabilitation of 517–519 East 11th Street. *(Alexander Garvin)*

Manhattan, 2012. After nearly four decades, the vacant lots on East 11th Street had been filled in with new row houses. *(Alexander Garvin)*

were then expected to move on to private-sector construction jobs armed with the training that they had received. Unfortunately, when the stipends ended, construction was far from finished and recipients had trouble finding jobs in the construction industry. When they lost their CETA stipends, they had no way of paying for basic necessities without abandoning the project. Since the loans covered interest payments only during an underestimated construction period, the project was in default even before the tenants moved in. Without a steady source of income, Interfaith-Adopt-a-Building had no choice but to stop paying the co-op's debt service.

As good an idea as it seemed to couple housing need with employment training, sweat equity on East 11th Street had simply been far too ambitious. Housing rehabilitation is difficult enough; it is impossible when the project is also expected to provide what society as a whole has failed to provide: training for the unskilled, jobs for the unemployed, income for the poor, and a sense of belonging for the rootless.

Although 517–519 East 11th Street remained occupied, other buildings on the block were abandoned by their owners, taken by the city for failure to pay real estate taxes, and demolished. During the 1990s, market conditions in the area improved considerably enough to attract private developers. Some previously empty lots were flourishing as community gardens (see Chapter 3) or as subsidized housing on sites reserved for infill development by the New York City Department of Housing Preservation and Development. Had these sites remained unused except as informal garbage dumps, the neighborhood would have gone further downhill or seemed attractive to private developers once market demand was restored. Thus, by the twenty-first century, when market activity had returned to the Lower East Side, this block on East 11th Street was again an integral part of an increasingly popular neighborhood.

Three Citywide Rehabilitation Programs

Most people think of rehabilitation in terms of individual projects. Rehabilitation, however, has a much broader role to play as part of a comprehensive, citywide housing strategy. The complexities of site acquisition, design, bureaucratic processing, construction, and marketing are such that few apartments will be added to a city's housing stock in any one year. Whatever new construction enters the market will inevitably rent or sell at higher prices than already existing apartments. More important, new construction does not deal with the inadequacies or deficiencies of the housing stock in which most city residents live. Any housing policy that concentrates on delivering new housing without addressing the maintenance of the existing housing stock will fail to ensure a supply of decent housing for most of the city's current residents.

Most governments effectively punish property owners who renovate their buildings by increasing the tax assessment on the property. Other, more incisive ones, however, exempt property owners from paying additional taxes as a way of forestalling disinvestment and eventual property devaluation; some even forgive a portion of the tax bill to compensate property owners for the investment that they have made in restoring their buildings. The alternative to using tax incentives to promote housing rehabilitation is to lend money directly to property owners who will renovate old buildings. Government lending of this sort for housing rehabilitation may make sense when banks are not willing to provide mortgages for old buildings (when they erroneously believe they have completed their useful life) or when the resulting rents are unmarketable and the lower interest rates that governments can charge will bring down the price sufficiently to attract residents.

New York City has operated three programs that encourage housing rehabilitation by the private sector. The earliest of them, the J-51 Program, reduced real estate tax payments to compensate for the increased debt service incurred to pay for rehabilitation. This was followed by the Municipal Loan Program, which also lowered debt-service payments by providing long-term mortgages at the city's cost of borrowing (usually two percentage points below conventional bank mortgage rates). Then, in the mid-1970s, the city began using its money as leverage to urge banks to invest in rehabilitation. Only the J-51 Program generated sufficient production, however, to have a major impact on the city's housing stock.

New York City's J-51 Program

New York City is the first municipality to establish a real estate tax incentive program that operates on a volume basis to encourage maintenance and rehabilitation of existing residential housing. This program, called J-51 after its section number in the city's administrative code, was enacted in 1955. Initially, it provided incentives for the elimination of unhealthy or unsafe housing conditions in tenements. Over the years, its scope was expanded to include upgrading of all existing multiple dwellings as well as the conversion of hotels, rooming houses, and nonresidential space into multiple dwellings. Between 1961 and 1988, J-51 provided benefits to more than 1.4 million apartments.[12]

J-51 provides two sorts of benefits for eligible buildings: tax exemption and tax abatement. The exemption is from any increase in real estate taxes resulting from improvements to the property, and it lasts 12 years. Consequently, property owners are not punished with increased real estate tax assessments when they improve their buildings. The abatement equals 90 percent of the reasonable cost of the improvements, but it cannot be taken at once. A limit of 8.33 percent of the certified reasonable cost of rehabilitation may be deducted from the property's real estate tax bill in any one year until either the abatement is exhausted or 20 years pass, whichever occurs first. In effect, it is a refund of 90 cents of every dollar spent on restoration (see Table 11.1).

Two elements have made J-51 effective: its certainty and its predictability. The law provides the *certainty* by ensuring that specific repairs that meet the requirements of the law will receive benefits, and *predictability* by defining the amount of those benefits. No public official may withhold approval. It is this nondiscretionary nature of the program that protects it from political pressure and graft. More important, banks can rely on J-51 benefits and will issue mortgage commitments based on these benefits.

Although J-51 has been successful, it is no panacea. If benefits are insufficient to reduce a building's economic rent to market level, owners will not renovate. Nor does J-51 necessarily provide housing for the lowest income group, which almost always requires further subsidy. In cities where market forces determine rents, exemption by itself will be

TABLE 11.1

MODERATE REHABILITATION WITH J-51 BENEFITS

Year	Tax without exemption or abatement, $	Tax after exemption before abatement, $	Abatement, $	Final tax bill, $
1	9,180	4,140	4,140	0
2	9,180	4,140	4,140	0
3	9,180	4,140	4,140	0
4	9,180	4,140	4,140	0
5	9,180	4,140	4,140	0
6	9,180	4,140	4,140	0
7	9,180	4,140	4,140	0
8	9,180	4,140	4,140	0
9	9,180	4,140	4,140	0
10	9,180	4,140	4,140	0
11	9,180	4,140	4,140	0
12	9,180	4,140	4,140	0
13	9,180	9,180	7,081	2,099
14	9,180	9,180	7,081	2,099
15	9,180	9,180	7,081	2,099
16	9,180	9,180	5,577	3,603
17	9,180	9,180	0	9,180
18	9,180	9,180	0	9,180
19	9,180	9,180	0	9,180
20	9,180	9,180	0	9,180
Total	183,600	123,120	76,500	46,620

Property: 19-unit, four-story walk-up apartment house
Preimprovement real estate tax assessment: $46,000
Preimprovement taxes: $4140 (.09 × $46,000)
City-issued certificate of reasonable cost of rehabilitation: $85,000
Maximum annual tax abatement: $7081 (.0833 × $85,000)
Maximum total tax abatement: $76,500 (0.9 × $85,000)
Postimprovement real estate tax assessment: $102,000
Postimprovement taxes without J-51 benefits: $9180 (.09 × $102,000)
Assumed tax rate for the duration of J-51 benefits: $9 for each $100 of assessed value

cost-effective. Without the inducement of tax exemption, however, a substantial amount of rehabilitation would not be likely to occur, and, thus, the city never would receive the additional taxes. The program can therefore be understood as a zero-cost subsidy because it does not cost the city anything to forgo taxes that it would not otherwise receive if the renovations were not conducted. Besides, when the exemption expires, the city collects additional taxes from any increased assessment, so in the long term the program generates not only additional, good-quality affordable housing but also an increase in tax receipts.

In cities where rent regulations virtually preclude landlords from increasing rents to recoup the costs of rehabilitation, tax abatement is the only way government can assist an owner to pay for restoration. Even then, though, the resulting increase in net operating income after taxes may not be enough. Owners also must be able to obtain financing to pay for rehabilitation. They will be able to do so only if banks and other lenders can count on fixed, nondiscretionary tax benefits large enough to justify a rehabilitation loan.

Municipal Rehabilitation Loans

New York City has also provided direct financing for housing rehabilitation through its Municipal Loan Program. The program was established in 1962, when the state legislature enacted Article VIII of the Private Housing Finance Law. Initially, Article VIII permitted municipalities to make mortgage loans to owners of multiple dwellings constructed prior to 1929. This requirement restricted lending to the city's oldest buildings with the worst apartment layouts.

Funds for the program came from serial revenue bonds authorized by the mayor. Bond proceeds were supposed to establish a self-sustaining pool of mortgage money to be recirculated as loans were amortized. The costs of operating the program and paying debt service on the bonds were to be paid from the proceeds of mortgage interest payments. Before the program was terminated in the mid-1970s, more than $125 million had been loaned for the rehabilitation of 7000 apartments. Fifteen years after the program was ended, virtually all of these loans had been foreclosed or were in default, not because the intentions or strategy were flawed but, rather, because the Municipal Loan Program was incompetently and fraudulently managed.

The law required 10 percent borrower equity, but it was often phantom equity. In some cases, owners of dilapidated properties had them appraised at unrealistically high values in order to show more than the required equity. That way, they could cover all possible expenditures with the municipal loan and walk away with cash.

Loan proceeds were advanced in stages as construction was completed. The officials who decided how much rehabilitation had been completed and how much money to release were underpaid. They were happy to augment their earnings with payments from borrowers who collected money for work that had not been done. Eventually, these crooks went to jail.

Ten years after it had begun, the city council determined that the Municipal Loan Program was "ill-conceived, badly executed, and fraught with corruption . . . [that] many persons associated with the program were either charlatans, thieves or incompetents . . . [and that it] was poorly thought out and foredoomed to failure."[13] New York City's experience with the Municipal Loan Program did not invalidate rehabilitation mortgage lending. Banks had a long history of successfully lending to existing property owners for rehabilitation. The Municipal Loan Program scandal simply highlighted the need for intelligent lending practices, competent management, and honest participants.

Joint Bank-City Rehabilitation Loans

New York City's fiscal crisis of the mid-1970s temporarily precluded further municipal lending for any purposes. The inability to continue lending was an opportunity to make additional reforms. The results were embodied in Article XV of the New York State Private Housing Finance Law, enacted in 1976.

Article XV established the Participation Loan Program (PLP), which was supposed to terminate the practice of making municipal loans directly. It authorized the city to join with financial institutions as a colender in mortgages for the rehabilitation of multiple dwellings. Prior to the 1980s, there was a clear separation between the savings banks and savings and loan associations that made long-term, permanent mortgages and commercial banks that made short-term, development loans. PLP eliminated city construction lending and, thereby, avoided the Municipal Loan scandals in which city inspectors had authorized payment for work that had not been done. The money from the city's portion of the mortgage was disbursed only when rehabilitation had been satisfactorily completed. The city provided its share of the permanent mortgage from the its annual federal Community Development Block Grant (CDBG), thereby avoiding a drain on the city's bonding capacity.

In a PLP project, the city lends CDBG money at 1 percent interest as part of a joint mortgage with a permanent lender who charges market interest rates. The proportion of 1 percent money and market-rate bank money is established so that the composite interest rate will produce the required economic rent (see Chapter 10, note 1). The city determines the amount of the loan based on an adequate level of rehabilitation and an interest rate that will enable the borrower to charge marketable rents that will cover operating expenses, real estate taxes, debt service, and a minimum return on equity, or collectively what has been defined earlier as economic rent.

Banks had not been willing to make rehabilitation loans because government rent regulation prevented adequate rent increases. Article XV allowed the city to restructure existing rents. Thereafter, the property, like most older buildings, is subject to rent regulation. Thus, from the start there was a presumption that participation loans would be made to occupied, not to vacant, buildings and that rents would have to be acceptable to existing tenants.

Even if rent increases were small, some tenants would not have been able to afford the new rents. Section 8 rent subsidies were set aside for those tenants who could not afford higher rents and who otherwise would have had to move (see Chapter 10). Only 15 percent of the tenants proved to need such subsidies.[14]

When the city and the participating financial institution issue a commitment letter for a permanent mortgage loan, the applicant is required to secure construction financing, either from the participating lender or from a government-supervised construction lender that is usually a commercial bank. The institution responsible for construction financing, using a licensed architect or engineer on a consultant basis, supervises construction and makes progress payments to the borrower. This lender certifies satisfactory completion of the work before the participation loan can take effect and the construction lender can recoup its money. Just as would happen in a completely private market development, the construction lender does not give out one penny unless the work is independently certified as having been satisfactorily completed. Upon completion of the renovation, the participation loan is serviced by the participating financial institution, thereby eliminating the need for a city debt-service collection bureaucracy and the possibility of political pressure against foreclosure for failure to repay the loan.

Manhattan, 2012. Thousands of apartments in Washington Heights were renovated between 1975 and 1990 with participation loans. *(Alexander Garvin)*

With both the permanent and the construction lenders checking the borrower's credit history, the validity of any refinancing, the scope of the work, the reasonableness of construction costs, and the marketability of projected rents, there is little chance of the abuses that plagued the Municipal Loan Program. This allows the city to concentrate on selecting projects and deciding how large its participation in the loan should be.

As of September 1988, $415 million had been lent for the rehabilitation of 635 buildings, containing 23,600 apartments. Of that, 55 percent came from private institutions and 45 percent from CDBG funds. Only a few loans are behind in debt-service payments or have been foreclosed. Over the next two decades, the focus of the program was completely shifted. HUD required that its funds be used entirely for the benefit of persons who met its definition of low income. Consequently, the city stopped using federal CDBG funds and used city capital funds instead. It also amended the legislation to allow its use in new construction and adopted regulatory constraints that make it inapplicable to most of the city's housing stock.[15]

Ingredients of Success

The rationale behind government housing rehabilitation programs should change with changes in each area's population, housing stock, and mortgage market. What may make sense during a period of rapid population growth will no longer make sense when population is declining. Similarly, a rehabilitation program that is ill-conceived for an area whose predominant building type is costly to convert to modern standards may be appropriate in another neighborhood with charming older structures. Whatever the rationale is behind government rehabilitation programs, success will depend on the same ingredients that determine the success of any program for fixing the American city.

Market

Demand for secondhand housing continually changes. When the elaborate ornament of the Victorian period is out of fashion, "tarting up" nineteenth-century structures is not likely to attract home buyers. During periods in which Victoriana is again in fashion, renovation of these same buildings is likely to be prevalent. Private-sector rehabilitation continually adapts to these changes in consumer taste. Government programs may change, but the objective will always be the same, the provision of adequate housing.

The best insulation from changes in fashion is for rehabilitation projects to offer lower-than-market rents. The New York City Housing Authority was particularly sensitive to this consideration at First Houses. If there were to be further public housing projects, the authority had to demonstrate that government-built, -owned, and -managed housing could successfully provide decent shelter on an ongoing basis. Consequently, it offered *at lower prices* a better product than was generally available in the surrounding area. Affordability alone is not enough. The Sweat Equity Project on East 11th Street avoided major capital investment, and so kept housing costs consistent with those of neighboring buildings. However, the project could not guarantee steady income or employment for its construction trainees and cooperative owners, who needed additional assistance to cover project costs that were otherwise market-based.

Location

The character of the buildings in any particular neighborhood is essential to any rehabilitation program's success. All three rehabilitation experiments discussed in this chapter involved the rehabilitation of Lower East Side tenements. Even before these tenements had been occupied, they were being described by Jacob Riis and other nineteenth-century reformers as slums.

As First Houses demonstrates, tenements can be altered to meet contemporary living standards. The cost of the extensive renovation required is substantial. Private-sector tenement rehabilitation will proceed as long as the resulting rents are marketable. In many tenement neighborhoods, such renovation is too expensive for most residents. Rehabilitating the buildings in these neighborhoods can occur only when substantial government subsidies are available. Such subsidies are not likely to be appropriated by any local, state, or federal administration. Moreover, since there is rarely much demand for public expenditures on buildings that do not require immediate attention, government assistance for moderate rehabilitation is also unlikely to be forthcoming. Thus, although housing rehabilitation is a cost-effective way to improve living conditions, buildings in many parts of the country will continue to deteriorate.

Design

The dimensions and arrangement of rooms within an apartment determine the degree of privacy provided to each occupant, the ease with which it can be furnished, and, as a result, its marketability. Any program involving extensive rehabilitation must balance the cost of altering the dimensions and arrangement of rooms with the rents that can be obtained for the altered apartments. Developers who have to compete for tenants are sensitive to these trade-offs. Public officials seldom pay much attention to this balance because subsidies allow them to avoid such choices. Unlike extensive rehabilitation, moderate renovation does not suffer from this problem because the tenants who remain in occupancy during the renovation will allow construction workers into their apartments only if they desire the improvements and consider postrehabilitation rents to be affordable.

Tax abatement and exemption programs spur privately financed rehabilitation projects that would not otherwise be cost-effective. Loan programs, on the other hand, tend to impose design requirements that inevitably result in extensive rehabilitation. Too often, new construction is more cost-effective. Consequently, rather than depend on publicly administered lending programs, we should provide banks with pools of below-market-rate-interest mortgage money and let them do the underwriting. This will direct subsidies largely to those items that would not otherwise be financially justified. More important, it will significantly expand lending for housing rehabilitation.

Financing

Publicly financed housing rehabilitation will never be a major government activity. There are too many competing demands for public expenditures. Even if there were enough money, too few public employees have the expertise to generate billions in secure mortgages. Nor should they. There is no need for government financing if safe loans can be made by private financial institutions.

Government assistance is required only where local banks are afraid to risk their depositors' money on housing rehabilitation or where developers cannot achieve marketable rents given prevailing rates of interest. The Participation Loan Program dealt with both situations. It provided the portion of the loan that covered the extra financial risk that lending institutions were unwilling to take and the lower interest rate that reduced rents to a level that most tenants could afford. Those few residents for whom this was a hardship were provided with additional government subsidies.

Government housing rehabilitation programs work best when borrowers and banks make market decisions and the public sector subsidizes desired results on a citywide basis. This approach is more equitable than tailoring the program to specific buildings. It also directs subsidies to those rehabilitation projects that are marketable as a result of public assistance. Therefore, if any city wishes to encourage housing rehabilitation it should establish a program, like New York's J-51, that lets the lenders and borrowers decide what changes are most likely to attract the tenants they need.

The best way to induce banks to increase lending for housing rehabilitation is to develop local joint-mortgage programs, like New York's Participation Loan Program, in which government provides a large enough portion of the gap between the cost of a rehabilitation project and what the bank would otherwise lend to justify a bank mortgage and an owner's equity investment.

Just as the housing stock of each city is different, so are lending practices. There is no way to develop a program that will be equally appropriate to every neighborhood. Thus, local government agencies will have to negotiate with local institutions the specific form of joint mortgage that best meets local conditions.

Time

All government rehabilitation programs must consider the time required for work to be completed. Some programs have reduced that time substantially. Moderate rehabilitation with tenants in occupancy usually provides enough revenue during construction to cover the cost of real estate taxes and debt service.

Maintenance and renovation of existing housing can take place only when lenders and borrowers are assured that they will not be adversely affected by business cycles. Five- or even ten-year exemption from any increase in real estate taxes due to rehabilitation is not enough. By extending the tax abatement period over 20 years, the J-51 Program increased cash flow and ensured a steady tax payment for the probable life of any mortgage used to finance the rehabilitation. The entire program was therefore aligned with time frames conventionally used in commercial financing, rendering it much more palatable to both developers and lenders. In addition, the longer time frame also helped to insulate property owners from changing market conditions.

Public officials must stop considering housing maintenance a short-term effort. Ongoing private housing rehabilitation requires a continuing role for government in creating stable financial conditions for the institutional lenders that provide rehabilitation mortgages. Anything less will guarantee continuing deterioration of the housing stock. Local governments have no way of ensuring a stable money market and stable interest rates. However, they can and should provide stable, predictable real estate taxes.

Entrepreneurship

Property owners will invest in housing rehabilitation only when they believe it is relatively safe and lucrative. J-51 created an investment climate of this sort. Once a property owner had complied with program requirements, the amount that real estate taxes would be reduced could be predicted from a published schedule of allowable costs for every expenditure on rehabilitation.

Even if the risk is low, developers will invest in rehabilitation only when most of the money comes from somebody else. J-51 provided banks with the certainty and predictability they needed to extend credit for housing rehabilitation. Initial loans were calculated as if there were no tax benefits. Once rehabilitation was completed and tax abatement was in place, loans were increased in proportion to increased project revenue. Therefore, developers made large equity investments only for the short period needed to complete rehabilitation, obtain J-51 benefits, and draw down the remaining portion of their mortgage commitment.

The Municipal Loan Program provided none of these conditions. The amount of the loan was left to the discretion of program officials. Once the loan was granted, progress payments were based on inspections by other government employees. Allowable rents were the responsibility of still others. Long-term property owners who avoided entanglement with government agencies kept out of the program. Instead, it attracted developers who were skilled at government processing or who found (often illegal) ways of obtaining government action. If we are to maximize the number of property owners and developers who rehabilitate older buildings, we must create conditions that minimize their equity investment, maximize mortgage lending, reduce development costs, and reduce the time required to obtain financing and complete the renovation. This can best be done by relying on local banks to provide financing and on government to provide the marginal funding needed to close the gap between the amount banks will lend without government assistance and the amount needed to make rehabilitation attractive to property owners and developers.

Rehabilitation as a Planning Strategy

Too many local governments pursue policies that discourage housing rehabilitation. They punish property owners who improve existing buildings by increasing real estate tax assessments. Eliminating this obstacle to improving the housing stock is easy: exempt all residential properties from any increase in taxes due to rehabilitation. This requires neither a new government program nor additional government employees. It may even reduce the number of people on the public payroll because there will be no need for tax assessors to calculate the value of private investment on housing improvements.

A small number of cities discourage housing rehabilitation by making it difficult to recapture expenditures in increased rent. Eliminating this obstacle to housing rehabilitation is also easy. In cities with rent regulatory systems that allow property owners to pass through the cost of renovation in the form of a rent increase, that increase should be automatic and permit a suitable period for the amortization of any expenditure (e.g., five years for building-wide improvements and three years for work done within an apartment). In situations where the municipal government wishes to avoid rent increases, it can, as New York City does with J-51 benefits, subsidize those improvements by abating real estate taxes.

Such policies will allow financial institutions to make loans for the rehabilitation of those buildings in which higher rents are marketable. In some cases, market rents will not be high enough to cover all the improvements that are needed. In such situations, government rehabilitation programs can close the gap. Below-market-interest loan programs, like the Participation Loan Program, can be established to cover the nonbankable portion of a rehabilitation loan while proportionately reducing debt service to a marketable level. If there is also a desire to avoid displacing the small percentage of tenants who cannot afford rent increases, a municipality can set aside special subsidies, as New York City did with Section 8, for those few residents who otherwise would have to move. But, whatever combination of housing programs a local government decides on, housing rehabilitation must be an important component. After all, most of the population inhabits existing housing; only a tiny portion will move to new housing in any particular year.

Notes

1. A *plat* is a plan for the actual or proposed territorial organization of a city or any of its parts, including the arrangement and dimensions of public spaces, streets, blocks, and building lots.
2. Historical and statistical material on New York City's building laws is derived from James Ford, *Slums and Housing*, Harvard University Press, Cambridge, MA, 1936; and Richard Plunz, *A History of Housing in New York City: Dwelling Type and Social Change in the American Metropolis*, Columbia University Press, New York, 1990.
3. Jacob Riis, *How the Other Half Lives*, originally published in 1890 and reproduced in *Jacob Riis Revisited*, Francesco Cordasco (editor), Anchor Books, Garden City, NY, 1968, pp. 30–31.
4. New York State Legislature, *Laws* (1867), ch. 980, sec. 17, pp. 2265–2273.
5. New York State Legislature, *Laws* (1901), ch. 334, sec. 17, pp. 889–923.
6. Charles Dickens, *A Tale of Two Cities* (originally published in 1859), Penguin Books, New York, 2010, p. 5.
7. In September 1933, Ohio was the first state to enact enabling legislation permitting the creation of municipal housing authorities. In February 1934, New York followed suit.
8. Historical and statistical information on First Houses is derived from Ford, op. cit., pp. 727–731; and Ira Robbins with Gus Tyler, *Reminiscences of a Housing Advocate*, Citizens Housing and Planning Council of New York, New York, 1984, pp. 27–30.
9. Matter of *New York City Housing Authority v. Muller*, 78 F 2d 684, 1935.
10. The Urban Homesteading Assistance Board, *Third Annual Progress Report*, U-HAB, New York, 1977, p. 3.
11. The grant was from the Community Services Administration to the 519 East 11th Street Cooperative Apartment Corporation.
12. New York City Department of Housing Preservation and Development, Office of Development, Division of Financial Services.
13. New York City Council, Committee on Charter and Governmental Operations, *Report on the Municipal Loan Program—Blueprint for Failure*, February 29, 1972, p. 1.
14. Barbara Leeds, Assistant Commissioner, Division of Financial Services, New York City Department of Housing Preservation and Development (1990–1994), interview with the author, March 8, 2000.
15. Barbara Leeds, Assistant Commissioner, Division of Financial Services, New York City Department of Housing Preservation and Development (1990–1994), interview with the author, March 8, 2000, and July 19, 2012.

12

Clearing the Slums

Bronx, 2012 *(Alexander Garvin)*

The rationale behind housing redevelopment is best captured in a single sentence by Jacob Riis: *The bad environment becomes the heredity of the next generation.*[1]

In common with many nineteenth-century reformers, Riis believed that rooting out tenements would eliminate a major impediment to safe, healthy family life. But replacing hovels with sanitary dwellings does not produce a good environment if the problems of surrounding slums soon engulf the new buildings.

Some reformers believe that surrounding slums can be prevented from overwhelming the new environment by creating separate enclaves whose critical mass precludes such intrusions. Others propose creating superblocks that supply all the elements needed for a healthy family life (schools, stores, recreation facilities, etc.) so that residents can spend almost all their time, should they want to, without leaving their chosen living environment.

Critics of both approaches argue that there is no way to isolate redevelopment projects from the surrounding city. One group of critics demands steady elimination of substandard housing. They believe that a growing supply of decent dwellings will inevitably draw customers away from unsound housing and eventually force its complete elimination from the market. Another group advocates government action to root out the pockets of blight that presently discourage private investment in an area. They believe that private market forces will create a decent living environment once government has cleared away any blighted areas. A third group relies primarily on multifaceted public action that goes beyond "salvation by bricks" alone and proposes that government implement a coordinated strategy that also includes social and economic programs for area residents.

Obviously, there is wisdom in each of these strategies; none of them is applicable to every situation. Wrongly applied, each has the potential of making things worse. Even when redevelopment is appropriate, there is unlikely to be enough government money with which to implement every desirable project. Consequently, public assistance should be directed only to those redevelopment projects whose benefits will spill over to improve housing conditions throughout the city. Then, instead of thinking in terms of housing units demolished or built, public officials will make redevelopment part of a citywide strategy for establishing the good environment.

The Bad Environment

Slum clearance has become part of an American tradition that dates back to the late nineteenth century, when Riis and other reformers went to war with one particular slum, Mulberry Bend, a notorious tenement district on the western edge of what is today Manhattan's Chinatown. He knew the area well, having covered its shootings and stabbings as a reporter for the *New York Tribune.* Riis argued that the only way to eliminate this infamous slum was for the city government to acquire the area's dilapidated tenements and level them. This strategy already had been tried in London, where the Common Lodging House Act of 1851 authorized the Metropolitan Board of Works to condemn and clear slum property for the purpose of providing replacement housing.[2]

In spite of Mulberry Bend's notoriety, the New York state legislature had difficulty justifying a public taking of private property. There had to be a public use for the land. Proponents of clearing Mulberry Bend had one in mind: a public park. They believed that playgrounds were essential to the battle with the slums (see Chapter 3). The legislature accepted this argument and, in 1887, approved the Small Parks Act, which authorized condemnation of privately owned land for the purpose of creating public playgrounds. Mulberry Bend thus became one of the nation's earliest public playgrounds and its first slum clearance project.

Riis heralded the success of Mulberry Bend's transformation, proudly boasting five years afterward that "not once has a shot been fired or a knife been drawn."[3] He used the example of Mulberry Bend to advocate further slum clearance, and for decades afterward reformers continued to make this argument by showing a correlation between slum clearance and a decline in criminal arrests, juvenile delinquency, tuberculosis, venereal disease, and other social or physical pathologies. In 1890, Riis crystallized his theories in the publication of *How the Other Half Lives,* a muckraking expose of life in the slums.

Techwood Homes in Atlanta provides a vivid example of this thinking, as manifest a generation later. Charles Palmer, a past president of the National Association of Building Owners and Managers, used to drive to work past the Georgia Institute of Technology. He was horrified by what he saw on his trip downtown: "crowded, dilapidated dwellings, ragged, dirty children, reeking outhouses—a human garbage dump."[4]

At the library, Palmer came across Jacob Riis's *How the Other Half Lives.* Riis's theories made such an impression that he went on to organize local support and obtain federal assistance to replace this slum with decent housing. Techwood Homes, the project that Palmer persuaded Washington to subsidize, opened in 1936. Where there had once been crowded, dilapidated dwellings stood an island of grass and trees containing two- and three-story redbrick buildings with 603 apartments and 109 dormitory units for Georgia Tech.

Riis, Palmer, and other advocates of slum clearance thought that replacing a slum and creating a desirable envi-

Manhattan, before 1887. Jacob Riis's photograph of living conditions in the "Bottle Alley" section of Mulberry Bend. *(Museum of the City of New York, Jacob A. Riis Collection)*

ronment for future generations was sufficient justification for the slum dwellers to lose their homes. What they ignored is that clearing slums also destroyed cheap housing and forced out residents who probably lived there because they could not afford anything better.

The national slum clearance effort began with enactment of the National Industrial Recovery Act of 1933. This legislation authorized the Public Works Administration (PWA) to build low-cost housing as part of its emergency action to create jobs. Techwood Homes was one of the first of 26 slum clearance projects that the PWA started before litigation terminated its program. The effort was resumed under the Housing Act of 1937, which subsidized local public housing authorities that sought "the eradication of slums" through clearance and construction of replacement housing for persons of low income (see Chapter 9).[5]

It soon became clear that there never would be enough money, entrepreneurial talent, or public support for municipal agencies to replace every slum with government-built, government-owned housing. Since the necessary resources could be found only in the private sector, state legislatures, starting in 1943, began to enact legislation that would permit local governments to use the power of eminent domain for the combined purpose of slum clearance and housing construction by private developers (see Chapter 11).

Stuyvesant Town, New York City

In 1943, Mayor Fiorello La Guardia and Parks Commissioner Robert Moses persuaded the Metropolitan Life Insurance Company (MetLife) to join with the city government in a major slum-clearance project. They chose the notori-

New York City, 1942. Tenements in the Gas House District that were cleared for the construction of Stuyvesant Town. *(Courtesy of MetLife Archives)*

ous Gas House District along the East River, north of 14th Street. Together, they lobbied the state legislature for passage of the Redevelopment Companies Law, which allowed the city to condemn blighted areas for resale to private developers who agreed to clear and reconstruct them pursuant to a government-approved plan. Once the legislation had been enacted, MetLife prepared a redevelopment plan for the Gas House District. It proposed clearing 18 blocks that were home to 16,000 residents and replacing them with Stuyvesant Town, a 75-acre superblock with 8800 apartments.[6]

The traditional street grid was replaced by an entire superblock of trees, flowers, grass, playgrounds, curvilinear paths, parking, and residential towers, but nonetheless lacking the schools, libraries, and other community facilities that were part of Le Corbusier's scheme (see Chapter 7). This towers-in-the-park scheme became a virtual cliché for all residential redevelopment (whether public housing projects or residential renewal projects) despite the fact that it was meant for office districts. None of them adopted the meandering buildings that Le Corbusier had proposed for residential areas.

Peter Cooper Village, a similar MetLife project directly to the north, followed Stuyvesant Town. But neither project was part of a continuous aggregation of green superblocks through which pedestrians were free to roam. They were separated from the rest of the neighborhood by wide streets. In fact, each had been specifically designed to be "quite independent of any development that might come around it."[7]

Both Stuyvesant Town and Peter Cooper Village responded to the need for a modern residential environment. Life in these projects was perceived to be so much better than anything around them that applicants clamored to get in. Even a half century after completion, hundreds of families remain on waiting lists for years in the hope of moving in.

These early redevelopment projects demonstrated that there was public support for redevelopment. They could not, however, prove that replacement of the most diseased quarters would result in city rejuvenation, because these projects involved only housing, parking, and stores and did not include the businesses that formed the city's economic base. That demonstration would come from the redevelopment of the Golden Triangle in Pittsburgh, discussed in Chapter 7.

1925. Model of Le Corbusier's Plan Voisin for the redevelopment of Paris. *(Courtesy of Artists Rights Society [ARS]/SPADEM, 1995, Paris)*

Manhattan, 1943. Model of the Metropolitan Life Insurance Company's proposal for Stuyvesant Town. *(Courtesy of MetLife Archives)*

Manhattan, 2005. Stuyvesant Town and Peter Cooper Village. *(Alexander Garvin)*

Manhattan, 2005. Stuyvesant Town and Peter Cooper Village. *(Alexander Garvin)*

The New Residential Environment

Proponents of redevelopment had good reason for thinking that some cities could not retain commercial tenants without major physical changes to their business districts. They were wrong, however, in believing that people were leaving the city only because they found inner-city residential districts filthy, congested, and inconvenient, or because existing apartments had inadequate plumbing and wiring or obsolete kitchens and bathrooms. Many residents were moving to the suburbs because they wanted to own their own houses, to give their children access to better public schools, or just to toss a ball around in their own backyards. Nevertheless, proponents of redevelopment tried to retain these people by creating modern residential environments within the city.

Most residential redevelopment projects were not generous superblocks containing all the necessities of a modern urban neighborhood. They were rarely large enough to justify much more than a convenience store. Nor were they part of a carefully integrated system of highways and garages that accommodated high-speed automobile traffic or even part of a vast, landscaped pedestrian park that provided a verdant setting for the leisure-time activities of the residents. With few exceptions, the projects that were built were mediocre caricatures of Le Corbusier's images of a brave new world (see Chapter 7).

Even redevelopment projects that were most closely patterned on Le Corbusier's vision (shorn of its elevated highways) had trouble matching the cost of living in less expensive suburban subdivisions. When residents moved to Stuyvesant Town, they paid $85 to $110 per month for their apartments. Homeowners in Levittown, who had purchased similar-sized residences with down payments of $400 to $800, paid monthly carrying costs of $50 to $60 per month.[8]

The Taking Issue

When the federal government adopted the housing redevelopment prescription with the Housing Act of 1949, opposi-

Washington, D.C., pre-1950. Southwest Urban Renewal Project prior to redevelopment. *(Courtesy of the Washington Post)*

tion was immediate and serious. While courts had decided that state and local governments could take private property for the purpose of providing government-owned housing to persons of low income, opponents insisted that the Constitution did not permit the federal government to take one person's property for the purpose of selling it to another (see Chapters 7 and 11). The issue was settled in 1954 by the Supreme Court in a landmark case known as *Berman v. Parker,* which involved the Southwest Urban Renewal Area (SWURA) in Washington, D.C.[9]

In 1946, Congress had enacted the District of Columbia Redevelopment Act, which determined that "substandard housing and blighted areas," such as SWURA, were "injurious to the public health, safety, morals, and welfare" and should be eliminated "by all means necessary and appropriate." It further determined that this could not be done "by the ordinary operations of private enterprise alone" and therefore required "comprehensive and coordinated planning . . . [and] the acquisition and assembly of real property and the leasing and sale thereof for redevelopment pursuant to a project area redevelopment plan." The act clearly stated that "redevelopment pursuant to a project area redevelopment plan . . . is hereby declared to be a public use."

The National Capital Planning Commission designated three areas as possessing potential threats to public health. One of these areas, SWURA, was located within eyeshot of the Capitol. Redevelopment of this strategic location, it asserted, would determine the future character of the city itself.[10]

The 560 acres known as the Southwest Urban Renewal Area had once been a prestigious residential neighborhood. By the time the area was designated for redevelopment, though, 76 percent of its 5600 dwelling units were judged substandard, 43 percent had outside toilets, 44 percent had no baths, 70 percent had no central heat, and 21 percent were without electricity. However poor the housing, its residents (more than 80 percent African American) were firmly rooted in the neighborhood, 65 percent having lived there for more than 10 years. What they lacked was decent shelter.

The planning commission hired landscape architect Elbert Peets to prepare a redevelopment proposal for the southwest section of the city. Peets proposed retention of the area's predominantly low-income population, rehabilitation of many of its traditional row houses, and selective replacement of the worst buildings with new low-rise structures. The Washington, D.C., Redevelopment and Land Agency, established to execute urban renewal plans, felt the plan was impractical on physical and financial grounds. It did not believe that the southwest area in question contained sufficient structures "susceptible to the Georgetown kind of rehabilitation." Moreover, it believed the area was so blighted that banks would be unwilling to finance anything unless the area were radically altered.

The redevelopment agency commissioned a second plan from Louis Justement and Chloethial Woodard Smith, two architects well known for their support of clearance and redevelopment. The new plan envisioned a modern residential district with few streets, ample open space, plenty of parking, apartment towers with handsome views of the Potomac, an area for low-rent public housing, and commercial buildings along the main thoroughfare. Eventually the redevelopment agency adopted a compromise that reduced the number of multistory apartment houses, included a substantial number of two- and three-story buildings, and paid lip service to the "historic and sentimental interest" of the area's street plan and architecture.

In 1953, the redevelopment agency began assembling land, demolishing buildings, preparing sites, and issuing invitations to developers to submit plans for redevelopment of southwest Washington. At this point, a major legal struggle began. A department store owner whose property was to be taken went to court claiming that his store was not a slum "injurious to the public health, safety, morals, and welfare." He argued that his property was being taken in violation of the prohibition against taking private property without due process of law established by the Fifth Amendment to the Constitution. Perhaps slum clearance would improve the health, safety, and welfare of the community, but how could there be a justification for taking a person's property merely to create a more attractive community?

The counterargument was that redevelopment would be impossible without a carefully considered plan for the entire renewal area. That plan assured developers and financial institutions that all slum properties would be eliminated and replaced by a suitable living environment. Without such assurances, developers would not be interested in building, nor would financial institutions readily provide the money. In *Berman v. Parker,* the Supreme Court decided that any property (blighted or not) that was required for a project could be taken. The decision was explicit:

> *When the legislature has spoken, the public interest has been declared in terms well-nigh conclusive. . . . This principle admits of no exception merely because the power of eminent domain is involved.*[11]

After this case was decided in 1954, there was no longer any question that legislatures could establish what constituted "public use" of land. What remained in doubt was whether redevelopment would provide new life to the country's aging cities and whether redevelopment would result in an improvement in the housing occupied by area residents. Neither could be ensured without financing for the replacement housing.

Rebuilding Southwest Washington, D.C.

Until 1949 there had been no way to reduce a developer's site costs sufficiently to result in the production of marketable new housing. As discussed in Chapter 7, Congress provided the mechanism in the urban renewal program established by Title I of the Housing Act of 1949. This legislation subsidized the difference between project cost and the resale price needed to make planned new construction financially attractive to private developers. Two-thirds of the subsidy came from the federal government, one-third from the locality.[12]

Very soon after enactment of the Housing Act of 1949, Congress realized that lenders were not providing mortgages in the federally approved urban renewal areas, nor were developers ready to risk the equity capital required by mortgage lenders skeptical about the wisdom of lending in these areas. To solve the problem, Congress adopted an approach that had worked during the Depression: mortgage insurance. It amended the Housing Act in 1954 to establish the FHA 220 Program, which insured bank mortgages that financed development in federally assisted urban renewal areas. These mortgages were insured for up to 95 percent of the "replacement cost" of newly built or rehabilitated housing.

Before Congress had established the FHA 220 Program or the Supreme Court had ruled in *Berman v. Parker,* the District of Columbia's redevelopment agency had already divided the Southwest project into three sections and invited developers to submit proposals for redevelopment. One small area was set aside for light industry. James Scheuer, later a congressman from New York City, and Roger Stevens, a successful Broadway producer who later became the power behind the Kennedy Center, submitted the successful proposal for the second area, known as Capital Park. It became a 30-acre complex of moderate-sized apartment buildings and row houses accommodating 1750 households. The third parcel, more than three-quarters of the renewal area, was assigned to William Zeckendorf Sr., one of the country's best-known and most adventurous developers.

Zeckendorf and his architect, I. M. Pei, conceived the

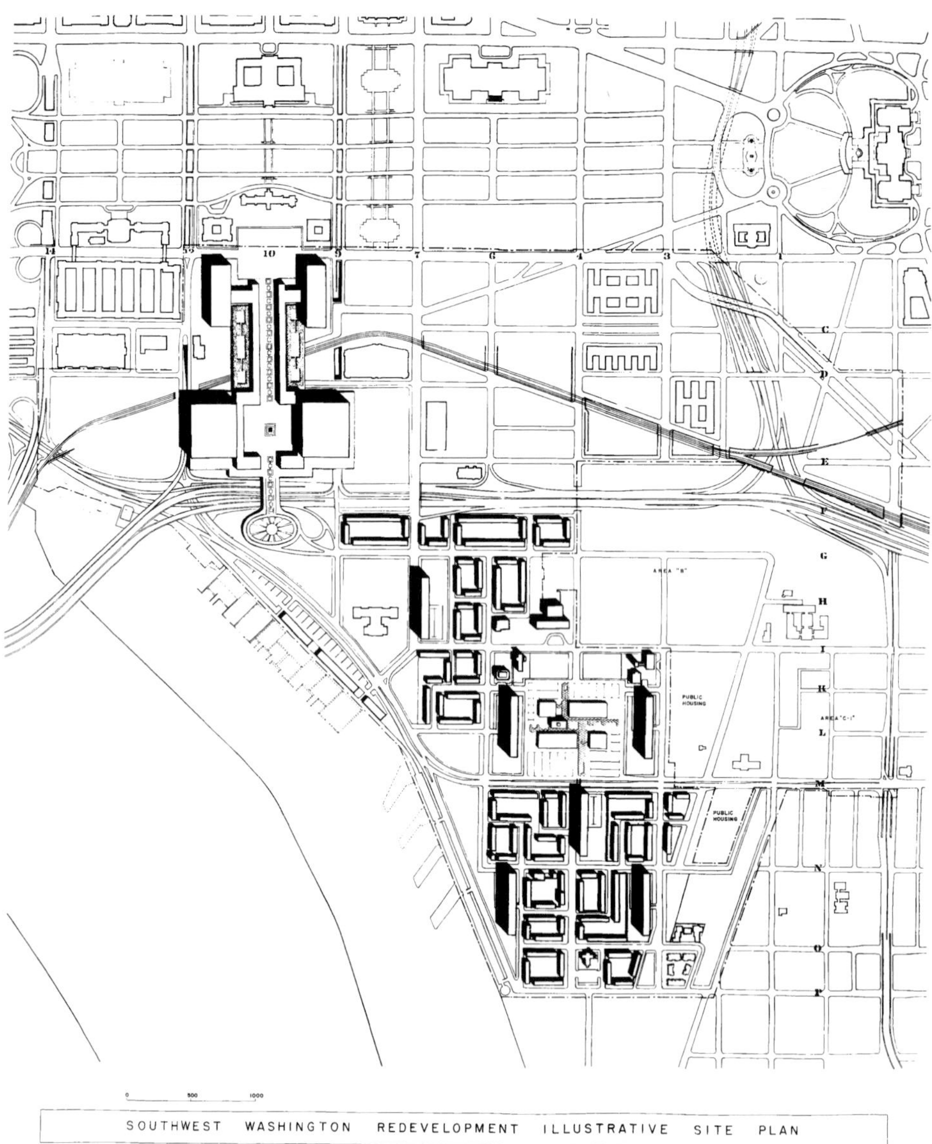

Washington, D.C., 1955. Redevelopment proposals made by I. M. Pei and William Zeckendorf for the Southwest Urban Renewal Area. *(Courtesy of Pei Cobb Freed & Partners)*

strategy that literally reconnected the Southwest project area with the rest of the city. As they saw it,

> *. . . the elevated tracks of the Pennsylvania Railroad, which rose just below [Independence Avenue] and cross the river to Virginia . . . form the visual and psychological boundaries of the area . . . separat[ing] and segregat[ing] Southwest Washington from the rest of the city.*[13]

Moreover, they felt the proposed eight-lane Southwest Freeway would add a second barrier, permanently disfiguring the neighborhood and eliminating any chance of attracting profit-motivated developers to rebuild the area.

Zeckendorf and Pei proposed to eliminate this barrier by building a 300-foot-wide mall that would stretch south from the Smithsonian Institution along 10th Street, bridge both the railroad and the freeway, and extend almost to the river. Along the mall they envisioned an office, cultural, and entertainment center named L'Enfant Plaza in honor of the author of the original plan for the national capital. At the riverfront they proposed a promenade, marina, and recreation area. For the rest of the area they planned row houses and elevator buildings containing 4000 apartments plus a "town center" with the retail stores necessary to serve all the residents of the southwest area

Although the project was constantly revised, took years to complete, and was largely built by others, Zeckendorf and Pei's vision is what determined the future of the Southwest area. New office buildings provided much-needed relocation space for federal agencies housed in temporary buildings

Washington, D.C., 2009. Wheat Row, completed in 1994–95 and retained as part of the Southwest Urban Renewal Project. *(Alexander Garvin)*

erected on the Mall during World War I as well as headquarters for the growing bureaucracies at the Department of Housing and Urban Development (HUD), the Department of Health, Education, and Welfare (HEW), Comsat, and the Department of Transportation. Developers built apartments for the growing market of government employees working in the area's new office buildings and throughout the city.

When it was finally completed in the 1970s, the Southwest Urban Renewal Project had eliminated 5600 "slum" dwellings, provided 5838 new apartments, and improved a critical section of Washington. It also improved the housing and living environment of the residents who had been displaced. Virtually everybody moved to significantly better accommodations. More than a quarter moved to public housing built at the time they were forced to move. The rest moved to better apartments in older buildings elsewhere in the city. This was possible because the flight of the middle class to the suburbs had increased the availability of such housing.[14]

Four decades later, the not-for-profit theater Arena Stage had become so successful that it had expanded. The Waterside Shopping Center had been replaced by entirely new, much larger commercial buildings that contained offices as well as retailers, and additional apartment buildings had been erected on the site. The project's public housing and predominant middle-income projects had remained in good condition.

The Southwest Urban Renewal Area (SWURA) helped to spur a revival of middle-class interest in other sections of the city. Rehabilitation of Washington's many row houses probably would have proceeded anyway, as would have construction of new apartment buildings. However, without the successful redevelopment of this very visible part of the city, it would have taken longer for the process to get started. Thus, the Southwest Urban Renewal Project provided both the constitutional basis for urban renewal and a model that could be copied on a building-by-building basis by developers throughout Washington. It did not, unfortunately, pro-

Washington, D.C., 1994. Arena Theater and Town Center Plaza in the Southwest Urban Renewal Project. *(Alexander Garvin)*

Washington, D.C., 2012. More than half a century later, the Southwest Urban Renewal Area is experiencing second growth. *(Alexander Garvin)*

vide a strategy that could reverse urban decay or be copied successfully by other cities.

HOUSING OCCUPIED BY SOUTHWEST RENEWAL AREA RELOCATEES[15]		
Condition	**Prior to relocation, %**	**After 5 years, %**
Good	22.2	85.7
Needs minor repair	26.4	14.3
Needs major repairs	19.9	0
Dilapidated or unfit	16.0	0
No answer	5.6	0

Washington, D.C., was not alone in considering subsidized redevelopment the path to a slum-free environment. As soon as Congress had enacted the program, Boston's Planning Board proposed housing redevelopment for the West End, a run-down Italian-American neighborhood on the edge of the downtown area. In the same year, the Newark Housing Authority identified the city's 16 most blighted areas and embarked on a renewal program that by 1957 included 395 blocks covering 25 percent of the city's residential areas.[16] In New York, Robert Moses, chairman of the newly created Committee on Slum Clearance, identified 9000 acres of slums requiring clearance and proposed five renewal areas. By 1960, when he resigned from New York City government, he had had been responsible for 32 percent of all Title I projects then in construction in the United States. Though that number is extraordinary, it does not reflect the equally extraordinary number of Moses failures. More than half of the 53 urban renewal projects he announced (often with elaborate brochures and commitments from major developers) had been rejected.[17]

Most cities, however, chose to concentrate on commercial redevelopment rather than residential reconstruction (see Chapter 7). They did not start serious housing redevelopment efforts until Congress amended the Housing Act in 1954. Ultimately, five redevelopment strategies emerged. Their success depended on the locations to which they were applied, the market in those areas, the appropriateness of redevelopment plans, the time they took to implement, and the willingness of developers to risk their money on local development objectives. However, they all required relocation of families and businesses from the sites selected for redevelopment, and they generated political opposition.

Strategy 1: The Self-Contained Enclave

Slum clearance had been advocated by Jacob Riis and other nineteenth-century reformers. Until the 1920s, however, private developers engaged in clearance on a scattered, lot-by-lot basis. The validity of redeveloping entire areas was demonstrated in 1928 when Fred F. French, a pioneering New York City real estate developer, completed Tudor City. He believed that only a giant, economy-size, self-contained enclave could survive negative pressures from surrounding areas and provide the necessary critical mass of new residents for community rebirth.

Manhattan, 2011. One of Tudor City's private parks. *(Alexander Garvin)*

Tudor City

French demonstrated his theory by assembling a large number of slum properties along East 42nd Street on the edge of the Manhattan business district. The site was sandwiched between the noisy Second Avenue Elevated Railway (the El) and the smelly slaughterhouses beside the East River. Few believed that housing at this location, although in walking distance of Grand Central Station, could attract the large potential market of people working in Midtown Manhattan. Others were scared off by the cost and complexity of assembling sufficient land to overcome surrounding negative influences.

French attracted customers by creating a residential enclave 30 feet above First Avenue and 42nd Street. There he built 3300 apartments in Tudor-style buildings that enclosed two 15,000-square-foot private parks and a private roadway bridging 42nd Street. The buildings turned their backs on the slaughterhouses along First Avenue, opening instead onto Tudor City's parks and streets.

From the beginning, Tudor City was popular with Midtown office workers, who eagerly rented apartments in the project. In the decades that followed, other builders erected apartment and office buildings in the area. Today, when the slaughterhouses have been replaced by the United Nations, the El has long been demolished, and east Midtown is a busy and expensive area, it is difficult to realize the enormous risk French had accepted in redeveloping the area.

Many of the first Title I projects in Washington, Boston, Chicago, and Newark accepted French's notion that since redevelopment took place in slum areas, it had to proceed on a sufficiently large scale to withstand the spillover of the surrounding slums that would not be directly impacted by redevelopment. This, plus the desire for economies of scale, undoubtedly contributed to the large size of the nation's first urban renewal projects. Robert Moses adopted this notion for New York City as the most effective way to generate massive amounts of new housing.

Harlem-Lenox Terrace

Despite the careful planning, self-contained apartment enclaves rarely revive the surrounding neighborhoods. Harlem-Lenox Terrace, one of New York City's first two redevelopment projects, illustrates this very well. The 15-acre site consisted of three full blocks between 132nd and 135th Streets and Fifth and Lenox Avenues. It was described in 1951 by the Committee on Slum Clearance, chaired by Robert Moses, as

> *. . . a group of buildings which are almost all ancient, poorly lighted, badly laid out, inadequately ventilated, and generally occupied by more families than they were originally designed to accommodate . . . 146 out of 164, or 89 percent, were classified as "run down . . ." 71 percent of the residential structures were . . . tenements built before 1901, with their excessive coverage of the lot, and inadequate courts and air shafts.*[18]

Manhattan, before 1951. Harlem-Lenox Terrace. *(Courtesy of the Citizens Housing and Planning Council, New York City)*

Manhattan, 2012. The Harlem-Lenox Terrace Urban Renewal Project became a "high-rise luxury oasis in the heart of Harlem." *(Alexander Garvin)*

Clearing these three blocks required demolition of 2068 apartments.[19] Most relocatees paid an average monthly rent of $29 for their "run-down" apartments and probably could not afford the projected monthly rent of $29 *per room* in the new buildings. The committee estimated that 1010 families were eligible for public housing, many of whom might go to the recently completed 1286-unit Abraham Lincoln Public Housing Project just across Fifth Avenue. Few people seemed to care that this project would probably not have enough vacancies to accommodate the relocatees. Nor did they wonder how public housing in other locations, given the low vacancy rate and high demand, could provide replacement housing for all the renewal projects Moses had recommended. As for the rest of the relocatees, they were on their own to find private housing.

When Lenox Terrace was completed in 1961, two streets had been closed to create a separate enclave that included a few retail stores, the already existing Harlem Boys Club and Playground, parking lots, and six 16-story redbrick apartment slabs financed with FHA 220 mortgages. The apartment buildings accommodated 1716 families, who paid monthly rents in excess of $50 per room, far more than the relocatees could have afforded.

Lenox Terrace immediately became a desirable residence for middle-class African Americans and remains so more than 60 years after Moses conceived it. It has become a "high-rise luxury oasis in the heart of Harlem," where

> *. . . doormen dress in spiffy blue uniforms and work around the clock. The private parking lot has trees and benches and glistening cars. And the long tenant list includes a roster of leading New York politicians, executives, lawyers and doctors.*[20]

It is part of a complex of six slum clearance projects extending along the Harlem River, only one of which, Abraham Lincoln Houses, is for families of low income. Each of these projects is designed as an enclave set apart from the rest of Harlem. The 6800 new apartments that these projects provide are certainly an improvement over the slums they replaced. But neither Lenox Terrace nor any of its neighboring enclaves have had much impact on the rest of the neighborhood. The adjoining blocks are still run-down—only more so. Lenox Terrace tenants now worry that the problems of neighboring "ghostly abandoned buildings and street-corner drug markets" may spill over into what one famous resident called "a substitute for moving to the suburbs."[21]

Harlem-Lenox Terrace did not generate further neighborhood improvements because poor residents who lived outside redevelopment projects could not muster sufficient rent to cover the cost of needed housing repairs, much less afford even more expensive redevelopment. Since they could not pay, private developers had no way of financing additional improvements.

Manhattan, 2012. Deteriorating tenements across Lenox Avenue from Harlem-Lenox Terrace remained unaffected for decades after the project was completed. *(Alexander Garvin)*

Lake Meadow and Prairie Shores, Chicago

Chicago, like many cities that built public housing, had been engaged in slum clearance since the beginning of the New Deal. The attempt to redevelop the South Side of Chicago, however, was not initiated by a government agency. The Illinois Institute of Technology (IIT, then called the Armour Institute of Technology) and Michael Reese Hospital were unwilling to relocate to another part of the city. But they were becoming increasingly concerned with the dilapidated condition of the surrounding neighborhood and the health of its primarily African-American population.[22]

When World War II ended, they were finally in a position to act. Two of the most prominent members of the IIT architecture faculty, Mies van der Rohe and Ludwig Hilberseimer, argued that large-scale intervention was the best way to deal with the situation. So, in 1946, IIT and the hospital joined with other area institutions and businesses to form the South Side Planning Board. Its objective was to obtain long-needed neighborhood improvements.

Over the next five years the South Side Planning Board identified the usual litany of slum problems: ramshackle buildings with overcrowding, inadequate plumbing, and inadequate recreational opportunities. In 1950, it commissioned planners from IIT, Harvard, and the University of Illinois to propose design alternatives for the area between 31st and 59th streets from Lake Michigan on the east to the Pennsylvania Railroad on the west. All three teams sought maximum light and air while providing what they considered to be minimum acceptable dwelling space for the area's future residents. Harvard's team, led by Walter Gropius, and the University of Chicago's team, led by Martin Meyerson, proposed high-density schemes, with the tallest buildings concentrated along the lakefront. The IIT team, led by Mies and Hilberseimer, also concentrated apartment towers along the lakefront, but their scheme proposed an average density that was one-quarter that of the other two schemes.

In 1946, while the planning board was studying the area, the Chicago Housing Authority (CHA) began the process of land acquisition and clearance of four blocks adjacent to Michael Reese Hospital for its expansion. The hospital paid the entire cost of acquisition and relocation, plus the CHA's expenses. These four blocks were the first part of what would become the city's largest urban renewal area. In time, it grew to include three privately financed housing developments, several public housing projects, and the expansion of IIT, Reese, and Mercy Hospitals.

The city established the Chicago Land Clearance Commission in 1947 to purchase and clear slum property and then sell that property at its reuse value to private developers and nonprofit institutions to implement an agreed-upon redevelopment plan. Thus, when Congress enacted Title I of the Housing Act of 1949, the Land Clearance Commission was already in place and ready to apply for the subsidies Title I provided to cover two-thirds of the difference between gross project cost and sale of property at its "reuse" value.

Chicago's first Title I project, Lake Meadows, was located in the midst of the South Side. At the time it was conceived, institutional lenders were skeptical about investing in an area that had been declared by local authorities and certified by the federal government as a "slum." Ferd Kramer, one of the members of the board of Michael Reese Hospital who had

Chicago, 2005. Lake Meadows provides middle-class apartment residents with the light, air, trees, and grass envisioned by Le Corbusier for his City of Tomorrow. *(Alexander Garvin)*

Chicago, 2005. Prairie Shores provided a living environment that was quite different from the older sections of Chicago's South Side. *(Alexander Garvin)*

been particularly instrumental in committing the hospital to redevelopment of the surrounding neighborhood, was to play a critical role in overcoming that skepticism. As president of one of the city's major real estate firms, president of the Chicago Metropolitan Housing and Planning Council, and past president of the Chicago Real Estate Board, he was one of the few people in a position to persuade lenders to invest in Chicago's South Side slum. In 1952 he convinced the New York Life Insurance Company that it was safe to offer a long-term mortgage for Lake Meadows, a 70-acre complex of 10 buildings containing 2033 apartments designed by Skidmore, Owings & Merrill (SOM).

The success of the first apartment buildings led Kramer and other friends of the hospital to invest in a second urban renewal project, Prairie Shores, which was completed in 1962. This 55-acre project consisted of five apartment buildings containing 1677 apartments designed by Loebl, Schlossman & Bennett, a local shopping center, and playground facilities. By then, government officials no longer considered redeveloping the South Side risky, and the Land Clearance Commission provided $6.2 million in Title I subsidies. Congress had by then amended the Housing Act to provide insurance to lenders who offered mortgages in urban renewal areas (see Chapter 2), so there was little difficulty in financing Prairie Shores.

After half a century, Lake Meadows and Prairie Shores are still fully occupied, largely by middle-class African Americans. They continue to provide desirable apartments at competitive prices in well-managed, well-maintained buildings. While they have continued to be oases of tranquility, they have had little or no impact on the rest of the South Side because, like Stuyvesant Town in New York, they were designed to be self-contained enclaves protected from the dangers of the surrounding slums. Nor did they have any impact on the problems of race and poverty because, like most urban renewal projects, they were designed to lower densities and provide decent housing without giving proper thought to how they might address social problems.

By reducing the number of site occupants at a time when the city was experiencing an influx of African Americans, the redevelopment of the South Side also reduced the supply of housing adjacent to downtown Chicago. As the NAACP explained, the South Side Planning Board, like the Chicago Plan Commission, "never made any realistic plans to accommodate the continuing and likely to continue Negro migration to Chicago."[23] In fact, 3416 lower-income families were moved from the site of Lake Meadows to make room for 2033 middle-income households. While 92 percent of these site occupants moved into standard housing that was usually an improvement over their previous residence, hardly any of those relocatees could afford the rents at Lake Meadows.

Strategy 2: The Superblock

At the start of the twentieth century, reformers argued that existing cities had become obsolete. They saw traffic hazards, noise, air pollution, disease, and crime as evidence that streets and blocks "built for an ancient pedestrian age" did not meet "the requirements of our motor age." As an alternative, Hilberseimer, already a well-known city planner in Germany before coming to Chicago, suggested building superblocks that included all the essentials of community life

Berlin, 1937. Ludwig Hilberseimer's idealized conception of a housing development for the Heerstrasse and the University of Berlin. *(Photograph copyright 1994, Courtesy of the Art Institute of Chicago)*

within walking distances that did not "exceed 15 to 20 minutes." These superblocks would be large enough "to support necessary communal, cultural, and hygienic institutions" and small enough "to preserve an organic community life."[24]

The superblocks, presented by their proponents in drawings, models, and photomontages, were diagrammatic at best. At worst they were anonymous rows of high-rise apartment buildings arranged in geometric patterns that ignored the complexity of daily life. Vehicular traffic was dispatched to regional highways so that the area devoted to local streets could be reduced to an absolute minimum and through-traffic eliminated from every superblock. Densities were kept to a minimum in order to prevent the overpopulation that was thought to be responsible for social, moral, and physical diseases prevalent in congested, obsolete cities. Space for day care, education, recreation, shopping, and institutional use was allocated by formulas that were intended to prevent wastage and optimize accessibility.

These superblocks remained theoretical abstractions until the Housing Act of 1949 made their realization possible. Public officials eagerly proposed them as the best way of eliminating the bad environment. Too often, the result of their efforts was another depressing dormitory stockade that could be ridiculed by anybody who opposed urban renewal. Only occasionally did such efforts result in projects like Lafayette Park and Portland Center, which approximate what proponents of the superblock had envisioned.

Lafayette Park, Detroit

Lafayette Park was conceived in 1946, before Congress had embarked on the national effort at urban renewal. As initially proposed, it was a city-sponsored redevelopment project on the edge of downtown Detroit. The 129 acres selected for redevelopment were to be cleared of 1953 families and 989 individuals, 98 percent of whom were African American, more than three-quarters of whom had annual incomes of less than $3500, and nearly a fifth of whom were receiving public assistance. Civic leaders believed that clearing this "slum" would remove a blighting influence on the central business district and that replacing it with a residential superblock would attract families who would otherwise leave Detroit for the suburbs.[25]

The project successfully overcame both community opposition and a taxpayers' lawsuit challenging the validity of public condemnation. But it was unable to overcome the need for a talented and experienced developer, a financial institution that would provide mortgage money, or substantial government subsidies to close the gap between project cost and any sales price that developers were willing to pay or banks to finance.[26]

The necessary subsidies became available when Congress enacted the Housing Act of 1949. Within a year, Lafayette Park, then known as the Gratiot Urban Renewal Project, was converted to an urban renewal project with $4.3 million in federal subsidies reserved for its execution. Finding a suitable developer was much more difficult. Local builders had never tackled anything as large or complex. When the site was put up for auction in 1952, nobody came forward. The following year a developer finally offered $1.27 million for the site, but had to drop out because he was unable to obtain financing (see Chapter 2).

Detroit, pre-1951. Site of Lafayette Park prior to clearance. *(Courtesy of City of Detroit Housing Commission)*

Banks and insurance companies were willing to finance housing development in Lafayette Park and dozens of renewal projects like it only after Congress amended the Housing Act in 1954 to include FHA mortgage insurance for urban renewal areas. The following year the city accepted a $1.17 million offer from Cities Redevelopment, Inc., a Chicago firm owned by Herbert S. Greenwald and Samuel N. Katzin.[27]

Greenwald and his design team proposed a 78-acre superblock that included 2000 apartments in six towers surrounded by clusters of one- and two-story row houses. Within a year, the first phase was in construction. When residents arrived in 1958, Lafayette Park seemed to announce the beginning of Detroit's transition into a new world "vertical to the sky, open to light and air, clear and radiant and sparkling." This brave new world died along with Herbert Greenwald in a plane crash a year later and there never were any other phases. Unfortunately, the project was completed by a different group of designers, whose work detracts from the ideal physical environment that Greenwald had attempted to create.[28]

Arriving at Lafayette Park by car, you first see the encircling trees and an occasional glimpse of the residences behind them. Then, turning down a cul-de-sac, you drive past one- and two-story row houses and park in either a driveway or a parking lot. There, safe from the intrusions of the sur-

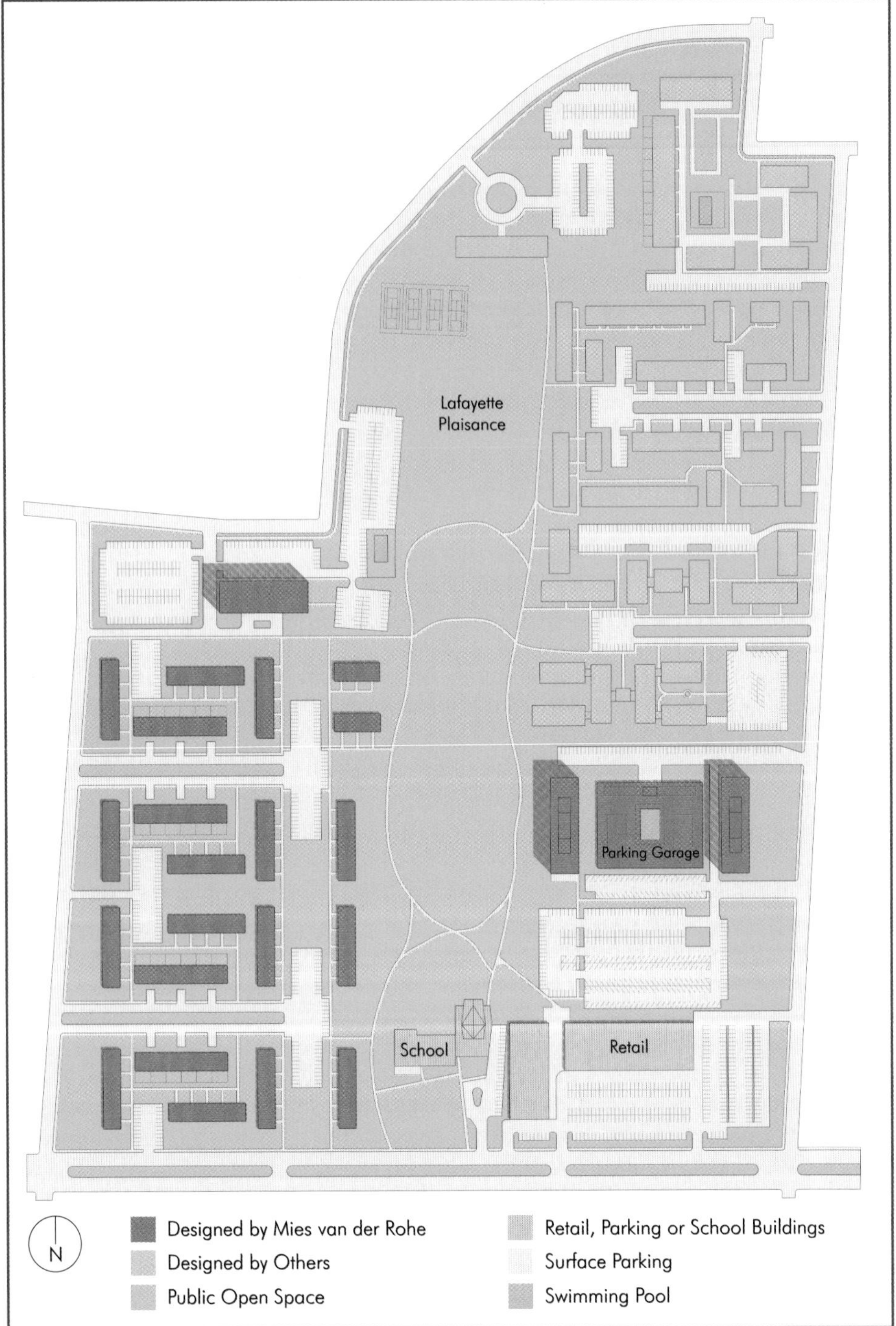

Detroit, 2001. Most of Lafayette Park was completed according to Mies van der Rohe's site plan. But only the buildings in dark blue were completed according to Mies van der Rohe's designs. *(Alexander Garvin and Ryan Salvatore)*

rounding city, you walk directly into an ideal living environment.

Lafayette Park is no isolated enclave. At its heart is a 19-acre park with broad expanses of grass, a clubhouse, and a swimming pool. Greenwald understood that its 2000 apartments could not provide enough customers to justify a full range of shops, schools, and other services within the project's perimeter, where they did not exist at the time. As a result, Lafayette Park includes a small shopping complex and an elementary school that also serve those who live outside Lafayette Park.[29]

Greenwald's design team created one of the nation's most urbane combinations of a natural and built environment. Despite Hilberseimer's insistence on spacing buildings to en-

Detroit, 1996. Cul-de-sac entry into Lafayette Park, with parked cars hidden from view in driveways 2.5 feet below garden level. *(Alexander Garvin)*

Detroit, 2001. The redevelopment of Lafayette Park was stalled for more than a decade until Congress provided banks with mortgage insurance in urban renewal areas, thereby creating a financial climate that attracted entrepreneurs to areas that had been officially designated as slums. *(Alexander Garvin)*

Detroit, 2008. Lafayette Park two-story row houses. *(Alexander Garvin)*

Detroit, 2001. Because the Lafayette Towers Shopping Center is located on the edge of the renewal area, it is a convenient shopping destination for both project residents and people from the surrounding neighborhood. *(Alexander Garvin)*

sure four hours of direct sunlight in every room, neither the spaces between buildings nor apartment layouts are determined by formula. Furthermore, the landscaping makes Lafayette Park a very special place. Even circulation is not simply a matter of connecting destinations: residents stroll past the front lawns and secluded yards of the low-rise residences and pass through intimate, tree-covered, common sitting areas on their way to broad expanses of grass with plenty of room to toss a ball or play touch football.

Although Detroit has continued to deteriorate and lose population, Lafayette Park has remained occupied and well maintained, perhaps because it is conveniently located only a few minutes' walk from the heart of the business district. Some of the buildings are owner-occupied. Others are inhabited by middle-income renters, who include a broad spectrum of racial and ethnic backgrounds. Lafayette Park is a unique island of stability, surrounded either by a combination of later, similarly located and financed but poorly designed housing projects and the abandoned wasteland of blocks untouched by redevelopment. Its benefits did not, however, spill over to improve housing conditions in the surrounding areas of Detroit. A similar, less architecturally stunning project in Portland demonstrates some ways in which a superblock can benefit the surrounding city.

Portland Center, Oregon

Civic leaders in Portland, Oregon, worried about the same problems that troubled their counterparts in other cities. Taxpaying residents were moving to the suburbs. Downtown retail sales were declining. They also saw the problem in terms of physical and functional obsolescence and decided the solution lay in redeveloping a 54-block "slum" on the edge of the business district. In 1950, the area was filled with junkyards, run-down hotels, dilapidated apartment buildings, and abandoned commercial structures. It also was home for 2300 people, 43 percent of whom were in single-person households.[30]

When redevelopment was first proposed, in 1952, the plan was to clear the entire 83.5 acres for a new civic auditorium and support services for the nearby business district. Four years later, when the area was rejected as the site for a civic auditorium, the project simultaneously lost its raison d'être and its developer. The Portland Urban Renewal Agency then proposed dividing what had become known as the South Auditorium project into three residential superblocks.

The market for downtown apartment towers in Portland had yet to be tested. Consequently, neither real estate developers nor financial institutions were willing to risk their time and money on the project. The necessary entrepreneurial element was supplied in 1958 when Portland voters approved the creation of a new public agency: the Portland Development Commission (PDC), which, together with the newly created Portland Center Development Corporation, accepted responsibility for reviving the ill-fated South Auditorium project.

In 1961, they hired the architecture firm of Skidmore, Owings & Merrill to redesign the project and Lawrence Halprin and Associates to devise suitable landscaping. Their plan for what became known as Portland Center still called for superblocks but retained the city's 200-foot street grid as a series of pedestrian walkways that defined building sites and recreation areas. However, it was redesigned as an integrated complex of apartment towers, low-rise buildings, offices, convenience retail stores, and generous, landscaped public spaces.

Today, residents of Portland Center still stroll past Halprin's landscaped walkways, fountains, and sitting areas on their way to apartments with impressive views of the city. Many continue to do their shopping within the superblock. More important, they live within walking distance of their downtown jobs. Their presence helps to keep the business district alive after the commuters have driven home.

Portland, Oregon, 2007. Residents of Portland Center enjoy a wide range of on-site recreational opportunities. *(Alexander Garvin)*

Portland, Oregon, 2007. Retail shops in Portland Center are used by project residents and those of the surrounding community. *(Alexander Garvin)*

There is no way of knowing whether Portland Center attracted residents who would otherwise have moved to the suburbs. What is clear is that it satisfied the demand for an orderly downtown residential environment with convenient shopping, trees, grass, and flowers.

Strategy 3: A Net Increase in Apartments

A different strategy for eliminating the bad environment is implicit in Fred French's other housing redevelopment project, Knickerbocker Village on Manhattan's Lower East Side. It calls for a two-pronged attack on a city's worst housing: first by clearing the worst slum housing and second by building more apartments than are eliminated. In this variant of the *filter-down theory,* the families who move into the new buildings vacate slightly less desirable apartments that are then occupied by other families, who in turn move from other less desirable apartments, and so on, until the city's least desirable apartments are left without occupants and eventually are removed from the housing stock.

New York City

In 1928, flush from the success of Tudor City, French embarked on an attempt to assemble 50 acres of the Lower East Side and to create an even bigger self-contained enclave of 40,000 apartments. He acquired one-third of the site before the Depression forced him to sell all but 5 acres of the infamous "Lung Block." Robert W. DeForest, commissioner of the city's Tenement House Department, had recommended that block for clearance in 1903:

> *I know of no tenement-house block in this city which is so bad from a sanitary point of view, or from a criminal point of view. . . . I understand that . . . these houses are permanently infected with the germs of tubercular disease, and that the only remedy and method of preventing the further spread of this disease from these houses is the destruction of the buildings.*[31]

There, with the help of a Reconstruction Finance Corporation mortgage and a 25-year real estate tax exemption,

Manhattan, 2000. Knickerbocker Village on the Lower East Side, completed in 1934. *(Alexander Garvin)*

Manhattan, 2005. Public housing built by the New York City Housing Authority has had as much as or more impact on the Lower East Side than Corlear's Hook, Robert Moses's first housing renewal project. *(Alexander Garvin)*

French erected a 13-story enclave that, like Tudor City, was organized around private open space, in this case two huge courtyards.

Knickerbocker Village replaced 1085 slum dwellings with 1593 well-designed new apartments. The net gain of 508 apartments, it was thought, might generate further neighborhood revival because even if the original-site tenants moved to tenement apartments without hot water or bathrooms, the families that moved to Knickerbocker Village would free up better apartments for other city residents living in worse conditions. They in turn would free up other, somewhat worse apartments, and so on, until 508 of the worst apartments would be filtered out of the city's housing stock.[32]

The economic dislocations caused by the Depression made it difficult to determine whether slum clearance combined with sizable net increases to the housing stock would, in fact, result in improved housing for families outside redevelopment areas. Far too little was built. After World War II, using Title I funding, Robert Moses conceived a massive slum clearance effort that he believed would work in just this fashion. By 1959, when he was forced to give up chairmanship of the Committee on Slum Clearance, Moses already had an approved redevelopment program that called for a net gain of 22,000 apartments. The number seems huge looking back from the twenty-first century, but it is dwarfed by the 84,819 public housing units erected during that same period by the New York City Housing Authority, operating completely independent of Moses.[33]

Despite Moses's massive redevelopment program and the enormous amount of public housing that was built, anybody living in New York City in 1940 or 1960 or 1980 could tell you that it was still suffering a major housing crisis. In those 40 years, although 1,050,351 housing units had been built and the population had *declined* by 383,356, there still was a terrible shortage of apartments. The shortage was intensified by a decrease in the average household size from 3.64 to 2.54 persons, producing an increase of 738,213 in the number of households that required shelter.[34]

New Haven, Connecticut

In New Haven, which also tried to achieve a net increase in apartments, the filtering process produced much more apparent results. Not only had the city pledged to "free itself of slums," it was more successful than any other American city in obtaining federal redevelopment subsidies. As of 1966, New Haven had received $745 in per capita federal urban renewal allocations, 20 times New York City's, 16 times Chicago's, and 7 times Philadelphia's. The nearest competitor was Newark, which had received $277.[35]

Like New York, Detroit, and Portland, New Haven had considered housing redevelopment long before passage of the Housing Act of 1949. The city's first comprehensive plan, jointly prepared in 1910 by architect Cass Gilbert and landscape architect/urban planner Frederick Law Olmsted Jr., proposed the redevelopment of the area between the railroad station and the downtown business district. Their proposal was not implemented. During the late 1930s the city had surveyed its blighted areas and proposed clearance for the purpose of building new public housing. In 1941–1942 it hired Maurice Rotival, a French city planner on the Yale faculty, to prepare another master plan. Like Olmsted and Gilbert's redevelopment scheme, it was not executed. However, both plans contributed the framework for the redevelopment projects initiated after World War II.

While there were areas of unsafe, inadequately heated, overcrowded housing, New Haven had nothing like the tenement districts of New York or Chicago. Most of the city's housing consisted of low-rise, low-density buildings whose renovation was neither complicated nor expensive. Furthermore, the relatively small quantity of housing that had to be replaced was within the financial capacity of Connecticut lending institutions. Thus, the promise of a slumless city was believable and realizable.

New Haven's redevelopment effort was conceived and directed by one of the best and brightest teams of municipal officials anywhere in the country. The man responsible for bringing it together and providing vision, leadership, and political moxie was Richard C. Lee, mayor from 1954 through 1969. He made the redevelopment proposals that had been discussed for years the cornerstone of his election campaign and 16-year mayoralty (see Chapter 7).

Lee needed talented administrators to develop imaginative but practical redevelopment proposals and then to shepherd them through the maze of bureaucratic stumbling blocks that could prevent their realization. As his first development administrator, he picked Edward Logue. Logue and the other members of this team produced plans for every neglected section of the city. Meanwhile, Lee worked with the city's political leaders to obtain the active local constituencies needed for acceptance by the Board of Aldermen. The mayor believed that when all these elements were in hand, Washington would have to provide the necessary financing—and Washington did.

The team's first undertaking, the Oak Street Area Redevelopment Project, obtained federal funding in 1956. The previous administration had already obtained conceptual approval for razing 15 acres. Lee and Logue tripled the area to be cleared and completely altered the redevelopment strategy. The centerpiece was a highway connector to the Connecticut Turnpike, 600,000 square feet of new office and retail space to be built on renewal land, and minor expansion by Yale University. The four new apartment buildings that seemed an afterthought were included because the newly elected mayor was so concerned about slum conditions in the area.

In 1955 there was as yet no way for the private sector to finance replacement housing for persons of low and moderate income. Nonetheless, public officials accepted demolition of 811 apartments because there was sufficient turnover of existing low-rent public and private housing to accommodate the relocatees. Moreover, construction of 812 new apartments guaranteed that there would be no net loss of housing. The officials also accepted displacement of the businesses occupying 62 commercial and 25 industrial structures. They believed that the benefits of slum clearance would surely outweigh the hardship of a move and thought that businesses would be able to carry the financial burden. Besides, redevelopment officials reported that 41 percent of the area's structures were substandard and required replacement.

Mayor Lee described the new apartment buildings in the

New Haven, 2011. Crown, University, and Madison Towers have taken on much greater prominence, because nobody remembers the massive relocation caused by the Oak Street connector and the commercial buildings that were part of the urban renewal project. *(Alexander Garvin)*

New Haven, 2011. Projects like the Dwight Coop maintained the small scale of the rest of the neighborhood. *(Alexander Garvin)*

Oak Street Project as "the most God-awful-looking things I ever laid eyes on."[36] He vowed there would be no more dull buildings and proposed to rebuild other areas with the help of some of the nation's best-known architects. He also shifted the redevelopment program away from large-scale clearance and toward selective redevelopment, housing rehabilitation, and affordable housing financed through federal programs.

Lee and Logue altered the original plan for the Wooster Square Urban Renewal Area to include a major housing rehabilitation effort, scattered clusters of new low-rise housing, small parks, and new community facilities (see Chapter 13). Tom Appleby, who succeeded Logue as development administrator in 1961, continued this approach. The next projects, Dixwell (247 acres) and Dwight Street (215 acres), covered whole neighborhoods still "basically in sound condition." Most of these areas were to remain untouched by redevelopment. Instead, small pockets of blight were replaced with small-scale buildings that blended into their surroundings.

Title I permitted a locality to make its one-third share of the project cost in the form of noncash credits rather than cash. Many cities used necessary expenditures for infrastructure improvements and community facilities as noncash credits. The leaders of New Haven turned this into a fine art. Not only did the city build 12 new schools in renewal areas as a way of covering its one-third project cost, it counted every allowable noncity expenditure. For example, when Yale University purchased city land for the construction of two new residential colleges, the sales price became a noncash contribution to the nearby Dixwell Urban Renewal Project. The acquisition cost for the site of the Yale School of Art and Architecture likewise became a noncash credit for the Dwight Renewal Project.

Prior city improvements that served the project area could also be counted. For this reason, New Haven regularly scoured old budgets for "catch basins, tree stump removals, even an abandoned public bathhouse," and other eligible items. Since the noncash credits in excess of the required one-third local contribution could be pooled and credited to other redevelopment projects, the city planned its renewal projects around much-needed local improvements and then used the surplus federal contributions to finance similarly needed improvements in other redevelopment projects. Thus, for New Haven, the Housing Act of 1949 could have appropriately been renamed the capital-budget substitution act.

No federal subsidy was left untapped. When Congress enacted the 221(d)(3) Program (see Chapter 10), New Haven rushed to use it for a series of moderate-income cooperative or nonprofit rental projects within already approved renewal areas. Columbus Mall Houses (72 apartments in the Wooster Square Renewal Area), Florence Virtue Houses (129 apartments in the Dixwell Renewal Project), and Trade Union Plaza (77 apartments in the Dwight Renewal Area) are typical in terms of scale and quality of architectural design.

When Richard C. Lee left office, almost every area of the city was involved in some effort at redevelopment. There was also major opposition to further government slum clearance, especially in the "Hill," a poor neighborhood with a substantial African-American and Puerto Rican population. Despite growing criticism of the redevelopment program, New Haven's housing market had improved markedly. The city's worst slums had been demolished. Most of the housing that remained was structurally sound and, with minor expenditures on rehabilitation, could have been expected to provide decent homes for many years to come. Since New Haven's population declined by 25 percent between 1950 and 1980, relocatees had been able to move to steadily improving accommodations. Although New Haven's attempts at downtown revitalization during this period were not particularly successful, the city's ambitious residential redevelopment program made great strides in providing a better environment for the next generation. If the process had continued, the revitalization of New Haven's residential neighborhoods might have become a model for the nation.

Strategy 4: Removing Frictional Blight

Another approach for eliminating slums calls for removing impediments to private investment. The argument is that if the unsightly structures, incompatible land uses, and noxious activities blighting an area are removed, neighboring property owners will make improvements, developers will purchase and rehabilitate or build, and banks will lend the money to pay for this. Therefore, government should acquire these blighted properties and resell them to developers who will execute the city's renewal plan.

Society Hill, Philadelphia

The strategy of eliminating blight that has a negative frictional effect on neighboring property is well illustrated by Philadelphia's Washington Square East Renewal Project, better known as Society Hill. This 120-acre area is located in the southeastern section of downtown Philadelphia, just west of the Delaware River (see Chapters 2 and 18).

Society Hill received its name from the Free Society of Traders, established in 1682, whose members originally purchased most of this land. The houses they built were called "Society houses." Although the society has been gone from the area for three centuries, the name persists.[37]

During the nineteenth century, the area was dominated by the expanding operations of the Dock Street Market, its wharves, warehouses, and traffic. Only poor families and transients moved into the midst of the noise, filth, and odors generated by the rat-infested market. By the mid-twentieth century, many of the area's formerly gracious Georgian, Federal, and Greek Revival town houses had been converted into storage facilities, manufacturing lofts, bars, rooming houses, and cheap tenements. Even Washington Square, one of Philadelphia's four original 8-acre park squares located in Society Hill, had become a favorite hangout for bums, perverts, and other "undesirables."

The Dock Street Market had become outmoded, and a new Food Distribution Center was being planned to replace it. During this time, Edmund Bacon conceived a strategy that used the relocation of the market as the basis for the revival of all of Society Hill. He also proposed the acquisition and removal of every incompatible land use, dilapidated building, and unsightly structure in the neighborhood, whether or not it was connected to the market. Approved for Title I funding in 1957, his plan called for preservation of the

Philadelphia, 2010. The carefully placed combination of new low-rise and high-rise buildings that augmented Society Hill's historic row houses attracted new residents to what had been thought of as one of the city's worst slums. *(Alexander Garvin)*

area's historic character and sizable stock of eighteenth- and nineteenth-century row houses; construction of new housing compatible in color, scale, and design with the buildings that were retained; and creation of the commercial, institutional, and recreational amenities needed in any healthy neighborhood.

FHA 220 mortgage insurance was the key to restoration of Society Hill's lovely houses. But by itself, insured bank financing was insufficient. There had to be property owners interested in and financially capable of rehabilitating their buildings, people willing and financially able to live there, and a method of being sure that the rehabilitation would be consistent with the rest of the neighborhood.

The Redevelopment Authority surveyed every structure in Society Hill. It found almost 700 buildings (more than 500 built before 1850) that could be economically rehabilitated. Then it established both general rehabilitation standards for the area and specific requirements for many of the buildings. Property owners were given 30 days to commit to restoration and an extended period to plan and start rehabilitation. If they failed to agree to renovate or to complete the rehabilitation, the Redevelopment Authority threatened to acquire their property through eminent domain.

In 1956, a group of civic-spirited business leaders established the Old Philadelphia Development Corporation to be the nonprofit implementation arm of the renewal effort. It acted as a general promoter of Society Hill, as a consultant to the Redevelopment Authority, and, where necessary, as interim developer for buildings and vacant properties that had to be acquired. It found purchasers for these sites, helped them to obtain necessary mortgage financing, and then led them through the maze of renewal requirements. The new owners were given at least one year to begin construction. If they failed, the Redevelopment Authority could take the property back and find another, more effective owner.[38]

During the 1950s, when millions of city dwellers were departing for what they thought to be more attractive suburban homes, some experts doubted that there would ever again be substantial demand for city residences. If Society Hill were to attract a market interested in living adjacent to the business district, potential residents had to be convinced that the area was going to improve and would soon have all the amenities needed to keep them from choosing a suburban alternative.

Once the Dock Street Market had moved, the Redevelopment Authority sought proposals for the site. The winning scheme was submitted by New York developer William Zeckendorf Sr. on behalf of his firm, Webb & Knapp. His team, which always included I. M. Pei as architect, had already been selected as a developer for Washington's Southwest Urban Renewal Area and for several of Robert Moses's redevelopment projects in New York City.

The Zeckendorf-Pei plan called for construction of 720 apartments in three towers, sited in a vast grassy area farthest away from the historic structures. It also included a supermarket, underground parking for 400 cars, and 14 three-story, brick, one-family town houses. The new town houses were intended to provide a transition between the modern apartment buildings and the historic brick rows.

Philadelphia, 1999. Society Hill's Greenways provide a public realm framework for the entire neighborhood. *(Alexander Garvin)*

The Society Hill Towers were concrete and glass. All the rooms had floor-to-ceiling windows with views of the Delaware River on one side or downtown Philadelphia on the other. These conveniently located apartments with splendid views and attractive rents (made possible by the subsidized land sale and FHA 220 mortgages) were just what was needed to attract new residents to the area.

The towers, completed in 1964, soon became a beacon announcing the revival of downtown living and physical proof of the area's revival for other developers eager to profit from the new market. Sometimes these developers built one-family town houses, filling the gaps between older row houses; sometimes, new groups of row houses; sometimes, apartment buildings. The town houses were often sited in combination with new open spaces and off-street parking. The infill structures, occasionally of radical modern design, were nevertheless sympathetic in scale and materials to the rest of the neighborhood. Contextual design was required by the design regulations written into the renewal plan. Moreover, each proposal had to be approved by an Advisory Board of Design.

Society Hill is unique among redevelopment projects because of the carefully designed system of handsome public open spaces conceived by Edmund Bacon. There are tree-lined brick and cobblestone streets and sidewalks lit by Franklin streetlamps, small parks designed to provide play

Philadelphia, 2009. The new buildings in Society Hill were designed to be compatible with its historic row houses. *(Alexander Garvin)*

areas for children and sitting areas for the elderly, and pedestrian greenways that connect the streets with the parks. These greenways were created by replacing vacant lots, run-down structures, and incompatible land uses with pedestrian ribbons that provide vistas to local landmarks.

Critics point to the decrease in the African-American population in Society Hill as evidence for the contemporary charge that urban renewal amounted to "Negro removal." African Americans (77 families and 63 individuals), however, amounted to fewer than 15 percent of the relocatees. Critics also denounce gentrification and point to the absence of low- and moderate-income housing. There is no doubt that the redevelopment increased the number of middle- and upper-income white residents in Society Hill. They had the money to pay for the area's new and rehabilitated housing and their money and spending habits affected the surrounding neighborhood by paying for further improvements to Society Hill that were not subsidized by the federal government.

Despite the relocation of 483 families and 551 individuals, the area's population increased from 3378 in 1960 to 4841 in 1970, and is even larger today. More than 1000 new dwelling units have been built and another 600 residential structures rehabilitated. Prior to redevelopment, Society Hill generated $454,000 in annual property tax payments. By 1974, this sum had climbed to $2.47 million, an increase of 444 percent. In 1950, approximately 3600 people had lived in Society Hill's 1700 housing units. At that time, a housing unit in Society Hill was approximately 40 percent more valuable than the median housing unit value in Philadelphia as a whole. By 2010, the area had nearly doubled in population, with approximately 6215 people living in 3875 housing units. Moreover, it was one of the city's most sought-after residences, with a median housing value of $490,000, nearly 3.5 times the median value of a housing unit in Philadelphia overall.[39]

The Washington Square East Renewal Project restored Society Hill to its former stature as a respectable and fashionable residential neighborhood. It sparked a revival of interest in simultaneously living and working in the center city. It also ended the redlining practices in which Philadelphia banks had engaged in the city's downtown residential areas. In less than 10 years, $180 million in private funds were invested in Society Hill. Tens of millions more have been invested since. Most important, it demonstrated to the country the effectiveness of removing frictional blight as a strategy for urban renewal.

Philadelphia, 2009. Headhouse Market. *(Alexander Garvin)*

New York City's Vest-Pocket Redevelopment

New York City also tried renewing neighborhoods by selectively removing unsafe, unsanitary, and incompatible structures in the West Side (1956–1994) and Bellevue South (1956–1976) Urban Renewal Projects in Manhattan. But the planning process and the results were quite different from those in Philadelphia. In New York, community groups were actively involved in selecting sites for clearance and determining their reuse. The beneficiaries were not just middle- and upper-income homeowners or profit-motivated private developers. Each renewal area included sites for middle-, moderate-, and low-income housing that were transferred for development to nonprofit organizations and to the New York City Housing Authority.

Of the 7800 new apartments in the West Side Urban Renewal Project, 800 were low-rent public housing, 4200 were

New York, 2005. Apartment houses erected on Columbus Avenue in the West Side Urban Renewal Area. *(Alexander Garvin)*

subsidized moderate- and middle-income dwelling units (15 percent of which had rents comparable to public housing), and only 2800 were fully tax-paying, privately financed apartments. Another difference was staging. Certain cleared sites were given construction priority so that relocatees from other redevelopment sites would not have to move from the neighborhood.

While both the Bellevue South and the West Side Urban Renewal projects are less visually charming and took far longer to complete than Society Hill, they were equally successful in attracting new middle-income residents. In each case, there was virtually unlimited demand for market-rate housing at these desirable locations a few minutes away from the Midtown business district. By removing blighted and incompatible structures from the neighborhood and providing financial incentives (below-market-interest-rate mortgages and real estate tax abatements) government eliminated impediments to private development. As in Society Hill, the market that these projects attracted spilled over to property not taken for redevelopment.

Mayor John Lindsay came to office in 1966 with a commitment to apply this same approach to other sections of the city. He believed that vest-pocket redevelopment was the appropriate alternative to Moses-style Title I renewal because it encouraged contextual architecture rather than immense housing projects. There was little support from Washington, however, until Congress adopted the same strategy of staged redevelopment for the Housing Act of 1968. This legislation established a Neighborhood Development Program (NDP) alternative to conventional Title I renewal. This program permitted housing redevelopment in designated areas to proceed on a year-to-year basis, based on the availability of financing, subsidies, and relocation resources.[40]

New York, like other cities, rushed to make use of these

Manhattan, 2012. By planning street-level retailing and retaining the street wall along 23rd Street in the Bellevue South Urban Renewal Project, the designers (Davis, Brody & Associates) were able to recreate New York City's active street life. *(Alexander Garvin)*

Bronx, 2012. Twin Parks.
(Alexander Garvin)

new funds. However, the Lindsay administration made sure that new redevelopment sites would include more new apartments than the number of units demolished and that these new developments could serve as relocation resources for the next stages of a comprehensive NDP.

Sections of Harlem, Mott Haven, and Twin Parks in the Bronx, and Coney Island and parts of Bedford Stuyvesant in Brooklyn were earmarked for NDP funding. In many cases, the designated builder-developer was the New York State Urban Development Corporation, whose first president and chief executive officer was Edward Logue, who had cut his redevelopment teeth in the renewal areas of New Haven. That experience had convinced him that mediocre design could ruin an otherwise successful redevelopment effort. The Lindsay administration was similarly committed to "good urban design." Together, they made the program a showcase for design by more fashionable architects (e.g., Richard Meier, James Polshek, Giovanni Pasanella, Prentice & Chan, and Davis, Brody & Associates).

The first NDP sites were the only ones completed. In January 1973, President Nixon unilaterally terminated the country's entire renewal program and declared a moratorium on additional subsidized housing. Further action in Twin Parks, Mott Haven, and other vest-pocket renewal areas was dependent on further subsidies, and without federal assistance, the city had no choice but to terminate further development.

The aborted vest-pocket renewal projects demonstrated that, by itself, selective replacement of dilapidated buildings is insufficient. For there to be further neighborhood improvement, the remaining property owners have to be able to obtain necessary financing. Residents of the areas selected by the Lindsay administration for vest-pocket renewal, unlike those of more prosperous areas of Manhattan or of Philadelphia's Society Hill, could not afford the necessary rent increases. In the absence of this middle-income market, neither developers nor lending institutions would initiate additional action. Thus, when Nixon terminated redevelopment subsidies, he also terminated further neighborhood improvement in the vest-pocket renewal areas.

Property owners had been abandoning their properties even before the Lindsay administration–initiated vest-pocket redevelopment program. Unable to finance rehabilitation of buildings that had deteriorated badly after years of deferred maintenance, much less cover the operating costs and taxes, there was nothing else for them to do. Abandonment continued unabated until large sections of the Bronx and Brooklyn were pockmarked with empty buildings and vacant lots. The administration of Mayor Edward Koch, however, began a program of selling these abandoned vacant properties for the nominal price of $500 per proposed new dwelling unit to nonprofit and developers of subsidized housing.

Some of these abandoned properties had become development opportunities. In 1982, the New York City Housing Partnership (a subsidiary of the New York Partnership and the New York Chamber of Commerce) perceived this opportunity. It initiated a program that selected sites it believed could be filled in with new buildings and sold to homeowners. It persuaded the city to grant an interest-free second mortgage of $10,000 to purchasers of two- and three-story row houses built by the partnership. This was supplemented by another interest-free second mortgage issued by the State of New York Mortgage Agency. Purchasers paid a cash down payment of 5 percent of the sales price. The remainder was financed with a first mortgage from an institutional lender.[41]

Since its creation, the New York City Partnership Housing Program has created more than 30,000 units of homeownership and rental housing using more than $4.2 billion in private-sector financing.[42] They differed from the Lindsay vest-pocket program in three ways: they were low-rise row houses rather than high-rise apartment buildings; they were resident owned and managed rather than owned and operated as private businesses; and they included mixed-income occupants (the owner and one or two renters whose incomes are below median income). The families who moved into

Bronx, 2012. Infill row houses in the Longwood section of the Bronx, built by the New York City Partnership. *(Alexander Garvin)*

partnership-sponsored housing took pride in their houses. They took care of streets in what were once dangerous, abandoned sections of the South Bronx. However, they are no different in their inability to generate a widespread and sustained private market reaction. Filling in scattered empty lots cannot provide the critical mass of customers needed to change much else in a neighborhood, unless, as in Society Hill, their replacement removes the only obstacle to a functioning market.

Strategy 5: Concentrated, Coordinated Government Action

Many urbanists believe physical redevelopment, by itself, is not enough. They argue that success also requires that residents' social, economic, and political ailments be treated. This approach calls for active program coordination among quite different government agencies, intense involvement by area residents, and enormous amounts of money. Moreover, taxpayers are rarely willing to foot the bill. Despite these difficulties, every so often Congress does try this strategy.

Model Cities

The nation's most ambitious attempt to integrate physical redevelopment with social, economic, and political action was embodied in the Demonstration Cities and Metropolitan Development Act of 1966. This legislation, popularly known as Model Cities, called for concentrated and coordinated government action (federal, state, and local) in specially designated *model neighborhoods*. In these model neighborhoods, the federal government paid 80 percent of the cost of planning, administration, and the nonfederal share of federal categorical grant programs, plus 100 percent of any new nonfederal programs.[43]

Cities everywhere wanted their share of Model Cities money. Community leaders, working with a wide range of professionals, prepared analyses of proposed model neighborhoods, established goals and objectives, and devised programs to achieve them. These were embodied in five-year strategies and one-year action plans that were submitted to Washington as the basis for inclusion in the program. By 1973, when the Nixon moratorium terminated Model Cities, $2.34 billion had been spent in 150 cities with designated model neighborhoods.[44]

Three huge sections of New York City were selected for the program and approved by HUD: Harlem–East Harlem in Manhattan, the South Bronx, and Central Brooklyn. By itself, the Central Brooklyn Model Neighborhood included 425,000 residents in three blighted neighborhoods: Bedford-Stuyvesant, Brownsville, and East New York.[45]

The Lindsay administration was ready to tap Model Cities money even before the legislation had passed. It appropriated "early action" money for acquiring sites and hiring "advocate planners" to work with community residents. The program they developed for East New York was probably the most carefully thought-out by its planners and the most strongly supported by community residents.

Planning for a Target Area: East New York, Brooklyn

East New York, a community of 100,000 located in the southern section of the Central Brooklyn Model Neighborhood, was experiencing rapid population transition. Its Jewish, Italian, Polish, and Lithuanian middle class began moving away in the 1950s. This exodus was accelerated when African Americans and Puerto Ricans, who had been displaced from their homes to make way for public housing projects in neighboring Brownsville, spilled over into the area.

Residents of East New York moved, on average, every 18 months. Rapid turnover in apartment occupancy led to strained landlord-tenant relations. Inadequate management and increasingly deferred maintenance exacerbated the deterioration of the area's already dilapidated buildings. By the time East New York was designated as part of the Central Brooklyn Model Neighborhood, it was filled with vacated, burned-out, and destroyed houses.

In November 1966, Walter Thabit, the advocate planner hired by the city, began a series of 17 Wednesday-night planning sessions with community leaders. Over 250 people came to at least one meeting of what was called the East New York Housing and Urban Planning Committee. Eighteen faithful members attended five or more sessions. In the end, more than 30 community organizations and churches had been represented. At these meetings, Thabit presented detailed analyses

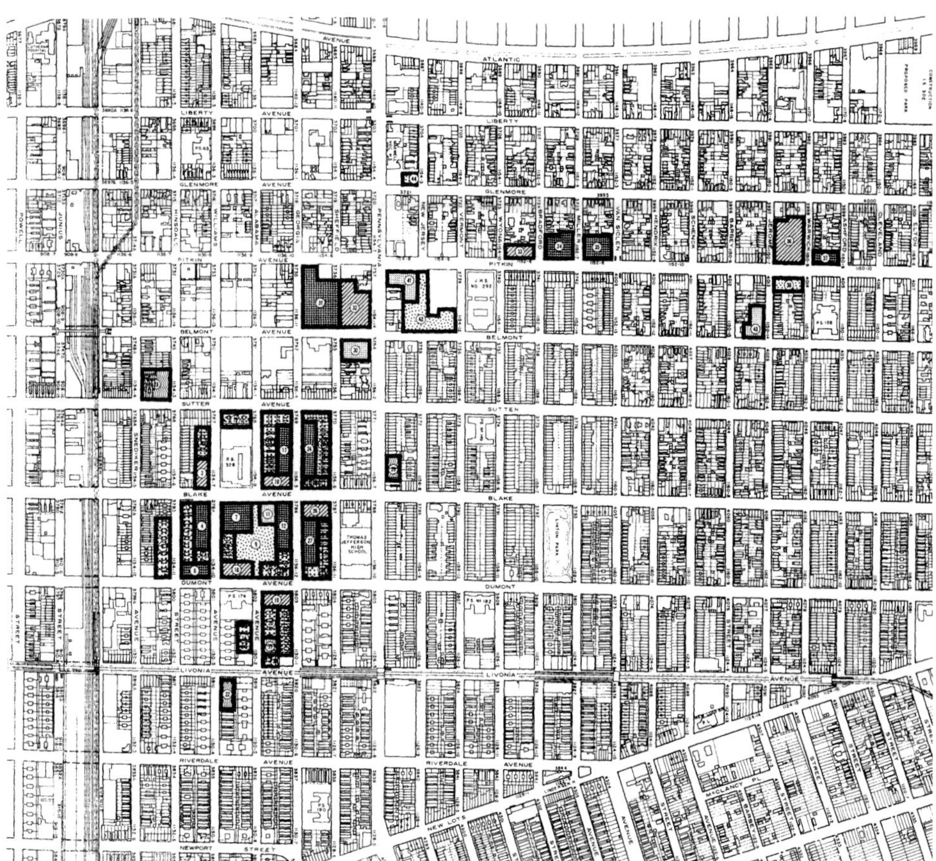

Brooklyn, 1967. Vest-pocket housing strategy prepared by Walter Thabit for the East New York Model Cities Program. *(New York City Housing and Development Administration)*

of industry, retailing, housing, transportation, population, community facilities, public services, and neighborhood organizations. By April, residents, municipal officials, and Thabit had completed a thorough analysis of the neighborhood.

The Model Cities Program in Central Brooklyn included proposals for everything: early childhood centers, after-school tutoring, bookmobiles, community service officers, fire prevention, sanitation, ambulance services, addiction treatment, college scholarships, industrial development, and job training. Its first-year budget totaled $25 million.

The individuals who came to the meetings in East New York encouraged the city's staff to concentrate these programs in a target area that had the most vacant buildings, fires, and crime. As Thabit explained, those who lived in the area wanted it to be livable, safe, and clean. Those who lived outside wanted the "welfare pesthole" removed. It was different from Mott Haven Model Cities, where initial housing development was followed by enough additional new construction and vest-pocket parks to restore the neighborhood.[46]

Brooklyn, 2012. Two blocks at the heart of the East New York section of the Central Brooklyn Model Cities Area were combined to create Unity Plaza, a superblock that integrated housing with a paved, open space designed and equipped for active recreation. *(Alexander Garvin)*

Brooklyn, 2012, Unity Plaza. New apartment buildings eliminated much of the neighborhood's active street life, while failing to attract it to the new open space at the center of the superblock. *(Alexander Garvin)*

There were disagreements about tactics, but the strategy was clear: target 800 to 1000 new public housing units and an equal number of substantially rehabilitated apartments in the "pesthole" and proceed from there. Local residents did not want high-rise buildings. Moreover six-story, semi-fireproof buildings could be built less expensively. Two blocks at the heart of the area were combined into a single superblock, and six-story buildings were sited along the perimeter of what they named Unity Plaza. Instead of open grass, however, common areas were paved and equipped for active recreation. The rationale behind this was similar to that of the self-contained enclave, except there was insufficient density to create a critical mass that would allow residents to be self-sufficient.

Unity Plaza opened in 1973, a few months after Nixon had terminated Model Cities with the aforementioned moratorium on further federal housing assistance. The development stood alone, surrounded by even more vacated, burned-out, and destroyed buildings. Facing fiscal disaster at that time, the city government was not able to replace federal Model Cities funding that had been terminated. Without money, there was no way to proceed with the ambitious program for East New York. Over the next four decades the city built some additional six-story subsidized housing and sold more than a dozen block fronts of vacant property for the Nehemiah development south of Unity Plaza. A few other city-owned, vacant sites were sold to the New York City Partnership, which built a few scattered one-family, resident-owned row houses. Those buildings eliminated the remaining burned-out and destroyed buildings—but, unlike the Mott Haven Model Cities Neighborhood or Bronx Nehemiah—not enough to have much effect on the surrounding neighborhood.

Neighborhood Strategy Areas

When federal funding was resumed via the Housing and Community Development Act of 1974, the burden of deciding the mix and level of funding for clearance, rehabilitation, economic development, public improvements, and other renewal activities was shifted from Washington to each recipient local government. Previously, each city had proposed specific projects that were approved by the Department of Housing and Urban Development (HUD) if they met the categorical requirements of specific federal programs. Now, federal grants were equitably distributed, funding levels were predictable, and local actions could satisfy local priorities. Never again could talented administrators in New Haven or

Brooklyn, 2012. Although some infill housing has been built where there were once empty lots and abandoned buildings, the social and economic problems of the area remain unsolved. *(Alexander Garvin)*

any other city get their citizens 19 times the money allocated to New Yorkers.

The new act created *block grants* for every locality, not just for those that successfully applied for *categorical grants*. The amount of the Community Development Block Grant (CDBG) was determined by a formula including population, extent of poverty, and degree of overcrowding. CDBG paid for 100 percent of the costs of housing rehabilitation, public improvements, open space, historic preservation, and economic development, not just redevelopment.

Most local governments faced with local political pressures and a multiplicity of local needs chose to avoid new slum clearance projects. They preferred using the CDBG grants to pay for such diverse items as playground renovation, sewer replacement, sidewalk repaving, street trees, and housing rehabilitation. Often the Community Development Block Grant simply paid for local budget items that might otherwise have been dropped.

The Carter administration decided CDBG would have greater impact if cities targeted funds based on a comprehensive community development program. In doing so it revived the Model Cities approach, but without the previous, exceptionally high level of funding. Localities were asked to designate Neighborhood Strategy Areas (NSAs) and concentrate and coordinate both CDBG funds and municipal-budget expenditures in those areas.

Many cities made a mockery of the process. In New York City, virtually every eligible section of the city was designated to become an NSA. For the first 10 NSAs, concentration and coordination consisted of listing all current and future government budget expenditures and transferring from the municipal budget those that were eligible for CDBG funding.

Just as NSA "targeting" was getting under way, the Reagan administration brought it to an end by reverting to the more pristine revenue-sharing philosophy that Congress had originally enacted in 1974. Localities could again obtain CDBG funds without complying with complex regulations. Lacking federal interference, most governments chose to ignore former federal stipulations requiring them to concentrate and coordinate their expenditures.

The Clinton administration revived the notion of coordination and concentration of the full range of social and economic programs for a few selected areas when it announced the Empowerment Zone Program. This program had even less chance of success because it proposed spending a fraction of the funds available for the Neighborhood Strategy Program or for the Model Cities Program.

Between its inception and 2011, the Community Development Block Grant Program has provided more than $275 billion for every possible expenditure that a city wished to make, provided that expenditure met the broadest interpretation of the program's requirements. Some cities, like New York, frequently used it to replace funds that would otherwise come for its capital budget. Many other cities used the money to try things that they otherwise never would have done.

The Critics

Given the enormous scale of most clearance and redevelopment projects, anger and opposition sprang up wherever one was proposed. In Boston, a "Committee to Save the West End" argued that while buildings might have needed repairs and apartments might have been overcrowded, the neighborhood was a good place to live. It managed to forestall final approval of the project until 1958. In New York, political opposition defeated nearly half of the projects Robert Moses proposed before retiring from the field in 1959. Opponents argued against both clearance and redevelopment.

Opposition to Clearance

There had been opposition to urban renewal from the very beginning. Challengers claimed that the area proposed for redevelopment was not blighted or that it would destroy the tight-knit sense of community within the neighborhood, that the project would eliminate jobs, or that it would rob the city of its ethnic heritage. In many cities the opponents to redevelopment could be ignored. In New York City, however, they were more successful and forced even the redoubtable Robert Moses to drop three of his first five urban renewal projects.[47]

A substantial number of Moses's projects never proceeded beyond the brochure stage or were altered significantly in response to opposition from site tenants and their political representatives. In 1950, when Moses proposed a project just north of Columbia University on Morningside Heights, residents objected to the characterization of the area as a slum. The residents organized the "Committee to Save Our Homes," which managed to stall approval for two years. In 1957, to obtain approval of his twelfth project, Lincoln

Bronx, 1958. Rendering of the Mott Haven Urban Renewal Project, one of Robert Moses's many unexecuted redevelopment schemes. *(Courtesy of the Citizens Housing and Planning Council)*

Manhattan, 2005. Morningside-Manhattanville Urban Renewal Project, approved after a two-year battle with area residents, and the public housing project (Grant Houses) that was completed too soon to provide housing for relocatees from the renewal project. *(Alexander Garvin)*

Center, Moses had to drop 4 of the 18 blocks he had proposed for clearance two years earlier (see Chapter 5).

Eventually, the ever-practical Moses began to avoid center-city locations, proposing instead sites that were in fringe areas where the relocation load would not be as serious. Nevertheless, criticism continued to mount. In response, Mayor Robert Wagner, in 1959, commissioned studies of the city's housing, relocation, and renewal programs. They were part of the mayor's strategy to oust Moses from his position as chairman of the Committee on Slum Clearance and to alter the character of the city's urban renewal program.[48] The strategy worked. When Moses resigned in 1960, 39 urban renewal projects were either still officially under way or in planning. The city eventually dropped 16 of them.[49]

Urban renewal was also under attack from widely read critics such as Herbert Gans, Martin Anderson, and Jane Jacobs. Gans argued that many renewal areas were not slums at all but vibrant neighborhoods whose major problem was the low income of their residents. Anderson insisted that one did not fix city neighborhoods by bulldozing them. Jacobs contended that the mistake lay in accepting Le Corbusier's vision of an ideal city. As she explained, "It was so orderly, so visible, so easy to understand. It said everything in a flash, like a good advertisement. . . . But as to how the city works, it tells . . . nothing but lies."[50]

Critics of clearance observed that you do not eliminate a slum by tearing it down and removing the people; that simply moves the problem elsewhere. In 1949, when the federal urban renewal program was enacted, there had been few "scientific" studies of the relocatees and where they went. Consequently, there was little evidence of the assertion that slum clearance simply moved the problem elsewhere.

One early study, though, a 1933 analysis of relocation from the site for the Knickerbocker Village, seemed to support the opposition. It found that, of 386 families that had occupied the tenements that were cleared, 83 percent moved back to Old Law tenements, 53 percent to apartments without toilets, 34 percent to apartments without hot water, and 38 percent to apartments without bathtubs.[51]

Studies of relocatees from federally subsidized urban renewal projects seem to prove the opposite. Analysis of relocatees from Washington's Southwest Urban Renewal Project found that

> *. . . nearly five years after the relocation, not a single family was in a home that was judged by a team of investigators as in need of major repair. . . . [There is] a significant increase in both the cleanliness and the orderliness of the dwellings. . . . [Today] only 1 percent of the families interviewed live in apartments that do not have bathtubs or showers with running water.*[52]

As for relocatees creating new slums, wall charts and pin maps showing where each family moved from the renewal areas of New Haven between 1956 and 1966 "demonstrated visually that no new ghettos were being created."[53] The only clustering was in public housing projects. A map of relocatees from Boston's West End produced similar results. Certainly, the relocatees brought their problems with them to new apartments. Whether these problems also spilled over with sufficient force to transform their new neighborhoods into slums is still debated.

By demolishing major amounts of unsanitary and unsafe housing, renewal projects reduced the opportunities for relocatees to move back into slum housing. In many cities this was less of a hardship than critics maintained. Relocatees were able to improve their housing because, during the 1950s and 1960s, these cities had declining populations. As a result, it was relatively easy to find housing that had been left behind by people moving to the suburbs.

In New York, where the drop in population came later, the experience was quite different. In 1954, the New York City Planning Commission found that many relocatees "doubled up or moved into furnished rooms or rooming houses."[54] The same thing happened in the West Side Urban Renewal Areas and later in redevelopment programs funded by the Model Cities Program.

Proponents of slum clearance argue that demolition removes the dangers of rats and other vermin, fire, trash, and other threats to a person's health and safety. They also argue that it removes a haven for criminals and thus reduces the risk of theft, assault, rape, and murder. Opponents respond by saying that the cost, in terms of disrupted lives, severed personal relationships, and destroyed identity, is not justified, especially given the vulnerability of slum residents. They point out that a person's home, on the most primal level, provides shelter from threats in the environment and that being forced to move from one's home, however inadequate, is a profoundly disturbing experience.[55]

It is difficult, perhaps impossible, to translate into dollars the psychological cost of being forced from one's home.

However, it is possible to estimate the economic impact of the move. Originally, the Housing Act of 1949 neither provided compensation for moving expenses nor recognized the need to compensate relocatees for any increased housing cost. As a result of Boston's West End Urban Renewal Project, for example, 86 percent of those moved were paying higher rents after relocation, and the median monthly rent had risen from $41 to $71. By 1964, Congress permitted reimbursement of up to $200 in moving expenses and up to $1000 over a two-year period to cover the cost of securing "a decent, safe and sanitary dwelling."[56]

In response to mounting criticism, Congress enacted the Uniform Relocation Assistance and Real Property Acquisition Policies Act of 1970. This legislation, which is still in effect, finally recognized the need for more generous compensation to the victims of slum clearance. It also established sufficiently high levels of payment to force municipal officials to consider the cost of relocation before deciding on further slum clearance.

Opponents also claim that areas chosen for renewal may have been characterized by badly maintained structures and poor residents but frequently were vibrant neighborhoods that provided a cohesive sense of community for the residents. The most persuasive advocate of this view is Herbert Gans, a sociologist whose book *The Urban Villagers* a study of Boston's West End, revolutionized public perception of areas chosen for renewal.[57]

The West End had been labeled "detrimental to the safety, health, morals, and welfare of the inhabitants" by Boston's renewal officials. Instead, Gans saw the West End as a "run-down area of people struggling with the problems of low income, poor education, and related difficulties." He observed that apartments "were usually in much better condition than the outside hallways," that the residents "could live together side by side without much difficulty," and "when emergencies occurred, neighbors helped each other readily."[58]

Gans's study came too late. A 48-acre section of the neighborhood was cleared and 7500 people forced to move. Today the West End is a pallid complex of modernistic boxes containing 2300 apartments, two office buildings, three shopping areas, two pools, a tennis club, and garages for 1200 cars. Except for one building containing subsidized apartments for the elderly, it is an enclave of the upper middle class.

Opposition to Redevelopment

Unlike opponents of clearance who consider the cost excessive, opponents of redevelopment believe the product unsatisfactory. Since new buildings in redevelopment projects had little ornament and few historical references, the initial criticism was from those who admired traditional architecture. Later it came from those who were unhappy with the new buildings. The problem they identified is best explained by architectural historian Vincent Scully, who observed that in a proposed redevelopment plan for the "Hill" neighborhood in New Haven, "purism and a distaste for life's messy multiplicity could go no further."[59]

There also was opposition from property owners and developers who argued that government-subsidized redevelopment constituted unfair competition with the private market. Owners of existing apartment houses and developers of conventionally financed new housing had reason to object. Not only were their competitors provided with land at subsidized prices, they also received better financing and paid lower real estate taxes.

These property owners had neither the know-how nor the ability to finance large projects, nor were they plugged into the political power structure. Consequently, public officials could dismiss their criticism as sour grapes while explaining that each proposed redevelopment project was a much-needed public action justifying temporary hardship

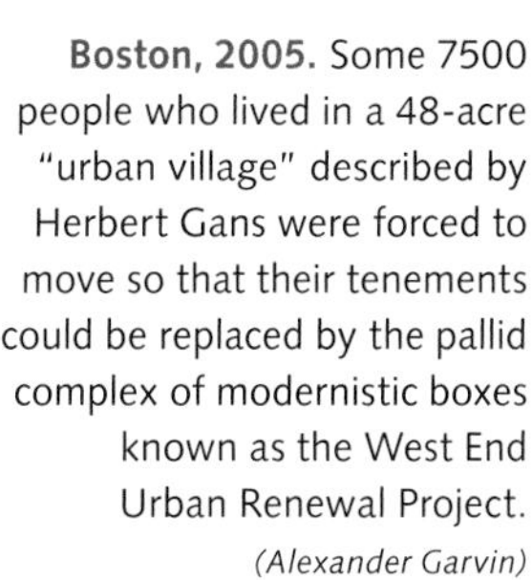

Boston, 2005. Some 7500 people who lived in a 48-acre "urban village" described by Herbert Gans were forced to move so that their tenements could be replaced by the pallid complex of modernistic boxes known as the West End Urban Renewal Project. *(Alexander Garvin)*

on the part of a very small number of greedy landlords incapable of participating in the urban renewal program.

Ingredients of Success

Until 1961, when Jane Jacobs published her eloquent book *The Death and Life of Great American Cities,* few critics challenged the effectiveness of redevelopment as a method of curing urban blight. Her indictment was loud and clear. Urban renewal was "a mirage, a pitiful gesture . . . to combat disintegration and instability that flow from the cruelly shaken-up city" that only created new slums.[60]

Instead of favoring large-scale clearance of aged and inefficient cities, she rejoiced in day-to-day neighborhood living and proposed that we emulate not some Utopian image, but the organic patterns of change common to our currently disorganized but vital cities. While she tried to diagnose the reason for failure and propose alternative action that would be more successful, she never differentiated among the various strategies for eliminating slums and blight, nor did she understand that there were circumstances in which one or more of these strategies might succeed.

Market

Redevelopment makes sense when the population of an area is declining. In such instances, there will be an increasing stock of available housing to which relocatees can move. Consequently, removing apartments from the market will not cause undue hardship. New Haven's massive redevelopment program, for example, eliminated the very worst housing during a period in which its population was declining. The result was an improved living environment along with an increasing vacancy rate. In New York City, on the other hand, even the net gain in available housing brought about by Robert Moses's redevelopment program could not satisfy the increasing demand for apartments.

It is more difficult to assess the effectiveness of residential redevelopment as a device for retaining population. Stuyvesant Town, Portland Center, and projects like them were meant to provide an alternative to the suburbs. They decreased densities, increased open space, and provided community amenities. Successful as they may have been in attracting largely white, middle-class residents, they could never have reversed the flight to the suburbs that had been under way since the start of the nineteenth century. Lake Meadows and Prairie Shores accommodated African Americans at a time when racial prejudice made their move to the suburbs quite difficult. By the end of the twentieth century, when many middle-class African Americans had overcome this obstacle, these well-located, well-maintained projects provided a most attractive alternative.

In New York City, where the residential vacancy rate was minimal, the urban renewal sites that Robert Moses cleared were virtually certain to be occupied. Although market demand was comparable in surrounding neighborhoods, his housing redevelopment projects were unable to generate further activity because competitors could not match this subsidized competition. They had neither Title I grants to reduce the cost of assembling and holding a site till it was ready for construction nor FHA mortgage insurance with which to obtain long-term, low-equity mortgages and thereby reduce debt service. Instead, developers built in less deteriorated, more expensive neighborhoods where they could charge enough to recoup their investment and earn a reasonable return.

Change in demand is not the only market factor that will determine the effectiveness of a redevelopment project. Income and taste play important roles. Removing blighted properties from an area like Society Hill attracts customers who can afford to pay for new and renovated housing. But it cannot revive neighborhoods like East New York, where residents cannot pay for repairs, much less full-scale rehabilitation or new construction, and where small changes in appearance are not enough to interest a large number of outsiders.

Location

Whether a renewal strategy is based on complete clearance and redevelopment or selective clearance and rehabilitation, the same approaches to location come into play: (1) exploitation of proximity to neighboring sections of the city that enhance marketability or screening out those sections that have a negative frictional impact, and (2) exploitation of area characteristics that are in high demand or introduction of new elements that will enhance marketability.

Fred French demonstrated the importance of proximity to employment, shopping, and entertainment when he built Tudor City on the edge of the Midtown Manhattan business district. He also demonstrated the importance of screening out the negative effects of traffic and uncongenial land uses. Renewal schemes as widely different as the Southwest Urban Renewal Project, Lafayette Park, and Society Hill had the benefit of locations that were similarly close to downtown Washington, D.C., Detroit, and Philadelphia. In addition, SWURA screened out the blighting influences of the highway and railroad by covering them with a platform, Lafayette Park entirely eliminated through traffic, and Society Hill replaced the Dock Street Market, thereby removing the source of the traffic.

Each of these projects dealt with the character of its location in a different way. Lafayette Park originally was flat terrain covered with substandard wooden dwellings in various states of dilapidation, often without running water, toilets,

Philadelphia, 2011. Federal slum clearance assistance, as used in Society Hill, restored hundreds of historic row houses that attracted new residents to what had been thought of as one of the city's worst slums. *(Alexander Garvin)*

and central heating. Detroit cleared this slum and started over again with a radically different living environment. Society Hill had the benefit of some of the nation's finest eighteenth- and early-nineteenth-century brick buildings. Bacon's solution exploited these lovely buildings. The Southwest Urban Renewal Area combined both approaches.

Design

All the components of a redevelopment project must be arranged in a manner that reinforces the particular renewal strategy. At Tudor City, Fred French exploited the drop in elevation between Second and First avenues to create two private parks and a private street that bridges over 42nd Street. These amenities contribute to the project's market appeal. This arrangement of buildings and open space created a place that has been distinctive enough to attract customers for 70 years.

Like Tudor City, Moses's urban renewal projects were giant-sized. Their distinctiveness, however, is not the result of a thoughtful arrangement of open space, streets, and buildings. Moses and his architects were more concerned with design standards than with the character of the living environment. New buildings are evenly spaced on flat open spaces of lawn or parking that provide each apartment with an equitable amount of light and air. But they do not provide the protected living environment of projects like Tudor City or Lafayette Park or the subtle integration into the surrounding city that characterizes Society Hill.

Large open areas are no more inherently incompatible with an exciting urban environment than new buildings. But when they take the form of the vast emptiness of many redevelopment projects, they deaden rather than revitalize the city around them. Light, air, and dramatic vistas are not enough. The trees and grass in many redevelopment projects help in providing color, scale, and the pleasure of foliage, but they, too, are not enough. Urban areas need open spaces that attract people and the activities that they bring with them. Where designers provide a program for this park-like environment, they bring these public spaces to life. At Stuyvesant Town, paved playgrounds for younger children, paved ball courts for teenagers, benches for watching parents, and tree-lined walkways for strolling adults provide the setting for social interaction. Portland Center created a verdant atmosphere conducive to leisurely strolling, tranquil meditation, and active recreation. In countless other residential redevelopment projects the vast stretches of dusty grass and parking lots provide only a setting for conflict between maintenance workers and "trespassing" tenants.

Society Hill takes a different approach to the arrangement of pedestrian circulation, public spaces, and buildings. Its new apartment towers were intended to announce the neighborhood's renewal. Customers interested in moving downtown drove toward the towers. Once they parked, pedestrian greenways and brick-paved sidewalks led them past the sitting and playing areas that their families would be able to enjoy if they chose to move either to a tower with windows overlooking Philadelphia or to one of the nearby row houses.

Lafayette Park uses architecture and landscaping to market a very different product. The generous, landscaped open spaces underscore the special character of the architecture, enhance the privacy of the residences, camouflage parked automobiles, and provide a sylvan setting for sitting, strolling, and playing ball. Consequently, customers have the benefit of a living environment quite different from the surrounding city.

Financing

The Housing Act of 1949 provided financial assistance that made redevelopment easy. Without the power of eminent domain, developers would have had to pay far more to buy properties from owners who were unwilling to sell. They also would have had to pay more to compensate tenants whose leases had not yet expired. More important, government paid the costs of planning, acquisition, holding property, and—frequently—relocation and demolition as well, until development could begin. At Lafayette Park and Harlem-Lenox Terrace, those activities extended for nearly a decade. The Southwest Urban Renewal Project took even longer. Few developers could have financed this activity without earning a penny during so long a period.

Housing redevelopment experienced the same problem prior to 1954, when Congress established the FHA 220 mortgage insurance program for residential buildings built or renovated pursuant to an approved urban renewal plan. A limited number of projects, like Lake Meadows, were able to obtain conventional long-term mortgages. After FHA 220 mortgage insurance became available, though, residential developers no longer faced this problem. Institutions were ready to provide mortgage loans for projects in areas that officially had been designated as blighted because they were insured by the federal government against loss.

Equally important, when sites were ready for development they were sold at a discounted price based on the reuse that the government sought for them and subsequently subsidized rather than on the cash invested or their true market value. Nevertheless, until 1954 most developers avoided redevelopment areas because they could not get financing. Banks were unwilling to make loans in areas the government had designated as blighted. Once financial institutions were protected from losses by FHA 220 mortgage insurance, they rushed to provide designated developers and property owners with permanent mortgages for new construction and rehabilitation.

Time

Title I eliminated financial risk during the development period; FHA 220 mortgage insurance ensured project financing for decades afterward. However, the program did not give adequate attention to the period of time people spent within a redevelopment area. Thus, success depended entirely on what developers and architects created.

Herbert Greenwald's planning team made circulation through Lafayette Park an aesthetic experience. Their approach to daily life in the superblock ensured a suitable living environment 24 hours a day, 7 days a week. Although completely different in conception and design, Edmund Bacon's scheme for Society Hill achieved the very same result. Pedestrian and vehicular circulation as well as shopping and recreation facilities were carefully arranged to accommodate everyday life. In most redevelopment projects in New Haven, New York, and elsewhere, such concerns were peripheral at best. No wonder Lewis Mumford could call such urban renewal "prefabricated blight."[61]

Queens, 1958. Rendering of Seaside Urban Renewal Project proposed by Robert Moses for Rockaway Beach, a site that required only the demolition of summer bungalows and was therefore presumed to be without serious relocation problems. *(Courtesy of the Citizens Housing and Planning Council, New York City)*

Most redevelopment projects were conceived in terms of slum clearance and measured in terms of new office space or new dwelling units. One reason was that, like Le Corbusier's City of Tomorrow, they were presented as towers in a park-like setting rather than as places in which people spent their waking hours. The planners of projects like Lake Meadows concentrated on the layout of its asphalt parking lots and the rental diagrams of its apartments in the sky. They neither explained nor made drawings of what happened as people made their way through a project. Shopping center developers would have used the walk as an opportunity to attract them to stores, restaurants, and perhaps even movie theaters. Projects like Portland Center might include shopping facilities and attractive recreation areas that made a walk through the renewal area worth the visit, but they were a decided minority.

Entrepreneurship

When the federal government initiated the urban renewal program, there were no developers with experience in redevelopment. The program made it relatively easy to get into the business. Nor were developers at risk for long. In many cases they did not take title to property until financing had been guaranteed. As a result, the risk extended only from the start of construction until apartments were rented. The tough period from project inception through to the entitlement to build was the responsibility of the local redevelopment agency.

The amount of cash at risk also was minimized. Conventional mortgages, had they been available, would have required at least 25 percent cash equity. An FHA 220 mortgage could be obtained with a nominal equity investment, a large portion of which came from mortgageable builder-sponsor fees rather than cash.

Thus, unlike many government programs, Title I had built-in incentives to encourage entrepreneurs to enter the redevelopment business. Any future program that seeks active entrepreneurial participation in redevelopment need only replicate the procedures of the federal urban renewal program.

Robert Moses thought he would avoid the entrepreneurship problem by first negotiating a deal with private developers and then announcing what he hoped would be a fait accompli. This was not always successful. From the beginning, political leaders were able to prevent some of his redevelopment schemes from obtaining the necessary public approvals. As the painful impact of relocation became increasingly evident, community groups also found that they could force delays and changes, and sometimes even defeat him. Eventually, Moses's disdain for his opponents interfered too much with the political feasibility of his projects. His increasingly unsuccessful public entrepreneurship became enough of a liability that Mayor Wagner was able to force him from office.

Housing Redevelopment as a Planning Strategy

As Jacob Riis has so graphically explained, redevelopment is intended to replace the bad environment that otherwise "becomes the heredity of the next generation." But redevelopment is not necessarily a way of increasing the housing supply or of improving the housing of slum dwellers. It eliminates housing (however dilapidated) and causes hardship for relocatees.

By itself, redevelopment will not eliminate slums. It must be accompanied by strategies for the survival of the new housing *still* surrounded by uncleared slums and for the improvement of untouched older housing *still* occupied by slum dwellers.

It has been years since anybody seriously proposed a major housing redevelopment project. It is time we recognized that redevelopment can revitalize appropriate sections of our cities. The errors of the past need not be repeated. In the right area, clearance and redevelopment can be a catalyst that triggers genuine urban renewal. But redevelopment is desirable only if the costs (in terms of disrupted lives and business) are low and if it truly results in a good environment. Then redevelopment can become a force for the improvement of living conditions throughout the city.

If, as in so many of Robert Moses's projects, new land uses, activities, buildings, and residents are set apart from

Detroit, 2008. Herbert Greenwald, Mies van der Rohe, Ludwig Hilberseimer, and Alfred Caldwell demonstrated that clearance and redevelopment provided the opportunity to create a version of the City of Tomorrow, which would remain an island of stability with little or no impact on the city of Detroit that continued to decay for more than six decades. *(Alexander Garvin)*

their neighbors, the result is not just segregation. This separation prevents project benefits from spilling over into surrounding neighborhoods and thus from stimulating further private activity. Instead, redevelopment must be integrated into the life of the surrounding city.

Eventually, opponents of redevelopment made the program extremely difficult and expensive to execute. Environmental protection and historic preservation statutes precluded indiscriminate use of the power of eminent domain. Relocation requirements greatly increased project costs. Most important, by making redevelopment one among many permitted uses of the Community Development Block Grant, Congress forced its proponents to compete with other demands for federal subsidies.

While redevelopment remains a viable prescription for some situations, further legislation is not needed. When and where cities wish to replace a functionally obsolete physical plant, the necessary enabling legislation is already in place. Projects like Portland Center can serve as models for future development.

Only a handful of cities will have good reasons to choose redevelopment. The experience, both good and bad, with Title I will stand them in good stead. The rest should avoid it, because there are very few situations in which government-subsidized redevelopment is the proper mechanism for fixing the American city.

We will never really know whether widespread residential redevelopment could have attracted the market that fled to the suburbs. The revival of Pittsburgh's Mexican War Street, Chicago's Printers Row, and the Ansonborough neighborhood in Charleston(discussed in Chapter 13), as well as countless other older urban neighborhoods, leads one to doubt that cities needed new residential districts "vertical to the sky, open to light and air, clear and radiant and sparkling" to keep their populations from moving to the suburbs.

Once we accept the fact that clearance does not necessarily eliminate slum problems and redevelopment and does not necessarily increase the supply of affordable housing, and once we understand that the primary utility of clearance is as a method for stimulating additional private development that would not otherwise occur, we can make housing redevelopment an effective device for fixing cities. Nothing more is needed because governments already possess all the powers they need for eliminating the bad environment.

Notes

1. Jacob Riis, *A Ten Years' War: An Account of the Battle with the Slum in New York*, originally published in 1910 and reproduced in *Jacob Riis Revisited*, Francesco Cordasco (editor), Anchor Books, Garden City, NY, 1968, p. 301.
2. Susan Beattie, *A Revolution in London Housing*, Greater London Council, The Architectural Press, London, 1980.
3. Jacob Riis, *The Peril and the Preservation of the Home*, 1903.
4. Charles F. Palmer, *Adventures of a Slum Fighter*, Tupper and Love, Inc., Atlanta, GA, 1955, especially pp. 7–30.
5. James Ford, *Slums & Housing*, Harvard University Press, Cambridge, MA, 1936, pp. 714–736.
6. Historical and statistical information on Stuyvesant Town is derived from Arthur Simon, *Stuyvesant Town USA*, New York University Press, New York, 1970.
7. Simon, op. cit., p. 31.
8. Rental ranges were obtained from Metropolitan Life Insurance Co. archives.
9. *Berman v. Parker*, 348 U.S. 26 (1954).
10. Historical and statistical information on the redevelopment of Washington's Southwest Urban Renewal Project is derived from Frederick Gutheim, consultant, *Worthy of a Nation: The History of Planning for the National Capital*, National Capital Planning Commission, 1977; *Berman v. Parker*, 348 U.S. 26 (1954); William Zeckendorf and Edward McCreary, *Zeckendorf*, Holt, Rinehart, and Winston, New York, 1970; Daniel Thursz, *Where Are They Now?*, Health and Welfare Council of the National Capital Area, 1966; and the District of Columbia Department of Housing and Community Development.
11. *Berman v. Parker*, 348 U.S. 26 (1954).
12. The subsidy was calculated by taking two-thirds of Net Project Cost (defined as Gross Project Cost minus proceeds from the sale of the property). Gross Project Cost was defined as including property acquisition, relocation, demolition and site preparation, and planning.
13. Zeckendorf and McCreary, op. cit., p. 207.
14. Critics of government housing redevelopment could well ask whether public action was necessary if sufficient improved housing accommodations were already available.
15. Thursz, op. cit., p. 28.
16. Harold Kaplan, "Urban Renewal in Newark," in *Urban Renewal: The Record and The Controversy*, James Q.Wilson (editor), MIT Press, Cambridge, MA, 1967, pp. 233–258; and Central Planning Board of Newark, *Re: New Newark*, 1961, p. 14.
17. Alexander Garvin, *The Planning Game: Lessons from Great Cities*, W. W. Norton & Company, New York, 2013, pp.156–163.
18. Committee on Slum Clearance, *Report to Mayor Impellitteri and the Board of Estimate*, New York City, January 1951, pp. 36–38.
19. Committee on Slum Clearance, *New York City Title I Progress*, New York City, July 20, 1959, pp. 24–25. I use the number of apartments demolished (2068) because Moses, in all the publications of the Committee on Slum Clearance, probably underestimated the number of relocatees as 1683 families.
20. Donatella Lorch, "Stray Bullet Makes Oasis in Harlem Fighting Mad," *New York Times*, June 13, 1990, p. B1.
21. Former state senator Basil Patterson, quoted in Lorch, op. cit., p. B1.
22. Historical and statistical information about Lake Meadows and Prairie Shores is derived from Kevin Harrington, "Hilberseimer and the Redevelopment of the South Side of Chicago," in Richard Pommer, David Spaeth, and Kevin Harrington, *In the Shadow of Mies: Ludwig Hilberseimer Architect, Educator, and Urban Planner*, Art Institute of Chicago and Rizzoli International Publications, New York, 1988, pp. 69–88; Miles L. Berger, *They Built Chicago*, Bonus Books, Inc., Chicago, 1992; Devereux Bowly Jr., *The Poorhouse, Subsidized Housing in Chicago 1895–1976*, Southern Illinois University Press, Carbondale, IL, 1978; and Harold M. Mayer and Richard C. Wade, *Chicago: Growth of a Metropolis*, University of Chicago Press, Chicago, 1969.
23. Faith Rich (chair of the Housing Committee of NAACP) to Ludwig Hilberseimer, letter, July 12, 1949.
24. Ludwig Hilberseimer, *The Nature of Cities*, Paul Theobald & Co., Chicago, 1955, pp. 192–193.
25. Historical and statistical information about Lafayette Park is derived from Robert Mowitz and Deil Wright, *Profile of a Metropolis*, Wayne State University Press, Detroit, MI, 1962, pp. 11–79; Roger

Montgomery, "Improving the Design Process in Urban Renewal," in Wilson, op. cit., pp. 454–487; City of Detroit Housing Commission, *Gratiot Redevelopment Project, Final Project Report,* Detroit, MI, June 30, 1964; Peter Carter, "Mies' Urban Spaces," in *The Pedestrian and the City,* David Lewis (editor), Van Nostrand, Princeton, NJ, 1965, pp. 11–26; and Pommer, Spaeth, and Harrington, op. cit., pp. 62–67.
26. *General Development Corporation v. City of Detroit,* 322 Mich. 459, 33 N.W. 2d 919.
27. Herbert Greenwald was an unusual developer who sought to combine the economic realities of real estate with a desire to create excellent architecture. Together with architect Mies van der Rohe, he was responsible for some of the finest apartment buildings in Chicago: Promontory (1946–1949), 860–880 and 900–910 Lake Shore Drive (1948–1951 and 1953–1956), and Commonwealth Promenade Apartments (1953–1956). In addition to Mies van der Rohe, the team he assembled to work on Lafayette Park included planner Ludwig Hilberseimer and landscape architect Alfred Caldwell.
28. Most of the low-rise buildings and three towers were completed as designed. The rest of the project was completed by other architects.
29. Mies van der Rohe was not the architect for these projects.
30. Historical and statistical information about Portland Center is derived from Carl Abbott, *Portland: Planning, Politics, and Growth in a Twentieth Century City,* University of Nebraska Press, Lincoln, NE, 1983; and Gideon Bosker and Lena Lencek, *Frozen Music: A History of Portland Architecture,* Western Imprints Press of the Oregon Historical Society, Portland, OR, 1985.
31. Ford, op. cit., p. 897.
32. Ibid., pp. 591–593, 706.
33. Committee on Slum Clearance, *New York City Title I Progress,* New York, July 20, 1959, pp. 24–25; and New York City Department of City Planning, *Public and Publicly-aided Housing 1927–1973,* New York, 1974.
34. New York City Department of City Planning, *New Dwelling Units Completed 1921–1972,* New York, 1973; and *Socioeconomic Profile,* New York, 1986.
35. Historical and statistical information on the redevelopment of New Haven is derived from Alan Talbot, *The Mayor's Game,* Harper & Row, New York, 1967; Robert A. Dahl, *Who Governs,* Yale University Press, New Haven, CT, 1961; L. Thomas Appleby, interview, November 21, 1988; City of New Haven Development Departments, *Oak Street Area Redevelopment Plan,* New Haven, CT, November 4, 1955, revised March 1959; City of New Haven Development Departments, *Church Street Redevelopment and Renewal Plan,* New Haven, CT, September 3, 1957, revised through May 27, 1964; City of New Haven Redevelopment Agency, *Dwight Renewal and Redevelopment Project Application for Loan and Grant,* May 1, 1962; City of New Haven Redevelopment Agency, *Dwight Renewal and Redevelopment Plan,* 1963; and Howard W. Hallman, *The Middle Ground: A Program for New Haven's Middle-Aged Neighborhoods,* New Haven Redevelopment Agency, September 1959.
36. Talbot, op. cit., p. 80.
37. Historical and statistical information on the redevelopment of Society Hill is derived from Philadelphia Redevelopment Authority, *Washington Square East Urban Renewal Area: Technical Report,* May 1959; Edmund Bacon, *The Design of Cities,* Viking Press, New York, 1967, pp. 242–271; Leo Adde, *Nine Cities: The Anatomy of Downtown Renewal,* Urban Land Institute, Washington, DC, 1969; Jeanne R. Lowe, *Cities in a Race for Time,* Random House, New York, 1967; and the Philadelphia Redevelopment Authority Public Information Office, 1979.
38. During its 12-year involvement with Society Hill, the Old Philadelphia Development Corporation participated in the rehabilitation or reconstruction of more than 325 properties. See Lowe, op. cit., pp. 345–347.
39. Garvin, op. cit., pp.181–188. Statistics for Society Hill in 2010 are taken from "Center City Reports: Leading the Way: Population Growth Downtown," Central Philadelphia Development Corporation, September 2011, and the U.S. Census Bureau, 2010 census data.
40. Unlike conventional Title I urban renewal, which required a comprehensive redevelopment plan including clearance of every blighted property, NDP permitted a city to time acquisition, relocation, and demolition to match available relocation resources and housing subsidies. Federal funds were allocated on an annual basis based on cash-flow requirements. In most other ways, redevelopment under NDP was similar to redevelopment under conventional Title I.
41. See www.nycp.org.
42. New York City Partnership, http://www.housingpartnership.com/index.php?option=com_content&view=article&id=68&Itemid=472.
43. For example, in urban renewal areas located in model neighborhoods, the federal government would cover 80 percent of the nonfederal portion of project cost, thereby increasing its share from 67 percent to 93 percent of project cost [i.e., $0.67 + (0.8)0.33 = 0.93$].
44. Historical and statistical information on the Model Cities Program is derived from Bernard J. Frieden and Marshall Kaplan, *The Politics of Neglect: Urban Aid from Model Cities to Revenue Sharing,* MIT Press, Cambridge, MA, 1975; U.S. Department of Housing and Urban Development, *The Model Cities Program—A Comparative Analysis of the Planning Process in 11 Cities,* U.S. Government Printing Office, Washington, DC, 1970; and HUD data published in Michael Cooper, "Cities Face Tough Choices as U.S. Slashes Block Grants Program," *New York Times,* December 21, 2011.
45. Historical and statistical information on the Central Brooklyn Model Cities and its program for East New York is derived from City of New York, Office of the Mayor, Model Cities Administration, *Partnership for Change,* New York, 1970; New York City Model Cities Committee (Donald Elliott, Chairman; Eugenia Flatow, Executive Secretary), *Central Brooklyn Model Cities Comprehensive City Demonstration Program,* vol. 1, New York, May 22, 1969; and Walter Thabit, *Planning for a Target Area—East New York,* New York City Housing and Development Administration, New York, October 1967.
46. Thabit, op. cit., p. 7.
47. Only Harlem–Lenox Terrace (see Chapter 11) and Corlears Hook (see Chapter 9) were built as conceived. The Williamsburg, Delancey Street, and Greenwich Village projects were eventually dropped.
48. See J. Anthony Panuch, *Relocation in New York City,* New York, 1959, and *Building a Better New York,* New York, 1960.
49. New York City Committee on Slum Clearance, *Title I Progress,* New York, January 29, 1960.
50. Jane Jacobs, *Death and Life of Great American Cities,* Random House, New York, 1961, p. 23.
51. Fred L. Lavenburg Foundation and Hamilton House, *What Happened to 386 Families Who Were Compelled to Vacate a Slum to Make Way for a Housing Project,* New York, 1933.
52. Thursz, op. cit., pp. 28–33.
53. Alvin A. Mermin, *Relocating Families: The New Haven Experience 1956 to 1966,* National Association of Housing and Redevelopment Officials, Washington, DC, 1970, p. 127.
54. New York City Planning Commission, *Tenant Relocation Report,* January 20, 1954.
55. Marc Fried, "Grieving for a Lost Home: Psychological Costs of Relocation," in *The Urban Condition,* Leonard J. Duhl (editor), Basic Books, New York, 1963, pp. 151–171.
56. Chester Hartman, "The Housing of Relocated Families," *Journal of the American Institute of Planners,* vol. 30, no. 4 (November 1964), pp. 266–286.
57. Herbert J. Gans, *The Urban Villagers,* The Free Press, Glencoe, IL, 1962.
58. Gans, op. cit., pp. 13–16.
59. Vincent Scully, *American Architecture and Urbanism,* Henry Holt, New York, 1988, pp. 250–251.
60. Jacobs, op. cit., p. 5.
61. Lewis Mumford, *From the Ground Up,* Harvest Books, New York, 1956, p. 108.

13

Revitalizing Neighborhoods

Milwaukee, 2012, Third Ward *(Alexander Garvin)*

If, as Jacob Riis believed, *"the bad environment becomes the heredity of the next generation,"*[1] then the rationale for government action to improve living conditions is obvious. It is not clear, however, that the most effective way to accomplish this is by relocating slum dwellers, demolishing the dilapidated buildings they occupy, and replacing them with safe, sanitary modern structures. As Chapter 12 describes, government agencies have spent hundreds of billions on slum clearance. Unfortunately, they have spent very little combating building deterioration. While there is every reason to concentrate resources where the need and situation are desperate, there is no reason to wait until that point, just as there is no reason to withhold minor medication until a patient requires major surgery.

Despite the pittance that has been spent on neighborhood revitalization, there is enough experience to demonstrate that abandonment can be prevented, deterioration can be reversed, and older neighborhoods can regain their health. Unfortunately, few neighborhood residents or public officials know much about effective neighborhood revitalization programs or how little money they require. As a result, cities continue to propose revitalization programs without being aware of what has worked and what has not. This is particularly disappointing because many of the most successful initiatives have been managed without government involvement.

In some cities, homeowners are attracted by the charm, architectural merits, and low prices of an older neighborhood. As more and more of them purchase and renovate small buildings, an area that was once thought of as an incipient slum is transformed into a vibrant neighborhood. In other places, households with less conventional lifestyles move into derelict lofts, garages, and warehouses, in the process transforming once-abandoned manufacturing districts into attractive residential neighborhoods. When either form of spontaneous revival appears unlikely, public action can provide the stimulus that revives the area.

This chapter begins with a discussion of the differences among neighborhoods and the reasons that, like cities, they are forever changing. It goes on to discuss the differences between the regular processes of neighborhood regeneration and the now commonplace conversion of former warehouse and manufacturing districts into vibrant residential areas. It then presents some government programs for reversing neighborhood deterioration and adapting abandoned manufacturing areas as residential communities, and evaluates what worked, what has not, and why. The chapter ends with a discussion of the comprehensive approach to neighborhood revitalization that successfully restored a section of New Haven, Connecticut, more than half a century ago.

Neighborhood Dynamics

Although their boundaries are frequently imprecise and continually changing, every city has well-known, identifiable neighborhoods. They may reflect physical appearance, social composition, cultural values, political interests, history, or any combination of these factors. Their physical conditions may be manifest as topographical features such as San Francisco's Russian Hill, Nob Hill, and Pacific Heights, or a particular building stock, such as Boston's Beacon Hill, South End, and Back Bay.

The identity of a neighborhood also may reflect the country of origin of its residents, as is the case in Los Angeles's Mexican-American, Vietnamese-American, Korean-American, Japanese-American, and other ethnic enclaves. Sometimes it arises from a neighborhood's social function. In New York City, Greenwich Village has long been a center for artists, writers, and a wide variety of nonconformist populations; the Lower East Side has sheltered successive waves of poor immigrants seeking the promise of a new world; the one-family-house sections of Queens have accommodated those seeking greater neighborhood stability. These neighborhoods take on their character by becoming congenial places for residents to share activities, lifestyles, and institutions.[2]

A neighborhood's cultural dimension is only in part a reflection of its social composition. New York City has many African-American neighborhoods. Only Harlem, which is neither its oldest, nor its most populous, nor its most depressed African-American neighborhood, is internationally known. And only because Harlem has a unique cultural significance to African Americans could Langston Hughes write:

> *I was in love with Harlem long before I got there and still am in love with it. Everybody seemed to make me welcome. The sheer dark size of it intrigued me. And the fact that at that time poets and writers like James Weldon Johnson and Jessie Faucet lived there, and Bert Williams, Duke Ellington, Ethel Waters, and Walter White, too, fascinated me. Had I been a rich young man I would have built musical steps up to the front door and installed chimes that at the press of a button played Ellington tunes.*[3]

Or Malcolm X:

> *Harlem was seventh Heaven! I hung around Small's and the Braddock bar so much that the bartenders began to pour a shot of bourbon, my favorite brand of it, when they saw me walk in the door.*[4]

The least discussed and perhaps most uncomfortable dimension of any neighborhood is political. Neighborhood residents share interests that may be at variance with public policies, surrounding communities, or accepted citywide standards. Chicago's Hyde Park–Kenwood neighborhood gained cohesion by opposing urban renewal projects; its Gold Coast reinforced its identity by restricting occupancy to certain "desirable" population groups; its Back of the Yards district became prominent by protesting environmental conditions and organizing labor representation.

The most admired and increasingly prominent dimension is historical. A good way to understand New Orleans is to get to know the history of its successively settled neighborhoods: the French Quarter, Marigny, Treme, the Garden District, and others. Sometimes identity is derived from a past that has little to do with the present. Laclede's Landing in St. Louis is no longer a riverside commercial district; New York's Ladies' Mile is no longer the home of major department stores; Kansas City's Westport District is no longer a pioneer settlement supplying the Santa Fe and Oregon trails. They are all nonetheless identifiable because of activities that took place in another era.

Patterns of Change

Neighborhoods continually change. These changes may reflect physical condition, social composition, patterns of consumption, and other internal neighborhood characteristics. They also may be caused by changing tastes, migration, cost and availability of credit, and other external factors. Sometimes these dynamics operate independently. San Francisco's Chinatown, for example, began to spill over into other areas during the 1970s because of a purely external factor: the Immigration Act of 1965 opened America, and San Francisco in particular, to increasing immigration from Hong Kong, Taiwan, and China. In Washington's Dupont Circle, external forces and internal characteristics interact. The combination of changing consumer demand (an external market factor) with a highly ornamented nineteenth-century building stock (an internal locational characteristic) was responsible for its decline after World War II and conversely responsible for its gentrification during the 1970s and 1980s.

Washington, D.C., 2009. Row houses on "S" Street in the Dupont Circle area that were renovated during the 1970s and 1980s. *(Alexander Garvin)*

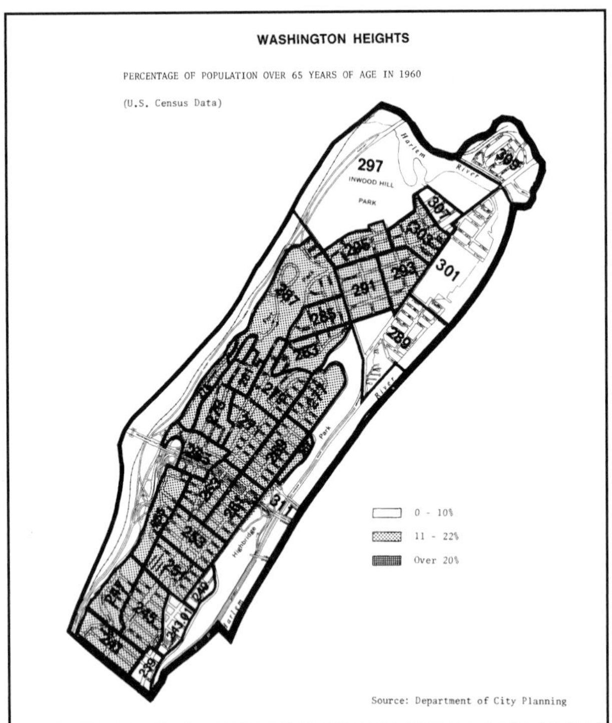

Manhattan, 1960. Washington Heights—percentage of population over 65 years of age. *(Courtesy of New York City Department of City Planning)*

Manhattan, 1970. Washington Heights—percentage of population over 65 years of age. *(Courtesy of New York City Department of City Planning)*

Dupont Circle's large, elegant mansions and more intimately scaled row houses were originally built for large upper-middle-class families that required a staff of servants. That lifestyle began to wane even before the stock market crash of 1929. With fewer families desirous and financially capable of living as their parents had, there was less demand for these ornate mansions. Inevitably, many once-elegant Beaux Arts buildings were converted into boardinghouses or multiple dwellings. Since Dupont Circle offered none of the suburban amenities so favored by young families of the day, its accommodations were not in great demand, and building owners spent as little as possible on maintenance. By the early 1960s, Dupont Circle was a sadly deteriorated neighborhood.

If changing tastes were responsible for the neighborhood's initial decline, they were equally responsible for its revival. When physical conditions had hit a nadir during the 1960s, many young adults chose to settle in the inner city rather than the suburbs. Dupont Circle was an easy walk for a large number of Washington's office workers. These relatively young city dwellers thought the area's older buildings had a charm and character that could not be duplicated. Given the condition of the neighborhood, they also could be purchased at low prices.

The "pioneers" who bought, rehabilitated, and usually reconverted the buildings to single-family occupancy often had difficulty obtaining bank financing for the work neces-

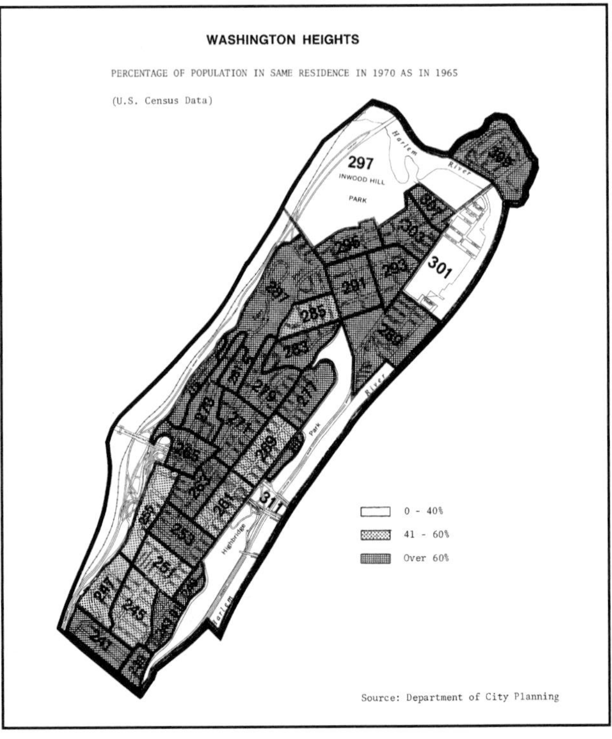

Manhattan, 1970. Washington Heights—percentage of population in the same residence in 1970 as in 1965. *(Courtesy of New York City Department of City Planning)*

sary to make needed improvements. Paying for restoration was easier for two-income households (often gay or lesbian couples) who had knowledge of and interest in design and were willing to put their sweat into renovation. Some pioneers chose to create one or two rent-paying apartments to help cover their costs. They set an example for more conventional developers who followed them into the neighborhood, converting the larger buildings into condominium and rental apartment houses. By the late 1980s, Dupont Circle was once again a fashionable middle-class residential enclave, one that had become too expensive for another generation similar to the one that pioneered its revitalization.

Boston, 1972. After 1950, many once-fine old buildings in the South End had been turned into rooming houses or abandoned. *(Alexander Garvin)*

Sometimes the rejuvenation of an area begins *because* it has deteriorated, gone out of fashion, or been abandoned. Occupancy declines to the point that property owners are willing to rent at almost any price. At this point, the regeneration of an area becomes the product of the courageous developers and pioneering building occupants who invest in what they believe are unique opportunities foolishly ignored by conventional realtors. In some cases, such as the historic districts of Charleston, Pittsburgh, and Atlanta, it requires an added push from a local civic organization.

The natural life cycle of an area's population (a purely internal factor) also may cause the neighborhood to change. In New York's Washington Heights, for example, the elderly population increased between 1960 and 1970. Since most people remained in the same residence during that period, it was clear that the increase in the number of elderly residents was not due to in-migration. The New York City Planning Department recognized an opportunity in the expectation that, as older residents retired, moved away, and died, apartment turnover would be inevitable. As a result of significant housing vacancies that would be occurring naturally, the department designated Washington Heights for inclusion in the city's just-announced Neighborhood Preservation Program.

The triggering mechanism for area rejuvenation also may be the legal or illegal reuse of old industrial and warehousing structures as artist's studios and residences. This is what happened in the 1960s and 1970s in New York's SoHo loft district (see Chapter 18), in the 1970s and 1980s in Chicago's Printers Row, in the 1990s south of Market Street in San Francisco, and most recently in the Third Ward in Milwaukee. Prices in these areas had declined to the point that they were obvious bargains. In all of these cities, government had little to do with this process. Government can, however, play an important role in encouraging conversion, as demonstrated by contemporary programs in the Pearl District of Portland, Oregon, and the Third Ward in Milwaukee.

Boston's South End

The South End was settled during the second half of the nineteenth century when landfill made this new territory available for development (see Chapter 14). The earliest residents were comfortable middle-class families who were drawn to the lovely row houses that filled its blocks. As other desirable areas opened up in Boston and in the suburbs, residents chose to move away, rents declined, and, with them, so did the condition of the buildings. By the Great Depression, 46 percent of its residents were on relief. While conditions had improved by 1940, more than one-fifth of its population was still on relief at a time when this was nearly five times the average for Boston as a whole.[5]

By the end World War II, the South End had become a skid row lined with bars. During 1961 alone, it was the site of more than 8000 arrests for drunkenness (one-third of all such arrests in Boston). Many neighborhood residents lived in rooming houses, 900 of which were licensed; hundreds more were not. Most important, living there was cheap.

Low prices and handsome nineteenth-century row houses (although usually dilapidated and often abandoned) attracted young people, many of whom were gays and lesbians. They could afford living in the South End and restore

Boston, 2006. By the twenty-first century, the South End had become one of Boston's most desirable residential neighborhoods. *(Alexander Garvin)*

the charming architecture by putting their own sweat into remodeling. This early activity triggered the neighborhood's gentrification.

The 15,000 residents of the South End in 1960 were at the bottom of the economic ladder, earning an average of $2000 per year when Boston's median income was $5500. In 2000, there were 16,000 people living in the neighborhood, earning $55,500 per year, 30 percent more than the average Bostonian.

The same patterns of change that transformed Dupont Circle took root in the South End for many of the same reasons. The housing stock, which had been well designed, could be easily restored. A new population found the low prices and charming old buildings attractive. Unlike in Washington, however, the Boston city government had initially tried to improve things by replacing some dilapidated buildings with not very attractive subsidized housing. Those efforts died with the Nixon moratorium in the early 1970s (see Chapter 12). Thereafter, government programs did relatively little to get in the way of the population that moved to the neighborhood; enterprising property owners who were able to get mortgages to pay for restoring the neighborhood's lovely old buildings transformed the South End.

New York City's SoHo

Today, the area named SoHo for its location (South of Houston Street), is known throughout the world as an art and retail center of unusual vitality, filled with trendy galleries, fashionable boutiques and restaurants, artist's studios, and expensive loft apartments. Few people remember that, for much of the twentieth century, SoHo had been a warehouse and manufacturing district. The industrial firms that remained in SoHo in the post-World War II period were marginal at best. Healthy businesses could not afford to operate in obsolete, multistory structures built before the turn of the century. In fact, by the middle of the twentieth century, the area was known for its dirty, neglected, half-empty buildings.[6]

SoHo also contains the largest concentration of cast-iron buildings in the country. These five- and six-story structures were erected during the second half of the nineteenth century. Many had ground floors with broad windows for the display of merchandise, middle floors for office work, and upper floors for storage. They are unique because of their self-supporting front walls and their interior columns that are built of prefabricated cast-iron pieces. These cast-iron

Manhattan, 2011. West Broadway in SoHo. *(Alexander Garvin)*

pieces are elegantly detailed components of unusual strength that permitted architects to design highly ornamented façades. Their large expanses of glass provide plenty of light to interior loft spaces supported by cast-iron columns rather than structural walls. They had been the showpieces of what had been during the second half of the nineteenth century one of the city's busiest commercial districts, well known for its hotels, theaters, and fine stores.

By the 1960s, it was amazing that these buildings were there at all. Robert Moses had proposed demolishing large sections of the district, first for the South Village Urban Renewal Project and then for a Lower Manhattan Expressway. Although the community had defeated both projects, the buildings nonetheless faced a much more powerful agent of destruction: neglect and abandonment.[7]

The city's zoning regulations precluded residential occupancy because planners mistakenly thought that they could retain manufacturing in Manhattan by restricting occupancy to industrial users. The zoning was easy to ignore. Artists with little money rented vast studio-residences in SoHo's neglected buildings at very low rates. Fortuitously, these large, empty spaces with high ceilings could also accommodate huge canvases and large sculptures. Despite the nonconformance with zoning, the restrictions on occupancy, and the prevalent noncompliance with building and fire codes, there was, by the 1960s, a large community of artists living in SoHo.

In the face of such widespread illegal occupancy, the City Planning Commission altered the zoning in 1970 to permit "joint living/work quarters for artists," provided that such quarters contained less than 3600 square feet. Although occupants had to be certified by the Department of Cultural Affairs as genuine "artists," the marketplace soon found ways around the restrictions. "Artists" managed to get certified for the sole purpose of converting whole floors into large unconventional apartments. Owners of buildings that were too large to qualify were happy just to get their space rented. They did not look too closely at the tenants to whom they rented. Nor did they or their tenants care whether renovation was consistent with the building's historic character or complied with building, fire, or zoning codes. Only after the city altered its J-51 tax exemption and abatement law in 1975, was there a cost-effective way to convert these buildings to legal residences that could be financed with insurance and bank mortgages (see Chapters 11 and 18).

During the 1960s, few people imagined that SoHo was the forerunner of similar, informal, tenant-initiated adaptive reutilizations of obsolete multistory industrial and warehouse structures. Fewer still understood that much less architecturally distinguished loft buildings would become similarly popular residential districts throughout the country.

Reversing Neighborhood Deterioration

It is amazing that so many revitalization efforts are as successful as they often are. Unfamiliar with revitalization efforts in other cities, each preservation group usually begins by reinventing the wheel. The transformation of Charlotte's Fourth Ward, for example, was the product of a few dedicated residents who were determined to live in a safe neighborhood within walking distance of their downtown jobs. When they began, there was little awareness that they were adopting a strategy similar to the one used successfully in New Haven's Wooster Square nearly two decades earlier.[8]

One reason for this ignorance is that, unlike downtown redevelopment or planned new towns, neighborhood revitalization is not glamorous enough to capture national attention. It is too narrow in scope, affects too little territory, and is too local in impact. Thus, success stories never get far beyond the local press. A second reason is that many people who see dilapidated buildings assume that aggressive code enforcement will coerce owners (especially negligent owners) to maintain their properties. When this approach fails, residents who are unaware of successful neighborhood revitalization programs abandon the effort to restore the neighborhood.

Once a revitalization effort has been successful, leaders who do not live in the area go back to their homes, enjoy their neighborhoods, and seldom move on to work in other deteriorating areas. This nearly universal tendency explains why, with the exception of New York City's Neighborhood Preservation Program, no city has developed the necessary cadre of trained technicians, proven programs, and political support to carry the work on to other deteriorating areas.

The only national institution that has for many years given assistance to groups interested in a coordinated neighborhood reinvestment strategy is Neighborworks, formerly known as Neighborhood Housing Services, Inc. (NHS). This nonprofit organization was established and funded by the Federal Home Loan Bank Board and HUD and has provided grants to neighborhood revitalization efforts throughout the nation that have generated more than $20 billion in reinvestment. These NHS-assisted local programs have varied greatly in scope and success.[9]

Some government neighborhood revitalization programs, like those of New York City, New Haven, and Charlotte, were developed in response to available federal assistance. New Haven's program for Wooster Square, for example, grew out of the neighborhood's opposition to federal highway and redevelopment projects. Whether they involve federal assistance or not, however, those programs that have been successful have certain common features:

- A neighborhood with attractive, basically sound housing stock that can be restored with relatively little effort and money
- Financial institutions that are prepared to make market-rate loans to property owners who can meet common underwriting requirements
- Residents and property owners who are willing and able to put time and money into improving the neighborhood
- Developers who are ready to invest in properties that will attract new residents
- A local government that invests in neighborhood infrastructure, community facilities, and public areas
- A local staff that provides property owners with assistance in dealing with city agencies and financial institutions

Crown Heights, Brooklyn

Owners of rental property are interested in making money and will do anything to avoid losing money. When the rent they receive allows them to make a return, responsible owners will keep their property in good condition. They have every incentive to keep doing so, because if they do not, their tenants are likely to move or stop paying rent. The new occupants will rarely be willing to pay as much for a deteriorating apartment. Others may be negligent, believing that they can allow their property to deteriorate without facing the consequences. There are, of course, unscrupulous landlords who will squeeze out as much revenue as possible by deferring repairs and taxes until they have little choice but

Brooklyn, 2012. Row houses on Sterling Place, Crown Heights. *(Alexander Garvin)*

to abandon the property. When enough buildings in a neighborhood start to become run-down, public pressure mounts to "do something about it." In an attempt to deal with these situations, Congress included code enforcement among the many programs that the federal government funds.

Congress enacted the Housing Act of 1964 as a way of expanding on initial neighborhood improvement experiences in the federal urban renewal program. Under Section 117 of the act, localities applied to HUD to qualify designated neighborhoods for Federal Area Code Enforcement (FACE). The act paid one-third of the cost of operating and staffing government offices and providing municipal improvements, such as paving, traffic signals, street lighting, and tree planting. Eventually, 171 cities were approved for FACE.[10]

Under Section 312, owners of one- to four-family houses in FACE areas could apply for mortgages that carried the below-market interest rate of 3 percent. These Section 312 mortgages, however, were available only to owners with incomes at or below the level permitted for public housing. Under Section 115, owner-occupants with incomes of less than $3000 per year also could apply for rehabilitation grants of up to $3500 for their one- to four-family houses. Few property owners were likely to meet these very low income requirements, so FACE consequently failed to generate major housing rehabilitation.

HUD approved New York City's application for FACE in 1967. It included a section of the South Bronx, just east of the Grand Concourse, and central Crown Heights in Brooklyn. The program had little applicability to the multiple dwellings of the East Concourse, but seemed well-suited to Crown Heights, which had a substantial stock of owner-occupied row houses.[11]

The 510-acre, 109-block section of Crown Heights designated for FACE contained 3734 buildings, almost half of which were one- and two-family dwellings. From 1820 to 1870, the neighborhood had been the home of a relatively prosperous African-American settlement known as Weeksville. This community persisted until white, middle-class families began moving into the handsome brownstone and limestone row houses that speculative builders erected during the closing decades of the nineteenth century. The neighborhood experienced a second building boom after 1920, when the IRT subway began service under Eastern Parkway. This time, developers erected four- and six-story apartment buildings in then-popular Tudor, Hispanic, and Colonial styles.[12]

Shortly after World War II, the neighborhood started to change. Rents had been frozen since the imposition of wartime price controls in 1942. As a result, landlords deferred nonessential maintenance. The state initiated the rent control program when the federal government lifted price controls in 1947. The continuing combination of frozen rents and escalating operating costs led to further deferred maintenance.[13]

Meanwhile, the population began to change. The Jewish families that had formerly constituted a significant portion of the Crown Heights population moved to the suburbs and were replaced by nonwhites, frequently middle-income immigrants from Jamaica, Haiti, and other Caribbean islands. By 1960, over half the population was nonwhite. The remaining Jewish population, especially its large Hasidic component, had trouble adjusting to ethnic change.

Although most rental properties were still well maintained, many nonwhite homeowners distrusted the apartment house tenants, especially the increasingly prevalent households receiving welfare who lived in the apartment buildings. While a drop in income did *not* accompany this change in ethnicity, it was accompanied by social tensions and, by the late 1960s, led the city planning commission to designate Crown Heights as an area that needed "preventive renewal." (See Table 13.1.)

During the mid-1960s, the planning commission had divided the city into areas that were considered sound, areas that required major action, and areas that required preventive renewal (see Chapter 19). In its *Plan for New York City,* it had advocated an allocation of about 30 percent of the city's housing resources to preventive renewal. Nothing approaching this amount was actually spent on transitional areas such as Crown Heights. However, the insistence on targeting some resources to transitional areas helped to persuade the New York City Housing and Development Administration (HDA) to apply to HUD for the FACE program.[15]

With the generous federal funding it received for FACE, HDA administered the program from a central office with eight professionals under the direction of an assistant commissioner. They also established a field office in Crown Heights that employed more than 20 professionals, including mortgage analysts, rehabilitation specialists, housing inspectors, plumbing inspectors, construction inspectors, community organizers, and even a sanitation patrol. They logged 7568 landlord interviews, 10,235 tenant interviews, 688 meetings, 1627 field visits, and 18,298 phone contacts.

The staff proceeded with the activities that HUD funded and the municipal government authorized, without any understanding of what needed to be done to revitalize the neighborhood. Statistically, their work resulted in the inspection of 2757 residential buildings, requests for 1476 court summonses, approval of 69 Section 312 loans and 14 Section 115

Table 13.1

CROWN HEIGHTS MEDIAN FAMILY INCOME IN 1959 DOLLARS[14]

Year	Crown Heights	Brooklyn	New York City
1959	$6078	$6245	$6554
1969	$6356 (+5%)	$7092 (+14%)	7671 (+17%)

grants, and issuance of 9 municipal rehabilitation loans. On top of that, $1.2 million was spent for sidewalk repair, new street name signs, new sanitation signs, new traffic signals, new streetlights, and new street trees. With the exception of the rehabilitation loans, the list of items funded is really a catalog of normal city expenditures that were selected because they were eligible for FACE reimbursement.

Half the buildings and 83 percent of the area's housing units were multiple dwellings that were *not* eligible for Section 312 loans or Section 115 grants. Nor were these programs likely to help the resident owners of the area's large stock of row houses. Crown Heights was not a poor neighborhood (see Table 13.1). In fact, most of the owners of its 1839 one- and two-family dwellings had incomes that were too high to be eligible for these programs. Consequently, over the five years that FACE was in operation, only 69 buildings qualified for Section 312 loans and 14 for Section 115 assistance.

Even though the neighborhood's multiple dwellings faced the greatest problems, they were ineligible for FACE assistance. In the absence of FHA-insured bank mortgages, the city offered municipal loans. However, the Municipal Loan Program did not have the necessary staff or funds to meet citywide demand for rehabilitation financing. Nor was there any mechanism for eliminating rent control. Besides, most property owners in Crown Heights wanted nothing to do with this scandal-ridden program (see Chapter 11).

Code enforcement was the least appropriate HUD-funded activity directed to Crown Heights. The only certain result of code inspection is an increase in the number of violations reported. Property owners may or may not make the repairs that will cure violations. Furthermore, the city is usually indolent in removing violations from its records, or it just fails to remove them. Thus, violations that may have already been cured may continue to show up on records.[16]

CROWN HEIGHTS—AREAS WITH SOCIAL AND PHYSICAL PROBLEMS

Areas requiring increased social services

Areas requiring limited building treatment

Areas requiring both increased social services and limited building treatment

Brooklyn, 1972. Crown Heights. *(Courtesy of New York City Department of City Planning)*

Brooklyn, 2009. Deteriorating apartment buildings in Crown Heights that did not qualify for federal assistance even though they provided housing for 83 percent of the residents of the Federal Area Code Enforcement Program. *(Alexander Garvin)*

Although code inspection in New York City is restricted to multiple dwellings and, therefore, did not apply to half the structures in Crown Heights during the FACE program, city inspectors somehow managed to gain access to 55 percent of the area's one- and two-unit structures. However, they were not authorized to issue violations. As if to make up for this, the Crown Heights FACE program completed cellar-to-roof inspections of all of its 1747 apartment buildings. Program records indicate that most of these buildings were brought into compliance with housing code requirements.[17] Despite this high level of code compliance, Crown Heights was still filled with deteriorating buildings. In 1972, the year that the FACE program was concluded, a HUD-funded study of the neighborhood found that conditions in Crown Heights had worsened.[18]

It is not fair to dismiss FACE because it was neither appropriate to Crown Heights nor successfully used there. In some cities with appropriate housing stock and ownership, and combined with significant additional activity, FACE was very helpful. By itself, however, FACE did not provide all the tools necessary to generate private-sector reinvestment, and New York did not supply the necessary additional ingredients or the strategy to make it work.

The Fourth Ward, Charlotte

The residents of Charlotte, North Carolina, used federal money for restoration rather than code enforcement. Their program lowered debt service on bank mortgages issued to landlords and homeowners who agreed to renovate their properties. During the 1970s, when neighborhood revitalization efforts began in the Fourth Ward of Charlotte, fire, unsafe building demolition, and increasing structural dilapidation had disfigured the area. Three dozen structures remained standing in the 20-block residential neighborhood that is beneficially located less than a five-minute walk from Charlotte's business district.[19] The Fourth Ward's increasing deterioration was especially startling because it had continued during a period when Charlotte's burgeoning economy had become a showpiece of the "new South." Furthermore, the city's residential market was expanding rapidly. Its population had grown from 134,000 in 1950 to 241,000 in 1960, and would be 315,000 in 1980.[20]

In 1975, a group of Charlotte residents came together to form an organization called Friends of Fourth Ward. They were united by their belief in the positive aspects of Charlotte's future, a commitment to living right in town, and a love of the dilapidated but charming Victorian buildings that remained in the Fourth Ward. James Dennis Rash, dean of students at the University of North Carolina at Charlotte (UNCC), became their informal leader.

Working with representatives of the city government and the business community, they developed a program that required action from both. The city agreed to close streets to through traffic; install new decorative brick paving, street lamps, street signs, and street trees; and acquire land for new neighborhood parks. They also persuaded the city to rezone the area and to create a historic district commission for the Fourth Ward that would review all development for compatibility with the area's historic character.

The Junior League took the lead by acquiring and renovating the John Newcomb House, one of the Fourth Ward's

Charlotte, 2001. By 2001, all the older buildings in the Fourth Ward had been renovated. *(Alexander Garvin)*

most significant structures. Duke Power Company paid for burying the Fourth Ward's utility lines. A consortium of seven local banks joined with the city government to create a below-market-rate-interest mortgage fund that lent money to residents ready to purchase, build, and renovate houses or apartments in the neighborhood. North Carolina National Bank (NCNB) established a wholly owned, nonprofit subsidiary to act as a development catalyst.

The mortgage fund was established by selling tax-exempt city revenue bonds to local banks. The lower rate of interest paid on tax-exempt bonds allowed the fund to lend money at below-market interest rates. During its six years in existence, the fund issued $25 million in mortgages to purchasers of new and old one-family structures, row houses, and condominium apartments. The interest rates on these loans were especially attractive because the fund had been established before the escalation in interest rates during the late 1970s and early 1980s that peaked with prime rates over 20 percent. Thus, they carried interest rates that were often less than half those available anywhere else in the region. Moreover, any purchaser could assume these mortgages, so reselling a house or condo in the Fourth Ward was easier than in many suburban areas. By the time the fund ran out of money in 1981, it had financed the restoration of 32 old houses, the construction of 10 new one-family homes, and the purchase of several hundred new and converted condominium apartments.

The role of NCNB Community Development Corporation (NCNBCDC) was no less critical than that of its mortgage fund. NCNBCDC acquired and renovated a 1929 apartment building and built new town houses. Altogether, it initiated four projects involving 112 apartments. These efforts triggered the interest of other developers, who went on to erect additional buildings.

During the nearly four decades since the Friends of Fourth Ward had initiated its revitalization efforts, the city's population had doubled and the neighborhood had become a popular alternative for Charlotte residents seeking attractive, in-town living. Real estate tax assessments in the Fourth Ward increased more than nine times between 1975 and 1981, the period during which the mortgage fund was in existence. During the next 25 years, more than 1100 dwelling units were built or rehabilitated. Current residents cannot imagine that the Fourth Ward was once a dilapidated neighborhood with refuse-strewn lots and burnt-out buildings.

Ansonborough, Charleston

Providing mortgage money may not be enough to spark a revitalization. Somebody has to want to live in a structure that at the time appears to be in hopeless condition and be ready, willing, and able to use the money to make it habitable. Preservationists in Charleston, Pittsburgh, and several other cities have devised a mechanism that deals with this problem: a revolving fund (see Chapter 18). The fund is used to purchase abandoned and dilapidated structures that were once charming residences. Building exteriors are spruced up, and any structural problems are repaired. The fund managers find a home buyer who is willing to purchase the building at their cost and to agree to a deed restriction guaranteeing that the

Charleston, 1974. Abandoned and deteriorating buildings acquired with money from its revolving fund by the Historic Charleston Foundation in Ansonborough. *(Alexander Garvin)*

exterior appearance will be maintained. The money is then recycled until all the buildings in the neighborhood have been dealt with. One particularly successful example is the Ansonborough neighborhood in Charleston, South Carolina.

By the 1950s, Ansonborough, the area just north of Charleston's Old and Historic District, was a victim of deterioration. Most of its 135 antebellum structures were vacant or dilapidated. The Historic Charleston Foundation, a local preservationist group, began a drive to preserve this decaying seven-block neighborhood. A first step came in 1966, when the foundation persuaded the city to enlarge its historic district to include Ansonborough. Next, it persuaded the city to install distinctive street signs identifying the area's special character.[21]

Decades of experience had demonstrated that landmark status, by itself, did not result in preservation. There also had to be a mechanism to get Ansonborough's fine old buildings into the hands of owners who would restore them. The foundation encouraged this process through the use of a revolving fund, which it utilized to acquire buildings. Once it owned them, it concentrated almost exclusively on their exterior appearance: occasionally they were so far gone that they needed to be demolished; other times their facades were restored. No matter what, however, it always resold them with protective covenants that ensured appropriate restoration and continuous exterior preservation. It was the entrepreneurial role the foundation played in obtaining new owners of these buildings that proved to be the critical element in saving the neighborhood for people who would take loving care of its buildings for decades thereafter (see Chapter 18).

By December 1972, the Historic Charleston Foundation had acquired and demolished seven buildings that were incompatible with the rest of Ansonborough. It also bought and moved 4 historic structures into the area and purchased another 54 buildings. Eight of these buildings were rehabilitated inside and out, five as rental properties and three for resale. Most important, the foundation persuaded banks to provide their ultimate owners with the mortgages they needed to finance acquisition and renovation.

In less than a decade, the revolving fund had been involved with one-third of Ansonborough's buildings. Their owners improved another third. As a result, between 1959

Charleston, 2011. Buildings throughout Ansonborough (now restored) were purchased and then resold with easements requiring restoration of their façades by the Historic Charleston Foundation. *(Alexander Garvin)*

and 1972, property values tripled, and by 1980, nearly every building in Ansonborough had been rehabilitated.

The Historic Charleston Foundation acted as the catalytic agent needed to generate market interest in Ansonborough and to persuade banks to lend money in the area. Many Charleston-area residents, both natives and new arrivals, perceived Ansonborough's potential. They were attracted by the comparatively low prices of its historic buildings and eager to settle in this extremely convenient inner-city location, but they were not willing to become pioneers unless a substantial number of its vacant and dilapidated buildings were renovated. Once a ready market of potential residents had been attracted and banks had been persuaded to provide financing, Historic Charleston used its revolving fund to keep the rehabilitation ball rolling. Restrictive covenants provided the necessary guarantees of continued maintenance and, thus, confidence in the area's future. Ansonborough's new residents did the rest.

Pittsburgh's Mexican War Streets

Neighborhood revitalization takes time, often more time that it took for the area to decline. Charleston's historic areas came back relatively quickly—over a couple of decades. Pittsburgh's different physical character and market conditions rendered its experience with the revolving fund different from Charleston's. Pittsburgh has charming older neighborhoods with charming nineteenth- and twentieth-century buildings. Few of them, however, are as exquisite as some of the antebellum houses of Charleston. More important, Charleston's historic preservation efforts coincided with a growing population and expanding economy. Between 1960 and 2010 the number of residents in Charleston nearly doubled. Pittsburgh's population, on the other hand, was cut in half, dropping to 305,000 in 2010. There might have been more demand for housing in this area if Pittsburgh had not been losing population and if there had not been plenty of inexpensive housing elsewhere, often in areas with just as much charm but which were thought to be safer. Moreover, many of the area's existing residents were poor African Americans, who could not afford to pay for more than basic patch-up.[22]

Despite these very different conditions, there was a similar increasing interest in historic preservation. In 1964, a group of dedicated citizens formed the Pittsburgh History and Landmarks Foundation (PHLF) to assist in meeting the city's diverse preservation needs. The PHLF issued publications on the architecture and history of the city and established a revolving fund to preserve distinctive older neighborhoods. By 1973, the revolving fund held nearly $500,000, making it the largest in the nation.[23]

The neighborhood popularly referred to as the "Mexican War Streets" was one of the areas selected by the PHLF for its revolving fund. Its streets had been laid out in 1848, when it was still part of Allegheny, an independent city across the river from downtown Pittsburgh. These streets were named for famous Mexican War battles and military officers, and,

Pittsburgh, 2012. The Mexican War Streets are a very convenient place for people to live who work downtown. *(Alexander Garvin)*

Pittsburgh, 2012. The charming character of the mid-nineteenth-century houses in Mexican War Streets attracted middle-class residents with other residential options. *(Alexander Garvin)*

within a few years, they were lined with modest, two- and three-story brick row houses with gingerbread wooden detailing.

By the early 1960s, the Mexican War streets had become "a mixture of young and old, white and black, home owners and slum rooming house dwellers, poor to middle-income residents."[24] The buildings they occupied had deteriorated badly. In an attempt to save old and historic structures and generate further area revival, the PHLF decided to concentrate the activity of its newly established revolving fund on the Mexican War Streets. The strategy was to buy out slumlords, reconvert rooming houses to family occupancy, relocate tenants "as appropriate, to public housing or other units in the area," fix up strategically located properties for continued use by low-income families, and thereby generate further investment by existing property owners and attract "new, particularly young, working residents" to the area.

During its first decade of operation, the PHLF purchased 24 separate buildings within the Mexican War Streets neighborhood. In order to keep housing costs at a level that was affordable to the area's existing low-income residents, the PHFL minimized rehabilitation and avoided costly replication of architectural details. When necessary, the PHLF used the revolving fund to subsidize tenants and rented the buildings at a loss. This kept poor families in the area. It also kept the buildings from being put on a sustainable economic footing and retarded further improvement.

As in so many other neighborhoods that have been revitalized, artists played important supporting roles. One of the earliest to come to the Mexican War Streets was Barbara Luderowski. In 1975, the same year that 27 acres of the Mexican War Streets Historic District were listed on the National Register of Historic Places, she purchased and moved into an old mattress factory that became a center for local artists, especially those interested in installation art. Seven years later the Mattress Factory opened its first exhibition in the building. By 2012, the Mattress Factory had grown into a complex of buildings that attracts more than 50,000 visitors annually.

At first, the revolving fund managed to save charming, inexpensive residences for households interested in living close to downtown Pittsburgh. For three decades, neither the revolving fund nor the Mattress Factory was enough to trigger much additional investment in the neighborhood. Nevertheless, there was a slow influx of new residents who fixed up vacant and dilapidated buildings.[25]

By the beginning of the twenty-first century, a critical mass of buildings had been restored and the area had begun to attract purchasers who perceived more than charming nineteenth-century buildings that could be acquired and transformed into uniquely attractive homes at relatively low prices. Residents could walk through the continually improving Allegheny Commons Park to reach the two sports stadiums, three museums, and several new office buildings and apartment complexes that had opened beside the Allegheny River. Downtown Pittsburgh was another five-minute walk across one of the area's three bridges. Their proximity had made living in the Mexican War Streets more competitive with Pittsburgh's other residential districts. Over the past half century, appreciation of the character of the area's architecture has also grown. In 2008 the National Register added 288 buildings to the 119 that had been listed 33 years earlier.

It may not have been easy and may have taken a long time, but the Mexican War Streets housing is again in good physical condition. Few empty buildings remain to be restored. Unlike other areas that had been revitalized over that period, however, many of the families who lived there in the 1970s are still there, living side by side with the somewhat more affluent newcomers who have helped to restore this lovely, affordable neighborhood.

New York City's Neighborhood Preservation Program

New York City's Neighborhood Preservation Program (NPP) is one of the few neighborhood revitalization programs created by city planners. Beginning in the 1960s, the City Planning Commission recognized the need for a balance between preservation and new construction. It insisted on including housing rehabilitation in most urban renewal projects, allocated money for housing rehabilitation when it issued the city's draft capital budget, called for a major commitment to preventive renewal in the city's master plan, and prepared a number of program proposals for mayors Robert F. Wagner and John Lindsay, both of whom were eager to develop successful strategies for neighborhood conservation.

When I entered city government in 1970, Donald Elliott, then chairman of the City Planning Commission, asked me to develop a neighborhood revitalization strategy that the Lindsay administration could implement. We spent the next three years fighting to have HDA try it. The city's housing establishment opposed any major allocation of funds to neighborhood revitalization because it did not produce *additional* housing. Besides, the establishment believed that the city's housing programs should be determined by federal assistance, which was largely devoted to new construction. Skepticism about neighborhood revitalization was further fueled by unhappy experiences with federally assisted code enforcement and city-assisted municipal rehabilitation loans. As long as Washington provided billions for redevelopment and new construction, there was little hope of changing this attitude.

In the wake of Nixon's moratorium on all federal housing assistance, Elliott's successor, John Zuccotti, persuaded Mayor Lindsay to issue Executive Order 80 in 1973. This law instructed the City Planning Commission to designate neighborhood preservation areas and the HDA to establish local offices "to coordinate governmental and community activities for neighborhood preservation . . . and to provide adequate public investment to support coordinated improvement programs."[26]

The City Planning Commission held public hearings and designated five neighborhood preservation areas: Washington Heights and Clinton in Manhattan, Crown Heights and Bushwick in Brooklyn, and West Tremont in the Bronx. HDA hired a staff of 100 and opened five neighborhood offices. However, it had neither a clear notion of what "neighborhood preservation" meant nor specific strategies for any of the very different areas that had been designated.[27]

As devised by the city planning department, NPP was directed to neighborhoods where the existing housing stock was still essentially sound and attractive but deteriorating and in need of only moderate rehabilitation. The idea was to target minimal amounts of city money and personnel in a manner that would induce the private sector to eventually resume full operation in these neighborhoods.

City money and personnel were used to create two new entities that, together, were intended to stimulate the necessary private-sector activity: decentralized housing offices and the New York City Rehabilitation Mortgage Insurance Corporation (REMIC). In response, 11 commercial banks and 23 savings banks established an independent, not-for-profit corporation, the Community Preservation Corporation (CPC), to provide rehabilitation mortgage financing.

Local offices were essential to program effectiveness. They were catalytic agents, on one hand stimulating property owners to spend small sums on building repair and maintenance, while on the other persuading fellow bureaucrats in city agencies to be more responsive to community needs. The offices provided a place in the neighborhood where both residents and property owners could come with their problems. The local offices were soon familiar with every community complaint, and they had to be responsive because they were judged by improvements in the neighborhood rather than by the amount of paper they processed. As a result, the staff did not wait passively for neighborhood business; it went out to generate it.

The city appropriated $7.9 million from its capital budget to start REMIC. Most of the money went into a mortgage insurance fund, initially limited to covering 20 times the amount of outstanding insured mortgage debt. Borrowers paid an annual premium of approximately 0.5 percent of the outstanding principal balance of each insured loan. The income from premium fees and investments was intended to cover REMIC's operating expenses and any mortgage insurance contract obligations.[28]

In the beginning, most bank lending officers knew little about NPP or REMIC. They were so worried about their existing loan portfolios that they avoided all but the very safest mortgages. Thus, a new lending institution for the five neighborhood preservation areas was essential. However, when CPC was established, its directors and officers were just as skeptical about lending in these areas. They began by restricting their lending to two neighborhoods the city government thought most appropriate: Crown Heights and Washington Heights.

In 1983, it extended its activities to all 13 of the city's designated Neighborhood Preservation Areas. Once the private sector grew more comfortable with lending for moderate rehabilitation in these areas, the administration of Mayor Edward Koch persuaded the Community Preservation Corporation to shift its emphasis to gut rehabilitation of the city's huge stock of vacant buildings that had been taken for failure to pay real estate taxes. As that inventory was depleted, it began to issue mortgages for new construction of apartments that would be rented or sold for "affordable" prices, expanded to the Hudson Valley in 1989 and to all of New

York State in 1995. As of 2010, it had issued more than $8 billion in mortgage loans on 144,000 affordable housing units.[29] As the subprime mortgage crisis intensified at the end of the decade, however, CPC cut back its lending activity and by 2011 was working to resolve problem loans on condominiums and other for-sale projects that had failed to perform."

REMIC insurance successfully curbed default risk and thereby encouraged lending institutions to provide mortgages. It also established standards that would be used for mortgages outside the preservation areas. During the 1980s, REMIC declined in importance because the state of New York established its own mortgage insurance program, SONY MAE. Its requirements were less rigid. More important, banks were more comfortable relying on the financial stability of the state. By the last decade of the twentieth century, REMIC ceased to exist. By 1990, however, REMIC had insured more than $102 million in mortgages for the rehabilitation of 291 buildings containing 14,284 apartments. Only $122,000 in insurance had to be paid—to cover a mortgage default on one building containing 36 apartments.[30]

In 1974, when I became HDA deputy commissioner, REMIC had only printed application forms, CPC had just opened its offices, and, in an indication of the city's fiscal crisis to come, I was instructed to terminate 15 percent of the staff. The next six months were spent reorganizing, training personnel, and devising preservation strategies for each area. Just as the neighborhood offices were beginning to generate owner-sponsored repairs and issue city-financed rehabilitation mortgages, New York City's fiscal crisis nearly terminated the program. With the city unable to issue additional bonds, there was no way to continue raising money to lend to property owners for rehabilitation. The budget crisis made it difficult to maintain enough staff to run five offices. Consequently, the program had to concentrate on Crown Heights and Washington Heights, the neighborhoods to which CPC had decided to restrict its operations. The other three offices became token operations.[31]

With virtually no city money and minimal personnel, HDA revised the Neighborhood Preservation Program to become more dependent on private-sector activity. It obtained state legislation altering the J-51 Tax Exemption/Abatement Program and establishing the Participation Loan Program (see Chapter 11). It shifted all construction lending to commercial banks. It began making joint bank–city rehabilitation loans at below-market interest rates. It replaced city bond proceeds with federal CDBG assistance as the source of funds for city mortgages. For the first time, the city was generating several dollars of private money for every public dollar lent for housing rehabilitation and doing so without charging a cent to the city budget.[32]

In order to avoid displacing neighborhood residents who could not afford higher rents, the administration asked Washington to set aside federal Section 8 subsidies for qualified tenants in buildings whose rehabilitation was financed with participation loans. These subsidies, which had not been available prior to 1974, permitted the program to subsidize neighborhood residents who could not afford increased rent and would otherwise have had to move.

Instead of more government spending or more government services, each preservation office directed existing government activity to critical sections of the neighborhood. The notion was that if a small number of problem buildings in each neighborhood were improved, local conditions would improve overall. In Crown Heights, for example, activity was targeted to large corner apartment houses. In Washington Heights, it was directed to small clusters of blocks.

Once critical buildings were identified, the preservation office sent rehabilitation specialists to ascertain what required immediate attention. With this list in hand, the office estimated the cost of repairs and negotiated with the owners to voluntarily make repairs. In exchange, the local office reached into the bureaucracy downtown to help owners obtain services to which they were entitled but which they had been unable to obtain due to inadequate representation or insufficient bureaucratic skills.

This selective effort to obtain voluntary repair agreements was far more effective in producing repairs than were official inspections, which cataloged code violations for the city's computer without obtaining improved living conditions. It was particularly effective when combined with mortgage financing for moderate rehabilitation. In Washington Heights, over the next 15 years, nearly *one-fifth* of the housing stock was renovated.

Washington Heights, Manhattan

Washington Heights includes the northern portion of Manhattan Island, extending from 155th Street to the Harlem River. In 1970, approximately 187,000 people lived in the neighborhood's 75,500 apartments: 15 percent were African American and 9 percent were Hispanic; 11 percent were on welfare, a figure lower than the city average. When Washington Heights was designated a Neighborhood Preservation Area, there was almost no abandoned housing. Its 3000 largely post-1929 structures were built to modern standards and contained better apartment layouts with larger rooms.[33]

Many of the neighborhood's predominantly five- and six-story apartment buildings were deteriorating. They required replacement or upgrading of basic mechanical systems, weatherproofing, and security systems, but this work had been deferred because banks were unwilling to make new mortgage loans and landlords of rent-regulated buildings, even if they could borrow the funds, were often unable to recoup money spent on improvements in the form of increased rents.

The social stress of adjusting to population changes only

Manhattan, 2011. Aerial view of Washington Heights, where two decades of effort by the staff of New York City's Neighborhood Preservation Program resulted in the renovation of 14,000 apartments. *(Alexander Garvin)*

complicated matters. Although large areas of Washington Heights continued to be German, Jewish, and Irish enclaves, there had been a significant increase in the Hispanic population. If population change was perceived as a downgrading of the neighborhood, property owners might be less attentive to building management and maintenance. Consequently, the preservation office targeted its activity in a manner that was intended to reassure older neighborhood residents that landlords were investing in property improvements. It also directed attention to those sections that were receiving increasing numbers of newcomers.

Responsibility for the program's success in Washington Heights lies with the directors of the neighborhood office, in particular, Michael Lappin and Barbara Leeds. Lappin became director in June 1974. For the next year and a half, he trained the staff, established an ongoing relationship with community leaders, and reached out to property owners to bring them into the program. Lappin became the Washington Heights loan officer for CPC, and in 1980 its president. Barbara Leeds succeeded him as director and remained in that position until 1984. In 1990, she became assistant housing commissioner for rehabilitation finance in charge of all the city's real estate tax- and mortgage-assistance programs for housing rehabilitation.

Leeds devised the strategy that targeted the post-fiscal-crisis programs to the critical sections of Washington Heights. It ignored the very strongest market areas, where little help was required. Instead, the office directed its initial attention to the rest of the neighborhood's fundamentally sound housing and later expanded into nearby weaker areas. Activity in the most deteriorated areas, which required massive subsidies, was deferred until suitable programs became available (as they did in the early 1980s, when the New York City Housing Authority and private developers were able to undertake gut rehabilitation with the heavy subsidies provided by the Section 8 program).

When Edward I. Koch became mayor in 1978, 1500 apartments had been rehabilitated under a variety of mortgage-assistance programs, 9300 apartments had been upgraded through landlord repair agreements, and 4000 apartments

Manhattan, 2012. Apartment buildings in Washington Heights that remained in good condition or were renovated with assistance from the Neighborhood Preservation Program. *(Alexander Garvin)*

had benefited from building-wide improvements assisted by J-51 tax exemption and abatement. The new administration added the one element of a true neighborhood revitalization effort that the city's fiscal crisis had precluded: additional funds for public services and community facilities. A task force that coordinated city activities in Washington Heights distributed the public services. In addition, the City Planning Commission designated Washington Heights a Neighborhood Strategy Area, in which it concentrated CDBG and capital budget projects. This resulted in a major program of park rehabilitation, sanitation services, infrastructure replacement, and assistance for the elderly.[34]

No neighborhood revitalization program anywhere in the country has resulted in as much private investment as was generated in Washington Heights. As of 1990, the Neighborhood Preservation Program had obtained more than 1336 voluntary repair agreements involving 49,370 apartments. By 1993, the Community Preservation Corporation and the City of New York had extended $92.2 million in joint mortgages for the rehabilitation of 193 buildings containing 7064 apartments, and more than 405 buildings with 13,898 apartments had been rehabilitated under various public programs.[35]

In those sections of Washington Heights north of the George Washington Bridge, for which the program had been devised and to which it was intensively directed, housing deterioration has been reversed and abandonment is minimal. In fact, dozens of buildings have been successfully converted to resident-owned cooperatives. In some spots developers have even resumed building small apartment buildings. Only south of the bridge, where moderate rehabilitation was insufficient and where residents were unable to afford higher rents, has the neighborhood continued to deteriorate.

Transforming Loft Districts into Residential Neighborhoods

The charming nineteenth-century residential buildings of Charleston and Savannah were easy to adapt to late-twentieth-century residential use. They needed owners who wanted to live there and lenders who would advance the money to restore them. But by the last third of the twentieth century, there were few new users for the old city warehouses and lofts, and few lenders prepared to provide the money to fix up what they thought of as obsolete structures. Nevertheless, there were declining business districts in Chicago, San Francisco, Cleveland, Minneapolis, and elsewhere that provided the seeds of a remarkable rebirth.

The explanation for empty buildings was not that demand for warehousing and manufacturing had disappeared. It was that the physical environment was obsolete: these activities were no longer as attractive in city locations as they had been during the nineteenth century and the multistory buildings that were once appropriate were no longer suitable for late-twentieth-century practices. In response to the creation of the Interstate Highway System, most companies shifted freight operations to truck, eliminating the advantage of a location along a railroad. Moreover, they wished to be rid of the added time and cost of shipping to and from buildings in traffic-congested city districts.

The SoHo success story provided a vivid model for what could happen in other declining business districts, but even public officials in New York City did not understand that it would not have gone very far without government action. To produce results more vigorous than illegal occupancy by artists or pioneers willing to live in unconventional residences, government would have to alter zoning, building, and fire codes, as well as real estate tax procedures. Without such action, there was no way to induce companies to make available building insurance, convince banks to provide the financing for conversion and rehabilitation, or attract conventional real estate development companies.

In the 1970s, Chicago, like New York, changed its laws and procedures to make possible the transformation of Printers Row into a hip, SoHo-like neighborhood. That, however, was not enough to create a similar loft district along Washington Avenue in St. Louis. Despite similar changes in law and investment in new street furniture, lofts that had vacancies in the mid-1990s still had vacancies more than a decade after they had been converted to residential use. Unlike New York City, whose population grew slightly from 7.9 million in 1950 to 8.2 million in 2010, St. Louis lost over 60 percent of its population, enormously circumscribing the market of potential renters. As in many other American cities, former St. Louis residents were moving to the suburbs, which in St Louis grew by more than 1 million residents during that 60-year period.

Although Chicago had suffered a similar population decline, from 3.6 million in 1950 to 2.7 million in 2010, market demand in the center of the city had remained high. Chicago, unlike St. Louis, had always had a large, downtown, residential population that was attracted to its lakeshore parks and used the extensive mass transit system to get all over the city. What the mass transit systems helped to make possible in New York and Chicago, however, was unlikely to work throughout the rest of the country, where potential loft tenants used automobiles to get to work, to school, and almost everywhere else.

Without major investment in the public realm, especially in transportation, it would be extremely difficult to generate serious market demand for the residences in loft districts outside New York and Chicago. Those districts could again become vibrant. But, as demonstrated by the revitalization of the Third Ward in Milwaukee and the Pearl District in Portland, Oregon, government had to provide a helping hand. That assistance, however, was a very different process from what had been necessary in the older residential areas of Charleston or Charlotte.

Chicago's Printers Row

The revival of Printers Row emerged from the vision and entrepreneurship of Harry Weese and Larry Booth, who were among Chicago's most prominent architects during the second half of the twentieth century. Where other people saw dilapidated and obsolete structures, they saw the opportunity to create an exciting, if unconventional, neighborhood. Harry Weese had heard how artists had resuscitated obsolete buildings in the SoHo section of New York City from his daughter, who lived there. He thought Printers Row, the loft district north of Dearborn Station, was ripe for a similar revival (see Chapter 6). In 1975, he persuaded a group of his friends to join him in acquiring and renovating the area's derelict buildings.

The first building they tackled was the 22-story Transportation Building opposite Dearborn Station. Converting this or any other building for residential occupancy when it had been used for bookbinding, printing, publishing, or warehousing was no easy matter. Many of these buildings had been stripped of marble panels, woodwork, and decorative details. The guts of the buildings had to be torn out in order to meet code requirements that had been conceived with new construction in mind. Compliance was not only prohibitively expensive; it eliminated the very characteristics that made the buildings so attractive. Worse yet, the buildings were assessed for taxes as though they contained functioning businesses. In many cases, owners had stopped

Chicago, 2009. Printers Row.
(Alexander Garvin)

paying real estate taxes, so anyone who purchased the buildings would have to pay off years of tax arrears.

Weese and his friends persuaded the city government to adjust the building code so that lofts could be reused as residences without having to be completely reconstructed to comply with specifications intended for new buildings. In 1983 they helped to persuade the state of Illinois to enact a law freezing property-tax assessments at prerenovation levels for eight years after rehabilitation of buildings in historic districts such as Printers Row.

They convinced the city to condemn the abandoned Transportation Building and sell the property to them for $150,000, free and clear of any tax liabilities. The building was already six years and $2.7 million in property-tax arrears. Under then-existing statutes, the city would have had to wait a full 10 years before taking possession for failure to pay taxes. By that time, further deterioration and vandalism would have precluded cost-effective renovation.

When the converted Transportation Building opened for occupancy in 1980, it contained 294 apartments. Within seven years, 15 buildings containing 1260 apartments had been renovated, at a cost of $110 million. Dearborn Station had been converted into a small-scale retail center. Where there had once been abandoned buildings there were now bookstores, restaurants, and even a hotel. In addition to acquiring and reselling the Transportation Building, the only role government had played was to alter the building code to permit the inexpensive residential conversion of loft buildings. The strength of market demand for the loft apartments became evident as soon as residents had moved into the Transportation Building. Thereafter, revitalization efforts were the product of other property owners, developers, and businesses.

Third Ward, Milwaukee

During the first half of the twentieth century, the Third Ward had the benefit of well-constructed four- and five-story, red-brick buildings, all of which had been erected after the area had been devastated by the fire of 1892. The businesses that occupied these structures had easy access to the rest of the city and, by rail and water, to markets throughout the Midwest. The value of those assets was greatly diminished with construction of highway I-794, which displaced a thriving network of local Italian-American businesses and residents and created a visual barrier between the Third Ward and the rest of downtown Milwaukee.[36]

Other Wisconsin locations had equally good or better access to national markets via the new interstate highway system. The one-story structures that were built on those sites could more easily accommodate the continuous production lines and forklift trucks that reduced the cost of production. As one might expect, occupancy in the multistory late-nineteenth- and early-twentieth-century urban lofts declined.

Milwaukee, 2012. Broadway has become an attractive Third Ward retail destination for residents of the entire city. *(Alexander Garvin)*

Businesses and property owners in the area mobilized to keep it from going the way of so many declining mercantile districts. The first step came in 1976 with the creation of the Historic Third Ward Association (HTWA), a private nonprofit membership organization established to promote the district. This was followed in 1984 with the designation of ten blocks of the area as "The Historic Third Ward District" by the National Register of Historic Places. Designation had great promotional but little economic impact. The first truly transformative action came in 1987 with the creation of a business improvement district (see Chapter 8). That was followed in 1992 with a $3.4 million investment in two arched gateways at the entrance to the neighborhood, new district lighting, a combination of angled parking with midblock parks in the center of two blocks of Broadway, and the relandscaping of Catalano Square.

Milwaukee, 2012. The Third Ward public Water Street Garage provides parking for the residents of nearby loft apartments without on-site parking. *(Alexander Garvin)*

Milwaukee, 2012. The Gateway brands the Third Ward as a special neighborhood. *(Alexander Garvin)*

These streetscape improvements began the visual transformation of the area into a neighborhood that would be welcoming to residents and shoppers—something that Pittsburgh's, Charlotte's, and Charleston's older neighborhoods did not need because they were already residential prior to the revitalization efforts. More important, the angled parking provided a visible destination for Milwaukee residents who drove everywhere. Those two blocks, however, did not provide enough parking to accommodate potential residential occupants of the increasingly empty loft buildings throughout the area. That changed with the construction of two city-sponsored parking garages, funded by tax increment financing similar to that used for Atlanta's BeltLine (Chapter 3) and Houston's Uptown District (Chapter 6). The first of these was a $5.5 million 500-car structure that opened in 1994. Another one followed in 2000 at a cost of $5.8 million and combining 436 parking stalls with 1700 square feet of retail space.

Once the parking and streetscape improvements were completed, the next issue was creating an even more welcoming environment for Milwaukee residents and retail customers. The city had been creating a downtown Riverwalk since the 1990s. In 2004, it put together a $17 million combination of federal, state, city, and private money to extend the walkway from downtown and pay for repairs to adjacent private docks and property walls. The Third Ward portion of the Riverwalk quickly became a magnetic amenity that sprouted bars, cafés, and restaurants. A year later, a BID-sponsored $10 million public market opened on the northern edge of the district.

At the start of the twenty-first century, there had been little more than 200 residents living in the Third Ward; a

Milwaukee, 2012. The Riverwalk connects the Third Ward with downtown Milwaukee. *(Alexander Garvin)*

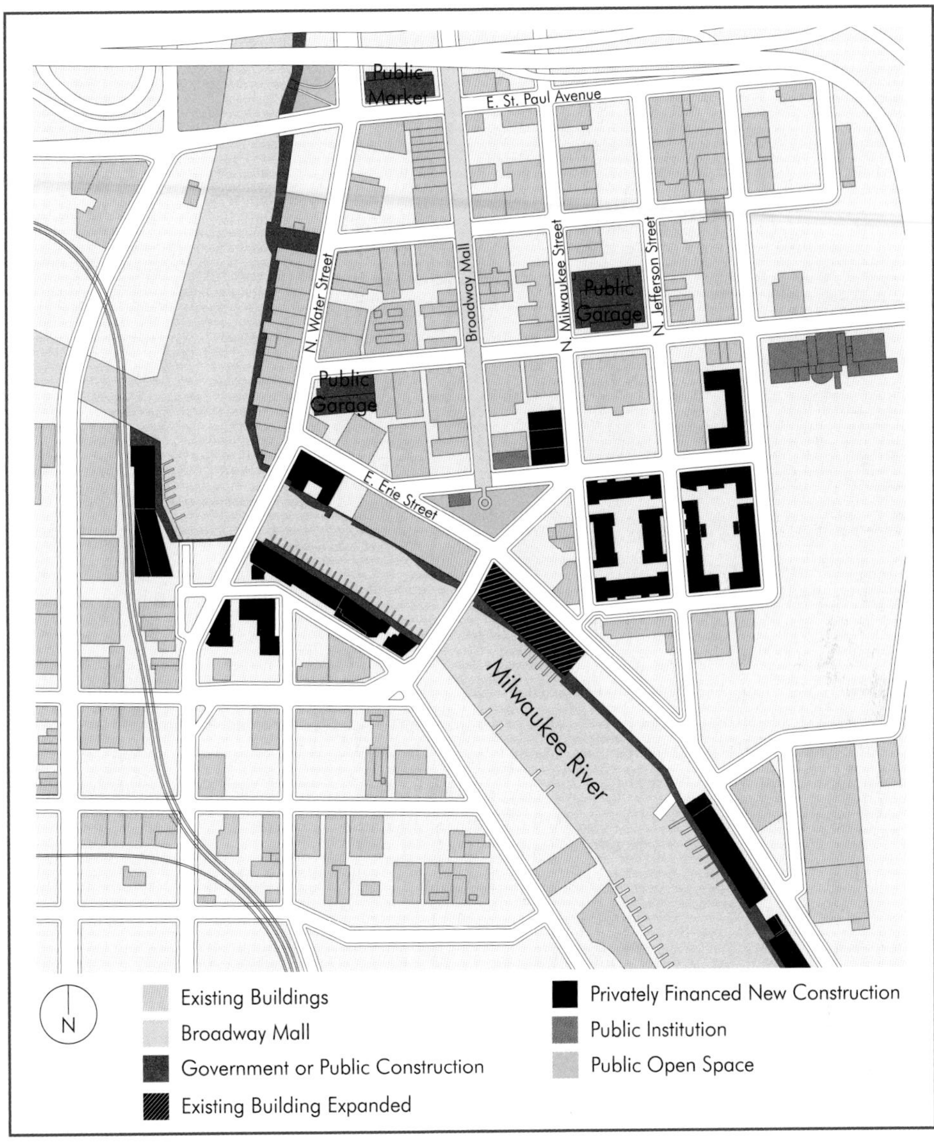

Milwaukee, 2012. Public investment in the Third Ward generated private real estate investment in the area. *(Alexander Garvin and Ryan Salvatore)*

decade later, more than six times that number occupied the area's many loft buildings. Moreover, the private market reaction extended beyond the boundaries of the Third Ward: hundreds more residents had moved into the new buildings erected just across the Milwaukee River in the Walker's Point Historic District.

In addition to the new residents, the Third Ward includes 400 businesses, shops, and restaurants, as well as 20 galleries and art studios, the Milwaukee Institute of Art and Design, the Off-Broadway Theatre, and the Broadway Theatre Center, which is home to the Skylight Opera Theatre and Milwaukee Chamber Theatre. As a result, the price of unimproved properties has escalated by 40 times its value 25 years earlier and total property assessment for the Third Ward has increased from $40 million to $450 million. This surely is an example of a widespread private market reaction to relatively modest public investments.

Portland's Pearl District

During the 1980s, the artists of Portland discovered cheap rents in deteriorating, partially vacant loft buildings just as the artists who moved into SoHo had made a generation earlier in New York. The same small café-restaurants that they enjoyed followed them, as did a few pioneering developers who wanted to repeat a similar successful loft conversion on Printers Row. There were, however, important differences. Although the district they discovered was about three-quarters of a mile from the center of Portland, unlike either SoHo or Printers Row, it did not have subway service.

Portland, 1998. Loft renovation began in the Pearl District at the end of the twentieth century. *(Alexander Garvin)*

Like Printers Row, it bordered on a vacant rail yard, and the privately owned buildings that had once depended on railroad access deterred investment there. But, unlike Chicago's Dearborn Station, there were no investors interested in creating a new community on the abandoned, 34-acre Burlington Northern Railroad property that was heavily contaminated (see the discussion of Dearborn Park in Chapter 15).[37]

Nobody knows exactly how this triangular, industrial district northwest of downtown Portland got its name. One explanation is that a resident of the area started to refer to it as Pearl's District in honor of a woman who liked to visit there because of hip people who were moving in. Another variant is that people thought "the old, crusty exteriors on the buildings [were] like the exterior of the oyster shell. But inside it's amazing: There are literally thousands of people inhabiting them, some illegally . . . not only painters and sculptors, but software-makers, wine distributors, poets and musicians."[38] That was late in the twentieth century, just before the city began making major investments in the area's public realm.

Public investment will not transform an industrial district into a residential neighborhood unless there already is enough market demand. A number of private developers perceived that market. They began acquiring obsolete loft buildings and converting them into condominiums. The first of these, the 27-unit Pearl Lofts Condominiums, opened in 1994, followed by Hoyt Commons, a 48-condo project. Once private developers had demonstrated a market, the city decided to accelerate the transformation of the neighborhood. In 1997, it entered into an agreement to acquire the rail yard and decontaminate the area. Two years later, the Portland Development Commission, the city's urban renewal agency, held a design competition for the property's reuse that was won by Peter Walker and Partners Landscape Architects. The plan that emerged after a couple of years of community meetings included the creation of two new parks on 200-foot-square

Portland, 2007. Public investment in Jamison Square created a recreational facility that is popular with both children and adults. *(Alexander Garvin)*

Portland, 2007. By the twenty-first century, the Pearl District had become a popular residential district. *(Alexander Garvin)*

blocks within the city's street grid and a 3-acre park between the district and the Willamette River. More important, starting in 2001 light-rail service on 10th and 11th Avenues connected the district with downtown Portland. The following year, the first of the two parks, Jamison Square, opened; the other block-size park, Tanner Springs Park, followed in 2005, and the groundbreaking for the Fields Park was held in 2012.

There was more public involvement with the Pearl District than in the Third Ward. The rail yard required excavation, decontamination, and safe disposal of debris. The capital cost and operating subsidies for transit service exceeded the one-time expenditure on Milwaukee's two public garages. The private investment in new construction, however, far exceeded anything in Milwaukee. Virtually the entire northern section of the Pearl District consists of new residential buildings. In addition, the district now includes more than 20 restaurants, cafés, and bars; two dozen art galleries; and 50 boutiques and retail stores. These metrics can be somewhat misleading, however. The greater amount of the private investment in the Pearl District is as much a reflection of the city's growth of the past half century as it is specific initiatives to revitalize the downtown. While Milwaukee's population declined from 637,000 in 1950 to 595,000 in 2010, Portland's grew from 374,000 to 584,000. The growth in its labor force is even more dramatic, increasing from 167,000 to 334,000 over the same 60-year period. What is clear, however, is that the city took proactive steps to ensure that its downtown would be appealing for both new city residents and significant growth in its metropolitan region. Growing market demand certainly makes neighborhood revitalization easier.

A Comprehensive Approach

It is ironic that the most successful neighborhood revitalization effort, that of Wooster Square in New Haven, Con-

Portland, 2007. Loft renovation in the Pearl District was so profitable that developers began erecting entirely new buildings. *(Alexander Garvin)*

necticut, is perhaps the oldest. Revitalization efforts in virtually every city discussed in this chapter occurred without knowledge of Wooster Square, and each time required reinventing the wheel. Moreover, with the exception of transportation-related investments, every technique used in these cities had been successfully pioneered in New Haven: insurance for banks that issued mortgages for housing rehabilitation; government-subsidized loans; capital spending for parks, landscaping, street improvements, and public buildings; spot clearance; modest infill new construction; and even the establishment of a local office to coordinate these activities.

Wooster Square, New Haven

Wooster Square is a 235-acre neighborhood centered around a lovely landscaped green created in 1825. Originally, it had been a resort fashionable enough to attract families from the South who traveled by boat from New Orleans and Charleston to spend their summers in New Haven. As the city grew, factories sprang up, and the resort was transformed into a working-class community—first primarily Irish- and later Italian-American. Single-family homes were converted into multifamily dwellings and rooming houses. Conditions deteriorated sufficiently that, in 1951, Wooster Square was one of nine areas identified by city planners as suitable for redevelopment.[39]

Maurice Rotival, consultant to the New Haven City Plan Commission during the 1940s and 1950s, advocated virtually complete clearance and construction of a heliport and an elevated highway to connect downtown New Haven to New York, Hartford, and Boston. In 1952, the Connecticut Highway Department released a plan for construction of a turnpike that would run through the western edge of the neighborhood

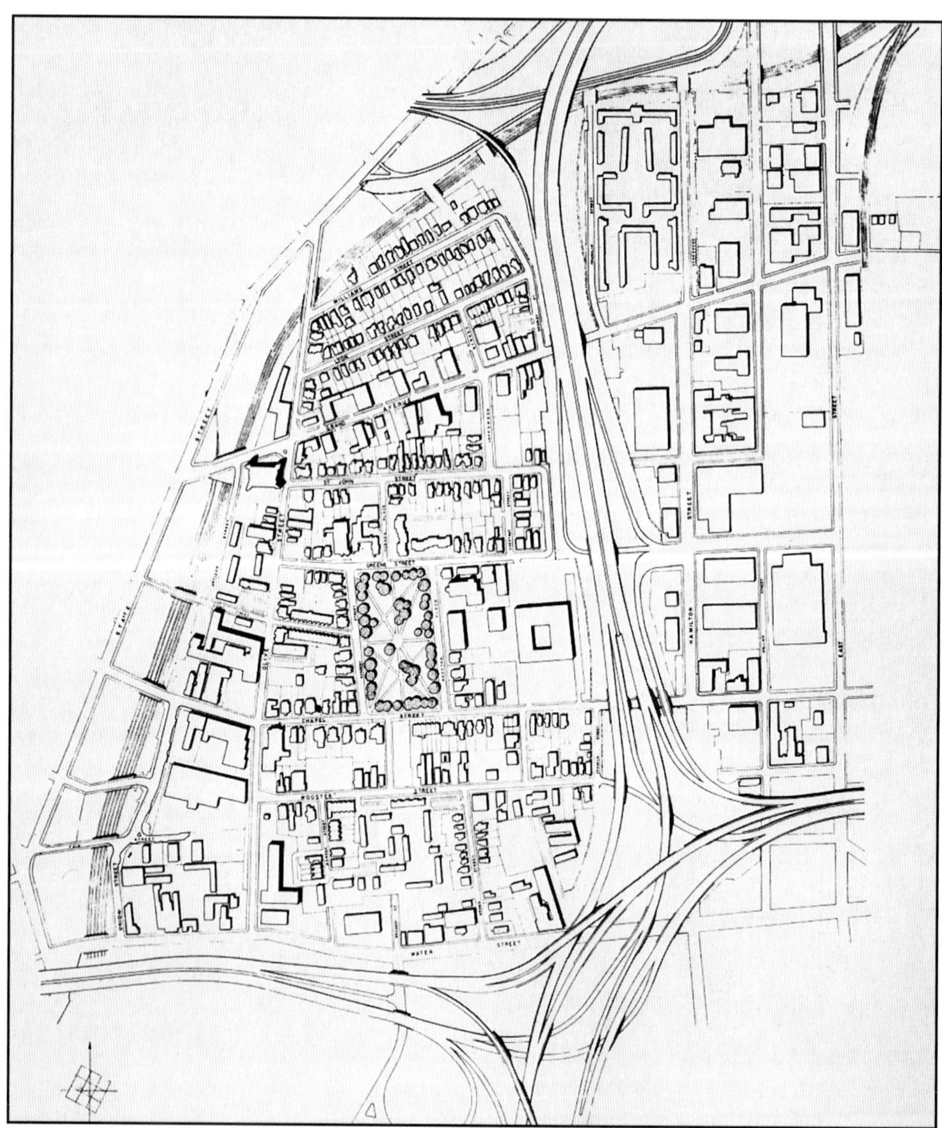

New Haven, 1965. Site plan for the Wooster Square Urban Renewal Area. *(Courtesy of City of New Haven Redevelopment Agency)*

New Haven, 1965. 10 Academy Street before renovation. *(Courtesy of City of New Haven Redevelopment Agency)*

New Haven, 2010. 10 Academy Street decades after renovation. *(Alexander Garvin)*

and forever separate it from downtown New Haven. The following year, the New Haven City Plan Commission published a different scheme that moved the highway (now Interstate Highway 91) several blocks east so that the area could still provide residences that were a short stroll from downtown. The revised plan also advocated complete clearance and redevelopment, proposing modern residential superblocks west of the highway and industrial parks on the other side.

Wooster Square's residents wanted neither superblocks with apartment towers nor superhighways. They liked their neighborhood and felt that all it needed was to be upgraded. They believed that once it was fixed up, those whom they considered undesirable would move elsewhere, and Wooster Square would again be a safe, comfortable place to live.

New Haven, 1965. Court Street before renovation. *(Courtesy of City of New Haven Redevelopment Agency)*

Few officials paid attention until 1955, when newly elected Mayor Richard C. Lee and his development director, Edward Logue, applied for Title I survey and planning funds for Wooster Square. After two years of active community participation in renewal planning, they proposed a strategy that included rehabilitation of 558 buildings, a new school, a new firehouse, numerous small parks and landscaped residential parking lots, 350 new street trees, and clusters of new low-rise housing. The approach was similar to that used in Society Hill in Philadelphia: spot clearance, scattered small-scale new construction, vest-pocket parks, and housing rehabilitation.

Many Wooster Square property owners were allocating a relatively small proportion of their income for housing and

New Haven, 2012. Court Street after renovation. *(Alexander Garvin)*

New Haven, 2010. Columbus Mall, one of the small-scale new construction projects in the Wooster Square Urban Renewal Project. *(Alexander Garvin)*

were ready to pay more for better accommodations. Their problem had been obtaining financing to pay for the improvements. Once Wooster Square was approved as a Title I project, however, mortgages became eligible for FHA 220 insurance. Consequently, banks began lending money for rehabilitation. By 1967, one decade after the plan had been unveiled, 353 residential structures had been or were in the process of being rehabilitated.

Public improvements qualified as noncash credits toward the city's one-third share of Title I project cost (see Chapter 7, note 14). Mayor Lee and Administrator Logue wanted to maximize these noncash credits, so they packed the Wooster Square Urban Renewal Project with every possible municipal improvement. Its new school and firehouse, which would have been built anyway, were counted as part of the local contribution. The small parks, street trees, and landscaped parking lots were simply the bonus that the neighborhood accrued as a result of clever financial planning.

New Haven, 2010. The Central Fire House, Wooster Square, was one of several non-cash credits New Haven received as part of the Wooster Square Urban Renewal Project. *(Alexander Garvin)*

Title I grants paid for a local office with a project director, a housing code inspector who could identify all essential repairs, an architect expert in property restoration, a rehabilitation specialist experienced in all areas of building practice, a mortgage adviser familiar with the intricacies of FHA processing and conventional bank financing, and a neighborhood representative skillful in working with property owners and neighborhood leaders. These specialists surveyed area buildings, prepared illustrative plans, made recommendations for property improvements, worked on mortgage applications, prepared lists of competent contractors, coaxed neighborhood residents into making minor improvements, cajoled property owners into major renovations, and smoothed the way with banks.

Within a decade of the project's approval, Wooster Square was again one of New Haven's most attractive neighborhoods. The Italian-American community remained. Its homes had been rehabilitated. Rooming houses had been reconverted to row-house apartments or single-family homes, and virtually all existing housing had been updated to modern standards. Ever since, it has remained one of the city's most desirable residential neighborhoods.

Ingredients of Success

Neighborhoods require periodic reinvestment. The money may not be available because the population is declining, because household income is declining, or both. Residents may be unwilling to spend additional money on property maintenance because they have other priorities. In those instances, code enforcement will be counterproductive. The money may not be available because institutional lenders have lost confidence in the neighborhood and are unwilling to lend money for rehabilitation (remedied in Charlotte by depositing money to be lent in a specific neighborhood). Creating a revolving fund to purchase and resell endangered properties (as happened in Ansonborough) may be sufficient. Private market activity can be triggered by carefully directed public investment in the public realm. In all of these instances, nonprofit organizations (like Historic Charleston) and government agencies (like New York City's neighborhood preservation offices or New Haven's Wooster Square Redevelopment Office) can play a role in changing the willingness of newcomers, existing residents, property owners, and banks to make such investments. Success, however, requires a strategy that is appropriate to the particular situation and an entity that can implement it effectively.

Market

Many cities have attractive but deteriorating neighborhoods. In the case of Dupont Circle, a new generation of homeowners discovered the neighborhood on their own. In other instances, government action may be needed to alter market conditions. Of course, if there is growing market demand, for a neighborhood like the Pearl District, it is difficult to avoid success. On the other hand, if, like Wooster Square, the area is losing population, government may be able to invest in public improvements that keep residents from moving away and even attract newcomers. If, as in Ansonborough, the area has already lost much of its market and needs to attract new residents, additional financial incentives may be needed. In all cases, however, successful revitalization programs will have to alter consumer perceptions and make the neighborhood competitive in price with its rivals.

The techniques for successfully altering market perception are the same for neighborhoods that need to retain residents as they are for neighborhoods that are seeking newcomers. Blighted structures must be renovated or, where necessary, replaced with attractive new residences that are compatible in size, scale, and character with the rest of the neighborhood. In both New Haven and Charlotte, pockets of blight were removed; once the physical condition had changed, so, too, did the negative impact that these dilapidated properties had on perceptions of the neighborhood. Targeted public investments, like the garages in the Third Ward, may be enough to alter market demand for buildings in the neighborhood that have no parking available.

There also must be visible evidence of investment. A neighborhood need not have special signs that designate it as being in a preservation area, as does Ansonborough. But streets and sidewalks must be kept in good repair and be well lighted. Street trees help, as do new park facilities. Without $8 million in municipal expenditures for street and sidewalk repaving and relighting, new vest-pocket parks, and new street trees, New Haven residents would have had little confidence in the continuing improvement of Wooster Square. Similarly, without the $4 million that Charlotte spent on repaving streets and sidewalks, creating and landscaping parkland, and installing new street signs and street lamps in the Fourth Ward, few city residents would have considered moving into the neighborhood. Investment in light-rail service to the Pearl District is perhaps the most cost-effective of the public expenditures described in this chapter. It increased market accessibility for everybody who worked in downtown Portland.

Reassurance can also come from visible investment in privately owned residential buildings. New York's Community Preservation Corporation requires that every rehabilitation mortgage include money for steam-cleaning building façades. In Ansonborough, Historic Charleston often restored building façades before putting houses up for sale and always put up prominent signs indicating that preservation was under way. Visible community investment, however, will not spur further rehabilitation if people are not willing or able to pay for it. A variety of subsidies can lower prices and thereby reach a larger market. Charlotte issued revenue bonds, which, because of their tax-exempt status, had a below-market interest rate. New York City allocated CDBG funds to be lent at 1 percent interest.

Location

Neighborhood revitalization can exploit locational advantages. Printers Row had the benefit of proximity to downtown business districts. Proximity, however, is less important than attractive housing stock. Wooster Square and Ansonborough had lovely wood frame houses. Washington Heights was filled with handsome masonry apartment buildings. There are areas, however, like the worst Lower East Side tenements described by Jacob Riis or the dilapidated shacks cleared for Techwood Homes in Atlanta, in which the housing stock had been so bad that it could not be made attractive enough to be part of a successful neighborhood revitalization program.

Successful revitalization strategies must be tailored to the area's specific characteristics. There is no point in targeting a program like FACE, which provides no assistance for multiple dwellings, to a neighborhood like Crown Heights, where most people live in multiple dwellings. In the same way, there was no reason to restrict activity in the Fourth Ward to preservation of the area's historic structures when the area's many empty lots called for a program that included scattered new construction.

Design

Different neighborhoods attract different residents, depending on the character of the building stock. SoHo, Printers Row, and the Pearl District appealed to artists, architects, and other professionals who needed lots of space but could not afford high prices. These migrants were attracted by brick walls, exposed beams, sleeping lofts, and unfinished materials that conventional households were unwilling to accept. Property owners were therefore initially able to attract them with raw space at low prices. Once the neighborhood was again filled with residents and shops, it developed a cachet. Lofts that artists had transformed into interesting residences were able to command prices that Printers Row pioneers were unable to afford.

The historic houses of Ansonborough attracted traditional families. Prior to the intervention of Historic Charleston, none of them would have moved in because they feared investing in a neighborhood filled with vacant and decrepit houses. The revolving fund, at very little expense, provided a method of quickly restoring building façades.

Fixing façades in Printers Row would have had as little effect as offering raw space in Ansonborough. The strategies used to revitalize these neighborhoods were successful because they were tailored to the character of the area, the demands of the market, and the design of the buildings. Copying a strategy that worked in one city neighborhood, without understanding the local building stock, is a recipe for failure.

Financing

Neighborhood revitalization is not possible unless lending institutions provide mortgage financing. Often, banks avoid involvement in deteriorating neighborhoods because lending money for rehabilitation is not as simple as underwriting new construction or issuing mortgages for the purchase of a one-family house. The process cannot be standardized. Buildings vary too much in design, condition, neighborhood setting, and marketability. In Charleston and Pittsburgh, preservation enthusiasts, community leaders, and public officials persuaded banks to provide mortgages for the historically significant, older housing stock in their cities. In Charlotte, they persuaded NCNB to sponsor the revitalization effort in one neighborhood, the Fourth Ward.

Some banks are willing to pioneer preservation areas. NCNB did so because it understood that the bank's success was "tied directly to the vitality of the economy where it does business [and that] if NCNB was to continue to thrive in Charlotte, the local economy must also continue to thrive."[40] Taking a risk on the Fourth Ward was quite profitable. NCNB became the prime construction lender in the neighborhood and obtained market interest rates on all these loans. The new residential market that was created also allowed it to become the neighborhood's leading permanent lender of conventional market-rate mortgages.

In New York, the city government established a corporation that insured institutional mortgages in neighborhoods designated for preservation by the City Planning Commission. New York's Community Preservation Corporation then concentrated on the designated neighborhoods, developed underwriting techniques that were appropriate to building conditions in those neighborhoods, and trained its staff to make the necessary loans.

Time

Changing a neighborhood takes time, and that means a long-term commitment. The FACE program was intended to last three years. In Crown Heights and many other neighborhoods, it was extended to five years—still not enough time to reverse the results of decades of deterioration. Public investment in the Pearl District began at the end of the twentieth century and continues to this day. We must, therefore, stop thinking of neighborhood revitalization as a temporary government function and make it an ongoing operation.

Revolving funds provided an initial spurt of housing renovation in Pittsburgh's Mexican War Streets. It took almost half a century, however, for the process of neighborhood revitalization to restore the district. Reversing years of deterioration usually takes at least as long as it took the area to decline. Nobody should begin such an effort unless they are prepared to persevere until the job has been accomplished.

Entrepreneurship

Neighborhood revitalization does not occur spontaneously. Somebody must own, renovate, and maintain the housing. In Charleston and Pittsburgh, local preservationists established nonprofit entities to purchase vacant and dilapidated buildings. Then they either sold the property to an owner who agreed to do the necessary work or they did the work themselves and then sold the property to a responsible new owner. The NCNB Community Development Corporation entered into joint ventures with local developers to purchase, build, and renovate houses and apartment buildings in the Fourth Ward.

One thing is certain: without a catalytic agent to start the ball rolling and generate the necessary level of activity, the only neighborhoods that will improve are the limited few like Washington's Dupont Circle, which improved on its own and needed no government assistance. The type of catalyst should reflect local conditions. In Ansonborough and the Fourth Ward, it was a nonprofit foundation. In Wooster Square, the urban renewal office performed this role. New York established and staffed neighborhood offices that reached out to building owners to offer them help in obtaining mortgage financing and then reached into the bureaucracy to speed processing by the city's own agencies.

In all cases, however, neighborhood revitalization requires vision and leadership. The developers who renovated the buildings in Printers Row knew that people would not move into new buildings next to vacant, rubbish-strewn lots or renovate dilapidated buildings unless the rest of the block was improved. Fortunately, Harry Weese and Larry Booth had the vision to insist that old loft buildings could become attractive residences, the willingness to convince the government to deal with tax delinquency issues, and the leadership skills to raise the money and find the contractors to renovate the first buildings in the area. Without their persistence, nothing would have happened.

Neighborhood Revitalization as a Planning Strategy

Many neighborhoods require virtually no public action. Property owners and residents do whatever is needed. They know what C. K. Chesterton understood: that "if you leave a white post alone, it will soon be a black post. If you particularly want it to be white, you must always be painting it again." That is the secret of neighborhood revitalization. The techniques, however, are no secret: tax policies that do not discourage property improvement, mortgage insurance, revolving acquisition funds, subsidized bank lending, capital investments in the public realm, and neighborhood offices to reach out to property owners and coordinate the application of all these techniques. They have been proven many times over in Chicago, Charleston, Charlotte, Pittsburgh, Milwaukee, Portland, New Haven, New York, and many other cities.

Continuing neighborhood deterioration only leads to expensive remedial action. The year before New York established its Neighborhood Preservation Program, it spent more than $28 million just for emergency-vacate relocation, unsafe-building demolition, and emergency repairs. Even if such remedial expenses are ignored, the replacement cost of the lost housing represents a capital investment larger than most cities will make over several generations. Adding to the cost of rebuilding the wide variety of lost community facilities, boarded-up stores, and abandoned institutions makes the price staggering. Duplicating the "community glue" that they provided takes far longer than just rebuilding them. Surely it is far cheaper and less time-consuming to spend a little now than vast sums over many years replacing whole neighborhoods.

The amazing thing about the failure to invest in neighborhood revitalization is that there is hardly any opposition to such expenditures. Suburban communities support such spending because it reduces pressure on the suburban market. Stable urban communities support it because it eliminates the threat of spreading blight. The only opponents are self-appointed advocates of the poor, who oppose gentrification and decry possible displacement sometime in the future when neighborhood deterioration has been reversed.

Displacement need not occur. As has been demonstrated in Washington Heights, existing low-income neighborhood residents who cannot otherwise afford renovated housing can be retained by targeting subsidies for exactly that purpose. Such directed expenditures are cheaper and less disruptive than allowing the neighborhood to deteriorate around residents who have to remain until unsafe conditions finally force them out.

There are many examples of spontaneous revival in loft districts and row-house neighborhoods. Much can be learned from these examples to help civic leaders devise successful revitalization strategies. Despite these successes, there remain deteriorating areas that cry out for attention. Public action has successfully revitalized older neighborhoods in cities as diverse as Charleston, Milwaukee, Portland, and New York City. These models can be adapted to similar situations throughout the country.

Notes

1. Jacob Riis, *A Ten Years' War: An Account of the Battle with the Slum in New York,* originally published in 1910 and reproduced in *Jacob Riis Revisited,* Francesco Cordasco (editor), Anchor Books, Garden City, NY, 1968, p. 301.
2. Walter Firey, *Land Use in Central Boston,* Harvard University Press, Cambridge, MA, 1947; Brian J. Godfrey, *Neighborhoods in Transition: The Making of San Francisco's Ethnic and Nonconformist Communities,* University of California Press, Berkeley, CA, 1988.
3. Langston Hughes, "My Early Days in Harlem," *Harlem: A Community in Transition,* J. H. Clarke (editor), Citadel Press, New York, 1964, p. 62.
4. Malcolm X, *The Autobiography of Malcolm X,* Grove Press, New York, 1964, p. 76.
5. Historical and statistical material on the South End is derived from Walter Firey, op. cit.; Langley Carleton Keyes Jr., *Rehabilitation Planning Game: A Study in the Diversity of Neighborhood,* MIT Press, Cambridge, MA, 1969; Robert B. Whittlesey, *The South End Row House and Its Rehabilitation for Low Income Residents, a* HUD-sponsored Report, Boston, 1969; and J. Anthony Lukas, *Common Ground: A Turbulent Decade in the Lives of Three American Families,* Vintage Books, New York, 1985.
6. Margot Gayle and Edmund V. Gillon Jr., *Cast-Iron Architecture in New York,* Dover Publications, New York, 1974; Margot Gayle and Robin Lynn, *A Walking Tour of Cast-Iron Architecture in SoHo,* Friends of Cast-Iron Architecture, New York, 1983; Margot Gayle, "America's Cast-Iron Heritage," in *Historic America: Buildings, Structures, and Sites,* Library of Congress, Washington, DC, 1983, pp. 159–182; James R. Hudson, *The Unanticipated city: Loft Conversion in Lower Manhattan,* University of Massachusetts Press, Amherst, MA, 1987.
7. Alexander Garvin, *Public Parks: The Key to Livable Communities,* W. W. Norton & Company, New York, 2011, pp. 155–156.
8. Karen Kollias with Arthur Naparstek and Chester Haskell, *Neighborhood Reinvestment: A Citizen's Compendium for Programs and Strategies,* National Center for Urban Ethnic Affairs, Washington, DC, 1977; Stephen A. Kliment (editor), *Neighborhood Conservation: A Source Book,* Whitney Library of Design, New York, 1975; and Real Estate Research Corporation, *Neighborhood Preservation: A Catalogue of Local Programs,* Office of Policy Research, Department of Housing and Urban Development, Washington, DC, 1975.
9. Neighborworks, www.nw.org/network/aboutus.
10. Find/SVP information services.
11. A third FACE area, Highbridge in the Bronx, received HUD funding from 1971 to 1973. Its housing stock was as inappropriate for code enforcement and Section 312 mortgages as was that of the East Concourse.
12. Historical and statistical material on Crown Heights is derived from Edward A. Gibbs (assistant commissioner of FACE), *Crown Heights Federal Code Enforcement Program, March 7, 1967–March 6, 1972,* City of New York Housing and Development Administration, New York, 1972, and Alexander Garvin (project director), *Crown Heights Area Maintenance Program (CHAMP),* vols. 1–3, New York City Planning Department, New York, 1972.
13. Between 1943 and 1971, when the New York state legislature instituted a "maximum base rent" formula permitting annual 7.5 percent rent increases, owners of continuously occupied rent-controlled apartments in New York City were permitted only two across-the-board rent increases—15 percent in 1953 and 8 percent in 1970.
14. For the evolution of the city's commitment to preventive renewal, see New York City Planning Commission, *New York City's Renewal Strategy/1965,* New York City Community Renewal Program, New York, 1965; *Between Promise and Performance . . . ,* New York City Community Renewal Program, New York, 1968; and *Plan for New York City,* vol. 1: *Critical Issues,* New York City Planning Commission, New York, 1969, pp. 138–143.
15. Garvin (project director), op. cit., vol. 3, p. 21.
16. Violations per se are meaningless. A building with recorded violations, even a myriad of violations, may be in very good condition. For example, failure to have the proper frame around a required document is a violation; so is a nonfunctioning boiler. One has no impact on living conditions; the other is fundamental.
17. Housing inspectors could only enter one- or two-unit buildings in which their visit had been authorized by the owner-occupant, a complaint had been filed, or a warrant had been issued.
18. Garvin (project director), op. cit., vol. 1, pp. 41–43.
19. Historical and statistical material on the Fourth Ward is derived from Christine Madigan, "The Revitalization of Fourth Ward, Charlotte, North Carolina," unpublished, 1990; James Dennis Rash, "Privately Funded Redevelopment in North Carolina," *Urban Land,* Urban Land Institute, Washington, DC, October 1983, pp. 2–7; Friends of Fourth Ward, *A Walk Through Historic Fourth Ward,* Loftin & Company, Charlotte, NC, undated; M. S. Van Hecke, "Cheap Loans Ending, But Fourth Ward Survives," *Charlotte Observer,* April 12, 1981; and correspondence and conversations with James Dennis Rash, October and November 1990.
20. U.S. Department of Commerce, Bureau of the Census, Statistical Abstract of the United States, 1978, p. 24, and Statistical Abstract of the United States, 1989, p. 33.
21. Historical and statistical material on Ansonborough is derived from Arthur P. Ziegler Jr., Leopold Adler II, and Walter C. Kidney, *Revolving Funds for Historic Preservation: A Manual of Practice,* Ober Park Associates, Inc., Pittsburgh, PA, 1975, pp. 56–61, and the Historic Charleston Foundation.
22. Historical and statistical information about Santa Barbara is derived from Ziegler, J Adler, and Kidney, op. cit.; mexicanwarstreets.org; www.mattress.org; and www.thenorthsidechronicle.com.
23. Ziegler, J Adler, and Kidney, op. cit., p. 78.
24. Ibid., p. 79.
25. Some will argue that this is neighborhood improvement without gentrification that can be emulated in other areas that wish to avoid forcing out lower-income residents. Indeed, many largely poor, African-American populations continued to live in the area, well into the twenty-first century.
26. Alexander Garvin (project director), *Neighborhood Preservation in New York City,* New York City Planning Commission, New York, 1973, pp. 49–82, 106–145.
27. The Neighborhood Preservation Program was made operational while Roger Starr was housing commissioner, during the mayoralty of Abraham Beame, and it became a major production program under Housing Commissioner Nathan Leventhal when Edward I. Koch was mayor.
28. REMIC insures qualified portions of first-mortgage loans by publicly regulated financial entities (i.e., savings banks, commercial banks, savings and loan associations, insurance companies, and pension funds). To be eligible for REMIC mortgage insurance, loans must be made to apartment buildings erected after 1901 within designated neighborhood preservation areas. They also must be at an interest rate not in excess of the ceiling imposed by the state banking board and be self-amortizing over terms ranging from 10 to 30 years. The insurance covers losses up to 90 percent of loans on the outstanding principal indebtedness incurred by rehabilitation and 20 percent of

the outstanding principal indebtedness on funds used to refinance existing debt or to finance acquisition. However, in no case can the insurance exceed 50 percent of the total of both.

29. http://www.communityp.com.
30. New York City Rehabilitation Mortgage Insurance Corporation (statistics cover the period from inception through October 31, 1990).
31. In 1979, the program was expanded to cover 13 neighborhoods.
32. When the city's fiscal crisis was over, during the 1980s, the Koch administration resumed using city capital funds for housing rehabilitation.
33. New York City Planning Commission, *Community Planning District Profiles*, New York City Planning Commission, New York, 1973.
34. Barbara Leeds, New York City's Neighborhood Preservation Program in Washington Heights, unpublished master's thesis, New York, 1981.
35. Community Preservation Corporation (statistics cover the period from inception through April 1993); New York City Department of Housing Preservation and Development (statistics for voluntary repair agreements cover the period from January 1975 through June 1989; these figures include repeat agreements over this 15-year period, so the actual number of buildings and apartments affected is lower than these totals).
36. Historical and statistical information on the Third Ward is derived from Megan E. Daniels, *Milwaukee's Early Architecture*, Acadia Publishing, Charleston, SC, 2010, and http://www.historicthirdward.org.
37. Historical and statistical information on the Pearl District is derived from Christopher S. Gorsek, *Portland's Pearl District*, Acadia Publishing, Charleston, SC, 2012; George Hazelrigg, "Peeling Back the Surface," *Landscape Architecture*, vol. 96, no. 4 (April 2006), Washington DC, pp. 112–119; http://www.pearldistrict.org; and www.inthepearl.com/Home/PearlDistrictHistory.
38. Margie Boule, see http://www.oregonlive.com/portland/index.ssf/2002/04/pearl_districts_namesake_was_a.html.
39. Historical and statistical information on Wooster Square is derived from Mary Hommann, *Wooster Square Design*, New Haven Redevelopment Agency, New Haven, CT, 1965; New Haven Redevelopment Agency, *1967 Annual Report*, New Haven Redevelopment Agency, New Haven, CT, 1967; and Alan Talbot, *The Mayor's Game*, Harper & Row, New York, 1967, pp. 107–109, 136–147.
40. James Dennis Rash, op. cit., p. 3.

14

Residential Suburbs

Tennessee, 2006. Collierville Tennessee, subdivision. *(Alexander Garvin)*

Americans want and can afford the privacy, security, and independence of homeownership. "To possess one's own home is the hope and ambition of almost every individual in our country, whether he lives in hotel, apartment house, or tenement. . . . Those immortal ballads, Home Sweet Home, My Old Kentucky Home . . . were not written about tenements or apartments . . . they never sing songs about a pile of rent receipts."[1] That is why two-thirds of the country lives in a home of their own, mostly in the suburbs. Yet many observers dislike the suburban landscape where those homeowners live. They think of suburbanization as a chaotic process that squanders both land and money.

James Rouse, the developer of Columbia, Maryland, is eloquent in his denunciation:

> *Relentlessly, the bits and pieces of a city are splattered across the landscape. By this irrational process, non-communities are born—formless places without order, beauty, or reason; with no visible respect for people or the land.*[2]

The alternative that he and many others have proposed is the planned new community. Although it is easy to believe that Rouse is correct when looking at some parts of the American landscape, a closer look reveals a different reality. As urban historian Kenneth Jackson explains in *Crabgrass Frontier*:

> *The theory that early suburbs just grew, with owners "turning cowpaths and natural avenues of traffic into streets," is erroneous. . . . Each city and most suburbs were created from many small real estate developments that reflected changing market conditions and local peculiarities.*[3]

These small suburban real estate developments grow out of a very rational human impulse: the attempt to escape the worst aspects of city life while simultaneously settling into an improved living environment. The combination of the *push* away from the city with the *pull* of something better is as old as city life itself. Pliny the Younger, writing nearly 2000 years ago about the commute from downtown Rome to his suburban home, tells a story that is repeated daily by tens of millions of Americans:

> *The place . . . is situated seventeen miles from [the city], so that after . . . having passed a constructive day, you come here to stay. It may be approached by more than one road. . . . [The roads] are difficult and long. . . . In one area the road is hedged in by woods and in another it opens up and spreads out in broad meadows.*[4]

Westwood, Los Angeles, 2004. Suburban residences "approached by more than one road . . . hedged in by woods . . . [and] broad meadows" [Pliny the Younger]. *(Alexander Garvin)*

In early twenty-first-century America, the commute is through a landscape that is filled with houses. The woods and meadows are gone because every American wants his or her own little bit of nature: a green lawn sweeping up to a house nestled among the trees and flowers, with blue sky and drifting clouds above—a vision that is not easily available in noisy, dirty, congested cities.

The concept of the suburb may go back to Roman times, when it simply meant settlements beyond the walls of the city. It has continued century after century since then, whether it was referred to as a *suburb* in England, *faubourg* in France, or *vorstadt* in Germany, for example. Until the nineteenth century, whenever a new wall was built farther out, what had been a suburb became an integral part of the city. City walls are not of major importance in the evolution of American cities. Thus, it seems reasonable to call everything built at a fairly low density *suburban*. This functional definition ignores municipal boundaries. However, whether a community is or is not within the political jurisdiction of a central city explains very little about its quality of life.

In America, profit-motivated developers have been a major factor in supplying this market for two centuries. They acquire relatively inexpensive land and hold it until they can sell lots to home builders at a price that covers the costs of carrying the property (e.g., maintenance, taxes, and debt service), subdividing it into building lots, and installing the necessary infrastructure (streets, water mains, drainage pipes, sewers, and utility lines), and also provides a return on equity that justifies their time, effort, and risk.

Shrewd developers often profit from cheap land in outlying areas by exploiting changes in transportation technology. In the nineteenth century, many developers invested simultaneously in land and the ferry, railroad, and mass transit systems that connected their land holdings with center cities. In the twentieth century, outlying territory was made accessible by government-financed highways. Whatever the transportation technology, once their property is sufficiently accessible to the growing market for new homes, developers profit from their investment by selling to home builders.

Another method of profiting from suburbanization is by supplying building lots that include amenities not available at other locations. Some developers market gated communities that underscore an area's privacy and special character. Others offer swimming pools, golf courses, or tennis courts.

It used to be that local governments determined the location and character of suburbanization by installing development infrastructure prior to development. This altered the sequence of suburbanization because, all other factors being even, developers favored sites with infrastructure already installed rather than land that still needed major investment. Not only have many local governments abandoned their responsibility for infrastructure, they also have abdicated their responsibility for providing public facilities, public transportation, access to nature, and places for community interaction. The only truly public component of the suburban landscape, its traffic arteries, is usually a confusing hodgepodge. Despite the widespread abandonment of active planning by local government, small, suburban real estate developments are not now and never have been "splattered across the landscape." They have to meet community standards and be approved by government agencies. Nor are they "formless." Although suburbs incorporate a wide range of combinations, they reflect three primary shapes: *rectilinear, curvilinear,* and (more recently) *clustered*. The rectilinear and curvilinear suburbs open onto public streets. Houses in the clustered suburb front onto open spaces (golf courses, clubhouses, playgrounds, swimming pools, etc.) that are intended for the occupants of the houses that surround them. All three can provide attractive living environments. Unfortunately, when they are combined, the resulting landscape can be quite disappointing, for there is rarely any overall pattern to provide a means of orientation.

We can change this pattern of suburbanization by giving it a public realm framework that will provide both residents and visitors with a powerful means of orientation, establish a series of places for community interaction, and lessen the fragmentation of suburban life. This can be accomplished in newly developing areas by requiring individual developers to set aside for *public* use not just roadways but also open space. Private yards and commonly used facilities would not qualify. It would have to be new open space available for general public use. A different open-space framework is needed for existing suburbs. It can be established by acquiring leftover and underutilized property and combining it into a continuous system of public places. Both prescriptions require local legislation that is based on an informed understanding of the three principal varieties of suburban development: rectilinear, curvilinear, and clustered.

Rectilinear Subdivisions

Some cities, such as Philadelphia, Chicago, and Detroit, expanded by filling in a preestablished plan. In states where settlement conformed to the rectilinear land surveys of the Northwest Ordinance or the Homestead Act, cities simply continued the legislated street grid. In other cases, they grew either by extending the existing street pattern to new territory or, like St. Louis, New Orleans, and Atlanta, by starting a new one, parallel and perpendicular to such topographic features as a bend in the river, a cliff, or a railroad line. Still others grafted rectilinear subdivisions onto existing regional roads. In every instance, expansion had to satisfy two needs: circulation and lot sales. The only requirements were binding rights-of-way and legal documents that could be used to

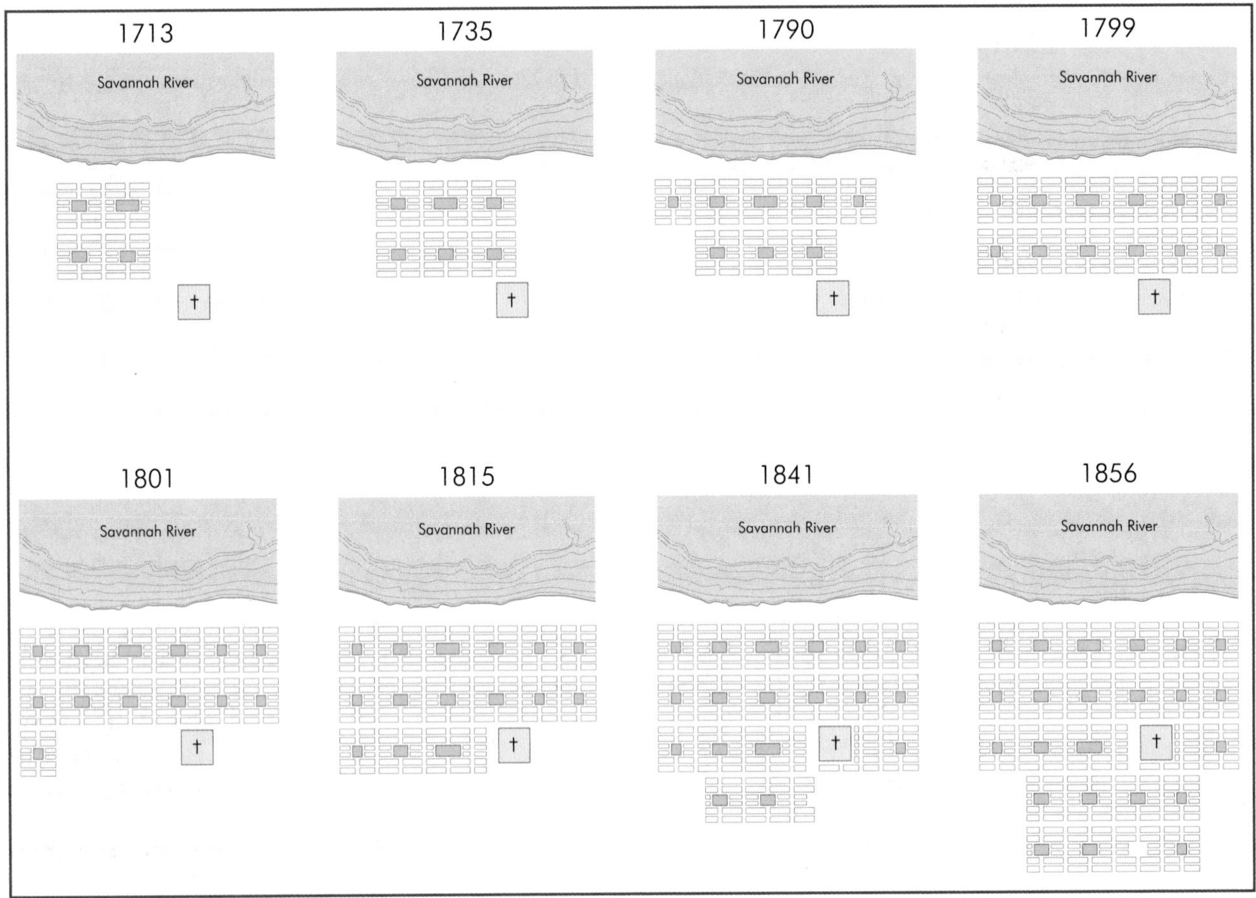

Savannah, 1733–1856. Suburban extension proceeded ward by ward in response to market demand.
(Alexander Garvin and Ryan Salvatore)

establish property boundaries and record ownership, easements, liens, and title to the land.

Savannah

Savannah is a unique example of a more civilized approach to suburbanization. Until the Civil War, each time population pressures required the opening up of new territory, the city fathers simply extended James Oglethorpe's initial grid by adding another ward. Every ward was centered on a landscaped public square bounded by eight rectangular blocks (see Chapter 3). Four of these blocks were set aside for churches, schools, or other public buildings. The other four were each bisected by a service alley and divided into ten 60- by 90-foot house lots.[5]

In 1856, when Savannah abandoned this approach to expansion, there were 26 such wards. Thereafter, developers accommodated new buildings by subdividing and recombining lots within existing wards (often replacing one-family homes with attached row houses) or by further extending a rectilinear pattern of streets and blocks, but without landscaped squares.

Brooklyn Heights and Prospect Park South

Brooklyn Heights was America's first commuter suburb served by public transportation. In 1819, when developer Hezekiah Pierrepont first advertised lots for sale, Brooklyn was an agricultural hinterland inhabited by no more than a few thousand people. They were clustered around a village that had only been chartered three years earlier. The lots were a short walk from the village. More important, they were directly across the East River from New York City, which already had a population of nearly 123,000.[6]

Hezekiah Pierrepont was a financial backer of Robert Fulton, the man who built and operated America's first commercially successful steamboat and who, in 1814, established the world's first steamboat ferry service. The steam ferry took a mere eight minutes to ply the waters between what is today Fulton Street in lower Manhattan and Fulton Street in Brooklyn. Pierrepont owned 60 acres of relatively cheap agricultural land, with a spectacular view of New York harbor. By investing in the ferry, he transformed this Brooklyn acreage into valuable residential real estate.

New York City from Brooklyn Heights, c. 1823. Hezekiah Pierrepont subdivided the undeveloped land in the foreground to create America's first suburb. *(Courtesy of I. N. Phelps Stokes Collection, Miriam and Ira D. Wallach Division of Arts, Prints and Photographs, The New York Public Library Astor, Lenox and Tilden Foundation)*

New York from Brooklyn Heights, c. 1836. In less than two decades, the eight-minute steamboat ferry ride between Brooklyn Heights and Manhattan had transformed Brooklyn Heights into a popular commuter suburb with houses lining the shore in order to benefit from the view. *(Courtesy of I. N. Phelps Stokes Collection, Miriam and Ira D. Wallach Division of Arts, Prints and Photographs, The New York Public Library Astor, Lenox and Tilden Foundation)*

Brooklyn, 2012. The distinctive posts designating the Prospect Park South community established its identity and underscored its desired privacy. *(Alexander Garvin)*

Pierrepont's plat for Brooklyn Heights consisted of 50-foot-wide streets and 200- by 200-foot blocks subdivided into 25-foot-wide house lots. His advertisements described "a place of residence combining all the advantages of the country with most of the conveniences of the city . . . *for a summer residence, or the whole year.*"[7] The first houses were built on sites made up of several lots. Within decades, many of these sites had been subdivided and filled in with masonry row houses. Today, when many row houses in turn have been converted to multiple-occupancy dwellings and others have been replaced by apartment buildings, it is hard to remember that Brooklyn Heights was once a low-density commuter suburb.

Prospect Park South is another Brooklyn subdivision developed for the express purpose of exploiting mass transit. The Flatbush Avenue trolley and the BMT subway, which had only recently extended across the Brooklyn Bridge into Manhattan, provided the opportunity.[8]

In 1899, real estate developer Dean Alvord purchased 50 acres just south of Prospect Park (see Chapter 3) where he, architect John Petit, and landscape gardener John Aitkin created a charming community in which "wife and children in going to and fro are not subjected to the annoyance of contact with the undesirable elements of society."[9] Privacy was emphasized at the entry points by brick piers decorated with a monogram formed from the letters *PPS*. Exclusivity was underscored by giving streets British names such as Albemarle, Argyle, and Buckingham. Protection from "undesirable social and moral influences" was provided by restricting construction to one-family houses costing more than $5000. Consistent design was guaranteed by deed restrictions requiring every lot to have a minimum street frontage of 50 feet, every house to be sited not less than 5 feet from its north lot line, and every yard to be open and unfenced.

Prospect Park South provides a park-like environment within the constraints of a rectangular street grid. Aitkin created beautifully landscaped islands in the middle of Buckingham and Albemarle Roads. He planted shrubs to conceal the subway line. Because trees stand at the front of each house lot rather than at the street curb, residents are less aware of the proximity of their neighbors. Visitors passing through think the tree-lined roadways of the district are wider than conventional streets—an illusion created by the landscaping.

When Pierrepont pioneered Brooklyn Heights, his customers could not have imagined that an international metropolis would soon engulf the houses they built on the edge of rural Long Island. Eighty years after Pierrepont had established his subdivision, Alvord's customers understood this only too well. To guarantee homeownership in an environment with privacy and a bit of nature, he had to do more than just provide a street grid within a short distance of a transit line. For this reason, he provided dedicated open-space islands and deed restrictions specifying the location and type of building for every lot. In the process, Alvord ensured that Prospect Park South would remain an enclave of one-family houses that would not be torn down for the construction of more profitable multiple dwellings.

The Garden District, New Orleans

In America, as in Europe, a common pattern evolved: as cities grew and attracted migrants seeking their fortunes, more prosperous residents sought green refuge in the suburbs. This desire for green refuge was particularly evident in New Orleans. In 1721, Leblond de la Tour and Adrian de Pauger established the plat of what is now known as the Vieux Carré (see Chapter 18). In 1800, when the plat had been filled in, *faubourgs* began to develop for French and Creole settlers and American newcomers alike.[10]

By 1830, the population of New Orleans had grown to 46,000, making it the fifth largest city in the country. The Anglo-Americans who were to become a solid majority by the 1850s were dominated by a growing number of energetic, opportunistic businessmen who made their fortunes in sugar, cotton, banking, insurance, shipping, wholesaling, and retailing. These increasingly affluent businessmen envisioned a home life surrounded by well-kept yards filled with greenery, to which they retreated after a busy day at work downtown. They were a ready market for a group of New Englanders led by Samuel Jarvis Peters, who in 1832 purchased a plantation on the suburban outskirts of the city, 2 miles south of the

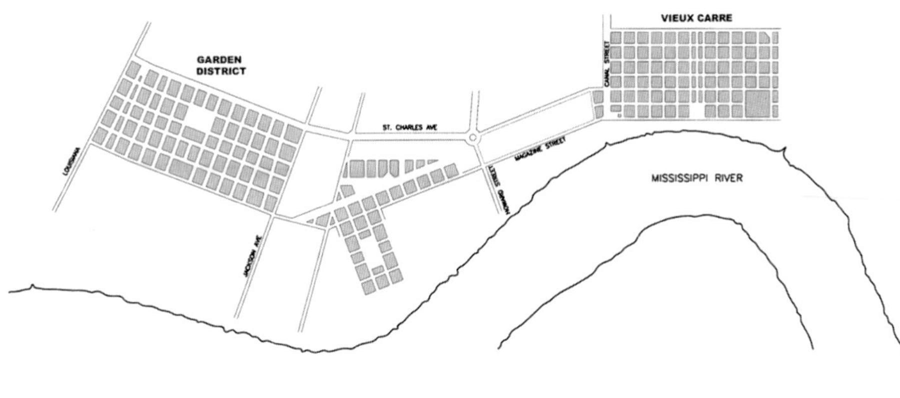

NEW ORLEANS – 1832

New Orleans, 1832. The Vieux Carré and initial plat of the Garden District. *(Alexander Garvin and Eric Clough)*

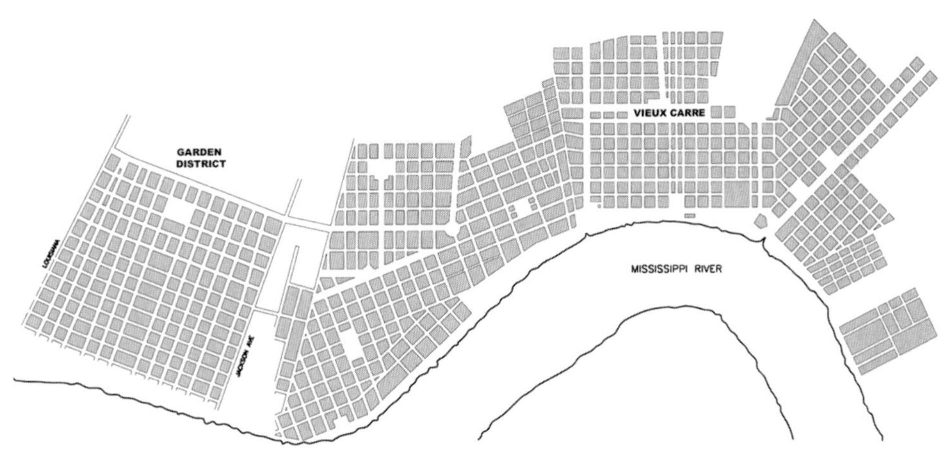

NEW ORLEANS – 2001

New Orleans, 2001. The Garden District, like all the faubourgs, has merged with the rest of the city. *(Alexander Garvin and Eric Clough)*

Vieux Carré. Peters hired Benjamin Buisson, who had served as an artillery officer under Napoleon, to lay out the new suburb, known today as the Garden District. Within one month Peters had subdivided and auctioned all the lots on the 66 blocks Buisson had laid out.

Many of the lots were purchased by speculators, who resold them to their ultimate occupants. Other purchasers held on till the Panic of 1837 forced them to sell. The houses, all of which were surrounded with greenery, were predominantly of three types: center hall villas, side-hall town houses, and center-hall two-story houses. The earliest of these were modest structures. By the middle of the nineteenth century, however, new houses had become quite grand.

New Orleans, 1998. The Garden District was one of America's first suburbs to provide a sylvan retreat from the noise and confusion of the city. *(Alexander Garvin)*

The Garden District was one of the earliest American suburbs to establish the model that was to be repeated ever after. City residents with enough money moved to a garden district, like the one in New Orleans, where they could live comfortably among a circle of people who shared their dream of an oasis on the edge of town.

The Private Places of St. Louis

St. Louis and Savannah are among the few American cities in which landscaped open space was central to the extension of suburban subdivisions. In St. Louis these subdivi-

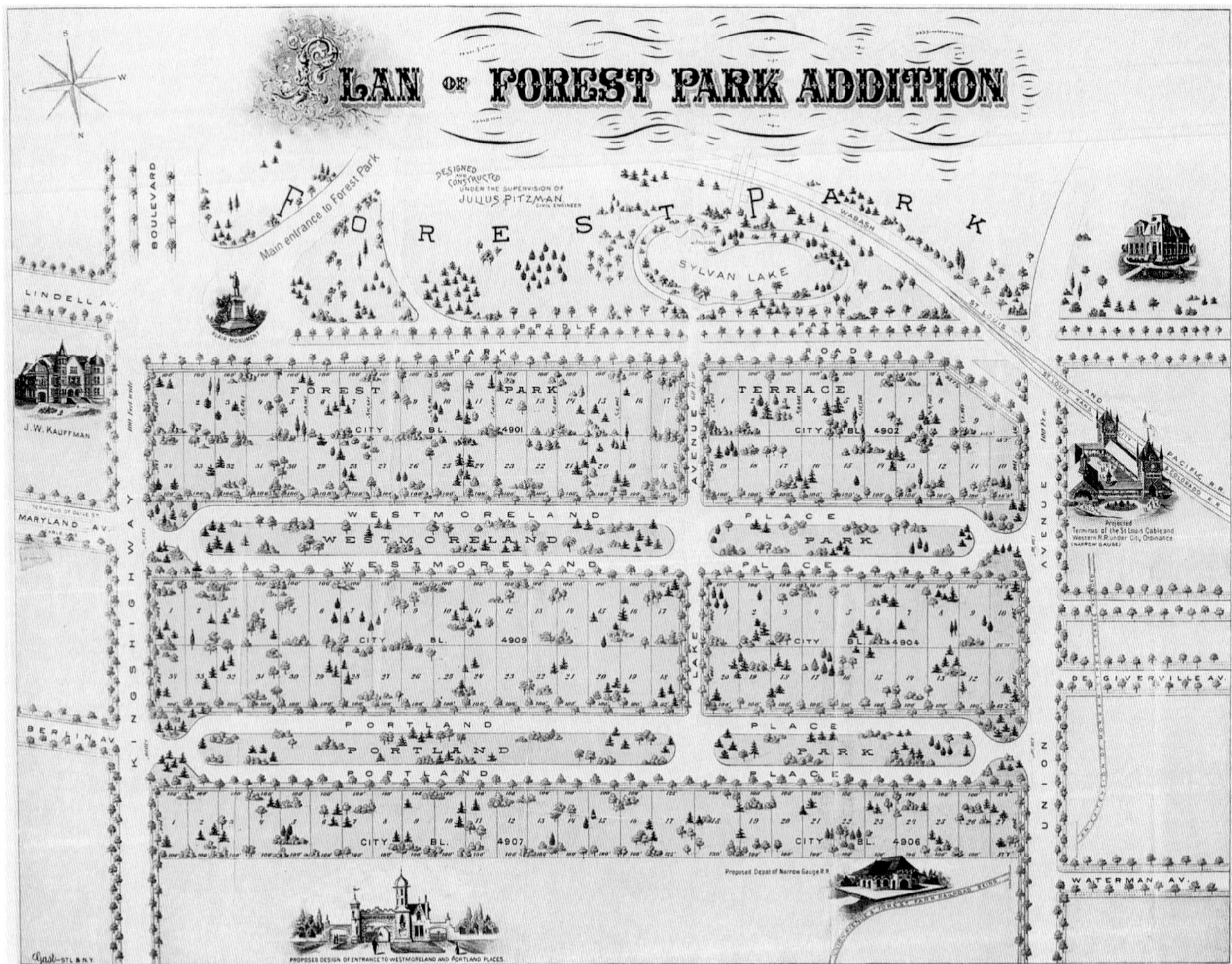

St. Louis, 1887. Prospectus plan of Forest Park additions (Westmoreland and Portland Places) showing the park islands that provided added privacy with landscaping that blocked the view of houses across the street. *(Courtesy of Special Collections, John M. Olin Library, Washington University in St. Louis)*

sions are called "private places" because, unlike Savannah's wards, they are private real estate ventures. The streets and park islands are owned and maintained by surrounding lot owners, not by the city. To emphasize this, most of the private places are defined by ornamental gates that also limit through traffic.[11]

Starting with Lucas Place in 1851, developers created more than 50 private places, mostly designed by city surveyor Julius Pitzman. The first private places were within walking distance of the business district. As the city became an increasingly important industrial metropolis, those citizens with enough money to select among different residential locations chose to move away from downtown congestion, noise, and grime. This move was accelerated during the 1880s by the introduction of mass transit. Whereas St. Louis had extended westward from the Mississippi River for a distance of only 1.5 miles when all transport was pedestrian-based, the trolley (well known to fans of the movie *Meet Me in St. Louis*) allowed settlement to extend along radial streetcar lines for a distance of about 6 miles. Developers lost no time in creating new private places on this newly accessible, cheaper suburban land.

While several private places include curving roads that adjust to topographical conditions or political boundaries, most are extensions of the city's rectilinear street pattern. Landscaped islands, 30 to 50 feet wide and hundreds of feet long, usually divide streets that pass through them. Residents of the houses that front on these park islands have the illusion that they live opposite a garden because the island's landscaping usually blocks the view of the houses facing their own.

The most impressive of these private streets were initially conceived in 1887 when the Forest Park Improvement

St. Louis, 2010. The ornamental gate of Flora Place establishes it as private property and limits through traffic. *(Alexander Garvin)*

St. Louis, 2010. The park island in the middle of Westmoreland Place provides greenery and privacy from residents across the street. *(Alexander Garvin)*

Association purchased a 78-acre site bordering the recently opened Forest Park (see Chapter 3). As usual, Julius Pitzman was hired to create a subdivision that was identifiable by the distinctive gatehouses at its entry. He laid out two private streets that were divided by park islands and connected by a cross street. The blocks were subdivided into large lots, 100 feet wide and 195 feet deep.

Within a year, the venture was sold to a syndicate of prominent business leaders who attracted other prominent and wealthy St. Louis residents to move there. They separated the property into two entities, Westmoreland Place and Portland Place, each with its own street association to which all property owners were required to belong and each of which was subject to property restrictions to which all property owners had to conform. The regulations restricted land use to residential purposes (but not boardinghouses), permitted only one house per lot, required all buildings to be set back 40 feet from the front lot line, obliged owners to spend at least $7000 in building their houses (none was built for less than $25,000), forbade fences or walls within the front 40 feet of any lot, prohibited the use of bituminous coal, and required association approval of all architecture. Changes to property regulations could be enacted only with unanimous consent of all the lot owners.

The only private places that have not survived are the first three. When their deed restrictions expired they were engulfed by St. Louis's rapidly expanding business district. At that time, land had become more valuable for nonresidential purposes. Consequently, these properties were sold. Subsequent private places have not faced this problem because their deed restrictions are self-perpetuating, thereby preventing use for anything but one-family houses. Indeed, since their initial development, Westmoreland Place and Portland Place have persisted as sites for some of the finest houses in St. Louis. They have retained their exclusive character because they are near Forest Park, offer an easy commute downtown, include distinctive gatehouses and landscaped park islands, are divided into unusually large lots, and observe deed restrictions to prevent intrusive development.

Back Bay, Boston

The government of Boston, unlike those of Brooklyn and St. Louis, chose to withdraw the process of suburban extension from the hands of private developers. This was not because of a belief in state planning or opposition to private land development. Rather, it was because pollution had become so serious that there simply was no alternative. The city could not wait for a developer who could raise the huge sums of money and take the extraordinary risks involved in reclaiming hundreds of acres that were quickly becoming a festering swamp west of Beacon Hill and the Public Garden.[12]

During the decade between 1840 and 1850, Boston's population increased from 85,000 to 114,000. Residents were confined to a small, hilly peninsula connected to the rest of Massachusetts by a narrow neck of land that is today Washington Street. The situation was somewhat improved as hills were gradually leveled and the resulting fill used to create developable property in small and isolated locations, but conditions became immeasurably worse during the second third of the nineteenth century when railroad lines cut off drainage in the Back Bay.

In 1814, the Boston and Roxbury Mill Company had been given riparian rights to more than 450 acres of the Back Bay. During high tide, the area filled up with brackish water. When the tide changed, the water drained back into the Charles River. Taking advantage of this tidal pattern, the Mill Company built a dam along the tidal Charles River. When the tide changed, the flowing water provided power for more than 80 mills. In 1834, the Commonwealth of Massachusetts authorized the construction of two railroad causeways crossing the Back Bay. These new rail lines impeded the flow of water, which resulted in litigation over the diminution in the Mill Company's ability to generate power.[13] More consequentially, they created a significant health menace because there was no longer a way for water to flow out of the bay. As a result, the Back Bay (into which city sewage flowed) became a stagnant swamp used by citizens as a dumping ground for

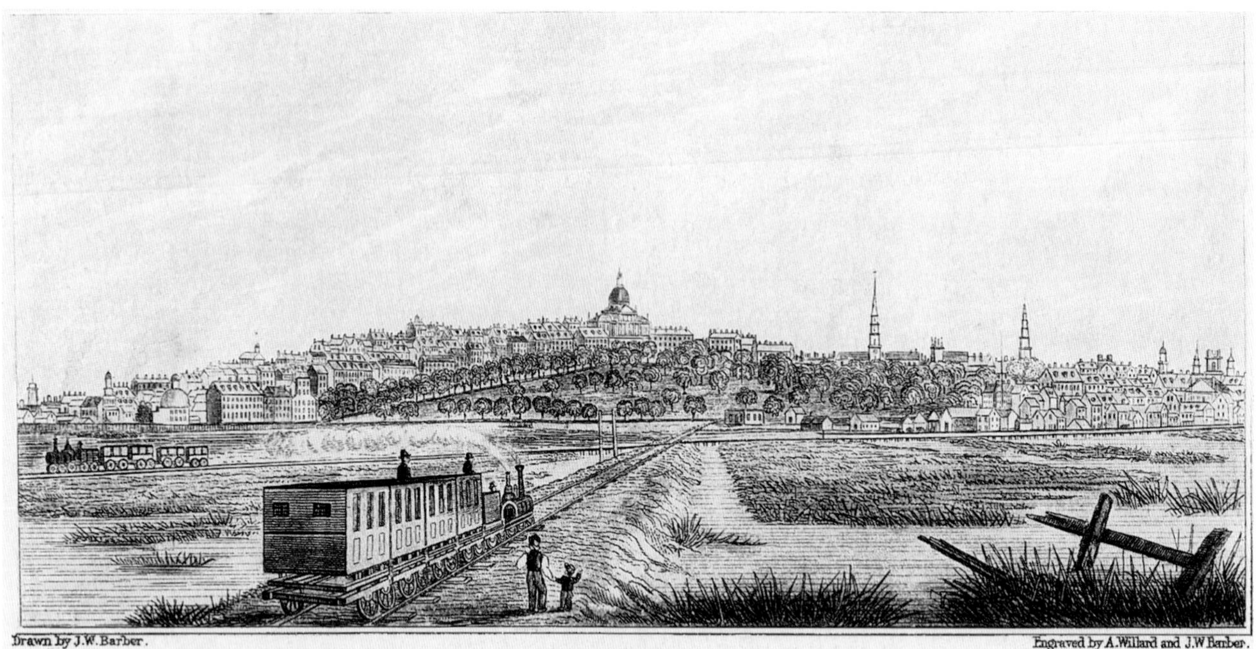

Boston, 1837. View of Boston from the Back Bay. *(Courtesy of Boston Athenaeum)*

Boston 1830. View of the "Neck" (Washington Street) connecting the South End of Boston. *(Courtesy of Boston Athenaeum)*

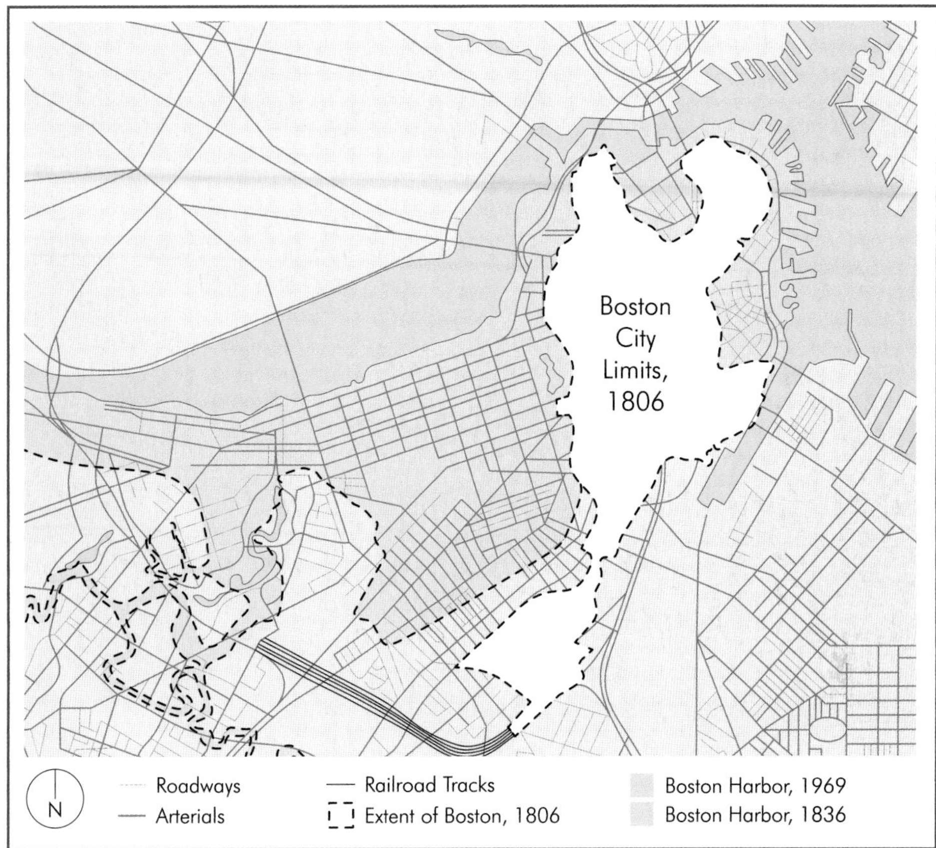

Boston, 1806 and 2012. 1806 map superimposed on a contemporary map of the city of Boston showing the extensive amount of the city that was created by landfill. *(Alexander Garvin and Ryan Salvatore)*

local garbage. In 1849, the Boston Health Department finally demanded that the area be filled.

At first, the only fill was for the purpose of widening the Washington Street neck leading downtown. While new streets were mapped, it was not till 1853, when horse-drawn omnibuses started to operate in the South End, that developers were able to exploit the rapidly expanding Boston market. Individual developers built row houses along the different street grids that emerged. They transformed the South End "into a region of symmetrical blocks of high-shouldered, comfortable red brick or brownstone houses, bow-fronted and high-stooped, with mansard roofs."[14]

An entirely different approach was taken for the expansion into the Back Bay. In 1852, the Commonwealth of Massachusetts created a permanent Commission on Public Lands.[15] Four years later, the commission agreed to the plat that forms the Back Bay. It consists of the five east-west streets running parallel to what had been the mill dam and the nine perpendicular streets that divide them into blocks. Each block is, in turn, bisected by a 16-foot east-west service alley

Boston, 2006. Row houses in the South End, one of the city's earliest suburban extensions. *(Alexander Garvin)*

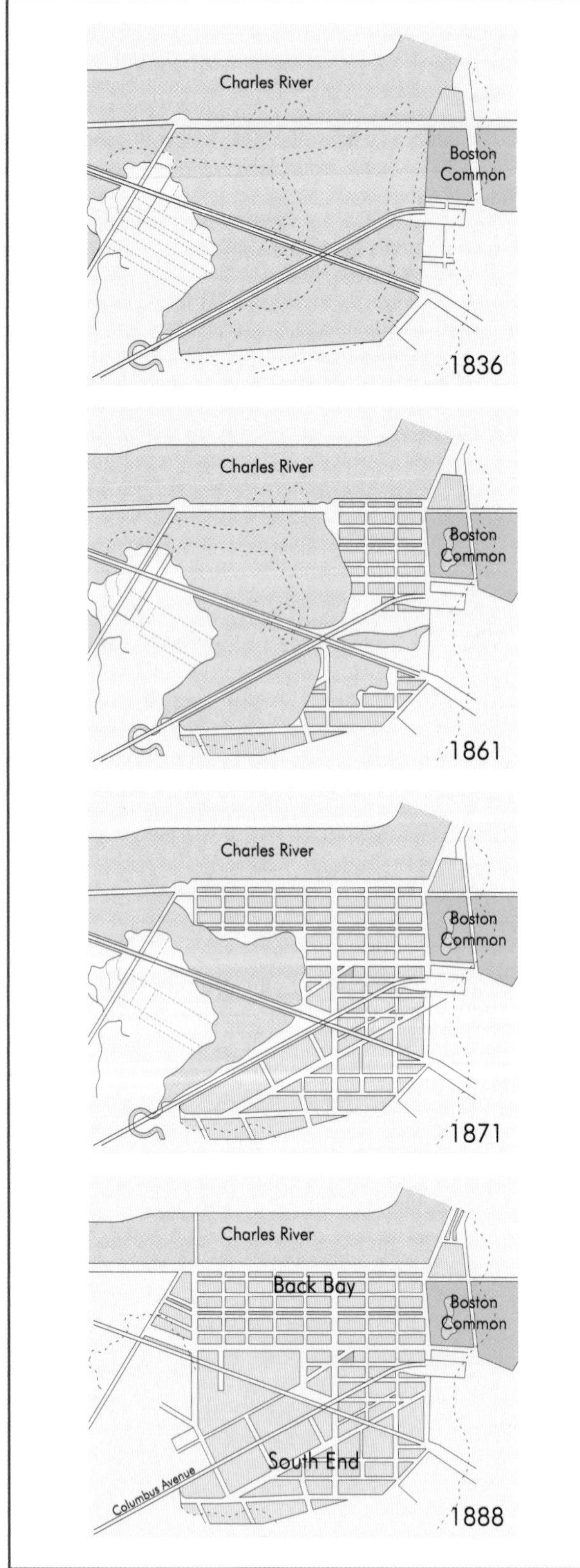

Boston, 1836–1888. Four maps showing the filling in of the Back Bay. *(Alexander Garvin and Ryan Salvatore)*

Boston, 1973. Aerial view of the Back Bay. *(Alexander Garvin)*

and divided into 25-foot lots. The dominant element of the Back Bay plat is its central east-west artery, Commonwealth Avenue, a 240-foot-wide boulevard with a landscaped central island that is broad enough to provide genuine parkland.

The commission contracted with a private company to supply fill (to an average depth of 20 feet) and to grade the site. A special railroad was built from the area to the gravel pits in Needham, 9 miles away, which, with the help of the recently invented steam shovel, supplied 3500 cartloads of fill every day. As the work was completed, each block was subdivided into lots and sold at public auction.

Building lots were not filled to street grade (17 feet above mean low tide). They stopped 5 feet lower in order to allow sewers and water mains to be installed below the service alleys. As a result, the typical Back Bay house has a functional basement service area, usually including the kitchen, 5 feet below street level. Reception and dining rooms are on the main floor, and bedrooms above. The below-street-grade service level had a fundamental impact on the design and consistency of the building stock in the area since it permitted Back Bay houses to keep the number of front steps to six, unlike the high-stooped buildings of the South End.

Other unusual features of Back Bay development were private deed restrictions and public laws regulating land use, building height, layout, and construction materials. They required that manufacturing uses be excluded throughout

Boston, 2008. The landscaped park island that runs down the middle of Commonwealth Avenue is wide enough to offer opportunities for adults to stroll, walk dogs, or sit under a tree, and still provide lots of room in which to play. *(Alexander Garvin)*

the district, commercial activity be prohibited on Commonwealth Avenue, buildings be set back not less than 20 feet from the sidewalk, structures be at least three stories high, buildings be constructed of masonry, and mansard roofs not exceed one story in height. These regulations produced a level of fire safety and structural soundness that would be matched only by twentieth-century building codes and a unity of architectural expression that has persisted despite the introduction of occasional high-rise buildings.

The project was completed in 1886 and produced a net profit of $3.4 million. Some of this return can be attributed to demand for development sites and to a location on the edge of existing city development. The rest was the result of three factors that distinguished the Back Bay from its suburban competition: the landscaped open space in the middle of Commonwealth Avenue; the system of sewers, service alleys, and water mains, which lowered the cost of preparing lots for development; and the building restrictions, which provided home buyers with a level of security not available in other developing areas. Each of these factors underscores the importance of the six ingredients of success.

Curvilinear Subdivisions

Despite prevalent and continuing use, the rectilinear subdivision began to wane in fashion during the second half of the nineteenth century. Public transportation (whether rail, streetcar, or subway) and then private automobiles made possible an entirely different pattern of suburban extension: the curvilinear subdivision. These new forms of transportation eliminated the need to connect new residential communities to any existing street system. Developers and commuters alike could leapfrog the edges of city development and move on to cheaper virgin land, unencumbered by previous development patterns. There they could obtain greater privacy, larger lots, and more generous landscaping, all at lower prices. Without the need to graft onto existing street patterns or limit developments to existing block dimensions, these new subdivisions could and sometimes did include elementary community facilities and landscaped open spaces.[16]

Developers could afford to pay prices that neither farmers nor country-estate owners could refuse. In between these early subdivisions and the cities from which their residents migrated, there remained farms, scruffy land uses, virgin forest, summer estates, and open territory. In time they, too, were replaced by residential subdivisions.

In contrast to common perceptions, this leapfrog pattern of suburbanization was neither casual nor rudderless. It was methodically planned for and carefully supervised by the real estate industry and by local governments. Landowners, brokers, mortgage lenders, insurance and utility companies, contractors, lawyers, and accountants all needed procedures that guaranteed accuracy and legality of title. Government agencies needed to provide the services residents expected. Accordingly, street layouts had to be adjusted to fit into probable traffic patterns and designed to accommodate delivery vehicles and fire engines. Similarly, water, sewer, gas, and electric lines had to conform to common engineering standards if the new subdivisions were to obtain service from regional utility companies. Water mains had to be large enough to supply the area and provide sufficient pressure to permit firefighting. In response to these needs, states enacted legislation requiring official approval and recording of all subdivisions.

California's laws regulating real estate subdivision are illustrative of the long history of government regulation and demonstrate how carefully supervised leapfrog suburbanization really is. The process of government regulation began in 1893 with a state law requiring that officially approved subdivision maps be legally recorded before anyone subdividing land could sell lots. In 1907, the state required developers to obtain approval by the local governing body of any streets dedicated for public use. Six years later, local governments were given the authority to establish layout standards for all streets. The 1915 Map Act required subdivisions to be submitted to the local planning commission (if one existed) or to the city engineer for consideration of their suitability in relation to the city's development plans. In 1921 and 1923, the act was amended to also include drainage, water supply, and other engineering features.[17]

After 1934, state requirements were augmented by those of the National Housing Act of 1934. The system of FHA mortgage insurance that grew out of this legislation provided a new financial basis for the American suburb (see Chapter 9). It also established common planning and design standards for developers around the country. These standards were contained in four publications—*Subdivision Development* (1935), *Planning Neighborhoods for Small Houses* (1936), *Planning Profitable Neighborhoods* (1938), and *Successful Subdivisions* (1940)—and provided the basis for property appraisal practice in the *FHA Underwriting Manual* and thus for essentially all bank lending. They covered design and engineering recommendations for subdivisions, streets, lot layout, utility installation, and landscaping. Properties that failed to meet these standards were not eligible for FHA insurance and thus had great difficulty obtaining bank financing.[18]

Olmsted and Company

America's suburban subdivisions, like urban neighborhoods, vary in size and quality. But they look very much alike because virtually all of them are imbued with the design philosophy of one man: Frederick Law Olmsted Sr.

Between 1857 and 1950, Olmsted's firm (first a partnership with Calvert Vaux; later with Henry Codman, Charles Eliot, and his nephew and stepson John Charles Olmsted; and eventually including his son and namesake, Frederick Law Olmsted Jr.) was involved in planning 450 subdivisions and new communities, 47 while the senior Olmsted was still

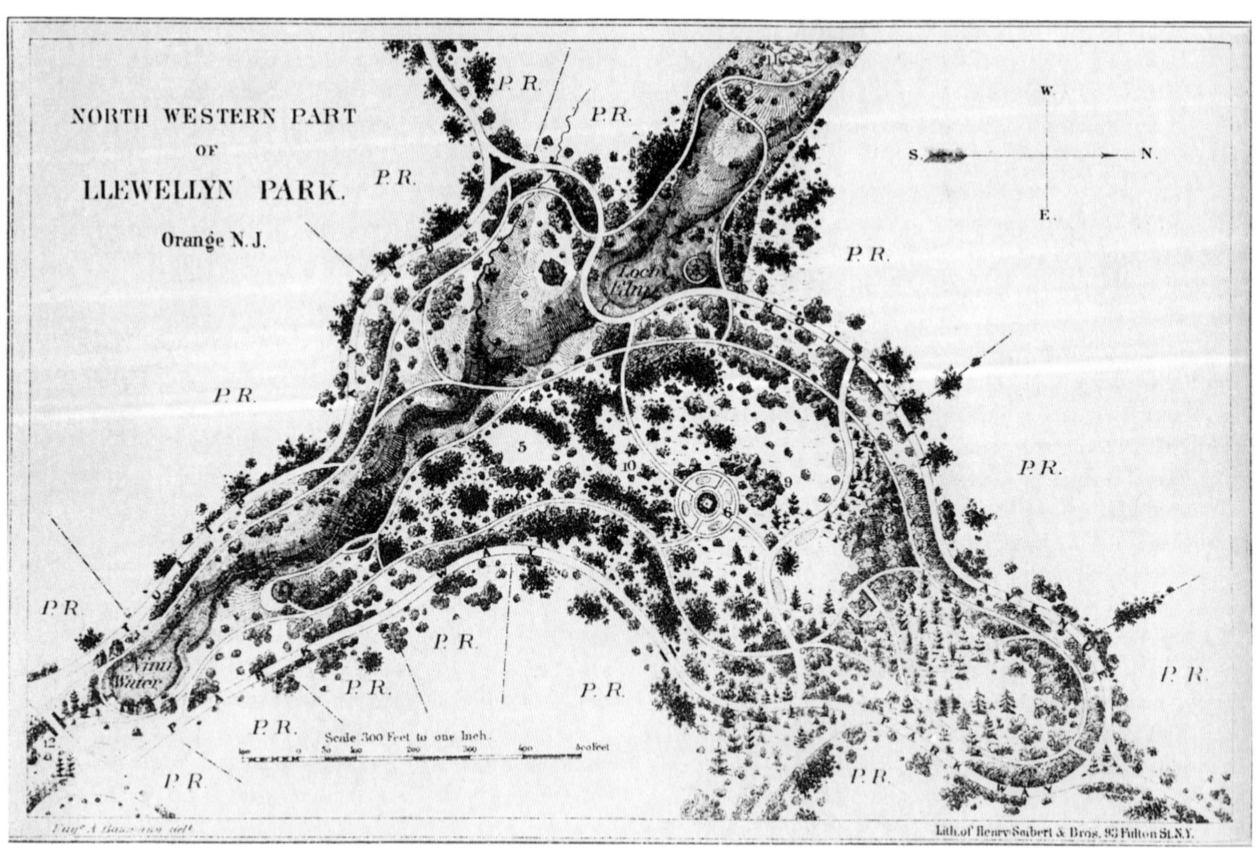

Llewellyn, New Jersey, 1859. Plan made six years after this community was founded by Llewellyn Haskell. *(From John W. Reps,* The Making of Urban America. *Copyright by Princeton University Press 1965, reproduced by permission of Princeton University Press)*

active. It actually prepared site plans for 270 communities. These suburban subdivisions, located in 29 states, Cuba, Bermuda, and the District of Columbia, along with thousands designed by former employees and students trained by its principals and tens of thousands designed by its imitators, still constitute the bulk of our suburban landscape.[19]

The senior Olmsted sat in on classes at Yale in 1844–1845 and tried to support himself as a farmer until 1854. Consequently, he was intimately familiar with horticulture, and this fact surely influenced his pragmatic approach to landscape design. During 1850 he toured Europe and probably visited some of the earliest suburban developments in England and the continent, later publishing a fascinating account of his trip entitled *Walks and Talks of an American Farmer in England.*[20]

While there are many eighteenth- and nineteenth-century English curvilinear precedents, landscape gardener Andrew Jackson Downing is usually given credit for contributing most to Olmsted's landscape design. Olmsted was an avid reader of Downing's magazine, *Horticulturalist,* which was first published in 1846. He corresponded with Downing before getting to know him during the two years before Downing's death in 1852, and he even met his future partner Calvert Vaux in Downing's home.[21]

Downing recommended "country villages" with "broad well-planted avenues of shade trees" and "a large central space always devoted to park or pleasure ground."[22] This park-centered approach was adopted for Llewellyn Park, New Jersey, a planned community with which Olmsted was certainly familiar. It was designed in 1853 for its founder, Llewellyn Haskell (hence its name), by the architect Alexander Jackson Davis, whom Olmsted also knew. Davis was one of the friends whom he consulted when he laid out his own farm on Staten Island.[23]

The 400 acres of Llewellyn Park were designed to be and remain today an exclusive residential compound with a romantic gatehouse to screen out unwanted visitors. Its 7 miles of winding roads and 3- to 5-acre house lots are organized around a 60-acre private park. Three successor trustees manage the park and roadways. Road maintenance and police protection are financed by fees assessed by a board of managers elected by Llewellyn Park's wealthy residents.

The Olmsted approach to the suburban landscape was both more functionally oriented and more democratic than Downing's. It involved neither gatehouses nor large central parks. As Olmsted explained, the principal requirements of a successful suburban community are "good roads and walks, pleasant to the eye within themselves, and having intervals of pleasant openings and outlooks, with suggestions of refined domestic life, secluded, but not far removed from the life of the community."[24]

By "good roads and walks," he meant clean, smooth surfaces and gracefully curved arteries without sharp corners, designed to accommodate several lanes of vehicular (carriage) traffic. Today this seems obvious, but paved streets were the exception when he started designing subdivisions. As late as 1890, not only were half the streets in America unpaved, they were used as much for dumping household garbage, industrial waste, and animal manure as for traffic.[25]

Louisville, 2008. Algonquin Parkway. *(Alexander Garvin)*

The curvilinear roads in Olmsted's subdivisions were more than vehicular arteries. They were lined with trees, and when the arch of trees grew large enough to engulf the roadway, it became so distinctive that the houses on either side receded into the background. These dominant, tree-lined roads created a public realm framework for "the life of the community," while simultaneously allowing the people in the houses to enjoy a refined and "secluded" domestic life.

In designing suburban communities, the firm's objective was a tranquil setting with plenty of grass and trees. It introduced turf and foliage by lining roadways with trees and setting houses back a sufficient distance from the lot line to allow for broad, open lawns. Trees provided the necessary seclusion for "refined domestic life." Open lawns ensured that residents would never be "far removed from the life of the community."[26]

Olmsted, Vaux and Company designed its first suburban community, Riverside, Illinois, in 1868–1869. While Riverside has many of the characteristics of the firm's residential subdivisions, it is really a new town and is considered separately (see Chapter 16). Among the earliest subdivisions for which the firm prepared designs were Tarrytown Heights,

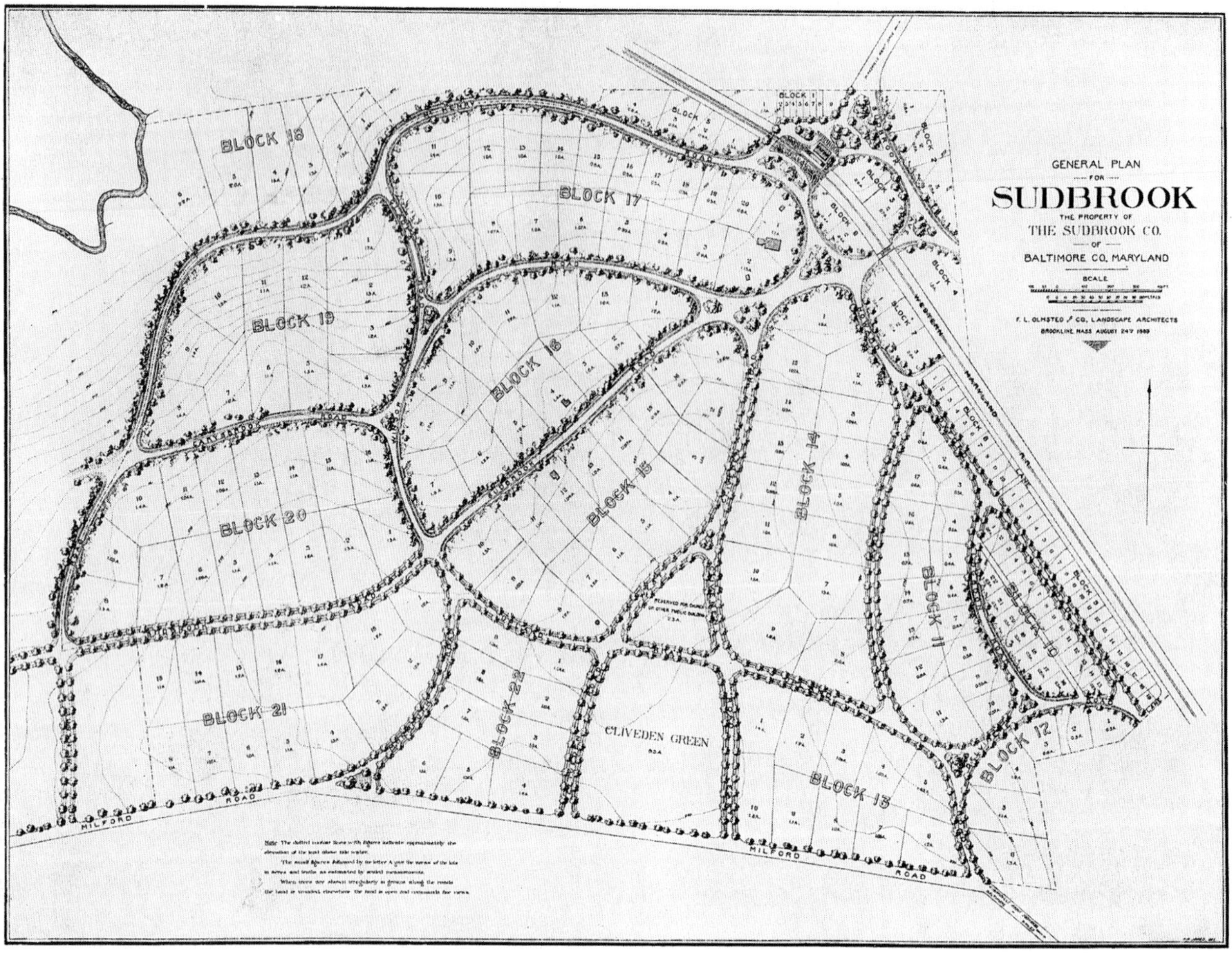

Sudbrook Park, 1889. Olmsted's general plan for Sudbrook takes advantage of the gently rolling landscape to create a charming suburban subdivision. *(Courtesy of the National Park Service, Frederick Law Olmsted National Historic Site)*

New York (1870–1872), Parkside in Buffalo (1872–1886), and Sudbrook Park outside Baltimore (1876–1892). They all had similar features: broad, gently twisting, tree-lined streets; curvilinear blocks with large lots; and small, irregular park islands at significant intersections.

Only Sudbrook Park was executed largely as designed. It was developed by a real estate syndicate that sought to exploit the recently opened Western Maryland Railroad station in nearby Pikesville. The syndicate purchased a 204-acre estate and commissioned the Olmsted firm to design a residential subdivision initially intended for summer residents. The strategy was to rely on the railroad for commuting to Baltimore and the nearby village of Pikesville for retail shopping.[27]

The firm's design subdivided the site into 1-acre house sites. Its gracefully curving, tree-lined roadways are superimposed onto the gently rolling topography. The plan limits the number of houses within each resident's angle of vision (thereby increasing the feeling of privacy) and heightens the sensation of nature by screening out surrounding structures. The resulting design makes each house seem to be an inevitable part of a "natural" landscape.

Sudbrook Park, 1997. Olmsted's formula of houses set back from a curving, tree-lined street set the pattern for suburban development throughout the country. *(Alexander Garvin)*

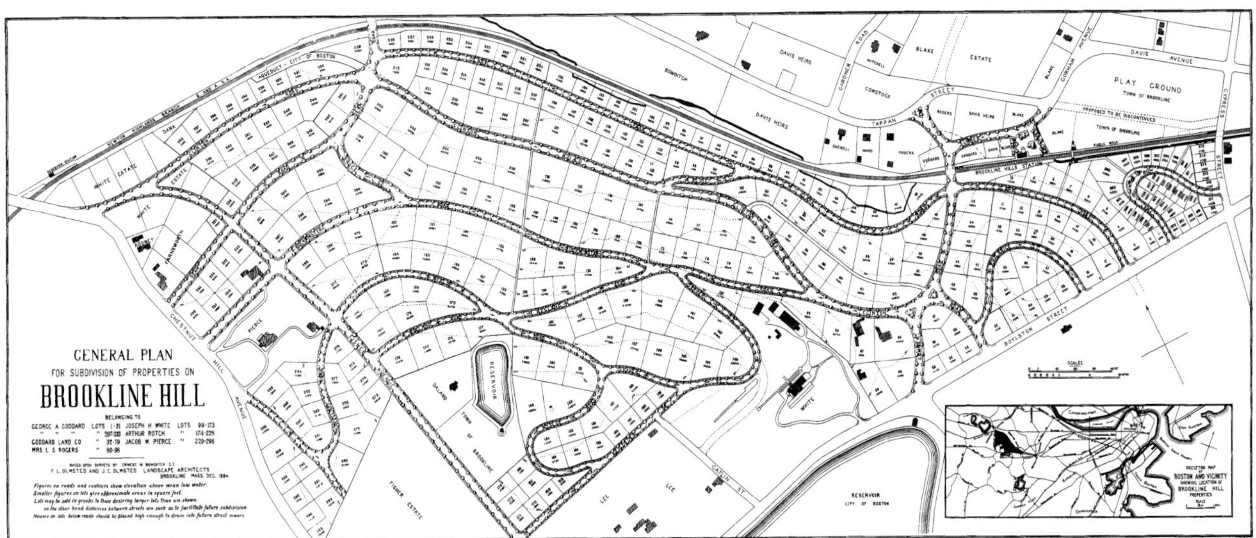

Brookline, 1884. By following hillside contours, Olmsted's plan for Brookline Hill (now Fisher Hill) minimizes steep slopes and provides many of the houses with wonderful views of Boston. *(Courtesy of the National Park Service, Frederick Law Olmsted National Historic Site)*

Similar curvilinear features were used in a very different way at Fisher Hill in Brookline, Massachusetts (1884–1892). Commuter access to Fisher Hill (originally Brookline Hill) was opened up by the Newton Highlands branch of the Boston & Albany Railroad (now part of the MBTA) and the streetcar lines on Beacon and Boylston Streets. The syndicate that developed Fisher Hill hired the Olmsted firm to design a modest year-round community on a steeply sloping site. Its plan consisted of lots, smaller than those at Sudbrook Park, fitted to the contours of the topography.[28]

From the street, Fisher Hill has the look of other Olmsted subdivisions. It is from the rear, however, that its exceptional features are revealed. The curving roadways of Fisher Hill were laid out to hug the contours of the site so that many of the houses can open out to the view below, which in most cases overlooks the nearby city of Boston.

At Druid Hills in Atlanta (1890–1908), Olmsted and Company used the usual tree-lined, curvilinear design in yet another way. Here, on a 1400-acre site, the firm proposed a broad, landscaped parkway intersected by winding roadways and landscaped waterways. The plan for Druid Hills, which was begun by the senior Olmsted and completed after his death, exploited the existing landscape. Steep slopes, ravines, and creeks were left undeveloped or became the focus of linear parks. To allow for natural drainage, roadways were located in shallow valleys and houses set back 50 to 75 feet on the slopes above. Wherever possible, the design preserved the majestic Southern pines, Spanish oaks, and evergreen

Brookline, 1995. The streets in Fisher Hill are located high enough above one another that the houses fronting them have views unobstructed by the roofs of their neighbors below. *(Alexander Garvin)*

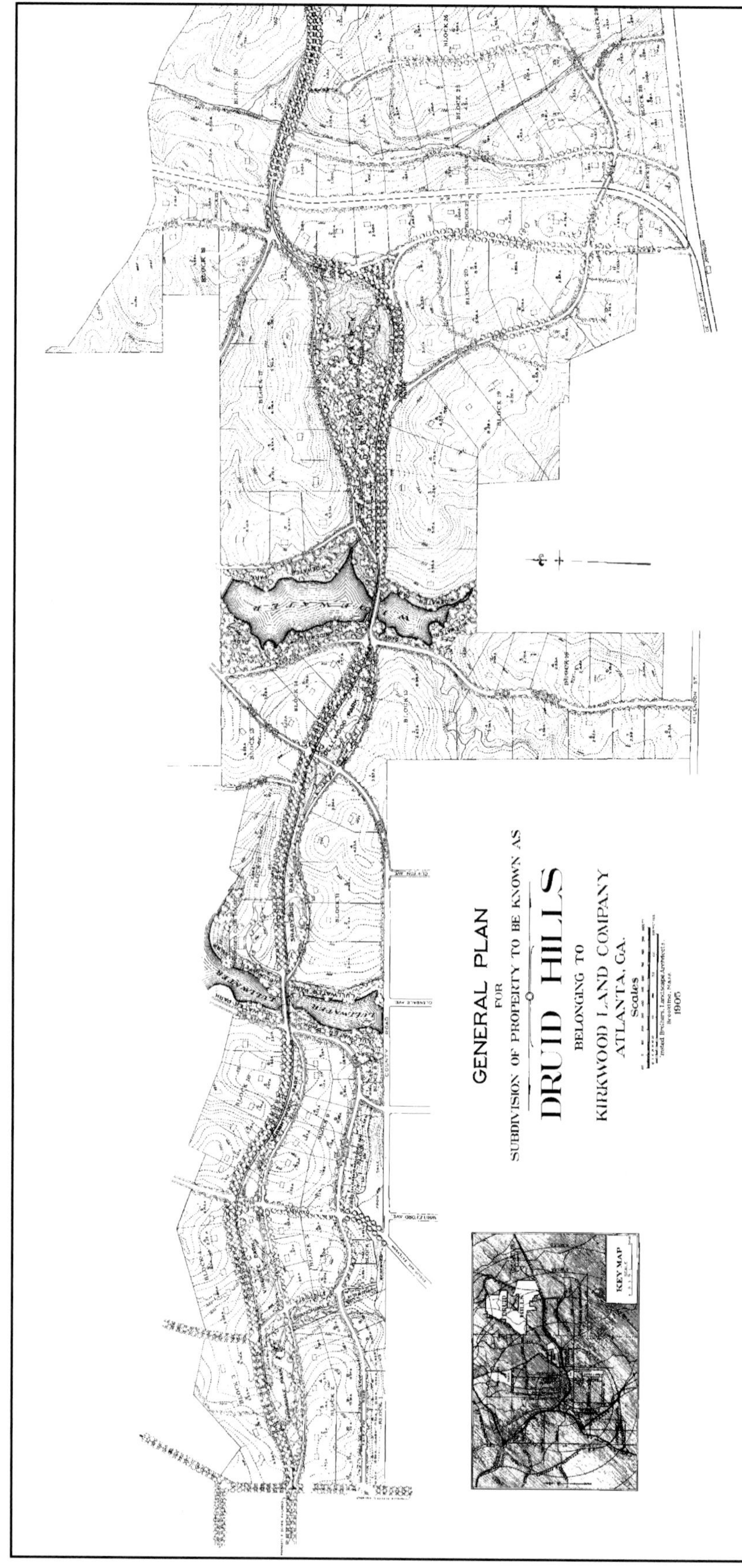

Atlanta, 1905. General plan of Druid Hills. *(Courtesy of the National Park Service, Frederick Law Olmsted National Historic Site)*

Atlanta, 2005. Rolling parkland in the middle of the flanking roadways of Ponce de Leon Avenue in Druid Hills. *(Alexander Garvin)*

magnolias that gave the landscape its distinctive regional flavor.[29] Although the initial developer, Joel Hurt, encountered financial difficulties and had to sell the property before the project could be completed, enough of the original Olmsted proposal was built to ensure that the balance of Druid Hills would be completed in keeping with the firm's design. That design is largely responsible for the continued appeal of Druid Hills as one of the most desirable residential sections of Atlanta.

The Olmsted firm's most famous subdivision is in Forest Hills, Queens, where, along with architect Grosvenor Atterbury, it laid out Forest Hills Gardens for the Russell Sage Foundation and, later, some of the surrounding area for the Cord-Meyer Development Company. The foundation sought to portray Forest Hills as an early demonstration of the effectiveness of enlightened community planning, and to this day it remains a popular reference for comprehensive subdivision planning.

The 175 acres that make up Forest Hills Gardens (1906–1911), largely designed by Frederick Law Olmsted Jr., were not developed as a speculative real estate venture. As explained by the Russell Sage Foundation, Forest Hills Gardens was intended "to create a suburb that would combine the beauty in arrangement of grounds with attractiveness and permanency of building . . . and at the same time to dispose of its property at prices that will give it moderate, but fair return for the money and time invested."[30]

Like so many other subdivisions, it was planned in conjunction with public transportation, in this case the Long Island Railroad, whose station was designed to be compatible with the rest of the community. A charming square that includes an inn and a few retail stores was created around the station. Two gently curving greenways radiate from the square to form spines of the community. Lots open onto tree-lined streets that provide urbane sites for one-family homes, row houses, and apartment buildings. The structures themselves are set back from the street and conform to the requirements of their deed restrictions, most visibly to the requirement that all the buildings have sloping, red clay tile roofs.

Forest Hills Gardens is not sufficiently distinctive to warrant the special notice it usually receives. Many subdivisions have been planned in conjunction with public transportation. The few stores around the Forest Hills railroad station have trouble competing with busy retail streets on the other side of the tracks. Deed restrictions may be responsible for an admirable unity of scale, color, and design, but the landscape is not more consistent than the private places of St. Louis, or many other communities with similar restrictive covenants. The gently curving streets create an agreeable living environment, but not enough to differentiate Forest Hills Gardens from other Olmsted subdivisions or to rival the exquisite landscapes of Druid Hills or Riverside.

As a proposed demonstration of the financial effectiveness of community planning, Forest Hills Gardens was a flop. In 1922, when the Russell Sage Foundation terminated its involvement, the loss on the investment totaled $360,800.[31] Speaking on behalf of the foundation, Clarence Perry, the planner who popularized the idea of the neighborhood unit, explained, "The cost of preparing the land, grading, electrical conduits, sewer systems, street lights, paving, and landscaping, while contributing greatly to the attractiveness of the development, was nevertheless unpredictably high."[32] There is no way of knowing whether a profit-motivated developer would have done better. However, most private developers would have been more careful to coordinate the installation of infrastructure with the pace and price of land sales.

Visiting Forest Hills or any of the Olmsted firm's subdivisions, one wanders along the curving roads, never sure where they will end and constantly surprised and entertained by some aspect of the design. This is the result of the firm's sensitivity to the landscape, achieved sometimes by adjusting to the topography and other times by inserting new elements to enliven its uniformity. It is also triggered by the firm's introduction of strategic openings along the roadways. Topographical features that were unsuitable for construction, such as waterways or steep slopes, may have dictated some openings while others were created by the intersections of the gracefully curving roadways; regardless, Olmsted designs exploited these physical conditions as design opportunities that could enrich the overall development rather than design liabilities to be avoided.

There is nothing exotic about the elements of an Olmsted

Queens, 2004. The houses at Forest Hills Gardens conform to deed restrictions that require all the buildings be set back from the street and have sloping, red clay tile roofs. *(Alexander Garvin)*

subdivision. These plans are basically networks of curving, tree-lined roadways bounded by houses set in the middle of open lawns. Sometimes, as at Riverside and Forest Hills, the curving roadways are used to enliven flat, undistinguished sites. In other places, like Brookline or Druid Hills, the roadways enhance topographic features. They may have become an ever-present suburban cliché, but in the hands of the Olmsted firm this combination produced places of unusual beauty. Even in the watered-down forms manifest in countless suburban communities designed by mediocre imitators, the Olmstedian subdivision is a scheme that usually transforms suburban lemons into lemonade.

Prairie Village, 2010. Olmsted-inspired curvilinear, tree-lined streets with houses set back from the property line. *(Alexander Garvin)*

Cluster Communities

Suburban developers seek to increase the number of residences per acre because this reduces land cost per dwelling unit and allows a project to supply amenities (swimming pools, tennis courts, children's play equipment, etc.) that homeowners could not otherwise afford. At the same time, they do not want to sacrifice the appearance of being in the country. In 1928, architects Clarence Stein and Henry Wright devised an outstanding design solution to this problem for Radburn, New Jersey. It was decades before their ideas took hold, and then only because of the growing popularity of a novel form of ownership, the condominium, and an innovative form of land use regulation, cluster zoning.

The Radburn Idea

When the senior Olmsted retired from active practice, in 1895, there were five automobiles in America. In 1928, when Clarence Stein and Henry Wright started laying out Radburn, New Jersey, there were 21.3 million. During those intervening 33 years, the automobile transformed daily life and made accessible vast new areas of cheap land. However, neither developers nor their designers departed much from the Olmsted formula. Instead of driveways leading to carriage houses, they built driveways that led to garages; instead

Queens, 2001. Sunnyside Gardens. *(Alexander Garvin)*

of simple roadways with a couple of traffic lanes, they built wider streets that allowed cars to park along either side.[33]

The most important subdivision to depart from the Olmsted formula was Radburn, a new form of community intended to answer the enigma: "'How to live with the auto,' or if you will, 'How to live in spite of it.'"[34] Radburn was developed by the City Housing Corporation (CHC), a limited-dividend company organized by realtor Alexander Bing. The CHC had been established for the purpose of building moderate-income housing, while simultaneously producing a modest 6 percent return for its investors. Its objective at Radburn was to create housing designed to the most advanced planning standards and "garden cities" based on the ideas with which Ebenezer Howard had been experimenting in England (see Chapter 16).

The plan for Radburn evolved from Stein and Wright's work on Sunnyside Gardens, a community of 1202 family units that they designed for a 77-acre site in Queens and which was a scant 15-minute subway ride from Times Square. Sunnyside Gardens is one of the most ingenious adaptations of the New York City 200-foot block. Each block consists of two-and-a-half-story, redbrick row houses enclosing a 120-foot-wide landscaped quadrangle.[35] Only 28 percent of the land is used for housing. The rest is landscaped open space. Buildings are slightly set back from the street to permit small front gardens and similarly set back from the inner quadrangle to provide each house with a small, private rear yard. At the time of its development, each inner quadrangle was held in common ownership by the residents of the block and protected by deed restrictions. Parking was provided in garage compounds on the edge of the site. Playgrounds, a baseball field, and tennis courts were provided at a 3.5-acre community-owned park. To avoid unnecessarily high initial expenditures, utility installation, street paving, and house construction were timed to meet market demand.

Sunnyside Gardens, unlike its more ambitious counterpart in Forest Hills, easily produced its projected 6 percent return and demonstrated the financial effectiveness of community planning. Upon completion in 1928, all unused land was sold for $646,000, three times its cost, including the initial purchase price, improvements, and carrying costs. Sunnyside Gardens' only flaw was revealed decades after completion: some of its residents preferred private gardens to common open space (see Chapter 17). "Fourteen years after the last of the original court easements expired, only 6 of the 15 center courts retain their original configuration; four are largely enclosed by fences along the property lines of the surrounding homes, though their pathways are unobstructed; three are completely enclosed."[36]

Flush with its success in Queens, the CHC decided to build a complete garden city for 25,000 residents. After examining 50 possible sites, it settled on 1 square mile of rolling farmland in Fair Lawn, New Jersey, a 10-mile drive from the George Washington Bridge, then under construction. The project, named Radburn, was again entrusted to Stein and Wright. Like Sunnyside Gardens, Radburn was not really a garden city such as Ebenezer Howard had envisioned. Land

Queens, 2005. The open space enclosed by the attached houses that line this block of Sunnyside Gardens is still owned by a homeowners' association. *(Alexander Garvin)*

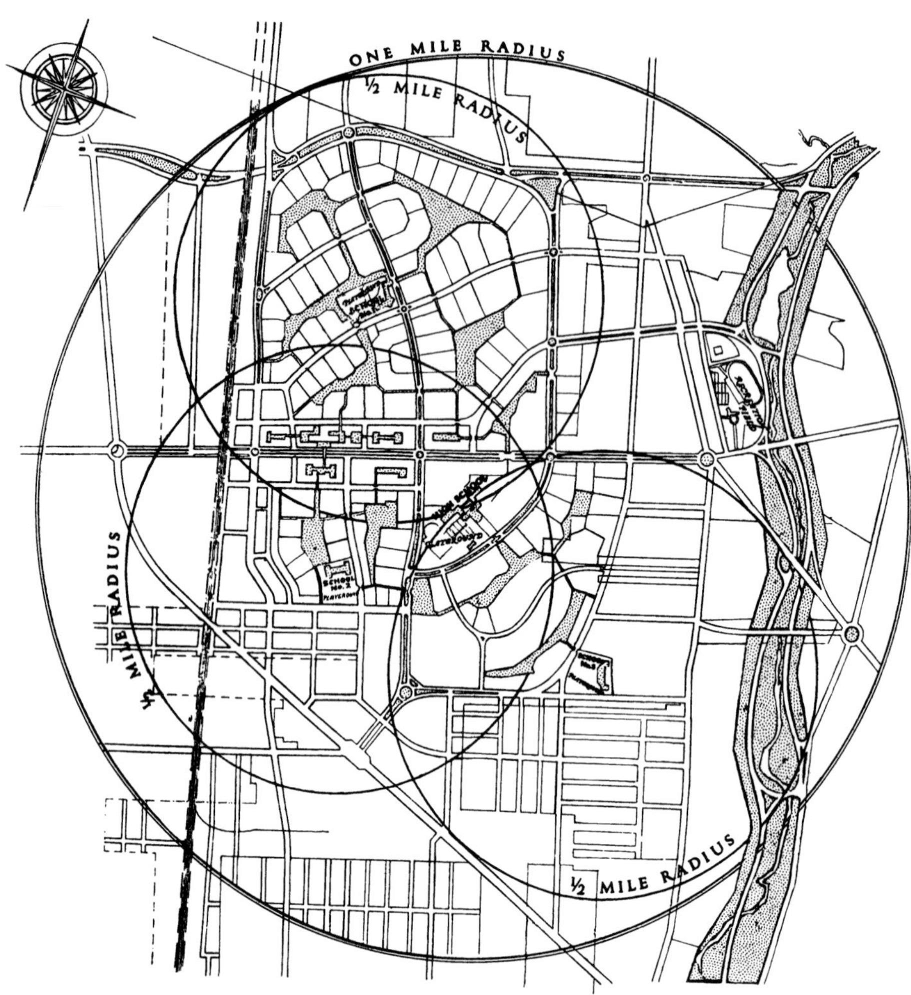

Radburn, 1929. General plan showing three school-centered neighborhoods, each ½ mile in radius. *(From Clarence Stein, New Towns for America, 1966; courtesy of MIT Press)*

was subdivided and sold to individual owners, not kept in common ownership for the long-term benefit of the community. There were no industrial land uses and, thus, no opportunities for residents to both live and work in Radburn. Instead, the CHC pioneered the design of a suburban subdivision for the "motor age."

With tens of millions of cars on the road, it was natural for Radburn's planners to want to eliminate traffic accidents. They did this by removing pedestrian traffic from the street, establishing an independent pedestrian circulation system, and planning an underpass wherever this pedestrian system intersected with vehicular traffic. Stein always credited the idea of separating pedestrian circulation from vehicular roadways to Olmsted and Vaux, who had pioneered the underpass in New York's Central Park (see Chapter 3) "almost half a century before the invention of the automobile."[37] Once Stein and Wright demonstrated its effectiveness, the underpass became the Radburn trademark and a commonly accepted symbol of good city planning.

It is ironic that the single underpass actually built became the popularly known trademark of Radburn, since other elements of the Radburn idea, especially the superblock, are so much more important. The Radburn superblock was probably conceived during what Lewis Mumford described as "the vivid interchange of ideas" that took place within the Regional Planning Association of America (RPAA).[38] As Stein explains: "A core of members [that included Stein, Henry Wright, Alexander Bing, Benton MacKaye, Catherine Bauer, and Lewis Mumford] met at least two or three times a week, sometimes more, for lunch or dinner . . . between 1923 and 1933."[39]

In 1927, when serious work on Radburn had begun, RPAA discussions took on a special focus. Their consideration of the number of people required for "a good elementary school" was one of the discussions that helped to give shape to the Radburn superblock. The principles behind the superblock were best articulated by Clarence Perry, one of the participants in these RPAA discussions. In his presentation of "the neighborhood unit" in volume 7 of the 1929 *Regional*

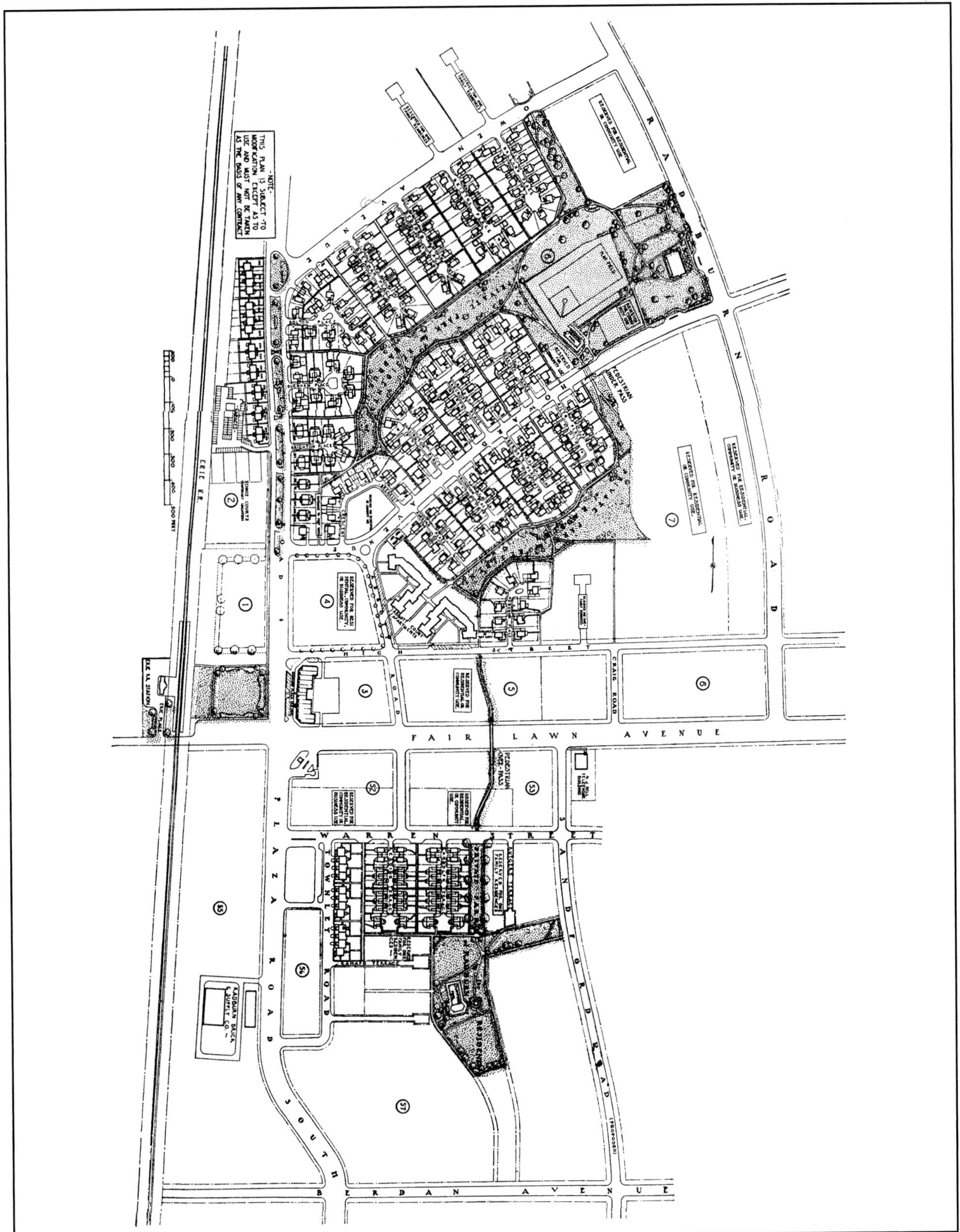

Radburn, 1930. Plan of development completed by 1930. *(From Clarence Stein,* New Towns for America, *1966; courtesy of MIT Press)*

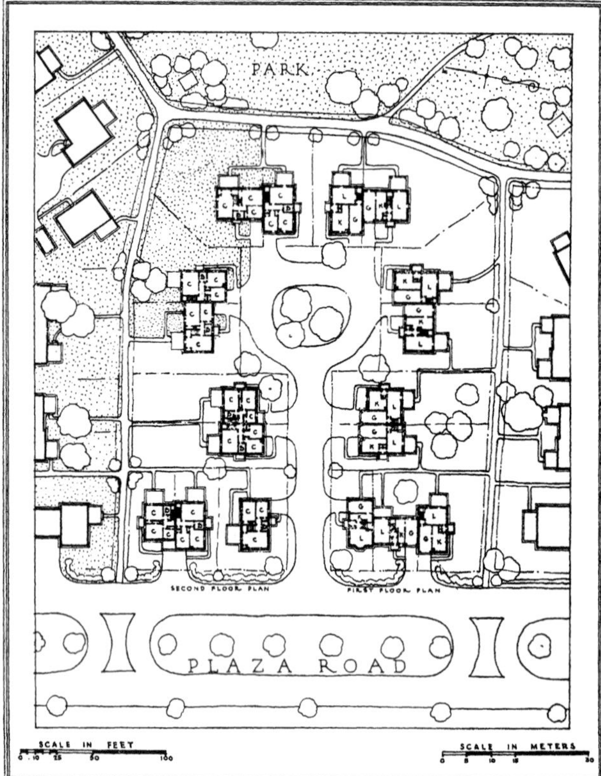

Radburn, 1929. Plan of houses grouped around a cul-de-sac (Burnham Place). *(From Clarence Stein,* New Towns for America, *1966; courtesy of MIT Press)*

Survey of New York and Its Environs (conceived and published by a contemporary organization, better known by its later title, the Regional Plan Association, whose only similarity to the RPAA was its name), he outlined these principles:

- *Size. A residential unit [should have a] population for which one elementary school is ordinarily required . . .*
- *Boundaries. The unit should be bounded on all sides by arterial streets, sufficiently wide to facilitate its by-passing by all through traffic.*
- *Open Spaces. A system of small parks and recreation spaces, planned to meet the needs of the particular neighborhood, should be provided.*
- *Institutional Sites. Sites for schools and other institutions having service spheres coinciding with the limits of the unit should be suitably grouped.*
- *Local Shops. One or more shopping districts adequate for the population to be served should be laid out in the circumference of the unit . . .*
- *Internal Street System. The street system . . . [should be] designed to facilitate circulation within the unit and to discourage its use by through traffic.*[40]

Each 30- to 50-acre Radburn superblock was conceived as a fully planned neighborhood for 7500 to 10,000 residents fitted within a radius of half a mile and centered around its elementary school and playgrounds. Shopping facilities were located on the periphery and were accessible both by foot and by car.

Radburn, 1993. Houses turned around to face the pedestrian path leading to common open space. *(Alexander Garvin)*

Radburn, 2006. Children at play in the common open space. *(Alexander Garvin)*

Planning based on pedestrian-traffic separation and superblocks required unscrambling "the varied services of urban streets."[41] They were reassembled in two intersecting systems that can be better understood when thought of as interlocking combs, one gray and the other green. The spine of the gray comb is an arterial street enclosing the superblock. Its teeth are cul-de-sacs, which provide vehicular access to a cluster of houses. These gray teeth alternate with green ones that provide pedestrian access to the houses and lead to the spine of the green comb, which Stein called the open-space backbone of the superblock. The diagrammatic expression of this design is quite different from what was actually created. By adjusting their diagram to the topography and carefully landscaping the resulting open space, Stein and Wright created one of America's most beautiful suburban subdivisions.

Radburn, 2006. Underpass. *(Alexander Garvin)*

The cul-de-sac, like the underpass, was based on precedent. As Stein later explained, he and Henry Wright had visited examples of superblocks with cul-de-sacs prior to planning Radburn.[42] Their models were the projects developed to demonstrate Ebenezer Howard's philosophy, in particular Hampstead Garden Suburb. However, unlike the conventional suburban houses fronting on the Hampstead cul-de-sac, Radburn houses were turned around to face the green pedestrian comb, thereby becoming the hinge that connected the two circulation systems. Kitchen and service rooms fronted on the vehicular cul-de-sac, while living rooms fronted on the garden and pedestrian walkway. Stein writes that this idea "was conceived by that imaginative genius Henry Wright" and that they had both wanted to turn the houses around at Sunnyside Gardens but were dissuaded by conservative opposition.

Not only was the first Radburn superblock beautiful and practical, it was economical. Eliminating through streets reduced the required length of the utility lines and paved streets. By serving fewer houses, streets also could be narrower and thus less expensive. This savings covered the cost of burying utility lines, "paid for the 12 to 14 per cent of the total area that went into internal parks, [and] also covered the cost of grading and landscaping."[43]

As at Sunnyside, Radburn properties were sold with deed restrictions protecting common open space. Given the size of the landscaped areas, the amount of pavement, and the sparse services provided by government agencies in semirural Fair Lawn, there also had to be a mechanism for disposing sewage, collecting garbage, lighting streets, and maintaining park areas and recreation facilities. To provide these amenities, the CHC established the Radburn Association, "a nonprofit, non-stock corporation to fix, collect, and disburse the annual charges, to maintain the necessary public services, parks, and recreation facilities, and to interpret and apply the protective restrictions." A self-perpetuating board of trustees, like the one established for Llewellyn Park, directed the association. Initially, there were no homeowners on the board. Perhaps in reaction to this form of taxation without representation, two months after the first family moved in, residents formed a Radburn Citizens' Association. It provided a forum for community opinion but had no real power. In response to its recommendations, in 1938 the Radburn Association was reorganized to provide residents more representation and democratic control, but also more responsibility.

The City Housing Corporation started Radburn just as the Depression caused its market to collapse. The project limped along until the CHC declared bankruptcy in 1935 and reluctantly sold the remaining land back to the farmers. As Stein explains, this was inevitable: "Continuous large-scale development is essential to the financial success of a new town such as Radburn. Otherwise the carrying charges on land, main highways, and utilities will soon devour possible profit and force the operating company deeper and deeper into debt."[44]

The completed portion of Radburn covers 149 acres. It contains 430 single-family dwellings, 44 two-family houses, a 96-unit apartment complex, 90 row houses, and 23 acres of parkland. The parkland currently includes two swimming pools, four tennis courts, three baseball fields, five outdoor basketball courts, numerous play areas, and a walkway system that led to the public school.

There is little demographic difference between Radburn and surrounding sections of Fair Lawn that were largely built following World War II. However, there is a world of difference in the patterns of daily life. Radburn's children actually play in its generous park facilities, not on the street as they do in neighboring areas. As a result, during its first 20 years, there were only two traffic fatalities in Radburn, both on surrounding main highways. Forty years after its completion, a research report documented in stark contrasts the differences between Radburn and the areas around it: 47 percent of Radburn's residents reported that they shopped for groceries on foot in contrast to the meager 8 percent of nearby residents that did. One-quarter of Radburnites used bicycles for utilitarian purposes while only 8 percent of nearby residents did.[45]

Fair Lawn, 1980. Children playing in the street lined with utility poles, directly across from Radburn. *(Alexander Garvin)*

Despite the obvious superiority of Radburn's planning, none of the developers in Fair Lawn chose to copy it. During the Great Depression, those private developers that were able to finance large subdivisions were unwilling to provide buried utility lines and other frills. Demand soared to such a degree after World War II that basic boxes sold like hotcakes. Consequently, developers created typical subdivisions with look-alike houses on streets with prominent utility poles. The overwhelming majority of America's developers have done the same. It is doubtful that 1 in 100 of these developers had even heard of Radburn. Only one component of the Radburn design crept into their work: the cul-de-sac. It reduced development costs, was endorsed by the FHA in its various publications, and became a familiar part of thousands of conventional and FHA-insured subdivisions when suburban development resumed after World War II.

If Radburn is unknown to the overwhelming majority of developers, it is revered by virtually all city planners, admired by many architects, and known to housing officials around the world. It is in their work that the Radburn idea has influenced the landscape. In America it became the model for some planned new communities built after World War II. But its greatest impact was in postwar Europe, where government had a more direct role in development. There, government planners made the cul-de-sac, the superblock, and pedestrian-traffic separation familiar elements of both inner-city redevelopment and suburban new communities. This is especially true of England, where the Radburn idea became an important element in government-planned communities. The new town of Stevenage, outside London, for example, includes superblocks, specialized road systems, an open-space backbone, and numerous pedestrian underpasses.

Condominium Communities

Radburn may not have become the prototypical motor age suburb. However, Stein and Wright were prescient in identifying a product the suburban market would soon demand: the relatively inexpensive, automobile-oriented subdivision with common landscaped recreation space.

During the first few years after World War II, developers had no trouble supplying the seemingly limitless demand for suburban houses. FHA and VA mortgages were extended to millions of home buyers. The Federal Aid Highway Act of 1956 provided 90 percent of the cost of interstate highways that made accessible vast areas of cheap land.

By the late 1960s, however, the cost of supplying suburban houses was climbing at a rate that exceeded the increase in consumer purchasing power. New suburban houses were becoming increasingly expensive because of the high cost of land, development, and financing. As land within reasonable automobile commuting distance of cities filled up with single-family detached houses, suitable sites for additional subdivisions became increasingly expensive. Moreover, construction costs were increasing faster than the general rate of inflation. Mortgage interest rates, which had fluctuated between 3 and 6 percent since the 1930s, suddenly went haywire. Between 1970 and 1974, the prime rate of interest changed 65 times, reaching a high of 12 percent. During the 1981–1982 recession, it changed 77 times, reaching the all-time high of 21.5 percent.[46]

The price of a large house lot on a meandering, tree-lined street became more than many first-time home buyers could afford. Simultaneously, an alternative, cheaper product, the cluster community, appeared on the market. Cluster development reduced costs by attaching residences to one another. It reduced the land cost per unit by fitting more single-family buildings onto the same site. It lowered construction costs by the value of the unbuilt exterior surfaces and it decreased utility installation and roadway expenses by the distance they no longer had to extend.

The cluster community may have been cheaper to produce, but it faced stiff consumer resistance. A site design had to be devised that would provide each resident with easy automobile access and an acceptable amount of private, landscaped open space. There needed to be a form of common ownership for roadways and open space. Most important, this pattern of common residential occupancy had to be made acceptable to lending institutions.

The obvious site design was similar to the Radburn superblock. Residents coming by car could turn off public arteries, drive along minimal commonly owned roadways, and park in their own driveways or garages. Once there, they could use commonly owned pathways to landscaped open spaces that satisfied their desire for a little bit of nature. The designers of cluster communities may have been familiar with Radburn. But the developers and bankers that built and financed them probably had never heard of it. For them, the Radburn idea was the natural product of the economics of development. Put alternatively and in the context of the six ingredients of success, the Radburn idea could proliferate only when a market for it could be sustained.

Any complex of individually owned residences with common roadways, open space, and community structures requires a form of common ownership. It also requires a legal entity with personnel and money to operate and manage commonly held parts of the complex. Developers ignored the Radburn approach to common ownership and management. They believed a board of successor trustees would be unlikely to appeal to the mass suburban market. Instead, they chose the condominium form of ownership.

The term *condominium* refers to that form of ownership in which individual title to a residence is coupled with an ownership interest in land and common areas (roads, walkways, landscaped areas, recreation facilities, etc.). The amount of this common ownership interest is determined by the ratio of square footage contained in the owner's unit to the total square footage of the project. This ratio also determines each owner's degree of control over condominium governance and the dollar amount to be paid for operating a condominium association and managing common areas within the project. The difference between a condominium association and a street association, like those established for the private places of St. Louis, and a community association, like those established for an Olmsted subdivision, lies in the form of common ownership and the specificity of the enumerated responsibilities of the condominium agreement.

The condominium concept was so new and different that prior to 1973, the U.S. Department of Housing and Urban Development did not publish annual statistics on the number of condominium units built. In 1975, more than 85 percent of the nation's condominium units were not even five years old.[47] Condominium ownership may have been slow in gaining acceptance, but it increased in popularity so quickly that by 1984, condo units represented 17 percent of national housing starts.[48]

Lack of lender acceptability was one of the reasons that condominium ownership was slow in gaining acceptance. The FHA refused to insure townhouses without direct street frontage until 1961, when it approved Hartshorn Homes, a 98-unit project in Richmond, Virginia.[49] That was also the year that Congress first empowered the FHA to provide mortgage insurance to condominiums. Within two years, the FHA published *Planned Unit Development with a Homes Association,* which established condominium eligibility requirements and explained the applicability of FHA Minimum Property Standards.[50]

Appraisal practice created another financing problem. Initially, the FHA, the VA, and most mortgage lenders refused to include the value of common areas and facilities in their appraisal of individual condo units. Hence, developers worried that purchasers would not be able to obtain sufficient mortgage financing. In practice, this was not a serious

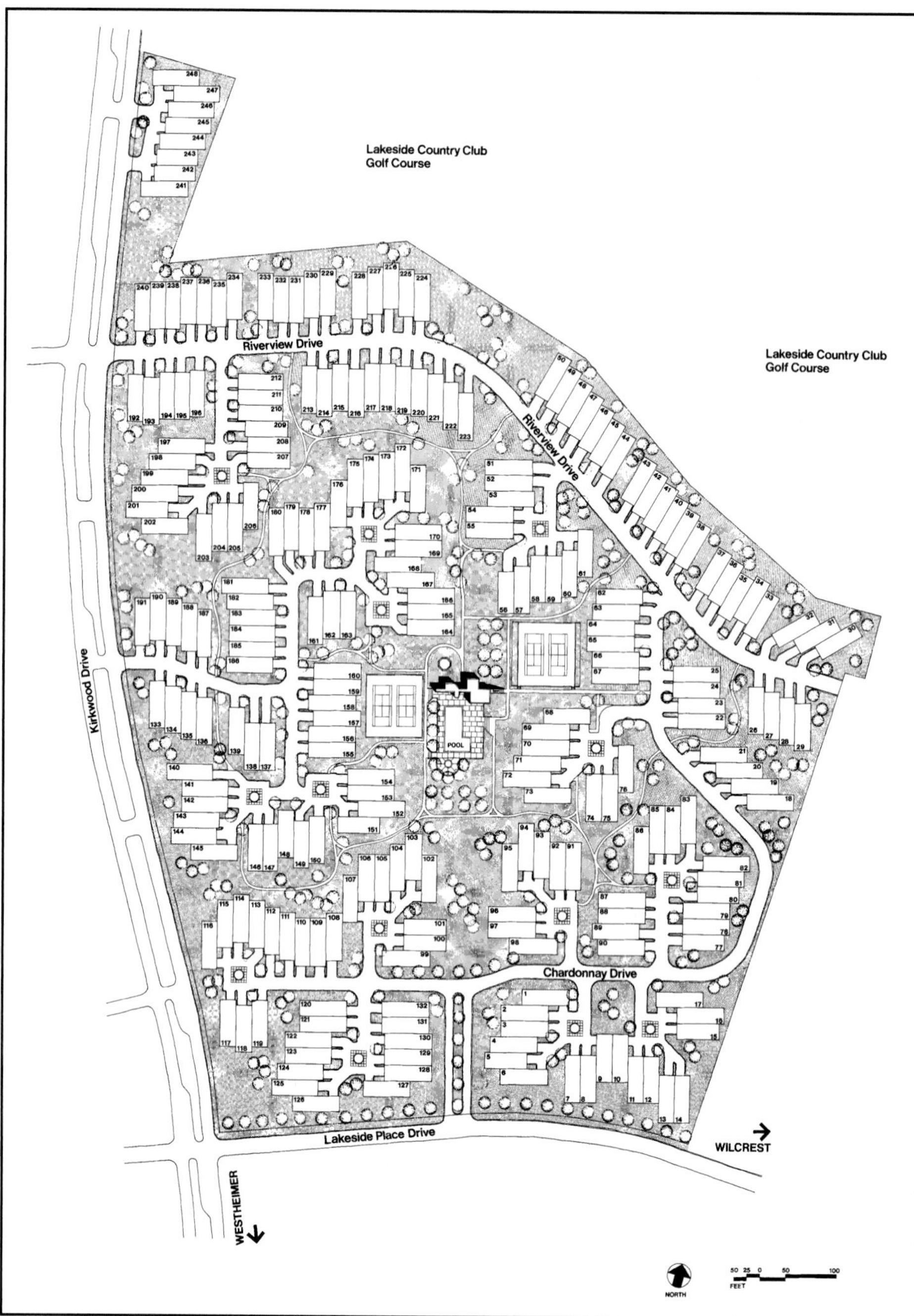

Houston, 1973. Subdivision plan for Epernay in which condominium units are grouped around cul-de-sacs and open onto greenways that lead to the swimming pool, tennis courts, and community center. *(Courtesy of Fisher-Friedman Associates)*

problem, however, as the market soon established a value that included these facilities, and mortgages reflected this reality.

The quality of condominium communities is as variable as the quality of earlier one-family house subdivisions. In the hands of Olmsted and Company these subdivisions become lovely residential communities. In lesser hands they become arid settings for "little boxes made of ticky-tacky . . . [that] all look just the same."[51] This also is true of condominium communities. Obviously, the quality of the architecture and the landscaping is critical. However, better condominium communities can be distinguished by their success in dealing with the two critical factors identified by Stein and Wright at Radburn: the automobile and open space. The critical importance of these factors is beautifully illustrated by Epernay and Ethan's Glen, two Houston cluster communities.[52]

Both projects were begun in 1973 by the same team: developer GreenMark Inc. (a subsidiary of Gerald D. Hines Interests), architect Fisher-Friedman Associates, and landscape architect Sasaki-Walker Associates. Despite differences in appearance, site plan, and landscaping, these projects

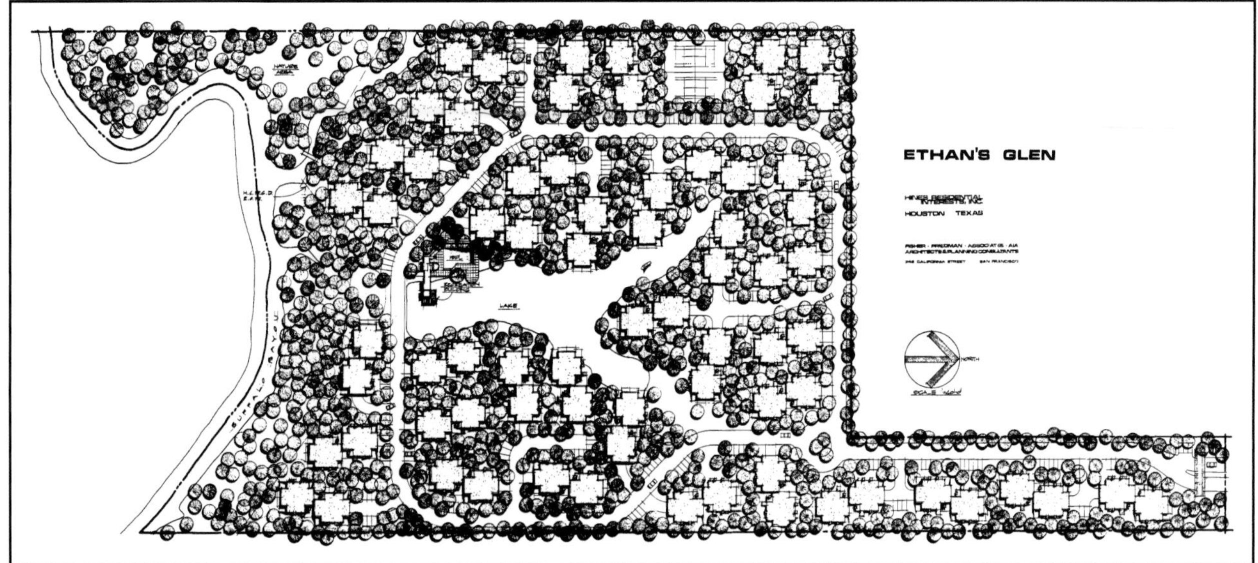

Houston, 1973. Subdivision plan for Ethan's Glen that retains the maximum number of trees by clustering eight units into one building and locating parking for 12 cars underneath. *(Courtesy of Fisher-Friedman Associates)*

achieve superior results because they both include generous, landscaped common areas and skillfully incorporate the automobile into their design.[53]

Epernay is a 248-unit project built on 43.5 relatively featureless acres next to the Lakeside Country Club and Golf Course. The design, like that of Radburn, is based on pedestrian-traffic separation within a superblock in which buildings are clustered around a vehicular cul-de-sac and residences face an open-space system with generous recreation facilities that include a swimming pool, clubhouse, and four tennis courts.

Epernay, however, is quite different from Radburn. It consists of clustered town houses, not single-family residences. The buildings include small landscaped courtyards that introduce a private bit of nature into the rooms within. Epernay separates pedestrian and vehicular traffic, but provides no underpass where they intersect. It is even more accommodating to automobiles. There are five parking spaces for each town house (two in resident-owned garages, two in the driveway, and one in the cul-de-sac cluster). Furthermore, each cul-de-sac provides more than just vehicular access. It is a carefully designed turnaround, specially paved to provide an urbane town house setting. Finally, excluding streets, 40 percent of Epernay is common open space, far more than Radburn.

Ethan's Glen is located in a 32-acre section of heavily wooded pine forest with a natural gully running through it. The site design exploits both features to produce a 288-unit condominium community distinctly different from its competitors. An earthen dam was built in order to create the 2-acre lake and establish a base for the loop road. In an attempt to preserve as much of the forest as possible, residences are gathered together into a freestanding structure that contains eight two-story units that are lifted half a level above grade. The half-level excavated beneath the cluster provides

Palo Alto, 2006. Peter Couts Village is one of numerous later-twentieth-century suburban subdivisions with houses grouped around common open space that residents use for recreational purposes. *(Alexander Garvin)*

12 parking spaces. As at Epernay, 40 percent of the site is common open space, including the lake, swimming pool, two tennis courts, and community center.

Epernay and Ethan's Glen may be more sensitively designed than many other cluster communities. Nevertheless, they illustrate that thoughtfully landscaped open space combined with intelligent handling of the automobile can provide suburbanites with amenities that they could not otherwise afford. This lesson was not lost on developers of rental housing. Once condominium clusters became popular, developers began creating rental communities that also integrated parking, included a swimming pool and clubhouse, and provided common open space. The difference between rental and condo communities is essentially the market they are trying to capture. Most rental communities are aimed at singles, recently married couples, and those without sufficient savings or income to buy their residence. The condo communities are usually directed at first- or second-time homebuyers.

Planned Unit Development, 1968. Site plan showing conventional street grid (with 1427 single-family house lots) that destroys the natural character of the site. *(Courtesy of New York City Department of City Planning)*

Cluster Zoning (Planned Unit Development)

Cluster development may be a popular marketplace phenomenon, but it is even more popular with those who wish to alter marketplace desires. In the attempt to encourage this alternative to cookie-cutter suburban subdivisions, they have persuaded most communities to add a cluster zoning or *planned*

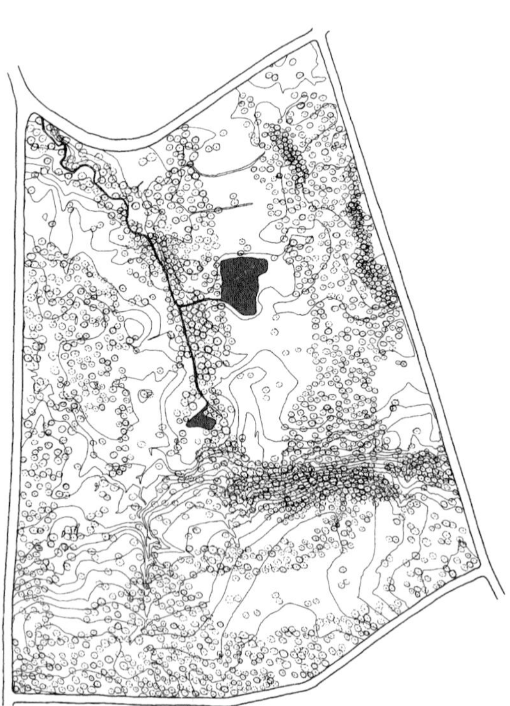

Planned Unit Development, 1968. Contour map showing existing features (trees, ponds, creek, and bounding road) of a 205-acre undeveloped site. *(Courtesy of New York City Department of City Planning)*

Planned Unit Development, 1968. Plan generated by using the cluster zoning approach would preserve the natural features of the site, provide for an elementary school and shopping facilities, and accommodate 1445 housing units in a a range of house types (detached and semidetached houses, town houses, and attached row houses). *(Courtesy of New York City Department of City Planning)*

unit development (PUD) alternative to conventional zoning requirements.

Radburn, Epernay, and Ethan's Glen required only subdivision approval. There were no zoning regulations to comply with. In most places, suburban development must comply with yard, setback, density, and a multitude of other zoning restrictions (see Chapter 17). Thus, the only way to create a cluster community with ample common open space, tailored to its site and natural features, is to do so outside traditional district-by-district zoning requirements.

Cluster zoning achieves this objective by overriding conventional lot-by-lot requirements, substituting instead general regulations that are applied to the site as a whole. It allows the same number of residences to be clustered in a manner that avoids costly construction on steep slopes or over waterways and permits attractive natural features to be retained without violating bulk, density, parking, or open-space requirements.

In its most straightforward form, cluster zoning permits density redistribution by relaxing zoning regulations for those projects that meet its requirements (e.g., minimum parcel size, minimum open space, and good site planning). A second approach also increases density for projects that provide specified site amenities. For example, Phoenix, Arizona, allows a 1 percent increase in density for each 4 percent of unimproved common open space or each 2 percent of improved common open space.[54] New York City provides bonuses for common open space, "good" site plans, community facilities (e.g., schools, day care centers, libraries), enclosed parking, and increased room sizes.[55]

One serious problem with cluster zoning is that the bonus is discretionary, not as-of-right. Since requirements such as "good site planning" are matters of opinion, the approval process can be abused by the reviewing entity. Some opponents fear that the additional density will lead to a decrease in housing values surrounding the development or that there will be an increased tax on municipal resources (sewer, water, schools) as a result of the increased density. As a result, poorly drafted PUD ordinances may result in bribery, political payoffs, or extra "amenities" demanded by adjacent (opposing) community groups. From a planning perspective, the most serious result may be scaring off reputable developers. One way to counteract these abuses and encourage cluster development is to specify minimal processing requirements and impose strict time limits. When Memphis, Tennessee, for example, limited the review period for PUDs to 14 days, it experienced "a dramatic increase in the number of PUD applications."[56]

The most important benefit of PUDs is not that they produce better suburban communities. That depends on the development team (entrepreneur, architect, landscape architect, and financial institution). The real benefit of cluster zoning is that it allows greater design flexibility, permits lower production costs, and produces a landscape with greater diversity in open-space arrangement and building distribution.

The New Urbanism

As historian Robert Bruegmann explains in *Sprawl: A Compact History,* critics have been unhappy with suburban development since the eighteenth century.[57] Clarence Stein's introduction of common open space was an early-twentieth-century attempt to restore a balance between community and privacy. At the close of the twentieth century, a group calling themselves *New Urbanists* initiated a different approach. They wished to reassert basic urban design principles they believed had been neglected or abandoned.[58]

The New Urbanism first emerged in the early 1980s as an outgrowth of Seaside, an 80-acre beach development on the Florida panhandle, between Pensacola and Panama City (see Chapter 16). Seaside was never intended to establish a whole new direction in suburban design. It began as a real estate venture that would maximize the return from a small parcel of Gulf of Mexico property. Robert Davis, the property owner, wanted the houses built on these particular lots to share a distinctive, regional appearance. Not long after beginning to work with Andrés Duany and Elizabeth Plater-Zyberk (DPZ), the designers of Seaside, Davis began a series of trips, some together with his architects and some without, to establish the common characteristics of traditional Florida Panhandle houses. Based on these trips, DPZ established the proportions, architectural details, and construction techniques that provided the basis for the design of the first two houses, for which ground was broken in 1981. The trips also provided the basis for the regulations that determined the appearance of all later construction.

At first, those few architects and planners who were aware of Seaside derided its conventional, easily understandable, emotionally accessible combination of picket fences, front porches, and gabled roofs. It was the customers who purchased the vacant lots, the contractors who built their houses, and the bankers who provided their mortgage financing who embraced what quickly became labeled as the New Urbanism. They quickly grasped the resale value of houses that included the best features of Florida's vernacular architecture. There was then and still remains today a huge market for an architecture that reflects traditional family values. The nearby, nondescript, "contemporary" bungalows that line Route 80 do not supply this market as well and thus cannot command the same high prices.

DPZ went on to design new communities that expanded on the principles they had first established at Seaside, and, within just ten years following their engagement for the design of Seaside, they had prepared plans for 10 new communities that would reach the development stage, including Kentlands (see Chapter 16) in suburban Washington, D.C., and Windsor, near Vero Beach, Florida. In 1993, they took the lead in establishing the Congress for New Urbanism, which would become a clearinghouse of information on the activi-

Seaside, 1990. The pink paving, crushed-shell parking area, gabled houses, and picket fences determine the character of this resort. *(Alexander Garvin)*

ties and projects of its members, an active proselytizer for their ideology, and a convener of annual meetings that bring together architects, planners, environmentalists, developers, public officials, academics, and a wide range of other people interested in reforming current development practices.

Architects may have been the first group to become interested in DPZ's work. However, they soon were followed by real estate industry professionals and by fascinated journalists. In 1992, the Urban Land Institute published a series of articles in its monthly magazine, *Urban Land.* Two years later, McGraw-Hill published Peter Katz's *The New Urbanism: Toward an Architecture of Community. The Atlantic, Smithsonian,* and an array of widely read periodicals printed stories on Seaside, DPZ, and DPZ's followers. *Newsweek*'s 14-page article devoted 7 pages to the 15 New Urbanist "ways to fix the suburbs."[59]

The growing popularity of New Urbanism is as much a function of Duany's determined advocacy in every part of the country as it is a function of Plater-Zyberk's equally determined effort to produce large numbers of intelligently educated professionals at the University of Miami, where she is dean of the school of architecture. The impact of their work extends far beyond their professional associates and students. In 1994, *Architectural Graphic Standards* (the book that is used in every American architect's office as a resource for the correct dimensions and construction details for virtually every aspect of architecture) first included a 13-page section on "site, community, and urban planning" prepared by Duany, Plater-Zyberk, and their colleagues.[60] In this way, New Urbanist planning concepts became an accepted urban design foundation for architectural work by draftsmen and designers who sometimes were completely unaware of the New Urbanism. The movement also affected traffic engineering. In 1997, the Institute of Transportation Engineers issued a manual entitled *Traditional Neighborhood Development Street Design Guidelines,* establishing design guidelines and standard dimensions for engineers involved with New Urbanist projects.[61]

New Urbanism claims to be a radical departure from conventional suburbanization. Its adherents favor mixed land use patterns that they hope will force residents to live, work, shop, and play in close proximity. This is more theory than reality. Mixed land use requires very high population densities and widespread public transit systems that are not present in most suburbs. Even in cities like Chicago and New York, where high densities and transit stations are common, few people actually live, work, and do their shopping all on the same block. They use the subway or the bus, or they drive. At four to eight houses per acre, the typical density of New Urbanist subdivisions, there are not enough customers within walking distance to support profitable retail activity. Consequently, retail districts develop at locations that are easily accessible by car to a large enough number of customers. As in most suburban subdivisions, houses are separated from these nonresidential uses because the residents wish to avoid intrusive truck deliveries, garbage collection, and other disruptive commercial traffic. Commercial uses, on the other hand, need to be accessible to as many custom-

Kentlands, 1993. Residents of this middle-class community purchased the nostalgic image of small-town America. *(Alexander Garvin)*

ers and employees as possible, not hidden amid residential streets.

Virtually all New Urbanists want to deemphasize the automobile. The best of New Urbanist–influenced subdivisions are definitely pedestrian-friendly. Yet none has been built around rail transit stations that would free residents from their dependence on the automobile. This form of development is unlikely to occur as long as local, state, and federal governments fail to make truly massive investments in new rail transit systems and substantially increase the allowable bulk and occupancy around the new transit stations. As a result, New Urbanist subdivisions have proven to be as automobile-dependent as other suburban subdivisions.

Like the modernist architects against whom many New Urbanists were rebelling, they believed that improvements to the physical environment would alter living patterns. In some ways they were right. Living in a house with a screened porch is different from living without one. However, just as the modernists failed to give birth to the brave new world of their dreams, so far the New Urbanists have failed to alter America's automobile-addicted suburban lifestyle.

It is design, not planning theory, that is responsible for the phenomenal market appeal of the best of the New Urbanist communities. Nobody seriously complains that New Urbanist communities suffer from sprawl. They tend to have definable edges, identifiable streets and districts, and clear distinctions between private property and the public realm. They have real streets, not loops or cul-de-sacs. Blocks are lined with sidewalks. Picket fences and front porches establish both where private property begins and where neighborly interaction can take place. There are no cars parked in the driveways because there are no driveways. Commuters on their way home turn down a narrow alley, park in a private garage, and pass through a rear yard (that often is far too small) before entering the house.

In nearly every market, there are now home builders who have copied one or another of the devices repopularized by the New Urbanism: rear-alley garages, row houses with front stoops, screened porches, and so forth. This is particularly evident in suburban Washington, D.C., where Kentlands (see Chapter 16) has set off a chain reaction. As the *Washington Post* explained:

> *They are proliferating in former farm fields and distant suburbs all around Washington, these clusters of brick row houses that look as though they were airlifted out of Georgetown. Some are imposing, New England style Victorians with wraparound front porches. Others are affixed with steeply angled stoops.*[62]

The explanation for the growing popularity of New Urbanist suburban design is its market appeal. Americans have been seeking instant roots since the nineteenth century. The first wave of suburbanization included Gothic farmhouses, Italian villas, Georgian manor houses, and French châteaux. At the beginning of the twentieth century, California invented a fantasy-Mediterranean architecture unknown in Spain, Italy, or France (see Chapter 18). New Mexico created a Santa Fe style that appeared to be a continuation of native American pueblo construction, but wasn't. No wonder the New Urbanists have been so successful with their combination of the best components of Charleston, South Carolina, and Cape Cod.

This market appeal is not simply a matter of supplying the demand for some connection to the past. Prior to the emergence of the New Urbanists, suburban developers were selling houses by providing ever more luxurious bathrooms and kitchens. With so much money going into private amenities, the obvious way to reduce development costs was to cut back any expenditures on the public realm. The real estate industry had gone too far in this direction. When the New Urbanists insisted on sidewalks and other public amenities, they were restoring the balance between investment in private property and the public realm. Consumer demand for this more "traditional" form of development quickly became obvious. No wonder the *Washington Post* could report that New Urbanist techniques were proliferating.

Consistent with the continued reference to the importance of design as one of the six ingredients for success, the most important reason for the popularity of the New Urbanism has been that so many New Urbanist communities are more beautiful than the characterless subdivisions that became common after World War II. Cookie-cutter suburbs have always lost out to more attractive competition when both are offered at similar prices. Duany and Plater-Zyberk may have repopularized traditional urbanism in their effort to respond to their clients' need for marketable communities. In the process, they started a revolution that has altered the public dialogue and is beginning to alter the future pattern of suburbanization. Like so many other revolutions, this one has made promises that cannot be fulfilled, but we must recognize that DPZ and their followers have provided marketable alternatives to many of the inadequacies of conventional suburban development. Their work is also likely to influence developers to provide the sidewalks, alley garages, houses with gables facing the streets, and other details that they have popularized.

Ingredients of Success

The history of residential suburban development seems almost easily predestined. The endless stream of customers flocked to a seemingly inexhaustible supply of land. Developers obtained financing, packaged the necessary elements, and offered their product for sale. Success only required doing it more effectively than the competition. The huge inventory of unsold suburban houses left in the wake of the subprime mortgage crisis is evidence that successful suburban development requires a deeper understanding of the ingredients of success.

Market

Residential suburbs attract customers by offering at lower prices a product that is equivalent to their competition, by offering amenities that are not available elsewhere, or both. Consequently, successful suburban developers are careful to keep their cost of production lower than the prevailing market price of neighboring properties or to offer a residential living environment that surpasses them.

Some developers are able to offer lower prices because the cost of land and infrastructure is spread over a large number of houses. Others find ways of lowering the cost of producing the houses themselves. The developers and designers of Sudbrook Park and Fisher Hill reduced site costs by exploiting local topography in a manner that minimized costly site work. Ethan's Glen reduced construction costs by combining eight residences into a single structure and minimizing exterior surfaces.

There are numerous ways of offering a better than average living environment. The developers of the private places of St. Louis provided enhanced privacy and landscaped islands. Sunnyside Gardens included carefully landscaped, commonly owned open space. Cluster communities add swimming pools, clubhouses, or other recreation facilities. Whatever the extras may be, they will help one project to outperform another.

Location

Accessibility is critical to the success of any suburban location. Boston's South End and Back Bay were sure to succeed because they were at the edge of the densely packed, rapidly expanding center city. Brooklyn Heights and Druid Hills became desirable because their developers exploited transportation systems that improved site accessibility, underscoring that location is a matter of time, not just spatial proximity.

Too many people erroneously believe that proximity has become less important because the automobile has made distant areas more accessible. Despite widespread dependence on cars and trucks, most of the American population remains concentrated on less than 2 percent of the land, and the increasing prevalence of traffic congestion has sustained the value of physical proximity. Since most development sites will continue to be located within commuting distance of relatively compact existing patterns of employment, shopping, and entertainment, we must find better ways of exploiting those locations.[63]

The success of any location is tied to the character of the area itself. Druid Hills had the benefit of ravines that provided natural drainage. Fisher Hill offered splendid views of Boston. Stands of fully grown pine provided a distinctive landscape for Ethan's Glen. In each case, the site plan amplified these assets.

Dull sites need something that will enliven them. Prospect Park South and Epernay were created on flat, characterless sites. Their success was dependent on skillful site planning and inventive landscaping. Too many residential suburbs fail because they offer an undistinguished location at prices that are similar to neighboring developments that offer something extra.

Design

The most common criticism of residential suburbs is their cookie-cutter appearance. Those whose design is better are able to exploit their competitive advantage. The developers of Sunnyside Gardens, for example, offered a better arrangement of streets, sidewalks, and open space than was available in neighboring sections of Queens. Lots on Portland Place and Westmoreland Place in St. Louis had the benefit of more imposing gatehouses, broader park islands, and larger house lots. The Olmsteds exploited the regional character of the Southern pines and evergreen magnolias that were native to Druid Hills. In all these instances, design was integral to the marketing strategy.

By altering the arrangement of lots, houses, and open space, the designers of the Back Bay and Radburn changed the pattern of daily life. The new combination of below-street-grade service alleys and broad streets in the Back Bay allowed builders to create more convenient row houses than were available elsewhere in the city. Stein and Wright accomplished a similar result in Radburn by altering the relationship between traffic arteries and houses, reorienting the service and living spaces of each house, and relocating and connecting outdoor play areas. As a result, Radburn's open spaces were more convenient and usable than those of competing developments. More important, the site plan established a new form of suburban living that had great appeal to home buyers who chose to live in Fair Lawn.

Dimensions are particularly important in determining the marketability of residential developments. For example, the relatively compact Radburn cul-de-sac allowed Stein and Wright to increase the amount of landscaped common open space without increasing project cost. The 25-foot minimum lot width in the Back Bay guaranteed that each building could fit four relatively wide rooms on each floor, two with windows opening onto the street and two more opening onto the rear yard. Row houses in competing areas were often too narrow to fit comfortably more than one room in front and another in back.

The "look" or "build" of a residential suburb is equally important. The gatehouses at Portland Place and Westmoreland Place distinguish them from more modest private places in St. Louis, just as the exquisite landscaping of Druid Hills distinguishes it from more conventional Atlanta subdivisions. Some designers exploit the existing landscape to provide their projects with distinctive character. The site plan for Ethan's Glen, for example, transformed existing pine trees into a design and marketing feature. On the other hand, there was nothing about the site of Forest Hills Gardens to differentiate it from the rest of Queens. Its distinctiveness was achieved by requiring all buildings to have red clay tile roofs.

Project failure is often a result of an unsightly design. In the attempt to avoid such failures, many communities adopt deed restrictions or enact subdivision regulations and zoning ordinances that determine lot size, building placement, allowable building materials, and landscaping. Duany, Plater-Zyberk, and their followers have succeeded in encouraging a number of suburban communities to rewrite their codes. We should be grateful for the marginal improvements that they have spawned. By themselves, however, such controls do not guarantee a convenient or beautiful suburban landscape.

Financing

All residential suburbs encounter financing challenges. Developers need capital to pay for property acquisition, planning, site development, infrastructure installation, and the expense of carrying the property till development costs have been recouped. Future residents need long-term mortgages to pay for the purchase of their houses. Thus, the availability and cost of financing is critical. Joel Hurt had to sell Druid Hills to a consortium that had better access to capital. Forest Hills Gardens and Radburn both faltered on their inability to provide even a limited return during periods of meager sales. In the Back Bay, however, the problem of development financing was eliminated by substituting the Commission on Public Lands for what is typically private-sector activity; that group front-ended the cost of planning, filling the site, and marketing individual lots and was able to pursue the development with a decreased sensitivity to the market than that to which a private developer would have been subjected.

In 1940, 15 percent of the American population lived in suburbs. Half a century later, that percentage had climbed to 46 percent.[64] Without low-interest, low-down-payment FHA and VA mortgages, this post-World War II suburbanization would not have been possible. More important, the availability of FHA and VA mortgages ensured that new suburban development would not be reserved primarily for the upper middle class.

During the 1960s, 1970s, and 1980s, local governments began to demand that developers donate land for schools, parks, and streets, and even that they help pay for water and sewer pipes, bury electric wiring, and provide extra landscaping and a variety of expensive amenities. The cost of these requirements was transferred to home buyers and threatened again to limit new development to affluent consumers who could afford these added costs. The only way to reduce development expenditures was to build row houses, apartment buildings, and condominiums. Without FHA acceptance of this form of ownership and FHA regulations establishing the requirements for condominium mortgages, such development would not have become widespread. So, once again, government mortgage insurance prevented suburban development from becoming the exclusive preserve of the affluent.

If we are to improve the character of our residential suburbs, we must reject national monetary policy that is dependent on roller-coaster interest rates. We must return to the relative stability of the two decades following World War II, when the Federal Reserve made sure that there was a stable supply of money that was large enough to finance landscaping, infrastructure, and community facilities—not just little boxes made of ticky-tacky.

Time

Some residential suburbs are designed in a manner that makes one's stay there a rich experience. The landscaping of an Olmsted subdivision, for example, enhances the experience of traveling home, spending the day on one's property, and mixing with the neighbors. Radburn provides playgrounds, swimming pools, landscaped recreation areas, and an elementary school, all gathered together in an environment free of the dangers of vehicular traffic. The promotional brochure for Epernay points out that:

> *You can enjoy relaxing around the pool or a quick game of tennis. A bicycle ride or jogging on the paths. Croquet in any one of a hundred spots in your green common yard. Or just enjoy your neighbors—an informal chat across the back-yard fence or cocktails in their patio.*[65]

Any of these examples illustrates that successful marketing of residential subdivisions requires more than a nice house at an affordable price. In addition, the trip through the project must be a pleasant experience, and living there must provide a wide range of activities 24 hours a day, 7 days a week.

The most difficult problem facing developers of residential suburbs, though, is financial security over the life of the project. Most residential suburbs require a massive investment in infrastructure and site work. The costs of carrying this investment cannot be charged to the first few lots that are sold; a steady stream of sales must retire them. Anything can go wrong: shortages of materials, strikes, sudden competition, unanticipated construction problems . . . The worst disaster is an economic downturn. If Radburn had not opened for sale just in time for the Great Depression, the City Housing Corporation might still be in business and Radburn might have become the first of many residential suburbs made up of superblocks with open-space frameworks and pedestrian-traffic separation. The subprime mortgage crisis is just the most recent example of how sensitive developers are to macroeconomic trends. Innumerable less-well-known developers went out of business between 2008 and 2010 because of the financial challenges that the crisis engendered.

Entrepreneurship

Some developers excel at coordinating the many activities and participants required to create a residential suburb. Others are better at finding innovative approaches to the development process. Public policies should make it easy for the developer to do both with minimum risk. The best way to effect this is by making it easier for developers, whether they are government agencies, limited-profit or nonprofit institutions, or private-market parties, to produce attractive residential suburbs.

FHA minimum property standards provide a demonstration of what intelligent government programs can achieve. Developers who depended on FHA financing were forced into producing residential suburbs that met higher standards. Once they established those minimum standards, developers of more expensive suburbs had to provide something better. Cluster zoning, for example, makes it easier for the developer to preserve natural features, provide amenities, and lower development costs.

Sometimes the cost, time, and risk involved in developing a residential suburb are too high to interest the private market. In those cases, the best option is to create and finance a public entrepreneur. Boston adopted this approach for the Back Bay. Because few areas involve such extraordinary resources or risk, the country continues to avoid public entrepreneurs and depends on the private sector.

Development by nonprofit and limited-profit institutions is often thought to be a more effective way of developing residential suburbs. Forest Hills Gardens, Sunnyside Gardens, and Radburn provide resounding evidence that they can produce projects that are more attractive than those of many contemporaneous, profit-motivated competitors. It is therefore tempting (but often wrong) to think that their projects are also less costly. Nonprofit developers still have to provide a reasonable return to their lenders. Furthermore, it is not at all clear that they spend less on personnel and overhead than their profit-oriented competition. In fact, Forest Hills Gardens was terminated because the Russell Sage Foundation could not compete with conventional developers' lower development costs.

The most serious problem with this model, though, is that only a handful of nonprofit entities are involved with real estate development. Like the City Housing Corporation, which developed Sunnyside Gardens and Radburn, they tend to go out of existence after a small number of projects. Without a large supply of experienced nonprofit developers, there is no way to sustain any serious level of production.

It is not the type of developer that is significant. It is the financial and regulatory context in which they operate that matters. Thus, the most promising route to improved suburban development is by way of laws that are easy to understand and enforce and do not depend on the caprice of any public official's taste, politics, or venality. The certainty and predictability of such laws reduces risk. More important, this type of legal framework establishes a regulatory environment in which developers are more likely to produce higher-quality, less costly, and more appropriate residential suburbs.

A Planning Strategy for the Suburban Landscape

The distinction between city and suburb withered away long ago. American cities engulf their suburbs almost as soon as they are created. Within a few decades of its inception, Brooklyn Heights was already part of the city of Brooklyn. Of the suburbs discussed in this chapter, only Riverside, Illinois, Sudbrook Park, Maryland, and Radburn, New Jersey, are not within the political boundaries of a large city.

As historian Robert Bruegmann has pointed out, the term *suburb* itself is not particularly useful. As he explains:

> *What was within the municipal boundary of any given city and what was outside it changed constantly and differed markedly from city to city. Contrary to the stereotypes created by many postwar "urban historians," suburbs were not at all homogeneous. There were affluent suburbs and modest ones, bedroom suburbs and industrial ones, white suburbs and black ones. The outer part of the metropolitan area was just as complex as the inner.*[66]

Even the notion of city limits has long ceased to have much meaning. H. G. Wells had it right in 1902, when he wrote that:

> *... "town" and "city" will be, in truth, terms as obsolete as "mail coach." For these new areas that will grow out of them we want a term, and ... we may for our present purposes call these coming town provinces "urban regions."*[67]

The urban regions he foresaw have not turned out to be single-city centered. For example, the Los Angeles metropolitan area (the second largest in the United States by population in 2010), covered 4850 square miles and included the cities of Long Beach, Santa Ana, Anaheim, Irvine, Glendale, Huntington Beach, and Santa Clara. The third largest, Chicago, which covered 10,856 square miles, included Gary,

Joliet, Evanston, Waukegan, Schaumburg, and a dozen others. Nor are central cities any longer the focus of life in these urban-suburban regions. As Christopher Tunnard explained in 1958, in his pioneering article, "America's Super-Cities":

> *The family . . . is not beholden to any one central business district either for shopping or for employment. Travel time is the chief factor here, not distance, and it is quite possible that husband and wife may go in opposite directions to get to their jobs.*[68]

Despite the complexity of these urban regions, their "suburban" residential districts are all variants of only three forms: rectilinear, curvilinear, and clustered. There is one significant difference between the clustered community and the other two. It is centered around common open space and therefore is as fundamentally private as is exclusive Llewellyn Park with its gatehouse and police force. The other two forms are dominated by public streets and sidewalks that, in the words of Frederick Law Olmsted Sr., ensure their residents are not "far removed from the life of the community."

Many of the better suburban subdivisions achieve their distinctiveness through deed restrictions that establish minimum development standards. The Olmsted firm, in its various incarnations, achieved such uniqueness by tailoring subdivisions to local topography, climate, and vegetation. Developers of more recent clustered communities have done so by building in a consistent (though often heavy-handed) architectural style.

Today, virtually all development is subject to local regulations. Unfortunately, regulation is not an automatic path to the production of lovely residential suburbs: government cannot legislate good design. The best course is to reduce to an absolute minimum the amount of time and money spent getting public approvals and let developers spend the savings on the projects themselves.

Even if every residential suburb were beautifully laid out and distinguishable from its neighbors, it would lack a mutually reinforcing relationship with neighboring districts. The gatehouses built for the private places of St. Louis provide visible evidence that residents wish to be protected from their surroundings. Stein and Wright's attempt to include within each Radburn superblock the basic components of daily life reflects a similar desire to insulate project residents from automobile traffic. The difference is that at Radburn, Stein and Wright envisioned a system of interconnected pedestrian spines that would tie its superblocks together into a single healthy community. But even if its internal circulation system had been completed, Radburn (like virtually all residential suburbs) would still lack a mechanism with which to promote interaction with other subdivisions and with the region as a whole.

Another difficulty with the suburban landscape is that visitors and residents alike rarely have any means of orientation. Except in those regions that remained wedded to the rectilinear structure provided by the Northwest Ordinance and the Homestead Act, or that are carefully oriented to the interstate highway network, or that are shaped by prominent topographical features, it is difficult to find one's way through the accumulation of loops, cul-de-sacs, and winding roads.

Observers frequently criticize residential suburbs for failing to provide shopping, employment, entertainment, recreation, and institutional and transportation facilities. This is unfair. Our urban regions do not suffer from a lack of shopping malls, drive-in retail establishments, office parks, or industrial districts. The private sector is ready to oblige wherever it identifies a ready market for such facilities.

If there is a lack of schools, hospitals, libraries, and other public institutions, it is because the public either does not want to pay for them or is unwilling to make the tough decisions needed to locate and build them. The same is true where there is insufficient public transit or parkland.

The common prescription for correcting all these problems is regional decision making. There is a vast literature on regional planning, but, as the landscape so graphically demonstrates, it has had very little impact. Nor is regional decision making likely to become more effective. The dense thicket of government jurisdictions is getting ever more crowded and the demand for local participation ever louder and more prevalent.[69]

If we are to correct the unsatisfactory nature of the suburban landscape, we must deal with two different situations. On one hand, we must augment already-built-up areas by providing adequate means of orientation and places in which the residents of its various subdivisions can interact with one another. On the other hand, we must see that developers build new residential suburbs with the framework of public places that has been missing in most existing suburbs.

An Open-Space Framework for Existing Suburbs

There is as yet no public support for reorganizing the hodgepodge of existing suburban communities. People may think it difficult to find their way through the loops and cul-de-sacs, but they will not support public action because they think this problem can be remedied only by creating new traffic arteries. Not only would this be expensive, it also would require condemning property and dislocating large numbers of people and businesses. Besides, most suburbanites would be opposed on the grounds that it would assist strangers to intrude into secluded communities. In fact, additional roads would not be of much help. Community interaction requires places that attract large numbers of people who want to do things together, not new traffic arteries that would only increase interaction among automobiles.

Palm Beach County, Florida, 2005. Much of the suburban landscape is a crazy quilt of small real estate developments built to satisfy changing market conditions and regulations. *(Alexander Garvin)*

Places of public interaction will not be forthcoming without changes in public opinion and political leadership. A good example is the open space acquired since 1967 by the city of Boulder, Colorado. When civic leaders wanted to preserve the remaining mountain backdrop and the entrances to the Boulder Valley, they did not create a regional government or publish a needs analysis; they imposed a 0.4 percent sales tax to be used for open-space acquisition.[70]

Boulder has acquired more than 40,000 acres. This acreage includes a system of greenways that was created out of runoff channels, drainage corridors, and streams. The money from Boulder's open-space tax allowed the city to expand, landscape, and transform these neglected corridors into bicycle/jogging trails, wildlife preserves, and recreation areas—not just storm drainage and flood channels. At the same time, the greenways connect new and old athletic fields, schoolyards, picnic areas, and playgrounds. Thus, Boulder's greenways perform the role of the open-space backbone that Stein and Wright had intended for Radburn.

While other communities may not have Boulder's extensive network of streams and creeks, they have other underutilized properties. These include unbuildable steep slopes, abandoned buildings (drive-in movies, isolated retail clusters, warehouses, schools, etc.), unpaved rights-of-way, once actively used roads, and empty lots. Such properties can be knitted together into a similar open-space framework.

The most serious obstacle to creating great suburban environments is the misconception that sprawl is a land development problem. The real problem can be found everywhere in already-built suburbs. It is only solvable by retrofitting existing suburbs with a great public realm framework.

Creating a Suburban Public Realm Framework

There are three simple actions suburban communities can take to begin creating this public realm framework. They involve the usual tools of the trade: regulation, incentives, and public investments. Local subdivision and zoning regulations must be amended to require *public* (rather than private or common) open space in any subdivision or condominium community. California, for example, allows local governments to create new parkland as part of the subdivision approval process. Its Subdivision Map Act authorizes cities and counties to require developers to dedicate parkland or to collect a so-called Quimby fee that will allow the jurisdiction to purchase land for parks. If every developer had to dedicate 30 percent of any site to public use, and if site occupants, whether owners or renters, were legally responsible for its maintenance, our suburban landscape would begin to have the public places needed for "the refined domestic life, secluded, but not far removed from the life of the community," that Olmsted recommended.

This strategy is so simple that public officials cannot stall while they check if a development meets legal requirements, developers cannot say it is too difficult to produce, and home buyers cannot fail to be attracted by the value of public amenities. Furthermore, it can be achieved without any appreciable increase in cost. Developers, whether building meandering subdivisions (with private open space) or condominium communities (with common open space), are already providing open space. All they need to do is locate it so that it is not separated for the exclusive use of their residents. Since most suburban developments already provide

DeKalb, Georgia, 2007. Existing traffic arteries. *(Alexander Garvin)*

DeKalb, Georgia, 2007. A proposal by AGA Public Realm Strategists that would create a public realm framework for the area. *(Alexander Garvin)*

substantially more than 30 percent open space, there will be plenty of land left over to provide each resident with appropriate accessory (private or common) open space.

Whether this new public open space takes the form of park islands, like those of the private places of St. Louis, or meandering linear parks, like those in Atlanta's Druid Hills, or some other form, it can transform noncommunities into places that are recognizably different from one another. As the landscape fills in, these new open spaces will become part of a sequence of public places that will provide a distinctive and varied setting for public activity, a connective tissue transforming independent developments into interdependent neighborhoods, and a means of orientation that can be understood and admired by residents and visitors alike.

The idea of incentivizing the private market is based on the plaza bonus in the New York City Zoning Resolution. If developers of large sites (e.g., 10 acres or more) provide 20 percent more public open space than is required, they could be rewarded with a bonus of 20 percent more interior floor area than would otherwise be allowed—but only if the open space is directly accessible from public streets and remains open to the public at all times. Between 1961 and 2000, New York's very similar open bonus resulted in the addition of more than 500 plazas covering 3.5 million square feet of privately owned, public open space.[71]

Third, every community should transform its traffic arteries into places that can be enjoyed, even when people are not in their cars. This means investing in sidewalks that are broad enough to accommodate more than one pedestrian at a time, bikeways that are protected from the traffic, and trees that line every public thoroughfare. This has already happened in Chicago where, beginning in 1996, the city installed planters in the medians of city streets and along its sidewalks and planted new trees throughout the city. By 2006, Chicago had more than 70 miles of newly planted medians and 300,000 additional street trees. It will be a matter of time before suburban communities, like Chicago, will be collecting additional real estate taxes from the houses that have gained value as a result of this public-sector investment and both additional real estate and sales taxes from the retail stores, restaurants, and cafés that open where zoning permits mixed land uses.[72]

Together, these three devices will begin the process of transforming placeless, first-growth sprawl into something worthy of the hopes and dreams of the people who live there.

Notes

1. Address of President Hoover delivered at Constitution Hall, Washington, DC, December 2, 1931, quoted in Charles Abrams, *The City Is the Frontier,* Harper & Row, New York, 1965, p. 254.
2. James W. Rouse, "The City of Columbia, Maryland," in H. Wentworth Eldredge (editor), *Taming Megalopolis,* vol. 2, Anchor Books, Garden City, NY, 1967, p. 839.
3. Kenneth Jackson, *The Crabgrass Frontier,* Oxford University Press, New York, 1985, p. 135.
4. Pliny the Younger (c. 61–114 A.D.) describing his villa at Laurentum outside Rome, quoted in J. J. Pollitt, *The Art of Rome c. 753 B.C.–337 A.D.,* Prentice-Hall, Englewood Cliffs, NJ, 1966, pp. 171–174.
5. The houses fronted on streets either 75 or 37.5 feet wide, and were serviced by a 22.5-foot alley.
6. I am persuaded by Christopher Tunnard and Kenneth Jackson that Brooklyn Heights is America's first planned suburb. See Tunnard, *The Modern American City,* Van Nostrand, Princeton, NJ, 1968, pp. 33–34, and Jackson, op. cit., pp. 25–32.
7. *Long Island Star,* December 24, 1823.
8. Historical and statistical information on Prospect Park is derived from New York City Landmarks Preservation Commission, *Prospect Park South Historic District Designation Report,* New York City, 1979.
9. *Prospect Park South,* Prospectus, Brooklyn, NY, 1900.
10. Historical and statistical information on the Garden District and the faubourgs of New Orleans is derived from Leonard V. Huber, *Landmarks of New Orleans,* Louisiana Landmarks Society, New Orleans, LA, 1991; S. Frederick Starr, *Southern Comfort: The Garden District of New Orleans, 1800–1900,* MIT Press, Cambridge, MA, 1989; and Roulhac Toledano, *New Orleans: The Definitive Guide to Architectural and Cultural Treasures,* Preservation Press and John Wiley & Sons, New York, 1996.

11. Historical and statistical information on the private places of St. Louis is derived from Charles Savage, *Architecture of the Private Streets of St. Louis,* University of Missouri Press, Columbia, MO, 1987.
12. Historical and statistical information on the South End and the Back Bay is derived from Walter Muir Whitehill, *Boston: A Topographical History,* Belknap Press, Harvard University Press, Cambridge, MA, 1968, and Bainbridge Bunting, *Houses of the Back Bay,* Belknap Press, Harvard University Press, Cambridge, MA, 1967, especially pp. 129–139, 250–255, 360–399.
13. For a discussion of the resulting lawsuit, see Stanley K. Schultz, *Constructing Urban Culture: American Cities and City Planning 1800–1920,* Temple University Press, Philadelphia, 1989, pp. 39–41.
14. Whitehill, op. cit., p. 122.
15. Between 1852 and 1855, the Harbor and Land Commission was known as the Commission on Boston Harbor and Back Bay Lands. After 1970, it became known as the Harbor and Land Commission.
16. Among the more interesting analyses of the impact of transportation on suburban growth are James E. Vance Jr., *Capturing the Horizon,* Johns Hopkins University Press, Baltimore, 1990; Sam B. Warner Jr., *Streetcar Suburbs, The Process of Growth in Boston 1870–1900,* Atheneum, New York, 1969; and David Brodsly, *L.A. Freeway,* University of California Press, Berkeley and Los Angeles, 1981, especially pp. 61–82, 151–160.
17. For a discussion of California's development regulations, see Marc A. Weiss, *The Rise of the Community Builders, The American Real Estate Industry and Urban Land Planning,* Columbia University Press, New York, 1987.
18. Jackson, op. cit., pp. 203–218; Weiss, op. cit., pp. 141–158.
19. Historical and statistical information on the Olmsted firm is derived from Charles E. Beveridge and Carolyn F. Hoffman, *The Master List of Design Projects of the Olmsted Firm 1857–1950,* National and Massachusetts Associations for Olmsted Parks, Boston, 1987; maps, drawings, and photographs maintained at the Frederick Law Olmsted National Historic Site in Brookline, MA, by the National Park Service; Charles Capen McLaughlin and Charles E. Beveridge (editors), *The Papers of Frederick Law Olmsted,* vols. 1–6, Johns Hopkins University Press, Baltimore, 1977–1992; Laura Wood Roper, *FLO: A Biography of Frederick Law Olmsted,* Johns Hopkins University Press, Baltimore, 1973; Elizabeth Stevenson, *Park Maker: A Life of Frederick Law Olmsted,* Macmillan, New York, 1977; Charles E. Beveridge and Paul Rocheleau, *Frederick Law Olmsted Designing the American Landscape,* Rizzoli, New York, 1995; and Witold Rybczynski, *A Clearing in the Distance: Frederick Law Olmsted and America in the 19th Century,* Scribner, New York, 1999.
20. I am unable to determine whether Olmsted knew or visited Blaise Hamlet (1811) or Park Village West (1830), Sir John Nash's pioneering, village-inspired suburban communities.
21. Charles Capen McLaughlin and Charles E. Beveridge (editors), *The Papers of Frederick Law Olmsted,* vol. 1, *The Formative Years 1822–1852,* Johns Hopkins University Press, Baltimore, 1977, pp. 74–77.
22. Andrew Jackson Downing, "Our Country Villages," *Horticulturist,* vol. 4, no. 12, New York, June 1850, pp. 537–541.
23. McLaughlin and Beveridge (editors), op. cit., p. 282.
24. Olmsted, Vaux and Company, *Preliminary Report upon the Proposed Suburban Village at Riverside, near Chicago,* originally published in 1868 and reprinted in S. B. Sutton (editor), *Civilizing American Cities,* MIT Press, Cambridge, MA, 1971, p. 299.
25. Schultz, op. cit., p. 175.
26. The idea of bringing the lawn right up to the house predates Olmsted, having been introduced during the eighteenth century by the English landscape gardener Capability Brown.
27. Most, but not all, of the streets of Sudbrook Park were completed as designed. Many of the original lots were later subdivided to accommodate smaller suburban houses. The railroad station was demolished. As a result, one can no longer see all of the community that Olmsted envisioned. Nevertheless, the bulk of Sudbrook Park remains as he designed it more than a century ago. Some sections of Parkside follow Olmsted designs. Tarrytown Heights was stillborn.
28. Historical and statistical information on Fisher Hill is derived from Cynthia Zaitzevsky, *Frederick Law Olmsted and the Boston Park System,* Harvard University Press, Cambridge, MA, 1982, pp. 114–117.
29. Historical and statistical information on Druid Hills is derived from Elizabeth A. Lyon, "Frederick Law Olmsted and Joel Hurt: Planning for Atlanta," in Dana F. White and Victor A. Kramer (editors), *Olmsted South,* Greenwood Press, Westport, CT, 1979, pp. 165–193; and maps, drawings, and photographs maintained at the Frederick Law Olmsted National Historic Site.
30. John M. Glenn, general director of the Russell Sage Foundation, *Statement,* dated September 4, 1911, quoted by Arthur C. Comey and Max S. Wehrly in *Urban Planning and Land Policies,* Urbanism Committee to the National Resource Committee, U.S. Government Printing Office, Washington, DC, 1939, pp. 104–105.
31. Minutes of the Sage Foundation Homes Company, April 15, 1924, quoted by Comey and Wehrly, op. cit., p. 108.
32. Comey and Wehrly, op. cit., p. 109.
33. Statistical and historical information on Radburn is derived from Clarence Stein, *Toward New Towns for America,* MIT Press, Cambridge, MA, 1966, and Eugenie Ladner Birch, "Radburn and the American Planning Movement," in Donald A. Krueckeberg (editor), *Introduction to Planning History in the United States,* Center for Urban Policy Research, Rutgers University, New Brunswick, NJ, 1983, pp. 122–151.
34. Stein, op. cit., p. 41.
35. Statistical and historical information on Sunnyside Gardens is derived from Stein, op. cit., and Franklin J. Havelick and Michael Kwartler, "Sunnyside Gardens—Whose Land Is It?" *New York Affairs,* vol. 7, no. 2, New York University, New York, 1982, pp. 65–80.
36. Havelick and Kwartler, op. cit., p. 73.
37. Stein, op. cit., p. 44.
38. Roy Lubove, Community Planning in the 1920s: The Contribution of the Regional Planning Association of America, University of Pittsburgh Press, Pittsburgh, PA, 1962.
39. Stein, op. cit., pp. 14–15.
40. Clarence Perry, *Regional Survey of New York and Its Environs,* vol. 7, Committee on Regional Plan of New York and Its Environs, New York, 1929, pp. 34–35.
41. Stein, op. cit., p. 47.
42. Ibid., p. 44.
43. Ibid., p. 48.
44. Ibid., p. 68.
45. "Radburn Revisited," *Ekistics,* March 1972, p. 200.
46. Colleen Grogan Moore, *PUDs in Practice,* Urban Land Institute, Washington, DC, 1985, p. 9.
47. Robert Engstrom and Marc Putman, *Planning and Design of Townhouses and Condominiums,* Urban Land Institute, Washington, DC, 1986, p. 5.
48. U.S. Bureau of the Census, *Statistical Abstract of the United States: 1989* (109th ed.), Washington, DC, 1989, p. 700.
49. Engstrom and Putman, op. cit., p. 5.
50. Moore, op. cit., p. 5.
51. Malvina Reynolds, "Little Boxes" (ASCAP).
52. Statistical and historical information on Epernay and Ethan's Glen is derived from Carla C. Sabala (editor), *Houston Today,* Urban Land Institute, Washington, DC, 1974, pp. 94–99.
53. Common areas and facilities are not managed by a condominium association but held in trust for the benefit of the homeowners by the Texas Commerce Bank.
54. Moore, op. cit., p. 17.
55. New York City Zoning Resolution, Chapter 8, Sections 78-34, 78-351-78-354.
56. Moore, op. cit., p. 19.
57. Robert Bruegmann, *Sprawl: A Compact History,* University of Chicago Press, Chicago, 2005, pp. 115–168.
58. Statistical and historical information on the New Urbanism is derived from *New Urban News* (a newsletter established in 1996 that is published six times a year); Lloyd W. Bookout, *Urban Land,* January 1992, pp. 20–26, February 1992, pp. 10–15, April 1992, pp. 18–25, June 1992, pp. 12–17, Urban Land Institute, Washington DC; Peter Katz, *The New Urbanism: Toward an Architecture of Community,* McGraw-Hill,

New York, 1994; Jerry Adler, "Bye-Bye Suburban Dream—15 Ways to Fix the Suburbs," *Newsweek,* May 15, 1995, pp. 40–53; John A. Dutton, *New American Urbanism: Reforming the Suburban Metropolis,* Skira, Milan, 2000; and Congress for the New Urbanism, *Charter of the New Urbanism,* McGraw-Hill, New York, 2000.
59. Adler, op cit.
60. Charles George Ramsey and Harold Reeve Sleeper, *Architectural Graphic Standards* (9th ed.), John Wiley & Sons, New York, 1994.
61. Institute of Transportation Engineers, *Traditional Neighborhood Development Street Design Guidelines,* Washington, DC, June 1997.
62. Alex Marshall: "Putting Some City Back in the Suburbs," *Washington Post,* September 1, 1996, pp. C1–C2.
63. Peter Wolf, *Land in America,* Pantheon Books, New York, 1981, pp. 24–25.
64. *New York Times,* June 1, 1992.
65. GreenMark, Inc., promotional brochure, *Epernay,* Houston, TX.
66. Robert Bruegmann, "Schaumburg, Oak Brook, Rosemont, and the Recentering of the Chicago Metropolitan Area," in John Zukowsky (editor), *Chicago Architecture and Design 1923–1933,* Prestel, Munich, 1993, p. 166.
67. H. G. Wells, *Anticipations,* quoted by Christopher Tunnard in "America's Super-Cities," in Wentworth Eldredge (editor), *Taming Megalopolis,* vol. 1, Anchor Books, Garden City, NY, 1967, p. 6. The Tunnard article is adapted from an earlier version originally published in *Harper's Magazine,* vol. 217, issue 1299, August 1958.
68. Tunnard, op. cit., p. 11.
69. A representative sample of this literature can be found in John Friedmann and William Alonso (editors), *Regional Development and Planning: A Reader,* MIT Press, Cambridge, MA, 1964.
70. Statistical and historical information on Boulder's open-space system is derived from Love & Associates, *Boulder's Stream Corridors—Design Guidelines,* City of Boulder Planning Department, Boulder, CO, 1989; City of Boulder, *Boulder's Greenways—Design Guidelines,* City of Boulder Planning Department, Boulder, CO.
71. Alexander Garvin, *Public Parks: The Key to Livable Communities,* W. W. Norton, New York, 2010, pp. 50–51; Jerold S. Kayden, *Privately Owned Public Open Space,* John Wiley & Sons, New York, 2000.
72. Alexander Garvin, *The Planning Game: Lessons from Great Cities,* W. W. Norton & Company, New York, 2013, p. 33.

15

New-Towns-in-Town

Chicago, 2012. Central Station. *(Alexander Garvin)*

As the nation's population exploded, the idea of creating new, planned communities within the existing urban fabric became increasingly popular. Lewis Mumford, one of the twentieth century's most ardent advocates of the idea, presented the case for such new, planned communities in 1967, arguing before Congress that:

> *. . . further increase of population in already congested centers should be met, not by intensifying the congestion in high-rise buildings, not by adding endless acres and square miles of suburbs, with ever-longer and more time-wasting journeys to work, but by building new, planned communities.*[1]

There would be no more messy, dirty, noisy, smelly, inconvenient neighborhoods! Only places that provided all the necessary educational, recreational, commercial, and community facilities: "an environment so rich in human resources" that, as Mumford put it, no one would willingly leave it even temporarily on an astronautic vacation."[2]

One way to avoid adding "square miles of suburbs" is to locate new communities within existing urban areas. This allows the growth areas to capitalize on sewers, water mains, and traffic arteries that are already in place. More important, such *new-towns-in-town* can provide the critical mass necessary for a full range of facilities that could never be justified by small, scattered real estate ventures by different developers.

Mumford thought these planned communities would be like Radburn: charming, low-rise, low-density communities "for the motor age." Le Corbusier, on the other hand, was convinced that the motor age required high-density, high-rise districts connected by superhighways. Others argued for extensions of existing street grids.

Whatever its appearance, a new-town-in-town requires a site of at least 30 acres. Contrary to conventional wisdom, urban sites of this size can be obtained without massive clearance and relocation. They are available wherever there are underutilized railroad yards, commercial waterfronts, industrial complexes, airfields, amusement parks, country clubs, and other facilities that could profitably move to less expensive locations. Parkchester, a new-town-in-town for 12,271 families in the Bronx, replaced the Society for the Protection of Destitute Roman Catholic Children. Fresh Meadows, a new community for 3287 families in Queens, replaced a country club and golf course. At Century City, in Los Angeles, an entire regional center with shopping facilities, office towers, and condominium apartments was built on the back lot of a movie studio.

Sites that are large enough for a new-town-in-town can only be purchased for huge sums. Even more money is needed to pay for streets, sewers, water mains, housing, schools, stores, parks, and the full range of community services. Mortgage financing on this scale can be supplied by only a few titanic financial institutions, corporations, labor unions, and government agencies. For this reason, many early efforts at new-towns-in-town were initiated and financed by insurance companies. Park La Brea in Los Angeles was financed by the Metropolitan Life Insurance Company; Fresh Meadows by the New York Life Insurance Company. Later new-towns-in-town, such as Brickell Key in Miami and Bat-

Bronx, 2012. 12,271 families live in Parkchester. *(Alexander Garvin)*

tery Park City in Manhattan, were financed by large corporations or state-chartered public authorities.

Timing is the most important decision facing developers of any new-town-in-town. In order to market their project, they must be able to provide prospective residents with neighborhood services as soon as they move in. This requires an enormous investment in infrastructure and community facilities. Since it will be years before all the residents will have moved in, successful developers have to schedule development so that project revenues will cover debt service on already installed infrastructure and community facilities. Moreover, the huge scale of these projects requires developers to be financially strong enough to weather the unavoidable macroeconomic changes that occur during any long-term development project.

During the 1940s and 1950s, when Parkchester, Park La Brea, and Fresh Meadows were created, most citizens welcomed such new communities. Necessary zoning changes were relatively easy to obtain. More recent projects have resulted in lengthy environmental reviews; political opposition from neighboring residents and property owners; costly, time-consuming litigation; and even demands that developers contribute "givebacks" in exchange for the right to proceed. As a result, only those few exceptional developers who are able to orchestrate all the participants, financial arrangements, and required approvals are able to develop a new-town-in-town.

The increased controversy over proposed new-towns-in-town is evidence that planned new communities may not be the panacea that Mumford envisioned. Some projects are successfully executed; others cannot be brought to completion. Some provide wonderful living environments; others are less than satisfactory. More important, some new-towns-in-town can be introduced with little or no adverse impact or even improve the surrounding city; others wreak havoc on the existing cities.

New-towns-in-town like Century City, Pentagon City, and RiverPlace, in Portland, demonstrate that easily accessible locations and sensible financing are essential to their success. However bland the design of these developments, high intensity and mixed use allow them to flourish while segregated, predominantly residential, large-scale *utopias around the corner* slowly lose their appeal to emerging generations that increasingly choose livelier urban settings for their businesses and residences.

There is now enough experience with new-towns-in-town to be able to determine when they will provide the rich environment for which Mumford argued so passionately and eloquently. That experience must be codified into project requirements with which developers will have to comply. Only then will opposition to large-scale, new development die down and increasingly well-planned new-towns-in-town appear within the clutter of our urban-suburban landscape.

Utopia Around the Corner

One of the most common arguments for developing new-towns-in-town is that they can provide thousands of families with decent homes and healthy living environments on sites that are virtually around the corner. Unlike redevelopment projects, they can be created with little or no displacement. Unfortunately, they often turn out to be colossal housing projects or shopping centers masquerading as mixed-use neighborhoods rather than vibrant communities.

Park La Brea in Los Angeles is a good example. MetLife built it in stages, between 1941 and 1948, on a 176-acre site. Superficially, Park La Brea appears to be a pallid copy of Le Corbusier's City of Tomorrow. However, it has neither the vast amounts of continuous parkland nor the many community facilities that Le Corbusier envisioned. Instead, Park La Brea includes 4200 apartments and plenty of parking, but little else. MetLife used the same formula at Park Merced in San Francisco to produce 3370 apartments on 200 acres.

Even if projects include more than just housing, they may still fail to become lively neighborhoods. This happens when the housing is so dominant that the project becomes little more than a dormitory. Despite their very different appearance and design, Fresh Meadows and Parkchester are good examples of how such dormitory communities can be developed with different features and different results.

Fresh Meadows is a watered-down, mid-1940s version of one of Clarence Stein's neighborhood units (see Chapter 14). Its 170 acres include a nursery school, a clubhouse, a small professional building, a small shopping center, 141 garden apart-

Los Angeles, 1990. Park La Brea offers 4200 families the open space and parking of Le Corbusier's "City of Tomorrow" but none of the commercial or community facilities that constitute a complete new-town-in-town. *(Alexander Garvin)*

Queens, 2006. Fresh Meadows. *(Alexander Garvin)*

ment buildings, and two brick residential towers. Despite its essentially low-rise character, short curving streets, and rolling lawns, the 3287 apartments at Fresh Meadows never coalesced into a lively, heterogeneous new-town-in-town because the apartments occupy almost all of the site and are carefully segregated from other land uses.

Unlike Fresh Meadows, Parkchester is more than an immense housing project. Begun in 1938, its 129-acre site includes a 100-store commercial center with a branch of Macy's and a 2000-seat movie theater. While the project's 171 elevator buildings and 12,271 apartments dominate everything, Parkchester succeeds at creating a series of identifiable places that, together, are understood as a community rather than simply a group of buildings. Visible changes in topography create unique building combinations and help define the spaces between each building as unique places. These individual open-space destinations have clear geometric shapes complemented by mature trees that offer a welcome contrast with the scale of the buildings. The stores along its busy retail streets are lined with colorful ceramic tiles that make them visibly different from the usual commercial outlets. Throughout the complex, buildings are decorated with unexpected small sculptures that individualize each residence, and at the center of everything is an oval park with decorative sculptures setting off a prominent fountain. Nobody would mistake Parkchester for a housing project whose buildings are evenly scattered within an undifferentiated landscape.[3]

Bronx, 2012. Parkchester is a new-town-in-town with more than 100 stores, a movie theater, and even a branch of Macy's. *(Alexander Garvin)*

Government officials happily encourage the construction of projects like Fresh Meadows or Parkchester because these developments can provide so many people with decent housing, especially when they are built on predominantly vacant sites, conveniently just around the corner from already-developed areas. There is an interesting history of government assistance for the development of such pseudo new-towns-in-town. The New York State Limited Dividend Housing Corporations Law of 1926, which provides real estate tax exemption to developers who limit their return on equity, is probably the earliest example of such assistance (see Chapter 10). However, it failed to deal with three major problems facing any developer of a new-town-in-town: the need for huge amounts of equity capital, the unavailability of very long term mortgage financing, and the high cost of mortgage money.

These problems were solved in the 1950s and 1960s, but only for a decade or two, when state legislatures established housing finance agencies with the ability to issue tax-exempt bonds, the proceeds of which could provide housing developers with mortgage financing (see Chapter 10). State housing finance agencies were able to extend the period of amortization for as long as 40 or 50 years. Since interest on the bonds is tax-exempt, they also were able to charge housing developers a rate of interest several percentage points lower than would be available at local banks. Most housing finance agencies reduce equity requirements to an almost nominal 5 or 10 percent.

Most government employees, however, are neither skilled at making investments nor able to withstand the political pressures that attend nearly any public investment in housing. It is simply too politically appealing, for example, to reallocate to operating costs monies actually intended for debt service; moves like this help limit increases in the monthly housing costs for tenants, but they also cripple the long-term success of housing projects. That is how projects like Co-op City in the Bronx nearly forced the State of New York into bankruptcy. Nevertheless, there continue to be insurance companies, pension trusts, and huge investment funds that have sufficient capital to make long-term investments in residential new-towns-in-town. Those big spenders often succeed in creating projects, like Brickell Key in Miami, that provide excellent financial returns but draw residents away from the rest of the city and, as a result, keep these local residents from spending time and money that could benefit the rest of the city.

Co-op City, the Bronx

The first project to obtain benefits under the New York State Limited Dividend Housing Corporations Law of 1926 was Amalgamated Houses, a union-sponsored project that involved staged construction of 1434 apartments spread over six blocks of the north Bronx.[4] It initiated a series of ever-larger planned residential communities that was brought to a halt in the early 1970s by the problems of Co-op City, still the nation's largest planned residential development (see Table 15.1).

The person behind these projects was Abraham Kazan, a union member who convinced the Amalgamated Clothing Workers of America that its members and their families should be able to obtain affordable housing. Kazan proposed raising equity capital in the form of nominal subscriber down payments, obtaining institutional mortgage financing (in the case of Amalgamated Houses, from the Metropolitan Life Insurance Company), and then building apartments for the subscribers. When residents moved out, they were required to sell their apartments back to the co-op for the amount of their original down payment plus the accumulated mortgage amortization for the period during which they owned the property.[6]

The federal government also encouraged these types of developments. The Limited Dividend Housing Corporations Law provided real estate tax exemption on all improvements for a period of 25 years if the return on the property were limited to 5 percent (later, 6 percent). Thus, occupants paid the low taxes that had been paid prior to the property's development. Since the project was a nonprofit cooperative, the only additional payments by the occupant covered the project's operating expenses and mortgage debt service.

When the first stage of the Amalgamated Houses opened in 1927, the *New York Times* called it "the finest and largest development of low-rent [sic] housing in the entire city." For a down payment of $500 per room and monthly payments of $11 per room per month, its occupants had a nice apartment in a modern building surrounded by trees, grass, and flowers. It was the first in a series of projects that provided more than 32,500 cooperative apartments.

By the time the final project, Co-op City, was conceived, Kazan had built up an effective development agency, known as the United Housing Foundation (UHF). Its only rival, in terms of number of apartments built or political influence, was the redoubtable Robert Moses. UHF projects were criticized for their banal, cookie-cutter designs; their overpowering, almost inhuman scale; and their insular site plans that carefully segregated project residents from the surrounding community. But, as Roger Starr explained at the New York City Planning Commission's public hearing on Co-op City,

> . . . *far from being inhuman, far from crushing the human spirit, far from presenting people with an inhuman envi-*

TABLE 15.1
UNION-SPONSORED COOPERATIVE HOUSING IN NEW YORK CITY[5]

Project	Borough	Number of apartments	Completion
Amalgamated Houses	Bronx	1,434	1927–1971 (in stages)
Amalgamated Dwellings	Manhattan	236	1930
Hillman Houses	Manhattan	807	1951
Corlears Hook	Manhattan	1,668	1958
Seward Park	Manhattan	1,728	1962
Penn Station South	Manhattan	2,820	1963
Warbasse Houses	Brooklyn	2,592	1965
Rochdale Village	Queens	5,860	1965
Co-op City	Bronx	15,389	1969–1972 (in stages)

The Bronx, 2011. Co-op City offers apartments at bargain prices in one of the nation's least inspiring new-towns-in-town. *(Alexander Garvin)*

ronment, inhumanly scaled, [they] have actually built developments which have been subscribed and oversubscribed, in which people have now been living for many, many years, and as nearly as we can make out—have been living successfully and happily.[7]

He also noted that residents were able to live in these projects at a price that was considerably lower than that of comparable new housing that was conventionally financed.

The original proposal for Co-op City called for the construction of 17,000 apartments on 300 acres of marshland that had previously been the site of Freedomland, an unsuccessful theme park.[8] City officials demanded and obtained changes. The project was cut to 15,389 dwelling units, 236 of which were in the mediocre town houses added to soften the height of the 35 apartment towers. The color of the brick was changed: buildings could be drab gray, not just drab red. Even the site plan was altered.[9]

Co-op City appears to be consistent with the stereotypes associated with typical UHF projects. There are, however,

The Bronx, 1994. The huge open spaces at Co-op City offer recreational opportunities that are not usually available in urban areas. *(Alexander Garvin)*

significant differences, in large part because the development's scale permitted more inclusions into the plan. This time, the UHF included shopping centers and a power plant that supplies residents with electricity. At city expense, it also built a separate educational park including two public elementary schools, two public intermediate schools, and one public high school. There even is express bus service to downtown Manhattan.

Despite these differences, though, Co-op City is still a housing project with conventional, drive-to shopping facilities that are set apart from the housing rather than a planned community for 50,000 residents. And therein lies the explanation of its failure: the rest of the Bronx lies on the other side of major highways and looks nothing like Co-op City. If, instead of arguing about brick color, the city had directed its attention toward minimizing the highway barriers to surrounding communities, Co-op City's separation would not be quite as stark. If, rather than demanding token town houses, the city had concerned itself with creating an internal street system in which residential structures, like those throughout the city, mingled with stores, schools, and other nonresidential uses, Co-op City could not be stigmatized as a project.

Kazan's conception of a working-class cooperative was in sharp contrast with the reality into which the occupants of Co-op City moved. Co-op City, like Kazan's other projects, was a resident-owned cooperative in name only. Residents understood that they were temporary occupants of dwelling units whose lower cost reflected public assistance. When residents of Co-op City gave up occupancy, they could only receive what they put in. Unlike owners of conventionally financed co-ops or single-family houses, they could not benefit from any appreciation in value. Their investment could never become a growing nest egg that would help finance the purchase of a new, even better residence or pay for their retirement in Florida.

From the beginning, the costs of building and operating Co-op City exceeded initial estimates. When co-op charges had to be increased, occupants refused to pay. Since they did not have any real equity interest in the project, they naturally blamed the "landlord," stopped paying "rent," and expected the government to step in. The Co-op City rent strike virtually terminated New York State's bond-financed, limited-profit housing program (see Chapter 11). A financial catastrophe was avoided only because Governor Hugh Carey, in 1975, persuaded the legislature to cover debt-service payments on the $390 million in bonds issued to finance Co-op City. Had the state's moral obligation to see that interest on the bonds was paid—at a time when New York City was in the midst of its own fiscal crisis—the country's financial markets would have collapsed. In the end, New York State had been forced into subsidizing middle-income citizens capable of paying for their own housing.[10]

The saddest result of what transpired at Co-op City is that $423 million had been spent without beneficial spillover elsewhere in the Bronx.[11] When the buildings were ready for occupancy, thousands of South Bronx households, especially from around the Grand Concourse, chose to move there. It can be argued that this provided a step up for both departing, primarily white, residents of Co-op City and for incoming, primarily Puerto Rican and African-American, residents who replaced them. On the other hand, it also can be argued that the sudden departure of thousands of stable families only weakened the neighborhoods they left behind. What cannot be challenged is that Co-op City increased the gulf between the more fortunate residents of the recently completed apartment complex and those who had to make do with the apartments they left behind.

Had Co-op City not been separated from the rest of the city by highways, had it not been designed to function independently of everything around it, and had it not been visibly different from the rest of the Bronx, there might have been some chance of its benefits spilling over into the rest of the area. Instead, 50,000 people moved into this new-town-in-town without bringing any centrifugal impact to their surroundings.

Brickell Key, Miami

While Co-op City may seem to be a man-made island set apart from the rest of New York, Brickell Key actually is a man-made island that is connected to Miami by a short bridge. It began as two islands created by Henry Flagler in 1896, when he dug a 9-foot-deep channel from the mouth of the Miami River to Biscayne Bay, creating what became known as Flagler's channel and, in its wake, two small islands. In 1943, Edward Claughton Sr., purchased them and then later acquired the bay between them. He proceeded to fill in Flagler's channel, thereby creating the 44-acre island popularly known today as Brickell Key. Three decades later, Swire Properties, a major international investment group

Miami, 2011. Brickell Key. *(Alexander Garvin)*

that develops and manages real estate throughout the world, purchased the island and began transforming it into an exclusive, upscale community.[12]

Swire Properties hired Brazilian architect Oscar Niemeyer to prepare an initial plan for the island's development, but the plan was quickly abandoned as impractical. Over the next four decades, the firm created a gated, automobile-accessible new-town-in-town that allows pedestrians unimpeded access. It contains 11 residential towers with more than 2900 condominium apartments, two buildings containing 207,000 square feet of offices, a 326-room Mandarin Oriental Hotel, and scattered restaurants, spas, and retail outlets serving building occupants. Depending on the building, each condominium tower offers its own combination of private swimming pools; beach access; gym; Jacuzzi; sauna; steam room; racquetball, squash, and/or tennis facilities; party rooms; lounge and multimedia rooms; valet parking; and maid, dry-cleaning, and other exclusive services. In addition, there is a shuttle bus that connects the island with the Brickell Metrorail/Metromover station and a number of intermediate stops along the way. It is a safe, secure place for those who can afford it.

Brickell Key proved to be a profitable investment for Swire Properties. It had the financing needed to supply a product for which there is market demand at a location with beautiful, unobstructed views of Biscayne Bay and Miami Beach. Brickell Key was designed to meet the expectations of its occupants at a time when nothing like it was available a few hundred feet away in one of the nation's most prosperous metropolitan areas. Residents have everything they could possibly desire on the island: fine restaurants, recreational facilities, a marketplace, and a wide range of services. It may be a successful commercial venture, but its impact as a planning initiative is far less convincing. It has little if any impact on the surrounding city because, once a resident has arrived, there is no reason to leave the island.

Dearborn Park and Central Station, Chicago

During the 1960s, Chicago's population declined by over 180,000. Businesses, not just residents, were leaving the city for the suburbs. Dearborn Station, destination of such legendary trains as the Santa Fe Chief from California, was about to close. Just north of the station, elevator buildings that had once been the center of Chicago's printing industry were now predominantly vacant (see Chapters 6 and 13). Retail sales on nearby State Street were still declining (see Chapter 6). Civic leaders feared that the southern section of the Loop, the city's central business district, might transform into the same sort of daytime-only downtown that was causing problems in Detroit, St. Louis, and other cities in the region that would soon be labeled the *rust belt*.[13]

The chairs of three of the city's major corporations (Sears Roebuck and Company, Commonwealth Edison Company, and Continental Illinois National Bank and Trust Company) were determined to prevent Chicago's central core from declining still further, since each had major business reasons to ensure that downtown Chicago remained vibrant. Sears had just built what was then the world's tallest building in the Loop. Commonwealth Edison was the biggest power company in the Midwest, with 6500 downtown employees. Continental Illinois, the Midwest's biggest bank, had hundreds of millions of dollars invested in downtown loans and mortgages. In early 1970, they and other downtown business leaders decided that the best way to prevent further decline was to replace the soon-to-be vacant rail yards south of Dearborn Station with a vibrant residential neighborhood. Since they were not experienced at real estate development, they sought professional assistance and raised $375,000 to pay for an area plan.

Ferd Kramer, the developer of the Lake Meadows and Prairie Shores urban renewal projects (see Chapter 7), and Philip Klutznick, the developer of the suburban planned community of Park Forest and the Old Orchard and Oakbrook shopping centers (see Chapter 4), were retained to spearhead the development of the neighborhood that that they named "Dearborn Park." The architecture firm of Skidmore, Owings & Merrill was hired to work with the Real Estate Research Corporation on a plan for the area. Their first proposal for a South Loop new-town-in-town was unrealistic: 250 acres covered with 40,000 dwelling units in numerous high-rise apartment buildings linked by glass-enclosed walkways to bridge city streets. As Kramer and Klutznick explained, there was no evidence that so many families wished to move downtown, that they wanted to live in a high-density, futuristic superblock, or that lending institutions were prepared to provide the huge amount of mortgage financing that would be necessary.

The scheme Kramer and Klutznick proposed was less ambitious but far more realistic. They suggested forming a limited-dividend corporation (Dearborn Park Corporation) whose equity would come from several local businesses, with the rest of the project's cost financed by conventional mortgages . The city government agreed to rezone the area for residential use and pay for the installation of whatever infrastructure and municipal facilities would be necessary.

After seven years and numerous problems, the Dearborn Park Corporation began construction on what would be 1671 dwelling units covering 51 acres, built in two phases. The project consisted of one-family houses, condominiums, and apartments interspersed with courtyards. When it was completed in 1994, Dearborn Park included a school, four small parks, off-street parking for 1940 cars, and the absolute minimum necessary convenience retailing. To provide resi-

Chicago, 2000. Dearborn Park is a secluded neighborhood within the city. *(Alexander Garvin)*

dents with the security that Kramer and Klutznick believed they would insist on, each phase was designed as a separate enclave accessible only by car from a single street.

The 21 businesses that formed the Dearborn Park Corporation in 1974 invested nearly $14 million. Twenty years later, when the corporation was dissolved, they received $5.7 million, representing a financial loss of $8.3 million excluding the opportunity cost of that money. In a myopic sense, the financial loss may have been unjustifiable, but the very existence of Dearborn Park meant that these corporations had protected their holdings in the Loop, provided housing for their employees and customers for their businesses, and enriched the city's tax base (thereby keeping their taxes down). The city had spent $18 million on streets, sewers, parks and other improvements for Dearborn Park. By 1995, the 51 acres that had paid virtually no real estate taxes 20 years earlier was contributing $5.1 million annually.

The spillover impact was even more significant. By 2001, nearby property owners had constructed or renovated apartment buildings containing more than 6235 apartments within a three-block radius of Dearborn Park. Moreover, Harry Weese and his partners had begun renovating build-

Chicago, 2000. Residents of Dearborn Park walk directly into downtown Chicago. *(Alexander Garvin)*

Chicago, 2012. Central Station is built on former railroad yards leading into downtown Chicago. *(Alexander Garvin)*

ings in Printers Row, immediately to the north (see Chapter 13). Building renovation there had transformed what had been a predominantly vacant loft district into a popular residential neighborhood. Scattered loft buildings to the south also had been converted to residential use.

In 1989, Central Station, a similar but more ambitious new-town-in-town, went into development on the site a few blocks east of Dearborn Park. What began as a 69-acre project on rail yards behind the shuttered terminal and freight complex of the Illinois Central Railroad and adjacent air rights has evolved into what will be 14 million square feet covering 80 acres. The project was conceived by Gerald Fogelson, whose firm, Fogelson Properties, teamed with Forest City Enterprises Inc. of Cleveland, Ohio, to build 7500 rental and for-sale apartments, small parks, offices, restaurants, and convenience retail stores.[14]

Like Dearborn Park, Central Station includes row houses, low-rise apartment complexes, and residential towers. One difference is that its high-rise buildings have phenomenal views of Lake Michigan and the downtown skyline. Another is its relationship to the city's street grid. Partially in response to requests from the city government, the developers donated the land needed to widen Indiana Avenue and extend Roosevelt Road in order to connect Lake Shore Drive with Michigan Avenue. They also donated property needed to widen and realign Lake Shore Drive and link it with Soldier Field and the Museum Complex that includes the Field Museum of Natural History, Shedd Aquarium, and the Adler Planetarium.

Unlike Dearborn Park, which protects its privacy by restricting vehicular entry to a few entry points, Central Station is accessible by all the east-west streets that cross Indiana Avenue, as well as from Grant Park and Roosevelt Road to the north. In most other respects, however, both projects are similar to Brickell Key. They provide a pleasant living environment for residents without offering anything that might attract people from elsewhere in Chicago. What makes them both different and particularly notable is that they have generated a widespread and sustained private market reaction, in addition to enhancing the contemporaneous revival of Printers Row. As of 2012, 75 new apartment buildings and 40 town houses have been built, and 60 loft buildings have been converted to residential use in the immediate surroundings of these two new-towns-in-town.

Chicago, 2012. Central Station. *(Alexander Garvin)*

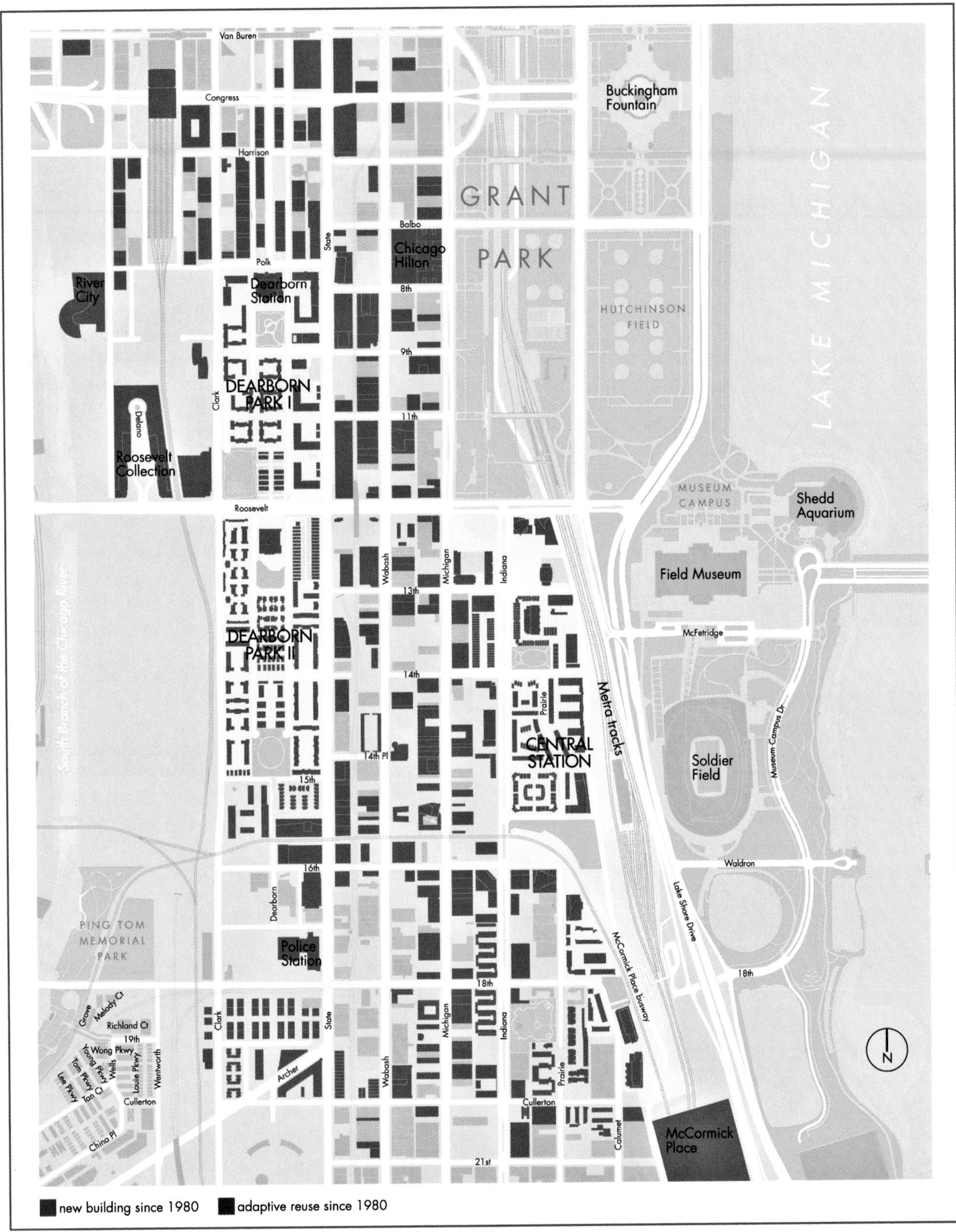

Chicago, 2012. The investments in Dearborn Park and Central Station generated widespread and sustained investment in nearby properties. *(Chicago Cartographics)*

The National Effort to Build New-Towns-in-Town

The federal government had a brief fling with residential new-towns-in-town. Title IV of the Housing and Urban Development Act of 1968 and Title VII of the Housing and Urban Development Act of 1970 provided a federal guarantee for up to $50 million in obligations issued by developers of approved new communities, as well as up to $20 million in interest-free loans (see Chapter 16). Only two new-towns-in-town, Cedar-Riverside in Minneapolis and Roosevelt Island in New York, were approved for funding.

Cedar-Riverside, Minneapolis

When Congress enacted legislation to encourage the development of well-planned new communities in 1968, it contemplated projects with a mix of "homes, commercial and industrial facilities, public and community facilities, and open spaces . . . [and] opportunities for innovation in housing and community development technology, and in land use planning."[15] Cedar-Riverside, in Minneapolis, was the first new-town-in-town to receive federal assistance.

Redevelopment of Cedar-Riverside had been under consideration for nearly four decades when it was finally approved by HUD for Title VII assistance. In 1934, the Works Progress Administration (WPA) recognized that the area was one of the most dilapidated sections in Minneapolis and suggested that it be cleared. In 1948, the city's Housing and Redevelopment Authority (HRA) designated it as a potential renewal site. A renewal plan was finally submitted 13 years later, but it was rejected by the Minneapolis City Council.

Minneapolis, 2005. Cedar-Riverside, an unfinished Title-VII-financed new-towns-in-town. *(Alexander Garvin)*

The area nonetheless started to change during the late 1950s. Two local hospitals began to expand into the area. A Lutheran seminary was transformed into a coeducational college. The HRA built 348 units of housing for the elderly. But the most significant change came when two interstate highways were constructed, because they separated Cedar-Riverside from the rest of the city.

Despite (and perhaps also because of) this activity, Cedar-Riverside's population fell from 8540 in 1940 to 4766 in 1960. The area changed from a neighborhood that consisted primarily of Scandinavian-American immigrants to one that reflected the nonconformist character of an emerging youth culture.

Investors, sensing the demand for housing generated by the aforementioned institutional expansion, began acquiring property in the area. The most important of these investors was a partnership consisting of a local doctor, his wife, and a university lecturer. By 1968, when the city council finally approved an urban renewal plan for Cedar-Riverside, this group had purchased nearly 200 parcels, had prepared a master plan for a new-town-in-town with a projected population of 30,000, had combined forces with several other local developers, and had approached HUD for funding under the recently enacted Title IV new-community legislation.

Architect Ralph Rapson and landscape architect Lawrence Halprin prepared an ambitious scheme that called for 12,500 apartments in six residential neighborhoods, 1.5 million square feet of commercial space, and 6 acres of cultural facilities. In the words of one of its promotional brochures, Cedar-Riverside was going to awaken a "new kind of urban life . . . filled with involvement. Filled with a vitality . . . depth, and breadth of experience."[16] What emerged was far less inspired: 11 reinforced concrete buildings containing 1299 apartments were organized around two quadrangles and connected by a system of walkways and pedestrian bridges. The sharp contrast between conception and reality at Cedar-Riverside was inevitable given the inexperience of its developers, the project's dependence on government financing, and the difficulties of operating within a complex political environment.

Government assistance ultimately proved to be the critical element both in initiating the development of Cedar-Riverside and in bringing it to a halt. Without local designation as an urban renewal project, the area's many property owners would have withheld their parcels from the project or demanded prohibitively high prices. Without the $24 million in federal new-community loan guarantees that were approved in 1971, the project's relatively inexperienced developers could never have obtained the necessary financing. Without new-community status, they could never have obtained the set-aside of scarce federal housing subsidies (117 units of leased public housing and 552 moderate-income units with FHA 236 subsidies) or the grants that helped to pay for the covered walkway system, the pedestrian bridge across Cedar Avenue, the plaza, or any of the other amenities that went into the project. In return for this government largesse, the developers had to accept complex and expensive regulations covering everything from equal employment opportunity to citizen participation.

Trouble began in 1973. First, the Nixon administration terminated all federally assisted housing and urban development programs. Then, residents of the area successfully initiated a suit claiming the project's Environmental Impact Statement was inadequate and that it violated the National Environmental Policy Act of 1969 (see Chapter 17). They succeeded in obtaining an injunction against continued development. Project tenants initiated a rent strike, and the urban renewal area's citizen advisory panel called for major changes in the redevelopment plan. The project's developers had neither the financial strength nor the political skills to continue. Predictably, Cedar-Riverside was brought to a halt.

Since Cedar-Riverside never evolved into the new-town-in-town that had been planned, there is no way of determining either the extent to which it would have awakened a "new kind of urban life" or generated a market reaction strong enough to vault over the highways to affect other Minneapolis neighborhoods. What is clear, however, is that federal grants and loan guarantees by themselves were not enough to ensure success.

Roosevelt Island, New York City

Roosevelt Island, the other new-town-in-town that received Title VII assistance, was specifically conceived to avoid the difficulties experienced by projects like Cedar-Riverside. There were no acquisition problems because the site was owned by the city of New York and no political problems because its developer, the New York State Urban Development Corporation (UDC), renamed the Empire State Development Corporation (ESDC) in 1995, was a superagency with truly amazing powers.

The UDC had been Governor Nelson Rockefeller's response to the riots that erupted in cities across the country during the mid-1960s. The statute creating this superagency was enacted in 1968. Written by Edward Logue, its first president and chief executive officer, the legislation provided the UDC with powers that could overcome every difficulty he had encountered earlier in his career, first as New Haven's development administrator and then as chairman of the Boston Redevelopment Authority (see Chapters 7 and 12). The UDC could condemn land, hire expert professional staff, ignore zoning and building regulations, and even issue tax-exempt bonds to finance development. Its most remarkable power was the ability to do all this without obtaining the approval of any other city, county, or state agency.

Logue needed a highly visible project with which to demonstrate UDC's effectiveness. That project came in the form of a proposed new-town-in-town for an underutilized 147-acre island in the East River, directly opposite the most expensive real estate in the world: Manhattan's fashionable Midtown office district and Upper East Side residential district.

Roosevelt Island, 2012
(Alexander Garvin)

Once known as Hog's Island and later Blackwell's Island, this valuable city-owned property—then called Welfare Island—was used for hospitals and nurses' residences. Over the years there had been proposals for high-density housing, parks, industrial development, and even an amusement park, like Copenhagen's Tivoli Gardens. When the Transit Authority announced it would build a subway tunnel that would connect Manhattan with Queens and run beneath the island, development seemed an obvious decision.

Mayor Lindsay appointed a blue-ribbon committee to determine the future of this underutilized asset. In 1969, based on studies prepared by the Development and Resources Corporation that was managed by David Lilienthal, the Welfare Island Planning and Development Committee proposed a new-town-in-town.[17]

No private developer could muster the funds necessary to execute the ambitious program envisioned for the island. When Logue announced that the newly created New York State Urban Development Corporation was ready to execute the plan, the mayor responded positively. Logue could get started immediately because the island had no organized community groups with whom to negotiate.

Lindsay had promised to build thousands of subsidized housing units in urban renewal areas throughout the city. In exchange for the chance to create this model new-town-in-town, Logue agreed to build and finance a large portion of the subsidized housing that the mayor had promised (see Chapter 12). He persuaded Governor Rockefeller to obtain special set-asides of housing subsidies and other federal programs from the recently elected Nixon administration. Thus, the island could be developed with funds that would not otherwise have been available to New York City.[18]

Logue renamed the project Roosevelt Island and hired Philip Johnson, a member of Lindsay's Welfare Island Planning and Development Committee, to prepare a master plan for the project.[19] One might have expected Johnson to produce a perfected version of the Garden City, the Radiant City, the Linear City, the Mega-City, or some other currently fashionable utopia. Instead he produced an innovative plan that responded to topography, landscape, and history.

Johnson proposed to preserve the island's historic structures. Existing hospitals (which were too expensive to move) were to continue operations. Private vehicular traffic was to be kept from the island. Cars would park at a "motor gate" from which there would be regular minitransit service. The island's waterfront was to become a continuous 4-mile promenade accessible to the public, and the buildings themselves would be gathered into high-density clusters separated by five new parks. At the core of this new-town-in-town, Johnson planned an exciting town center with two large public spaces, a glass-roofed retail arcade, a 300-room hotel, and at least 200,000 square feet of office space. All together, there were to be 5000 apartments divided into two neighborhoods that would be built on either side of the town center.

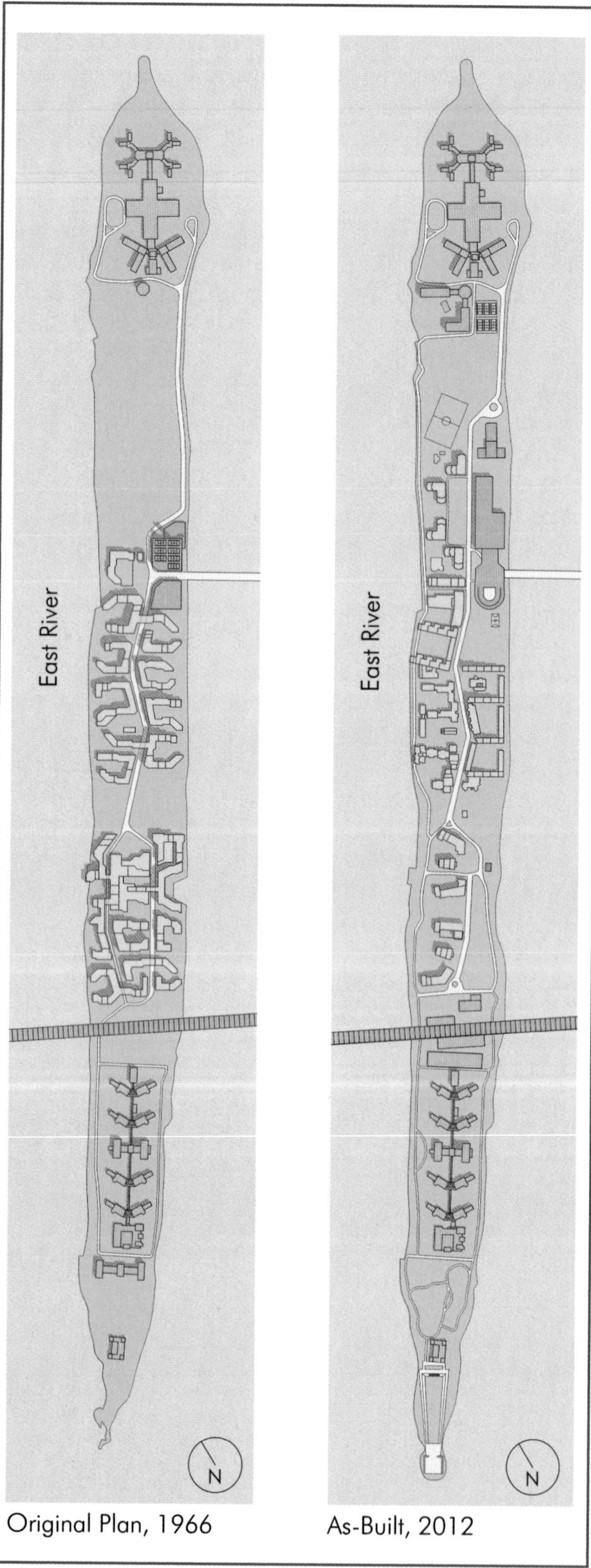

Roosevelt Island, 2012. Philip Johnson's unexecuted master plan for Roosevelt Island in which Main Street was interrupted every 100 feet by a 60-foot-wide view of the river. *(Alexander Garvin and Ryan Salvatore)*

The master plan organized these elements along a central spine in a unique manner intended to maximize proximity to the riverfront on both sides of the island. Twelve-story apartment buildings, shaped like the letter C, were staggered so that their flat back ends defined a 50-foot-wide street that led to an east-west town center that was the "main street" of the scheme. Pedestrians strolling along this spine would enjoy river views opening every hundred feet alternately to the west and then to the east. There were also river views for building residents whose apartments faced the open ends of these C-shaped buildings and thus the waterfront beyond. In this manner, Johnson's innovative plan simultaneously defined the street, provided changing vistas of the water, and created a new, dynamic relationship between the island and the rest of the city.[20]

The UDC hired more "practical" architects, who misunderstood the master plan and designed more conventional buildings. The unique, river-oriented apartment buildings were replaced by mediocre quadrangles, one side of which blocked the water views. The 12-story height limit was ignored and the 50-foot-wide north-south spine became the island's actual Main Street because the town center was eliminated.

While Roosevelt Island did benefit from minor grants made possible by its status as a Title VII new community, it was financed without federal guarantees. The UDC provided below-market mortgage money by issuing tax-exempt bonds. The new apartment buildings also were effectively exempt from New York City real estate taxes. These two subsidies allowed the UDC to lower rents sufficiently to attract people who otherwise would not have been able to afford living on the island. In addition, the federal government provided FHA 236 subsidies for 1003 apartments, thereby ensuring that the Roosevelt Island residents would represent a range of income levels.

Roosevelt Island, 2012. The main and only street of Roosevelt Island became a high-density corridor unlike the lively retail area built over a new subway station that had been envisioned in the original master plan. *(Alexander Garvin)*

The plan for Roosevelt Island did offer a few innovations. Anticipating delay in subway construction, Logue persuaded the city to subsidize construction of an aerial tramway connecting the island with midtown Manhattan. No other planned new community is served by a tramway or by any other form of mass transit with such spectacular views. In addition, rather than using noisy garbage trucks, all buildings were provided with the Disney World Automated Vacuum Collection System. When the first apartments were occupied, resident children attended day care centers and minischools that were located within the island's residential buildings, so that "children, parents and teachers" could "interact with each other and develop into an active community unit."[21]

Like Cedar-Riverside, development at Roosevelt Island was brought to a halt in 1973 by the Nixon moratorium. Unlike Cedar-Riverside, though, the halt was temporary. The UDC had already financed and built much of the infrastructure for the rest of the project. As a government agency, it could wait until conditions changed before continuing with housing construction. However, development did not resume for more than a decade because unsubsidized rental, co-op, or condo apartments could not be produced at prices that were low enough to attract additional residents.

Enough construction has been completed to evaluate the success of Roosevelt Island as a planned community. When tramway service began in 1976, the development comprised 2148 apartments, waterfront promenades, lawns with children tossing Frisbees, glass-enclosed swimming pools, minischools, and even a "Rompers Roost" preschool. Eighteen months later, the *New York Times* reported:

> *The "shops, restaurants, and a range of services" that planners envisioned have not materialized yet, and as a result the "excitingly urban" Main Street mentioned in the Roosevelt Island brochure has not materialized yet either. There is no hardware store, household appliance store, shoe store, ice cream fountain, barber shop or beauty parlor... although the stationery store does sell T-shirts proclaiming that Roosevelt Island is "The Little Apple."*[22]

Nearly three decades had passed between the time when Johnson first prepared his plan and when the long-promised subway station that had triggered the island's redevelopment

Roosevelt Island, 2012. View of Manhattan's Upper East Side from the riverfront esplanade at Roosevelt Island. *(Alexander Garvin)*

finally opened in 1990. During that time, 3025 apartments had been completed, but the population that they housed was too small to support a viable retail district. Moreover, without additional land uses to attract off-island visitors, Roosevelt Island essentially remained a dormitory suburb.

The Roosevelt Island Operating Corporation (RIOC) replaced UDC as the island's developer in 1984, and in 1995 RIOC initiated a planning process aimed at devising a more attractive future and recapturing the long-lost spirit of innovation that had characterized the project's inception.

In 2001, RIOC selected a developer to complete the island's development. In the decade since then, an additional 2500 apartments have been plunked down in no particular order between the original Main Street corridor and the Queensborough Bridge. The first three of the six new buildings were financed by medical/educational institutions (Cornell Medical College, New York University, Memorial Sloan-Kettering) and are occupied by graduate and medical students, and staff and visitors to those institutions. Unlike previous buildings on the island, none of the new residences are subsidized. As a result, the composition of resident population is slowly shifting to people who can afford market rents, but, in general, neither the new buildings nor the new occupants have had any dramatic impact on the island. The waterfront esplanades and recreational opportunities remain wonderful amenities enjoyed by the islands' residents, and retailing on Main Street remains inadequate.

Just when the promise of a brave new world seemed to have been forgotten, several activities on Roosevelt Island began to contribute to the prospect a much more vibrant future. In 2012 the long-promised Roosevelt Memorial, designed by architect Louis Kahn for the southern tip of the island, was completed and opened to the public. The architecture critic of the *New York Times* predicts that this "virtual walk-in sculpture" will "more than honor the 32nd president"—it will "bring to life a neglected but symbolic

Roosevelt Island, 2012. The structures erected at Roosevelt Island during the first decade of the twenty-first century were conventional apartment buildings. *(Alexander Garvin)*

stretch of prime shorefront" and attract thousands of visitors from around the world.[23]

The more significant change will be manifest over the next two decades as a consortium of Cornell University and the Technion–Israel Institute of Technology complete what will be a 2.1 million-square-foot campus serving 2500 students, 280 professors, and countless other occupants of the area between the Queensborough Bridge and the Roosevelt Memorial. When everything is finished, the southern portion of the island will have become the lively part of New York City that Philip Johnson once imagined would develop over the subway stop in the middle of the island.

Privately built utopias like Co-op City, Brickell Key, Dearborn Park, and Central Station, as well as government-sponsored ones like Cedar-Riverside and Roosevelt Island, raise the inevitable question of whether a residential new-town-in-town can also be a planning success. As the following discussion of Century City, Pentagon City, and RiverPlace demonstrates, success requires that the development be conceived as a new city district with land uses that attract residents from the rest of the city, and it also requires that the plan be designed as an integral part of the surrounding city.

New City Districts

The most persuasive public policy rationale for new-towns-in-town is that they can benefit the surrounding city. One way is by generating additional employment and taxes. Based on this argument, Co-op City, Cedar-Riverside, and, at least to date, Roosevelt Island are flops. Despite utopian rhetoric, they simply moved city residents from one part of town to another without bringing any new business to the city.

Another way in which a new-town-in-town can be of benefit is by generating further development around its periphery. The self-contained planning of new-towns-in-town, like Atlantic Station in Atlanta, Georgia, prevents this from occurring. This is even more obvious at Brickell Key and Roosevelt Island, which are separated from the surrounding city by water; Co-op City is cut off by highways. As a result, essential commercial, community, and recreational facilities had to be provided within these projects, further reducing the reasons for them to spill over into surrounding neighborhoods.

New districts, like RiverPlace, can also improve the character of city life if they supply services (retail stores, recreation facilities, cultural establishments, etc.) that would not otherwise be available to surrounding neighborhoods. Once again, projects like Brickell Key and Cedar-Riverside, which have little to offer except apartments, are unlikely to attract many outsiders.

There are, however, mixed-use new-towns-in-town that do provide these benefits. RiverPLace includes retail and recreation facilities that attract people from the surrounding metropolitan region. Century City in Los Angeles and Battery Park City in lower Manhattan provide thousands of jobs and generate millions of dollars in taxes.

RiverPlace, Portland, Oregon

RiverPlace, in Portland, Oregon, began as a public effort to clear blighted and underutilized warehouses and industrial buildings along the Willamette River. There had long been disagreement over the proper reuse of the waterfront among advocates for vehicular and rail transportation, riverfront commerce, and recreation. The controversy was resolved in 1974–1976 with the replacement of the multilane Harbor Drive by the mile-long Governor Tom McCall Park.[24]

Civic leaders were not satisfied with a beautiful swatch of green along the downtown waterfront. They wanted McCall Park to become a popular destination for residents from all over the metropolitan region. To accomplish this, they introduced retail destinations that would attract visitors to each end of the park. Accordingly, the city established the Saturday Market under the Burnside Bridge at the northern end of the park. The market, which operates Saturdays and Sundays from March through December, brings thousands of customers who wish to acquire wares from local artisans and craftspeople. A 73-acre site that had been occupied by a plywood mill and steam plant at the south end of the park was entrusted to the Portland Development Commission (PDC). It issued $6 million in bonds to finance environmental remediation at the site, to construct a new marina, and to make improvements to streets, sidewalks, pedestrian esplanades, and other infrastructure.

As a result of a national design competition, the PDC awarded the private market portion of the project to the Cornerstone Development Company based in Seattle. The first phase of Cornerstone's project included a 74-room hotel and restaurant, an athletic club, a small office building, 190 condominium apartments, a 200-boat marina and floating restaurant, 15 retail stores (mostly food service), and a single level of covered parking for 500 cars. The first condominium apartments were completed in 1988, during a period of slack demand. Fortunately, Cornerstone had the financial wherewithal to offer apartments for rent and wait till it could sell them. The hotel, restaurants, stores, and marina, however, were an instant success. They attract astonishingly heavy use for such a modest development. (The stores occupy only 26,000 square feet.) Consequently, the esplanade gets quite crowded, especially in the evenings and on weekends between April and October. This success spurred construction of phase two, which included another 290 apartments, six stores, a 300-car parking garage, and the eight-story headquarters of the Pacific Gas Transmission Company. In 1995, a third section that contained 182 townhomes was added by Trammel Crow Residential.[25]

Portland, Oregon, 2007. The riverfront esplanade and Marina at RiverPlace. *(Alexander Garvin)*

The design of RiverPlace by the Bumgardner Architects, Gaylord, Grainger, Libby, O'Brien-Smith Architects, and the Zimmer Gunsul Frasca Partnership is appropriate to its site on the edge of downtown Portland. None of the buildings is higher than five stories. They are topped with a variety of gabled roofs that attract the attention of motorists passing across the Marquam Bridge, just to the south, and workers in the office buildings to the west. While radically different from the masonry, metal, and glass buildings downtown, the vernacular wood construction at RiverPlace was not only more affordable to construct but was also consistent with traditional northwestern American architectural style, for which there was demand in the market.

The best aspect of the design is its clever balance between the privacy for downtown residential developments and the busy public spaces needed in order for the waterfront to develop into an active part of the downtown. The occupants of the office, hotel, and apartment buildings enter from the city side, without having to pass through the retail portions of RiverPlace. At the same time, esplanade customers can stroll along the riverfront without disturbing the privacy of other site occupants.

Because of their location along the waterfront esplanade, these stores appear to be a major destination, when in reality they are only a small part of this 73-acre residential new-town-in town. Thus, RiverPlace is thought of as a popular destination for visitors from all over the metropolitan area, rather than an $84 million residential new-town-in-town.

Century City, Los Angeles

Century City takes its name from a 263-acre site that was originally the back lot of 20th Century Fox studios. It is strategically located in the middle of the rapidly growing western corridor between downtown Los Angeles and the Pacific Ocean. Directly to the north are the exclusive residential and commercial sections of Beverly Hills; to the west, Westwood; and to the south, the Hillcrest Country Club and the Rancho Park Golf Course.[26]

Spyros Skouras, president and chairman of the board of Fox, understood that the lot was much more valuable for real estate development than for movie production. In 1958, Skouras hired architect Welton Beckett to prepare a master plan for a complex of office towers, hotels, retail stores, and apartment buildings. He hoped to sell the concept and the property for $100 million. Unfortunately, no buyers materialized because the cost to develop the site—between half a billion and a billion dollars—was simply not financeable.

After nearly a year, William Zeckendorf Sr. agreed to purchase the property for $56 million, but only after he had devised a scheme that reduced the cash required; 20th Century Fox would lease back 75 acres for $1.5 million a year in rent. As he explained his intentions to sell off the revenue stream from the leaseback, "I was actually paying only $31 million, because that $1.5 million in rent, at six-percent interest, would be worth $25 million to some insurance company."[27]

Even with the reduced equity requirements, Zeckendorf had real trouble raising the money. He eventually persuaded the Aluminum Company of America (Alcoa) to enter into a joint venture with him despite its previous policy of investing only in aluminum production or sale. When they closed the deal in 1961, the price had been reduced to $43 million. Eventually, Alcoa bought out Zeckendorf and developed the project on its own.

The site was divided by traffic arteries into superblocks, which were in turn subdivided into large individual sites. Thus, Alcoa could schedule development to meet market demand, occasionally building something itself, but usually selling large parcels to those office, hotel, and residential developers who were able to obtain the necessary financing.

Century City has long provided a glitzy setting for the active business world in Los Angeles. This was not always true. When the first buildings opened during the 1960s, the place had the look of a ghost town, especially at night when the office workers drove home. But as additional buildings were completed, activity increased and it became a quintessentially twentieth-century development—the dream of auto-dependent Los Angeles in aluminum and glass. Re-

Los Angeles, 2013. Avenue of the Stars—the main artery of Century City, a new-town-in-town that includes more than 9 million square feet of office space, 2000 condominium and town house apartments, 3 major hotels, a nine-story hospital, a 17-acre shopping center, multiplex movie theaters, and one of the city's largest legitimate theaters. *(Alexander Garvin)*

gional traffic sped past the project along Santa Monica, Olympic, and Pico boulevards, turning off these arteries to breeze down Avenue of the Stars, a monumental north-south artery with a median strip for trees, flowers, fountains, and sculpture that extends for nearly a mile. Workers, residents, and visitors drive directly into below-ground or hidden parking structures and then travel by elevator or escalator up to an office, apartment, or retail destination, without ever having to emerge into the landscaped plazas, malls, and walkways that surround the buildings.

Fifty years after Alcoa began development, Century City included nearly 10 million square feet of office space, two major hotels, a nine-story medical center, the 20th Century Fox Studios, a 17-acre shopping center, and parking for thousands of cars. Its southern half remains predominantly residential, containing 2100 condominiums and apartments that house more than 2600 residents, over 94 percent of whom work within Century City itself. Carefully walled off from the rest of the city, these residential projects have consistently remained among the city's most expensive addresses.

Century City did not turn out to work the way its planners had expected, but nonetheless can be considered a success. When the project was originally developed, occupants and visitors were intended to remain in superblocks and cross vehicular traffic on pedestrian bridges. Several pedestrian bridges remain, but due to a combination of the mild weather in Los Angeles and the inconvenience of driving from building to building, office workers and residents going to activities in a different superblock usually walk along the vast boulevards that were designed for fast-moving cars.

The Westfield Group acquired the shopping center in 2002 and has aggressively modernized, expanded, and improved it with a keen focus on unique retailers, destination restaurants, and evening uses. The mall's new shops, recently expanded movie theaters, supermarket, and dining terrace have been hugely successful at drawing upscale customers. The final phase of the mall's modernization will include a new retail anchor store, several more small boutiques, more parking, and a mixed-use tower containing a hotel and residences.

Not all of Century City's original developments were successes. The Century City Hospital, St. Regis Hotel, and the Shubert Theater were all casualties of changing tastes and markets, which ultimately caused them to be replaced by medical offices, condominiums, and 2000 Avenue of the Stars, a 790,000-square-foot office building. In many respects, 2000 Avenue of the Stars marked a turning point for development in Century City. Although the mammoth office building sits atop a giant parking deck, it is easily accessible on foot and contains a mix of restaurant and retail uses, a public park, and the Annenberg Center for Photography. Like 2000 Avenue of the Stars, a number of the more recent office developments have begun to engage the pedestrian realm and offer a mix of land uses.

Several projects planned before the economic collapse of 2008 are likely to proceed into construction. Unlike the first generation of development in Century City, however, these projects are all urban in character, oriented toward the street and designed to be accessible to pedestrians. When the long-planned Metro Purple Line subway station finally opens, Century City will become among the most high-intensity, mixed-use, walkable new-towns-in-town in the United States.

The "hard-bitten, super-commercial urbanity" of the "bland and blank, completely faceless high-rise buildings" in Century City would seem to preclude success as a

Los Angeles, 2013. One of the gated residential enclaves in Century City. *(Alexander Garvin)*

new-town-in-town.[28] Consequently, city planners and architects alike ignore it. However, as the ever-observant architect Charles Moore noticed three decades ago, "What is thoroughly puzzling . . . is why so many people write about it as such a success in all the other ways that matter, and why they see it as such a positive addition to the Los Angeles scene."[29] It should not be puzzling. Despite its mediocre architecture, Century City is a success because it is as much a regional employment, shopping, and entertainment destination as it is a residential community and because it is fully integrated into the traffic and land use patterns of its surroundings.

The gated residential superblocks of Century City will most likely never develop the character and charm of a lively city neighborhood. Nor will the carefully landscaped but antiseptic business superblocks offer the variety and richness of commercial activity on Wilshire Boulevard or Rodeo Drive in nearby Beverly Hills. Although recent developments have marked a change in direction, Century City's residential and commercial superblocks are fundamentally too self-contained and too isolated from one another to generate the vibrant pedestrian activity that is fundamental to conventional, pedestrian-based environments. While Century City may itself become more accommodating to pedestrians, it is unlikely that its workers, residents, and visitors will ever spill over onto neighboring streets of nearby communities, close as they may be. They would have to get back into their cars to drive there.

Whatever one's opinion of Century City's architectural design or urban presence, it is a notable real estate success story. Seventy-five acres of the 280-acre property were sold back to 20th Century Fox. In 1986, Alcoa sold its remaining interest in the project for $620 million to JMB/Urban Development Company.[30] Alcoa made a fortune on this one property because the rapid transformation of Los Angeles into the nation's second-largest metropolitan area provided an unmatched market for new office, retail, and residential space. Furthermore, the site was located in the path of the city's major corridor of expansion. Alcoa did not have to pay debt service on a mortgage covering property acquisition and development costs. More important, the company had deep pockets and, therefore, was able to pay real estate taxes and operating expenses without immediately having to cover these costs out of land sales. Most important, the site plan subdivided each superblock into individual parcels that could be developed whenever the market was ripe. Consequently, JMB/Urban Development and Westfield Companies continue to earn profits from the project by making better and better use of one of the most accessible destinations in the region.

Battery Park City, Manhattan

Battery Park City, may have been planned, financed, and developed by a state agency, but its similarity with Roosevelt Island stops there. The new-town-in-town on Roosevelt Island was created for entirely opportunistic reasons: housing without relocation and better utilization of scheduled transportation improvements. At Battery Park City, these same opportunistic consequences were the by-product of a bold urban planning strategy intended to revitalize Manhattan's downtown business core.

Lower Manhattan had been losing jobs to office districts in midtown Manhattan, Connecticut, Westchester, and New Jersey in the post–World War II period. The downtown business community tried to counter this decline by encouraging construction of new office space. Its effort began with the opening of the Chase Manhattan Building in 1960, and continued under the leadership of David Rockefeller (president of the Chase Manhattan Bank) when the Downtown Lower Manhattan Association pressed successfully for mapping a new interstate highway along the West Side and for construction of a World Trade Center over the railroad tracks of the Port Authority Trans-Hudson (PATH) train to New Jersey.

The proposed new highway offered an opportunity for restructuring business patterns throughout Lower Manhattan, while the fill excavated for the foundation of the proposed trade center could be deposited along the shoreline to extend Manhattan hundreds of feet into the Hudson River. Together, they supplied the rationale for an intelligently planned new-town-in-town on the western edge of Lower Manhattan. This project, which came to be known as Battery Park City, created sites for office expansion and walk-to-work residences that could take advantage of Lower Manhattan's underutilized infrastructure. What better way to improve the working environment, diversify the character of downtown business life, and enhance the city's economic and tax base?

These objectives were extended to the entire downtown business district by the *Lower Manhattan Plan* prepared for the New York City Planning Commission in 1966. This document articulated for the first time the need for carefully planned, large-scale, mixed-use development, includ-

Manhattan, 1969. An early version of Battery Park City showing the projected multilevel parking structure, hexagonal office buildings at the southern end of the site, and waterfront promenade. *(Courtesy of New York City Department of City Planning)*

ing thousands of new apartments, along the periphery of Lower Manhattan. Building in the Hudson and East rivers guaranteed that no one would be displaced, nor would existing services be interrupted. Furthermore, these riverfront projects would be able to take advantage of the tremendous existing transportation infrastructure in Lower Manhattan at night and on weekends, when the area would otherwise be deserted.[31]

Battery Park City may have started with the compelling public policy rationale of reinforcing downtown Manhattan. However, its initial master plan, prepared at the behest of Governor Nelson Rockefeller by architect Wallace K. Harrison, would have delivered the exact opposite: an independent zone boldly separated from downtown Manhattan. The design called for 92 acres of newly created land plus 24 acres of air rights over a depressed West Side Highway that would extend 1.2 miles from the Battery on the south to Chambers Street on the north. A 32-foot-high parking structure was planned for the Hudson River side of the new interstate highway. It was to be covered with a platform that extended the entire length of the development and contained an enclosed shopping mall. The southern end of the platform was dominated by three hexagonal towers, 40, 50, and 60 stories high, providing 5 million square feet of office space. Marching northward two-by-two was a series of high-rise apartment slabs containing 18,000 apartments.

Lower Manhattan Plan, 1969. The Lower Manhattan Plan adapted to include the contemporary master plan for Battery Park City. *(Courtesy of New York City Department of City Planning)*

In the face of enthusiastic advocacy by Governor Nelson Rockefeller, brother to the aforementioned president of Chase, and the entire downtown business establishment, the New York state legislature voted to proceed with Battery Park City. In 1968, it created the semiautonomous Battery Park City Authority with the power to issue the bonds necessary to finance the plan. No capital expenditures were required from the city or state, other than the cost of depressing the proposed West Side Highway, better known by its later name, Westway. (See Chapters 3 and 17.) Once the bonds were amortized, the project was expected to generate an additional $16 million per year in real estate taxes. Under the terms of the lease the authority negotiated with the city, one-third of the apartments were to be market-rate housing, one-third were to be middle-income housing, and one-third were to be low-rent housing. The necessary housing subsidies were supposed to be generated by the profits from the commercial space and market-rate housing, or be provided by the federal government.[32]

The authority issued $200 million in bonds to pay for planning, landfill, site improvements, administration, and debt service until revenues would be generated by the development. By 1976 the landfill was in place, but further work was stalled by a fiscal crisis at both the city and state levels. Project planning continued because bond proceeds could continue to be used for improvements, administration, and interest payments until 1980, when the first amortization payments were due to bondholders. FHA insurance was obtained for a 1712-apartment complex called Gateway Plaza. However, it became clear that neither Gateway Plaza nor any other revenue-producing component of Battery Park City would be completed in time to cover bond amortization requirements.[33]

The three-person board of directors of the Battery Park City Authority was composed of members who had been appointed for fixed terms by Governor Rockefeller. They steadfastly refused to alter the plan. When Governor Hugh Carey finally was able to replace two members in January 1979, the authority was folded into the New York State Urban Development Corporation. That group's president and chief executive officer, Richard Kahan, used the possible bond default as an opportunity to prevent a planning disaster. He commissioned an entirely new master plan, which Alexander Cooper and Stanton Eckstut completed within a few months.

The architects discarded everything that had been planned except Gateway Plaza, which was about to start construction. They chose instead to base their scheme on traditional New York City development patterns. As explained by Alexander Cooper, they set out to re-create "the best of New York," settling on such models as the East River esplanades at Carl Schurz Park and Brooklyn Heights, the neighborhood surrounding Gramercy Park, and the apartment houses along Central Park West. These places represented a form of urbanization that was radically different from the gigantic superblocks in Harrison's scheme.[34]

The new plan projected an eventual population of 30,000 residents and 31,000 workers. It eliminated development over the proposed highway. The remaining 92 acres of landfill were to be devoted 42 percent to residential uses (14,000 housing units); 9 percent to a commercial center relocated to be as close as possible to the subway and PATH stations at the World Trade Center; 30 percent to parks, open space, plazas, and a stunning riverfront esplanade; and 19 percent to streets and avenues. Two pedestrian bridges were planned to pass over West Street and the proposed highway.[35]

The new design also reduced initial infrastructure costs. Installation of more than half a mile of trunk sewer lines, water mains, and utility conduits was avoided by moving the proposed office buildings from the southern tip of the project to the middle of the site, where these lines entered

Manhattan, 1992. The World Financial Center at Battery Park City seen from the Hudson River. *(Alexander Garvin)*

the site. Elimination of the platforms that Harrison had designed saved millions more. Most important, the traditional block-by-block and lot-by-lot approach to development allowed infrastructure to be installed when dictated by market demand for building sites.

Instead of superblocks built by a few superdevelopers, Cooper and Eckstut chose to create separate neighborhoods made up of moderate-size city blocks that were, in turn, divided into moderate-size building parcels that could be built at different times by different developers. Instead of a continuous platform 32 feet above the rest of Lower Manhattan, they chose to create streets and squares that were woven into the pattern of existing city streets and which extended existing pedestrian views of the waterfront. Instead of separating Lower Manhattan from the Hudson River, they designed a riverfront esplanade that would become a favorite downtown destination. Instead of isolating office workers away from mass transit in three towers at the southern end of the project, they proposed to create a commercial center across from a major subway interchange, right at the heart of Battery Park City. Best of all, instead of a fiscal disaster, their design made possible a financial triumph.

When Richard Kahan renegotiated the terms of a new agreement with the city, the federal government no longer offered the generous housing subsidy programs that had been available when the original plan had been approved. He was therefore forced to eliminate all subsidized housing and establish a pay-as-you-go basis for all development. The first parcels hit the market in the early 1980s, when New York City was entering a period of rapid economic growth. There was active competition for virtually every site that was initially opened for development. Bids were unexpectedly high. The success of these first projects generated additional interest in the second group of sites when they were ready for development.

Contrary to expectations, the best bid and schedule for the four separately offered office sites was from a bidder for *all* the proposed office buildings: Olympia and York, a Canadian-based real estate firm, which at that time had plenty of equity capital and ready access to secure, long-term financing. It transformed the projected 6 million square feet of office space, 280,000 square feet of retail space, the 18,500-square-foot glass-enclosed public Winter Garden, and the 3.5-acre public plaza into a single, unified World Financial Center designed by Cesar Pelli and Associates. Thus, all the lucrative, revenue-producing commercial sites were quickly brought to completion, guaranteeing, in one fell swoop, that Battery Park City would be able to meet its bond obligations on schedule.

Battery Park City generates hundreds of millions of dollars in revenues beyond these obligations—revenues that in 1986 the state legislature voted could be spent on construction and rehabilitation of low- and moderate-income hous-

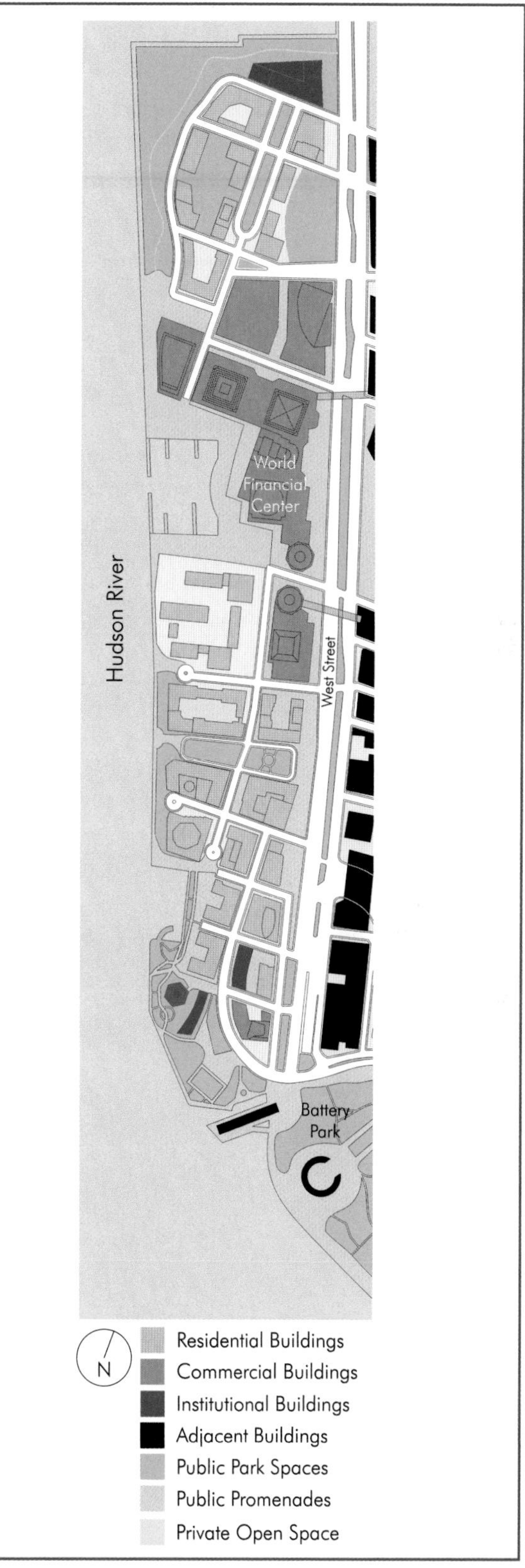

Manhattan, 2012. The Cooper/Ecktut plan for Battery Park City evolved over the following 30 years. *(Alexander Garvin and Ryan Salvatore)*

ing throughout New York City. Mayor Koch and Governor Cuomo envisioned using this money for the renovation and construction of 60,000 low- and moderate-income dwelling units by issuing $1 billion in bonds backed by the revenue streams from Battery Park City. As of the end of fiscal 1992, the New York City Housing Development Corporation had issued $455 million in bonds for the renovation of 3685 apartments in the South Bronx and Harlem. When the Giuliani administration took office in 1994, it chose not to earmark its share of the Battery Park City surplus for any particular purpose, using it instead to fund the general city budget, a large portion of which went to housing. By then it had become clear that many of the apartment buildings erected at Battery Park City after 1994 would be 80/20 Low-Income Housing Tax Credit (LIHTC) projects in which 20 percent of the units would be set aside for persons of low income (see Chapter 11).[36]

There was little activity at Battery Park City during the recession of the early 1990s, except for completion of its substantial public parks and construction of the Museum of Jewish Heritage. Because infrastructure improvements had been financed during the previous decade from revenues paid by the initial office buildings, the corporation was able to hold out until the market shifted and sites again could be sold for development. It also made changes to the design of the north residential neighborhood that allowed temporary ball fields to become permanent facilities. Battery Park City is now complete. In addition to the World Financial Center (whose major corporate tenants include American Express, Dow Jones & Co, and Deloitte), Battery Park City also includes the Goldman Sachs World Headquarters, an intermediate school, one of the city's premier high schools, 12 rental and 16 condominium apartment buildings providing residences for more than 13,000 people, 2 hotels, a multiplex movie theater, more than 350,000 square feet of retailing, a marina, a ferry station, and 13 parks, including a stunning esplanade along the Hudson River.

The eventual abandonment of the Westway had left a vast right-of-way between Battery Park City and the financial district, and the destruction and subsequent construction activity at the World Trade Center site has separated Battery Park City still further from the rest of Lower Manhattan. It remains to be seen whether future planning will close this unfortunate separation. However, the patina of time is beginning to set in, and Battery Park City has already lost its image as a new community, recognized now as one of New York's many unique, established neighborhoods.

Pentagon City, Virginia

The Cafritz-Tompkins Group acquired the 116-acre site of what is today Pentagon City during the 1940s, when the area was largely undeveloped and notable only for its proximity to the Pentagon. In those days, few developers foresaw the tre-

Manhattan, 2012. The riverfront esplanade at Battery Park City. *(Alexander Garvin)*

mendous growth of government or the dramatic changes that growth would bring to the national capital and its metropolitan region. Furthermore, most developers expected this growth to occur in conjunction with the projected suburban beltway, not within eyeshot of the Pentagon.[37]

For years, the Cafritz-Tompkins Group remained satisfied with the site's low-density industrial and commercial uses. This outlook changed when the Washington, D.C., Metropolitan Area Transit System (more commonly known as the Metro) opened in 1976. By then, the cluster of offices and apartment buildings at Crystal City, the next Metro stop from Pentagon City, was in the process of becoming one of Washington D.C.'s most important regional subcenters. Anticipating the tremendous potential market that would accompany the opening of the Metro station and, by extension, Pentagon City, the site's owners hired Dewberry, Nealon, and Davis to prepare a master plan for a whole new city district. They used this master plan to persuade Arlington County to rezone the area for high-density, mixed-use development: 1.25 million square feet of offices, 800,000 square feet of retailing, 2000 hotel rooms, 5900 housing units, a 300-bed nursing home, 300 units of subsidized elderly housing, an 11-acre public park, and thousands of much needed parking spaces.

Cafritz-Tompkins began the project with the nursing home, 300 apartments in a low-rise apartment complex, and 600 apartments for the elderly, all at the southern end of the site. It quickly became apparent that they would not be able to proceed with a project of this scale and complexity without a skilled developer who had the ability to package the project, market the space, and attract the necessary capital. In 1977, they entered into a joint venture with Rose Associates, a major New York City–based developer, for the 99 acres that had yet to be built.

Rose Associates had been established in 1928. It already had financed and built thousands of apartments and millions of square feet of office space that it owned and managed. It also had long experience with that most complicated of politico-financial markets, New York City. Thus, there were few if any difficulties that the firm had not encountered and overcome.

Daniel Rose, the partner in charge of Pentagon City, devised the strategies which led both to Cafritz-Tompkins selecting his firm as the developer and to the project's financial success. The deal Rose negotiated transformed time into an asset rather than a continuing debt-service liability. Instead of Rose Associates making loan payments on the ever-increasing cost of the project, Cafritz-Tompkins agreed to hold each section of the property until it was ready for development, then sell it at the market price plus a 15 percent interest in the venture. It was a productive compromise for everyone involved: Cafritz-Tompkins benefited by capturing the site's appreciation in value, while Rose Associates benefited by avoiding any cash payment or debt service until the project was ripe for development.

Rose also found a way to profit from intermingling uses at the site rather than permitting the complexities of the development to become a stumbling block. He avoided building the retail heart of the project until the subway station had been completed, a hotel deal had been negotiated, and office space either had been occupied or preleased. This delay allowed the occupants of the office, retail, hotel, and apartment buildings to simultaneously provide customers for one another.

The Cafritz-Tompkins-Rose joint effort began in 1980 with construction of 562,000 square feet of office space in two structures that were built for and subsequently sold to MCI Communications. This was followed in 1985 by the sale

Arlington, Virginia, 2012. A hotel, shopping mall, apartment houses, and numerous office buildings line Hayes Street at Pentagon City. *(Alexander Garvin)*

of land to Lincoln Properties for an additional 516,000 square feet of offices. By 1984, there was enough office space to attract Melvin Simon and Associates to become a joint-venture partner in the creation of an 860,000-square-foot Fashion Centre Mall that included Nordstrom's, Macy's, and another 170 stores; a 362-room Ritz-Carlton Hotel; a 172,000-square-foot office tower; and a 4534-car parking garage. In 1988, Rose Associates started another joint venture with the Sumitomo Corporation of America to build the 16-story, 299-unit Parc Vista Apartments. It has been followed by the Metropolitan (a 15-story 325-unit apartment building), Pentagon Row (a 1.5-million-square-foot mixed-use development with 504 apartments, shops, restaurants, and galleries), and the conversion of Pentagon City Hospital into a 183-unit apartment house.

Like Century City, Pentagon City demonstrates three of the ingredients of a financially successful new-town-in-town: the presence of a rapidly growing market, a location conveniently accessible to that market, and a development strategy that synchronizes capital expenditures with a dependable revenue stream. However, it is more than a popular shopping facility surrounded by separate office, hotel, and residential superblocks. Thanks to the Metro station and nearby Interstate 395, Pentagon City acts as a centripetal force attracting hundreds of thousands of people to a once-empty section of the Capital region. Residents from surrounding blocks filter through Pentagon City on their way to the subway and to their workplaces, miles away. The Fashion Centre at Pentagon City is the largest shopping complex that is within two Metro stops from the heart of downtown Washington, D.C. As a result, when the subway is not used for the journey to work, it is used by people from all over the metropolitan area who come to Pentagon City to shop, dine, and find entertainment. By 2012, it had become the pulsating heart of a dense and busy regional subcenter, generating activity in surrounding areas. The low-density warehouse and industrial buildings between Crystal City and Pentagon City have been replaced with big-box retail outlets (including one of the busiest Costco facilities in the nation) and popular new apartment buildings, with even more projects approved for construction.

Arlington, Virginia, 2012. The Fashion Centre Mall at Pentagon City attracts customers from the entire Washington, D.C., metropolitan area. *(Alexander Garvin)*

Atlantic Station, Atlanta

The new city districts that went into development during and after the 1990s had little of the "brave new world" idealism of earlier new-towns-in-town. They were usually conceived as real estate ventures that could profit from an available site, financing, or both, with little or no consideration of any impact on the city as a whole. Atlantic Station, for example, is built on the brownfield site of the former Atlantic Steel Mill in Atlanta. It made use of a government-approved Tax Allocation District (TAD) to cover the cost of public facilities and infrastructure (streets, sidewalks, sewers, and water mains), similar to the TAD that was later used by the city to finance the BeltLine (see Chapter 3). During the first 25 years in every TAD, any tax revenue in excess of the real estate taxes paid annually prior to approval of the TAD is allocated only to that district and goes first to pay debt service on the bonds issued for any initial public improvements. If there is anything left over, it is used to finance further capital investments exclusively within the district. In addition to remediation and project infrastructure, TAD proceeds were used to finance the extension of Seventeenth Street and a new eight-lane-wide bridge over Interstate I-75/85, thereby connecting Atlantic Station with the thriving Midtown section of Atlanta to the east and the lower-density residential districts to the west.

Atlantic Station was acquired in 1997 by the Jacoby Development, a land development company that completed most of the project by 2005. It consists of five parts: garage parking

Atlanta, 2007. Atlantic Station.
(Alexander Garvin)

under a platform that supports the ground plane for a commercial district; a low-rise, multifamily residential district that encircles a water retention pond; a second set of low-rise, multifamily residential buildings opposite IKEA; and row houses south of Sixteenth Street. The apartment complexes all include substantial multistory garages. The commercial district consists of office towers and a hotel that are located on one side of a retail complex occupied by national chains, a multiplex movie theater, and a supermarket erected on a conventional system of streets and sidewalks policed and maintained by project personnel. The water retention pond is bounded by landscaped edges with walkways, crossed by a pedestrian bridge, and anchored at one end by an imposing triumphal arch. It masquerades as a central park that functions well for resident dog walkers. Nobody else is able to use it because street-side parking is not permitted.

Atlantic Station does attract thousands of people, who come there to shop at IKEA, Target, and other stores, or to go to the movies. When they are done, they return to their cars and drive home. Although Atlantic Station is a magnet attracting people from all over the metropolitan region, there is no spillover activity to the surrounding parts of Atlanta. Midtown is east of the highway and too far away to walk to. A neighborhood comprising single-family homes separates Atlantic Station from the Georgia Tech campus, three-quarters of a mile south. There is no reason for people to go to the largely industrial/warehousing districts to the west and north. Thus, this major Atlanta destination has generated a modest private market reaction similar to that of the isolated and primarily residential new-towns-in town Brickell Key and Co-op City rather than Battery Park City or Pentagon City.

Ingredients of Success

The success of Dearborn Park, Century City, and Battery Park City is no accident. Each was able to tap a huge market because it offered an easily accessible product that its customers desired and could afford. Despite very different designs, development within each of these new-towns-in-town could be slowed without significant financial consequences when market demand slackened and then resumed when conditions improved. Most important, major investments in infrastructure and community facilities could be delayed until there were enough project revenues to justify the additional debt service.

Market

Some people believe that creating new-towns-in-town in rapidly expanding regions like post–World War II Los Angeles or Washington, D.C., is not difficult. However, too many planned new communities have failed for anybody to believe that success is simply a matter of selecting a site within a growing market area. The projects that succeeded did so by offering a better deal than the competition.

Some new-towns-in-town offer a better deal by marketing a good product at lower prices. Abraham Kazan's projects, for example, minimized the down payment and lowered monthly housing expenditures. They entirely eliminated profit and received exemption from a portion of local real estate taxes. Later projects also obtained long-term, below-market-rate-interest mortgages. Cedar-Riverside tried a similar strategy. However, its lower prices depended on federal subsidies. As soon as these subsidies were no longer available, the project lost its market and came to a halt.

Projects like RiverPlace and Pentagon City attract their market by offering amenities that are not readily available elsewhere. RiverPlace provides a marina and a waterfront esplanade. Pentagon City includes a Metro station. Dearborn Park offers a degree of privacy that is difficult to find elsewhere in downtown Chicago. Roosevelt Island offers its own public school and island-centered municipal services.

If the surrounding city is to benefit from the market attracted by a new-town-in-town, however, there must be a reason for interaction. Not all the customers who come to the Fashion Centre Mall at Pentagon City return home without checking out what is across the street. Nobody from outside Roosevelt Island, on the other hand, patronizes its stores. These retailers have to survive from purchases made exclusively by the island's residents.

Location

New-towns-in-town often occupy leftover sites that previously were not sufficiently attractive for development or have been discarded by previous users. Thus, site characteristics rarely can be exploited for marketing purposes. In a few cases, though, they have contributed to project success. Roosevelt Island, RiverPlace, and Battery Park City, for example, all exploit the possibility of dramatic waterfront views.

If RiverPlace were far from downtown Portland or Pentagon City far from downtown Washington, there would be less interest in living there. Mere proximity to an expanding market, however, is not sufficient. New-towns-in-town also must be easily accessible.

The pedestrian network of Battery Park City is a case in point. Until the original project design was abandoned, office workers would have had to hike nearly half a mile from the subway to get to their jobs in the office towers at the southern tip of the project. Access was radically improved by relocating these buildings to a site directly across from the PATH and the intersection of subway stations in downtown Manhattan. The pedestrian bridges, which were erected as part of the final scheme, allow office workers to avoid the heavy vehicular traffic on West Street. These bridges also provide easy access for visitors from all over Lower Manhattan who come to stroll along its traffic-free riverfront esplanade, wander among its retail stores, and make use of its restaurants during their lunch hour. Perhaps the least anticipated linkage has been by bicycle. The 5-mile-long bike lane that is part of the Hudson River Park (see Chapter 3), brings people from all over the city to enjoy the splendid parks and esplanade that provide stupendous views of the river and New Jersey.

Design

The design of a new-town-in-town is largely dependent on its size, the arrangement of its components, and their relationship with the surrounding city. Both Roosevelt Island and Brickell Key, for example, are separated by water from the rest of the city. Roosevelt Island's problems arise from a design that ignores this reality. To provide a satisfactory living environment, its design would have to adjust to that separation by providing enough residents to support a broad range of commercial and community facilities. When the "town center" that Philip Johnson had envisioned was eliminated, the project was divided into two sections. They are far enough from one another that neither will have enough residents to support a full range of retail stores. The residents of Brickell Key can simply walk across a bridge to reach the rest of the city.

Unlike these islands, Co-op City did not have to be separate from the rest of the Bronx. Its arteries could have been connected with neighboring communities, thereby providing a critical mass of 50,000 customers able to support a wide array of services. Moreover, its use of independent shopping centers rather than continuous shopping streets deprives citizens on both sides of the highway of the safety and vitality of 24-hour activity. The result is a living environment that is as dull as the 35 drab brick towers that dominate the project.

Century City and Battery Park City, on the other hand, are megaprojects that are *not* designed to be independent districts. Not only do they include thousands of apartments, they also contain millions of square feet of office space. Century City and Pentagon City also include regional shopping malls with department stores. As a result, each of these new-towns-in-town has become a major city destination with people coming and going 24 hours a day.

Financing

Hundreds of new-towns-in-town never see the light of day because stable, long-term financing is unavailable. Others, like Cedar-Riverside, are terminated when money dries up. Financing for new-towns-in-town is based on the same principles that determine any real estate venture. However, because they require such huge sums, obtaining the necessary capital is no easy matter.

There are many ways to reduce the need for capital. Dearborn Park raised $14 million in capital by persuading 21 Chicago businesses to contribute to a fund that, at best, promised a limited return on that investment. It eliminated the cost of infrastructure by persuading the city of Chicago to pay $18 million for items that would otherwise have been the developer's responsibility. Roosevelt Island involved no acquisition financing because it was constructed on government-owned land. Century City avoided major expenditures on traffic arteries and infrastructure by exploiting the existing Los Angeles street grid.

The key to a successful financing strategy, however, is intimately related to time and lies in avoiding substantial borrowing until there is an obvious source of revenue to cover debt-service payments. Rose Associates deferred the cost of carrying Pentagon City until tenants had already been signed up. Financing the infrastructure at Atlantic Station was provided by the TAD there. At Battery Park City, a state authority issued tax-exempt bonds whose interest was prepaid until the project had attracted sufficient occupants. Once Olympia and York agreed to build the World Financial Center, Battery Park City was able to cover debt service and generate millions in additional revenues for other public purposes. On the other hand, the Roosevelt Island Operating Corporation has been running a deficit for two decades because it cannot recoup infrastructure and community facility costs from residents who are not yet in place.

Time

Passing through Battery Park City is a delightful experience for residents, workers, and visitors. Along the way, one can stop at a restaurant in one of the courtyards of the World Financial Center, purchase something in one of its shops, pause for a sandwich on a bench overlooking the marina, stroll along the riverfront esplanade, or wander through one of the small parks. There are no amenities like these to attract people as they pass through Park La Brea (now a gated community) or Cedar-Riverside. Nobody but the occupants and their guests go to Brickell Key.

The single most important ingredient in the success of a new-town-in-town is the ability to survive over very much longer periods of time. It may take several decades to complete a planned new community. During that time, the project will have to survive continually changing consumer tastes and several economic cycles. Even a financially successful project will have to avoid development during serious economic downturns. Consequently, it must be conceived in a manner that will enable it to survive on the cash flow from already developed portions of the project. At Century City that was relatively easy because the scheme required little in the way of infrastructure. As a result, the cost of carrying the property was largely a matter of paying real estate taxes and minimal operating expenses till land sales could resume. The design of Battery Park City solved this problem by covering debt service and most other expenses out of revenues from the World Financial Center and charging residents for additional services as such services become necessary when residents moved in. Roosevelt Island has yet to balance project expenses with revenues. The TAD financing of Atlantic Station and the Mitchell Lama financing of Co-op City eliminated the problem of timing investment in tandem with the arrival of tenants.

Entrepreneurship

There are two simple models for developing new-towns-in-town. Like Park La Brea, a project can be financed and built by the same entity. This model has become less and less popular because few financial institutions are willing to risk hundreds of millions of dollars on one project or to include within their organization the necessary staff of real estate developers. Alternatively, a property owner like Zeckendorf/Alcoa can commission a plan to be filled in by individual developers in response to market demand. This model has also become more difficult to execute because there is a decreasing number of sites for new-towns-in-town that can be developed without huge up-front investments in infrastructure.

Relying on the nonprofit sector is not possible. Nonprofit developers like the United Housing Foundation require years to develop a staff with the necessary development experience. More important, they need a continuing source of revenue to support the organization. Few nonprofit organizations are able to fund such operations, and fewer still are able to sustain the necessary revenue stream. Those that engage in real estate development have the added burden of policies that try to minimize monthly housing costs for residents of their projects. Consequently, they are unwilling to charge the generous fees that will keep them in operation. As soon as anything goes wrong, as it did at Co-op City, they are unable to stay in operation. That explains why there are few nonprofit housing developers that propose new-towns-in-town and none (now that the United Housing Foundation is out of the development business) with the resources to develop them.

The results tend to be variable when government acts as the developer of a new-town-in-town. As long as Ed Logue was in charge of Roosevelt Island and Richard Kahan directed the development of Battery Park City, their projects moved forward relatively swiftly and with admirable design and financing results. They remained in charge for a few years. After they left, the various agency officials in charge took more than three decades to bring the projects to completion. RiverPlace is one of the few government-developed new-towns-in-town that achieved its initial objectives relatively quickly and with enviable results.

Thus, the entrepreneurship needed for the creation of new-towns-in-town is most likely to come from profit-motivated developers who find ways of minimizing risk or from publicly created and financed development entities that need not consider risk. Dearborn Park is unique in obtaining entrepreneurial talent without paying for it. Ferd Kramer and Philip Klutznick, two of the nation's most successful developers, were willing to devote much of the last 20 years of their lives to making the project a reality because they thought it important to revive a dying part of their home town. The same was true of Gerald Fogelson, who devoted considerable time and effort obtaining support for Central Station from all the area's stakeholders prior to getting city approval of the project.

New-Towns-in-Town as a Planning Strategy

Lewis Mumford admired new-towns-in-town such as Fresh Meadows, believing that "if all New York were designed on the same principles, the roads leading out of it would not carry such a weekend load of desperate people looking for a spot of green or a patch of blue or a pool of quiet—the mirage of the great metropolitan desert."[38] He was wrong, however, in believing that new-towns-in-town were the alternative to "intensifying the congestion in high-rise buildings" or endless additional "square miles of suburbs." Neither New York nor any other city is likely to grow primarily by the incremental addition of such new-towns-in-town. There is no way to mass-produce new-towns-in-town because they require too much patient capital, because too much of every metropolitan area is already filled in, and because political opposition will prevent all but a very few from reaching completion.

Mumford was right, however, in believing that new-towns-in-town could provide attractive alternative residential neighborhoods and enhance the quality of life in adjacent areas. To do so, they must be conceived as more than oversize housing projects with the minimum necessary overlay of community facilities. They must be transformed into destinations that can attract the additional users. Then they will support amenities that a city could not otherwise afford to install or continue to maintain. Most important, if considered as strategic citywide investments, they can, like Battery Park City, enhance the very character of city life.

Governments everywhere are faced with proposals for new-towns-in-town from private builders, public agencies, and civic organizations. Typically, once a scheme gains momentum, it causes problems for everyone. Community groups are urged to accept a plan they find unsatisfactory. Developers struggle to save their investments and commitments. Government tries to mediate, but its power to recommend alternatives is limited. The situation is conducive to neither sensible planning nor the reconciliation of differences. What government, community groups, and developers all need is a planning framework that allows desirable projects to proceed.

Any planning framework for the evaluation of new-towns-in-town must separate privately sponsored from publicly assisted projects to ensure that, at a minimum, publicly assisted projects have a public policy justification. However, since both privately sponsored and publicly developed communities will require rezoning, they should be evaluated in a similar manner and meet the same minimum standards.

New-towns-in-town will affect traffic patterns, sewage-treatment capacity, water supply and drainage systems, school loading patterns, and myriad other resources. If a particular project does not have any significant negative environmental impact and does not require significant additional municipal construction or expense, there is every reason to rezone. Similarly, if the developer (private or public) is prepared to mitigate its negative impact, the site should be rezoned.

The difficulty with such simple consideration on the merits is that too often a dispute over the degree of negative impact becomes the basis of divisive political controversy and costly, time-consuming litigation. One way to avoid this is to publish standards enacted by the appropriate legislative bodies, which, if met by the project, result in automatic rezoning. These standards could include: primary and secondary processing of all sewage; a minimum level of supply of potable water per square foot of development; available school seats for every type of apartment created; a minimum number of parking spaces per square foot of office, retail, and housing; and so forth. Such standards should be quantifiable and measurable. A specific agency (e.g., the city planning department) would be given a specified period of time to verify that the standards have been met. Thereafter, development would be as-of-right.

A similarly straightforward approach should apply to design. The issue here is not the amount of light and air, the compatibility of land uses, or the appropriateness of the proposed density. Zoning and building regulations cover such matters. The issue is impact on surrounding areas. No new-town-in-town should be designed, like Co-op City, in a manner that attracts thousands of residents while draining vitality from and failing to benefit surrounding areas. These problems can be avoided if the barriers (e.g., highways, waterways, ravines) between the proposed new-town-in-town and the surrounding neighborhoods are bridged by streets and buildings, thereby tying the new district to the surrounding neighborhood; if the streets and land use patterns of the project are designed as extensions of surrounding arteries; and if the proposed new-town-in-town provides additional land uses that attract people who would not otherwise be there. Whatever criteria are selected, they should be specified and enacted into law so that desirable new-towns-in-town can reduce time-consuming bureaucratic review or expensive litigation and provide a sense of real estate certainty to private market developers.

Publicly assisted projects cannot be justified simply on the basis that they will have no negative impact on the surrounding city and will be designed in a desirable manner. There must be a defensible public-policy rationale to warrant government assistance, whether it is assistance for assembling sites, or financing the prior installation of necessary infrastructure and community facilities, or providing long-term, low-cost mortgages for residential development. Nor is it enough, as happened at Atlantic Station, to build a much-needed bridge or road that connects the project with the rest of the city without also directly linking to contiguous districts that could profit from spillover from the people and businesses who go to and from that new-town-in-town.

Government assistance should be forthcoming only where, as at Dearborn Park and Central Station, it will stimulate investment that would not otherwise occur and where, as at RiverPlace, it will generate a desirable market reaction in adjacent areas. The problem with this simple standard is that there must be a way of separating parochial interests (whether powerful developers, domineering public agencies, or selfish community groups) from the interests of the region as a whole, without resorting to litigation.

The resolution of such controversy is to be found in the need for the massive amounts of patient capital. If a public agency believes a specific new-town-in-town is important enough to the future development of a metropolitan region to justify public financing, it should be ready to put that financing up to a vote. The vote should be on the bond issue required to finance the project, any contemplated public subsidies, and all other government approvals. Truly desirable new-towns-in-town will generate the votes needed to override parochial opposition. Once a project has been voted on, there will be many fewer opportunities for expensive, time-consuming agency processing or litigation.

So far, only one city has tried to develop published standards by which to judge proposals for new-towns-in-town. That effort failed. In 1973, the New York City Planning Commission inaugurated a series of workshops and publications as part of its ongoing planning process. One of these workshops was devoted to large-scale development. It examined 20 typical sites ranging in size from 10 to 300 acres. In fact, the agency had identified more than 50 potential sites but was afraid that talking about them would generate unnecessary community opposition. Even if all the potential sites had been openly discussed, nothing would have happened, because the agency abandoned the effort a few months after it had started.[39]

Many of the new-town-in-town sites identified in 1973 kept reappearing. Those that took shape as serious proposals developed into major controversies that were eventually resolved based on the political power of the participants with little or no attention to a project's impact on the city as a whole. For example, every consideration of a proposed new-town-in-town for the West Side Rail Yards in Manhattan, known now as Riverside South, focused on opposition from adjacent residents, not on any intrinsic reasons for development in that location. Similarly, when a coalition of government agencies proposed a planned new community along the East River waterfront in Long Island City, eventually called Queens West, they failed to disclose the need for well over $2 billion in public expenditures for infrastructure and community facilities, money that clearly was not available. At the same time, opponents made expensive, unrealistic lists of changes and amenities they felt might mitigate the negative consequences of the proposal, without being able to focus discussion on whether the proposed project would achieve any of the city's planning and development objectives. In both cases it took years to reach acceptable solutions—in great measure because the people who would eventually live in these projects were not represented in the discussion.

The course of these controversies in New York City has provided a lesson that its citizens have yet to learn: cities need clearly articulated standards and strongly presented policies by which to evaluate any proposed new-town-in-town. In their absence, every city is doomed to divisive battles and ad hoc decisions that may or may not produce healthy communities, intelligent consumption of precious, underdeveloped land, or desirable patterns of urbanization.

Notes

1. Lewis Mumford, *The Urban Prospect,* Harcourt, Brace & World, New York, 1968, p. 211.
2. Lewis Mumford, *The Highway and the City,* The New American Library, New York, 1964, p. 235.
3. In 2000, an investment subsidiary of the Community Preservation Corporation (see Chapter 13) completed the conversion of Parkchester from a rental project to a residential condominium and extended a $130 million loan for the rehabilitation of the 8200-unit south condominium.
4. Elliott Metz (project director), *Public and Publicly Aided Housing in New York City 1927–1973,* New York City Planning Commission, New York, 1974. The total of 1434 dwelling units includes 263 units subsequently demolished for Amalgamated Houses extension.
5. Metz, op. cit. *Completion* is defined as the issuance of the project's final certificate of occupancy. Tenants may have moved in earlier, upon the issuance of a temporary certificate of occupancy.
6. Eugene Rachlis and John E. Marqusee, *The Landlords,* Random House, New York, 1963, pp. 131–163.
7. Roger Starr, statement at the public hearing held by the New York City Planning Commission on April 28, 1965, as reported by the commission's stenotypist.
8. William Zeckendorf Sr. with Edward McCreary, *Zeckendorf,* Holt, Rinehart, & Winston, New York, 1970, pp. 291–292.
9. Ada Louise Huxtable, *Will They Ever Finish Bruckner Boulevard?,* Collier Books, New York, 1972, pp. 77–80.
10. New York State Division of Housing and Community Renewal, *Statistical Summary of Programs,* New York, March 31, 1978, p. 81.
11. Total project cost is estimated by the New York State Division of Housing and Community Renewal at $422,699,700 (excluding $46 million in construction financing for the Co-op City Educational Park).
12. Historical and statistical information on Brickell Key is derived from ULI-the Urban Land Institute, *Miami, Florida Metropolitan Area . . . Today,* ULI-the Urban Land Institute, Washington, DC, 1983, pp. 61–63; Allan T. Shulman, Randall C. Robinson Jr., and James F. Donnelly, *Miami Architecture,* University Press of Florida, Gainesville, FL, 2010, pp. 80–81; http://www.swireproperties.com; and http://www.brickellkey411.com.
13. Historical and statistical information on Dearborn Park is derived from Lois Wille, *At Home in the Loop,* Southern Illinois University Press, Carbondale, IL, 1997, and Dennis McClendon, Chicago CartoGraphics.
14. Historical and statistical information on Central Station is derived from Gerald W. Fogelson, Central Station: *Realizing a Vision,* Rancom Communications, Evanston, IL, 2003, and Deborah Johnson, "The New Face of Chicago's South Loop," *Urban Land,* ULI-Urban Land Institute, Washington, DC, April, 2002, http://www.centralstationsouthloop.com/.
15. Housing & Urban Development Act of 1969, Sec. 402.
16. Historical and statistical information on Cedar-Riverside is derived from Judith A. Martin, *Recycling the Central City: The Development of a New-Town-in-Town,* Center for Urban and Regional Affairs, University of Minnesota, Minneapolis, MN, 1978.
17. Welfare Island Planning and Development Committee (Benno C. Schmidt, Chairman), *Report of the Welfare Island Planning and Development Committee,* submitted to John V. Lindsay, Mayor, City of New York, February 1969.
18. In addition to Welfare Island, Logue also sought to develop a series of new-towns-in-town over the Penn Central Railroad right-of-way that extended along the Harlem River in the Bronx. Only one of

them, the 1654-unit Park River Towers project, was completed. It was developed in conjunction with Roberto Clemente Park, the first state-financed park within the city of New York.

19. The idea of renaming the island after Franklin D. Roosevelt had been suggested by the *New York Times,* in an editorial dated February 15, 1969.
20. Philip Johnson and John Burgee, *The Island Nobody Knows,* New York State Urban Development Corporation, New York, October 1969.
21. New York State UDC, *Roosevelt Island Schools P.S.-I.S. 217,* undated brochure, p. 2.
22. Matthew I. Wald, "Many Roosevelt I. Shops Still Empty," *New York Times,* December 25, 1977, sec. 10, p. 1.
23. Michael Kimmelman, "Decades Later, a Vision Survives," *New York Times,* September 13, 2012, A1.
24. Carl Abbot, *Portland: Planning, Politics, and Growth in a Twentieth-Century City,* University of Nebraska Press, Lincoln, NE, 1983, pp. 215–216 and 224–225; Robert Moses, *Public Works: A Dangerous Trade,* McGraw-Hill, New York, 1970, pp. 758–765.
25. Historical and statistical information on RiverPlace is derived from *RiverPlace,* ULI Project Reference File, vol. 18, no. 3, Urban Land Institute, Washington, DC, January–March 1988; www.terrain.org/Issue_7/UnSprawl/RiverPlace/riverplace.html; and Simmons B. Buntin, "Sustaining the City through Sustainable Development," Unsprawl Case Study #7, Planetizen Press, 2000, http://www.terrain.org/unsprawl/7.
26. Historical and statistical information on Century City is derived from Cheryl G. Cummins (program director), *Los Angeles Metropolitan Area . . . Today,* Urban Land Institute, Washington, DC, 1987, pp. 110–124, and Zeckendorf with McCreary, op. cit., pp. 245–254; http://cityplanning.lacity.org; and Century City Chamber of Commerce.
27. Zeckendorf with McCreary, op. cit., p. 247.
28. Charles Moore, Peter Backer, and Regula Campbell, *The City Observed: Los Angeles,* Vintage Books, New York, 1984, pp. 211–212.
29. Ibid.
30. The transaction included five parcels of undeveloped land, plus half-ownership in the twin 44-story Century Plaza Towers, plus the Century Plaza Hotel and Towers.
31. Wallace, McHarg, Roberts, and Todd Whittlesey, Conklin, and Rossant, Alan M. Voorhees & Associates, Inc., *The Lower Manhattan Plan,* New York City Planning Commission, New York, June 1, 1966.
32. There was never any hope of generating sufficient revenues from the new office buildings and market-rate housing to subsidize so large a quantity of apartments. The proponents of Battery Park City, like so many other advocates, had cynically proposed something they were ready to eliminate when it was proven infeasible.
33. Luckily, bond proceeds had been invested at a rate of interest that was higher than that paid to the bondholders.
34. Alexander Cooper, interview, August 1990.
35. Battery Park City Authority, *Battery Park City Vital Statistics,* undated brochure.
36. David L. A. Gordon, *Battery Park City—Politics and Planning on the New York Waterfront,* Gordon and Breach Publishers, Amsterdam, 1997.
37. Historical and statistical information on Pentagon City is derived from *Pentagon City,* ULI Project Reference File, vol. 20, no. 4, Urban Land Institute, Washington, DC, 1990, and an interview with Daniel Rose, October 3, 1990.
38. Lewis Mumford, *From the Ground Up,* Harcourt, Brace & World, New York, 1956, p. 12.
39. See New York City Department of City Planning, *Large-Scale Development in New York City,* New York City, New York, October 1973.

16

New-Towns-in-the-Country

Celebration, Florida, 2006 *(Alexander Garvin)*

Some planned new towns, including Chestnut Hill in Philadelphia, Beverly Hills, California, and Seaside, Florida, began as vacation refuge new towns. The second group, including Lake Forest, Illinois, Shaker Heights, Ohio, and Reston, Virginia, were intended to be suburban satellites of large and growing cities. Some new towns, such as Riverside, Illinois, and Palos Verdes Estates, California, are places of rare beauty. Others, such as Levittown, Long Island, and Lakewood, California, were profitable business ventures that also produced huge quantities of inexpensive houses. Still others, like the bulk of the Title VII new towns started during the Nixon administration, were flops and are now forgotten.

Thousands of planned new towns in America may have been intended as vacation refuges or as utopian satellites of congested cities. However, from the beginning, some of their proponents have also seen them as devices for solving other pressing problems. Some argue that land is a scarce resource that, once consumed, can be reclaimed only at considerable expense and disruption. They believe that carefully planned new towns preserve this scarce resource, maximize open space available to nearby populations, and site land uses so that they interact with infrastructure and community facilities in a more efficient and economical manner.

Others argue that planned new towns are cheaper because they can be created on large properties that do not require costly, time-consuming site assemblage. Since infrastructure, community facilities, and municipal services can be provided more efficiently, the assumption is that development costs will be lower.

Some advocates even believe that there will be less crime, juvenile delinquency, and other social ills in well-planned new communities. Their rationale is not preposterous. New towns usually provide convenient access to community facilities and municipal services that are not as easily available elsewhere. Their village-like organization may reduce conflict and increase social interaction. During the nineteenth century, these ideas led to the establishment of utopian new communities by Shakers, Rappites, Icarians, Owenites, and Fourierists. Such communities continued to be founded into the twentieth century, though far less frequently.[1]

More recently, in reaction to post–World War II suburbanization, architects and planners have designed communities that invoke the image of small-town America. They have evolved into two groups, both of which advocate compact subdivisions with defined boundaries in which the activities of daily living are within walking distance. The *New Urbanists* and their followers call for high-density districts served by mass transit, rather than what they decry as vacuous sprawl accessible only by automobile (see Chapter 14). The *Neotraditionalists* accept the fact that the electorate is not prepared to invest in transit corridors designed to organize entire metropolitan regions. They advocate planned communities that respect historical precedents and include a rich and substantial public realm, not just one-family houses.

In all these cases, there is a utopianism that extends to a political rationale for new towns. Somehow, the promise of improved social interaction gets translated into a racially and economically integrated society. Orderly industrial location gets translated into living and working in the same community.

Even excluding utopian communities, company towns, and seats of government, all these claims can be demonstrated by some planned new towns. Palos Verdes Estates and Columbia offer residents unusually generous open space and recreational facilities. Sea Ranch, California, preserves thousands of acres of meadowland that would otherwise have been torn up for little boxes made of ticky-tacky. Celebration and New Albany provide pedestrian-accessible recreation facilities that are unavailable in nearby subdivisions. The designs of Riverside and Shaker Heights integrate transportation and utilities in an efficient and attractive manner. Houses in Levittown and Lakewood were a better buy than contemporaneous nearby tract development. The quality of life at Seaside and Beverly Hills is more pleasant than in most suburbs. These same planned new towns can also be used to demonstrate flaws in the physical, economic, social, and political rationales propounded by their advocates: Levittown is not prettier; Shaker Heights is not more convenient; Beverly Hills is not cheaper; Palos Verdes Estates is not healthier socially; Reston is not politically more vigorous than neighboring communities.

Some new-town advocates will admit the debatable nature of these rationales. They may even admit that the underlying reason for their continuing faith in planned new communities is a distaste for existing cities. Naturally, they assume that everybody feels the same way and, when provided with this much-desired alternative, will happily leave outmoded, unsatisfactory cities. Like the English social visionary and political activist Ebenezer Howard, they believe that once new towns have become sufficiently pervasive, consumers no longer will have any reason to remain behind in "obsolete" cities, which will then inevitably wither away.

Developers have been creating new-towns-in-the-country for nearly two centuries. Some of them began as vacation refuges and became self-sufficient communities that function 365 days a year. Others began as utopian satellites that are now quite independent of the cities that spawned them. Yet neither variety of new town has led people to abandon cities.

Nor are new-towns-in-the-country an exceptional phenomenon. In 1992, the Urban Land Institute issued a directory of projects that were then under way. It included 529 planned communities of more than 300 acres that contained "a substantial amount of housing units," a "significant recreational amenity package," and often "significant commercial uses as well."[2]

Nevertheless, well-intentioned reformers continue to insist that the new town is the antidote to both urban blight and suburban sprawl. On three occasions they even persuaded the federal government to get into the new-town building business. None of these efforts demonstrated the superiority of government-assisted new-town development over more modest forms of urbanization. In fact, the loveliest of America's new-towns-in-the-country—Riverside, Palos Verdes, Sea Ranch, and Seaside—were started at different times by different developers that conceived of them as for-profit real estate ventures.

The new-town-in-the-country, like the rectilinear subdivision, the curvilinear subdivision, the clustered community, and the new-town-in-town, is simply a form of urbanization. Like them, it should be allowed to flourish wherever there are developers, investors, and lenders that are willing to risk their money. Rather than saddle these new towns with responsibility for fixing our urban-suburban environment, government should ensure that they do not damage it, and step out of the way.

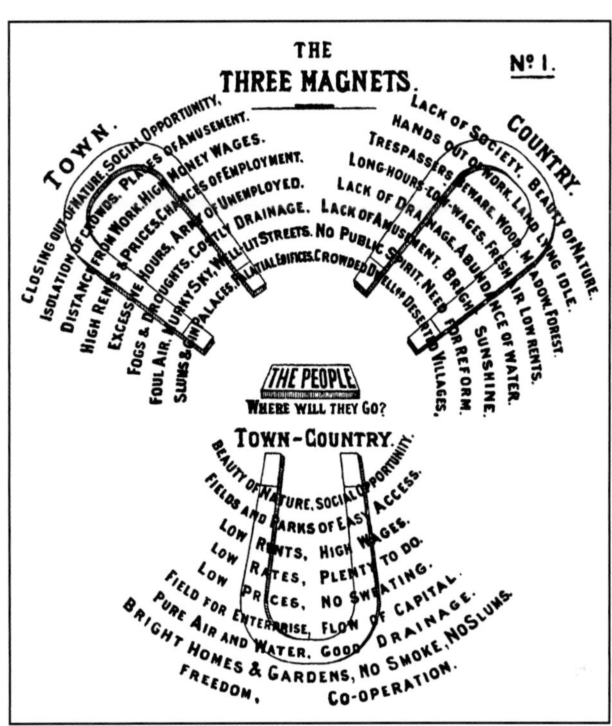

Ebenezer Howard, 1898. The best of "town and country must be married and out of the joyous union will spring a new hope, a new life, a new civilization." *(From E. Howard,* Garden Cities of Tomorrow, *MIT Press, Cambridge, MA, 1965)*

Ebenezer Howard and the Garden City

In 1871, at the age of 21, Ebenezer Howard emigrated from London to Nebraska. After a year as a farmer, he moved to Chicago and began a career as a shorthand reporter. In 1876, Howard returned to England, where he earned a living as a stenographer, worked on a series of mechanical inventions, and participated in organizations interested in land reform, urban poverty, and other public issues. His interest in social reform led, in 1898, to the publication at his own expense of *Tomorrow: A Peaceful Path to Real Reform* (revised and reissued in 1902 under its better-known title, *Garden Cities of Tomorrow*).[3]

This influential book identified the problems of living in existing cities (closing out of nature, overcrowding, high rents and high prices in general, air pollution, slums, etc.) and of living in the country (lack of society, unemployment, long hours at low wages, lack of cultural and recreational facilities, etc.). Howard argued for "a third alternative, in which all the advantages of the most energetic and active town life, with all the beauty and delight of the country, may be secured in perfect combination."[4] This third alternative, which he called a *garden city,* would combine beauty of nature with social opportunity, entrepreneurial opportunity with the flow of capital, accessible parks with slumless residential areas, pure air and water with plenty of things to do, and so forth.[5]

The suburban utopia that Howard proposed had many of the features of English eighteenth-century suburban communities, where villas were built among open fields and groves of trees and where children could grow up sheltered from the dangers, cruelties, and immorality of the crowded, filthy streets of nearby London. However, Howard's conception went beyond withdrawing one's family from the evils of the city in order to settle in a house surrounded by trees and grass and flowers.[6]

Howard wanted to create a completely new entity. As he explained, "Town and country must be married and out of this joyous union will spring a new hope, a new life, a new civilization."[7] This attempted marriage of town and country was probably influenced by his experiences in America. During the four years that he lived in Chicago, an increasing number of city residents moved to new towns built in conjunction with suburban railroad stations. Howard must have been familiar with these Illinois communities, particularly Lake Forest and Riverside. The 1871 promotional brochure for Riverside boasts, in language very reminiscent of Ebenezer Howard's later writings, that it will "combine the conveniences peculiar to the finest modern cities, with the domestic advantages of the most charming country."[8]

Lake Forest and Riverside, Illinois

In 1856, members of Chicago's First and Second Presbyterian churches decided to found a Presbyterian school. They acquired a 2300-acre irregular, wooded site along Lake Michigan, on the recently completed Chicago and Milwau-

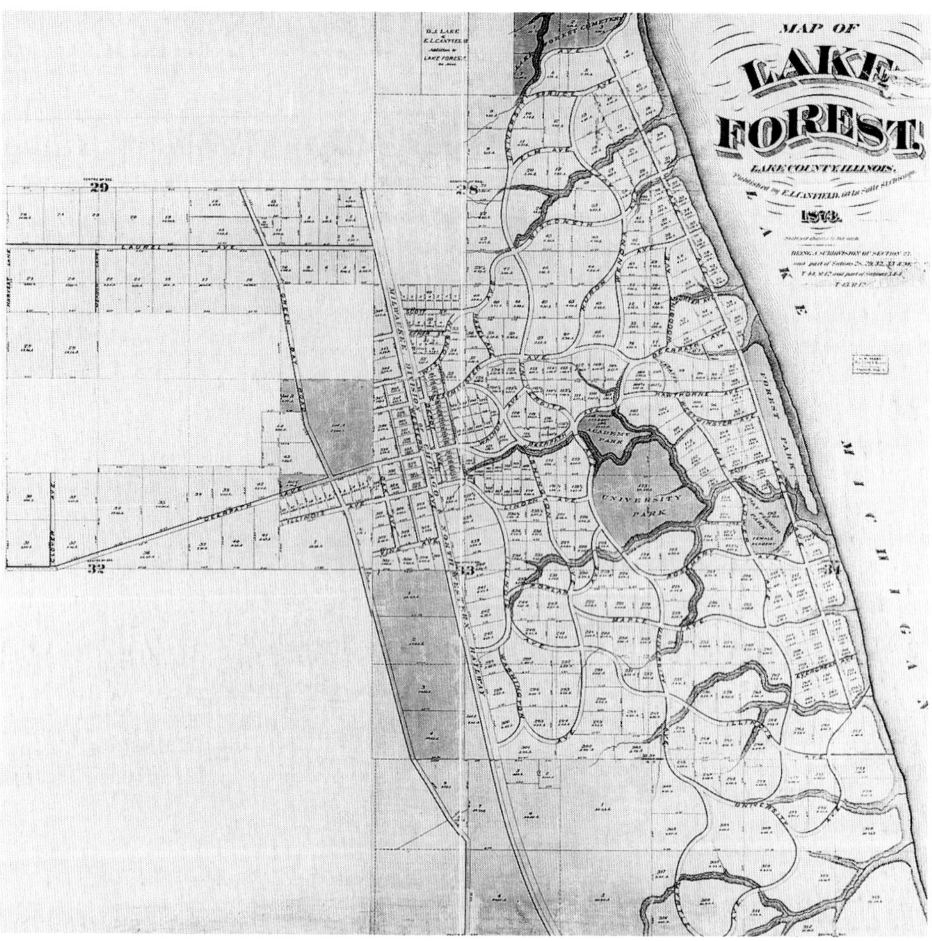

Lake Forest, 1873. Plat of Lake Forest in which the streets radiate from the railroad station to Lake Michigan, following routes that minimize topographical problems. *(Courtesy of Chicago Historical Society)*

kee Railroad (later, the Chicago and Northwestern Railroad), 25 miles north of the city. The site's attractions included a lovely shoreline along Lake Michigan, a varied, rolling landscape quite different from the surrounding prairie.[9] The Lake Forest Association, formed to develop the site, hired landscape architect Almerin Hotchkiss to prepare a plan for the new community. Little is known about the association's design intentions. Its business intention, however, was to auction every other lot and use the money to establish Lake Forest Academy and, later, Lake Forest College on a central 62-acre site.[10]

The lot sale that was held in 1857 was a success. To generate additional demand for its property, the Lake Forest Association built a resort hotel. When it opened the following summer, hundreds of wealthy Chicagoans could see for themselves the luxurious house sites that had been made available. The community's popularity spread quickly, and by 1861, when Lake Forest was incorporated as a city, it was already one of "the most aristocratic of all Chicago's suburbs."[11]

The winding roads and irregular blocks of the Hotchkiss plan reflect the popular romantic, picturesque esthetic of the period (see Chapters 3 and 14). But the design itself was determined by the need to cut through thicket-covered hills and ravines to connect the railroad station on the western edge of town with the lakeshore along its eastern edge. The vehicular roadways that emerged exploit natural drainage patterns, minimize the number of ravine crossings, and radiate in an irregular manner from the railroad station and town center eastward toward the house lots stretched out along the lakeshore.[12]

As a fledgling suburb of Chicago, then only beginning to be recognized as a national center, Lake Forest would not become known outside the region for some time. Furthermore, its unique topographic features made it an unlikely model for other sites. That role would be played by Riverside, the Chicago suburb designed by Frederick Law Olmsted.[13]

Like Lake Forest, the site of Riverside was made accessible by railroad, gained popularity once its 124-room hotel opened in 1870, and soon thereafter became a fashionable upper-middle-class suburb. The similarity ends there. Riverside was a business venture, not a semicharitable land operation.[14]

Eastern entrepreneur Emery Childs understood that the Chicago, Burlington, and Quincy Railroad station that opened in the middle of the David Gage farm in 1864 greatly increased the site's proximity to downtown Chicago. He established the Riverside Improvement Company for the express purpose of acquiring this 1600-acre farm 9 miles west of Chicago, subdividing it into house lots, and offering them for sale.

The "low, flat, miry, and forlorn" site was located on the prairie, far from Lake Michigan. Its only distinguishing feature was the shallow Des Plaines River, which slowly wound its way through the farm. Childs needed a design that would attract customers to this now accessible but forlorn location.

He hired Olmsted, Vaux and Company, whose work on Central Park in New York City and Prospect Park in Brooklyn had made its reputation as the premier landscape architecture firm of the day. Instead of a picturesque design similar to that of Llewellyn Park (see Chapter 14) or Lake Forest, Olmsted proposed a scheme that could be copied (frequently quite badly) anywhere in America.

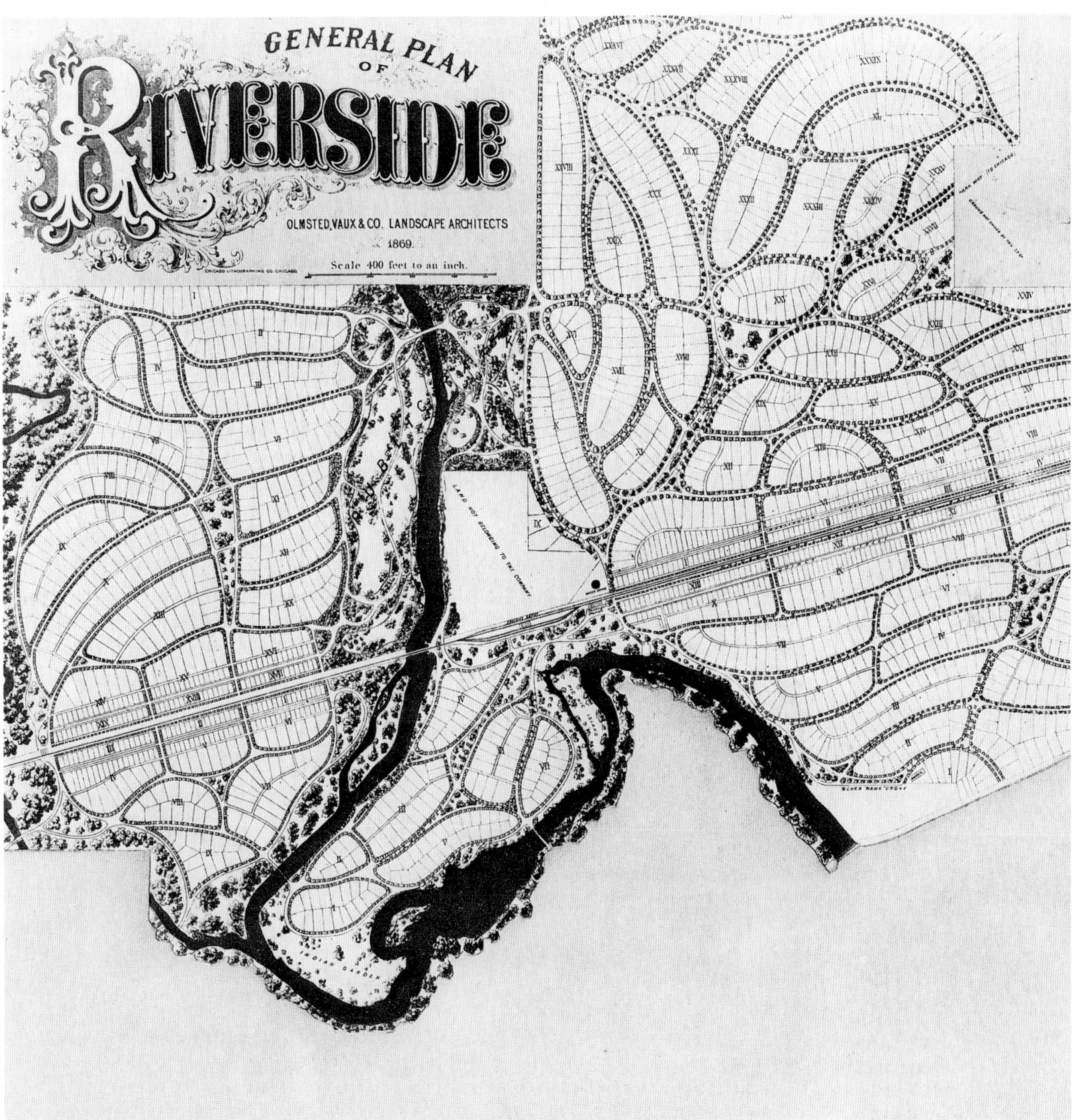

Riverside, 1869. Olmsted and Vaux's curvilinear plan was intended to enliven this flat, prairie-like site.

(Courtesy of the National Park Service, Frederick Law Olmsted National Historic Site)

Riverside, 2004. Tree-lined open lots with houses set back from the property line as a method for creating a "refined domestic life, secluded, but not far removed from the life of the community." *(Alexander Garvin)*

Olmsted used traffic arteries and landscaping to create an Arcadian refuge for the middle class. Seven hundred of its 1600 acres were set aside for roads, borders, walks, and parks. Houses were located about 150 feet from one another and set back at least 30 feet from the roadways. Fences were prohibited. Thus, the first impression of Riverside is one of airy expanses of open space, or what Olmsted referred to as "a sense of enlarged freedom" (see Chapter 3).

Unlike Lake Forest, where home builders had to cut through the existing landscape to open roads and clear house sites, the curvilinear streets of Riverside were intended to enliven the forlorn prairie. Olmsted designed 40 miles of carriage road and 80 miles of walks that followed curving routes with seemingly endless glimpses of what was just around the bend.

Between 1869 and 1871, the Riverside Improvement Company planted "47,000 shrubs, 7,000 evergreens, and 32,000 deciduous trees . . . some of them 19 inches in diameter . . . and 80 feet high . . . used 50,000 cubic yards of MacAdam stone, 20,000 cubic yards of gravel . . . and over 250,000 cubic yards of [excavated] earth."[15]

In the process, Olmsted demonstrated how clever landscaping could transform even the bleakest site into marketable house lots. He explained, "We cannot judiciously attempt to control the form of the houses which men shall build, we can only, at most, take care that if they build very ugly and inappropriate houses, they shall not be allowed to force them disagreeably upon our attention."[16]

Olmsted's solution was to combine the 30-foot front yard with two trees planted between the street and the house. This combination of open front lawns, houses set back from the street, and tree-lined roadways was a cheap, easy way of ensuring privacy. The secluded houses in Lake Forest were hidden by hedges, walls, and bushes, which precluded the neighborly interaction across the front lawn that took place on the open lawns of Riverside. Once the Riverside's trees matured, they formed an arch of foliage that became a green, public realm corridor, while the houses set back across their

Riverside, 1993. Gracefully curving arteries that are "pleasant to the eye within themselves" but also with "pleasant openings and outlooks." *(Alexander Garvin)*

open lawns became less and less prominent. The Riverside model quickly became the design formula for innumerable suburban subdivisions (see Chapter 14).

The plan provided all the up-to-date conveniences of city life: paved streets and sidewalks, water and sewer mains, fireplugs, a gravity drainage system, a gas works, individual gas hookups, and street lamps. In addition to the hotel, there was a block of stores and offices near the railroad station. The only planned convenience that was not executed was Olmsted's proposed 150-foot-wide parkway extending all the way into Chicago.

The Riverside Improvement Company went bankrupt in 1873. Too much of its initial $1.5 million investment had been spent on land acquisition, infrastructure, landscaping, and operating expenses. Lot sales, while adequate at first, plummeted during the panic of 1873. Without a revenue stream that exceeded expenditures, there was no way to pay for further improvements, to pay debt service on a mortgage that could finance them, or to repay the money that had been borrowed. Consequently, only 60 percent of the Olmsted and Vaux plan was completed as designed.

Riverside may have been established for the purpose of selling residential building lots for profit. However, what Olmsted had conceived and the Riverside Improvement Company had built was no suburban subdivision; it was an entire new town. Anything that remained to be done could be accomplished by an established government with the power to raise the money needed to pay for public services or the upkeep of the recently completed streets, utilities, and community facilities.

In 1875, only six years after the start of development, Riverside was incorporated as a village with a charter enabling it to provide fire, police, and other municipal services similar to those provided by the incorporated city of Chicago. This was an easy step because, unlike most residential land development projects of that era, Riverside had been established with its infrastructure already in place. Nearly a century and a half later it remains an independent community with a population of approximately 8900 people.

The Garden Cities Movement

Like Riverside, the garden city that Ebenezer Howard described was "to be planned as a whole, and not left to grow up in a chaotic manner." It would include every modern urban convenience: water, sewer, surface drainage, gas, power, streets, sidewalks, streetlights, and so forth. Houses were to stand on "their own ample grounds." Howard's description of the landscape even sounds like Olmsted: "ample space for roads, some of which are of truly magnificent proportions, so wide and spacious that sunlight and air may freely circulate, and in which trees, shrubs, and grass give to the town a semi-rural appearance."[17]

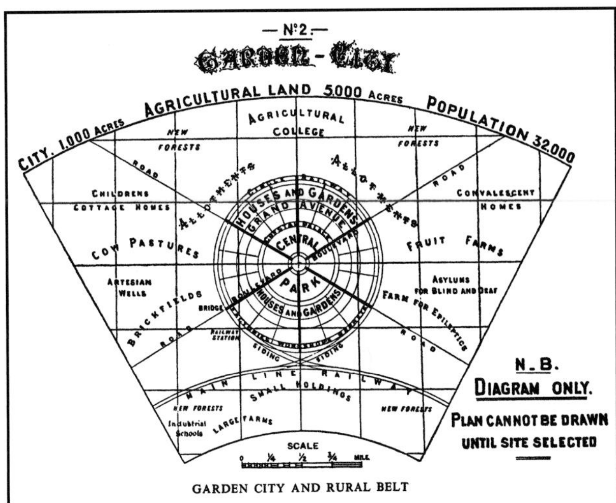

Ebenezer Howard, 1898. The garden city was intended to be a place "in which all the advantages of the most energetic town life" would be "secured in perfect combination with all the beauty and delight of the country." *(From E. Howard,* Garden Cities of Tomorrow, *MIT Press, Cambridge, MA, 1965)*

Despite apparent similarities, the garden city that Howard conceived was quite different from Riverside or Lake Forest. Its idealized shape was presented by Howard as circular. However, none of the new town projects with which he was involved took this form. As he later explained, "The diagrams in my book were never more than diagrams."[18]

The "new life" that Howard envisioned was to take place in 1000-acre, 30,000-person garden cities surrounded by a 5000-acre, 2000-person agricultural greenbelt. This was no dormitory suburb. Manufacturing firms located right in town would provide residents with employment. Farms located in the greenbelt would supply all the agricultural products needed by the residents.

Each garden city would own all land in perpetuity. Individual sites would be leased to their occupants at rents "based on the annual [market] value of the land."[19] Thus, as land appreciated in value, often as a result of community investment in infrastructure and public facilities, this increase accrued to the community rather than to the property's current occupant. Any earnings in excess of the cost of infrastructure and public facilities could then be used for the further benefit of the community.

Garden cities would be governed by boards of management whose activities were paid for by ground rents on municipally owned land. These boards would consist of a central council that was to be responsible for the planning, development, expenditures, and overall policy, and departments that were to be responsible for day-to-day administration. There were to be three groups of departments: Public Control (i.e., finance, assessment, law, and inspection), Engineering (i.e., roads, water, sewers, drainage, transit, light,

Letchworth, 1972. A garden city that looks like many other residential sections of suburban London. *(Alexander Garvin)*

and power), and Social Purposes (i.e., education, baths, music, libraries, and recreation). The members of the central council and the heads of the departments were to be elected from among the community's occupants.[20]

Howard's method for curing London's ills was to keep establishing new garden cities and filling them with the residents and businesses that moved from overpopulated London, until the city could be entirely abandoned. This same prescription was to be employed wherever there were obsolete, overcrowded cities.

In 1899, to promote the ideas presented in his writings, Howard founded the Garden City Association (now the Town and Country Planning Association). No garden cities were forthcoming. Consequently, four years later he decided to demonstrate the validity of his ideas, and he formed a joint stock company to build the first genuine garden city: Letchworth. The company acquired a 3822-acre site 35 miles north of London and held a design competition that was won by architect-planners Barry Parker and Raymond Unwin.

Their scheme combined formal geometry with the more organic forms of the traditional English village. A central square was established at the highest spot on the site. Radial avenues extended from the square to provide access to the rest of the town and to the surrounding greenbelt. The only distinctive feature of the design was the use of the "close" to eliminate disturbing through traffic and reinforce a sense of community. Parker and Unwin further developed the close (better known as the *cul-de-sac*) at Hampstead Garden Suburb.[21]

The proposed 5/1 acreage ratio of greenbelt to garden city turned out to be too costly. Furthermore, the two-thirds of Letchworth's area that was devoted to the greenbelt was a park, not an agricultural settlement. The notion that greenbelt residents would provide the agricultural products needed by garden city residents turned out to be a romantic illusion, especially once efficient, speedy freight movement became common and cheap.

Howard's vision of a self-contained workforce also had little in common with the realities of the marketplace or the desires of garden city residents. Industrial firms did move to Letchworth. By 1949 there were 65 manufacturers, ranging in size from a handful of employees to one firm with over 1000. However, many of the workers traveled to Letchworth from surrounding areas, while many Letchworth residents worked outside the town.[22]

Neither Letchworth nor Welwyn, Howard's second garden city, started in 1919, was successful in demonstrating superiority over conventional, developer-initiated suburban development then under way on smaller sites around greater London. Neither garden city was physically different from these less-ambitious dormitory suburbs. Nor did they exhibit the lovely combination of town and country of Lake Forest, Riverside, or some of the better developer-initiated new towns in America.

Despite the failure of so many of Howard's theories, they continued to have enormous influence. During the 1920s and 1930s, reformers, intellectuals, and politicians throughout England demanded that the government institute a program of new-town development as an antidote to urban overcrowding and wasteful suburbanization. In 1946, Parliament finally enacted a New Towns Act that over the next 26 years produced 22 satellite new towns and expanded countless existing towns.[23]

Howard's influence extended far beyond England. After World War II, countries throughout the world experimented with government-planned new towns. Among the most noteworthy were the new-town programs of Sweden, Finland, France, and Israel.

While American real estate developers built suburban communities that combined the best aspects of city and country, this was not because Howard proposed it. It was embodied in America's earliest suburbs and elaborated by Frederick Law Olmsted long before Howard wrote his book. The rest of the prescription had been irrelevant and was ignored from the start.

There is no American municipality that owns all the land within its boundaries and rents that land to its occupants. Howard proposed municipal land ownership as a way of capturing the value of community investment in infrastructure and public facilities. In America, the increase in property values that results from community investments is captured quite effectively in the form of increased real estate tax assessments.

The idea that garden city residents should purchase their food from farmers who occupy surrounding agricultural greenbelts had always been ridiculous. Why would anybody forgo the chance to pay the lower prices made possible by more efficient production techniques in more appropriate locations and climates?

The notion that people should live and work in the same

small community was unrealistic. Most people don't want to be trapped in company towns. They *want* the ability to move from job to job. They *want* to choose their residence based on their personal assessment of schools for their children and available shopping and recreation opportunities. They *want* to be able to sell their homes and move elsewhere whenever they desire.

The reason Howard's ideas persisted in the United States is that they were given persuasive physical form by Clarence Stein and were popularized by Lewis Mumford and the other members of the Regional Planning Association of America (see Chapter 14). However, it was not Howard's philosophy that influenced most American new towns. From the beginning of the nineteenth century, there was general unhappiness with city life and consumer demand for an alternative. This combination led inevitably to an attempt by profit-motivated real estate developers to supply that alternative, whether as Olmsted's nineteenth-century or Stein's twentieth-century conception of a garden suburbia.

The Vacation Refuge New Town

Many Americans have a profound yearning to escape the city for a healthier life in "natural" surroundings. Since most people are unable to entirely abandon the city, the next best thing is to get away as often as possible. Consequently, those families that can afford it get out of town whenever they can, sometimes moving to the country for the entire summer. During the eighteenth century, there were not enough such wealthy families to warrant development activity. However, during the nineteenth century, cities grew so rapidly that a large market developed, and smart investors acquired attractive country real estate.

The typical sequence began with a railroad or roadway that brought a beautiful landscape within easy access of a prosperous, growing city. Developers, often with business connections to the railroad, would seize the opportunity to profit from the newly accessible site by building a hotel for vacationing city dwellers. They would also subdivide nearby property into building lots. As the area became popular, vacationers would seek an alternative to the relatively short, expensive hotel stay. They would buy the building lots and erect summerhouses. In time, many of the houses would be converted into year-round residences.

As the nearby city prospered it also expanded, frequently enveloping what turned out to be a temporary refuge. Eventually, Chestnut Hill, Beverly Hills, and countless other vacation refuge new towns, which had been established to exploit the lovely countryside, ended up as distinctive sections of vast metropolitan areas. In some places, the increasing demand for space and higher prices induced developers to convert larger houses into multiple dwellings. In other places, they demolished and replaced them with tract houses or condo communities.

During the twentieth century, as the workweek grew shorter, leisure time increased, and disposable income rose, more and more families tried to get away from the city. Increasingly cheaper air travel and an extensive interstate highway network opened more distant sites. The communities established at these locations reflect much deeper desires for ideal living environments. As a result, Sea Ranch in California, Seaside in Florida, and other vacation refuge new towns have begun appearing in beautiful but previously untrammeled natural areas. These new communities are often more varied in character than those built closer to established cities. They also exhibit a much more obvious attempt to create an alternative to conventional city and suburban living.

Chestnut Hill, Philadelphia

Most visitors to Chestnut Hill assume it is quite separate from the city of which it has been a part for nearly a century and a half. But Chestnut Hill was not developed as a residential extension of Philadelphia or even as a speculative subdivision for commuters. It began as an alternative to spending the summer in the congested city.[24]

In 1854, when this section of northwest Philadelphia was incorporated into the city, the Reading Railroad initiated rail service connecting it with the downtown business district, 11 miles away. At that time, Chestnut Hill was a verdant, rolling landscape dominated by the steep valleys etched by Wissahickon and Cresheim Creeks, which later became part of Fairmount Park. Prior to the railroad, the only access to the

Philadelphia, 2009. The Wissahickon Inn (now Chestnut Hill Academy) provided a refuge for vacationing families. *(Alexander Garvin)*

Philadelphia, 2009. French Village in Chestnut Hill, designed in the mid-1920s by Robert McGoodwin. *(Alexander Garvin)*

city was along Germantown Avenue. The railroad opened the area to country excursions by day-trippers from Philadelphia but brought few other changes.[25]

Chestnut Hill's transformation into a summer resort began with the extension of the Pennsylvania Railroad to Chestnut Hill in 1872. Henry Howard Houston, the general (or head) freight agent of the railroad, lived in the area. He decided to profit from the new line by purchasing 3000 acres of hilly farmland overlooking the picturesque Wissahickon Valley, now only half an hour by rail from downtown Philadelphia. He hired architects G. W. and W. D. Hewitt to prepare a conventional, rectilinear plat for the area. The plan allowed him to sell lots at increasingly high prices, while building houses for rent on much of the remaining land.

The critical marketing device was the Wissahickon Inn (now Chestnut Hill Academy), built in 1884. Whole families left town for a summer stay at the Wissahickon Inn, from which fathers could easily commute to work in town. Houston also donated the land for the Philadelphia Cricket Club (a country club) and St. Martin-in-the-Fields Episcopal Church. Once the hotel, country club, and church were in place, Chestnut Hill became a fashionable resort for the Philadelphia elite. Upper-middle-class families became acquainted with the area from their visits to the Wissahickon Inn. Houston sold them sites on which they built large houses. They transformed Chestnut Hill into an elite suburb with tree-shaded streets, broad lawns, and big houses set back from the street.

After Houston died, his son-in-law, Dr. George Woodward, continued building houses for rent in Chestnut Hill. The most interesting of these are a series of charming residential developments in Norman, Cotswold, and French vernacular styles. Some are groups of one-family houses. Others are attached clusters that heighten the community's village-like appearance.

The 30-minute train ride made Chestnut Hill too convenient. It quickly evolved from a summer resort into a year-round commuter suburb. In fact, Chestnut Hill is so convenient that during the latter decades of the twentieth century it began to experience second growth. Developers have been subdividing the larger estates into house lots and condo communities, sometimes even converting the larger manor houses into multiple dwellings. As a result, Chestnut Hill is slowly losing its role as an alternative to conventional city life.

Beverly Hills, California

If a tourist hotel was an effective device for marketing Chestnut Hill, it was crucial to Beverly Hills. Americans had been streaming to the West Coast by the thousands in search of their vision of Arcadia. The farmland that would become Beverly Hills was located midway between the rapidly growing city of Los Angeles and the Pacific Ocean. It was an almost perfect embodiment of that Arcadia: gentle plains, lying just below the rolling foothills of the Santa Monica Mountains, warmed by a semitropical sun, cooled by soft ocean breezes.

Land acquisition for a new community was relatively simple, because initial colonization by the Spanish had divided the area into enormous ranchos—in this case, the Rancho Rodeo de las Aguas, named after the "gathering of waters" from Coldwater and Benedict Canyons during the rainy season. Easy access had been provided since 1887 by a steam railroad that ran along Santa Monica Boulevard, connecting Los Angeles and Hollywood with the Pacific Ocean. The line was electrified in 1896, sold to the Southern Pacific Railroad in 1909, and merged into the Pacific Electric interurban railway system. Despite a benign climate, picturesque landscape, strategic location, and transit service, the area spawned only unsuccessful real estate ventures. Instead of new homes, it remained occupied by lima beans.[26]

In 1900, a group led by Burton E. Green decided to drill for oil in the area. They formed the Amalgamated Oil Company and purchased what was then the 3608-acre Hammel and Denker Ranch. Instead of oil, they struck water. Green simply shifted gears and, in 1906, formed the Rodeo Land and Water Company, which purchased the site in order to resell the land as individual house lots. He hired Wilbur Cook, a New York landscape architect, to design a luxury community with broad, tree-lined streets and large, attractive house lots. The plat, officially recorded in 1907, discarded the north-south Los Angeles street system, replacing it with a gently undulating street grid of long blocks, running roughly at a 45-degree angle to Wilshire Boulevard. Green named his model community Beverly Hills, partly in memory of Beverly Farms, Massachusetts, and partly in recognition of the hilly landscape at the northern end of the site.

The new town of Beverly Hills was advertised in Los Angeles newspapers, but as one 1907 visitor explained, there was nothing there: "We got off the Pacific Electric car at the station and looked around. Very young trees, uniform in variety and spacing, had a sort of merry, hopeful look."[27] Streets were paved, but sewers, gas lines, and streetlights had yet to be installed. As a result, hardly anyone bought. During the next three years only six new residences were built north of Santa Monica Boulevard.

Beverly Hills, 1915. North Crescent Drive, the newly built Beverly Hills Hotel, and the many vacant building lots that, when filled in, would become one of America's most prestigious new-towns-in-the-country. *(Courtesy of Bison Archives)*

Beverly Hills, 2013. No longer a vacation retreat, Beverly Hills is now filled with homes of the rich and famous. *(Alexander Garvin)*

The situation changed dramatically in 1911, when Green decided to build the Beverly Hills Hotel. He persuaded Margaret Anderson, then manager of the popular Hotel Hollywood, to take a similar position at the Beverly Hills Hotel, but only by offering her an option to buy the hotel if it was successful. When it opened the following year, she brought her reputation, her staff, many of the furnishings, some of her clients, and, most important, instant success. Green happily sold her the hotel.

Vacationers from all over the country began coming to the Beverly Hills Hotel. Many fell in love with the area, establishing homes and businesses there. Local residents also settled in the area. By 1914, there were enough residents for Beverly Hills to become an incorporated city, quite separate from nearby Los Angeles. Within a few years, however, it had merged with the rest of the region's automobile-dominated landscape, distinguishable only because of its different street grid and more expensive houses. In 1950 its population had grown to 29,000 and would grow to 34,000 in 2010.

The Beverly Hills Hotel had been the critical element needed to attract customers to this previously unsuccessful new community. Once there, they bought the vacant lots, built the houses, filled the empty streets, and transformed the remaining fields of lima beans into an international mecca for the rich and famous: a "fabulous clime . . . Where the rain does not rain. It just drizzles champagne . . . Where romance is the theme of the day . . . [and] everything is tremendous! titanic! stupendous!"[28]

Sea Ranch, California

Sea Ranch, the country's most dramatic vacation refuge, is a 10-mile-long stretch of splendid desolation on the Pa-

Sea Ranch, 2011. Wooden houses nestled into the meadows on the cliffs overlooking the Pacific. *(Alexander Garvin)*

cific coast, 100 miles north of San Francisco. Its 5200-acre site was shorn of redwood, fir, and pine forests during the 1880s and 1890s. All the loggers left behind was windswept, grassy meadowland, interrupted by an occasional logging cabin, barn, or shed. Later, Monterey cypress hedgerows were planted to act as windbreaks.[29]

The site was acquired for $2 million by Oceanic Properties, a real estate subsidiary of Castle & Cooke of Hawaii, in order to develop it as a low-density, second-home community named Sea Ranch. In 1964, it hired the landscape architecture firm of Lawrence Halprin and Associates to prepare a master plan and the architecture firms of Joseph Esherick & Associates and Moore, Lyndon, Turnbull, & Whitaker to prepare architectural prototypes with which to begin development.

Halprin applied ecological principles to capitalize on the area's awesome, natural beauty. The idea was to design:

> *a place where wild nature and human habitation could interact in a kind of intense symbiosis. . . . We derived lessons from analyzing the wind erosion of cypress trees and studied how they were shaped into specific slopes and pitches. We realized that they were models for establishing roof slopes in buildings, which could extend their protection from the wind. We also looked carefully at drainage patterns and studied ways by which natural drainage could be channeled to maximize run-off in a non-destructive way.*[30]

Soil, climate, wind, and days of sunlight were analyzed to determine which areas were appropriate for single-family houses, condominium structures, active community use, roads, and supplemental cypress windbreaks. As a result, half of Sea Ranch has been set aside as undeveloped meadows and woods that are in common ownership, deeded to a landowners' association, which is responsible for its maintenance. One-quarter is private land belonging to individual homeowners. The remaining quarter is "restricted" private land that belongs to the adjoining lot owner, but cannot be built on or planted with nonindigenous vegetation.

Sea Ranch is bisected by California Highway One. Houses in the meadows west of the highway are clustered around cul-de-sacs that tend to be perpendicular to the coast. The land in between has been left in a natural state and is overgrown with grasses and wildflowers. There are neither roads connecting the cul-de-sacs nor sidewalks. The only concession to human activity is a trail that runs along the edge of the cliff, offering residents spectacular views of the surf and an occasional glimpse of passing whales and sea lions. The land east of the highway slopes upward, quickly turning into forested hillsides. Here, houses are hidden in the trees or clustered into groups of condominium units.

Architectural design was subjected to the same rigorous ecological analysis that had been applied to the land. Wind studies revealed the need to shelter houses from the northwest wind. Consequently, most houses must be oriented toward the south, with gardens and parking on the leeward side. They also must have slanting roofs to direct the wind up and over adjacent open areas. All exterior finishes must be natural. Reflective surfaces and bright colors are not permitted. These requirements are embodied in a series of printed restrictions intended to generate buildings similar in size, scale, color, and material. To be absolutely sure that these restrictions have the intended effect, a three-person design committee must approve all construction plans.

The initial buildings by Esherick and by Moore, Lyndon, Turnbull, & Whitaker established a common architectural idiom inspired by the simple country barns and leftover logging structures that were already in place: shed roofs without overhangs for the wind to flutter, rough vertical redwood siding, and large windows placed low enough to profit from the splendid views. Virtually all of their buildings have been felicitous additions to the landscape. Later structures designed

Sea Ranch, 2011. Houses in an open landscape owned in common by its residents. *(Alexander Garvin)*

by others generally followed their example, but "as the years have passed . . . principles have been altered by succeeding people—owners, builders, and designers . . . whose motivations . . . are less pioneering."[31]

Despite the disappointment with more conventional, later development patterns, the only real threat to Halprin's design came from government. In 1972, voters approved a referendum that established a California Coastal Zone administered by six regional commissions (see Chapter 17). Amazingly, the North Central Commission discarded Halprin's ecologically based plan because it felt the scheme did not provide the public with adequate views of or access to the ocean. The commission imposed a moratorium on further construction until Oceanic Properties, acting on its own behalf and that of the property owners in the Sea Ranch Association, agreed to establish public parking areas and public access ways to the beach, to cut down many of the Bishop pines that blocked motorists' views, and to monitor septic tank operation. The commission also demanded that the number of building lots be reduced from 5200 (an average of one house per acre) to 2329.

Most property owners objected to this form of extortion. The association tried to negotiate. Oceanic tried to litigate. Every step they took was futile. Finally, in 1980, the California state legislature enacted compromise legislation (the Bane Bill) that exempted all of Sea Ranch from the Coastal Commission's permit process and paid $500,000 to settle all litigation and to obtain five public access ways to the beach. Development quickly resumed. By 1992, there were houses on half the building lots. In 2012, there were still only 1300 permanent residents.

Every structure erected at Sea Ranch reduces by one more little bit the splendid desolation of its landscape. As a consequence, in a very real sense, this is an example of no-win planning. On the other hand, in comparison to what would have happened had Oceanic Properties adopted more conventional real estate practices, Halprin's ecological approach is a triumph. The master plan ensured that thousands of acres of irreplaceable meadows and woods would be set aside for future generations. It helped to make hundreds of houses (but not all) as much a part of the landscape as the hedgerows that were added decades earlier.

Despite the moratorium, Sea Ranch did not turn into a financial disaster, largely because so little had to be invested in infrastructure. Nevertheless, few developers have tried to emulate its planning. Perhaps this is because there are so few sites that are as beautiful. Perhaps it is because higher-density development generates more cash. As a result, Sea Ranch remains an appealing refuge for vacationers and weekenders that is unlikely to be replicated in more populated sections of the country.

Seaside, Florida

Seaside, Florida, the vacation refuge with even fewer residents in its 326 houses than Sea Ranch, has become the model for numerous early-twenty-first-century planned communities. It represents a radical rejection of previous planning theory. It is small: 80 acres, compared with 5200 acres for Sea Ranch or 3600 acres for Beverly Hills. It is proudly urban, compared with the windswept meadows at Sea Ranch or the tree-shaded lanes of Chestnut Hill. It is designed for continuous pedestrian interaction, compared with the broad vehicular boulevards of Beverly Hills or the splendid desolation of Sea Ranch. Although Seaside is so different from these profitable new communities, it has been no less financially successful.

Seaside began when architectural enthusiast Robert Davis inherited 2800 feet of Gulf of Mexico beachfront. The bulk of the site, on the Florida panhandle, between Pensacola and Panama City, was separated from the 30-foot-high dunes and broad white beach by Route 30A, a county road connecting scattered bungalows, motels, and nondescript villages. Rather than sell this undeveloped land for a few thousand dollars per acre, Davis decided to create an exclusive beach

Seaside, 1990. The beach. *(Alexander Garvin)*

resort and sell his land by the square foot at much, much higher prices.[32]

In 1980, Davis hired architects Andrés Duany and Elizabeth Plater-Zyberk to develop a plan for a vacation refuge that would function simultaneously as a real town. Even though this was the firm's first urban planning project, Duany and Plater-Zyberk embarked on what has become a mission to reintroduce traditional American town-planning practices into the suburban marketplace (see Chapter 14).

On paper, the Duany and Plater-Zyberk design looks more like the new-town plans of the years just before and after World War I than a later-twentieth-century beach resort. Its main elements include the beach, Route 30A, which slices off an imaginary second half of the plan, a demioctagonal central square that functions as a town center, several broad avenues radiating from the town center, residential streets with vistas that terminate in colorful public structures, picturesque beach pavilions on axis with residential streets perpendicular to Route 30A, and pedestrian paths that bisect the residential blocks and provide secondary access to all the houses.

Because of its street network, block and lot pattern, and the scale, texture, and color of the buildings, Seaside has none of the academic formalism of its geometric plan. The key feature is not the axial vista; plenty of awful streets have axial vistas. It is the reduction of every roadway, sidewalk, and path to a size that is dominated by *people,* not cars.

Seaside was planned as the first new community of a "postmotor age." Rather than standard one-way arteries with several 12-foot-wide automobile lanes mandated by every traffic-engineering textbook, Duany and Plater-Zyberk designed two-way, 20-foot-wide pink-brick streets. Rather than concrete sidewalks, driveways, garages, and sections of street that are carefully marked for parallel parking, they designed narrow shoulders of crushed shell that are comfortable for pedestrians and just large enough to accommodate cars.

Seaside also eliminates repetitive block and lot patterns. Most town grids have standard dimensions that are seldom followed. As time passes, lots are combined and recombined to fit the dimensional requirements of different buildings, built at different times for different purposes. This produces the rich texture we so admire in our oldest city neighborhoods. Duany and Plater-Zyberk achieved this from the very start by varying the widths, depths, and shapes of the building lots and by introducing diagonal streets that break the grid of blocks. This variety of lot location, orientation, size, and price naturally allows variation in household size and income that most new towns develop only decades after initial settlement.

The witty elements in Seaside's buildings are the purely decorative notes that the developer has added to the streetscape: pergolas, beach pavilions, and even a water tower, brought from its former location in Norfolk, Virginia. These charming structures, which pretend to have functional attributes, fool nobody. They are there to add a note of delight to what would otherwise be a conventional streetscape.

Duany and Plater-Zyberk's most conventional (but very

Seaside, 1994. Aerial photo of the almost-completed 80-acre beach resort for 326 houses. *(Alexander Garvin)*

Seaside, 1990. Tupelo Street on axis with the pavilions at the entrance to the beach. *(Alexander Garvin)*

effective) tool for achieving Seaside's distinctive look is a set of regulations that establish eight building types and a common design vocabulary: wooden siding painted in pastel colors, white or cream-colored wood trim, vertical windows, metal gable or shed roofs (either corrugated, crimped, or standing seam), screened porches and verandas, picket fences, and so forth. These regulations are presented in a diagrammatic table of performance standards and in more detailed written esthetic and construction codes. To ensure that the results are consistent in character, a town architect (selected by its developer and planners) must review and approve all building plans.

Now that Seaside has been built out, the validity of its planning has become clear. The beach with its glorious dunes remains the star attraction. To it, Davis has added the colorful beach pavilions, Per-spi-cas-ity (an open-air gift and clothing market), a popular upscale bistro, a tiny bookstore, and a sandwich shop. Even without the projected 200-room hotel, the beachfront has the panache of a much larger resort.

The plan was so effective that lot prices have climbed to amazing heights. The typical vacant lot in Seaside (about 4100 square feet) sold for $18,800 in 1982, when the first lots were put up for sale. This price, about $4.50 per square foot, was far higher than could be obtained for inland lots in any nearby community. By 1996, Seaside's remaining smaller lots (about 2200 square feet) were going for $343,333, about $156 per square foot. Most purchasers come from cities in nearby Alabama, Georgia, Louisiana, and Florida. One reason for the increasingly high prices is that purchasers do not have to spend every weekend at Seaside. The Seaside Community Development Corporation rents cottages for owners when they are not in residence. During 1996, the 233 units that were offered for rent had an occupancy rate of 52 percent, at an average nightly rate of $260.[33]

The central square is a long way from becoming a town center. A decade after the project was started, it still included only three small retail establishments (a gourmet grocery, a yogurt shop, a house-garden-gift store) and a doll-sized post office. Nor will it include much more until Seaside attracts enough retail customers from other communities along Route 30A.

Even without enough customers to justify so imposing a town center, Seaside is already more than a holiday refuge. Tourists from nearby hotels and residents from nearby houses and villages do come to Seaside. The town's minimal retail outlets could not survive on just the occupants of a couple of hundred cottages. Furthermore, Seaside has attracted a rental market many times larger than the market provided by its few homeowners. Thus, it is not just a successful residential development. Seaside is a chosen destination for thousands of people from all over the country because relatively inexpensive air travel has made it so accessible.

Seaside has not been financially successful just because of its design or its sales strategy. Robert Davis kept his initial cash investment minimal. He installed street paving and a packaged sewage treatment plant only after the first houses were up. Even the roadways were minimal: pink pavers for the street surface and crushed shell for the cars parked along the shoulders. Davis invested in infrastructure only in small increments, timed to coincide with lot sales. In doing so, he achieved what every community builder hopes to achieve. He balanced cash outflow with revenues from lot sales.

Seaside represents one possible paradigm for a post-motor age. Its designers were well-intentioned in seeking a happier balance between a community and its automobiles. More important, they demonstrated that high density can lead to a most humane living environment. Nevertheless, this densely packed, second-home community built on small lots is not any more affordable for the bulk of America's automobile-dependent, middle class than Sea Ranch, with its open meadows and dark woods. They both provide (quite different) idyllic environments for the few who can afford to pay for the luxurious setting they offer.

The Utopian Satellite New Town

Satellite new towns evolve less from the attempt to find refuge from city life than from the attempt to satisfy the demand for an ideal living environment. As early as 1871, the developers of Riverside, Illinois, were promoting it as a place with:

> *. . . splendid improvements, found in no other suburb . . . [A place where] wives and children live a more quiet and satisfactory life, and one of far greater and more varied enjoyment than can possibly be attained in any city, and can do it without sacrifice of urban conveniences.*[34]

Unlike many vacation refuge new towns, suburban utopias are intended to be and remain satellites of large cities that provide them with consumers for their land and houses, employment for their residents, and even the services without which they could not exist. Also unlike the vacation refuge new towns, these satellite communities usually do not offer stunning beaches, pine forests, a semitropical climate, or many of the other natural attractions of an alluring resort. Virtually every suburban satellite has chosen to provide its little bit of country by adopting the Olmstedian esthetic: curvilinear, tree-lined roadways and broad lawns with houses set back from the street.

Sometimes, as with Mariemont, Ohio, and Columbia, Maryland, the sites had to be assembled in bits and pieces; sometimes, as with Lakewood and Irvine, California, they already were a single property. Whatever the method of acquisition, the difference between financial success and failure has proved to be the developer's ability to carry the debt service on the investment in land, infrastructure, and facilities until cash flow from sales exceeded debt service, operating expenses, and the other costs of carrying a huge inventory of unsold building lots.

Similarly, some satellite towns, like Levittown and Lakewood, attracted their customers by providing decent shelter at a price affordable to the vast majority of the population. Others, like Mariemont and Palos Verdes Estates, did so by offering amenities not available elsewhere. But, whatever the marketing strategy, sales to customers depended on the availability of mortgages with affordable debt-service payments.

Satellite new towns may differ in their financing and marketing strategies, but every one was established on open land that initially was far outside city limits and then was engulfed in a rapidly suburbanizing landscape. Some satellite new towns are different from these surrounding suburbs. Mariemont, Reston, and Columbia, for example, can be distinguished by substantial open space and recreation facilities. Palos Verdes Estates can be distinguished by the character of its architecture and the unusually successful integration of the built environment with the topography. Despite the utopian rhetoric, careful planning, and extra amenities, life in most satellite new towns is not very different from that in the surrounding suburbs. Only in a very few cases can one find what Ebenezer Howard dreamed of: town and country "secured in perfect combination."

Shaker Heights, Ohio

The exclusive Cleveland suburb that is today called Shaker Heights is the creation of two bachelor brothers: Oris and Mantis Van Sweringen. The Van Sweringens entered the real estate business in 1900. After five years and a few costly mistakes, they bought some land in the eastern suburbs of Cleveland. The quick profits they made on the first 200 acres helped them convince investors to put up the $1 million needed to buy the rest of a 1400-acre property bordering a park.[35]

The North Union Society of the Millennium Church of United Believers, better known as the Shakers, had acquired the land they purchased in 1822 for a utopian settlement. Their settlement was disbanded in 1889. Three years later, the property was sold to a group of Cleveland businesses called the Shaker Heights Land Company, in recognition of the former settlement and the site's elevation, 400 to 600 feet above Lake Erie. Before completing the community's first roads, the company sold the land to a Buffalo syndicate that failed to sell or lease anything until 1905, when the Van Sweringens took it over.

The Van Sweringens understood that they also would be unable to sell much land until the site had ready access to downtown Cleveland. By agreeing to subsidize part of the fare, they persuaded the Cleveland Railway Company to extend its streetcar line three more miles to Shaker Heights. When service began in 1907, the Van Sweringens finally began to profit from lot sales. Unfortunately, the streetcar stopped at the western edge of the site. The company refused to extend service any farther. As a result, the Van Sweringens went into the transit business. In 1911, they established the Cleveland and Youngstown Railroad Company, which built a profitable four-track railroad running from Cleveland, through Shaker Heights, all the way to Youngstown. The success of this project eventually led them to acquire the Nickel Plate Railroad and, in 1930, to build Terminal Tower, Cleveland's 708-foot-high combination railroad station and retail-hotel-office center.

While the Van Sweringens were making millions in the railroad business, they were also making millions in Shaker Heights real estate. The brothers assembled 4034 acres, hired the Pease Engineering Company to prepare a plan for their new town, and devised deed restrictions and design standards that specified allowable land uses, architectural styles, and construction materials. The community that emerged includes one-family houses, "English-style" apartment buildings, a colonial-style shopping complex (octagonal Shaker Square), and several private country clubs built, like the community's schools, on land donated by the Van Sweringens. There is no industry.

Shaker Heights is divided into separate residential neighborhoods by its Y-shaped system of parks, waterways, and ravines that provide natural drainage and by two major east-west boulevards with rapid transit lines running down the middle. On paper, the design is a confused mixture of naturalistic curves, parallel streets, and elliptical figures. It looks more reasonable when driving through Shaker Heights's rolling landscape because of the skillful adjustments to topographic features, the common district lot sizes, which help to differentiate the various neighborhoods, and the property restrictions, which generate a uniform look.

Many of the property restrictions are tied to lot width: the distance houses must be set back from roadways, the distance they must be set back from neighboring lots, minimum house width and depth, and even porch dimensions. Houses

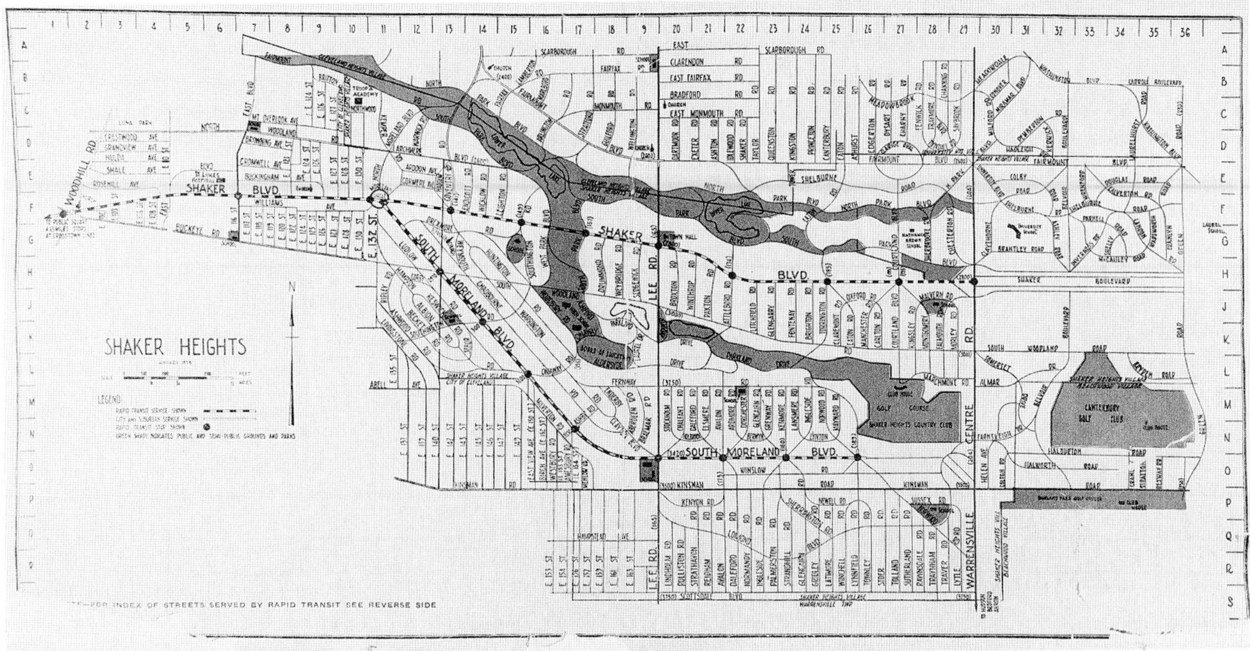

Shaker Heights, 1928. Map showing transit lines and parks. *(Courtesy of the collection of the Shaker Heights Historical Society, Shaker Heights, Ohio)*

must be two stories high, and garages, which the Van Sweringens considered unsightly, must be at the rear of the lot. Only specific periods of English, French, or Colonial architecture are allowed. The "Shaker Village Standards" even specify appropriate color schemes and materials for each of these styles.

The uniformly upper-middle-class, WASP character of the Shaker Heights population began to change after World War II, when, for the first time, Jews were allowed to purchase homes in the area. In 1955, an African American moved into the Ludlow section of Shaker Heights. At first there was panic and blockbusting. However, within a couple of years white and African-American residents formed the Ludlow Community Association (LCA) to stabilize the neighborhood. The LCA asked homeowners to let it handle sales. It established a second-mortgage fund to help purchasers with their down payments and organized cocktails and dinners for prospective purchasers of all ethnicities and races. Within a few years, fewer houses were being sold and prices had begun to increase.

By the 1960s, Shaker Heights's population had reached its peak of 36,600. During the middle of that decade, Ludlow's population stabilized at 45 percent white. Since then, several other of its neighborhoods have been integrated, mostly in its southern sections, resulting in a 1980 population that was 74 percent white and 24 percent African American. Over the next half century, the population of Shaker Heights declined to just above 28,000, 55 percent white and 37 percent African American.[36]

Palos Verdes Estates, California

If there is one satellite new town with a truly ideal living environment, it is Palos Verdes Estates. Its site is a spectacular peninsula jutting out into the Pacific Ocean. This location provides it with the only consistently pure air in the Los Angeles basin. It also provides easy access to one of the most economically successful and culturally diverse metropolitan areas in the world.

The plan of Palos Verdes Estates, by Frederick Law Olmsted Jr. and Charles Cheney, amplifies the attractions of the site by fitting curvilinear roads to the steep topography of the peninsula, carefully nestling house lots into its slopes, and providing spectacular views of the Pacific coast. In spite of all these assets, the 26-year attempt to create a new town on the Palos Verdes Peninsula very nearly failed four times before Palos Verdes Estates was incorporated as a city in 1939.[37]

Like Beverly Hills and so many other Southern California communities, the 16,000-acre Palos Verdes Peninsula belonged to a succession of ranchos until 1882, when it was acquired by Jotham Bixby, a wealthy California landowner. Bixby planted eucalyptus trees and leased some land to Japanese-American farmers, but did little with the peninsula's barren hillsides and terraces, finally selling it for $1.5 million in 1913. The purchaser defaulted, and the land was resold to a syndicate put together by Frank A. Vanderlip, president of the National City Bank of New York.

Vanderlip, who had never seen the site, believed its stra-

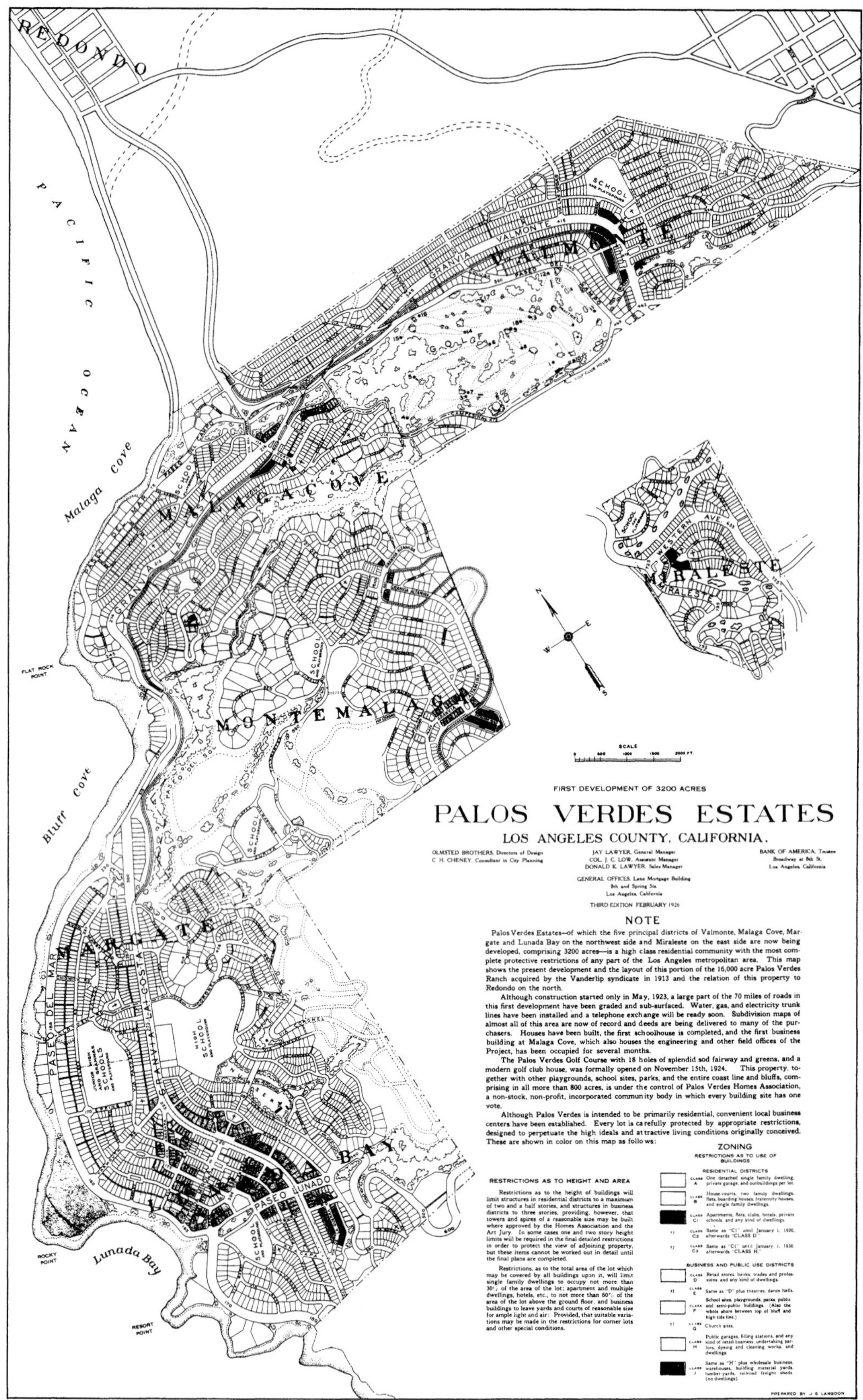

Palos Verdes, 1926. Olmsted and Cheney's plan locates streets along terraces that previously had been eroded by the Pacific Ocean. *(Courtesy of the National Park Service, Frederick Law Olmstead National Historic Site)*

tegic location, 23 miles southwest of Los Angeles, guaranteed success. It took but one journey to Palos Verdes for him to decide how to develop its potential. As he later explained:

> *[The] first sight of Palos Verdes Ranch was . . . one of the most exciting experiences of my life. Before me lay a range of folded hills, miles and miles of tawny slopes patched with green, thrusting themselves abruptly from the Pacific. Above me were broad natural terraces, with here and there a little farm, backed by a range of taller hills. Wherever the road passed over a hillcrest I could see the shore-line of the ranch as a series of bold headlands spaced off by gleaming crescent beaches. . . . The most exciting part of my vision was that this gorgeous scene was not a piece of Italy at all but was here in America, an unspoiled sheet of paper to be written on with loving care.*[38]

Inspired by the Amalfi coast in Italy, Vanderlip decided to create his own version of a Mediterranean town. Starting in 1914, he hired a series of architects and planners to make proposals. All he built, however, was his own part-time residence.

In 1920, a real estate firm secured an option on the property, which it sold two years later to developer E. G. Lewis for $5 million. Lewis wanted to build a $35 million city with a regional harbor and airport and a population of 200,000. The money was to come from individual investor-partners. Although he claimed to have interested 7000 subscribers, Lewis was unable to put the pieces together. In 1923, Vanderlip stepped in for the second time, forming the Commonwealth Trust Company to take over the project. Although Lewis's investors were offered their money back, 4000 of the original backers stayed with the project. The salvaged money allowed the best-located and most beautiful 3200 acres to be acquired for what came to be called the Palos Verdes Project. Vanderlip held on to the rest of the property, selling some sections and developing only small parcels, until his death in 1937.

By 1924, 6 miles of storm drains, 14 miles of gas mains, 2 miles of underground conduits, and 12 miles of overhead electric lines had been installed, and more than 100,000 trees and shrubs had been planted. This was simultaneously the reason for the project's marketing success and its financial problems. Subscribers were delinquent in making promised capital contributions. Land sold more slowly than had been anticipated. Consequently, expenditures on development and marketing exceeded sales revenues and capital contributions. Fortunately, the project only sold land. Had it also built houses for sale, even more capital would have been needed.

To cover cash-flow requirements, the project issued $1 million in bonds. When the downturn in the real estate market at the end of the decade turned into a general collapse, the project's trustees were unable to meet interest payments, and again the new company nearly failed. They precluded foreclosure by reorganizing as a corporation (Palos Verdes Estates, Inc.) and transforming the bonds into stock.

Two years after Vanderlip's death, the new town ran into trouble for yet another time. This time, the residents stepped

Palos Verdes Estates, 2013. Houses, with a low profile along the street, nestle into the landscape as if they were a natural feature of the topography. *(Alexander Garvin)*

Palos Verdes Estates, 1988. The view opening onto Santa Monica Bay and the Los Angeles Basin. *(Alexander Garvin)*

in to save the project. From the beginning, Palos Verdes Estates had been managed by its Homes Association, which maintained the streets, parks, and water and sewer systems; provided other municipal services; and paid Los Angeles County taxes. County taxes were particularly high because the Homes Association, which was not a government entity, was taxed on all its land, which constituted more than 25 percent of the community and included parks, playgrounds, a golf course, and 4.5 miles of seashore. The money for all these expenses came from an annual assessment collected from each property. Many of the property owners, hard hit by the Great Depression, stopped paying the assessments. By 1938, the Homes Association was $62,500 in debt, more than half of which was in the form of back taxes owed to the county.

The situation was complicated by the position of Palos Verdes Estates, Inc., which still owned more than half the undeveloped lots and was behind in both its Homes Association assessments and its county taxes. The corporation's proposed solution, in lieu of paying back taxes, was to donate large sections of shoreline (owned by both the Homes Association and Palos Verdes Estates, Inc.) to the county for use as public parks. However, residents were opposed to losing control of their most scenic lands to "the gay throngs of picnickers" and the "NOT SO GAY throngs of loafers, of prowlers, of drunkards" expected to come with them.[39] Since the Homes Association made decisions by majority vote of its 5155 lots (a large number of which were still held by Palos Verdes Estates, Inc.), the developer's proposal could very well have been accepted, despite overwhelming homeowner opposition.

When residents realized that the "solution" would still leave them with county tax bills for future years, they adopted a different strategy. First, they obtained passage of state legislation permitting establishment of a park and recreation district to which they could deed park land, and thus avoid further taxes. Then, in 1939, they voted to incorporate as the City of Palos Verdes Estates. Forever after, no matter how many vacant lots were owned by developers, decisions affecting the community would only be made by resident voters. In part as a result of these actions, in part as a result of the improving real estate market, Palos Verdes Estates, Inc., settled its debt to the Homes Association and, in 1940, sold its remaining property at a huge public auction.

Palos Verdes Estates had the benefit of a strategic location and a spectacular site. Its design capitalized on them to create an almost ideal version of the living environment sought by the Los Angeles market, so ideal that by 1982 the Los Angeles County assessor valued peninsula land at $3.2 billion. Two aspects of Olmsted and Cheney's planning were responsible for creating this highly valued living environment: the street system and the building restrictions.

Palos Verdes Estates was one of the first planned communities in the Los Angeles region that did *not* depend on the Pacific Electric Railroad. In fact, it was one of the nation's first new towns that was (in words that Clarence Stein would later use in describing Radburn) designed for the motor age. The plan included a scenic circumferential roadway that not only afforded drivers spectacular views but also allowed through traffic to bypass strictly residential sections. The remaining streets were all designed for easy driving and, where possible, views.[40]

Olmsted's design was based entirely on the physical character of the site. The peninsula included 13 major terraces that had been formed by wave action and erosion. Instead of spending money to move thousands of tons of earth, he located the major traffic arteries, commercial centers, golf course, schools, and playgrounds in these generally flat areas. Canyons, steep hillsides, and shoreline cliffs were left as parkland. The remaining sloping land was made accessible by secondary roads, thereby producing choice residential sites with spectacular views. These sites allow houses to be set back from the roadway enough to accommodate driveways and garages. More important, the landscape remains dominant because many houses appear to be only one story high, when they really are multistory structures built into the hillside so that the occupants are able to enjoy the unobstructed panorama below.

Olmsted and Cheney understood that the site plan by itself did not guarantee that Palos Verdes Estates would be built as they envisioned it. They also knew that any consumer had to "be sure when building his home there that his neighbor will have to build an equally attractive type of building."[41]

Their solution was to prepare a set of protective restrictions that applied to all construction. Its objectives were to preserve the views, increase the natural beauty of the landscape, and create the ideal living environment demanded by the consumer. A seven-member art jury initially appointed by the developer enforced the protective restrictions. It had to approve any plans for property subdivision, building construction, artwork, fences, sidewalks, and awnings.[42]

The protective restrictions, in a manner similar to a zoning ordinance, specified land use, height limits, setbacks, and yard requirements. Three types of architecture were established and mapped for specific sections of town. Type I had to have plaster or stucco walls and roofs with slopes no greater than 35 degrees that were covered with reddish-colored tiles. This meant building in the California-Mediterranean style popular at that time (see Chapter 18). Type II allowed roofs as steep as 45 degrees and greater latitude in color and appearance. Type III permitted roofs as steep as 60 degrees and darker colors and materials. Whatever the type of architecture, materials and colors had to be used "honestly, actually expressing what they are, and not imitating other materials."[43] In practice, most buildings approved by the art jury, even the gas station, were in a California-Mediterranean idiom.

The protective restrictions also included the statement that "there are the usual restrictions prohibiting Negroes, Asiatics, and people of other than white or Caucasian race, except in the capacity of domestic servants."[44] Time and Supreme Court decisions have made this the only part of the restrictions to be discarded. By 2010, 23 percent of the more than 13,000 people living in Palos Verdes Estates were non-Caucasian.[45]

Frank Vanderlip's dream of building an American version of the Amalfi coast along the Palos Verdes Peninsula is now reality. The stucco and red-tiled houses, which seem to grow naturally from the peninsula's terraced hillsides, give one the impression of the same centuries-old partnership between man and nature that is typical of the Italian coastline. The nonresidential sections of Palos Verdes have a similarly Mediterranean look. Malaga Cove Plaza, the community's major business center, completed in 1925, includes the arcades of an Italian piazza. It even has a copy of Bologna's Neptune Fountain in the middle of its parking field. Palos Verdes' fantasy-Mediterranean appearance has certainly helped it to exploit its lucrative market and spectacular location. But its success was even more dependent on a patient developer with pockets deep enough to cover debt service on the money spent for land acquisition, planning, and development (streets, sewers, water mains, lighting, schools, libraries, landscaping, etc.) during the periods of slow sales.

Mariemont, Ohio

At the time Frederick Law Olmsted Jr. was conceiving Palos Verdes, John Nolen, one of his former students at Harvard's School of Landscape Architecture, was designing a very different new town: Mariemont, Ohio. In 1903, at the age of 34, Nolen entered Harvard to begin a career as landscape architect and urban planner. He went on to complete more than 400 projects, including plans for 50 new towns.[46]

Nolen was hired by Mrs. Mary Emory, a wealthy Cincinnati widow, to design Mariemont, a new town 10 miles east of the city. Mrs. Emory had become convinced that a model town for working people would "furnish better housing" and

> *. . . illustrate intelligent and sane town planning, by following a preconceived plan along scientific lines, including proper road building, harmonious house construction and the relation of homes to industrial areas and public service facilities.*[47]

She decided to demonstrate the validity of building a model community and in so doing also establish a fitting memorial to her late husband.[48]

Between 1910 and 1920, Mrs. Emory acquired 18 farms along the Pennsylvania Railroad in an elevated area overlooking the Little Miami River. The Wooster Pike, which provided easy automobile access to downtown Cincinnati, ran right through the site. It was a perfect spot to demonstrate that good planning "could be duplicated wherever initiative, capital, and social planning could be combined to support building of new towns and suburbs."[49]

Ebenezer Howard and the garden cities movement influenced Nolen and Emory. As a result, Mariemont was intended to be fundamentally different from such profit-motivated real estate ventures as Shaker Heights and Palos Verdes Estates. It was conceived as a small, self-contained community, very much like Letchworth, with a source of employment right in town. The plan allocated 250 of the site's 485 acres for the town: 62 acres along the railroad for industry, 32 acres for a retirement center, 59 acres for parkland, and 82 acres for playgrounds. Land and buildings were intended to remain in the hands of its philanthropic developer until the town was completed, so that its residents, like the leasehold residents of Letchworth, could enjoy its benefits without being priced out by land speculators.

Mariemont, 1988. Affordable residences that would not be out of place in one of Ebenezer Howard's "garden cities." *(Alexander Garvin)*

Despite its aspirations, Mariemont is not very different in appearance from other residential suburbs built during the 1920s. Like so many such communities, Mariemont has a Tudor-style central square containing shops and public facilities. As at Riverside, Shaker Heights, Palos Verdes, and most other new towns planned before World War II, deed restrictions cover the type and use of structures, the cost of dwellings, setback and yard requirements, architectural approval, and racial occupancy. Its only distinguishing characteristics are the land set aside for industry, the unusually generous recreation facilities, and buried utility lines.

Mariemont's tree-lined streets are dominated by the multilane Wooster Pike, which runs right through the middle of town. It handles regional traffic and acts as the town's main thoroughfare. Local streets are laid out to fit a geometric Beaux Arts design. The resulting residential blocks enclose large, central, open areas that can be used for delivering goods, storing vehicles, or recreation. The excessive size of these open areas unnecessarily increased the amount of land allocated to each dwelling unit and the lineal distance of infrastructure installation.

When Mariemont's first buildings were ready for occupancy in 1923, all necessary utilities (including enough conduits to supply three-quarters of the building sites with steam from a central plant) were already installed. Initial infrastructure and land development expenses were too great to be recouped from revenues generated by the community's residential buildings. Consequently, before Mariemont reached one-fifth of its planned occupancy, development had to be curtailed. One-family houses were sold as private residences. Multiple dwellings were sold as conventional real estate investments. To nobody's surprise, the community that was planned for wage earners eventually gentrified.

In-town manufacturing employment proved to be as illusory as continuing community ownership of the land. As Nolen and so many other city planners have discovered, designating land uses and employment patterns is futile without market demand. That demand was never forthcoming. As a result, Mariemont's industrial land has never been occupied.

Mariemont was a noble experiment. Unsightly utility lines do not disfigure its tree-lined streets. It provides wonderful recreation facilities. However, unlike Radburn or Sea Ranch, its design did not reduce initial infrastructure requirements. No effort was made to time its cash expenditures on infrastructure, community facilities, and landscaping with revenues from houses and stores. Thus, it does not present the argument for the superiority of planned new towns over conventional developer-built suburbs that Mary Emory and John Nolen intended.

Levittown, Long Island, and Lakewood, California

Despite the experience of Mariemont, many new towns, especially those built right after World War II, have been quite effective in providing working people with decent affordable homes. Working people constitute the largest single market in the United States. When Congress established FHA mortgage insurance and VA-guaranteed mortgages, it opened that huge market to developers (see Chapter 9). All that was necessary was a standard product that met FHA and VA requirements and a method of supplying a steady stream of those products at prices that working people could afford. Abraham, William, and Alfred Levitt were among the first builders to demonstrate how a planned new town could achieve these objectives.[50]

Levitt & Sons was established during the Great Depression. Abraham Levitt, a Brooklyn attorney, foreclosed some mortgages on Long Island property. His son, Alfred, who had formal training in architecture, designed a house for the site. His other son, William, sold it at a profit. Over the next four years, Levitt & Sons, the company they established, built and sold another 600 houses, mostly at prices below $20,000. The company went on to create Strathmore-at-Manhasset, a 500-unit subdivision on the north shore of Long Island. Work ceased during World War II, but resumed shortly afterward with 1000 conventional, two-story plus basement and garage, five-and-a-half-room houses, selling for around $10,000. Their ambition, however, was to make a fortune selling inexpensive, mass-produced houses.[51]

They began by assembling Long Island potato fields (at $3600 an acre) into what eventually became the site of a 4700-acre new town. There, beginning in 1947, Levitt & Sons built standardized, single-story, 750-square-foot houses on 60- by 100-foot lots. The first houses were "Cape Coddages," which were identical in plan but came in five exterior variations. Inside were a living room, two bedrooms, a bathroom, an unfinished attic, a fully equipped kitchen, and (as a marketing extra) a Bendix washing machine.

To bring sales prices down to $7990, houses were built on concrete slabs into which pipes for radiant heating were embedded. Levitt used standard, prefabricated components in-

Levittown, 1991. Olmstead's formula of tree-lined, curving streets, open front lawns, and houses set back from the street minimized the monotony of standardized, mass-produced "Cape Coddages." Successive improvements and remodeling give each of the houses its own personality. *(Alexander Garvin)*

stalled in 26 standard operations by 80 subcontractors, who paid union scale but used nonunion labor (see Chapter 10). As *Time* described:

> *Every 100 feet the trucks stopped and dumped identical bundles of lumber, pipes, bricks, shingles, and copper tubing—all as neatly packed as loaves from a bakery. Near the bundles, giant machines with an endless chain of buckets ate into the earth, taking just thirteen minutes to dig a narrow four-foot trench around a 25-by-32-foot rectangle. Then came more trucks loaded with cement and laid a four-inch foundation for a house in the rectangle. After the machines came the men. On nearby slabs already dry, they worked in crews of two and three, laying bricks, raising studs, nailing lath, painting, sheathing, shingling. Each crew did its special job, then hurried on to the next site. Under the skilled combination of men & machines, new houses rose faster than Jack ever built them; a new one was finished every 15 minutes.*[52]

Like automobile manufacturers, Levitt & Sons made model changes every year. For 1950 it was ranch-style houses with built-in television sets. When Levittown was finished in 1951, 17,442 houses had been built.

Unlike most new towns, Levittown did not start with a completed plan. It was designed in stages that were determined by the pace of land acquisition. Nevertheless, the plan adhered to the usual Olmstedian esthetic: tree-lined, curvilinear streets with houses set back from the roadway. Perhaps because the site was not assembled when development began, Levitt & Sons built scattered neighborhood centers that included convenience retailing, a gas station, a swimming pool, a children's playground, and a community recreation building with bowling alleys and a restaurant. The firm also donated land for a variety of small neighborhood parks, two high schools, two intermediate schools, six elementary schools, and six Protestant, two Catholic, and two Jewish congregations.

The stores in these centers were laid out in a shallow strip bounded on either side by streets that had been widened to allow for curbside, angle parking. At first they flourished. As Long Island's population soared, shopping malls and retail outlets were built along the nearby Hempstead Turnpike. These competitors attracted customers from Levittown. Many of Levittown's retailers were unable to survive without this resident market. The vacant stores they left behind are a continuing problem for each of Levittown's community centers.

Like the Levittowns that were built later in Pennsylvania and New Jersey, Levittown, Long Island, is no longer starkly uniform in appearance or narrowly homogeneous in population. The trees have matured, and the landscaping now varies from lot to lot. So do the houses, which, after more than 40 years, have been remodeled, expanded, resurfaced, and repainted many times. Not one has retained its initial appearance. They now sell at more than 20 times their original purchase prices. Many first-time home purchasers have retired and moved away, and the predominantly lower-middle-class white families with young children have been replaced by a community of 65,000 residents with a broader range of age, income, and race.

A year after Levitt & Sons began its first planned new town, the Lakewood Park Corporation initiated a similar venture on a 3500-acre site, 23 miles south of downtown Los Angeles. The corporation had been formed by three prominent southern California real estate developers: Ben Weingart, Louis Boyer, and Mark Taper. Like the Levitts, they had decided that a planned new town could reduce development costs enough to offer returning veterans a dream house for no money down plus $50 a month.[53]

The site had been part of the Rancho Los Cerritos until 1866, when it was sold to a consortium including the Bixby family. They farmed portions and used the rest as grazing land for vast herds of sheep until 1897, when they sold it for $500,000 to a Montana company that used it as farmland. In 1949, it sold the property to the Lakewood Park Corporation for $8.8 million.

Developing this new town was easy. The Lakewood Park Corporation simply extended the Los Angeles north-south/east-west street grid across its level farmland. This elementary site plan was largely the work of Louis Boyer and architect Paul Duncan. It deviated from the grid by providing "landscaped parkway panels" to insulate the purely residential streets from regional traffic on wider boulevards. Several blocks were set aside for neighborhood schools and their playing fields. A 255-acre site at the center of the grid became the Lakewood Shopping Center, which for a few years after its completion in 1951 was the world's largest shopping center.

Like Levitt & Sons, the Lakewood Park Corporation standardized its product (there were only seven model homes), organized construction along assembly-line procedures, supplied marketing extras (stall showers and electric garbage disposal units), and provided sidewalks, trees, lawns, and shrubbery. The Prudential Life Insurance Company provided the financing for both the residential and (once the May Company had agreed to sign a long-term lease for a department store) commercial development. The houses, which ranged in price from $7500 to $9000, were so popular that in 1951 *Time* reported that "on what was once an old sugar beet Field . . . 30,000 people stampeded one day last week to purchase houses in Lakewood Park."[54] By the time the project was completed, more than 17,000 houses had been built.

Lakewood, 1990. Despite successive alterations, the flat aspect, skimpy landscaping, and rectilinear block and lot pattern continue to emphasize the uniformity in building design. *(Alexander Garvin)*

By 2010, Lakewood included 9.5 square miles and 80,000 residents. Eighty percent of its 27,000 dwelling units are single-family homes that, like those in Levittown, have been sold and remodeled numerous times. The centrally located Lakewood Shopping Center, unlike the scattered retail strips in Levittown, has remained a busy and profitable enterprise because it is designed to serve the town's 80,000 residents *plus* customers from surrounding areas.

When the first houses in Levittown were offered for sale, critics howled that its ticky-tacky little boxes would spawn a generation of bored, alienated conformists and in very short order become the "slum of the future." Had they known about Lakewood, their howl would have been even louder and even less well founded. The residents of both Levittown and Lakewood eagerly grabbed the "best buy for the money" and went about their lives. They could recognize the importance of what the critics were blind to: affordable housing.

Both Levittown and Lakewood were initially populated largely by first-time, working-class home buyers. As time passed, they became communities with a range of age profiles, household sizes, income levels, and ethnic and racial groups. They have begun to exhibit the variety and patina that come only with age. Nevertheless, Levittown and Lakewood are difficult to distinguish from the suburban fabric of either neighboring Nassau and Suffolk counties on Long Island or Los Angeles and Orange counties in California. Their importance does not lie in similarity to conventional suburbs or differences from more ambitious utopian communities. It lies in their demonstration that mass-produced new communities can provide working people with affordable homes in a living environment that has the amenities that they desire.[55]

Reston, Virginia, and Columbia, Maryland

The new towns of the 1960s and 1970s were conceived in reaction to the criticism of Levittown, Lakewood, and their ilk. Rather than rows of identical inexpensive houses, planners proposed a variety of house sizes and styles; rather than mass-produced dwelling units built by the same developer, they proposed to let different sites be built by different developers; rather than simple plats that could be built cheaply and quickly, they proposed elaborate plans that included a wide variety of open spaces, retail stores, employment centers, and community facilities. The more ambitious their proposals, the less likely that the proposed new towns would be financially successful.

Of all the planned communities started during those years, Columbia, Maryland, and Reston, Virginia, seemed the most promising. Both were located within one of the nation's fastest growing metropolitan areas, Washington, D.C. Both were initiated by enlightened developers eager to demonstrate an alternative to suburban sprawl. Both had financing from large corporations more interested in long-term appreciation of value than short-term profits. Nevertheless, both experienced financial difficulties.

Reston is named for its developer, Robert E. Simon (REStn). Simon bought 7400 acres, 23 miles west of downtown Washington, in 1961, for $13 million. The site was a well-drained, pleasantly rolling, tree-covered landscape with a lake at the southern end of the property. It was strategically located in a portion of Fairfax County that lay astride the highway linking Washington with Dulles International Airport and was expected to become one of the region's major growth corridors. Building a planned new town on this site was an attractive enough investment for Simon to obtain $20 million from the Hancock Insurance Company and $25 million from the Gulf Oil Company (Gulf guaranteed $18.8 million in loans from others and gave $3.7 million as a long-term loan and $2.6 million as a construction loan).[56]

Reston, 2010. The plan of Reston ties together its various communities with a system of open-space sinews, which simultaneously provides natural drainage and a natural habitat for a wide variety of flora and fauna. *(Alexander Garvin)*

Simon hired A. D. Little & Company to prepare a development strategy and architects Whittlesey & Conklin to give it three-dimensional shape. The new town they devised was intended as a "serious experiment in city planning . . . to discover what should be done to create a quality environment."[57]

They projected a population of 75,000 living in 22,000 dwelling units of the widest variety: detached houses, town houses, lake houses, hilltop houses, and apartment houses. The design included seven residential villages, each with accessory shopping, schools, and recreation facilities, two town centers, and a 1300-acre industrial park that was intended "to provide employment opportunities for a large portion of Reston residents."[58]

People who live in company towns often work in the same community. In Reston and most other towns, people live and work in different places. Consequently, its industrial areas do not provide employment for many Reston residents. Most of their employees come from outside Reston.

The villages are separated from one another by a system of open-space "sinews" that preserved the natural forest and ground cover as much as possible. The sinews, like Radburn's open-space system, were designed to provide pedestrians with uninterrupted access to village shopping centers, schools, and the full range of neighborhood facilities. As at Radburn, there were underpasses separating pedestrian and vehicular traffic. However, the open-space sinews were quite different in conception. The idea was to create "something of the busy life and character of a fine city street, with all of its visual and social interest, without its problems of automobile traffic."[59]

The open-space sinews make lovely bicycle paths and jogging trails that also connect Reston's villages into a single new town. But they never could provide the "busy life and character of a fine city street." Reston is built at densities that are far too low for active street life. Moreover, the sinew system provides seemingly endless recreational opportunities for residents without interfering with the ecosystem that sustains a wide variety of flora and fauna. As of 2010, this system consisted of 1350 acres of open space, 55 miles of pathways, 16 outdoor swimming pools, 48 outdoor tennis courts, 2 golf

Reston, 1976. Radburn-inspired underpasses protect pedestrians using its open-space sinews. *(Alexander Garvin)*

Reston, 2010. The artificial lake at Lake Anne Village provides a charming setting for recreation activities that are scarce in neighboring suburban subdivisions. *(Alexander Garvin)*

courses, and 64 recreation areas with facilities for ice-skating, volleyball, basketball, baseball, picnics,and other activities.

At the core of each village there was to be a 15-acre center that would include housing, churches, community buildings, and restaurants, not just convenience shopping. Thus, as at Levittown, each village center was really a civic space providing a social focus for the surrounding population.

Unlike Levitt, Simon did not start with model houses from which he made sales that covered development expenses as they were incurred. His marketing strategy was to start with an 18-hole golf course; an artificial lake large enough for swimming, fishing, sailing, and all manner of water sports; and other facilities unavailable elsewhere in the Washington suburbs. This required installation of storm drainage, sewer, water supply, and road systems from the beginning. More important, it required massive amounts of cash, up front, with low financial returns.

Residential sales always take time. At Reston they were slowed because it had no access connecting it to the area's only limited-access highway, which at that time ran uninterrupted from the District of Columbia to Dulles Airport. Even if everything had sold quickly, it still would have been difficult to recoup development costs from initial property sales. Simon ran out of money in 1967 and had to surrender control to his major financial partner, Gulf Oil.

Gulf retained much of the original planning but abandoned Simon's commitment to architectural quality. Although the architectural press objected, the market response was positive, and by 1971 Reston experienced its first annual positive cash flow. Nevertheless, Gulf's board of directors decided to concentrate solely on the oil business. In 1975, it sold Reston's 3700 acres of undeveloped land (with most of the infrastructure already installed) to the Mobil Land Corporation for $33.4 million. Two factors allowed Mobil to greatly accelerate land sales. It shifted to residential development by home builders who marketed houses at different price points to a wider range of customers. More important, in 1983 it succeeded in getting an interchange on the highway to Dulles Airport.

In 1990, when 19,000 households lived in Reston, Mobil opened the first phase of a 460-acre "town center" that was to include office buildings, retail space, restaurants, an 11-screen multiplex movie theater, and more than 1400 condominium and rental apartments. This town center was to bring the vitality and variety that Simon hoped would make Reston a "quality environment" as different from homogenized Levittown as it was from small suburban subdivisions.

Mobil also eventually divested itself of Reston, selling the remaining undeveloped land in 1996 as part of a package of 18 planned communities that were acquired by Westbrook Partners for $324 million. The town center, office buildings, and retail space were sold separately.

In some ways, Reston evolved as initially planned. As of 2010, it included 58,000 residents (70 percent Caucasian) living in five villages rather than seven, 914 acres of land set aside for light manufacturing, 1.5 million square feet of retail space, and 19.3 million square feet of office space. Its planners, however, did not anticipate that northern Virginia would become the site of one of the largest concentrations of

Reston, 2010. The Reston town center became one of Washington, D.C.'s busiest suburban shopping destinations. *(Alexander Garvin)*

office space in the United States. Consequently, they could not have imagined a town center of the sort that has evolved, nor that Reston would be the site of 60,000 jobs (more than two jobs per Reston household). They also did not anticipate that competition from big-box retailing, power centers, and shopping malls in the Virginia suburbs of Washington, D.C., would make it difficult for Reston's small village retail centers to remain major shopping destinations for its residents. They did, however, succeed in making Reston a distinctive community that became an integral part of suburban Washington.

Columbia, Maryland, evolved from developer James W. Rouse's belief that building a well-planned new town was more profitable than splattering bits and pieces across the landscape. Rouse had the expertise and reputation to borrow the money he needed to prove his contention. The Baltimore-based Rouse Company, which he headed, was a successful real estate business devoted largely to shopping centers. It had financed $750 million of development and operated projects with more than $100 million in annual revenues. This record of real estate success persuaded the Connecticut General Life Insurance Company to invest $23.5 million and the Chase Manhattan Bank and the Teachers Insurance & Annuity Association of America to provide another $25 million. Debt-service payments on the money from Connecticut General were deferred for 10 years, during which time the project's limited initial cash flow was expected to help offset development and operating costs.[60]

Land acquisition began in 1962. Rouse purchased 165 farms and land parcels in 140 separate transactions. He was able to buy these 15,600 acres for the low price of $1500 per acre because he used dummy purchasers to keep the project secret. Next, he persuaded officials in Howard County, Maryland, to approve the necessary rezoning and tax-exempt bonding authority and to extend the necessary roads and public utilities to the edge of the site.

The plan for Columbia was developed by Rouse and a group of professionals working under the direction of William Finley (the vice president in charge of the project) and Morton Hoppenfeld (its planning and design director). From the beginning, they worked with national experts in government, family life, health, education, psychology, sociology, housing, transportation, and recreation. These experts played a role that was similar to the one played by the Regional Planning Association of America in designing Radburn (see Chapter 14). They met twice a month over a period of six months, and later on an ad hoc basis, to evaluate alternative approaches to planning and development. The scheme that emerged was based on existing land features (drainage patterns, land slopes, tree masses, historic and other valued sites, vistas, etc.); water, sewer, and transportation requirements; ideas about community life; and a retail shopping hierarchy that reflected the Rouse Company's experience in shopping-center development. The Columbia they envisioned was to have a population of 110,000 living in 30,000 residences distributed among nine villages and a town center.

Columbia, 2001. Oakland Mills Village, one of Columbia's nine community retail centers. *(Alexander Garvin)*

Housing clusters, built by developers who did not need Rouse financing, were the plan's most elemental building block. These housing clusters were aggregated into groups of 700 to 1200 residences that could support a *neighborhood center* with a day care center, an elementary school, a convenience store, a multipurpose meeting room, a swimming pool, and playgrounds. Three to five of these neighborhoods (about 3500 residences) were intended to support a *village center* that would include an intermediate school, a library, a worship center that could be shared by different religious groups, a supermarket, a drugstore, a laundry, a barber and beauty shop, a restaurant, and a few other stores. These nine villages (about 30,000 residences) were planned to provide a sufficient market for a *town center.* Consequently, Columbia's town center includes a 70-acre, air-conditioned shopping mall, plus office buildings and some entertainment facilities.[61]

There is a similar hierarchy for transportation: cul-de-sacs, pedestrian and bicycle paths, neighborhood streets, collector streets, and, finally, the widened and landscaped Columbia Pike (U.S. Route 29), which bisects the town. This hierarchy is also reflected in Columbia's political makeup. Residents belong to a village association that, in turn, provides one member of the citywide Columbia Association.

Each neighborhood and village was designed to be visibly defined by surrounding open space. Thus, as in so many new-towns-in-the-country, the landscaping transforms scattered housing and retail centers into a distinctive new town. Approximately 1800 acres were set aside for industrial parks that attracted large employers to Columbia. As at Reston,

Columbia, 2001. Nearly 30 percent of this planned community is devoted to an open-space system that provides residents with all sorts of recreational opportunities. *(Alexander Garvin)*

they are romantic relics of the garden cities movement. Most Columbia residents work in other communities.

Like Reston, Columbia ran into financial difficulties. In 1973, the land sales, which had averaged $24 million per year during 1971 and 1972, suddenly dropped to $6.5 million. The decline in sales could not have come at a worse moment, because the Rouse Company no longer had the benefit of 10-year, interest-free financing. Major investors, including the Connecticut General Life Insurance Company, had no alternative but to refinance $155 million, delaying debt-service payments until 1978, and in the process obtaining 80 percent of the controlling stock (which could be bought back by the Rouse Company with dividends from property sales). In 1982, Connecticut General merged with the Insurance Company of North America to form Cigna, which became Columbia's only remaining lender of the original five. Three years later, the Rouse Company bought out Cigna (which allegedly had written off $50 million in losses) for $120 million.

By 2010, Columbia had grown from 15,600 to 18,000 acres. It included 35,000 residences (rather than the planned 30,000) with a population of nearly 100,000 people (rather than the planned 110,000) living in the nine planned neighborhoods outside the town center; 2500 local businesses employed more than 60,000 people. Nearly 30 percent of Columbia was open space. These 5300 acres included 3 lakes, 19 ponds and natural open-space areas, the 40-acre Symphony Woods with the Merriweather Post Pavilion (the site of numerous summer concerts by the Baltimore and Washington symphony orchestras and other nationally known performers), and more than 83 miles of walking, bicycling, and jogging trails.

Unlike Reston, the commercial hierarchy of convenience and village centers remained healthy despite competition from big-box retailers and shopping malls outside Columbia. The Rouse Company made sure that a trip from the residential sections of each of its villages to this competition took enough time to be inconvenient. As the owner-operator of the village centers and the 1.26-million-square-foot Columbia Mall, it also was careful to lease stores to retailers whose merchandise would appeal to nearby residents. At the turn of the twenty-first century, Columbia Mall was enlarged and remodeled. As of this writing, it includes five department stores, 230 shops and restaurants, a multiplex movie theater, and parking for 7000 cars.

The town center along Columbia's lakefront, unlike the Reston town center, has failed to develop into the exciting urban center that was envisioned during the 1960s. The 31,000 town center workers and 100,000 community residents ought to constitute the necessary critical mass of customers that would be needed to support a genuine town center. However, by locating a 1.2-million-square-foot shopping mall across a broad parkway, Columbia's planners lost the benefit of thousands of customers who drive to the mall to do their shopping or just hang out, as well as the office workers who cross the broad-span footbridge on their way to the Columbia Mall, rather than remaining in what could have been a vibrant lakefront town center.

The planning practices used at Reston and Columbia established a pattern for new town development that persists to this day: a central-place hierarchy (particularly with regard to retailing), defined business and industrial park areas, and recreation opportunities. This pattern satisfies the requirements of home builders as well as retail, office, and industrial developers. Most important, it satisfies the pervasive market demand for combining an active town life with the beauty and delight of the countryside.

Irvine, California

The largest and financially most successful new town in America is Irvine, California.[62] Its 83,000 acres occupy 17 percent of Orange County and could accommodate 26 Palos Verdes Estates, 52 Riversides, or 1038 Seasides. The project was so successful that in 1977, 17 years after development of the new town began, the Irvine Company was sold for $337 million, and then in 1983 resold for $518 million.[63]

Unlike Riverside, Palos Verdes Estates, or Seaside, Irvine is neither a serious experiment in city planning nor a model that can be copied by other new town developers. It is too big. Moreover, the site, which was purchased by James Irvine in 1867, remained in the hands of the same family until it was sold 110 years later. Even when development began, not one cent had to be advanced to cover purchase of the property or debt service on its acquisition.

Irvine is located 40 miles southwest of Los Angeles. It stretches from the Pacific coast, just inland of Newport Beach, for 22 miles, to the Santa Ana Mountains. James Irvine Jr. inherited the property from his father in 1886. Eight years later, he formed the Irvine Company, which transformed it into a sprawling, combination cattle ranch and citrus operation. To prevent the breakup of his property after his death, James Jr. willed 54.5 percent of the Irvine Company to a foundation controlled by handpicked life trustees. Thus, his heirs, who owned the rest, had the benefit of a luxurious income but could not sell any property.

During the 1950s, Orange County began to experience the development pressures that would soon make it one of the fastest-growing and wealthiest parts of the country. As a result, the Irvine Company faced continually increasing real estate taxes. Rather than sell property piecemeal to pay taxes, the trustees decided to go into the development business.

Architect William I. Peireira prepared Irvine's first master plan in 1960. With so much property to plan for, Peireira had the luxury of conceiving a real city, which he projected would reach its target population of 430,000 in the year 2020. It included the 1000-acre Irvine campus of the University of California, plus another 560 acres for student, faculty, and staff housing. It also accommodated the San Diego Freeway, the Santa Ana Freeway, the Orange County Airport, and a U.S. Marine Corps air base, all of which were completely independent of the Irvine Company. Major employment centers were located at the eastern, western, and southern edges of the site. Scattered in between these huge land uses were Irvine's residential areas and the retail, recreation, and community facilities that serve them.

By themselves, Irvine's industrial and business parks occupy more land than all but one of the new towns discussed in this chapter. Unlike industrial areas in the other communities, those at Irvine were not romantically conceived to provide resident employment. They were intended to be self-sufficient facilities and quickly provided more jobs than there were town residents.

As of 2010, Irvine contained 20.7 million square feet of space devoted to manufacturing and wholesale uses, 13.7 million square feet devoted to research and development, and 24.2 million square feet of office space. Most of this space was located at the 5000-acre Irvine Spectrum (the largest master-planned commercial and technological headquarters in the world), the 2700-acre Irvine Business Complex, and 183-acre University Research Park. The North American headquarters of 11 international companies are located in Irvine. It is the home of 10 universities, including the 1000-acre Irvine branch of the University of California. All together, this included more than 13,000 firms employing 191,000 people.[64]

Newport Center, developed on Irvine land, is technically in the city of Newport Beach. By 1987, this 622-acre business center included eight office towers, a three-building medical center, three hotels, a design center, numerous low-rise buildings, and the Fashion Island shopping center (with 5 department stores, 200 retail shops, and the Irvine Ranch Farmers Market).

In 1971, to protect Irvine from annexation and additional taxes, the central portion of the site was incorporated. The resulting City of Irvine has a sphere of influence (including all areas of possible future incorporation) of 53,000 acres. Its extraordinarily heterogeneous population in 2010 included 212,000 people (45 percent non-Hispanic White, 39 percent Asian, 9 percent Hispanic, and 2 percent African American) living in 84,000 residences.

As at Reston or Columbia, residents live in neighborhoods with their own recreation center, swimming pool,

Irvine, 1989. The third largest concentration of office space in California. *(Alexander Garvin)*

Irvine, 1989. Irvine Spectrum, located at the intersection of interstate highways I-405 and I-5, is one of the premier shopping destinations in southern California. *(Alexander Garvin)*

and community facilities. Independent developers have built some, but many are Irvine Company projects. These residential areas include 45 retail shopping centers ranging in size from small village clusters to the Irvine Spectrum Center, Orange County's premier shopping and entertainment center (see Chapter 4), more than 80 schools, 485 acres of public parkland and 5200 acres of protected open space, 40 miles of off-street hiking and bicycle trails, and health clubs. In addition, there are two lakes, four golf courses, and four major tennis clubs.

Irvine is surely not the utopian community that most new-town proponents have in mind when they argue for the superiority of planned new communities over conventional suburban development. It is indistinguishable in physical appearance from the rest of Orange County. With the exception of university students and staff, it is similarly indistinguishable in social composition. It is unique among planned new towns, however, in its ability to satisfy market demand, exploit its location along southern California's busiest freeways, and benefit from the financial strength of its sponsor. Nor has any other new-town-in-the-country grown more quickly, going from a population of 62,000 in 1980 to 212,000 in 2010, when it ranked ninety-sixth in size among all American cities.[65]

National Policy Initiatives

Despite thousands of privately developed new towns, various city planners, social critics, and public officials periodically demand that government get into the new-town business, often as a way of increasing housing production. They have been able to convince Congress to fund new-town development three times, but in each case only in response to some crisis.[66]

The first governmental new-town program provided shelter for workers involved with military-related production during World War I. It was promoted to Congress on the basis that war-production efficiency demanded sound housing and a suitable living environment for workers and their families. Since overwhelming amounts of capital were

Irvine, 1989 *(Alexander Garvin)*

needed for the war effort, the argument went, government had to provide the funds for additional housing. The program ended with the armistice.

Another group of government-subsidized new towns was built in response to the Great Depression. This time Congress was told that the "resettlement of destitute and low-income families from rural and urban areas" required "the establishment, maintenance, and operation . . . of [planned new] communities in rural and urban areas." The courts terminated this new-town program.[67]

The last group of government-supported new towns was an outgrowth of the "urban crisis" of the 1960s, when angry demonstrations led Congress to seek better ways of providing "community services, job opportunities, and well-balanced neighborhoods in socially, economically, and physically attractive living environments." This time, new-town proponents persuaded Congress that planned new towns would help solve virtually every aspect of the urban crisis. New towns were oversold as devices for reducing "inefficient and wasteful use of land," preventing "destruction of irreplaceable natural and recreational resources," providing "good housing," increasing "unduly limited options for many of our people as to where they may live, and the types of housing and environment in which they may live," increasing "employment and business opportunities," supporting "vital services for all . . . citizens, particularly the poor and disadvantaged," reducing "separation of people within metropolitan areas by income and by race," and even increasing the "effectiveness of public and private facilities for urban transportation."[68] This effort was ended by the 1973 Nixon moratorium on housing and development programs.

Despite dubious claims for the curative effects of new-town development, all three federal programs initiated valuable experiments in urban planning and created some lovely communities. More important, they also provided insights into the ingredients of successful new-town development, as well as its irrelevance as a device for fixing the American city.

World War I Villages

During 1917, Congress appointed three committees to examine whether full war production depended on construction of additional housing for workers in war industries. All three had little doubt that it would be impossible for private capital to supply the necessary shelter. In response, the Wilson administration established two housing-production programs, one operated by the Emergency Fleet Corporation (EFC) of the U.S. Shipping Board and the other by the U.S. Housing Corporation (USHC), organized by the Department of Labor. Together, these two agencies helped produce 15,183 family dwellings and 14,745 accommodations for single workers.[69]

The EFC was organized in 1917, under the Shipping Act of 1916, to acquire, maintain, and operate merchant vessels. In 1918, its authority was extended to provide housing for shipyard employees and their families. It did so by providing 10-year mortgages to realty companies incorporated by shipbuilding concerns. The mortgages covered 70 to 80 percent of the cost of development and carried an interest rate of 5 percent. However, these loans, totaling $67 million, were expected to be repaid at the end of the war, in an amount reduced by 30 percent to allow for inflated war costs.

The staff and consultants hired to run the program included some of the country's most prominent architects, housing specialists, and urban planners, among them Frederick Law Olmsted Jr., John Nolen, and Henry Wright. Before it finished its work in 1919, the EFC had been involved with 28 projects that produced 9185 dwelling units in 15 states and 23 cities. As directed by Congress, by 1924 it had disposed of all its loans at a net loss of about $42 million.

The USHC shared the same personnel with the EFC and had a similar but even more illustrious group of advisers, including Olmsted, Nolan, Grosvenor Atterbury, and Lawrence Veiller. Despite the congressional requirement that the housing be temporary in character, the advisers recommended construction of permanent housing in a series of industrial villages. The USHC followed that recommendation. How-

Bridgeport, Connecticut, 1998. Seaside Village remains one of the most charming of the government-planned communities erected during World War I. *(Alexander Garvin)*

ever, instead of making loans to shipbuilding concerns, it built these industrial villages itself, spending $52 million for 55 projects, finishing 27 developments in 16 states and the District of Columbia. After the war, the USHC sold off its 5998 family dwellings and 7181 single-worker units for a net loss of $26 million.

Yorkship Village (now known as Fairview) in Camden, New Jersey, is one of the largest and most interesting of these war villages. It was financed by the EFC and developed by the Fairview Realty Company, established by the New York Shipbuilding Company. Fairview was designed by architect Electus D. Litchfield in collaboration with Pliny Rogers. The project covered 225 acres. Only 90 acres were subdivided into lots for 1438 dwelling units (1021 row houses, 300 semidetached houses, 56 apartments, and 61 detached houses). There were also a few stores and a theater.[70]

Yorkship Village has none of the grandeur and scale implied by its Beaux Arts design. The plan, which like Letchworth had a central square with avenues radiating from it, also includes a variety of geometrical open spaces and landscaped formal boulevards. These 48 separate squares, parks, parkway islands, and playgrounds are individually small. Together, they occupy barely 36 acres. Its two-story buildings are also small.

The 78 World War I villages long ago merged into their urban or suburban surroundings. Today, they are distinguishable only by their relatively small buildings and period architecture. With the exception of Yorkship Village, which is admired by Duany and Plater-Zyberk, none of these war villages has had much impact on new-town planning. However, they themselves are what new community planning and development was all about. They also provided a model for the later housing-subsidy formulas of the PWA (see Chapter 10).

The New-Deal New Towns

When the Roosevelt administration came to office in 1933, one-third of the nation was unemployed and ready to try any prescription for relief. Among the prescriptions with which it experimented was one that promised to achieve a better world by resettling the jobless into planned new communities. The $108 million spent on this experiment produced 99 planned new communities with 10,938 dwelling units.[71]

While these planned new communities sprang from a common attempt to create a new utopia and usually housed fewer than 100 families, they were established to serve widely different purposes. There were farm colonies, industrial settlements, colonies for stranded workers, subsistence gardens for city workers, cooperative associations, and forest homesteads. Only Greenbelt, Maryland, Greenhills, Ohio, and Greendale, Wisconsin, were genuine, planned new towns (Table 16.1). However, unlike the others, which have long since faded into obscurity, these three new towns continue to serve as object lessons in the use of public open space and community facilities to create superior living environments.

TABLE 16.1
THE GREENBELT TOWNS

Town	Area, acres	Dwelling units	Cost
Greenbelt, Maryland	3,600	890	$13,701,817
Greendale, Wisconsin	3,510	640	$10,638,466
Greenhills, Ohio	5,930	737	$11,860,628
Total	13,040	2,267	$36,200,901

The Greenbelt towns were the brainchild of Dr. Rexford Guy Tugwell, a member of the Columbia University economics faculty, who became a Roosevelt adviser during the presidential campaign of 1932 and undersecretary of agriculture in 1933. Tugwell believed that a national program established to build intelligently located and planned new towns would guarantee the efficient use of land and resources while simultaneously providing jobs, housing, and a healthy living environment for the unemployed. When Congress enacted the Emergency Relief Appropriations Act of 1935, he thought he had the necessary legislative sanction for such a program and convinced Roosevelt to sign an executive order establishing it within the newly created Resettlement Administration in the Department of Agriculture. Naturally, he was appointed its director.

Tugwell and his staff began by studying the demographic and economic trends of more than 100 cities to determine promising locations for the new towns. They selected 25 for further study. Roosevelt approved $68 million for eight new towns. The Resettlement Administration eventually settled on five, two of which were later dropped from the program.

The new towns that Tugwell envisioned were to be "demonstrations of the combined advantages of country and

Greenbelt, Maryland, 1969. "Moderne" design gives the town center a distinctive New Deal character. *(Alexander Garvin)*

city life for low-income rural and industrial families."[72] He sought the best professional advice on how to create such towns, settling on Clarence Stein to provide general guidelines for the architects and planners that the Resettlement Administration hired to design them. Stein made sure that the planning reflected the principles established by Ebenezer Howard for his garden cities, by Clarence Perry for his neighborhood unit, and by Stein himself for Radburn.

The siting criteria that Stein established were intended to minimize capital expenditures and operating costs. In that way there would be enough money left over to also maximize infrastructure and community facilities. More important, as Stein explained, while the Resettlement Administration could cover initial capital costs, the "projects were to be unified, self-supporting communities in which a tenant's monthly charges must cover *all* costs."[73]

The necessary cost savings were achieved by avoiding prevalent through streets lined with buildings and attached garages. Instead, they relied on cul-de-sacs, clustered parking, and residences that were near public open space. Stein established criteria for house design that ensured maximum furnishability at minimum cost. This resulted in relatively compact residential plans, straightforward construction, and a common vocabulary of simple materials: brick, concrete block, stud walls, and standard double-hung windows.

Eighteen months into Tugwell's effort to pioneer a new pattern of rural-urban industrial life, the program was brought to a halt by the U.S. Court of Appeals in the District of Columbia. It declared Greenbrook, the fourth and largest of the Greenbelt projects, which was to have been built between New Brunswick and Princeton, New Jersey, to be unconstitutional. Dean Acheson, counsel for the surrounding township, had argued that Greenbrook would radically alter its character, unreasonably increase local expenditures without supplying a corresponding increase in its tax base, depress existing property values, and leave existing residents in danger of losing home rule. The court not only accepted these arguments, it also found the Emergency Relief Appropriations Act to be invalid. The Roosevelt administration decided not to appeal the decision to the Supreme Court. Nevertheless, it proceeded with the three projects already under way.[74]

Greenbelt, Maryland, is 13 miles northeast of Washington, D.C., far enough away at the time to represent a serious commute. In an attempt to provide nearby employment opportunities, it was built in conjunction with the 8659-acre Department of Agriculture National Research Center. As is usually the case with employment areas designated to supply jobs for new towns, this proved to be an appealing illusion. The employees of the Research Center commute from all over the metropolitan region, not just from Greenbelt, which is a dormitory suburb for people who work throughout the Baltimore–Washington area.

Of the 3600 acres acquired for Greenbelt, only 217 acres were used for the town; 500 acres were reserved for future expansion, 250 for parks, and the bulk of the rest for the greenbelt. As a result, Greenbelt is the beneficiary of extraordinarily generous amounts of open space. Its plan consists of two parallel, crescent-shaped main roads that wrap around residential superblocks and a town center with retail stores, a school, and recreation facilities. This crescent is divided every 1000 feet into 14-acre superblocks by a series of cross streets that connect the main roads. As at Radburn, the superblocks are bounded by attached residences or garden apartment buildings and served by an internal system of pedestrian walkways. However, the similarity to Radburn is illusory. Greenbelt's open space is undefined by the surrounding residences, shabbily landscaped, and in many cases has been fenced off by its residents. The walkways neither are continuous nor provide true separation from vehicular traffic. Except in two instances where they connect to underpasses that lead to the town center, the walkways all end at the street.

Greendale, which only provided a 10-acre site for light industry, made no pretense of providing resident employment. Since it is only 7 miles southwest of Milwaukee, it is also a shorter commute. Nevertheless, Greendale, like Greenbelt, reflects the romanticism of the garden cities movement. Approximately 1830 of its 3510 acres were set aside for 13 full-time dairy farms and 53 subsistence farms, most of which have either been subsumed into the surrounding greenbelt or subdivided as suburban residences. Only 170 acres were used

Greendale, Wisconsin, 2012. A creekbed provides the corridor for this community's greenbelt. *(Alexander Garvin)*

Greendale, Wisconsin, 2012. Affordable houses in a generously landscaped setting. *(Alexander Garvin)*

for its residences. As at Radburn, there are cul-de-sacs with houses turned around onto pedestrian walkways that connect to a landscaped open-space system with a creek running through it.

Greenhills also made no pretense of providing industrial employment, and, being 11 miles north of Cincinnati, its 5930-acre site was even more agricultural. In fact, a great deal of the land had to be leased back to its farmers. Only 1300 acres were used for the town. The bulk of the remaining land was dedicated as public open space. As a result, Greenhills is the only one of the three new towns completely enclosed by a permanent greenbelt.

While Greenhills does have cul-de-sacs, most of the houses (primarily small, redbrick Cape Cod cottages and attached four-family residences) line its curvilinear streets. The difference from most suburbs is that they also back onto large, landscaped open spaces. At the heart of Greenhills are its shopping center, school, Olympic-size swimming pool and recreation center, public buildings, newspaper, credit union, savings and loan, and golf course.

Better than any of Tugwell's other projects, Greenhills captures the essence of what he had in mind when he wrote:

> *I really would like to conserve all those things which I grew up to respect or love and not see them destroyed. I grew up in an American small town and I've never forgotten it. No one was very rich there, but no one was very poor either.*[75]

If any child now living in Greenhills grows up to administer another governmental new-town program, he or she will have no trouble repeating these very same sentences.

Tugwell's experiment in public planning, construction, ownership, and administration of new communities was terminated in 1949 when Congress enacted Public Law 65. It required the negotiated sale of the Greenbelt towns, preferably to nonprofit organizations, cooperatives, or veterans' groups. Greendale was sold to its tenants, and its greenbelt was largely eliminated by the Milwaukee Community Development Corporation, which had purchased it for additional housing development. Nevertheless, over 1000 acres remained in use as park and recreation facilities. Greenbelt's houses, plus 708 acres suitable for development, were sold to a veterans' group. Its apartment houses and shopping center were sold to separate investors. The greenbelt was transferred to the Department of the Interior. After a few years, the veterans' group ran into financial trouble and sold several large undeveloped parcels to commercial developers, who built the usual mediocre tract homes. The town of Greenhills plus 600 acres of vacant land were sold to a nonprofit, cooperative homeowners corporation. Some of the surrounding land went to the Cincinnati Park Service and to the army, and the rest to a development corporation that created the more expensive subdivision of Forest Park.

The Title IV and Title VII New Towns

During the late 1960s and early 1970s, when government went into the new-town business for the third time, it did so pursuant to legislation specifically designed for the purpose. The program, which evolved from Title IV of the Housing and Urban Development Act of 1968 and Title VII of the Housing and Urban Development Act of 1970, had none of the previous sense of urgency, experimental fervor, or utopian romance.[76] Everybody concerned understood and was prepared to provide for:

> *(1) the large initial capital investment required to finance sound new communities, (2) the extended period before initial returns on this type of investment can be expected, (3) the irregular pattern of cash returns characteristic of such investment.*[77]

As a result, this new-town program was tailored to the requirements of the nation's financial markets. It provided federal guarantees of up to $50 million on bonds, notes, or other obligations issued by developers of approved new communities and 15-year, interest-free loans up to a maximum of $20 million.[78]

Government officials assumed that new-town development would also utilize funding from the Urban Mass Transportation Act, Land and Water Conservation Fund Act, the Higher Education Facilities Act, and other government programs. As in the Model Cities program, supplementary cash grants were made to state and local agencies. The grants could cover the local share of other government programs, provided they did not exceed 20 percent of project cost and the total federal share did not exceed 80 percent.

Congress wanted the new towns to reflect both national and community standards. Developers had to secure all required state and local reviews and approvals, meet area-wide planning requirements, provide a substantial amount of low- and moderate-income housing, comply with all applicable civil rights laws, and adopt an affirmative-action program for equal opportunity in employment, housing, and business enterprise.

TABLE 16.2
TITLE IV- AND TITLE VII-ASSISTED NEW COMMUNITIES

Community	State	Developer	Type	Acres	Result
Cedar-Riverside	Minnesota	Private	In-town	100	Partial
Flower Mound	Texas	Private	Separate	6,156	Aborted
Gananda	New York	Private	Separate	5,847	Aborted
Harbison	South Carolina	Private	Separate	1,734	Revised
Jonathan	Minnesota	Private	Separate	4,884	Revised
Maumelle	Arkansas	Private	Separate	5,319	Revised
Newfields	Ohio	Private	Separate	4,032	Aborted
Park Forest South	Illinois	Private	Separate	8,136	Revised
Radisson	New York	Public	Separate	2,800	Aborted
Riverton	New York	Private	Separate	2,125	Aborted
Roosevelt Island	New York	Public	In-town	147	Completed and augmented
St. Charles	Maryland	Private	Separate	6,980	Profitable
Shenandoah	Georgia	Private	Separate	7,250	Aborted
Soul City	North Carolina	Private	Separate	5,287	Aborted
The Woodlands	Texas	Private	Separate	16,939	Profitable

Fifteen projects were approved prior to Nixon's unilateral 1973 termination of every federal housing and renewal program. These approved projects were allegedly selected based on their compliance with HUD's performance standards that required "a financial plan or program demonstrating that the project is and will be financially sound."[79] (See Table 16.2.)

Despite the rhetoric, political criteria appear to have played a greater role in project selection. Soul City was to be built in a completely undeveloped part of North Carolina by an African American–owned company organized by civil rights leader Floyd McKissick. HUD approved Roosevelt Island while Nelson Rockefeller was governor. It provided supplemental funding for the New York State Urban Development Corporation. Along with Cedar-Riverside, Roosevelt Island provided HUD with a chance to fund inner-city development, rather than just suburban sites (see Chapter 15).

Every project but The Woodlands and Roosevelt Island defaulted on its HUD-guaranteed obligations, and only The Woodlands and St. Charles have proceeded largely according to plan (Table 16.2). Most projects had unrealistic cash-flow projections. Some were located beyond the range of serious market demand or ran into opposition from local interest groups.

The Woodlands is a project of the Mitchell Energy and Development Corporation, a *Fortune* 500 company with assets in excess of $2 billion. Its 16,939-acre site is 28 miles north of Houston. In the 1970s, when development began, Houston was one of the most rapidly urbanizing sections of the country. The Woodlands was readily accessible to this large and expanding market. Property began to be assembled in 1964 by George P. Mitchell, after his firm purchased a lumber company that owned 50,000 acres of timberland in four counties north of Houston. To this, Mitchell added several large sites, but avoided 12 holdouts in order to reduce the land cost to the relatively low price of $1688 an acre.[80]

Mitchell needed experienced consultants to deal with this essentially flat, poorly drained site with low-quality vegetation. He selected architect William Pereira, the designer of

The Woodlands, 2007. Residents enjoy the lakes so prevalent in planned new communities. *(Alexander Garvin)*

The Woodlands, 2007. 50,000 people work at the more than 1750 businesses in The Woodlands. *(Alexander Garvin)*

Irvine, and landscape architect Ian McHarg. The plan they arrived at, like those of Columbia and Reston, preserved the most attractive woodlands. It minimized hydrological disruption by using a natural drainage system of floodplains, small ponds, 39 lakes, and numerous ditches. As a result, 3909 acres (23 percent) of the site is public open space.

The target population for The Woodlands was 150,000 people (estimated to reach 117,000 in 2916) in 47,375 dwelling units (15 percent low- and moderate-income). The residences, like those of Columbia and Reston, were grouped into 19 neighborhoods, 6 villages, and a metropolitan center that were all, in turn, defined and connected by the open-space system. The plan also called for 2000 acres in industrial parks; a 335-acre commercial, conference, and leisure center; a university; trade, research and medical centers; community facilities; and public services.

Despite setbacks caused by business cycles and changing local market conditions, The Woodlands proceeded without significant difficulties. By 1980 it included 170 businesses employing 3000, 12 churches, 6 public schools, and 8800 residents in 2951 dwelling units. In 1983, HUD released the Mitchell Energy Company (which had made timely payments on its $50 million in HUD-guaranteed obligations) from its role as a Title VII new-town sponsor. Development continued throughout the decade, although somewhat retarded by the impact of the energy crisis. By 1990, more than $2 billion had been spent developing The Woodlands. Its 2010 population exceeds 100,000 people, and more than 50,000 people work there.

Neotraditional New Towns

During the last quarter of the twentieth century, a varied group of designers who were unhappy with post–World War II suburbs proposed designs that appealed to both developers and a great number of their customers. In some cases, like those of Jaquelin T. Robertson (architect/planner of New Albany, Ohio, and master planner/architect of Celebration, Florida) and Robert A.M. Stern (master planner/architect of Celebration), their designs were a logical evolution of their belief in an urbane architecture that dated back to their work as students at the Yale School of Architecture during the 1950s and 1960s. For Andrés Duany and Elizabeth Plater-Zyberk (who also graduated from Yale, but in 1974), it began with their work on Seaside, Florida, and Kentlands, Maryland, and evolved into a passionate commitment to change the very character of American suburban development (see Chapter 14). For others, neotraditional design established simple principles that made it easy to provide developers with a marketable product.

Whatever the motivation, neotraditional designers share a commitment to what they believe to be intelligent urbanization. They decry the ragged agglomeration of suburban subdivisions, proposing instead compact, understandable street patterns with clearly identifiable destinations. They believe that too many suburban subdivisions are placeless concentrations of resident-owned houses, proposing to augment these houses with a substantial public realm. They dislike the pared-down, characterless appearance of many suburban buildings, proposing an alternative based on indigenous or historical architectural styles. Their hopes and dreams are beautifully articulated in an undated advertisement for Celebration, the new town built by Disney:

> *There once was a place where neighbors greeted neighbors in the quiet of summer twilight. Where children chased fireflies. And porch swings provided easy refuge from the care of the day. The movie house showed cartoons on Saturday. The grocery store delivered. And there was one teacher who always knew you had the "special something." Remember that place? Perhaps from your childhood. Or maybe just from stories. It held a magic all its own. The special magic of an American home town. Now, the people at Disney—itself an American family tradition—are creating a place that celebrates this legacy. A place that recalls the timeless tra-*

ditions and boundless spirit that are the best parts of who we are.

New Albany, Kentlands, and Celebration are among the most important of the many neotraditional communities created by these designers. New Albany combines an ecologically sensitive approach to landscape design with definable destinations and buildings of a consistent Virginia Georgian style of architecture that has proven to be commercially successful. The design of Kentlands evolved out of a seven-day event in which designers, developers, public officials, and residents of Gaithersburg, Maryland (a suburb of Washington, D.C.), came together to plan a new community. Celebration is the product of the Disney Company's desire to promote the very best in new-town planning.

New Albany, Ohio

Columbus, Ohio, is at the center of one of the fastest-growing metropolitan areas in the country, increasing 14.5 percent in the 1990s to a population of 1.54 million in 2000 and another 21.4 percent to 1.87 million in 2010. Les Wexner wanted to live in the country in the rapidly growing northeastern section of Columbus. Wexner is the president and CEO of The Limited, Inc., a company that began in 1963 with one store in Columbus and has grown to include over 5300 stores and nine retail brands, among which are Express, Lane Bryant, Structure, and Victoria's Secret. He soon realized that in order to prevent his newly adopted domain from becoming endless miles of suburban ticky-tacky, he would have to assemble land and transform it into a well-designed, new, and different kind of community. In 1986, soon after he began purchasing property, he asked Gerald McCue, then dean of the Harvard School of Architecture, landscape architect Laurie Olin, and initially Robert Stern, Thierry Despont, and Roger Sahli to come up with alternative development ideas. Later, Jaquelin Robertson was added to the group, replacing Stern. Based on their work, Wexner decided to develop the area as a planned residential community called New Albany, which would be characterized by good schools, ample civic spaces, and widespread opportunities to explore creek beds, walking trails, quiet sitting areas, and other recreational open spaces.[81]

Wexner was acutely aware of the need to purchase a critical mass of land at low enough prices to permit the development of a first-rate planned community. But because he decided to do so without purchasing overpriced holdouts, the team had to find an inexpensive way to identify the substantial but discontinuous property assemblage that was to become New Albany. They settled on a familiar iconic image: open, white wooden fences that are common in horse country. These New Albany fences still enclose the golf course and other major open spaces.

New Albany, 2001. By encircling the golf course with jogging trails, bicycle paths, and vehicular roadways, residents and visitors alike enjoy the benefits of this substantial, landscaped open space. *(Alexander Garvin)*

In 1996, Wexner decided to diversify residential New Albany. He asked the Georgetown Company (see Chapter 4) to join him as his development partner. Wexner and Georgetown understood that New Albany, like Irvine and other successful planned communities, needed a commercial tax base to help cover the cost of its schools and public services. Consequently, they began acquiring land for commercial purposes. As of 2012, the 3000-acre New Albany Business Park included six campuses occupied by 31 companies with more than 11,000 employees.

The business park generates 40 percent of the revenues for New Albany's operating budget. Its financial success and the desire for additional tax revenues have led Wexner and Georgetown to develop a 700-acre science and technology park adjoining the business park. Its health care campus alone employs more than 1100 people who occupy a surgical hospital and four medical office buildings, making it a major medical hub for Columbus's major growth corridor.

In 2012, New Albany's 7800 residents had begun to fill in its projected 46 neighborhoods, each slightly different in character, but each designed in a Virginia Georgian vernacular. This residential portion of the new town is expected to cover 2750 acres, 35 percent of which will be open space. In addition to the small original village of New Albany and the residential neighborhoods, it includes a village center, public schools, a public library, six country clubs with golf courses, a convenience retail center, the business park, and the science and technology park.

New Albany is no ordinary golf-course community. Its first 300-acre golf course is part of an open-space system shaped by existing creeks, wetlands, and drainage basins. Unlike most golf-course communities, the links are not hidden behind houses. They are enclosed by the New Albany wooden fence and by parallel pedestrian and jogging paths and vehicular roadways. Thus, people will have the pleasure of seeing large expanses of landscaped open space as they stroll, jog, or ride by. There is a financial rationale to this design. The longer perimeter on the far side of the roadway

New Albany, 2001. The design of the high school and library was inspired by the University of Virginia. *(Alexander Garvin)*

creates more sites that front on the golf course and therefore greater revenue from lot sales than would have been available from the smaller number of sites that would have opened directly onto the golf links.

New Albany looks markedly different from other post–World War II suburbs, even faux-Georgian suburbs. The school is inspired by Jefferson's University of Virginia. The country club could be mistaken for an eighteenth-century Palladian villa. Even individual houses would not look out of place in Virginia. Jaquelin Robertson, the person most responsible for this appearance, grew up in Richmond in a neo-Georgian house designed by William Lawrence Bottomley. At the time, he was dean of the School of Architecture at the University of Virginia. The first four large houses in New Albany were designed by Robertson as an homage to Bottomley. Knowing that local builders would put up most houses, he took a planeload of Central Ohio contractors to Richmond to learn about its brickwork. Robertson also helped to develop New Albany's design guidelines, has been intimately involved with approving building designs, and assisted in the design of civic and commercial buildings throughout the community.

The residents of New Albany represent a variety of household sizes and income groups. They do not live in structures that combine different land uses, or in residential enclaves that mix different housing types, nor are they separated from one another in gated communities. This is the result of the development strategy adopted by the New Albany Company (NACO) and the Georgetown Company.

When parcels are ready for development, the Georgetown Company, as development manager, examines market conditions to determine the prices at which demand for new residences is most intense and what kind of residences would best satisfy that demand. It works with Robertson, McCue, and local architects to arrive at an appropriate site plan. The package then becomes the basis of what Wexner and Georgetown ultimately decide to do. At that point, Georgetown works out a business arrangement with a builder who specializes in that type of residence.

Each neighborhood takes on a character of its own that reflects the overall plan and its builder's experience with that particular product. The economies of scale that come with repetition allow the builder to offer houses at competitive prices. The care that is lavished on site planning and landscaping provide the consumer with a setting that is far superior to that of any similar house within the market area.

The distinctive neighborhoods that result are tied together by a system of continuous open spaces and leisure paths that connect them with one another, in the same way, but not in the same form, as Clarence Stein's open-space backbone for Radburn (see Chapter 14). The character of each area also is reinforced by its open spaces. In some places, neighborhoods are organized around boulevards with landscaped center islands. In other places, open-space ovals are lined with houses. Many neighborhoods contain small sitting areas that exploit scenic vistas.

When New Albany is completed, like all healthy communities, it will consist of different neighborhoods built at different times by different companies for different residents who share common characteristics. However, unlike many

New Albany, 2001. Edge-of-woods is just one of many charming, Georgian residential complexes. *(Alexander Garvin)*

cities and suburbs, its continuous public realm and its common architectural vocabulary will unify its quite distinctive parts into a single, beautiful, and very special community.

Kentlands and Clopper's Mill, Maryland

The plan for Kentlands evolved out of an intense, seven-day planning period, which Duany Plater-Zyberk called a *charrette*. The term originated at the École des Beaux Arts in Paris. *Charrette* is the French word for the horse-drawn carts that would come to the various architectural studios around the city to pick up student projects and deliver them to the Beaux Arts for review. Most drawings were unfinished. The students would get on the cart and finish their work in a mad rush while "*en charrette*."

The work produced at the Kentlands and other New Urbanist charrettes is not produced by a single individual. It is the work of a panel of experts meeting regularly with community residents, public officials, and developers, coordinated by a single design firm—at Kentlands, by DPZ. Panels of expert visitors have been used for decades to certify already agreed-upon projects or to stop them. In some instances, expert panels have recommended innovative and effective programs of action. The Urban Land Institute, for example, organized and administered the expert panel that recommended the location and planning of Denver's Coors Field.

A DPZ charrette is not just an opportunity for experts to present their vision of an ideal future. The participants discover and explore proposals that have a serious chance of coming to fruition. The discipline of listening to and exploring ideas allows participants to resolve contradictions, establish feasibility, determine cost, and rally support. It also enhances the opportunity for designs that reflect local conditions to be discussed and adopted. Once accepted, the resulting plan establishes an appropriate context for future development. The predictability that this context provides reassures property owners and developers. The certainty that all future action will fit this context provides the security that is necessary if private investment is to take place.

Another reason DPZ charrettes differ from other expert panels is that, rather than reviewing completed work or recommending future work, they can complete the needed design work in the presence of the experts and the community. In order to do this, the firm brings a large staff of architects who prepare initial building designs for specific sites, which they revise during the week-long planning process. Consequently, at the conclusion of the process, DPZ produces preliminary building designs that can be used to demonstrate the physical and financial feasibility of the participants' recommendations.

In 1987, Joseph Alfandre & Company, Inc., purchased an option on a 352-acre property in Gaithersburg, Maryland, 23 miles northwest of Washington, D.C., and hired DPZ to manage a charrette, at which the firm developed its initial plan for what became known as Kentlands. The following year, Alfandre purchased the site for $40 million. DPZ's plan called for 48 acres of roads (14 percent), 91 acres of retail and commercial uses (26 percent), 20 acres for civic uses (6 percent), 56 acres of common open space (15 percent), and 137 acres of housing (39 percent). This required $20 million in infrastructure and site improvements.[82]

Construction and sales began in 1990. Unfortunately, Alfandre did not have the same flair for real estate as Robert Davis, the developer of Seaside. Since Kentlands was meant to be a place of permanent residence rather than just a vacation resort, Alfandre had to provide better initial amenities. Buyers of houses in a resort community might tolerate unpaved roads for a few years, but people intending to live in a place full-time demand pavement and other basic facilities. Alfandre was unable to balance the cash outflow for debt service and taxes on his $60 million investment in land, infrastructure, and buildings with cash income from sales. Consequently, the lender, Chevy Chase Savings Bank, took the property from its defaulting borrower in 1991. The situation might have been different if land sales had not begun during the recession of the early 1990s. Even so, covering the soft costs on a $60 million investment would have been difficult. Had Alfandre, like Davis, delayed spending on infrastructure, the situation might have been different.

In some respects, the physical appropriateness of the plan proved to be as illusory as its financial appropriateness. Kentlands ought to illustrate the principles enunciated in 1993 in the *Charter of the New Urbanism*, which states:

> *Many activities of daily living should occur within walking distance, allowing independence to those who do not drive, especially the elderly and the young. Interconnected networks of streets should be designed to encourage walking, reduce the number and length of automobile trips, and conserve energy.*[83]

This is true for those residential sections of Kentlands that are near one particular nonresidential land use. However, because there is a clear separation between institutional, commercial, and recreational areas, and because these land uses are just far enough away from one another to discourage walking, people in Kentlands drive to church, drive to the tennis courts, and drive to the supermarket. This is neither bad nor different from any other suburban community built to a density of 4.7 houses per acre. (Lakewood has 4.9 houses per acre, Levittown 3.7, and Reston 3.1.) At this density, there is no way to support the mixed land uses and intense pedestrian activity that are called for by the New Urbanists—exclusively from purchases by Kentland's residents.

DPZ's original 1987 plan for Kentlands included a

Kentlands, 2012. The traditional design of the houses in Kentlands is enhanced by picket fences, front porches, and pitched roofs. *(Alexander Garvin)*

thoughtfully designed retail district with apartments above street-front retail space. Parking was located behind the buildings. Combined with thousands of residents who lived within a five-minute drive of Kentlands, the planned 1655 residential units provided enough customers to test DPZ's clever design. Unfortunately, the bank could not find anybody who was ready to invest in the proposition that suburban shoppers coming by car from nearby communities or from already-built sections of Kentlands would drive to parking lots they could not see while driving their cars. Consequently, the developer to whom the bank sold this portion of the commercial district scrapped the original plan in favor of the usual strip mall, whose only concession to neotraditional design was some Palladian motifs decorating its anchor stores.

In 1999, the remaining section of the retail district was finally completed as "Market Square." No one could mistake it for a strip mall. Stores, restaurants, and the cinema front on real streets lined with sidewalks. They are built to the property line. There even is an intimation that people might live above the stores. Of course, there are huge parking fields on either side of Market Square. The retailers attract plenty

Kentlands, 2012. Market Square. *(Alexander Garvin)*

Kentlands, 2012. Row houses at Kentlands are similar to those being built everywhere in suburban Washington, D.C. *(Alexander Garvin)*

of activity. However, the florist, furniture outlets, and other retailers depend on regional customers who drive to the parking field visible from the road and walk to their ultimate destination.

One way in which Kentlands is a most successful demonstration of New Urbanist design is its block and lot pattern. The *Charter of the New Urbanism* calls for "a broad range of housing types and price levels" that "can bring people of diverse ages, races, and incomes into daily interaction, strengthening the personal and civic bonds essential to an authentic community."[84] Most suburban developers subdivide their property into blocks whose lot sizes are all identical. They do so because that is an easy way to subdivide their property, and it simplifies legal documents. The uniform lot size makes the lots easier to market. It reassures purchasers that their neighbors will be similar to themselves. But it also produces a cookie-cutter landscape.

At Kentlands, as in most of the firm's other designs, DPZ varied lot sizes. Consequently, houses of quite different sizes are built next to one another. Single-family houses and row houses are mixed together. Only low-rise apartment buildings occupy a separate enclave. The New Urbanist rhetoric insists that this will produce desirable social heterogeneity. The implied mix of class, education, and income, however, is more theoretical than real. It is not the social engineering that is important, it is the physical results. Varying lot sizes avoids the raw newness and monotony that is typical of standard suburban subdivisions. When construction is completed, communities like Kentlands almost instantly take on the patina of older communities whose lots have been subdivided and recombined enough times to guarantee variety within a unified structure.

This variety comes at a cost. There are no sites large enough for those builders who are seeking the economies of large-scale production. Consequently, Kentlands houses must be sold at prices that are higher than those of similar houses built by national home builders in neighboring communities. On the other hand, it can be argued that the premium paid for houses at Kentlands is evidence of the value consumers place on living in a community designed according to New Urbanist principles.

While Kentlands succumbed financially to the recession of the early 1990s, Clopper's Mill, a nearby development in Gaithersburg, survived. Clopper's Mill is a 318-acre

Clopper's Mill, 2012. Row houses at Clopper's Mill are similar to those being built everywhere in suburban Washington, D.C. *(Alexander Garvin)*

planned community that includes a church, a high school, a 130,000-square-foot shopping center, and 1039 dwelling units. Like Kentlands, there are large and small one-family houses, row houses, and low-rise apartment buildings. However, unlike Kentlands, each building type is concentrated in one area rather than intermixed with others. The seven home builders who erected one-family houses developed large tracts of land because they wanted to profit from the economies of large-scale construction on a single site and the easy and inexpensive marketing from a single location it made possible. In doing so, they were able to sell at prices that were 30 percent lower than those of comparable units at Kentlands. The density of development at Clopper's Mill (3.3 units per acre) is lower than at Kentlands, because a larger portion of the site, 62 acres (20 percent), is set aside for open space that includes tennis courts, a basketball court, six playgrounds, 7000 feet of exercise paths, and two swimming pools.[85]

Elm Street Development, Inc., the developer of Clopper's Mill, placed the property under contract in 1984, but unlike Alfandre, it did not take title to the property all at once. Elm Street took title when each portion of the site was ready for development, thereby minimizing debt service and taxes during the development period. The planning process began in 1987, and it took two more years to obtain all the approvals. Unlike at Kentlands, the approved plan minimized site expenses and synchronized them to accompany land purchases. Development began in 1992. Nevertheless, the recession prevented the developer from balancing expenses with revenues. Unlike at Kentlands, there was no single creditor. No one was in a position to get their money back because each portion of the project was dependent on the other creditors. David Flanagan, the president of Elm Street Development, had built strong relationships with lenders, contractors, landowners, and other project participants. Consequently, he was able to delay payments until conditions had improved. When market conditions began to change, Elm Street began consummating land purchases, selling large parcels to major builders, and satisfying creditors. By moving quickly to exploit improving market conditions, the developer was able to decrease debt service and build momentum for further sales. By 1994, the project was generating positive cash flow. The closing on the purchase of the last portion of the site was completed in 1995, when home builders had demonstrated substantial demand for the residences already completed.

Kentlands and Clopper's Mill are the same size and were developed at the same time in the same market area. At Clopper's Mill, initial diversity was the product of a set-aside (required by Montgomery County) of 15 percent of multifamily units for people with low incomes, unlike at Kentlands, where there was merely the possibility of diversity growing out of variation in house or lot size. By the second decade of the twenty-first century, however, Kentlands sales and rental prices had substantially increased, frequently rising to well over $ 1 million, at a time when it was still easy to find houses in Clopper's Mill selling for less than $500,000.

If the business acumen shown at Clopper's Mill had been applied in the development of Kentlands, this New Urbanist community might have been a resounding financial success. Conversely, if the planning principles devised for Kentlands had been adopted by the developer of Clopper's Mill, property values there might have risen as much as they did at Kentlands. But as long as more conventional planned suburban communities like Clopper's Mill are able to make substantial profits, neotraditional developments like Kentlands, regardless of their aesthetic appeal, will not meet with widespread acceptance by the real estate industry.

Celebration, Florida

Celebration, like Irvine, is the product of circumstances that cannot be repeated. The Walt Disney Company acquired the site in 1964–1965, while it was assembling 28,000 acres to create Disney World outside Orlando, Florida. In addition to the amusement park, Walt Disney himself intended to build a utopian planned community called *Epcot* (Experimental Prototype Community of Tomorrow). After his death, the idea was dropped.[86] A nonresidential Epcot was completed in 1982. This 9600-acre portion of the site remained fallow until the early 1990s. During that time, the population of the Orlando metropolitan area mushroomed, growing by 2.5 times from 1970 to 1990. By 2000 it would climb to 1.6 million, making it the sixteenth largest such area in the United States.

When it became clear that this portion of the property would not be needed for Disney World's expansion, the company decided to use it to provide its 55,000 area employees with a broader range of housing opportunities and to insulate the amusement park from conventional tract development that had reached its borders. Consequently, Disney entered the real estate business as the developer of a new community.

Two entirely new companies were established: the Celebration Company, to design and develop a 4900-acre new town surrounded by a 4700-acre protective natural area, and Celebration Realty, to manage it. In typical Disney fashion, a number of well-known architects were selected to prepare plans for a small portion of the site. Finally, in 1989, Disney decided to proceed with two firms, Cooper, Robertson & Partners and Robert A.M. Stern Architects as master planners and town center architects. They were asked to design a pedestrian-friendly community. The plan, which was approved for development by Osceola County in 1994, permits the development of 8065 dwelling units, 810 hotel rooms, 325 time-shares, 3 million square feet of office space, 2 million square feet of retailing space, 1.7 million square feet for industrial uses, and a 5000-seat performance center. Pursuant to Florida law, two Community Development Districts (for

Celebration, 2006. Downtown Celebration includes a spectacular waterfront promenade. *(Alexander Garvin)*

the residential and commercial sections, respectively) were established to install and provide water, sewer, and other services.

Although undertakings of this magnitude do not develop overnight, when the first residents move in they expect to be able to use community facilities characteristic of a mature town. Celebration is unique in trying to meet these expectations. In 1996, when the first residents moved in, they were able to patronize a downtown retail district, a business park, health facilities, and a school, long before there was really a large enough population to support such amenities.

Downtown Celebration is built along the shore of Lake Evalyn and the one-block-long Market Street that leads to it. There is a town hall, a bank, a post office, a 515-seat cinema, a 115-room hotel, and small three- and four-story office and residential buildings with stores and restaurants at ground level. These buildings, designed by such signature architects as Philip Johnson, Cesar Pelli, Robert Venturi, and Denise Scott-Brown, make it a virtual museum of late-twentieth-century architecture.

The 36-acre education complex, operated by the Osceola County School District, is only a block away, making it very much a part of downtown. It includes a school for 1000 students (kindergarten through the twelfth grade), a "teaching academy" that receives visitors who come to observe Celebration's teaching activities, a full range of sports facilities, and an environmental study center. The teaching academy served as a temporary school in 1996, when the first residents moved to Celebration. The school itself was ready about a year later. Disney promised a "state-of-the-art school" for students

Celebration, 2006. Market Street provides Celebration with a genuine urban center. *(Alexander Garvin)*

Celebration, 2006. The office park at Celebration is integrated with an extensive system of public open spaces. *(Alexander Garvin)*

from all over the county. It subsidized the physical development of the school by donating land, design services, and other support worth $11 million and also provided another $9 million for teacher training and operating enhancements. The school's program was developed with educators from Johns Hopkins, Stetson, Harvard, and other universities. The design, by architect William Rawn, is more ambitious than that of most schools in Florida. Given Disney's promises and money, as well as the presence of so many children from outside Celebration, it is not surprising that the education complex proved to be controversial.

The 60-acre Celebration Health Campus opened at about the same time. It, too, is special. While the Health Campus encompasses a full range of medical, dental, and laboratory services, including a 60-bed acute-care facility, it can be better described as a "wellness center." The medical services are combined with swimming pools, sauna and steam rooms, a day spa, gymnasiums for basketball and volleyball, and a children's sport center.

Public open spaces abound at Celebration. As at Sea Ranch, its designers began by setting aside sacrosanct natural areas—in this case, wetlands and drainage corridors rather than open grasslands. There is an 18-hole golf course, a downtown lake, miles of walking paths and nature trails, and numerous squares, small parks, and playing fields. The area around Lake Evalyn and much of the surrounding greenbelt includes freshwater marshes, forested wetlands, pines, scrub, palmetto prairie, and dozens of species of animals. Disney is committed to environmentally sensitive restoration and management of these areas and of the additional 8500 acres of wilderness preserve it had to acquire in order to obtain the necessary regulatory approvals for Celebration. However, unlike so many other planned new communities, people are encouraged to spend time in the natural areas and enjoy them. They are neither fenced nor gated. Indeed, there are pedestrian walkways leading through many of them and benches for people who wish to linger and enjoy the scenery. Even the signs at the entry points to these facilities welcome visitors and establish the rules for their use.

Houses in Celebration are intended to have the look of those in a small southern town. In part, this is a natural outgrowth of planning each of the residential neighborhoods around distinctive landscaped public spaces and tree-lined streets. These neighborhoods provide comfortable settings for one-family houses, town houses, and apartment buildings. In part, it is the result of requiring home builders to

Celebration, 1998. Residences that recall the timeless traditions of small-town America. *(Alexander Garvin)*

comply with a common set of architectural principles set forth in the *Celebration Pattern Book*, prepared for the Celebration Company by UDA Architects, a Pittsburgh urban design firm. The pattern books that were common during the nineteenth century presented three-dimensional images illustrating an approach to construction that homeowners, designers, and builders accepted as exemplary. Like those predecessors, the *Celebration Pattern Book* presents alternative techniques; unlike them, property owners are required to use the patterns.

In preparing the *Celebration Pattern Book*, its many contributors studied more than 30 small towns and city neighborhoods. From them, UDA abstracted the characteristics of estate lots (90 feet wide by 130 feet deep), village lots (70 feet wide by 130 feet deep), cottage lots (45 feet wide by 130 feet deep), and town house lots (22 feet wide by 130 feet deep or 28 feet wide by 100 feet deep). The section of Celebration that contains each type of lot is carefully marked on a map. The *Pattern Book* then illustrates the yard, driveway, setback, façade, porch, and fence requirements for each type. Based on its study of traditional residential architecture, Celebration's planners decided on six styles: Classical, Victorian, Colonial Revival, Coastal, Mediterranean, and French. The *Pattern Book* also illustrates those styles' requirements for massing, materials, and architectural details.

Celebration's pattern book residences are very much in demand. In 1998, houses sold every other day at an average sales price of $377,000, by far the county's price leader. The six Florida and four national home builders that erected these houses sold them at prices that were "42 percent higher than sales at other central Florida development."[87]

Celebration is halfway to being completed. As of 2010, 7400 residents (82 percent Caucasian and 11 percent Hispanic) lived in its 4100 housing units. Despite its thoughtful planning, charming appearance, and obvious marketability, Celebration cannot be a template for other planned new towns. Few developers can afford the time and cost the Disney Company devoted to its design. Assuming they could, it is unlikely they would be able to afford the front-end costs of its extensive public realm of streets, squares, parks, and recreation areas, or be able to spend more than $20 million subsidizing a school for an entire county long before there would be anywhere near enough new-town children to fill its classrooms, or justify building an entire downtown years before there would be enough customers to support it. Nevertheless, in so many other ways, its planning and design is so good that it will provide ideas that can help other developers to produce better planned new communities.

Ingredients of Success

Riverside, Mariemont, Reston, Kentlands, and every other new town that has run into financial difficulties has suffered because the initial investment in land, infrastructure, and community facilities could not be recouped from property sales. Their developers and designers understood that completion would take years, but failed to adjust their planning so that the pace of development could match cash flow. Chestnut Hill, Levittown, Seaside, and the other planned communities that avoided financial difficulties did so because development and financing could respond to periodic changes in demand.

The roles played by location, design, and entrepreneurship are equally important. Shaker Heights may have been in the path of Cleveland's expanding suburban market, but until the Van Sweringens bought a railroad and extended transit service to their property, it was not easily accessible to that market. Railroad service may have made Riverside accessible, but without Olmsted's design, this prairie location would have been far less alluring. Seaside may have had a wonderful beachfront site and a unique design, but without Robert Davis's skillful marketing, rental rates could never have justified the extremely high prices he charged for development sites.

Market

New towns succeed when people and businesses want to move there. They cannot be attracted just by designating land areas for residential, industrial, commercial, or agricultural uses. The planners who assigned land uses to specific sites at Mariemont or the Greenbelt towns confused their desires with reality, just as the designers of Kentlands failed to understand that their dream of a pedestrian-oriented shopping district was impossible given the number of consumers within walking distance and the far larger number of consumers needed to support those stores. Celebration was able to overcome this market reality because Disney was prepared both to minimize and to subsidize downtown retailing.

Land uses inevitably reflect demand. Lakewood's shopping center was built because the May Company had identified a large market area without a major department store. When air-conditioned malls became popular, the Lakewood Center was reconfigured and air-conditioned so that it would continue to attract shoppers. Reston's long-promised town center became a reality when there were enough customers in the Virginia suburbs of Washington, D.C. However, as retail patterns shifted and large chain stores began to dominate the market, its village shopping districts had trouble competing with more up-to-date regional centers.

Market income is as important as market size. Levittown would not have been successful if its customers had not been able to afford its low FHA mortgage rates. The real trick, however, is attracting customers to an affordable planned community. Burton Green built a hotel to introduce people to Beverly Hills. The Levitts attracted customers by providing appliances and landscaping not offered by the competition. Emery Childs used Olmsted's design to distinguish Riverside from other suburban communities then being built on the prairie outside Chicago, just as Les Wexner depended on the Virginia Georgian style of architecture to differentiate New Albany from competing tract development in suburban Ohio.

Location

Palos Verdes Estates and Sea Ranch have the benefit of stunning oceanfront locations. The Woodlands avoids merging into suburban Houston by retaining critical portions of its forested landscape. At Reston, Columbia, Celebration, and other recent new communities, semi-artificial lakes perform a similar role.

Other planned communities with less attractive locations have to compensate for this disadvantage. Olmsted described the site of Riverside as "low, flat, miry, and forelorn." His curvilinear, tree-lined roadways with houses set back on open lawns enlivened this uninteresting site. It became a formula that functioned as successfully in Levittown and countless other communities. Kentlands achieves a similar result by varying lot size. Its variety of house designs is the product of trying to fit the different characteristics of each lot. The *Celebration Pattern Book* produces a distinctive appearance on uninteresting, flat sites by differentiating among estate, village, cottage, and town house lots within each district.

At locations with an inflected topography, like Palos Verdes Estates, the Olmsted firm used a similar curvilinear formula to exploit site characteristics. Roads were placed along ledges, on the crests of hills, or in valleys as a way to ensure good drainage and easy travel. The Olmsted firm determined the width of roadways based on their function. Streets had to be wide enough to carry carriage, delivery wagon, and, later, motor vehicle traffic. These vehicles did not have to mount steep slopes, because the firm avoided locating streets in those areas of a site. The curving roadways the firm designed (whether fitted to the topography or placed in a manner to create interesting views) guaranteed that very few lots were of the same size and shape. The Olmsted firm avoided monotonous, repetitive siting and provided for a variety of house and household sizes. These streets are still convenient today when they are used for FedEx deliveries, recyclable garbage pickups, school bus routes, and automobile commutes.

Proximity is even more important than site characteristics. Lake Forest, Riverside, and Chestnut Hill were entirely dependent on the railroad to bring residents downtown; Mariemont, Palos Verdes, and Irvine, on the other hand, depend on automobile access. Reston almost failed because customers were unable to use the highway to Dulles Airport. Seaside is unimaginable without the airplane.

Design

The lots, blocks, and streets of Riverside are large in order to create what Olmsted described as a feeling of seclusion that is "not far removed from the life of the community." These ample dimensions allowed Childs to market Olmsted's "sense of enlarged freedom." Similarly, the lots, blocks, and streets of Seaside are small because Duany and Plater-Zyberk were intent on restoring the dominance of residents over their cars. These small dimensions make Seaside so distinctive that Davis can sell land at prices that greatly exceed those of nearby locations along the Gulf of Mexico.

Tugwell wanted to put "houses and land and people together in such a way that . . . our economic and social structure will be permanently strengthened."[88] In most suburbs, churches and synagogues perform a part of that function, as do nearby ball fields, jogging trails, and swimming pools. Mariemont, Greenhills, Reston, Columbia, Irvine, and most other planned communities allocate large amounts of territory for open space and recreation that mix "houses and land and people" in just the way Tugwell intended. At New Albany, small sitting areas and landscaped squares and ovals provide more intimate settings for social interaction.

Shopping is another activity that can bring people together. Corner stores depend on high-density living patterns that are difficult to achieve in most new towns. Thus, the arrangement of retail outlets plays a critical part in determining the character of life in planned communities. At Lakewood, where a large amount of retail activity is relegated to major arterials, what social interaction there is takes place in its vast shopping mall. At Columbia and Reston, interaction is repackaged in the form of easily accessible neighborhood and village retail centers that include libraries, day care centers, and other community facilities.

Landscaping can change the character of any planned community. As Olmsted predicted, roadside trees and houses set back on open lawns soften a Spartan design, even one as dull as Levittown. Leaving thousands of acres of meadowland untouched allows Sea Ranch to retain a feeling of splendid desolation. People drive three hours just to enjoy this landscape.

People who live in Palos Verdes Estates cherish the Hispano-Mediterranean buildings that emerged from its design regulations. They also benefit from the way its buildings fit into the landscape. Many houses are built on lots that are on the downside of a slope. From the street, they appear to be modest, one-story structures. Many are really multistory houses that open out to the view below. Building regulations and lot patterns produce similarly distinctive, but very different, buildings at Sea Ranch, Seaside, and Celebration. Those who prefer natural materials, shed roofs, and large expanses of glass opening onto dramatic vistas will choose houses at Sea Ranch. Those who prefer front porches, painted clapboard siding, and small, cozy rooms build second homes in Seaside. Celebration offers a composite, if nostalgic, vision of the Old South.

Financing

The developers of Irvine and Seaside had the benefit of inherited land. They are exceptions. Most new-town sites are purchased at considerable expense. Consequently, they begin with a commitment to pay a return on any equity investment and debt service on any borrowing. As more and more money is spent for planning, infrastructure, community facilities, and marketing, that commitment increases. The Chevy Chase Bank foreclosed on Kentlands when debt service on those costs grew to a magnitude that could not be justified by land sales.

This financing requirement forces new-town developers to devise ways of reducing debt service, especially during the early years, when development expenses are high. Connecticut General and Teachers Life Insurance, the institutions that financed initial development at Columbia, allowed the Rouse Company 10 years of accrued debt service because they correctly assumed that during the initial years of development, there would be little or no cash flow from sales. Nonetheless, neither they nor the Rouse Company expected a serious downturn in the economy to further delay payments on their loans. Clopper's Mill was able to ride out the market fluctuations that nearby Kentlands could not because the site was an agglomeration of different properties that did not have to be acquired all at once. Consequently, site work and infrastructure investment could be delayed until market conditions had improved and demand had reached a level that justified construction.

The developers of Lakewood avoided extensive debt service commitments by preselling model homes and using this money on a pay-as-you-go basis. They were able to avoid large outlays because they used a simple lot and block pattern that could be extended or withheld in response to market demand. The complex designs for Reston and other more recent planned communities preclude this form of minimalist financing.

Residents also face financing problems. They need long-term mortgages with which to finance home purchases. The Levitts solved this problem by preprocessing FHA and VA mortgages on their standardized houses. But, like thousands of other developers, they had to accept FHA site-planning recommendations. Home builders in Irvine use a similar procedure when they market model homes. But, because the Irvine Company sells land with approved subdivision plans to different home builders, there is a greater variety of house designs and price levels.

Time

The time spent passing through or finding one's destination in a planned new community can be critical to its success.

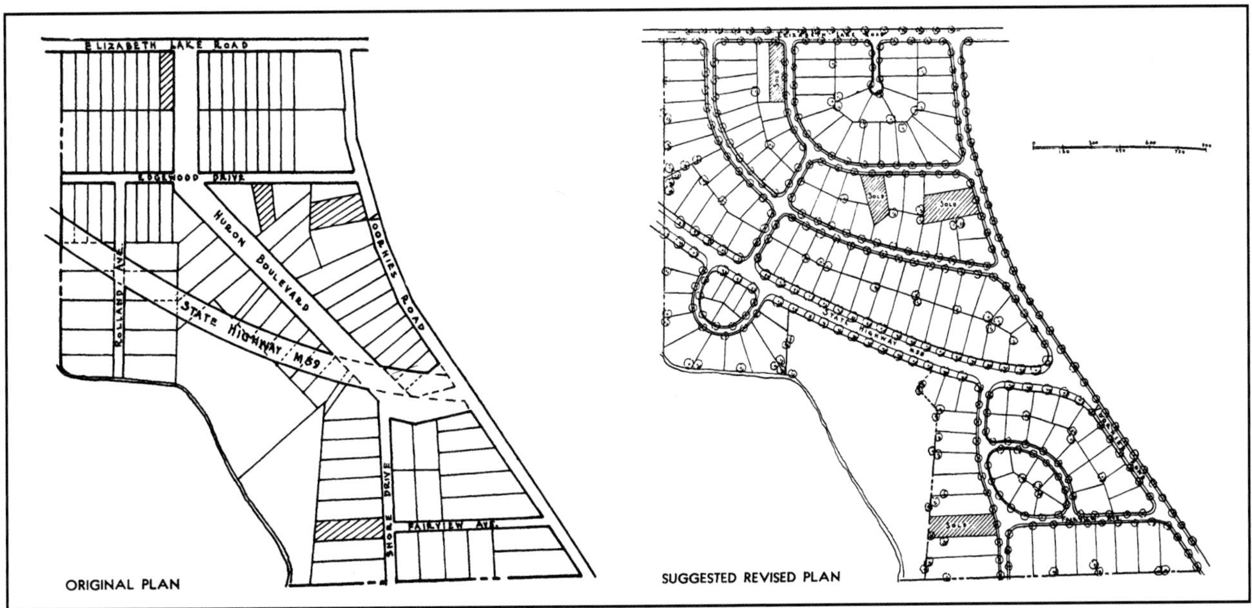

FHA, 1939. Diagram of the undesirable "original plan" and the preferred curvilinear street pattern recommended for subdivisions receiving FHA mortgage insurance. By providing mortgages to developments that followed the government's preferred plan, institutional lenders changed the face of America's residential suburbs. *(From Federal Housing Administration,* Planning Profitable Neighborhoods, *Washington, D.C., 1938)*

Nobody driving along the major state roads that pass through Palos Verdes Estates, Sea Ranch, or Seaside can fail to notice that they are distinctive. In each case, that drive is part of the marketing strategy. Taking the roads through Chestnut Hill or Mariemont, on the other hand, gives little indication of anything special.

The range of experiences encountered when driving through New Albany is considerably more varied than that encountered in Lakewood or Levittown. Each village is different from the next, and they, in turn, are different from the open spaces of the golf course or the village center or the business campus. The streets of Levittown, Lakewood, and New Albany are all lined with similar houses occupied by households that are similar in size and income. That, however, is their only similarity. There is nothing but houses in the residential sections of Levittown and Lakewood. In New Albany, one encounters formal, oval open spaces, secluded sitting areas, small parks with scenic vistas, and creek beds that meander through a naturalistic landscape, following long-established drainage patterns.

Suburban subdivisions rarely provide for their residents' needs and activities 24 hours a day, 7 days a week. New towns are supposed to. Whether they do depends on their size and density. Irvine is large enough to include an airport, three industrial and business parks, more office space than any city in California except Los Angeles and San Francisco, concert halls, theaters, shopping malls, hospitals, and so forth. The 80 acres of Seaside will never include anything like that, even when all 326 house lots have been filled in. Similarly, there are enough customers around Rodeo Drive in Beverly Hills to support a range of retail stores that is the envy of much of the world. Yet, even after three decades, Sea Ranch does not have a supermarket and is unlikely ever to support more than convenience retailing. Thus, Seaside and Sea Ranch cannot provide their residents with the range of living experiences one expects in completely successful new communities.

New towns usually take decades to complete. Lakewood and Levittown are among the few exceptions. Even in these cases, it has taken them decades to develop the patina of a complete community. Kentlands solves this problem by aggressively mixing different lot sizes and low-density residential structures. The most serious time problem, however, is presented by the changes in market conditions that accompany business cycles. Palos Verdes Estates survived only because Frank Vanderlip had deep pockets and provided the patient capital to see the project through. Riverside, Reston, and Columbia nearly failed because their developers could not time the pace of cash outflows to changes in pace of sales.

Entrepreneurship

New-town developers, more than most real estate entrepreneurs, must have real vision. Burton Green believed Beverly Hills could replace hundreds of acres of lima beans. Frank Vanderlip imagined his own version of the Amalfi coast built on the grasslands, farms, and cliffs of the Palos Verdes Peninsula. Robert Davis envisioned an upscale community of architecturally distinguished second homes on 80 nondescript acres along the Gulf of Mexico. Each of these developers perceived a market opportunity that others had missed. These sites would have been developed anyway, but they would have been indistinguishable from surrounding communities.

These developers were willing to risk their capital. Green built a hotel when others thought he was throwing good

Seaside, 1982. Street paving and other infrastructue improvements at Seaside were installed only after there were enough lot sales to pay for those improvements. *(Alexander Garvin)*

money after bad. Every time Palos Verdes ran into financial trouble, Vanderlip put up the cash to save his vision. Davis was ready to risk his land on a venture that had never been tried before.

Coordination also pays a more important role than in other forms of real estate development. This is particularly evident in the assembly-line production techniques used by the Levitts. But it was no less important a century earlier when Childs had to coordinate the financing, design, and marketing of Riverside. Over a period of three years, he oversaw the construction of 40 miles of carriage road, 80 miles of pedestrian walkway, a 735-foot-deep artesian well, 108-foot-high water tower, as well as the planting of 32,000 deciduous trees, 7000 evergreens, and 47,000 shrubs.

Entrepreneurship requires more than identifying an opportunity others have missed, devising a way of exploiting that opportunity, and coordinating all the participants needed to bring a project to fruition. Developers have to be flexible enough to deal effectively with uncertainty and risk. Davis understood this when he minimized investment in Seaside's infrastructure until he had generated the necessary level of market demand. When Clopper's Mill faced market uncertainty, David Flanagan was able to use his long-standing business relationships to reassure his creditors and maintain their participation until the real estate market revived. Joseph Alfandre did not have those skills. Consequently, the Chevy Chase Bank foreclosed and replaced him as developer of Kentlands.

The New Town as a Planning Strategy

Writing nearly a century ago, Ebenezer Howard insisted that well-planned new towns would demonstrate how to better provide:

> *opportunities of social intercourse . . . beauties of nature . . . higher wages . . . reduced rents . . . abundant opportunities for employment . . . admirable sanitary conditions . . . beautiful homes and gardens . . . and cooperation . . . by a happy people.*[89]

Howard and other new-town advocates have oversold their product. Planned new communities, even those that make creative use of the six ingredients of success, do not guarantee a better physical environment, or a social utopia, or even a more economical daily life.

The Woodlands, Sea Ranch, New Albany, and many other planned new towns do preserve valuable open-space resources that would otherwise have been lost. But preplanning public open space in new towns is not the only way to provide communities with good recreation facilities. Many cities still purchase land for public parks. Minneapolis and Boulder have park systems that rival those of any planned new towns. Other communities enact land use regulations that require private developers to provide common open space.

Many new towns have buried utilities and efficient infrastructure layouts. But, as with public open space, there are alternatives. The municipality can install utilities prior to urbanization. Furthermore, subdivision regulations can be just as effective in controlling utility installation by individual home builders.

Similarly, new towns are not the only communities that make efficient use of public transportation. Los Angeles during the turn of the twentieth century and Atlanta in the 1970s and 1980s provided mass transit service to areas that had not yet fully developed. In both cities, this resulted in more effective land use patterns around the newly opened subway stations.[90]

As the Title VII new towns so graphically illustrate, carefully planned new communities may be more costly than opportunistic development of scattered sites.[91] Ironically, new-town proponents were so aware that huge amounts of up-front, patient capital would be necessary that they used this as a justification for enacting the Urban Growth and New Communities Development Act of 1970.

Greenhills, Celebration, and many other planned communities are visibly superior to their competition. Riverside, Palos Verdes Estates, Sea Ranch, Seaside, and New Albany include oases of rare beauty in a desert of mediocrity. However, planning a large piece of land as an entire new-town-in-the-country may not produce better results than dividing it into smaller sites that are planned and built by different designers and developers.

It is understandable that architects, planners, social critics, developers, and government officials all want to create their own version of utopia and sell it to a ready public as the best alternative to our urban-suburban mess. In 1935, when Tugwell began building the New Deal new towns, he thought he had found that alternative. As he explained, "My idea is to go just outside centers of population, pick up cheap land, build a whole community and entice people into it. Then go back into the cities and tear down slums and make parks of them."[92]

Building such miniature utopias has nothing to do with fixing cities. People like Jane Jacobs even contend that new towns siphon resources and talent away from cities. She therefore concludes that they are a recipe for "undermining their economies and killing them."[93] Our great cities may be facing a myriad of problems, but the nation's many new-towns-in-the-country are not responsible. For two centuries, while developers were creating planned new communities around New York, Los Angeles, Chicago, Philadelphia, and every other large American city, many of these cities have remained very much alive. If we are to fix our cities, we should avoid this debate, let new-town developers go about what is a very difficult business, and concentrate on the many devices that will improve our troubled cities.

Notes

1. Dolores Hayden, *Seven American Utopias,* MIT Press, Cambridge, MA, 1976; Charles Nordhoff, *The Communistic Societies of the United States,* Schocken Books, New York, 1965.
2. Urban Land Institute, *Planned Communities, New Towns, and Resort Communities in the United States and Canada,* ULI, Washington, DC, 1992.
3. F. J. Osborn, preface to Ebenezer Howard, *Garden Cities of Tomorrow,* MIT Press, Cambridge, MA, 1965, pp. 9–28.
4. Howard, op. cit., pp. 45–46.
5. Stanley Buder, *Visionaries and Planners: The Garden City Movement and the Modern Community,* Oxford University Press, New York, 1990.
6. Robert Fishman, *Bourgeois Utopias: The Rise and Fall of Suburbia,* Basic Books, New York, 1987, pp. 18–72.
7. Ebenezer Howard, *Garden Cities of Tomorrow,* MIT Press, Cambridge, MA, 1965, p. 48.
8. Riverside Improvement Company, *Riverside in 1871 with Description of Its Improvements Together with Some Engravings of Views and Buildings,* originally printed by D. & C. H. Blakely, Chicago, 1871, and reprinted by the Frederick Law Olmsted Society of Riverside in 1981, p. 6.
9. Historical and statistical information about Lake Forest is derived from Michael H. Ebner, *Creating Chicago's North Shore: A Suburban History,* University of Chicago Press, Chicago, 1988, especially pp. 27–35, 68–77, 195–209.
10. Hotchkiss's first name and exact identity have yet to be discovered. See Ebner, op. cit., pp. 27–28.
11. Journalist James B. Runnion writing in 1860, quoted by Ebner, op. cit., p. 30.
12. The town was later extended to the west side of the railroad.
13. Calvert Vaux was in Europe during 1868 when Olmsted prepared the plan of Riverside.
14. Historical and statistical information about Riverside is derived from Olmsted, Vaux and Company, "Preliminary Report upon the Proposed Suburban Village at Riverside, Near Chicago," Sutton, Browne and Company, New York, 1868, reproduced by David Schuyler and Jane Turner Censer (editors), *The Papers of Frederick Law Olmsted,* Johns Hopkins University Press, Baltimore, 1992, pp. 273–289; Riverside Improvement Company, op. cit.; Walter L. Creese, *The Crowning of the American Landscape: Eight Great Spaces and Their Buildings,* Princeton University Press, Princeton, NJ, 1985, pp. 221–240; and Ann Durkin Keating, *Building Chicago: Suburban Developers and the Creation of a Divided Metropolis,* Ohio State University Press, Columbus, OH, 1988, pp. 61–63, 73–74, 86–87, 174, 184.
15. Riverside Improvement Company, op. cit., pp. 17–18.
16. Olmsted, Vaux and Company, op. cit., p. 286.
17. Howard,*Garden Cities of Tomorrow,* pp. 54, 67, 76–79.
18. Ebenezer Howard, "The Relation of the Ideal to the Practical," *The Garden City,* London, February 1905, pp. 15–16.
19. Howard, *Garden Cities of Tomorrow,* op. cit., p. 51.
20. Ibid., p. 90.
21. Frank Jackson, *Sir Raymond Unwin: Architect, Planner and Visionary,* A. Zwemmer Ltd., London, 1985.
22. C. B. Purdom, *The Building of Satellite Towns,* J. M. Dent & Sons, London, 1949, pp. 110–118.
23. Ray Thomas and Peter Cresswell, *The New Town Idea,* Open University Press, Milton Keynes, 1973.
24. Historical material on Chestnut Hill is derived from Willard S. Detweiler, *Chestnut Hill: An Architectural History,* Chestnut Hill Historical Society, Philadelphia, 1969, and David R. Contosta, *Suburb in the City: Chestnut Hill Philadelphia* 1850–1990, Ohio State University Press, Columbus, OH, 1992.
25. Philadelphia City Planning Commission, *Northwest Philadelphia District Plan,* Philadelphia, 1966.
26. Historical material on Beverly Hills is derived from Fred E. Basten, *Beverly Hills: Portrait of a Fabled City,* Douglas-West Publishers, Los Angeles, 1975; Karl Stull (editor), *Beverly Hills: An Illustrated History,* Windsor Publications, Los Angeles, 1988; Pierce E. Benedict (editor), *History of Beverly Hills,* Cawston and Meier, Beverly Hills, 1934; and brochures published by the Beverly Hills Chamber of Commerce.
27. Quoted in Basten, op. cit., p. 27.
28. E. Y. Harburg (lyrics) and Jerome Kern (music), "Californ-i-ay," T. B. Harms Company, 1944.
29. Charles Moore, Gerald Allen, and Donlyn Lyndon, *The Place of Houses,* Holt, Rinehart & Winston, New York, 1974, pp. 31–48; "Ecological Architecture: Planning the Organic Environment," *Progressive Architecture,* Reinhold Publishing, Cleveland, May 1966, pp. 121–137; Richard Babcock and Charles Siemon, *The Zoning Game Revisited,* Oelgeschlager, Gunn & Hain, Boston, 1985, pp. 235–254; Donald Canty, "Sea Ranch," *Progressive Architecture,* Reinhold Publishing, Cleveland, May 1993, pp. 86–91; Lawrence Halprin, "Revisiting the Idea," *Progressive Architecture,* Reinhold Publishing, Cleveland, May 1993, pp. 92–93; Donlyn Lyndon, "Lyndon's Assessment," *Progressive Architecture,* Reinhold Publishing, Cleveland, May 1993, pp. 93–95.
30. Halprin, op. cit., p. 92.
31. Ibid., p. 93.
32. Historical and statistical information on Seaside is derived from *Seaside,* ULI Project Reference File, vol. 16, no. 16, Urban Land Institute, Washington, DC, October–December 1986; Beth Dunlop, "Coming of Age," *Architectural Record,* McGraw-Hill, New York, July 1989, pp. 96–103; Alex Krieger with William Lennertz, *Andres Duany and Elizabeth Plater-Zyberk: Towns and Townmaking Principles,* Harvard Graduate School of Design, Cambridge, MA, 1991; promotional brochures printed by the Seaside Community Development Corporation; and lectures by and correspondence with Andrés Duany.
33. Seaside Development Corporation.
34. Riverside Improvement Company, op. cit., p. 15.
35. Historical and statistical material on Shaker Heights is derived from Ian S. Haberman, *The Van Sweringens of Cleveland,* Western Reserve Historical Society, Cleveland, OH, 1979; Eugene Rachlis and John E. Marqusee, *The Landlords,* Random House, New York, 1963, pp. 60–86; Eric Johannesen, *Cleveland Architecture 1876–1976,* Western Reserve Historical Society, Cleveland, OH, 1979, pp. 57–59, 131–133, 167–183; and Mark A. Dettlebach, "Shaker: A Suburb That Hit the Heights," unpublished, 1983.
36. Dettlebach, op. cit., pp. 23–28; and U.S. Bureau of the Census.
37. Historical and statistical information on Palos Verdes Estates is derived from Deland Morgan, *The Palos Verdes Story,* Review Publications, Palos Verdes Estates, CA, 1982; Augusta Fink, *Time and the Terraced Land,* Howell North Books, Berkeley, CA, 1966; Arthur C. Comey and Max S. Wehrly, *Urban Planning and Land Policies,* Urbanism Committee to the National Resource Committee, U.S. Government Printing Office, Washington, DC, 1939, pp. 85–89; and John Bacon, "Palos Verdes Estates: A Formative History," unpublished, 1982.
38. Frank A. Vanderlip, *From Farm Boy to Financier,* New York, 1935, quoted in Morgan, op. cit., p. 8.
39. Morgan, op. cit., p. 143.
40. Not only was Frederick Law Olmsted Jr. his firm's partner in charge of Palos Verdes Estates, he was also one of the first people to build himself a house there.
41. Palos Verdes Homes Association, *Protective Restrictions,* Palos Verdes, 1923, p. 1.
42. Among the first members of the art jury were Frederick Law Olmsted Jr., Charles Cheney, and Myron Hunt, a prominent Los Angeles architect.
43. Palos Verdes Homes Association, *Protective Restrictions,* op. cit., p. 34.

44. Ibid., p. 4.
45. U.S. Census Bureau.
46. John L. Hancock, "John Nolen," in *American Landscape Architecture,* William H. Tishler (editor), Preservation Press, Washington, DC, 1989, pp. 70–73; "John Nolen: The Background of a Pioneer Planner," in Donald A. Krueckeberg (editor), *The American Planner: Biographies and Recollections,* Metheun, New York, 1983, pp. 37–57.
47. *Cincinnati Inquirer,* April 23, 1922.
48. Historical and statistical material on Mariemont is derived from Comey and Wehrly, op. cit., pp. 92–97, and Robert B. Fairbanks, *Making Better Citizens: Housing Reform and Community Development in Cincinnati, 1890–1960,* University of Illinois Press, Urbana, IL, 1988, pp. 49–55.
49. The Mariemont Company, cited by Fairbanks, op. cit., p. 54.
50. Historical and statistical material on Levitt & Sons and its new towns in Long Island, Pennsylvania, and New Jersey is derived from Rachlis and Marqusee, op. cit., pp. 228–256; Herbert Gans, *The Levittowners,* Pantheon Books, New York, 1967; and John T. McQuiston, "If you're thinking of living in Levittown," *New York Times,* November 27, 1983.
51. A few years after Abraham Levitt died, William Levitt bought out his brother and went on to build a multi-million-dollar business.
52. *Time,* July 3, 1950, p. 67.
53. Historical and statistical material on Lakewood is derived from Lakewood Living Corporation, *Lakewood Living: 35th Anniversary Edition,* Lakewood, CA, 1989, and *Lakewood Service Guide,* Lakewood, CA, 1989; Creative Network, *Lakewood 35th Anniversary Magazine,* Santa Ana Heights, CA, 1989; and Bensel Smythe, "Lakewood Park, Lakewood Center, 'Twin Miracles' in L.A. Metropolitan Area," *Los Angeles Daily News,* November 7, 1951.
54. *Time,* April 17, 1950, p. 99.
55. Most, but not all, residents of Levittown and Lakewood are contented with their communities. What dissatisfaction there is comes primarily from teenage, elderly, minority, and nonconformist residents. See Gans, op. cit., pp. 153–304.
56. Historical and statistical material on Reston is derived from Simon Enterprises, "Reston Virginia," May 1962, unpublished; James Bailey (editor), *New Towns in America: The Design and Development Process,* John Wiley & Sons, New York, 1973; "A Plea for Planned Communities," *Architectural Record,* McGraw-Hill, New York, December 1973; Ann Mariano, "Reston's Town Center Soon to Be a Reality," *Washington Post,* June 30, 1990; Tom Grubisich and Peter McCandless, *Reston: The First Twenty Years,* Reston Publishing, Reston, VA, 1985; Joshua A. Olsen, *Planned Communities: Are They Worth It?,* unpublished, New Haven, CT, 1999; Marc Hochstein, "Planned Communities at Middle Age," *Grid,* New York, June 2001, pp. 66–71; Alan Ward (editor), *Reston Town Center: A Downtown for the 21st Century,* Academy Press, Washington DC, 2006; Charles A. Veatch (with Claudia Thomspon-Deahl), *The Nature of Reston,* Charles A Veatch Company, Reston, VA, 1999; and Susan E. Jones, *Reston, Virginia: A New Town,* The Reston Historic Trust, Reston, VA, 2010.
57. Robert E. Simon, quoted in Bailey, op. cit., p. 14.
58. Simon Enterprises, op. cit., pp. 6–7.
59. Ibid., p. 3.
60. Historical and statistical material on Columbia is derived from James W. Rouse, "The City of Columbia, Maryland," in H. Wentworth Eldredge (editor), *Taming Megalopolis,* vol. 2, Anchor Books, Garden City, NY, 1967, p. 839; Morton Hoppenfeld, "A Sketch of the Planning-Building Process from Columbia, Maryland," *Journal of the American Institute of Planners,* vol. 33, no. 5 (November 1967), pp. 398–409; Kathy Sylvester, "Columbia Birthday," Associated Press, July 21, 1977; Robert M. Andrews, "Developers Battle Economic Downturns, Lack of Interest; Planned Towns Face Hard Road to Success," *Los Angeles Times,* August 31, 1986; Olsen, op. cit.; Brett Rubin, *Columbia, Maryland: Learning from the Past to Develop a Successful Retail Center,* unpublished, New Haven, CT, 2000; www.therousecompany.com; Hochstein, op cit.; and Robert Tennenbaum (editor), *Creating a New City: Columbia, Maryland,* Partners in Community Building and Perry Publishing, Columbia, MD, 1996.
61. It quickly become evident that not every village was able to support a separate library and interfaith center.
62. Historical and statistical material on Irvine is derived from Martin J. Schiesl, "Designing the Model Community: The Irvine Company and Suburban Development," in Rob Kling, Spencer Olin, and Mark Poster (editors), *Postsuburban California,* University of California Press, Berkeley, CA, 1991, pp. 55–91; Cheryl G. Cummins, *Los Angeles Metropolitan Area . . . Today 1987,* Urban Land Institute, Washington, DC, 1987, pp. 306–333; "A Plea for Planned Communities," *Architectural Record,* McGraw-Hill, New York, December 1973, pp. 112–117; G. Christian Hill, "Real-Estate Heiress Fights to Get Control of California Empire, "*Wall Street Journal,* March 28, 1977; Paul Goldberger, "Coast 'New Town' Has Old Familiar Look," *New York Times,* March 29, 1978; Pamela G. Hollie, "Social Patterns Are Forming as Irvine Ranch Expands," *New York Times,* May 4, 1980; Gladwin Hill, "Big but Not Bold: Irvine Today," *Planning,* American Planning Association, Chicago, February 1986; U.S. Census Bureau, *Census 2010;* City of Irvine, City Manager's Office; http://www.cityofirvine.org; and City of Irvine, California, *Comprehensive Annual Financial Report,* June 30,2011.
63. The original deal was contested by Irvine heiress Joan Irvine Smith, which resulted in a sale to a group that included Alfred Taubman, Henry Ford II, Donald Bren, and Joan Irvine Smith. In 1983, the rest of the group was bought out by Donald Bren in a sale that was also contested by Joan Irvine Smith. See Hill, op. cit.
64. City of Irvine, California, *Comprehensive Annual Financial Report,* June 30, 2011.
65. U.S. Department of Commerce, Bureau of the Census.
66. The federal government also has built residential communities in conjunction with military installations and nuclear facilities. Among the more famous are Oak Ridge, Tennessee, Hanford, Washington, and Los Alamos, New Mexico.
67. Select Committee of the House Committee on Agriculture, "Hearings on the Farm Security Administration," 78th Congress, First Session, 1943–1944, p. 966.
68. Housing and Urban Development Act of 1970, Title VII—Urban Growth and New Community Development, Sec. 701(b).
69. Historical and statistical material on the World War I villages is derived from William J. O'Toole, "A Prototype of Public Housing Policy: The USHC," *Journal of the American Institute of Planners,* vol. 34, no. 3 (May 1968), pp. 140–152, and Robert M. Fisher, "Origins of Federally Aided Public Housing," in *Federal Housing Policy and Programs Past and Present,* J. Paul Mitchell (editor), Center for Urban Policy Research, Rutgers University Press, New Brunswick, NJ, 1985, pp. 231–235.
70. Historical and statistical material on Yorkship Village is derived from Comey and Wehrly, op. cit., pp. 64–67, 134.
71. Historical and statistical material on the Greenbelt towns is derived from Paul K. Conkin, *Tomorrow a New World: The New Deal Community Program,* Cornell University Press, Ithaca, NY, 1959; David Myrha, "Rexford Guy Tugwell: Initiator of America's Greenbelt New Towns, 1935–1936," in Krueckeberg (editor), op. cit., pp. 225–249; Stein, op. cit., pp. 118–187; K. C. Parsons, "Clarence Stein and the Greenbelt Towns," *Journal of the American Planning Association,* Chicago, vol. 56, no. 2 (Spring 1990), pp. 161–183; and Lloyd W. Bookout, "Greenbelt, Maryland, A 'New' Town Turns 50," *Urban Land,* Urban Land Institute, Washington, DC, August 1987, pp. 7–11.
72. U.S. Senate, Resettlement Administration Program, Document 213, 74th Congress, Second Session, 1936, p. 7.
73. Stein, op. cit., p. 122.
74. *Franklin Township v. Tugwell,* 85F. 2d 208 (D.C. Cir. 1936).
75. Rexford Tugwell quoted in Myrha, op. cit., p. 245.
76. This legislation is also referred to as the New Communities Act of 1968 and the Urban Growth and New Communities Development Act of 1970.
77. Housing and Community Development Act of 1968, Title IV, Sec. 403.
78. Historical and statistical material on HUD's New Communities program is derived from *Planning New Towns: National Reports of the*

U.S. and the U.S.S.R., U.S. Department of Housing and Urban Development, Office of International Affairs, Washington, DC, 1981; Bailey (editor), op. cit.; and "A Plea for Planned Communities . . . New Towns in America—with Lessons from Europe," *Architectural Record,* McGraw-Hill, New York, December 1973, pp. 85–144.

79. Samuel C. Jackson, "Toward an Urban Growth Policy," in Bailey (editor), op. cit., pp. 138–140.
80. Historical and statistical material on The Woodlands is derived from George T. Morgan Jr. and John O. King, *The Woodlands: New Community Development 1964–1983,* Texas A&M University Press, College Station, TX, 1987, and http://thewoodlands2.reachlocal.com.
81. Historical and statistical material on New Albany is derived from unpublished materials supplied by the New Albany Company, the Georgetown Company, and Cooper Robertson & Partners, architects, as well as interviews with executives of those firms; www.newalbanycompany.com; and interviews with Edgar Lampert, November 19, 2012, and March 7, 2013.
82. Historical and statistical material on Kentlands is derived from *Kentlands,* ULI Project Reference File, Urban Land Institute, Washington, DC, 1994; Olsen, op. cit.; and Rubin, op. cit.
83. Congress for the New Urbanism, *Charter of the New Urbanism,* McGraw-Hill, New York, 2000.
84. Ibid.
85. Historical and statistical material on Clopper's Mill is derived from David Flanagan, *Clopper's Mill,* ULI Case Study (unpublished), Urban Land Institute, Washington DC, 1998, and Olsen, op. cit.
86. Historical and statistical material on Celebration is derived from Andrew Ross, *The Celebration Chronicles,* Ballantine Publishing Group, New York, 1999; UDA Architects, *Celebration Pattern Book,* Walt Disney Company, Pittsburgh, PA, 1997; Beth Dunlop, *Building a Dream,* Harry N. Abrams, New York, 1996; Michael Pollan, "Town Building Is No Mickey Mouse Operation," *New York Times Magazine,* December 14, 1997; and Celebration Company, *Celebration Fact Sheets,* Celebration, FL, 1998.
87. Lyn Riddle: "At Celebration, Some Reasons to Celebrate," *New York Times,* March 7, 1999, sec. 11, p. 5.
88. Rexford Tugwell, quoted in Myrha, op. cit., p. 227.
89. Howard, Garden Cities of Tomorrow, pp. 48–49.
90. Edwin H. Spengler, *Land Values in New York in Relation to Transit Facilities* (first published in 1930), AMS Press, New York, 1968; Leon S. Eplan, "Transit and Development in Atlanta," in *New Urban Rail Transit: How Can Its Development and Growth-Shaping Potential Be Realized?,* Subcommittee on the City of the Committee on Banking, Finance, and Urban Affairs, House of Representatives, 96th Congress, First Session, U.S. Government Printing Office, Washington, DC, 1980, pp. 129–144.
91. Richard Peiser, "Does It Pay to Plan Suburban Growth," *Journal of the American Planning Association,* American Planning Association, Chicago, Autumn 1984, pp. 419–433.
92. Rexford Tugwell (diary, March 3, 1935), quoted in Myrha, op. cit., p. 236.
93. Jane Jacobs, *The Death and Life of Great American Cities,* Random House, New York, 1961, p. 21.

17

Land Use Regulation

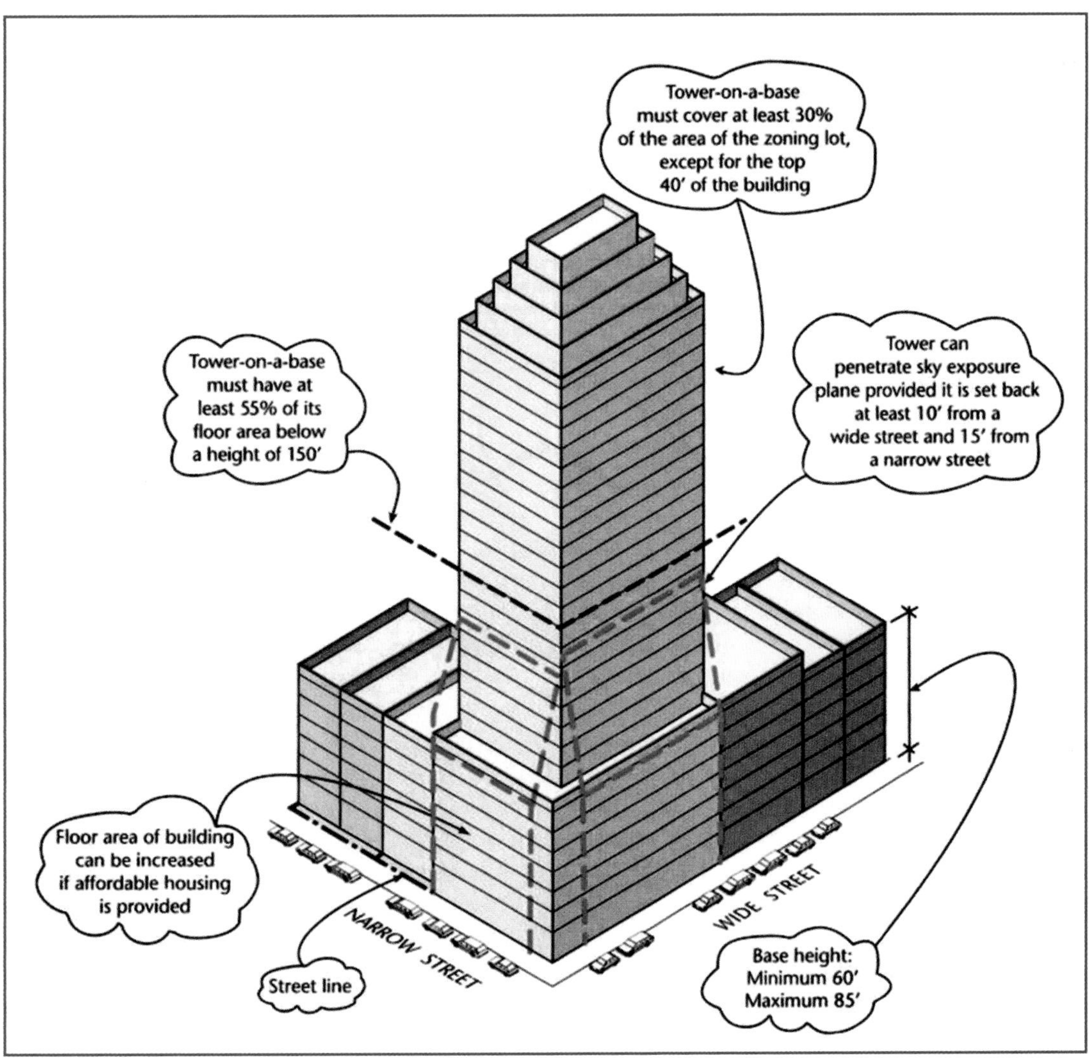

Land use regulation *(Courtesy of New York City Department of City Planning)*

For centuries, municipal governments have specified what property owners may and may not do with their land. The earliest American land use regulations were scattered ordinances preventing property owners from harming other citizens or damaging their property. In 1672, for example, Boston enacted legislation that required structures to be built of fireproof materials such as brick and stone. Twenty years later, it restricted the location of slaughterhouses, stills, and tallow manufacturers. By the end of the nineteenth century, every major city had similar land use statutes.[1]

A second group of regulations was modeled on the nation's first comprehensive zoning resolution, enacted by New York City in 1916. These regulations combined into a single ordinance a series of restrictions that concerned land use, height, bulk, siting, and (later) density requirements.[2]

During the second half of the twentieth century, a third set of laws provided inducements for property owners to establish plazas, arcades, atriums, covered parking, day care facilities, "affordable housing," and other facilities that were thought to be in the community's interest. This approach shifted the cost and responsibility for providing public space and facilities from government agencies to individual property owners.

The most recent group of laws is intended to manage growth and prevent inadvertent destruction of scenic vistas, natural habitats, unpolluted waterways, and other irreplaceable resources. These laws come in two varieties. One attempts to predetermine the location and sequence of development and avoid any unsuitable impact on the natural or human-made environment. The other avoids all planning and mandates specific review of the widest range of possible impacts that might be caused by proposed changes in land use.

The rationale behind each of these approaches is that common action can achieve results that could not be produced as quickly, efficiently, or inexpensively if they were mandated by the government, and, moreover, might not be produced at all by the market operating independently. Economists call these situations *externalities*. They may be caused by individuals acting in their own perceived self-interest, but doing so out of ignorance, shortsightedness, or lack of concern for the rest of society. The demand for intervention may also arise from society's need to provide some publicly accepted goal that can be achieved only by common action and cannot be withheld if some beneficiaries refuse to pay. Whether the aim is to prevent the undesirable or to stimulate the beneficial, these are situations that individual citizens find difficult to deal with on their own; as a result, addressing these circumstances in a manner that puts first and foremost the common good above all else requires public intervention.[3]

By the end of the twentieth century, land use regulation had ceased being purely a method for preventing property owners from causing harm to their neighbors and had grown into one of the most popular techniques for improving cities and suburbs. Many communities that employed land use regulations to achieve a broad range of goals began to apply them in a manner that was so broad in scope, specific in content, and discretionary in application that they clashed with such other societal objectives as fairness, economy, and efficiency. For example, the attempt to achieve one objective, such as protection of a lovely natural landscape, might clash with another objective, such as lowering housing costs; the resulting conflict might cause additional problems, such as discrimination against minority groups.

In some areas, ordinances have become so complex that property owners who only employ architects and lawyers are no longer able to determine what they can and cannot do with their land. They must also employ one set of consultants (usually former municipal employees) to advise them on their initial proposal and another set of consultants (also usually former municipal employees) to expedite the processing of their requests. Inevitably, the additional time and costs associated with these added personnel are passed on to consumers, who are increasingly unable to afford the benefits. In the process, not only is any improvement of the physical environment seriously retarded, but the cost of living escalates, as well. Some people became so frustrated with the situation that at the end of the twentieth century a reform movement arose that advocated the elimination of use, density, and environmental regulations. Instead of comprehensive zoning with all its considerations, they support zoning that considers only form.

Whatever method of regulation may be in place, the difficulty and cost of changing existing regulations has now become so burdensome that it is tackled only for major projects, by developers with deep pockets, politicians with powerful connections, and civic groups with sizable constituencies. As a result, unsatisfactory and ineffective land use controls remain in place long after they can be justified.

We must clear away the existing thicket of confusing, conflicting, and counterproductive statutes. There are plenty of workable regulations that are fair, inexpensive, easy to administer, and responsive to local conditions and priorities. Wherever they exist, they should remain in place. They should also be extended to those areas where they have not been heretofore applied. Whatever the combination of statutes, they should be altered to harmonize with one another. Only then will we be able to protect and improve both the natural and the built environment.

Private Land Use Regulation

Private citizens can protect themselves by using deed restrictions. Deed restrictions may be initiated by the government entity or property owner who established an initial system of streets, blocks, and lots, or by a later property owner. The deed (with restrictions) is then formally and permanently recorded with the relevant government agency, usually the county clerk. Such deed restrictions were common during the nineteenth century. From its inception in 1857, for example, Boston's Back Bay was encumbered with requirements that governed height, construction materials, siting, and even such appendages as steps, porches, balconies, and bay windows. In 1869, when the developer of Riverside, Illinois, began selling lots for development, the sale included deed restrictions that prohibited fences and required houses to be set back at least 30 feet from any public thoroughfare. The private places of St. Louis usually included deed restrictions that required a minimum cost of construction and design approval by the street association (see Chapter 14).

Deed restrictions have serious drawbacks. Only those properties that have restrictive covenants get the benefits of regulation. The amount of land affected therefore varies with the locality. Unfortunately, traffic regulation, water supply, pollution control, and many other city planning objectives cannot be accomplished unless the entire region is included in the regulatory framework.

Another drawback is that private regulatory action is often limited in its duration and effectiveness. The minimum construction cost requirements of the private places of St. Louis, for example, were rendered irrelevant by inflation. Moreover, deed restrictions can expire. When the time is up, property owners may choose to scuttle previous decisions. Individual dwellings at Sunnyside Gardens were organized around the edges of city blocks, leaving common open space in the middle. Each block had a restrictive covenant preserving the common open space. When the covenants expired in the mid-1970s, some blocks chose to subdivide the open space, turning it into small individual backyards.[4]

Private land use regulations are often difficult, time-consuming, and expensive to enforce. Property owners have trouble preventing their neighbors from breaking the rules. Homeowners in Levittown, Long Island, for example, are all required to abide by restrictive land covenants, among which are requirements that "Laundry must be hung only in the rear on a revolving portable dryer, which must be taken down when not in use" and "Lawns must be mowed and weeds removed at least once a week between April 15 and November 15." These rules may be trivial—too trivial for court action. But even if they were not, litigation still might not be practicable. Enforcing covenants is difficult, costly, and time-consuming because it requires legal counsel. Furthermore, the only individuals with standing to sue are property owners affected by the same covenants.

More pernicious covenants have been used to restrict occupancy to specific racial and ethnic groups. In 1948, the Supreme Court, in the case of *Shelley v. Kraemer,* ruled that the equal protection provisions of the Fourteenth Amendment precluded state courts from enforcing this form of contractual discrimination. Nevertheless, scattered cases of exclusion persist wherever the cost and complexity of proving discrimination have deterred legal action.[5]

Even the process for altering restrictions can be cumbersome and unfair. Unless otherwise specified, unanimous consent of all the affected properties is required. Even when changes can be made if agreed to by a simple majority, it usually means a majority of the lots affected. This can cause unanticipated difficulties, such as were faced by the residents of Palos Verdes Estates when they sought to avoid paying county real estate taxes on commonly owned parkland (see Chapter 16).

Most localities stopped regulating land use solely by private agreement because the time, cost, and complexity of separately and individually drafting, executing, amending, and rescinding these agreements proved to be too great. It seemed better to prescribe actions by legislation rather than by written agreement among the property owners affected, to make decisions by majority vote rather than by unanimous consent, and to enforce them by administrative agencies rather than by the courts.

Houston, Texas, is the only major American city that continues to regulate land use largely by private agreement. While Houston's deed restrictions are pervasive, they have been augmented by laws that establish minimum lot sizes for areas with and without sewers and minimum setbacks for buildings on local streets and major thoroughfares, as well as by an ordinance that allows public enforcement of private covenants. The city's attempts to combine its land use legislation into a single comprehensive zoning ordinance were rejected by referendum in 1948, 1962, and 1993.

The Police Power

The power to govern implies the power to establish suitable regulations protecting the order, health, safety, and morals of the community, or—in nineteenth-century parlance—the power to "police." The extent to which localities can regulate land use, however, is determined by the degree to which this *police power* is circumscribed by the Constitution, and, specifically, the Fifth and Fourteenth Amendments. The Fourteenth Amendment, approved in 1868, was intended to protect civil rights and included the prohibition against state legislation depriving "any person of life, liberty, or property without due process of law." Such language clouds the constitutionality of local property regulations, because any restriction will deprive citizens of the use of their property to some extent.[6]

The extent of the police power was defined by Justice John Marshall Harlan in the case of *Mugler v. Kansas* (1887).

The case arose when Mugler, a brewery owner, challenged a Kansas law that forbade the manufacture or sale of alcohol, arguing that the law had rendered his property valueless by eliminating its use as a brewery. Justice Harlan explained that legislation that prohibits activity that is "injurious to the health, morals, or safety of the community," but "does not disturb the owner in the control or use of his property for lawful purposes nor restrict his right to dispose of it," is not a "taking" of the owner's property. It is "only a declaration by the state that its use by anyone, for certain forbidden purposes, is prejudicial to the public interests."[7]

The critical case in determining the point at which regulation becomes a *taking* in violation of the Fourteenth Amendment came in 1922 in *Pennsylvania Coal Company v. Mahon.* During the previous year, the legislature enacted the Kohler Act in an attempt to protect hundreds of thousands of residents of northeastern Pennsylvania from mine subsidence. Land above a large number of mine shafts had subsided, leaving sags and humps in once-level streets, sometimes breaking water, sewer, and gas mains, and even opening cemeteries, so the legislature attempted to solve this problem by enactment of the new law. The Kohler Act prohibited mining of anthracite coal in designated areas where such mining could cause the subsidence of streets, bridges, railroads, conduits, cemeteries, public buildings, factories, stores, and private residences.[8]

Much of the land in that part of Pennsylvania had been sold by coal companies that retained mineral rights below the surface of the property. The deeds to these properties often included the stipulation that the purchaser waived future claims for property damage or personal injury due to mine subsidence against those who retained subsurface mining rights.

In September 1921, the Mahon family was notified in writing that within 10 days the Pennsylvania Coal Company would start mining operations that could cause the subsidence of their house. They started proceedings to enjoin the action as a violation of the Kohler Act. The coal company countered, saying that the legislation was not an exercise of the police power by the legislature, but, rather, an unconstitutional taking of their property (the right to mine coal under the Mahon's house) without due process of law or compensation.

When the case reached the Supreme Court, Justice Oliver Wendell Holmes, writing for the majority, found that "to make it commercially impractical to mine certain coal has very nearly the same effect for constitutional purposes as appropriating or destroying it," and was therefore a taking. He further enunciated the principle that established the constitutional limits of government land use regulation: "while property may be regulated to a certain extent, if regulation goes too far it will be recognized as a taking."[9] Holmes's principle fails to answer the question: *what is too far?* It is the question that should always be paramount in the mind of anybody trying to regulate land use.[10]

The Role of Federal, State, and Local Government

The words *police power* cannot be found in the Constitution. It is among the many powers retained by the states. Nevertheless, because of a variety of other constitutionally established responsibilities, the federal government does regulate land use. For example, the Constitution gives the federal government the primary role in policing interstate commerce and, thus, responsibility for ensuring the navigability of the country's waterways. By implication, it must police land use along the water's edge. To better carry out these functions, Congress passed the *Rivers and Harbors Act of 1899.* It gave the Corps of Engineers specific authority to regulate construction of dams, dikes, piers, wharves, bulkheads, and other structures. The law also allows the Corps to establish pierhead and bulkhead lines and to regulate excavation and landfill extending into navigable waters. Another example is the *Federal Water Pollution Act of 1972,* which required "zero discharge" after 1985 of any refuse into U.S. waterways other than liquids flowing from streets or sewers. It allowed the federal government to establish performance standards for new industries locating along navigable waterways. The *Wild and Scenic Rivers Act of 1968, Clean Air Act of 1970,* and *Coastal Zone Management Act of 1972* are only a few of the many other federal laws with land use implications that flow from other constitutionally authorized activity.

The most potent device in federal land use regulation does not derive directly from the Constitution but, rather, from federal expenditures. Any time Congress appropriates funds, it has the right to demand that the recipient comply with specified requirements that can include anything from preventing discrimination to complying with prevailing wage rates. The most important of the funding-derived statutes that affect land use, the *National Environmental Policy Act of 1969* (NEPA), does so by requiring that every major federal action as well as every federally funded project prepare detailed statements concerning any major effects they could have on the quality of the environment (see the subsections "The National Environmental Policy Act of 1969" and "NEPA's Impact" that appear later in this chapter).

Local authority to police land use derives from state constitutions. City, county, and regional governments are considered to be "creatures" created by the states and are entitled to carry out only state-authorized functions. For this reason, when comprehensive zoning became the favored form of land use regulation, Secretary of Commerce Herbert Hoover appointed a committee to draft a model state-enabling statute. The resulting Standard State Zoning Enabling Act was issued by the Government Printing Office in 1924. Within four years, it "had been used wholly or largely in zoning laws enacted in some thirty states."[11]

Early zoning ordinances were narrow in scope. They dealt with simple questions of design and land use, ignoring such other issues as drainage, air and water quality, or noise. These first zoning laws could not have been sufficiently inclusive because the individuals who wrote them had neither the knowledge nor the experience to draft truly comprehensive regulations. Additionally, political opposition would have been so virulent that they never would have been adopted. Even if they had been adopted, no government agency could have hired the staff to enforce them because trained personnel did not exist. Equally important, they applied only in governmental jurisdictions that had adopted such ordinances. As a result, vast stretches of territory remained unregulated.

For at least half a century, large segments of the planning profession continued to duck environmental issues. The staffs of most planning and zoning commissions did not include individuals trained to consider environmental issues. The vacuum was filled by enacting environmental regulations that involved such issues as the siting of power plants, supervision of surface mining, protection of agricultural land, conservation of water resources, and the management of floodplains, wetlands, and shorelands. These regulations usually applied to property outside city limits. Proponents of such regulation were asking for the same things that had been obtained decades earlier by urban reformers demanding zoning. They were just filling the jurisdictional void caused by the absence of city governments and traditional zoning.

California's coastal-zone management is probably the best known and surely one of the most far-reaching examples of such state land use regulation. It began as the *California Coastal Zone Conservation Act of 1972,* a citizen-sponsored initiative, better known as Proposition 20. The act established six regional commissions and one 12-member state commission with the responsibility of preparing a coastal zone plan for California's entire 1100-mile oceanfront. That plan had to include policies on land use, transportation, conservation, public access, recreation, the siting of public facilities, ocean minerals, living resources, maximum population densities, and educational and scientific uses.[12]

For the next four years, while the California Coastal Commission prepared its plan, there was a virtual moratorium on all development. The completed plan was largely rejected by the state legislature, which instead passed the *California Coastal Act* and *State Coastal Conservancy Act of 1976.* Under this new legislation, local governments regained the power to adopt land use plans and zoning ordinances. However, a reconstituted Coastal Commission was responsible for approving their actions, as well as virtually all new development within a broadly defined coastal zone that varied in width from 1000 yards to 5 miles. These awesome responsibilities required a substantial administrative apparatus. Within six years, the commission had a full-time staff of 188 and an annual budget of $10.3 million. By 2010 its staff had dropped to 125.[13]

During the 1970s, Florida, Oregon, Vermont, Maine, and Wyoming took the lead in adopting statewide land use programs, which, like California's coastal-zone management program, attempted to regulate areas of critical concern as well as projects that had major regional impact. Essentially all this legislation was aimed at safeguarding the environment while simultaneously protecting such favored activities as farming, fishing, hunting, logging, and tourism. Whether aimed at the California coastline or the Wyoming mountains, however, these acts were generated as much by unhappiness with the character of new development as by the desire to protect scarce resources that were particularly vulnerable to new development.[14]

Land Use Regulation Strategies

Early-twentieth-century urban reformers who fought for city zoning ordinances, and later-twentieth-century environmentalists who fought for state and federal legislation to protect the countryside, share a common distaste for entirely market-driven land use decisions. They are not willing to wait till the damage is done, till the stream has been polluted, the traffic is in gridlock, or the landmark has been demolished. Instead, they would augment the private market either with legislatively specified requirements, such as codes and ordinances, or with administrative review, such as discretionary action by a government agency. However, both the proponents of entirely market-driven land use and the proponents of government regulation are burdened with the same unrealistic expectation: that results inevitably will be satisfactory.

Unlike arithmetic problems, land use decisions never have one and only one correct solution. Their success depends on the value systems of the people who examine them. Predominant among these values are the desires to protect individuals from harm, to ensure the continuing utility of private property, to encourage orderliness, and to ensure stability.[15]

There has been since the earliest colonial settlements a continuing stream of regulations aimed at protecting private property. Prior to the twentieth century, fires that destroyed whole city districts were common. Starting with laws such as Boston's 1672 statute requiring that structures be built of fireproof materials, proponents of government land use regulation have fought for increasingly detailed construction requirements, all intended to protect property from fire. Consequently, by the end of the twentieth century, construction in virtually every American city was subject to building

Washington, D.C., 2000. Continuous cornice line on 15th Street established by the height limit enacted by Congress in 1899. *(Alexander Garvin)*

and safety codes to minimize the possibility of property destruction in citywide conflagrations.

Orderliness—society's attempt to keep pigs out of its parlor—is another of the values behind land use regulation. In 1647, it was even put in similar terms by a New York City ordinance that attempted to avoid a "disorderly manner . . . in placing pig pens and privies on the public roads."[16] It was not until 1908, however, that any major American city tried to find a place for everything and, by extension, put everything in its place. In that year, the Los Angeles city council enacted ordinances that established seven industrial districts in which mills and factories were permitted and three residential districts from which they were excluded.[17]

In his book *Icons and Aliens,* John Costonis identifies the desire for stability and reassurance in the face of change as one of the values behind land use regulation. He explains that the environment is "an assemblage of natural and built features, many of which have become rich in symbolic import. The 'icons' are features invested with values that confirm our sense of order and identity. The 'aliens' threaten the icons and hence our investment in the icons' values." Land use regulation, Costonis explains, can be better understood as society's attempt to preserve "icons" and exclude "aliens."[18]

The attempt to protect "icons" from the threat of "alien" intrusion is graphically illustrated by 1899 legislation in which Congress established height limits for the District of Columbia. Among the purposes of these height limits was the desire to ensure that the Capitol dome (a genuine icon) would dominate every section of the city and that the city's streets and avenues would be protected from intrusive tall structures (aliens?) that would deflect attention from the Capitol.[19]

In its less attractive form, this desire for stability and reassurance becomes an excuse for ethnic and racial discrimination. A good example is San Francisco's attempt, during the latter part of the nineteenth century, to exclude (possibly alien) public laundries from many sections of the city. This effort was directed at 240 businesses, many of which also were de facto social centers owned by (definitely alien) subjects of the emperor of China or by Chinese Americans.[20]

In 1880, the city and county of San Francisco enacted legislation making it unlawful to operate a laundry in nonmasonry buildings without having first obtained the consent of the board of supervisors. At that time, nine-tenths of San Francisco was built of wood. There were about 320 laundries in the city and county of San Francisco, of which all but 10 were housed in wooden structures. Three-quarters of all laundries were Chinese-owned and -operated. Over the next five years, more than 150 "subjects of China" were arrested on the charge of carrying on business without having consent of the board of supervisors, while all 80 of the Caucasian-owned laundries remained in business, unaffected by the legislation.

In 1886, in the cases of *Yick Wo v. Hopkins* and *Wo Lee v. Hopkins,* the Supreme Court found this legislation to be a violation of the equal protection guaranteed by the Fourteenth Amendment. It upheld the right of the board of supervisors to regulate activities "which are against good morals, contrary to public order and decency, or dangerous to public safety." But it also found that the danger of fire was no more significant in three-fourths of the territory affected (10 miles wide by 15 miles long) than in "other farming regions of the State." The court held that while there was a danger from fire in built-up portions of the city, "a fire, properly guarded, for laundry purposes, in a wooden building, is just as necessary, and no more dangerous, than a fire for cooking purposes or for warming a house." Furthermore, the court observed, "clothes washing is certainly not opposed to good morals or subversive of public order or decency."[21]

Comprehensive Zoning

Although there were a vast number of land use regulations already in effect by the beginning of the twentieth century, the movement for such regulations often is erroneously thought to have been initiated with the zoning resolution enacted by New York City in 1916. This law was indeed innovative, but it was the comprehensive scope of the resolution more than its attempt to assign specific land uses to different sections of the city that was new. For the first time, a land use ordinance simultaneously specified permitted land use, building height, and building placement for the entire city. It did so by providing three bulky sets of *zoning maps,* which designated the regulations that applied to every block and lot within the city limits, and a *zoning text* that explained them.

The 1916 resolution set off a chain reaction in the rest of the country. Within five years of its passage, 76 communities had adopted similar statutes. By 1926, that number had grown to 564. Thus, when the Supreme Court validated the constitutionality of comprehensive zoning in the seminal case of *Ambler Realty Company v. Village of Euclid,* the overwhelming majority of American cities had adopted comprehensive zoning patterned on the law New York had pioneered a decade earlier.[22]

The same course of action reoccurred in 1961 when New York City completely revised its resolution. At that time, it eliminated height limits, replacing them with more flexible bulk regulations; it also added both density controls and incentives to encourage an increase in the amount of public open space. Once again, the techniques innovated in New York were taken up by cities across the country.

New York City Zoning Resolutions of 1916 and 1961

Although both of New York City's trailblazing zoning resolutions tried to deal comprehensively with every section of the city, they did not emerge from a comprehensive planning process. Nor were they based on previously adopted, comprehensive city plans.[23] The 1916 resolution was enacted because a group of powerful business leaders and good-government reformers were unhappy with existing real estate activity and sought legislation to protect their property, ensure the orderly development of the districts they frequented, and establish stable land use patterns for those areas.[24]

Lord & Taylor, Saks, Tiffany, and many other fine stores objected to manufacturing firms moving to the "famous retail sections" of the city. So did the Waldorf-Astoria Hotel, the University Club, the Union League, and many other exclusive establishments. In response, they started an advertising campaign that complained about "factories making clothing, cloaks, suits, furs, petticoats, etc."[25]

The intruding land uses generated pushcarts, trucks, and workers who mingled with patrons of expensive stores, hotels, and private clubs, making it difficult for these establishments to retain customers. Consequently, the stores, hotels, and clubs were forced to move, often giving up substantial investments in properties that had lost their former value. A growing number of business leaders, led by the Fifth Avenue Association, demanded action that would alter the terms of private market competition by keeping neighboring property owners from freely renting or selling to "incompatible" neighbors.

These activists found ready allies among good-government reformers who had a long history with legislation that guaranteed minimum levels of light and air for apartment buildings (see Chapter 11). In 1887, they had persuaded the state legislature to amend the old tenement law to limit the height of all future multiple dwellings.[26] Their first twentieth-century success, the New Tenement Law of 1901, specified a height limit of 1.5 times the width of the widest street on which any apartment house was to be built.[27]

Spurred on by success in regulating apartment construction, reformers moved on to commercial buildings, to which the tenement laws did not apply. Perhaps in response to fires like the one in 1911 that destroyed the Triangle Shirtwaist Company and resulted in the death of 146 workers, reformers gained increasing support for regulation of commercial buildings. They allied themselves with design professionals, who were increasingly enamored of continuous cornice lines.[28]

In 1912, the borough president of Manhattan appointed a Heights of Buildings Commission, whose proposed height limits were never adopted. In 1914, the city's Board of Estimate appointed a Commission on Building Districts and Restrictions. It was not until the completion in 1915 of the Equitable Building on lower Broadway, however, that the effort to regulate commercial buildings attracted sufficient political support to obtain legislation. The 1.2-million-square-foot Equitable Building rose 540 feet, uninterrupted, from the sidewalk to its roof. Its floor area was almost 30 times that of the lot on which it was built. No building of such enormous bulk had ever been seen in New York, or anywhere else. Suddenly the public was in an uproar over the possibility that all of Manhattan could be covered by similar buildings that would darken sidewalks and generate serious pedestrian and vehicular traffic congestion.[29]

In 1916, the coalition of business leaders and political reformers finally obtained approval of a zoning resolution that segregated land uses by district and limited the bulk and placement of the buildings that could contain those uses. They understood that property owners would still want the highest possible return on their investments. The new law was intended to prevent landowners from maximizing return by erecting buildings whose occupants would be "incompat-

SHALL WE SAVE NEW YORK?

A Vital Question To Every One Who Has Pride In This Great City

SHALL we save New York from what? Shall we save it from unnatural and unnecessary crowding, from depopulated sections, from being a city unbeautiful, from high rents, from excessive and illy distributed taxation? We can save it from all of these, so far at least as they are caused by one specified industrial evil—the erection of factories in the residential and famous retail section.

The Factory Invasion of the Shopping District

The factories making clothing, cloaks, suits, furs, petticoats, etc., have forced the large stores from one section and followed them to a new one, depleting it of its normal residents and filling it with big loft buildings displacing homes.

The fate of the sections down town now threatens the fine residential and shopping district of Fifth Avenue, Broadway, upper Sixth and Madison Avenues and the cross streets. It requires concentrated co-operative action to stem this invading tide. The evil is constantly increasing; it is growing more serious and more difficult to handle. It needs instant action.

The Trail of Vacant Buildings

Shall the finest retail and residential sections in the world, from Thirty-third Street north, become blighted the way the old parts of New York have been?

The lower wholesale and retail districts are deserted, and there is now enough vacant space to accommodate many times over the manufacturing plants of the city. *If new modern factory buildings are required, why not encourage the erection of such structures in that section instead of erecting factory buildings in the midst of our homes and fine retail sections.*

How it Affects the City and its Citizens

It is impossible to have a city beautiful, comfortable or safe under such conditions. The unnatural congestion sacrifices fine residence blocks for factories, which remain for a time and then move on to devastate or depreciate another section, leaving ugly scars of blocks of empty buildings unused by business and unadapted for residence: thus unsettling real estate values.

How it Affects the Tax-payer

Every man in the city pays taxes either as owner or tenant. The wide area of vacant or depreciated property in the lower middle part of town means reduced taxes, leaving a deficit made up by extra assessment on other sections. Taxes have grown to startling figures and this affects all interests.

The Need of Co-operative Action

In order that the impending menace to all interests may be checked and to prevent a destruction similar to that which has occurred below Twenty-third Street:

We ask the co-operation of the various garment associations.
We ask the co-operation of the associations of organized labor.
We ask the co-operation of every financial interest.
We ask the co-operation of every man who owns a home or rents an apartment.
We ask the co-operation of every man and woman in New York who has pride in the future development of this great city.

NOTICE TO ALL INTERESTED

IN *view of the facts herein set forth we wish to give publicity to the following notice:*—We, the undersigned merchants and such others as may later join with us, will give the preference in our purchases of suits, cloaks, furs, clothing, petticoats, etc., to firms whose manufacturing plants are located outside of a zone bounded by the upper side of Thirty-third Street, Fifty-ninth Street, Third and Seventh Avenues, also including thirty-second and thirty-third Streets, from Sixth to Seventh Avenues.

February 1st, 1917, is the time that this notice goes into effect, so as to enable manufacturers now located in this zone to secure other quarters. Consideration will be given to those firms that remove their plants from this zone. This plan will ultimately be for the benefit of the different manufacturers in the above mentioned lines, as among other reasons they will have the benefit of lower rentals.

B. ALTMAN & CO.
ARNOLD, CONSTABLE & CO.
BEST & CO.
BONWIT TELLER & CO.
J. M. GIDDING & CO.
GIMBEL BROTHERS
L. P. HOLLANDER & CO.
LORD & TAYLOR.
JAMES McCREERY & CO.
R. H. MACY & CO.
FRANKLIN SIMON & CO.
SAKS & CO.
STERN BROTHERS

The undersigned endorse this movement for the benefit of the City of New York

Vincent Astor
University Club
Union League Club
Criterion Club
Ritz-Carlton
Hotel Biltmore
Hotel McAlpin
A. A. Vantine & Co.
Mark Cross Co.
Astor Estate
Waldorf-Astoria
St. Regis Hotel
Hotel Gotham
Hotel Belmont
Hotel Manhattan
Hotel Netherland
Hotel Lorraine
Charles Thorley
Astor Trust Co.
Columbia Trust Co.
Fifth Avenue Bank
Guarantee Trust Co.
Harriman National Bank
M. Knoedler & Co.
H. W. Johns-Manville Co.
Yale & Towne Mfg. Co.
Scott & Fowles Co.
Tiffany & Co.
Gorham Co.
Black, Starr & Frost
Theodore B. Starr, Inc.
Dreicer & Co.
Marcus & Co.
E. M. Gattle & Co.
Charles Scribners' Sons
Maillard's
W. & J. Sloane
Aeolian Company
C. G. Gunthers' Sons
A. Jaeckel & Co.
Tiffany Studios
Higgins & Seiter
Davis Collamore & Co.
The Edison Shop.
Frank L. Slazenger
Brooks Brothers
Knox Hat Co.
Theo. Hofstatter & Co.
James McCutcheon & Co.
Cammeyer
J. & J. Slater, Inc.
De Pinna
Kennedy & Company
Fred'k Keppel & Co.

We ask Citizens, Merchants and Civic bodies to co-operate and send letters endorsing this plan to the committee, care of J. H. Barton, chairman, 267 Fifth Avenue.

New York, 1916. This advertisement in the *New York Times* demanding action to save the city from incompatible land uses played a major role in persuading the city to enact comprehensive zoning.

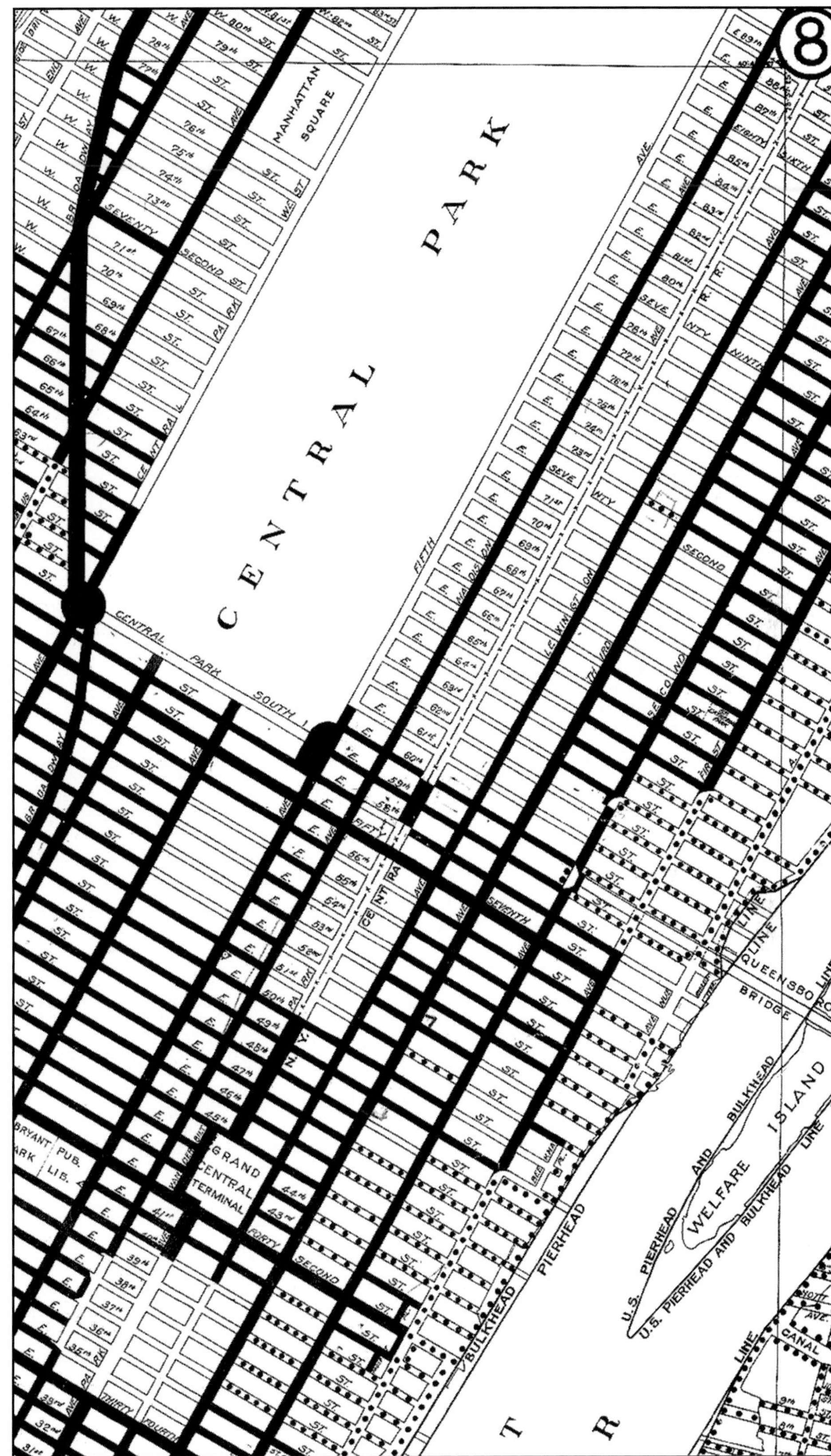

Manhattan, 1916. Land uses permitted by the zoning resolution of 1916. Streets and avenues left blank were zoned for residential use only. Streets and avenues marked in black were zoned for either commercial or residential use. Streets and avenues lined with black dots were zoned for unlimited land use. *(Courtesy of New York City Department of City Planning)*

Manhattan, 1893. Fifth Avenue north of 42nd Street was entirely residential prior to zoning. *(Courtesy of Museum of the City of New York)*

Manhattan, 2008. Fifth Avenue north of 42nd Street became entirely commercial after it was zoned for commercial use. *(Alexander Garvin)*

ible" with their neighbors, or by building bulky behemoths blocking sun and sky.

The new zoning resolution determined the future of every part of the city in which development was to take place, but nowhere was it more effective than on Fifth Avenue. When the ordinance took effect, Fifth Avenue was lined by the mansions of the wealthy along virtually the entire stretch between 42nd and 96th streets. The new ordinance specified that the blocks opposite Central Park could be used only for residential purposes, while those south of the park could be used for either commercial or residential purposes, but not for manufacturing. Residential uses disappeared from the blocks below Central Park, where commercial uses were permitted, because office and retail tenants paid higher rents

Manhattan, 1893. Fifth Avenue north of 65th Street was entirely residential prior to zoning. *(Courtesy of Museum of the City of New York, The Byron Collection)*

Manhattan, 2011. Fifth Avenue north of 65th Street remained entirely residential because it was zoned exclusively for residential use. *(Alexander Garvin)*

than apartment residents. However, opposite the park, apartments for the wealthy replaced the mansions of the wealthy, because there the zoning resolution permitted only residential land use. Indeed, most Fifth Avenue town houses were eventually sold to developers, who could make more money by replacing them with elevator buildings.

The impact of height regulations was no less dramatic. Wherever buildings were erected, they had to comply with the new height limits. Between enactment in 1916 and even through to new structures built today, Park Avenue between 50th and 96th streets, has been outlined by a 210-foot-high street wall because the zoning resolution forbade façades that exceeded 1.5 times the width of the avenue.

Given the speed with which the zoning resolution had been adopted and the complete lack of experience with such legislation, demands for revision were inevitable. By 1961, when the 1916 resolution was supplanted, 2500 amendments had been approved. The pressure for change was seldom directed at market-driven real estate activity. Instead, it was aimed at the regulations themselves, which were thought to be inadequate, inflexible, and unnecessarily confusing.[30]

Manhattan, 2005. The uniform height of buildings along Park Avenue is the result of the height limit imposed by the zoning resolution of 1916. The single tall building in the distance was built pursuant to the 1961 Resolution. *(Alexander Garvin)*

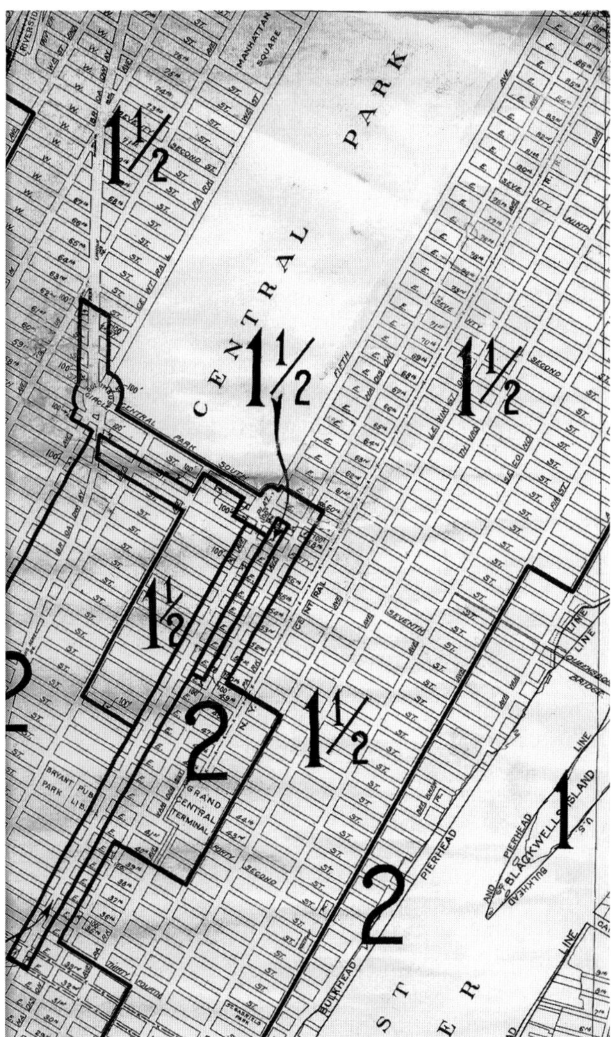

Manhattan, 1916. Height limits established by the zoning resolution for the blocks along Park Avenue north of 50th Street set building height at 1.5 times the width of the street. *(Courtesy of New York City Department of City Planning)*

The 1916 resolution had been adopted without much consideration of existing or likely future development patterns. At full site utilization, it permitted a city of 55 million residents and 250 million workers. Moreover, to a great extent, both the use and bulk restrictions were unrelated to what then existed. Over half the city's population inhabited districts that were not zoned exclusively for residential use. Of the areas that were zoned exclusively for residential development (largely outside Manhattan), more than half were zoned for large apartment buildings that were more appropriate to Manhattan's exclusive Upper East Side.

The regulations themselves had been developed with little concern for construction practices or enforcement procedures. Buildings were required to fit specific envelopes without regard to site, use, structure, or cost of construction. Worst of all, consulting three sets of maps and numerous written regulations was cumbersome even for the city officials with the responsibility of interpreting and enforcing them.

Despite all these deficiencies, the reform movement became effective only in 1947, when Robert F. Wagner Jr. was appointed chairman of the City Planning Commission. He understood that zoning had to be appropriate to existing conditions. The city had changed significantly since its pioneering zoning resolution had been adopted. At that time, supermarkets, chain stores, and shopping centers had been unknown. Industrial firms had not yet begun to abandon their multistory lofts for suburban locations where there was adequate land for horizontal production lines. Cars, trucks, and planes had not yet become the dominant forms of transportation.

Wagner commissioned the architectural firm of Harrison, Ballard, and Allen to prepare a new ordinance that would be appropriate to the second half of the twentieth century. Before it was completed, he was elected borough president of Manhattan, a position from which he was unable to continue his efforts for a new zoning resolution, but he made zoning reform a major objective of his administration when he was elected mayor in 1953.[31]

Wagner waited until there was sufficient political support before proceeding. Zoning reform would affect every business, neighborhood, and voter in New York City. For this reason, having a committed constituency was a political necessity. Conditions had ripened sufficiently in 1956 for the mayor to appoint James Felt, a successful real estate agent, to be chairman of the City Planning Commission. Felt led the fight for modernizing the zoning resolution. He hired a second architectural firm, Voorhees, Walker, Smith, and Smith, to draft the new resolution. In 1961, after years of consultations with architects, lawyers, real estate agents, builders, community groups, and civic organizations, after more than 850 changes to the proposed text, and after dozens of formal and informal hearings, the city adopted a final version of the new zoning resolution.[32]

The 1961 resolution consolidated the three sets of maps into a single map system. Like the 1916 resolution, it indicated use, bulk, and building placement requirements. To these were added density and parking regulations. The text was also simplified by presenting many of the requirements in tabular form. The most important revisions, however, were in content.[33]

For the first time, the city had density controls that established the number of people likely to occupy every development site. Ostensibly, that density was related either to the existing capacity of infrastructure (water, sewer, streets, transit, etc.) and community facilities (schools, libraries, parks, playgrounds, etc.) or to future capacity, once planned capital construction was completed. Since New York had yet to prepare a comprehensive city plan, this was more theory than fact (see Chapter 19).

Another major change was the replacement of height limits with flexible bulk regulations that, district by district, specified the maximum allowable floor area as a multiple of lot area. Each building's shape would now depend on the regulations for that zoning district.[34] The new resolution also sought to provide additional space, light, and air for pedestrians, particularly in densely built-up areas, like Lower Manhattan. In selected areas of the city, for every additional square foot of sidewalk or plaza, projects were entitled to increase the amount of floor area beyond what would otherwise be allowable.

As in 1916, New York's innovations were immediately copied. Cities everywhere adopted the single-map system. They junked height limits, instead specifying a ratio of floor area to lot area. Some adopted the bonus for plazas and arcades. Many also went on to develop their own variants.

Cities include densely built-up areas and vacant land; highly prized districts that are not likely to change and obsolete areas that certainly will be rebuilt; complex, heterogeneous neighborhoods and districts whose consistency verges on the monotonous. Despite these variations, comprehensive zoning resolutions apply the same techniques to all sections of all cities, techniques that regulate allowable land uses, density, bulk, building placement, and open space.

Land Use

New York City's 1916 zoning resolution and all zoning ordinances patterned on it established three land use categories: exclusively *residential, commercial* (in which residential uses are permitted), and *unlimited* (to which manufacturing is relegated). This system recognizes that most built-up city districts include complex combinations of land uses that, when zoned *unlimited,* can coexist.

Following the precedent of New York City's 1961 zoning resolution, most cities dropped the unlimited category, replacing it with *manufacturing* zones in which certain commercial, but not residential, uses are permitted. The logic behind this segregation of manufacturing is that the traffic, noise, fumes, and industrial waste that are generated are inappropriate to most residential areas. As the number of manufacturing firms in American cities has declined, these regulations have assumed a new role: now they are as much for the protection of the industrial firms. Nevertheless, as the number of "clean" industries and businesses with fewer than 10 employees increases, this logic is becoming obsolescent.

Zoning ordinances usually are adopted after cities already have been built up. Thus, no matter how carefully use categories are mapped, a substantial number of existing uses inevitably will fail to conform to the approved zoning. The idea is for these *nonconforming* uses eventually to pass away and be replaced by those that will conform with the zoning. Since after-the-fact prohibition can be considered a taking that would require compensation, nonconforming uses are usually permitted to persist, so long as the degree of nonconformity is not increased.

The disparity between actual land uses and zoning sometimes results in unfortunate situations. Occupants of a nonconforming building may find it difficult to obtain insurance, mortgage financing, or building permits for remodeling. The resulting disinvestment may spur the deterioration of the surrounding area. In fact, the very existence of nonconforming uses may render irrelevant any land use policy established by a zoning resolution, for by the time such uses are replaced, the zoning itself may have become obsolete.

Some critics question even the desire to separate land uses. Jane Jacobs, in her pioneering book, *The Death and Life of Great American Cities,* for example, argues for mixed land use as a way of ensuring safety and neighborhood vitality, pointing out that cities need "people who go outdoors on different schedules . . . for different purposes, but who are able to use many facilities in common."[35] The only way to generate activity 24 hours a day is to allow for buildings that will contain the widest variety of uses. Starting in the 1970s, more in response to entirely local situations than to Jacobs's cogent analysis, cities began redesignating zones for mixed use.

Density

The 1961 resolution introduced regulations which, for residential construction, specified *lot area per room* and later *lot area per dwelling unit.* The intent was to establish, on a district-by-district basis, a maximum number of households that could be adequately served by local water, sewer, transportation, hospital, school, library, and park systems.[36]

There are serious flaws in the rationale for controlling density. Regulations are applied without reference to a zoning district's demographic characteristics and remain in place although those characteristics are continually changing. Consequently, the same lot-area-per-room requirement applies to houses in which there are widely different occupancy characteristics and service requirements. This is difficult to justify if neighborhoods with large concentrations of low-income residents (in which the number of people per room and the amount of public services on which they depend tends to be high) have the same lot-area-per-room requirement as wealthy neighborhoods (in which there is a greater number of residents who belong to country clubs, own second homes, and take frequent vacations). It is also difficult to justify when the same lot-area-per-room requirement remains in place long after the proportion of single-person households has changed, thereby changing loading patterns for schools, recreation facilities, and transit systems.

Density regulations reflect an implicit assumption that higher density results in greater crowding. As Jane Jacobs so eloquently points out:

> *High densities mean large numbers of dwellings per acre of land. Overcrowding means too many people in a dwelling for the number of rooms it contains. . . . It has nothing to do with the number of dwellings on the land, just as in real life high densities have nothing to do with overcrowding.*[37]

The equation of density with crowding also results in a bias against high-density districts. As a result, zoning regulations consign lower-density districts to forever lack populations large enough to support the eating places, clothing shops, hardware stores, and other services and amenities that should be close at hand in any urban neighborhood. They also force increasing dependence on—and time lost in—automobiles.

Bulk

There are two main techniques for regulating bulk (the three-dimensional presence of a building). The first, *height limits,* predates comprehensive zoning. New York City limited building height in 1887, Washington in 1899, Los Angeles in 1904. The rationale was different in each case. New York wanted to ensure adequate light and air for the occupants of apartment houses; Washington wanted a symbolic setting for the national Capitol; Los Angeles wanted to minimize damage from earthquakes. New York's 1916 resolution consolidated height limits into five districts, in each of which the maximum permitted building height was a different multiple of the width of the street on which a building was to be erected. The buildings that complied with the height limits tended to be fat and squat, cover most of their site, and leave only minimal open space.

Attitudes about building height vary. Most potential building occupants will want to be as high as possible. They recognize that the higher they are, the less they will be disturbed by street noise and fumes, the more light and air they will get, and the better their view will be. For these reasons, both apartment dwellers and office occupants pay higher rents for higher floors, with the highest rents on the top floor. Pedestrians, on the other hand, like buildings to be as low as possible. They want to see the sky and feel the sun as they wander through the streets.

Nor do attitudes about building height remain constant. During the first half of the twentieth century, communities prized skyscrapers as evidence of prosperity and technological superiority. Chicago and New York even thought it important to compete to erect the world's tallest building. More recently, San Francisco, Seattle, and other cities have sought to prevent further skyscraper construction.

The second device for regulating bulk, the *floor area ratio* (FAR), relates a building's bulk to the lot on which it is to be built. The floor area ratio is defined as the total floor area (either already built or to be built) on a zoning lot divided by the area of that zoning lot. The allowable floor area is therefore determined by multiplying the allowable FAR by the lot area. The maximum allowable FAR depends on city policy. New York, for example, has 10 residential districts for which the allowable FAR ranges from 0.5 to 10 or, with a bonus, 12.

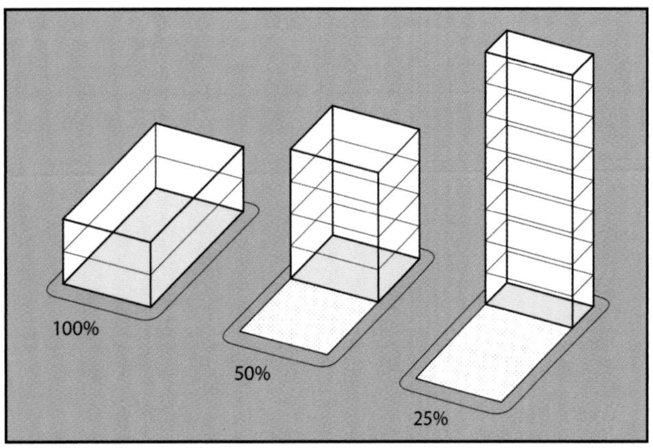

Three buildings with the same floor area ratio and equal bulk: one covering 100 percent of the site, one covering 50 percent of the site, and one covering 25 percent of the site. *(Alexander Garvin and Owen Howlett)*

One argument for a floor area ratio as opposed to a height limit is that it allows the developer and architect to select a building configuration that is less costly to produce, more marketable, more appropriate to the uses the building will contain, and less constrained to fit a particular architectural esthetic. The allowable floor area remains the same no matter what design is selected. For example, a 5000-square-foot lot with an allowable FAR of 2 has a maximum floor area of 10,000 square feet. The building, however, can cover the entire lot and rise two stories, or cover half the lot and rise four stories, or cover one-quarter of the lot and rise eight stories, or any other combination that produces a building with no more than 10,000 square feet of floor area. A second argument in favor of an FAR approach is that height limits force property owners to squeeze in as many floors as possible, thereby keeping ceiling heights within the building to the absolute minimum. Moreover, activities which require high ceilings (e.g. theaters, basketball and badminton courts, exhibition halls, etc.) may not fit within the allowable height.

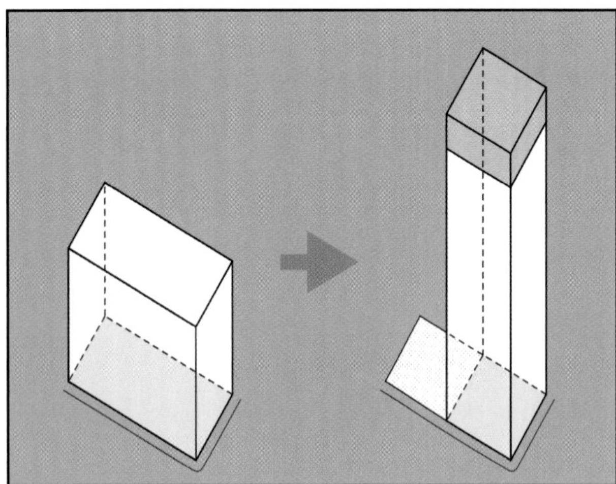

Manhattan, 1961. The plaza bonus allows the property to include 2 more square feet of interior floor area than would otherwise be allowed by the zoning, for every square foot of public open space that is provided on the site. *(Alexander Garvin and Owen Howlett)*

Half a century of experience with FARs has demonstrated that, while they afford greater design flexibility, they also produce a chaotic cityscape, with buildings that vary in height and are set back differing distances from the property line. These setbacks can interrupt pedestrian movement, causing a decline in retail activity.

Building Placement

In addition to mapping use and height districts, ordinances that were patterned on the 1916 resolution established "area districts." The term *area* is a misnomer, since these area districts specified neither building area nor lot area, but, rather, the number of feet a building had to be set back from the front, rear, and side lot lines. The resulting yards were intended to provide minimal levels of light, air, and privacy for building occupants (especially in detached and semidetached houses).

Yard dimensions were unrelated to potential uses. Sometimes the leftover spaces that they produced were transformed into off-street parking spaces for automobiles or off-street loading areas for trucks. In other cases, their dimensions made them almost unusable. This was especially true of side yards. Some were so narrow that they could only be used as alleys in which to collect trash.

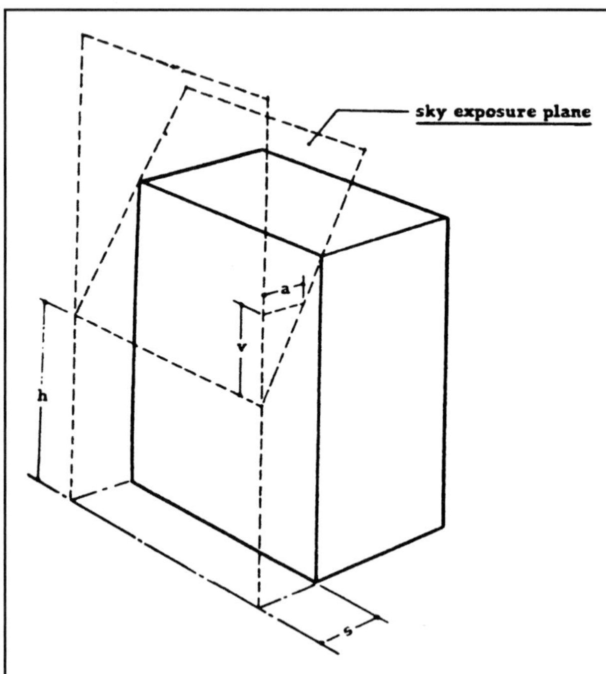

The sky exposure plane. *(Courtesy of New York City Department of City Planning)*

Manhattan, 1985. A conventional office building designed to fit the sky exposure plane. *(Alexander Garvin)*

A second widely applied building placement technique is the *sky exposure plane*. A sky exposure plane is an imaginary inclined plane that extends from the property line. It establishes a building envelope beyond which nothing can be built. Its purpose is to guarantee adequate natural light (and frequently direct sunlight) on both street and sidewalk.

Of all the techniques of comprehensive zoning, building placement regulations have tended to have the most stultifying effect. These well-intentioned requirements tend to become a straitjacket into which architects and developers automatically fit their buildings.

Manhattan, 2013. The IAC building on 11th Avenue, designed by Frank Gehry, demonstrates that interesting architecture can be created within the confines of a sky exposure plane. *(Alexander Garvin)*

Open Space

Zoning ordinances increasingly have sought to encourage private property owners to set aside space for the general public. This space has taken the form of wider sidewalks, plazas, arcades, publicly accessible galleries, and even interior passages connecting streets. New York City's 1961 resolution did so by introducing two techniques: the *open-space ratio* and the *plaza bonus*.

The open-space ratio (OSR) for a lot is defined as the square footage of open lot area divided by the floor area. Each zoning district includes a range of permissible OSRs, which, together with other regulations, determine the allowable floor area. The architect and developer select that combination of open space, building coverage, floor area, and height that best meets their needs and also complies with the zoning. However, if they wish to maximize buildable floor area by taking advantage of available bonuses, they are forced to use larger lots and erect slimmer structures that leave a high proportion of open space.

Manhattan, 2011. The plaza bonus often is little more than a sidewalk enlargement. *(Alexander Garvin*

Manhattan, 2008. Sometimes the plaza bonus makes a major addition to the city's public realm. *(Alexander Garvin)*

The plaza bonus is an incentive for developers to provide the public with accessible open areas around their buildings. Such plazas presumably increase pedestrian space in densely built-up areas and sun and sky in areas with a preponderance of tall buildings. In compensation for this public benefit, developers receive a bonus of additional rentable floor area.

The 1961 resolution established zones in which every square foot of exterior public plaza entitled the developer to an additional 2 square feet of interior floor area. This bonus "proved almost embarrassingly successful. Between 1961 and 1973 some 1.1 million square feet of new open space was created."[38] Its effectiveness, however, varied significantly. In lower Manhattan, with its narrow, twisting streets and busy sidewalks, the plaza bonus produced miraculous results. It freed pedestrian traffic, provided sitting areas, opened vistas to the sky, and even allowed the sun to light up small portions of ground. But opposite parks and along wide avenues, the newly created open space was superfluous. In other places, it produced "sterile, empty spaces not used for much of anything except walking across."[39] In 1975, in an attempt to improve the character and utility of these plazas, the New York Planning Commission adopted guidelines for seating, planting, and street furniture, but as with all such guidelines, their success is dependent on the quality of the design.

Variances

Besides the few devices discussed previously, zoning resolutions have so many other provisions that they may require several volumes. Given the plethora of requirements, many structures that predate the zoning will not meet the terms of one or more of them. When this refers to land use, it is called *nonconforming*; when this refers to bulk, it is called *noncomplying*. Just as with nonconforming uses, noncomplying conditions are permitted to persist, as long as the degree of noncompliance is not increased. However, especially in the case of major renovation, this may not be possible. There also may be problems in erecting new structures that comply with the zoning. For this reason, there has to be a procedure for authorizing variances from the allowable zoning.

In a process that has parallels in all other municipalities, New York City established an appointed Board of Standards and Appeals. The board can authorize a variance, which is an approved deviation from the applicable zoning regulations, provided that (1) there are unique physical conditions, (2) these physical conditions make it impossible for the property owner to earn a reasonable return if the proposed building completely conforms to the zoning, (3) the variance, if granted, will not seriously alter the character of the surrounding neighborhood, (4) the hardships justifying the variance have not been created by the property owner, and (5) the variance granted is the minimum needed to afford relief.

While these are reasonable standards, in New York and elsewhere there are no clear, quantifiable guidelines for determining compliance with such findings. As a result, some of the individuals charged with providing quasi-judicial relief from unreasonable requirements deviate from both the letter and the spirit of the required findings. In the absence of either certainty that a variance will be granted or predictability of the degree of relief, property owners frequently hire attorneys who specialize in obtaining variances or appeal to "influential" friends. A few make contributions to "deserving" recipients. If they fail to obtain relief, they go to court.

Village of Euclid, Ohio v. Ambler Realty Co.

From the beginning, comprehensive zoning was attacked as an unreasonable taking of private property. It was not until 1926 that the Supreme Court, in the seminal case of *Euclid v. Ambler*, established that citywide comprehensive zoning was a constitutional exercise of the police power. Since then, disputes have centered on specific provisions of particular zoning laws.[40]

The village of Euclid occupies 16 square miles along the coast of Lake Erie, just north of Cleveland. In 1922, when it adopted a comprehensive zoning resolution patterned on that of New York City, Euclid was a village of farms and scattered suburban houses, with a population of 10,000. Two railroads and three broad arteries, including fashionable Euclid Avenue, provided easy access to a burgeoning metropolitan region.

With Shaker Heights already in development six miles

to the south, Euclid seemed on the brink of change. Unlike Shaker Heights, however, Euclid had neither a development plan nor an owner-developer to shape its future (see Chapter 16). Instead, it chose comprehensive zoning with a stated intent "to keep Euclid Village as free from unsanitary conditions as possible, and to locate those unsanitary conditions in [a] segregated district."[41]

In 1911, the Ambler Realty Company began assembling an undeveloped 68-acre site between Euclid Avenue and the Nickel Plate Railroad. While its purpose in purchasing this site cannot be verified, land speculation is the obvious explanation. It also seems clear that Ambler Realty thought proximity to the railroad and to Euclid Avenue eventually would result in a sale to an industrial user who would gladly pay several times the land value that a residential developer would.

When Euclid introduced comprehensive zoning, it divided Ambler's property into three zones. The largest and northernmost strip of land that extended along the railroad was, for all practical purposes, zoned *unlimited*.[42] The southernmost strip, extending back 150 feet from Euclid Avenue, was zoned exclusively for one- and two-family houses. The 40-foot-wide strip in between was zoned for apartment buildings and community facilities as well as one- and two-family houses. This strip was intended to act as a buffer between residential Euclid Avenue and the probable industrial users in the unlimited zone along the railroad. However, its 40-foot width and elongated configuration made construction of apartment buildings and community facilities difficult.[43]

Ambler Realty believed that, by restricting what it could do with its property, the village of Euclid had deprived it of a significant portion of the land's value. Ambler took its case to court, claiming that property had been taken without compensation or due process of law, in violation of both the Ohio Constitution and the Fourteenth Amendment of the U.S. Constitution. U.S. District Judge David Courtney Westenhaver agreed. He also argued that Euclid had imposed a legislatively sanctioned but transitory idea of beauty, which was not an appropriate exercise of the police power. Worse yet, it was a subterfuge "to classify the population and segregate them according to their income or situation in life . . . [thereby] furthering such class tendencies."[44]

The village of Euclid appealed the decision, and by the time the appeal reached the Supreme Court, the Euclid case had taken on national significance. On the surface, the issue was whether Ambler had sustained a decrease in the value of its land and whether that decrease had gone too far and was therefore a taking. However, with similar zoning ordinances in cities across the country, Euclid became a test case for the constitutionality of this ten-year-old prescription for fixing cities.

There was incontrovertible testimony that land in Euclid would sell for higher prices if it could be used for industrial purposes than if its use were restricted to one- and two-family houses, and for still higher prices if its use were entirely unrestricted. Ambler's brief, therefore, questioned the reasonableness of the government action that reduced value. It pointed out that the most desirable land for residential use, the wooded hillsides along Lake Erie, had been zoned to exclude all but those who could "maintain the more costly establishments of single-family houses." The "men, women, and children who, for reasons of convenience or necessity, live in apartment houses or in more restricted surroundings of two-family residences" and who are "most in need of refreshing access to the lake or the better air of the wooded upland" were segregated to less desirable inland sites. Ambler further argued that by segregating industrial uses to the land around the railroads and relegating those who lived in apartment buildings and two-family houses to territory adjacent to it, zoning imposed one group's notion of order upon the entire community. It even challenged the very philosophy of zoning by arguing that no legislature could "measure, prophetically, the surging and receding tides by which business evolves and grows" and therefore could never "foresee and map exactly the appropriate uses" or "the amount necessary for each separate use."[45]

The case was argued before the Supreme Court in January 1926. In the months that followed, the National Conference of City Planning, the National Housing Association, the Massachusetts Federation of Town Planning Boards, and a variety of other interested parties filed briefs in support of the constitutionality of comprehensive zoning. The Court scheduled a re-argument for October.

In a 5-to-3 decision written by Justice George Sutherland, the Supreme Court found that comprehensive zoning in general, Euclid's ordinance in particular, and its application to Ambler's property constituted a valid exercise of the police power. In explaining the majority opinion, Sutherland wrote that:

> *. . . the segregation of residential, business, and industrial buildings will make it easier to provide fire apparatus suitable for the character and intensity of development in each section; that it will increase the safety and security of home life; greatly tend to prevent street accidents, especially to children, by reducing the traffic and resulting confusion in residential sections; decrease noise and other conditions which produce or intensify nervous disorders; preserve a more favorable environment in which to rear children, etc.*[46]

Once the Supreme Court had spoken, every major American city—except Houston—that had not yet enacted a comprehensive zoning resolution adopted one.

Incentive Zoning

When New York City included a floor area bonus for developers who provided public plazas in its 1961 zoning resolution,

Manhattan, 1997. The Plaza for 140 Broadway provided an expansion of the public realm. *(Alexander Garvin)*

Manhattan, 2005. When obstructions to potential terrorist attacks were installed on the plaza of 140 Broadway, the public lost the amenity it had paid for by allowing the property owner additional interior floor area that attracted additional people to the site. *(Alexander Garvin)*

it crossed the boundary between regulation intended to ensure minimum standards and regulation intended to induce desirable public amenities at no immediate dollar cost to government. The rationale for the plaza bonus is that while the general public benefits from additional light, air, and room for pedestrians, the chief beneficiaries of the bonus are those who have to bear any additional cost (i.e., occupants, visitors, and owners of the property). While extra expenses may be incurred in providing this amenity, those expenses are more than offset by the additional revenues that are derived from the building's additional floor area. Most important, while the additional floor area generates additional pedestrian and vehicular traffic, increases utilization of infrastructure and community facilities, and casts longer shadows, this additional burden is offset by the benefits of the public plaza. Since property owners can build a conventional structure and are not obliged to take advantage of the bonus, the taking issue is eliminated.

The plaza bonus inevitably led to other forms of incentive zoning. The earliest bonuses were largely for such pedestrian-oriented amenities as arcades, through-block walkways, and small parks. Some cities went beyond pedestrian amenities, enacting bonus provisions for preservation of historic structures, cultural and entertainment facilities, public art, nurseries and day care centers, low-income housing, and a variety of other "public benefits."[47]

Incentive zoning requires a *base FAR* and a *bonus ratio*, but not necessarily a *bonus cap* and an *FAR cap*. The base FAR establishes the bulk and density that can be accommodated in every location. Presumably, any increase in site utilization beyond this base FAR results in an additional load to the area's infrastructure and community facilities, an increase in noise and pollution, and a diminution of light and air. This negative impact, however, will be offset by the use, improvement, or facility for which there is a bonus.

The *bonus ratio* indicates the amount of additional floor area that can be obtained in exchange for providing the desired use, improvement, or facility. For example, Hartford has a bonus ratio of 1:4 for transient parking. Thus, for every square foot of parking that is provided but not specifically required for the subject development according to the zoning regulations, the developer is entitled to 4 additional square feet of rentable floor area above the base FAR.

Since there is a point at which the bonusable amenity no longer mitigates the additional burden, some zoning resolutions cap the amount of the benefit. Hartford's transient parking bonus, for example, has an FAR cap of 2. As a result, no matter how much parking is provided, a property's FAR base cannot be exceeded by more than 2 FAR. Furthermore, to ensure that the combined effect of various bonus provisions does not result in unacceptable bulk and density, some cities also cap the total possible FAR for different sections of the city.

In 1961 New York City's zoning bonus for what Jerold Kayden has dubbed "privately owned public space" (POPS) applied exclusively to plazas and arcades.[48] Over the next 22 years it was extended to elevated plazas; through-block arcades, connections, and gallerias; covered pedestrian spaces; open-air concourses; and sidewalk widening. As of 2000, there were 503 such POPS providing the city with more than 3.5 million square feet (the equivalent of 82 acres or 30 New York City blocks) of additional open space on the most expensive properties in Manhattan. These POPS ranged in size from as little as 174 square feet to 3 acres.[49]

At first there were no specific requirements for obtaining the bonus. Over time, the City Planning Commission ad-

Manhattan, 2006. The privately owned public open space now known as Zuccotti Park is the result of plaza and arcade bonuses for 1 Liberty Plaza. *(Alexander Garvin)*

Manhattan, 2011. For several months, Zuccotti Park was the center of the Occupy Wall Street Movement. *(Alexander Garvin)*

opted specific requirements, including the requirement that there be a prominent sign stating the hours particular plazas had to be open to the public. After the terrorist attack on the World Trade Center on September 11, 2001, property owners (without authorization) began installing bollards, planters, and other obstructions to motor vehicles and people that might be carrying bombs, thereby to some degree obstructing "public" access without giving back any of the interior rentable space that had been obtained from the plaza bonus. The issue of whether POPS is truly public became controversial on the tenth anniversary of the 9/11 attack, when the POPS provided by 1 Liberty Plaza (later renamed Zuccotti Park) was taken over by members of the "Occupy Wall Street" movement for a number of weeks. Clearly, the plaza bonus was quite effective in increasing the amount of space available to the public, but not as effective in ensuring convenient public access.

For bonus provisions to be attractive to property owners, the value of the additional floor area must be greater than the cost of providing the bonusable amenity. Otherwise, no developer will provide it. Determining a suitable bonus is quite difficult. There is no way to predict with certainty the cost of supplying the bonusable amenity. Not only do development and service costs vary, so does the ability of the developer. Nor is there an accurate means of predicting the value of the bonus. Its value will depend on regional economic conditions, citywide demand, district-wide vacancy rates, project location, site dimensions, building configuration, and even the developer's financial situation. The only predictions that can be made are that if the bonus formula is insufficiently generous, nothing will happen, and if it is overly generous, the public will distrust its validity. Unhappiness with Seattle's incentive zoning, for example, became so intense that in 1989 a referendum forced substantial curtailment.

Seattle, the Citizens Alternative Plan (CAP)

Of all American cities, Seattle has the most sophisticated incentive zoning and the most experience with citizen opposition to its effects. The city first adopted incentive zoning in 1963, providing bonus provisions for pedestrian amenities such as plazas and arcades, but without specifying any maximum floor area ratio. As time passed, pressure mounted for zoning incentives that were more sensitive to social issues. By the late 1970s, city officials had begun work on their response: a comprehensive downtown plan and new zoning provisions for both the city's residential neighborhoods and its central business district.[50]

It was the completion in 1984 of the 76-story Columbia Seafirst Center, however, that generated enough political support for city council action. This new office tower had managed to combine enough zoning bonuses to reach an FAR of 28. To prevent further construction of truly giant office towers, in 1985 the city council adopted a *Land Use and Transportation Plan* and new downtown zoning provisions.

The 1985 rezoning divided downtown Seattle into 12 zoning districts. The density for each district was based on its function and character as well as on projected traffic and transit capacities established by the downtown plan. The number of bonusable items was expanded from 5 to 28 (Table 17.1). This time, however, the rezoning established maximum areas that each public benefit feature could award as a bonus, as well as FAR caps for each zoning district. There were also view-corridor, setback, street-wall, retail, and height requirements for each district, with a maximum height of 240 feet in the retail core and 400 feet in the office core.

Bonusable "public benefit features" included (1) *pedestrian amenities* (parks, wider sidewalks, plazas, and "hill-

Table 17.1
BONUS FOR PUBLIC BENEFIT FEATURES: CITY OF SEATTLE ZONING CODE[51]

Public benefit feature	Bonus ratio*	Maximum eligible area
Human service unit:		
New structure	7	10,000 square feet
Existing structure	3.5	10,000 square feet
Day care:		
New structure	12.5	10,000 square feet
Existing structure	6.5	10,000 square feet
Cinema	7	15,000 square feet
Shopping atrium	6 or 8	15,000 square feet
Shopping corridor	6 or 7.5	7,200 square feet
Retail shopping	3	0.5 times lot area (not to exceed 15,000 square feet)
Parcel Park	5	7,000 square feet
Rooftop garden:		
Street-accessible	2.5	20% of lot area
Interior-accessible	1.5	30% of lot area
Hillclimb assist (escalator)	1.0 FAR	Not applicable
Hillside terrace	5	6,000 square feet
Sidewalk widening	3	As required to meet the required sidewalk width
Overhead weather protection	3 or 4.5	10 times the lot's street frontage
Sculptured building top	1.5 square feet per square foot of reduction	30,000 square feet
Small-lot development	2.0 FAR	Not applicable
Short-term parking		
Above grade	1	200 parking spaces
Below grade	2	200 parking spaces
Performing arts theater	12	Subject to Public Benefit Features Rule
Museums	5	30,000 square feet
Urban plaza	5	15,000 square feet
Public atrium	6	5,500 square feet
Transit station easement	25,000 square feet	2 per lot
Transit station access:		
—Grade-level	25,000 square feet	2 per lot
—Mechanical	30,000 square feet	2 per lot
Housing	Subject to Public Benefit Features Rule	Maximum bonus: 7 times lot area

*Ratio of additional square feet of floor area granted per square foot of public benefit feature provided.

climb assists" [a.k.a. escalators]), (2) *land use preferences* (performing arts theaters, cinemas, shopping facilities, and museums), (3) *social services* (affordable housing and, day care, health, and drug treatment facilities), and (4) *design features* (rooftop gardens, sculptured building tops, and atria). Some bonuses were directly related to a building's increased occupancy and its impact on the immediate surroundings (parking, overhead weather protection, and transit station access). Others (affordable housing, museum space, and performing-arts theaters) were intended to have citywide impact.

The bonus provisions were usable in a manner that could double the base FAR. In Seattle's most intensely developed office district, for example, a site zoned for an FAR of 10 could—with bonuses—reach an FAR of 20. One has to wonder whether the underlying base FAR was correct to begin with, if a site's bulk and density could be so easily doubled. Were the infrastructure and facilities downtown really able to support such increased utilization, either concentrated at one particular site or cumulatively, if the entire district were developed with the maximum combination of bonuses?

The citizens of Seattle soon experienced the impact of the proposed new downtown zoning because the developer of the Washington Mutual Savings Bank Building chose to comply with the zoning while it was being revised. The building that emerged under the 1985 regulations combined sufficient bonuses to add 28 stories to the 27 permitted by the underlying zoning. The new 55-story tower galvanized an opposition fed

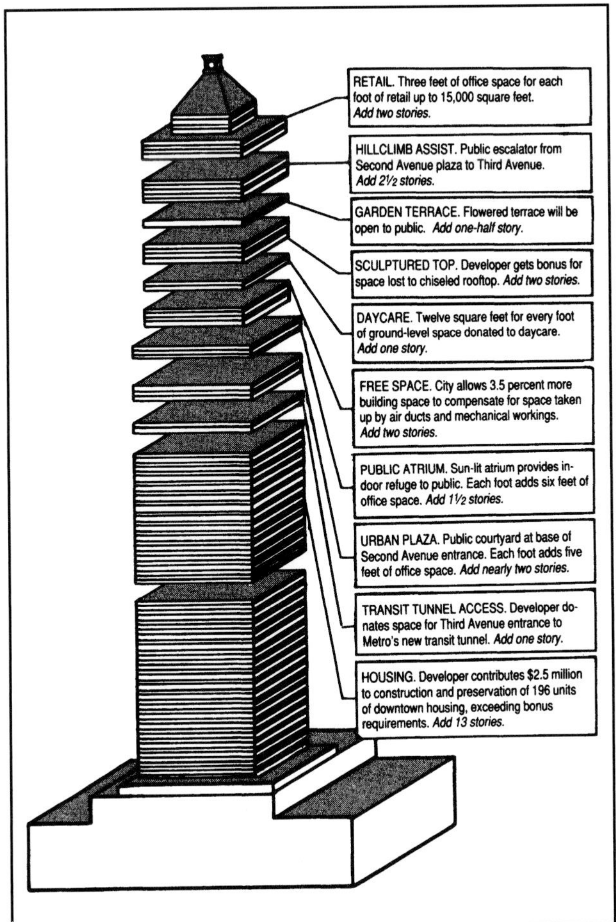

Seattle, 1985. Diagram showing additional bulk accumulated for the Washington Mutual Savings Bank Building by using FAR bonuses made available by rezoning. *(Courtesy of ULI—The Urban Land Institute)*

Seattle, 1990. At first the 55-story Washington Mutual Savings Bank building (extreme right) seemed to tower over the street. *(Alexander Garvin)*

up with "runaway" growth, "out-of-scale" skyscrapers, and "dangerous levels" of noise, fumes, and congestion. They were inspired by the example of San Francisco, which, in 1986, had approved a voter initiative called Proposition M that capped at 950,000 square feet the maximum amount of new office construction that could be authorized during any single year.[52]

Opponents of the new downtown zoning put together a Citizens Alternative Plan (CAP), which limited new downtown office construction to 0.5 million square feet per year

Seattle, 2010. Over time, the Washington Mutual Savings Bank Building blended into the Seattle skyline. *(Alexander Garvin)*

until 1995 and 1 million square feet through 1999, and lowered both the base FAR and the height limits in the city's retail and office core. CAP's proponents secured sufficient signatures to place the proposal on the ballot in 1989, and it was approved by 62 percent of the voters. CAP proved to be irrelevant. Market conditions precluded construction of this much office space. Over the next quarter of a century, other high-rise buildings were erected in Seattle. As a result, the 55-story tower no longer seems "out of scale." Whether the building's sculptured top, atrium, and other amenities justified doubling the number of people and vehicles coming to the site, is debatable. What is not debatable is that the added traffic did not place an undue burden on this city's infrastructure. Thus, one can ask whether the original zoning was justifiable, given the greater carrying capacity of that infrastructure. However, that question may be answered, the effort to CAP further development is sure to be duplicated wherever voters think incentive zoning has gone too far.

Exactions

Some communities adopt policies that make compensatory benefits a requirement of project approval. They perceive property development in terms of the additional load to infrastructure, community facilities, and public services. Rather than just tax future site occupants for this public burden in the same manner that they do current residents, they ask developers to provide "compensatory" amenities. As a result, instead of paying for amenities either with budget expenditures or with a floor area bonus, the community transfers the cost to developers, who pass it on to consumers in the form of higher prices.

Such exactions are common where property development is not as-of-right. In these cases, the developer submits a development proposal for approval either by agency officials or a public board. Subdivision plan approval is perhaps the most common example. The proposed block and lot layout is reviewed for its impact on traffic, drainage, sewage, waste disposal, water supply, and so forth. Since standards for approval are left to the discretion of the reviewing entity, there is an opportunity to negotiate for items that will mitigate prospective negative impact. Project approval then becomes contingent on relocating vehicular access, maintaining wetlands, preserving distinctive landscape features, or other changes. In this manner the desires of the reviewer supplant the printed requirements of the underlying zoning.

There is a clear police-power rationale wherever the amenities that are demanded are directly related to the added burdens caused by that particular development. For example, since factories generate truck traffic, it is reasonable to require that they provide sufficient off-street loading facilities. Similarly, since shopping malls generate automobile traffic, it is reasonable to require that they provide sufficient space for off-street parking.

When an agency demands an amenity that is not clearly related to a specific development, property owners may legitimately claim it has been exacted from them only because they cannot proceed without agency approval. That is what happened when the California Coastal Commission established the goal of increasing public access to Pacific Ocean beaches and made the issuance of development permits conditional on the property owner's agreement to provide the public with a path to the beach.

James and Marilyn Nollan applied for a permit to enlarge their beachfront house in Ventura County. The California Coastal Commission agreed to grant them a permit, provided that they officially recorded an easement for the public to pass through a portion of their property on its way to the beach. The ostensible rationale for this exaction was that the larger house would decrease the view of the ocean and prevent the public "psychologically . . . from realizing [that] a stretch of coastline exists nearby that they have every right to visit." The commission also stated that the house would "burden the public's ability to traverse to and along the shorefront."[53]

The case of *Nollan v. California Coastal Commission* reached the Supreme Court in 1987. The Court found that public access to the Pacific Ocean would not be materially affected if the Nollans built a larger residence on their property and, thus, that the purpose behind the Coastal Commission's action was "quite simply, the obtaining of an easement to serve some valid governmental purpose, but without payment of compensation." The opinion went on to explain that a "legitimate state interest" justifying the exercise of the police power could exist only if there were an "essential nexus" between the regulation and the additional public burdens caused by the development. In the absence of this "essential nexus," the action of the Coastal Commission was deemed to be an unconstitutional taking of private property for public use in violation of the Fourteenth Amendment.[54] Seven years later in *Dolan v. City of Tigurd, Oregon,* the Supreme Court also explained that the nature and extent of any imposed regulation must be roughly proportionate to its impact.[55] Florence Dolan, wanted to enlarge her plumbing and electric supply store and to provide additional parking. The City had previously enacted a Community Development Code that required any person seeking a building permit to provide 15 percent of the lot as landscaped open space and a transportation plan that allocated a portion of the lot for a pedestrian/bicycle path. Dolan was denied permission to proceed with construction unless she deeded one-tenth of her property to the City of Tigurd for a storm drainage control system and a 15-foot strip for the pedestrian/bicycle path. Mrs. Dolan objected, saying that this was a taking of her property. She took her appeal all the way to the Supreme Court, which agreed, saying that the transfer of property to city ownership was not proportionate to the amount of work she had asked permission to do.

Even when exactions may have a reasonable relationship with the construction proposed by a property owner, they frequently produce circumstances in which less development than would otherwise be permitted exclusively by the zoning is ultimately approved, thereby diminishing rental or sales revenue. They also usually increase project cost by requiring a more expensive design, additional construction, or some other contribution. As a result, no developer can be certain that projects that comply with the requirements will also be financially feasible.

The prospect that a project that complies with zoning regulations may still not be approved or the unpredictability of what additional requirements may be exacted can lead to unintended and undesirable results. The most obvious of these is increased project cost. During the weeks or months that an agency negotiates an acceptable project, the developer is forced to spend additional sums to carry the project (debt service, taxes, maintenance, etc.). This not only increases consumer prices, it also scares away both potential developers and the financial institutions that ultimately finance the projects. Neither of those parties is willing to invest time and money in situations whose outcome is unclear.

Another undesirable consequence is the influence peddling by individuals who present themselves as specially able to obtain necessary approvals. In addition to increasing project costs, the increased complexity of the development process excludes from it all but those who are sufficiently well connected to obtain project approval. The more obvious and easily dealt with problem is graft. If it can be demonstrated that project approval is contingent on payoffs, the guilty parties can and should be prosecuted.

A subtler problem is created by exactions extorted by community groups. Neither venality nor illegality is involved. Project opponents simply agree to drop their opposition in exchange for "mitigating" amenities. In the words of a committee established by the Association of the Bar of the City of New York to examine this issue, such amenities "at bottom constitute taxes, which are not levied evenhandedly on the basis of neutral principles but are required from developers on a case by case basis" and by extension "cast government in an unjust and therefore untenable role."[56]

Special Districts

There will always be districts in which conditions are sufficiently special that, without additional provisions, citywide zoning regulations are unable to accomplish the desired results. Charleston, South Carolina, was the first city to come to grips with this problem when, in 1931, it included in its zoning ordinance an Old and Historic Charleston District, which included special safeguards for the area's unique historic structures (see Chapter 18). Other cities slowly adopted the same approach, establishing either special historic or special design districts. The use of special zoning districts for specific purposes, however, became prevalent only after New York City established a Special Theater District in 1969.[57]

Special district zoning took New York by storm. Four years after adopting the Special Theater District, the city had a dozen special districts. Twenty years later there would be 33, many of which had no impact whatsoever, and one of which was even expunged from the zoning resolution. The reason for the many special districts in New York, San Francisco, Santa Monica, and other cities is that they have real appeal to public officials faced with community demands. Officials can simultaneously demonstrate responsiveness and provide special interests with special zoning districts without spending one penny.

The effectiveness of any special zoning district is dependent on the appropriateness of its special regulations. This is vividly demonstrated by the provisions of two special districts intended to spur street life and retailing. The first, Manhattan's Special Yorkville–East 86th Street District, was a flop. The other, Santa Monica's Third Street Mall District, has helped to revive what might otherwise have become one of the worst planning mistakes on the West Coast.

New York City's Special Theater District

New York's 1969 comprehensive plan observes:

> *The theater is a key to a whole complex of communications activities of which New York is the acknowledged center. The hotel and restaurant business also depends to a large extent on the theaters; in fact, the presence of so many legitimate theaters is a very important reason that Manhattan is a major tourist center.*[58]

For this reason, whenever the theater industry is threatened, the city government responds. Radio, talking pictures, and television had in succession threatened the market for live theatrical performances. These were threats over which city government had little control. During the 1960s, however, when New York's theater industry faced a more fundamental threat, elimination of its capital plant, the City Planning Commission responded by establishing a special zoning district.[59]

At that time, demand for new office space in Manhattan was so great and the availability of development sites so small that developers began searching outside the traditional office districts in Lower and Midtown Manhattan. The theater district, located west of fashionable Fifth Avenue, provided a perfect hunting ground. The Port Authority Bus Terminal

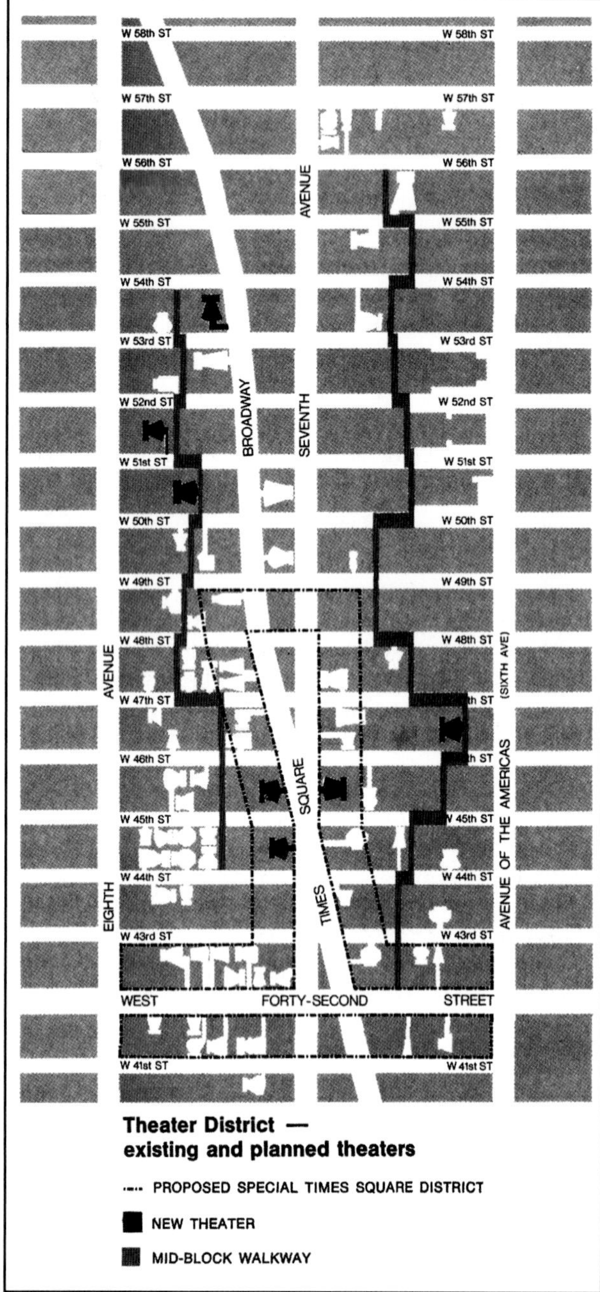

Manhattan, 1969. Map of the Theater District showing sites large enough on which to build modern office structures. The theaters in black were scheduled for demolition. Five of these structures were eventually built using the special zoning bonus for structures including new theaters. *(Courtesy of New York City Department of City Planning)*

was located in the southwest corner of the area. Grand Central Terminal and Pennsylvania Station were half a mile away. The many subway lines that converged around Times Square had excess capacity. For developers, though, its most attractive feature was the existence of large sites, whose buildings (i.e., theaters) used a fraction of the bulk permitted by the zoning. Moreover, these sites could generate substantially greater revenues as office buildings.

The chairman of the City Planning Commission, Donald Elliott, recognized the graveness of the situation. He wanted to engineer a way in which zoning could alter the probable result of market competition for land in the theater district. The Planning Department's urban design group, under the leadership of Richard Weinstein, devised a floor area bonus that would make it financially and architecturally feasible to include theaters within new office buildings at no net cost to the developer. Elliott, however, did not believe that the developers would be interested in extra floor area if it required them to go through the uncertain, time-consuming process of applying for a zoning change. More important, as an attorney, he knew that any spot change to the underlying zoning would be challenged successfully in court if it provided benefits only to a few fortunate property owners.[60]

Working with agency staff, Elliott devised a solution: (1) designate the theater district as a special area in which generic zoning was supplemented by special provisions intended to support the theater industry, and (2) allow all property owners in the area who were willing to include theaters within new buildings to apply for a *special permit* to erect a structure 20 percent larger than those that did not include new theaters. The proposed Special Theater District did not try to eliminate the pressures of the real estate market. In a manner similar to that of the generic plaza bonus, it harnessed market pressure, but in Elliot's adaptation, it was leveraged to produce new theaters rather than public open space. Its bonus provisions could be obtained quickly because the City Planning Commission, having already set forth its land use policy for the Special Theater District, did not need to spend months studying each application. Since every property in the district was eligible for a theater bonus, there could be no court challenge from property owners claiming to have been denied their Fourteenth Amendment right to equal protection of the law. Most important, by making provisions elective rather than mandatory, special-district zoning also avoided the taking issue.

The district's bonus provisions were used in building five theaters, the first new theater construction since the Great Depression.[61] One of the resulting buildings, the government-subsidized Marriot Marquis Hotel, required demolition of the Helen Hayes and Morosco Theaters. While a large new theatre was incorporated into the hotel, the furor caused by the loss of these two beloved theaters led the City Planning Commission, in 1982, to add a theater rehabilitation bonus, which provided benefits to projects that incorporated theater renovation and a special "demolition permit" for any of district's 44 listed legitimate theaters.

In 1982, the commission adopted a Special Midtown Dis-

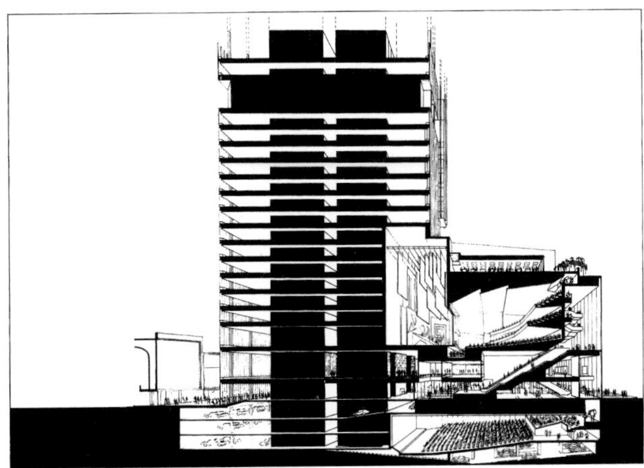

Manhattan, 1969. Illustration showing how a new theater could be included within the structure of a new office building. *(Courtesy of the New York City Department of City Planning)*

trict, in which the Special Theater District was reestablished as the Theater Subdistrict. Six years later, it replaced the theater construction bonus with a theater retention bonus. This provision allows developers who build within the district to purchase up to 1 FAR (but no more than 50,000 square feet of floor area) from any of the 44 existing legitimate theaters, provided that the theater owner records a deed restriction that requires proper building maintenance and use in perpetuity as a legitimate theater.[62]

The effectiveness of each of these theater incentives is dependent on market demand and processing time. Obviously, during periods in which there is little demand for new office space, developers will not be interested in erecting new buildings, and, thus, there will be no market for a theater bonus. That is surely one reason that no buildings were erected that utilized the theater retention bonus during the real estate downturn of the early 1990s.

The lack of interest in the theater construction bonus during the real estate boom of the 1980s, however, was a product of the time and risk involved. Ironically, the reason for this was the special permit that Elliott devised to speed up zoning changes for the theater district. In 1977, the mayor issued an executive order establishing the City Environmental Quality Review (CEQR) requirement. Among its provisions, CEQR required an environmental review prior to any discretionary act taken by the City Planning Commission. Since the issuance of a special permit was deemed to be discretionary, the theater construction bonus suddenly required CEQR analysis.[63]

Manhattan, 2011. Three decades after signage requirements were added to the Special Theater District, Times Square is ever more ablaze with activity. *(Alexander Garvin)*

The cost of a CEQR study reduced the value of the theater construction bonus, but not sufficiently to prevent its use. The more important problem was that it could take as long as 24 months to obtain agency certification that the environmental study was complete and satisfactory. At that point, the project entered the city's Uniform Land Use Review Process (ULURP). This meant another six months of public hearings and public controversy during which there was no certainty of ultimate approval.[64]

Even if developers had been guaranteed approval, too much time was required. Two years of additional debt service, real estate taxes, and design and processing expenses could easily exceed the value of the bonus. And even if the benefits had been worth the added costs, when the process was completed, market conditions could have changed sufficiently to make the project financially infeasible. For these reasons, there were no developers willing to apply for only 20 percent more floor area than they were entitled to as-of-right.

In 1998, the city planning commission decided to modify its regulations and extend the boundaries of the Special Theater Subdistrict so that this zoning incentive would once again generate improvements to the theater district. These revisions restored a special permit mechanism that granted additional floor area for new theaters and allowed the transfer of available development rights from listed theaters "in exchange for retaining, preserving and maintaining" them "for legitimate theater use." At the time, the most important change appeared to be permitting the transfer of development rights anywhere within the Theater Subdistrict.[65]

The more lenient transfer rules, however, proved to be less important than the impact of an unfunded mandate approved by the City Planning Commission and Board of Estimate in 1988, and extended still further when the transfer of development rights was approved in 1998. The mandate required any property owner, seeking to remodel an existing building or to build a new one, to provide a minimum of one illuminated sign with a surface area of not less than 1000 square feet for each 50 linear feet of street frontage.[66] The mandate was vigorously opposed by many of the area's property owners who argued that it would reduce the value of their properties. In fact, illuminated signs could be rented for huge sums, far more than their cost to property owners. As this market-driven reality became more patent, property owners naturally rushed to provide even more than the required minimum signage. As a result, they made a great deal of money from the "unfunded mandate," and the Great White Way, already famous for its neon lights, acquired so much bright, colorful advertising, that street lamps became entirely unnecessary. With the bright lights came even more tourists and spending. Not only had the threat to the theater district that seemed imminent in the late 1960s evaporated, but the economic activity attracted to the district by all that neon began to overshadow the theater industry.

Manhattan's Special Yorkville–East 86th Street District

The Special Yorkville–East 86th Street District, sometimes referred to as the "Anti-Gimbels District," was approved in 1974. It arose ostensibly in response to demands from the residents of Manhattan's exclusive Upper East Side, who were determined to preserve the "Middle European charm" that was fast disappearing from the area. The 86th Street neighborhood had long been characterized by its German-American hofbraus and konditoreis, as well as by a wide variety of small stores catering to Central European immigrants. The impetus behind the special district, however, was more closely aligned with the customers attracted by a department store and movie theater. In the words of a planner residing in the area: "When Gimbels was built and the movie theater opened in the Gimbels, replacing the old RKO . . . [they] attracted consumers from Harlem and the Bronx, and my wealthy neighbors objected to blacks coming into the neighborhood. And there began a real pressure to do something to keep further development of this sort from the neighborhood."[67]

The staff of the city planning department perceived the problem to be similar to the competition between theaters and office buildings. In this case, chain stores, fast-food restaurants, and discount outlets that could pay higher rents were squeezing small stores out. They proposed to change the terms of that competition by establishing a special district that restricted each retail establishment to a maximum frontage of 25 feet. This and other provisions were incorporated into the Special Yorkville–East 86th Street District that was approved by the City Planning Commission. The only dissenting vote came from Sylvia Deutsch, who later served as the commission's chair. She predicted its failure, saying that "fast-food establishments can still flourish and small specialty stores will not come in, regardless of frontage size, unless there is a market they can serve."[68]

As she predicted, the special district only accelerated the decline of the area's older retail establishments. They could not afford escalating rents. Fast-food outlets, high-turnover discount stores, and chain stores, however, could pay through the nose for stores of even limited frontage. Consequently, instead of preventing the demise of a gemütlich 25-foot-wide sauerbraten spot along 86th Street, the new zoning spawned a 25-foot Burger King. In 1989, after most of the ethnic businesses had moved away, this ill-conceived special district was finally voted out of existence.[69]

Santa Monica's Third Street Mall District

The stores along Santa Monica's Third Street Pedestrian Mall also faced competition. It came from Santa Monica Place, the city-subsidized shopping arcade that opened in 1980 (see

Chapter 4). Three blocks of Third Street, bounded by Broadway and Santa Monica Place on the south and Wilshire Boulevard on the north, had been pedestrianized in 1965 at a cost of $703,000. Ninety percent of this cost was paid for by local businesses in the form of an assessment. Naturally, there was considerable unhappiness among local businesspeople when it became apparent that Santa Monica Place was attracting their customers. By the mid-1980s, the mall had become a seedy strip of third-rate retailers, hurting for business.

In 1986, the city adopted the Third Street Mall Specific Plan, which set forth a cumulative mix of actions to reverse deterioration and transform Third Street into a thriving shopping street. They included: (1) establishment of the nonprofit Third Street Development Corporation to supervise the mall's revitalization, (2) redesign and reconstruction of pedestrian areas, (3) construction of additional parking, (4) implementation of peripheral street and alley improvements, (5) adoption of the Third Street Mall zoning district, (6) creation of design guidelines to implement the new zoning, and (7) establishment of an Architectural Review Board to interpret the design guidelines. Together, these actions were intended to transform the mall into a Third Street Promenade that operated "as a unified whole rather than a series of individual shops." The new promenade, it was hoped, would extend active use into nighttime hours and reconfigure merchandising to "be better able to compete with the newer shopping centers which were designed as a single unit."[70]

The special zoning district includes the six blocks bounding the Third Street Promenade. It established a height limit, a base FAR of 3, a group of permitted first-floor

Santa Monica, 2012. Third Street Promenade after new development that made use of the FAR bonuses of the Special Zoning District. *(Alexander Garvin)*

"public inviting" uses along the promenade, and a series of bonuses to encourage improved business conditions. The bonuses included: 0.5 FAR for hotel and mixed-use ("design-entertainment") structures that provide off-street parking, 0.5 FAR for incorporating passageways connecting the public parking garages on each of the six surrounding blocks, and 1 FAR for housing.[71]

Since the adoption of the Third Street Plan, a large number of new buildings have been erected within the zoning district. The most important are three multiplex movie theaters with 16 screens and mixed-used buildings on the corners of the avenues crossing the promenade. A typical example is Janns Court, a 131,000-square-foot building that includes underground parking, 2 restaurants, 1 four-screen multiplex theater, 3 floors of offices, and 32 apartments.[72] The new customers these buildings have brought to the promenade include office workers and their clients, downtown apartment dwellers, and hordes of evening revelers from all over the metropolitan area, who fill its theaters, bars, and restaurants. They have also spurred the renovation of some of the older buildings along the promenade. The special district's bonuses, especially for "design-entertainment" uses, worked so well that the Third Street Promenade has become the destination of choice for people coming from communities in Western Los Angeles County, especially young people. It has become so popular that when the Macerich Company purchased Santa Monica Place and completed a repositioning of it in 2011, it removed the skylight and air-conditioning and turned the shopping center into an extension of the Promenade.[73]

Growth Management

Most zoning ordinances only regulate how property can be used and what can be built there. Some also try to prescribe where and when development can take place. The rationale for assigning control over the location and timing of development to government is that an unregulated marketplace can make costly mistakes. Presumably, without government-initiated growth management, developers will build in areas whose infrastructure, community facilities, and public services cannot handle the impact, either because these community resources will be overloaded or because they are not yet in place.[74]

The explanation for this market failure is that in making location and timing decisions, developers choose to minimize the additional marginal cost to themselves rather than to the community as a whole. The apparent solution is for government to determine the location and timing of development. This, proponents argue, will result in orderly development that begins where there is excess capacity and continues as government installs the necessary infrastructure and facilities. Eventually development will reach service capacity within the geographic fringes of the community. In this way, government will not have to spend capital funds or incur operating expenses before there are sufficient users to pay for them.

Some observers argue that growth management by government is unnecessary because the real estate market already includes a mechanism that minimizes development in locations that are not properly serviced or are not yet ready for settlement. They contend that the consumer will pay more to live where there are good schools, adequate water and sewer capacity, good traffic conditions, and so forth. For that reason, developers will more readily choose to build in those areas. Others criticize growth management as "gangplank" planning: once existing residents have moved in and can profit from an area's properly supplied public services, they pull up the gangplank so that others cannot move in and overload the community.

When growth management programs were first enacted by Ramapo, New York, in 1969 and Petaluma, California, in 1972, these were contentious issues. Ramapo chose to supplement its zoning with growth management based on technical analysis of the load on municipal services. Petaluma chose an annual growth cap and a political process to determine which projects would go first. Since there were no precedents for either approach, neither community was able to determine in advance if it would obtain better results than neighboring areas that depended entirely on the conventional zoning nor who would be harmed and who would benefit from such growth management. There is now sufficient experience with both approaches to evaluate the results.

Ramapo, New York

The unincorporated town of Ramapo is located west of the Hudson River, about 30 miles north of New York City. It consists of a 49-square-mile area that is not very different from the rest of surrounding Rockland County. When the Tappan Zee Bridge was completed in 1955, the area became easily accessible to the rapidly suburbanizing population of the New York metropolitan area and to home builders eager to profit from this new market.[75]

Residential construction in Ramapo soared from 332 dwelling units per year in the last half of the 1950s to 785 dwelling units per year between 1960 and 1964. The population more than doubled, rising from 20,410 in 1960 to 47,711 in 1970. In response to the sudden influx of population, the town adopted a series of planning initiatives, many of which were financed with federal funds. They included a sewer district (1966), a master plan (1966), an official map (1967), a housing code (1968), subdivision regulations (1968), and a six-year capital improvement program (1968). The most controversial measure, however, came in 1969, when Ramapo amended its zoning ordinance to include a section establishing a special permit procedure intended to synchronize housing construction with the availability of infrastructure, community facilities, and public services.

Ramapo established a system that assigned points for different degrees of drainage capacity and sewer service, as well as for the distance from an improved road, park or recreation facility, and firehouse. Proposed developments that could aggregate at least 15 of 23 possible points were entitled to obtain a special permit allowing them to proceed. If the town's capital improvement plan called for items that would bring the point total to 15, a project could be deferred and given a building permit for the year the necessary service would be in place. Property owners without the necessary 15 points could apply for a variance. Alternatively, they could agree to "buy out" the necessary points by providing at their own expense whatever facilities would bring their total to 15.

The impact of the growth management was dramatic. Residential construction for the period 1970 to 1974 dropped to 233 dwelling units per year, a level lower than prior to the opening of the Tappan Zee Bridge. As a result, the population increased by less than 6500 people between 1970 and 1980. Ramapo had wanted to manage growth, not stunt it, but the impacts of its growth management initiatives were simply too stifling. Consequently, in 1983, the town voted to terminate its growth management program.

Ramapo's growth management plan clearly slowed development. It also failed to determine location and timing. This was no accidental failing. It is a fundamental flaw of all growth management plans. They have no way of predicting or adapting to economic cycles.

An equally fundamental explanation is the improbability of any community rigidly following either a comprehensive plan or a capital improvement program. Too often, such programs are wish lists—to be revised in the face of changing construction costs, changing estimates of available tax revenue, and changing demands for public expenditures. In Ramapo, the capital improvement plan had to be altered to deal with unanticipated emergencies (major damage from two hurricanes), changes in federal and state assistance programs, inflation, and unanticipated project costs. As a result, Ramapo could not offer sufficient construction sites where the requisite 15 points were available.

Another explanation is regional competition. Developers were not willing to pay for Ramapo's capital improvements when they could build in surrounding areas without adding the cost of these improvements to the cost of development. The same diversion of development activity to surrounding areas occurred in Petaluma and other cities with growth management programs.

Petaluma, California

Petaluma is located in Sonoma County, California, about 40 miles north of San Francisco. It was founded in 1833 and became a truck-farming community that was prosperous enough to generate a lively array of late-nineteenth- and early-twentieth-century architecture. Development pressure began to grow after the relocation and widening of Interstate 101 in 1956. This improved artery significantly decreased commuting time to San Francisco and opened vast tracts of Sonoma County farmland to residential development. By the end of the decade, Petaluma had grown 37 percent, reaching a population of 14,085. Steady growth continued and by 1970 Petaluma was a city of 24,870. Suddenly, the production of new housing mushroomed, soaring from an average annual increase of 346 units per year during the last half of the 1960s, to 591 in 1970, and 891 in 1971.[76]

The city of Petaluma responded in 1971 by adopting a temporary freeze on development while it sought a suitable way to control what was thought to be runaway growth. It wished to retain its small-town character and to create a permanent greenbelt that would maintain the distinction between city and country. Petaluma also wanted to encourage infilling of undeveloped land in central areas prior to new development along peripheral areas that required additional municipal expenditures on community facilities and infrastructure, and to balance development among all sections of the city (recent construction had been almost exclusively east of Interstate 101). After a year of public discussion and considerable controversy, the city adopted a five-year Environmental Design Plan that capped development at 500 dwelling units per year and a Residential Development Control System that established the mechanism for determining where that development would take place.

The criteria established by the Residential Development Control System included the capacity of local water-supply, sewer, drainage, road, fire-protection, and school systems, "quality of design," and the goal of producing 8 to 12 percent low- and moderate-income dwelling units every year. These criteria were to be applied annually to all proposed development projects. A 17-member Residential Development Evaluation Board, chosen by the city council, then voted on the projects that were allowed to proceed.

After five years of experience, Petaluma revised its Environmental Design Plan. The growth cap was changed to an annual increase of 5 percent, which by the early 1980s meant 700 dwelling units per year. In addition, infill projects of less than 5 acres, projects with fewer than 10 units, and housing for low-income, elderly, and handicapped persons were exempted from the cap. In 1981, another set of revisions replaced the appointed members of the evaluation board with the city's architectural review committee and planning commission

Petaluma, 1991. Turn-of-the-century buildings that growth management was intended to preserve. *(Alexander Garvin)*

and shifted to a rating system that operates year-round. It also increased the exemption to projects of fewer than 15 units. Including all city review and permit procedures, acceptable projects take 6 to 18 months to obtain approval.

It can be argued that Petaluma's growth management plan has helped the city to retain its charm, architectural distinctiveness, low density, and small-town character. However, the older sections of Petaluma had never really been threatened. It had always been cheaper for developers to build on fresh sites than to assemble lots with old houses and tear them down.

Growth management helped Petaluma to reduce leapfrog development and to grow in a fairly compact manner that minimized additional government spending for infrastructure and community facilities. Development has taken place west of Interstate 101, not just on the eastern side of town. Moreover, despite the predominance of single-family house construction, some attached dwellings and cluster developments have also been built.

Once Petaluma introduced growth management, virtually all new development included buried utilities, paved sidewalks, and all those good things that planners recommend. Nevertheless, there is little appreciable difference in appearance between suburban development in Petaluma and in the rest of Sonoma County. The differences lie in the character of housing construction in Petaluma. The year before growth management was initiated, more than half the new houses built in Petaluma were sold at or below $25,000 (constant 1970 dollars). At the end of the first five-year plan, only 3 percent were sold at or below $25,000 (constant 1970 dollars).

Petaluma, 1991. Typical residential development that complies with growth management legislation. *(Alexander Garvin)*

During the same period in Santa Rosa, a city just to the north that had no growth management program, the number of newly built houses selling at or below $25,000 (constant 1970 dollars) remained at 37 to 38 percent. During that same five-year period, the percentage of new houses of less than 1900 square feet in Petaluma dropped from 72 percent to 40 percent, while in Santa Rosa it dropped less than 4 percent, to 81 percent.

From the beginning, developers failed to fill Petaluma's annual growth quota. Even after restrictions were loosened in 1977, at least half the programmed housing construction failed to materialize. Growth management proved to be sufficiently cumbersome, time-consuming, and, therefore, costly to discourage some local developers from building in Petaluma. Instead, they chose to build in more hospitable nearby communities. Thus, given the steadily increasing demand for housing in the area, the value of Petaluma's existing dwellings increased, while the variety (in terms of cost and size) of new housing produced has been reduced.

Smart Growth and Sustainable Development

During the 1990s, the rhetoric changed. Too much opposition had developed to Ramapo-like statutes that obstructed or delayed inevitable change. Instead, civic leaders, public officials, city planners, environmentalists, and developers advocated "sustainable development" and "smart growth." These terms often had different meanings, depending on who used them. For some people, they are no different from no growth or slow growth. For others, they are a sophisticated way of moving growth elsewhere. In the main, however, these terms refer to development that meets present needs without compromising the ability of future generations to meet their own needs.[77]

Advocates of sustainable development usually are in agreement that cities and suburbs should evolve in ways that enhance everybody's quality of life, are economically "sound," and produce environmentally "friendly" results. They demand cleaner air and water, mass transit opportunities, affordable housing, and ample public open spaces. The disagreements occur when local officials have to appropriate funds or levy taxes to pay for these community benefits. Demands for mixed land use, attractive buildings that are compatible with neighboring structures, and preservation of historical buildings engender similar disagreements. Local officials have difficulty determining how to revise zoning and other land use regulations to make those results possible. The good news is that many of the relevant players are talking to one another and using similar phraseology.

Environmental Review

The Standard Zoning Enabling Act of 1924 and every state law based on it require that the zoning power be exercised "in accordance with a comprehensive plan." The rationale for this is brilliantly explained by Lewis Mumford: "Zoning without city planning is a nostrum; and city planning without not merely an initial control of the land—which every municipality has in its unplanned areas—but a continuous supervision over its actual development and uses is merely a branch of oratory or mechanical drawing."[78] The problem was that the authors of most original zoning statutes thought of these texts as devices for achieving a desirable end result, while their successors changed them largely in response to new and pressing demands.

Few communities maintain up-to-date comprehensive plans, capital improvement programs, or zoning ordinances that deal comprehensively with the interplay of their physical, functional, economic, social, political, cultural, and environmental issues (see Chapter 19). This absence of serious city planning leads citizens to seek legislation that will deal with issues untouched by comprehensive zoning. That is one reason that Ramapo and Petaluma enacted growth management plans. It is also the reason for land use regulations independent of zoning, which protect landmarks, historic districts, and environmental quality. If zoning had been truly comprehensive, there would have been no need for supplemental statutes covering historic preservation or environmental protection.

The earliest efforts to supplement "comprehensive" zoning were those of preservationists. Charleston, which in 1931 established the nation's first historic district, chose to make preservation a component of its zoning ordinance (see Chapter 18). Preservationists, while trying to persuade other communities to create historic districts, directed their main effort at the federal government. The first step taken was the establishment during the 1930s of the Historic American Buildings Survey and the National Register of Historic Places, which listed designated national landmarks. The most important step came when Congress enacted the Historic Preservation Act of 1966. This law required an examination of the impact of actions by federal agencies or federally financed programs on landmarks listed in the National Register. It was this requirement of an impact study that in 1969 prevented the construction of a highway through the Vieux Carré in New Orleans (see Chapter 18). The preservationist reliance on review and study prior to permitting development to proceed rendered preexisting zoning regulations far less important. As time passed, zoning in many areas remained unchanged despite changes in population, new market conditions, and activities that had not been conceived when the zoning was enacted. In response, citizens demanded project review prior to approval (even when development could proceed as-of-right). The most important demands for review came from environmentalists who correctly insisted that environmental concerns had not been considered when the zoning was enacted.

The National Environmental Policy Act of 1969

Preservationists were not alone in objecting to the narrow scope of supposedly comprehensive zoning ordinances. Environmentalists shared their view. They also chose federal legislation as a way of introducing environmental considerations into land use regulation and adopted the impact study technique pioneered by preservationists. The National Environmental Policy Act of 1969 (NEPA), which grew out of their efforts, is the most significant land use regulation since the enactment of comprehensive zoning in New York City in 1916. Like that early resolution, NEPA has spawned similar statutes in about half the states of the Union.

NEPA sets forth national environmental policy and establishes the Council on Environmental Quality to supervise implementation. Its most important provision is Section 102(2)(c), which provides that:

> *. . . all agencies of the Federal government shall . . . (c) include in every recommendation or report on proposals for legislation and other major Federal actions significantly affecting the quality of the human environment, a detailed statement by a responsible official on—*
>
> *(i) the environmental impact of the proposed action,*
> *(ii) any adverse environmental effects which cannot be avoided should the proposal be implemented,*
> *(iii) alternatives to the proposed action,*
> *(iv) the relationships between local short-term uses of man's environment and the maintenance and enhancement of long-term productivity, and*
> *(v) any irreversible and irretrievable commitments of resources which would be involved in the proposed action should it be implemented.*[79]

Thus, NEPA forces federal agencies considering any action whatsoever to determine whether that action is major. If it is, the agency must determine if it "significantly" affects the environment, what the alternatives are, and, if a detailed *environmental impact statement* (EIS) is required, what should be done to see that it is prepared in a procedurally and substantively correct manner.

"Major federal actions" include more than establishing an army base, erecting a post office, demolishing a landmark listed on the National Register, or other actions directly undertaken by federal agencies. The term also includes federal decisions to approve, fund, or license activities carried out by others (e.g., designating a route for inclusion in the Interstate Highway System, funding an urban redevelopment project, licensing utility companies to build nuclear power plants). As a result, NEPA opened a huge number of previously private activities to public review.

It is extremely difficult to establish a threshold for determining whether an action "significantly" affects the environment. As a result, agencies play it safe. Even activities that are thought to be of minor import are no longer permitted without prior environmental review, if only to see whether it is possible to avoid producing a full EIS. Since the federal government is involved in a vast array of funding, licensing, and approval actions, NEPA has made environmental review a virtual necessity for myriad nonfederal activities that were not previously subject to regulation.

The Council on Environmental Quality established the guidelines for preparing and approving an EIS. These involve specific actions that may take years to complete. The relevant agency must decide whether an action requires an EIS. If not, it issues a *negative declaration* and proceeds with the action. When an EIS is required, the agency must first publish in the *Federal Register* a notice of intent to prepare one and, when more than one agency is involved, indicate which is the lead agency with supervisory responsibility.

The EIS must consider both direct impact and secondary effects (e.g., additional traffic congestion, air pollution, and noise). In addition, it must assess all reasonable alternatives to the proposed action and their primary and secondary effects. Finally, the relevant agencies must give serious consideration to any detrimental environmental consequences caused by the proposed actions that cannot be practicably mitigated.

Once a draft EIS has been prepared, the responsible agency must publish in the *Federal Register* (and in suitable local newspapers) a notice of its availability and seek comments from the public and from other agencies with expertise or jurisdiction over the action under consideration. All these comments must be considered before the agency prepares a final EIS. This final EIS then must be published prior to any federal agency review, decision making, or course of action, no matter what environmental protection requirements are imposed as part of the approval process.

Given the increasing obsolescence of most land use regulations, there may have been little alternative except to mandate an informed response to every pressing situation. But this has meant adopting a reactive approach that is the very opposite of planning. *Planning* requires deliberation by a legislative body that balances conflicting goals and reflects a future that is desired by the society as a whole. *Impact analysis* transfers decision making to the judiciary, which has the much narrower role of deciding whether a particular result has been adequately considered.

NEPA's Impact

NEPA was intended to be a method of encouraging federal agencies to consider the environmental consequences of their actions. While this desirable result has been achieved, too often the process becomes an elaborate, time-consuming, and extremely expensive justification of decisions that already have been made. It has also spawned a bonanza for clerical workers, testing laboratories, engineers, scientists, planners, and lawyers, all engaged in preparing, analyzing, or challenging environmental impact statements.

More important than its role in the decision-making processes of federal agencies is that NEPA provides a statutory basis for private lawsuits. Controversial projects now can be stalled for years while opponents pursue litigation against the agency responsible for issuing an EIS. The basis for private legal action can be procedural, substantive, or both. As a result, some minor flaw can result in major delays while the relevant agency takes the necessary curative action. In the process, the agency is forced to deal with one particular concern instead of balancing all significant interests.

The demise of New York City's Westway (see Chapter 3) presents a particularly vivid example of the way in which single-function interests can dominate environmental reviews and warp the decision-making process. To satisfy NEPA requirements, the U.S. Army Corps of Engineers dutifully conducted the mandated environmental review, producing a study of the highway and various rebuilding alternatives. Almost immediately, the study came under legal attack from environmental protection groups. The Hudson River Fisherman's Association asserted that it failed to consider properly Westway's effect on the spawning habits of the striped bass (a major element of New York's fishing industry). The court agreed and ordered another study.

The second review, which adhered to a strict interpretation of the law, focused on the impact of Westway's traffic flows, the different alternatives for building a highway on Manhattan's West Side, and its effect on the environment surrounding its route. Yet Westway raised a huge number of other issues: Should such large sums be spent on a transportation project rather than other city needs? If so, should the money be used to build a highway rather than to repair existing city arteries or to improve mass transit? How would this new roadway affect existing traffic conditions? Was the proposed plan (parks, manufacturing, housing, etc.) the right way to develop 178 acres of created land? How would so large a project affect the neighborhoods bordering Westway?

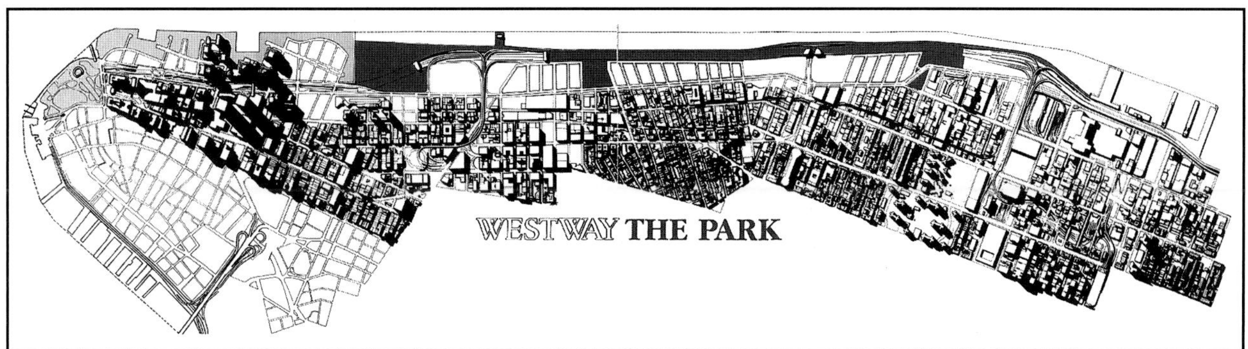

Manhattan, 1985. Plan of the defeated proposal for Westway—an underground highway and landfill project extending along the West Side of Manhattan. *(Courtesy of New York City Department of Parks and Recreation)*

What would be its impact on the city's population, employment patterns, tax base, air quality, and so forth? The EIS barely considered these issues.

Ultimately, *none* of them played a crucial role in determining whether to proceed. Westway was brought to a halt by litigation initiated by a variety of environmental and community interest groups. The critical issue proved to be that the EIS had not properly examined the effect that demolishing Hudson River piers would have on striped bass that spawned below them. Striped bass were no endangered species. They did, however, contribute an estimated $200 million and 5600 jobs to the commercial and recreational fishing industries in 1980. Consequently, determining the effect of displacing bass from the waters under the piers, where they paused during their winter migration to the Atlantic Ocean, was no frivolous question.

In response to the concerns raised by the Hudson River Fisherman's Association, the Army Corps of Engineers spent $3.1 million over the next two years to catch 57,429 fish (11,415 of which were striped bass). In the words of one of the men in charge of the study, the entire effort was devoted to a determination of "what fraction of the striped bass we are willing to displace to build the project."[80] The results were insufficiently conclusive to prevent further litigation.

Accordingly, in 1985 government officials canceled Westway not because they had evaluated whether Westway was the best or most cost-effective way to use federal money in New York City, or even because it was the correct action to take in response to serious environmental concerns. As explained in Chapter 3, it was canceled because public officials were afraid that the city would lose its ability to obtain a $1.7 billion federal grant if Westway faced further delays. The holdup imposed by the environmental review proved simply too intense and too time-consuming for the proponents of the project to withstand.

During the quarter of a century since the demise of Westway, environmental review has become a standard component of any land use decision. It is, however, a decision-making process that is seriously flawed. The review focuses on the environment as if it is otherwise in no way subject to human actions. Moreover, it relies on an approach that emphasizes the singularity of certain functions to the exclusions of all others; in the case of Westway, this was the impact of transportation, as if nothing else were happening. A narrow analysis of this sort may be acceptable in rural Nebraska, where the highways are generally flanked by single-function farmland. But it does not make any sense in complex cities like New York, Chicago, or Los Angeles, where the territory through which proposed highways will run is affected by a myriad of forces other than transportation.

State and Local Environmental Review

Once NEPA had been signed into law, environmentalists obtained similar legislation on the state and local level. Unfortunately, they failed to perceive the fundamental difference between the situation facing the federal government and that faced by local communities. The federal government has no comprehensive mechanism for controlling land use. Local governments, on the other hand, already have land use regulations, which often have been in place for centuries.

In 1970, California became the first state to adopt legislation similar to NEPA. Four years later, 11 states had NEPA-like statutes. By 2012, 41 states had "some requirement for agencies to assess the environmental impact of their actions."[81] These statutes make it impossible to make land use decisions simply by complying with published comprehensive zoning resolutions. Consequently, rather than make decisions based on a community's explicit land use policy, property owners in areas with NEPA-like statutes are forced to make decisions based on an assessment of the probability of lawsuits challenging the validity of environmental impact studies and the cost and time required for both preparing the studies and fighting possible lawsuits.

Few state or local environmental-review laws are as rigorous as those of California. The California Environmental Quality Act of 1970 (CEQA) applies to all state agencies, boards, and commissions, to "the legislative bodies of all cities and counties," and (since 1972) to "all other local governmental agencies." In the case of *Friends of Mammoth v. Board of Supervisors,* the California Supreme Court interpreted CEQA to apply to any agency granting or denying a permit, including permits for actions taken in compliance with local zoning resolutions and building codes. The court observed that these regulations and codes had been enacted without the benefit of rigorous environmental impact analysis.[82]

As a result of this decision, environmental reviews have become necessary for any action by any agency anywhere in California, except when (1) it is a purely ministerial action that involves no discretion by public officials but only the application of fixed standards and objective measurements, or (2) it is specifically exempted by state legislation. Environmental review is thus required prior to the issuance of building and grading permits, sewer- and water connection permits, franchises, leases, subdivision plans, redevelopment projects, variances, zoning changes, and a variety of other entitlements.

As in the case of NEPA, if the environmental impact is not significant, the relevant agency can issue a *negative declaration* eliminating the need for an environmental impact report (EIR). CEQA, however, is more stringent than its national progenitor. NEPA only requires that federal agencies "consider" the potential significant adverse environmental impacts presented in an EIS. The California version forbids agencies from approving projects that might have significant adverse impacts when feasible alternatives or mitigating measures could lessen such impacts.[83]

Unlike California, most states have adopted environmental protection statutes that are more closely patterned on NEPA. Nevertheless, the proliferation of environmental review requirements can restrain development. This is true not only for situations in which a full EIS is required. Even when a project is entitled to a negative declaration, the money and time required to obtain agency action retard development and raise costs.

The chilling effect of agency processing is well illustrated by the time required by the New York City Planning Commission to make minor changes to special zoning districts. In early 1987, a property owner decided to build an apartment house in one of the city's many special districts. His site was a vacant corner that had been zoned for residential use decades before the special district had been enacted. The project was feasible only at the area's originally zoned floor area ratio. Despite the fact that the zoning map still displayed this FAR, special zoning district regulations permitting that bulk had lapsed 10 years earlier.[84]

The owner met with agency staff, who assured him that the change was not only minor, it was also consistent with agency policy. Three months later, he filed a City Environmental Quality Review statement. Over the next 18 months, various city agencies requested studies whose relationship to the zoning change remains difficult to discern. They included a survey of manufacturing firms within 400 yards of the site (for which no change from its zoned residential use was requested), a compilation of all fire department permits for area gas stations, an analysis of the ethnicity of the surnames of area residents enumerated by the censuses of 1850 and 1900, and a study of the site's possible archaeological resources.[85]

Eighteen months after the initial CEQR submission was received by the agency, the city issued a negative declaration; 28 months later, the City Planning Commission voted unanimously for the zoning change; 30 months later, the zoning change was unanimously approved by the Board of Estimate. Construction plans for a 12-story apartment house were prepared. Unfortunately, by the time they were approved by the New York City Buildings Department, market conditions had changed sufficiently that the project was no longer financially feasible.[86]

The Community Board, City Planning Commission, and Board of Estimate all held public hearings and completed their work within the required statutory 6-month period. The other 24 months of review were caused by civil servants who wanted information that was of interest to them and that might possibly be needed to measure or mitigate the project's impact. There being no specific printed requirements for CEQR submission, they had been given a hunting license to uncover anything that might have a deleterious impact. Conscientious reviewers did their best to cover every possible angle, without reference to the impact of any added cost or delay. The developer, who had to protect the project from lawsuits challenging the completeness of the environmental review, had no choice but to respond to every request for additional information, no matter how frivolous.

Nobody involved with reviewing this request for a *negative* declaration was responsible for completing that review for a specific budgeted sum of money, nor within a fixed period of time. Neither was anybody charged with determining the relative importance of any of the information requested. This is the very antithesis of comprehensive zoning, which is based on a printed document with explicit requirements. Zoning resolutions are prepared by consultants and civil servants who must produce them for a specific budgeted sum and then submit them to elected officials for approval. Unlike environmental reviews, the very process by which a zoning resolution is produced and ratified guarantees that it chooses among conflicting goals and establishes priorities.

Most jurisdictions have adopted a more reasonable approach. Hudson County, New Jersey, for example, excludes entire categories of projects from review and routinely processes its few environmental assessments within a period of weeks. Nevertheless, both NEPA and NEPA-like local legislation increase the time required for, and the cost of, development. They generate direct costs (preparing reports and engaging in litigation) and indirect costs (debt service, taxes, and maintenance on a property during the period of agency review). Both increase uncertainty and thereby restrict development to those who have the time, money, and experience to obtain agency approval. There is one serious difficulty, however, that is unique to *local* environmental-quality-review statutes. They invalidate local zoning ordinances. For, if every project has to be examined individually, there is no reason for communities to develop carefully considered land use policies that balance conflicting goals. Nor is there any reason for them to maintain a written set of land use regulations that predetermine what land use activity will or will not be authorized.

Form-Based Zoning

During the last decade of the twentieth century, a group of architects, led primarily by Andrés Duany and Elizabeth Plater-Zyberk, started a movement to upend comprehensive zoning. In preparing plans for Seaside (see Chapter 16), they sought human-scaled, pedestrian-oriented communities. Communities of this sort cannot be mandated by conventional zoning ordinances, which rely on listing what is prohibited. Conventional zoning cannot prescribe a specific floor area, building height, or building placement. Consequently, proponents of form-based zoning jettison the notion of permitting property owners to build anything that does not *exceed allowable limits.* They oblige anybody who builds to create a structure that includes *explicit features.* At Seaside, those features were based on a careful examination of the traditional architecture of the Florida panhandle. For the communities that came later, the characteristics were based on a similar respect for traditional construction in that section of the country.

The result was an entirely new *form-based* zoning that mandated building types (bungalows, row houses, etc.), frontage conditions (stoops, porches, arcades, shop windows, etc.), building heights and placement, open spaces (plazas, courtyards, patios, etc.), and street standards (maximum block lengths, midblock pedestrian connections, alley patterns, etc.). This new zoning also stipulated building massing, orientation, roof shapes, materials, details, and other aspects of building design. It therefore became impossible to build huge mansions in neighborhoods without them. In bungalow neighborhoods, property owners had to build similar bungalows; in row house neighborhoods, they had to build similar row houses.[87]

Besides the mandates, there were two other aspects to any form-based code. It prohibited radical shifts in scale between districts and it minimized the text and drawings needed to convey the requirements. The Duany/Plater-Zyberk progression of scales, called *transects,* ranged from rural to high-density urban core. These transects were presented in an evolving handbook entitled *Smart Code,* first published in 2003. Each transect area contained building typologies that reflected the relationships among streets, yards, and buildings that were common to that district. Thus, the transects determined the density and character of each district to be developed.[88]

At Seaside and some of the earliest New Urbanist communities, their form-based codes were laid out on a single page, listing all the requirements by zone. This minimization of text and detail was particularly appealing to ordinary citizens, who could no longer decipher the complex, comprehensive zoning resolutions that had taken decades to evolve. The emphasis on walkability meant higher than usual levels of

Miami, pre-2009. Typical development permitted prior to Miami 21, the city's comprehensive rezoning initiative. *(Courtesy of Duany Plater-Zyberk & Co.)*

Miami, post-2009. Development based on the form-based zoning principles that require transitional, midrise buildings between high-rise, corner buildings and one-family houses on the rest of the block. *(Courtesy of Duany Plater-Zyberk & Co.)*

density. This attracted support from developers, who wished to reap profits from properties that would support additional buildings if New Urbanist regulations were adopted.

Form-based zoning has spread beyond its Florida roots. Denver has inserted form-based "Main Street zones" into its existing code, initially for Colfax Avenue and eventually for its low-, medium-, and high-density commercial corridors. After 500 community meetings and four years of discussion, Miami exchanged its old code for form-based zoning that covers the entire city.[89]

Despite their simple origin, most form-based codes are becoming increasingly complex and can no longer be laid out on a single page. By concentrating on form alone, they still fail to deal with issues that are routinely included in comprehensive zoning texts. The most obvious of these are land use and density—which in turn determine the load carried by local schools, community facilities, and infrastructure as well as the traffic and noise that penetrate city neighborhoods. Perhaps the least appreciated is the deeply conservative attitude of form-based zoning. It insists that whatever is, is right and, accordingly, all new development should emulate existing buildings within the district. It is as if Thomas Jefferson had never explained that "as new discoveries are made, new truths discovered and manners and opinions change, with the change of circumstances, institutions must advance also to keep pace with the times," and that "we might as well require a man to wear still the coat which fitted him when a boy as civilized society to remain under the regimen of their barbarous ancestors."[90]

Ingredients of Success

During the second third of the twentieth century, communities developed and maintained standardized comprehensive zoning regulations that allowed citizens to predict the future character of any area. This reduced the risk of incompatible activity and stabilized market conditions. Once assured of a predictable future, financial institutions, real estate developers, and community residents could more confidently invest in real estate.

Market conditions began to change in the late 1960s. The primary causes were increasing rates of inflation, rising interest rates, changing income tax formulas, new banking procedures, and other factors that had little to do with land use regulation. Maintaining existing land use regulations could *not* have prevented real estate prices from rising. However, instead of maintaining existing land use regulations to help steady the investment climate, governments did precisely the opposite. Heady with the success of comprehensive zoning, they introduced incentive formulas, special zoning districts, growth management plans, historic preservation requirements, and environmental-quality-review legislation. These innovative land use regulations increased the cost of complying with government regulations, the complexity of obtaining institutional financing, and the uncertainty of government approval. However, like the simpler statutes of the middle third of the twentieth century, they succeed only during periods of market demand. After all, there is nothing to regulate when nobody wants to build.

Market

There is a false notion among proponents of land use regulation that their statutes will inevitably affect the real estate market. In fact, they may be irrelevant. No matter what zoning provisions had been adopted for Manhattan's 86th Street, for example, the number of German and other middle-European retail outlets would have declined in direct proportion with the decline in the area's immigrant population.

In many cases, land use regulation does affect the market. Improperly conceived land use regulations can stifle development pressure. Ramapo's growth management plan, for example, curtailed demand for what otherwise would have been viable development sites.

Location

There are two components of every zoning resolution: text and maps. Success lies in applying appropriate regulations (text) to the appropriate locations (map). The nation's first comprehensive zoning resolution provides a dramatic illustration of the successful combination of map and text. Its authors wanted to guarantee the city's most powerful commercial enterprises an auspicious environment in which to do business. By mapping Fifth Avenue below Central Park as a district in which manufacturing was forbidden, they successfully manipulated market pressure to a location that the business community wanted.

New York City's 1961 zoning resolution includes an equally dramatic illustration of the importance of location, one that failed. Its authors assumed continuing demand for manufacturing space and zoned numerous sections of the city for manufacturing, including enormous sections of Lower Manhattan. In fact, manufacturing employment in New York City, as in most large cities, was already declining. Between 1956 and 2012, the number of manufacturing jobs plummeted from 532,000 to just over 74,000.[91]

When the 1961 zoning resolution was adopted, multistory industrial lofts in many locations of Lower Manhattan were already partially vacant. As demand for the uses specified by the zoning declined, so did prices. In some areas, SoHo and Tribeca in particular, the availability of cheap space attracted uses quite different from those established by the zoning. Eventually, "illegal" tenants forced the City Planning Commission to legitimize their occupancy by creating special zoning districts for SoHo and Tribeca (see Chapter 18). By continuing to zone other locations for manufacturing despite insufficient demand for such uses, public officials prescribed for these areas a future of increasingly vacant and underutilized structures. This resulted in occupants finding ways to circumvent the zoning, public officials rezoning those areas at a later time, or both.

Design

There are a few communities, such as Santa Barbara, California, and Santa Fe, New Mexico, where appearance is crucial. These cities have nurtured sizable tourist industries by requiring all new construction to comply with an idealized regional esthetic (see Chapter 18). In most cities, however, the design component of land use regulation is not a matter of architectural style. It is an expression of land use policy. The placement of buildings along a street is a good example. If the goal is to improve pedestrian circulation, buildings should be required to set back a minimum distance from the property line, thereby obtaining wider sidewalks. On the other hand, if the goal is to increase the continuity of retail shopping, setbacks from the property line should be forbidden, thereby preventing any interruption in window-shopping.

The trick is to stick with simple design requirements that will produce a suitable and consistent physical environment. Washington, D.C., provides an apt illustration. Nearly a century ago, Congress enacted height limits intended to strengthen the symbolic character of the national capital, while simultaneously ensuring plentiful light and air. These height limits were related to the width of streets and their role within the city. Neither architects nor developers could misconstrue this simple device, nor could they subvert it. It helped Washington to develop into a gracious city that is particularly consistent in appearance and in which the relative importance of each thoroughfare is reflected by the character of adjacent buildings.

Architects and developers frequently terminate the design process when they have complied with the letter of the law. This transforms minimum requirements into standardized designs that make a mockery of land use regulation. New York City's sky exposure planes are a good example. They were intended to—and did—increase light and air for both pedestrians and building occupants. Unfortunately, too many architects and developers devised schemes that were *primarily* fitted to the relevant sky exposure plane. As a result, New York is filled with hundreds of office and apartment buildings built in the shape of stepped pyramids that may increase the availability of light and air but contribute nothing to the functioning of the city or to the coherence of its appearance. On the other hand, form-based zoning can make quite desirable adjustments to the built environment. The design requirements adopted in Miami's form-based zoning, for example, have softened the impact of large new buildings erected next to low-rise buildings, in the process reducing opposition from nearby residents.

Manhattan, 2012. Seagram Building Plaza on Park Avenue. *(Alexander Garvin)*

Public officials not only disfigure cities by allowing minimum design requirements to be transformed into standard market products, they also do so by pursuing momentary fashions rather than fundamental design principles. A good example is New York City's plaza bonus. It was adopted in an attempt to encourage development like the extraordinarily beautiful Seagram Building, with its inviting public plaza on Park Avenue. The design, by Mies van der Rohe and Philip Johnson, paid no attention to setback regulations, height limits, or sky exposure planes. In fact, it could not have been built without zoning modifications. The glorious pedestrian environment it creates is a product of the designers' concern for the way in which people walk in, out, and past the building: the approaches along Park Avenue, 52nd Street, and 53rd Street; the diagonal movement up steps onto the site; the interplay of visitors on the plaza; the changing views of the city as one passes into the glass-enclosed, 24-foot-high elevator lobby.[92]

Dozens of buildings erected after 1961 made use of the plaza bonus. Not one creates a pedestrian environment that equals what Mies and Johnson achieved. In Lower Manhattan, the additional pedestrian space has been a genuine asset. Elsewhere, the plaza bonus has resulted in some of the nation's most expensive but least necessary open space. These plazas could have been avoided if public officials had not unthinkingly adopted a fashionable design and instead had applied it only to situations in which it would be beneficial.[93]

Financing

Land use regulations mandate expenditures for planning and processing, as well as specific configurations of use, bulk, and density, and public amenities that would not otherwise be provided. They all increase the cost of development, affect the availability of financing, and cause consumer prices to increase. In some cases, these added costs are justifiable even to the property owner (e.g., off-street parking for retail stores, off-street loading facilities for factories, ticket-holder waiting space for movie theaters). However, too often they are adopted without any consideration of affordability.

In the case of both scattered early land use restrictions and later comprehensive zoning, the community as a whole carried the cost of planning. It paid these costs because individual property owners had no interest in or reason to develop citywide policies and could not be forced to do the necessary planning. Communities also did so because they thought it only fair that these costs be shared by everybody who benefited. More recent discretionary forms of land use regulation, such as state environmental review statutes, transfer this cost to individual property owners, who, in turn, pass them on to their customers.

Discretionary reviews, such as those required by some growth management programs, NEPA, and NEPA-like statutes, add uncertainty and the cost of litigation over the validity of decisions. Their major impact, though, comes from adding the cost of carrying the property (debt service, taxes, maintenance, and equity capital) during the review period. In the case of New York's theater-construction bonus, this was sufficient to preclude further theater construction, despite a potential 20 percent increase in buildable floor area.

The most pernicious form of increased cost is one that does not cost the consumer or the government any cash. Such situations are caused by regulations that induce developers to pay for public facilities that elected officials are not ready to fund through normal budgetary procedures. Seattle's pre-CAP zoning provides a good example. The local government did not appropriate money to pay for sculptured building tops, skylit atriums, or escalators. Instead, it offered downtown property owners added floor area as compensation for providing these amenities. The electorate would not tolerate this prostitution of zoning and voted to "CAP" development.

Zoning can also be a boon to property owners seeking to finance their properties. The mandated signage requirements in the Times Square Special Zoning District, for example, made possible considerable additional revenue from advertising. That revenue could be used to justify mortgages in excess of what lending institutions would have approved prior to the rezoning.

Time

Time is usually the forgotten component of land use regulation. When it becomes a conscious ingredient, however, it can bring significant improvement to city life. Land use regulations affect the time in which an individual passes through an area, the 24-hour period during which people use the area, and the period of years during which the character and utilization of that area can change significantly.

The land use regulations for Santa Monica's Third Street Promenade illustrate how successful land use regulation can shape the first two periods of time. These regulations include specified ground-floor retail uses along the promenade and FAR bonuses for projects that provide direct pedestrian access to the promenade from public parking structures and for projects that include hotels, department stores, and design and entertainment centers. Together, they combine to shape the time spent walking from parking, past impulse-shopping facilities, on to more substantial retail destinations. They also encourage land uses that extend patronage beyond weekdays to nights and weekends. Once the Third Street Special Zoning District was enacted, developers responded by remodeling or rebuilding many of the sites along the promenade. In short order, this seedy strip of stores that used to close by 6 p.m. became one of the Los Angeles region's liveliest shopping areas. Its movie theaters and restaurants attract crowds into the wee hours of the morning, seven days a week.

Growth management plans are an example of land use regulation that extends the conception of time beyond hours or days, to a period of years. Essentially, they are a form of timed zoning. When conceived to extend over several seasons, growth management can reduce the load on overburdened facilities and prevent government from spending extra sums to play catch-up. For a growth management plan to be successful, however, it must include a capital improvement program that outpaces privately financed real estate development and an annual threshold for additional construction that exceeds market demand. Petaluma's growth management strategy provides an excellent demonstration of these principles. It established a growth threshold that was double the number of housing units actually produced. As a result, Petaluma was able to minimize additional capital expenditures without unduly restraining construction.

Entrepreneurship

The beauty of comprehensive zoning is that it minimizes the time, effort, and money needed for property development. Its requirements are readily available in a published document that can be purchased by community residents, property owners, and their architects. Most important, it provides lending institutions, developers, and their consumers with *certainty* that all development will comply with preestablished rules and *predictability* as to the future character of the environment. Without such government guarantees, lending institutions would be far less likely to provide the credit needed for any community to prosper.

There is a second, equally important result. The certainty and predictability of preestablished rules reduces risk. This, in turn, increases the number of people willing to get involved with real estate development. It explains why development during the middle third of the twentieth century was dominated by small real estate entrepreneurs. The resulting competition (together with New Deal banking reform, federal income tax incentives, and highway construction) helped to keep the price of real estate reasonable and allowed the national rate of homeownership to rise from one-third to two-thirds of the population (see Chapter 9).

As the regulatory environment has grown more complex, the rules less certain, and the cost of doing business greater, the number of players in the development game has been curtailed. Increasingly, real estate has become the preserve of sophisticated, well-financed entrepreneurs, well-connected consultants with specialized knowledge of regulatory requirements, and seasoned veterans able to interpret and manipulate government operations. Sadly, the landscape they are creating is neither appreciably more convenient nor more beautiful. It is only more expensive and more likely to generate community opposition.

There is another, even less attractive, result. The lack of certainty and predictability has increased the danger of influence peddling. Developers who require negative declarations, special permits, or any other discretionary government action are inevitably tempted to hire advocates who they believe (often erroneously) can influence the decision. They also are tempted to make deals with community organizations and special interest groups whose assistance they (also, often erroneously) believe can influence the decision.

In this increasingly discretionary environment, some developers expedite favorable action by paying off government employees. No one should tolerate such graft. Those making or receiving payoffs belong in jail. Unfortunately, communities across the country unwittingly increase opportunities for corruption by increasing the number of land use regulations that require discretionary action by government officials.

Land Use Regulation as a Planning Strategy

We have created a crazy quilt of land use statutes and discretionary procedures that needlessly increases the price of housing, places of employment, and virtually everything else. These obsolete and often conflicting statutes disfigure the landscape with inappropriate structures, prevent large territories from being used in economically productive ways, generate angry citizen protests, and require for their operation an ever-increasing army of bureaucrats and consultants. A second unfortunate consequence is unjustifiable discrimination. There may be a justifiable public policy rationale for treating most theaters around Times Square differently from theaters everywhere else in New York City. It is difficult to explain why people wishing to build in a few other special districts must comply with unusual requirements that do not apply to other properties within the city. But it is impossible to justify excluding new stores on lots that are wider than 25 feet from on three blocks of 86th Street in Manhattan. It is time to replace this crazy quilt with a single, easy-to-understand statute that truly balances competing interests, treats property owners equitably, and nurtures the development of attractive, healthy, affordable communities.

The first step in doing so is for every political jurisdiction to combine into a single statute all land use regulations: zoning, historic preservation, growth management, environmental protection, and whatever other local laws are pertinent. Without a single combined statute of this sort, there can be no truly comprehensive planning. The task is not as easy as it was in 1924, when Secretary of Commerce Herbert Hoover commissioned the Standard State Zoning Enabling Act. At that time, there were relatively few communities with comprehensive land use regulation, fewer court decisions defining the requirements of such regulation, and even fewer "experts" in the field. Today, the authors of any land use statute face a powerful array of interests represented by existing local and state statutes. They must consider a huge quantity of complex and frequently conflicting objectives. (Environmental protection alone requires consideration of topography, air quality, water supply, waste disposal, soils, drainage, floods and floodplains, vegetation, natural habitats, etc.). The professionals who write our land use regulations have to confront the objections of lawyers, architects, engineers, city planners, and other consultants whose careers are dependent on the existing pattern of land use regulation.

The objectives for their text should be clarity, simplicity, and brevity. This is no small task, because any new statute presents not just a problem of substance but also a problem of language. Too many laws and the administrative regulations instituted to implement them consist of negatives, double negatives, exceptions, and qualifications. There is no need for any intelligent person to face a zoning resolution such as that of New York City, which is replete with language like this:

> *However, no existing use shall be deemed non-conforming, nor shall non-conformity be deemed to exist solely because of . . . (c) The existence of conditions in violation of the provisions of either Sections 32–41 and 32–42, relating to Supplementary Use Regulations, or Sections 32–51 and 32–52 relating to Special Provisions Applying along District Boundaries, or Sections 42–41, 42–42, 42–44, and 42–45 relating to Supplementary Use Regulations and Special Provisions Applying along District Boundaries.*[94]

The length and complexity of many zoning resolutions is so great that these statutes are indecipherable not only to property owners and community leaders, but also to nonspecialized, professional architects and lawyers. The New York City Zoning Resolution, for example, is contained in three bulky volumes that include 35 pages of General Provisions, 406 pages covering 61 Special Districts, 38 pages covering 16 Manufacturing Districts, 32 pages covering 61 Commercial Districts, 160 pages covering Residential Districts, 18 pages of special regulations, 15 pages on Administration, and 6 Appendices.

Land use regulations should aim for equity, efficiency, and economy of operation. The best way to achieve this aim is to eliminate all discretionary action and make the statutes self-enforcing. This can be done by transferring the permitting process from government into the hands of licensed professionals. Only professional engineers, architects, and lawyers licensed by a designated state agency would be allowed to issue building permits. To be licensed, these professionals would have to successfully complete a training program operated by the designated state agency and subsequently pass a licensing examination. Anybody licensed to issue permits who did so improperly would lose the right to issue permits, have his or her professional status revoked, and be subject to substantial fines.

Local land use statues are usually written by appointed planning commissions and approved by elected legislatures. Restricting the role of a planning commission to managing land use regulations will result in the same ineffective growth management devices as those of Ramapo and Petaluma. Sustainable development can be achieved only when capital improvements (whether paid for by government or by project developers) are made in conjunction with the businesses, institutions, and residents that will eventually occupy the buildings affected by the regulations. Otherwise, residents (for good reasons) will oppose zoning that allows development that the existing water supply system, sewer system, street grid, or schools cannot handle. At the same time, property owners (for equally good reasons) will claim that if the zoning precludes construction, they will be deprived of the use of their property without compensation. Thus, the commission should also be made responsible for drafting the community's capital improvement program.

Virtually all existing land use regulations suffer from a common flaw. They envisage an ideal end state without providing for changing market conditions or changing community priorities. For this reason, any consolidated land use statute must establish triggering mechanisms for changing regulations as well as criteria for determining the character of those changes. The triggering mechanism could be a regular event (e.g., the decennial census) or a specific condition (e.g., a threshold vacancy rate). Under no circumstances, however, should it be left to the discretion of local officials. Otherwise, the quantity of land assigned to inadequately productive uses will continually increase.

It makes sense to require an environmental review of any large projects that will, by their very nature, greatly affect the community. It makes no sense to require such a review for discretionary actions by any local agency. Consequently, any consolidated land use statute should establish fairly high thresholds that trigger the preparation of a review to determine whether an environmental study is needed and much higher thresholds that trigger actual preparation of that review. Otherwise, the costs of doing business and providing government services will become prohibitive.

There should be no required environmental impact studies for changes in land use regulation approved by a planning commission. Any changes in these regulations are by their very nature comprehensive, rather than reactive, and require balancing a wide range of societal goals, some of which will be in conflict with one another. Moreover, the cost and time required for such studies will result in avoiding action, except when it is particularly pressing. That, in fact, is why communities with local regulatory agencies whose actions are subject to such studies end up with so many obsolete laws.

Similarly, any government action requiring an environmental review must require publication of an estimate of the impact on consumer prices of any regulatory changes. Otherwise, public officials will continually add individually desirable mandates that are ostensibly paid for by property owners without any information about their likely cost to a property's occupants or to the community as a whole.

There is no way to guarantee that suitable land use statutes will be developed. Nor is there any guarantee that they will be widely accepted or even be successful. If we do not try, however, cranes will continue to erect intrusive and unsightly new buildings; unwanted land uses will continue to disfigure otherwise delightful neighborhoods; needless actions will continue to befoul land, sea, and air; and all of these results will be accompanied by escalating prices and increasingly bitter political controversy.

Notes

1. Fred Bosselman, David Callies, and John Banta, *The Taking Issue*, Council on Environmental Quality, U.S. Government Printing Office, Washington, DC, 1973, pp. 51–104.
2. Seymour I. Toll, *Zoned American*, Grossman Publishers, New York, 1969.
3. For an economist's explanation of the rationale behind government intervention into the marketplace, see Otto Eckstein, *Public Finance*, Prentice-Hall, Englewood Cliffs, NJ, 1967, pp. 8–13.
4. Franklin J. Havelick and Michael Kwartler, "Sunnyside Gardens: Whose Land Is It Anyway?" *New York Affairs*, vol. 7, no. 2, New York University, New York, 1982, pp. 65–80.
5. *Shelley v. Kraemer*, 334 U.S. 1 (1948).
6. For a discussion of the police power and nineteenth-century court interpretation of its applicability, see Stanley K. Schultz, *Constructing Urban Culture, American Cities and City Planning* 1800–1920, Temple University Press, Philadelphia, 1989, pp. 33–91, and Bosselman, Callies, and Banta, op. cit., pp. 105–123.
7. *Mugler v. Kansas*, 123 U.S. 623, 8 S. Ct. 372 (1887).
8. Bosselman, Callies, and Banta, op. cit., pp. 124–138.
9. *Pennsylvania Coal Co. v. Mahon*, 260 U.S. 393 (1922).
10. Justice John Paul Stevens, writing for the majority in the case of *Keystone Bituminous Coal Association v. De Benedictus*, U.S. 107 S. Ct. 1107 (1987), took a somewhat different approach, explaining that the regulation of mining rights was appropriate when it applied to all property (not just that of coal companies) and balanced the "private economic interests of the coal companies against the private interests of the surface owners."
11. Theodora Kimball Hubbard and Henry Vincent Hubbard, *Our Cities To-day and To-morrow: A Survey of Planning and Zoning Progress in the United States*, Harvard University Press, Cambridge, MA, 1929, p. 21.
12. Elaine Moss (editor), *Land Use Controls in the United States: A Handbook on the Legal Rights of Citizens*, National Resources Defense Council, New York, 1977, pp. 117–118.
13. www.coastal.ca.gov.
14. Hawaii is the sole exception. There, land use regulation has always been a state function. A statewide comprehensive zoning ordinance was adopted in 1961.
15. Jed Rubenfeld, "USINGS," *The Yale Law Journal*, vol. 102, no. 5 (March 1993), pp. 1077–1163.
16. *The Records of New Amsterdam*, I, 4, quoted in James Ford, *Slums and Housing*, Harvard University Press, Cambridge, MA, 1936, p. 28.
17. Marc A. Weiss, *The Rise of the Community Builders*, Columbia University Press, Cambridge, MA, 1987, pp. 79–106.
18. John J. Costonis, *Icons and Aliens: Law, Aesthetics, and Environmental Change*, University of Illinois, Chicago, 1989, pp. xv–xvi.
19. This explanation of height limits in the national capital does not appear in Costonis's book. It is entirely my own, as is the discussion of the Chinese laundry cases following.
20. *Yick Wo v. Hopkins* and *Wo Lee v. Hopkins*, 118 U.S. 356 (1886).
21. *Yick Wo v. Hopkins* and *Wo Lee v. Hopkins*, 118 U.S. 356 (1886).
22. Hubbard and Hubbard, op. cit., pp. 162–163.
23. Some attempts at comprehensive planning had been initiated in 1914, when the city established a Committee on the City Plan. However, its report, *Development and Present Status of City Planning in New York City*, had only a tangential relationship to the 1916 zoning resolution. Nor did the 1961 ordinance evolve from a comprehensive plan. Although the city established a planning commission in 1938 and required it to produce a comprehensive plan, none was produced until eight years after the 1961 resolution had been enacted (see Chapter 19).
24. Toll, op. cit., pp. 173–183.
25. *New York Times*, advertisement, March 5, 6, 1916.
26. The height limits were 70 feet for streets less than 60 feet wide and 80 feet for wider streets.
27. Alexander Garvin (project director), *Neighborhood Preservation in New York City*, New York City Planning Department, New York, 1973, p. 137.
28. Toll, op. cit., pp. 143–157.
29. The maximum bulk for any office building permitted by the 1961 zoning ordinance is 18 times the lot area on which it is erected or, with special bonuses, 21.6.
30. Harrison, Ballard, and Allen, *Plan for Rezoning the City of New York*, New York City Planning Commission, New York, 1950.
31. Interview with Robert F. Wagner, Jr., October 16, 1991.
32. Carter B. Horsley, "The Sixties," *Planning the Future of New York City*, New York City Planning Commission, New York, 1979, p. 29.
33. Voorhees, Walker, Smith, and Smith, *Zoning New York City*, New York City Planning Commission, New York, 1958.
34. The switch from height limits to a floor area ratio had been proposed by Harrison, Ballard, and Allen in 1950 and had already taken place in Seattle, which had adopted many of their proposals as part of its comprehensive rezoning of 1957 (Harrison, Ballard, and Allen, op. cit., pp. 44–45).
35. Jane Jacobs, *The Death and Life of Great American Cities*, Random House, New York, 1961, p. 152.
36. One has to wonder how this could be done in the absence of a regularly updated, comprehensive city plan and capital improvement program.
37. Ibid., p. 205.
38. William H. Whyte, *City*, Doubleday, New York, 1988, p. 233.
39. Ibid., p. 234.
40. *Village of Euclid v. Ambler Realty Co.*, 272 U.S. 365 (1926); and Toll, op. cit., pp. 213–253.
41. Charles X. Zimmerman, mayor of Euclid, quoted by Toll, op. cit., p. 215.
42. It was not entirely unlimited because there was a class of uses that was prohibited altogether.
43. *Village of Euclid v. Ambler Realty Co.*, 272 U.S. 365 (1926).
44. At 297 Fed. 316, quoted by Toll, op. cit., p. 215.
45. *Village of Euclid v. Ambler Realty Co.*, "Brief and Argument for Appellee," pp. 78–79, 83, quoted by Toll, op. cit., pp. 232–233.
46. *Village of Euclid v. Ambler Realty Co.*, 272 U.S. 365 (1926).
47. Terry Jill Lassar, *Carrots and Sticks: New Zoning Downtown*, Urban Land Institute, Washington, DC, 1989.
48. Jerold S. Kayden (with the New York City Department of City Planning and the Municipal Art Society of New York), *Privately Owned Public Space: The New York Experience*, John Wiley & Sons, New York, 2000, pp. 43–60.
49. Ibid, p. 48.
50. Historical and statistical material on zoning in Seattle is derived from Lassar, op. cit., especially pp. 20–25, 31–33, 81–82, 121–122; Terry Jill Lassar, "Seattle's Zoning Saga," *Urban Land*, September 1988, Urban Land Institute, Washington, DC, pp. 34–35; T. Richard Hill, "Seattle's Voters CAP Downtown Development," *Urban Land*, August 1989, Urban Land Institute, Washington, DC, pp. 32–33; and Robert Stevenson, "Downtown Seattle: To CAP and Beyond," 1991, unpublished.
51. Abstracted from Lassar, *Carrots and Sticks*, op. cit., p. 18.
52. Lassar, *Carrots and Sticks*, op. cit., pp. 70–76.

53. *Nollan v. Coastal Commission,* 107 S. Ct. 3141 (1987).
54. Ibid.
55. *Dolan v. City of Tigurd, Oregon,* W.L. 276693 (1994).
56. The Special Committee on the Role of Amenities in the Land Use Process, "The Role of Amenities in the Land Use Process," reprinted from *The Record,* vol. 43, no. 6, Association of the Bar of the City of New York, New York, 1988, p. 7.
57. Richard F. Babcock and Wendy U. Larsen, *Special Districts: The Ultimate in Neighborhood Zoning,* Lincoln Institute of Land Policy, Cambridge, MA, 1990.
58. New York City Planning Commission, *Plan for New York City,* vol. 4, Department of City Planning, New York, 1969, p. 57.
59. Jonathan Barnett, *Urban Design as Public Policy,* Architectural Record Books, New York, 1974, pp. 17–27; and Richard Weinstein, "How New York's Zoning Was Changed to Induce the Construction of Legitimate Theaters," in *The New Zoning,* Norman Marcus and Marilyn W. Groves (editors), Praeger, New York, 1970, pp. 131–136.
60. Interview with Donald Elliott, October 17, 1991.
61. They are the American Place Theater in the Stevens Building, the Gershwin and the Circle in the Square theaters in the Uris Building, and the Minskoff and the Marquis theaters in buildings of the same name.
62. Interview with Con Howe (executive director of the City Planning Department during the period in which these changes were made), October 14, 1991.
63. Executive Order 91, August 24, 1977.
64. New York is divided into 59 community districts. Each district has a 50-person *community board* jointly appointed by the borough president and city council members that represent the affected district. The ULURP process requires public notice and hearings by the affected community board, the city planning commission, and the city council (until September 1990 this role was taken by the Board of Estimate, which was abolished as a result of charter reform).
65. That is what I thought when, as a member of the city planning commission, I voted for those changes to the Times Square Special Zoning District.
66. Interview with Con Howe, September 2, 2012.
67. Babcock and Larsen, op. cit., pp. 99–100.
68. New York City Planning Commission, *Minutes of Meeting,* New York, April 3, 1974.
69. New York City Planning Department, Special Yorkville–East 86th Street District: A Rezoning Proposal, New York, 1989.
70. City of Santa Monica, *Third Street Mall Design Guidelines,* Santa Monica, CA, undated, p. i.
71. Third Street Development Corporation, *Third Street Mall Specific Plan,* City of Santa Monica, Santa Monica, CA, 1986, Appendix, pp. A1–A15.
72. Urban Land Institute, *Janss Court,* ULI Project Reference File, vol. 21, no. 3, Urban Land Institute, Washington, DC, January–March 1991.
73. For a more detailed explanation, see Alexander Garvin, *The Planning Game: Lessons from Great Cities,* W. W. Norton & Company, New York, 2013, pp. 58–63.
74. Randall W. Scott (editor), with the assistance of David J. Brower and Dallas D. Miner, *Management and Control of Growth,* vols. 1–3, Urban Land Institute, Washington, DC, 1975; and Douglas R. Porter (editor), *Growth Management,* Urban Land Institute, Washington, DC, 1986.
75. Historical and statistical material on Ramapo is derived from Rockland County Planning Board, *Rockland County Data Book,* County of Rockland, New York, 1965, 1969, 1976; *Golden v. Planning Board of Ramapo,* 285 N.E. 2nd 291 (1972); Israel Stollman, "Ramapo: An Editorial & the Ordinance as Amended," in Scott, Brower, and Miner, op. cit., pp. 5–14; Manual S. Emanual, "Ramapo's Managed Growth Program," *Planners Notebook,* vol. 4, no. 5, (October 1974), Urban Land Institute, Washington, DC; and Robert Guenther, "Ramapo, N.Y . . . is seeking to encourage growth, a sign of the economic times," *Wall Street Journal,* August 31, 1983, p. 23.
76. Historical and statistical material on Petaluma is derived from Frank B. Gray, "The City of Petaluma: Residential Development Control," in Scott, Brower, and Miner, op. cit., pp. 149–159; *Construction Industry Association of Sonoma County v. City of Petaluma,* 522 F. 2d 897 (9th Cir. 1975), *cert. denied,* 424 U.S. 934; Warren Salmons, "Petaluma's Experiment," in Porter, op. cit., pp. 9–14; and Seymour I. Schwartz, David E. Hansen, and Richard Green, "The Effect of Growth Control on the Production of Moderate-Priced Housing," in Porter, op. cit., pp. 15–20.
77. One of the earliest publications to present these concepts is known as the Brundtland Report. See World Commission on Environment and Development (Gro Harlem Brundtland, chairman), *Our Common Future,* Oxford University Press, Oxford, UK, 1987.
78. Lewis Mumford, "Botched Cities," *American Mercury,* no. 18 (October 1929), p. 147.
79. 42 U.S.C. § 4331 (2) (C).
80. Col. Fletcher H. Griffis, Army Corps of Engineers district engineer, quoted by Sam Roberts in "For Stalled Westway, a Time of Decision," *New York Times,* June 5, 1984, p. B1.
81. Donald G. Hagman, "NEPA's Progeny Inhabit the States—Were the Genes Defective?," *Urban Law Annual,* vol. 7, School of Law, Washington University, St. Louis, MO, 1974, pp. 3–56; Jeffrey T. Renz, "The Coming of Age of State Environmental Policy Acts," *Public Land Law Review,* School of Law, University of Montana, Missoula, 1984, pp. 31–54; and http://www.wildlifelawnews.com/statbio/pb-bst14.html.
82. *Friends of Mammoth v. Board of Supervisors,* 8 Cal. 3d 247, 502 P.2d 1049, 104 Cal. Rptr. 761 (1972).
83. Michaeil H. Remy, Tina A. Thomas, and James G. Moose, *Guide to the California Environmental Quality Act (CEQA),* Solano Press Books, Point Arena, CA, 1991.
84. New York City Planning Commission, Calendar Items 53 and 54, September 20, 1989.
85. Dresdner Robin & Associates, Inc., *Project Data Statement: Special Little Italy Zoning District, Houston Street Corridor (Area B),* Jersey City, NJ, May 1987; supporting documents filed in connection with CEQR No. 87–311M; and Grossman & Associates, Inc., *An Archeological and Historical Sensitivity Evaluation of Surviving Open Areas Within the Little Italy Special Zoning District (Area B), Blocks 508, 509, and 521 (CEQR No. 87–311M),* New York, October 1988.
86. The developer was forced to spend tens of thousands of dollars for the various tests and studies required by city personnel reviewing the CEQR submission. There was only one significant additional obligation that was not already required by the zoning. The property owner agreed to remove any contaminated soil that might be found on the site. It was a gratuitous requirement for which no special environmental review was needed, because every lending institution in the city would have required the same commitment in conjunction with its construction loan.
87. Donald L. Elliott, Ratthew Goebel, and Chad Meadows, *The Rules That Shape Urban Form,* Planning Advisory Service Report Number 570, American Planning Association, Chicago, 2012.
88. Andrés Duany, Jeff Speck, and Mike Lydon, *The Smart Growth Manual,* McGraw-Hill Companies, Inc., New York, 2010.
89. Manny Diaz, *Miami Transformed: Rebuilding America One Neighborhood, One City at a Time,* University of Pennsylvania Press, Philadelphia, 2012, pp. 164–180; and Garvin, op. cit., pp. 34–35.
90. Thomas Jefferson, letter to Samuel Kercheval, July 12, 1816.
91. Edgar M. Hoover and Raymond Vernon, *Anatomy of a Metropolis,* Doubleday Anchor Books, Garden City, 1962, p. 248; U.S. Department of Commerce, Bureau of the Census, *County Business Patterns,* Washington, DC, 1958, http://censtats.census.gov/cgi-bin/cbpnaic/cbpsect.pl,;and New York City Department of City Planning.
92. Philip Johnson, "Whence and Whither: The Processional Element in Architecture," *Perspecta 9/10,* School of Art and Architecture of Yale University, New Haven, CT, 1965, p. 168.
93. Whyte, op. cit., pp. 103–131.
94. New York City Planning Commission, *Zoning Resolution,* Article V, Chapter 2, Section 52–01.

18

Preserving the Past

Charleston, 2011 *(Alexander Garvin)*

Sometimes a nostalgic interest in the past motivates people to preserve old buildings. Other times a dislike of the newly built environment produces a movement for preservation. During the 1960s, Christopher Tunnard, one of the pioneers of the preservation movement, proposed a more inclusive rationale for preserving the past. He believed it was society's responsibility to cultivate what he called the "cultural patrimony."[1]

Cultural patrimony means inheritance from one's ancestors. It includes places of historical significance (George Washington slept there), esthetic prominence (Frank Lloyd Wright designed it), social import (Native Americans live there), cultural significance (Tin Pan Alley, where American pop music was born), public importance (festivals take place there), and scenic distinction (guidebooks say it is worth the visit). As Tunnard conceived it, society's responsibility is not simply to preserve what we receive from our forebears and pass it on to future generations. It is to select what is significant, nurture and enhance it, make improvements, and pass on to the next generation an environment that is richer, more fulfilling, and more beautiful.

Selecting what is significant is not easy. Tin Pan Alley is a world-famous component of America's cultural patrimony. Yet, as the Cy Coleman and Joseph McCarthy song explains:

> *There aren't many songs about it*
> *And what's even more strange to me*
> *Not a map has a dot, to distinguish the spot*
> *Where that hurly-burly street used to be.*[2]

Most Americans, even New Yorkers, have no idea that 28th Street between Broadway and Sixth Avenue, once the heart of the country's popular music industry, is Tin Pan Alley. How could they? There is no sign there telling them about it. Irving Berlin songs are no longer tried out upstairs on pianos in music publishers' offices. Besides, the publishers left decades ago. Nevertheless, every second of the day somebody remembers, plays, or sings what was produced on this block of 28th Street. After all, cultivating the cultural patrimony does not require conferring landmark status on a building or designating a historic district. In the same way, declaring something a landmark does not mean that it will be lovingly maintained or even remain in existence.

As Justice William Brennan explained in the Supreme Court decision that declared New York City's landmark preservation law constitutional: "Structures with special historic, cultural, or architectural significance enhance the quality of life for all . . . represent the lessons of the past and . . . serve as an example of quality for today . . ."[3] Preserving that legacy can provide tremendous benefits to the surrounding city: economic benefits from the tourists it attracts, social benefits from a more heterogeneous population seeking a broader range of living environments, and cultural benefits from its enhanced setting for artistic activity. These additional benefits often offset the costs of preservation. A good example is the privately financed recreation of eighteenth-century Williamsburg, the colonial capital of the Dominion of Virginia. Tourist spending since this restoration was completed probably has generated enough economic impact to pay for Williamsburg and a dozen other reconstruction projects around the country.

Manhattan, 2011. The buildings that were once occupied by world-famous "Tin Pan Alley" are all still there. The music business has moved elsewhere. There is not even a sign to inform passersby that this is part of the nation's cultural heritage. *(Alexander Garvin)*

Approaches to preservation vary markedly. Some people think preserved buildings should be displayed like works of art. In 2012, more than 15,000 house museums in America were doing just that.[4] Others think that preserved buildings should continue to be used in the same manner as they were when they were initially constructed. Since that may not be possible or even cost-effective, many preservationists favor adapting old buildings for new activities. Their conception of adaptive reuse keeps older buildings more or less physically intact. When that is not feasible, some people believe that keeping the building's original façade in place is sufficient, or at least preferable to tearing down the entire building. *Façadomies* of this sort are increasingly common. The approach advocated by Tunnard, cultivating the cultural patrimony, requires a more flexible and nuanced attitude to the past. Many of its proponents prefer to intermingle still usable parts of old buildings with new structures that meet contemporary needs.

There is not enough money to create mini-Williamsburgs wherever there are properties that are worth preserving, nor are there enough tourists to visit these places or curators to maintain them. Moreover, there is every reason to keep from

Washington, DC, 2009. A genuine façadomy. *(Alexander Garvin)*

transforming the American landscape into one continuous, institutionally owned and operated museum. Instead, we should find ways to make the cultural patrimony an integral part of daily life. That means retaining landmarks without destroying their continuing utility as residences, offices, stores, warehouses, schools, police stations, and so forth.

When Tunnard proposed his rationale, the preservation movement was in its infancy. Federal, state, and local governments owned and maintained for public view some well-known monuments. Charleston, New Orleans, and a handful of smaller cities had public bodies with the legal authority to supervise privately owned but publicly designated landmarks and privately owned property in publicly designated historic districts. Since that time, more than 2000 communities have enacted legislation to protect landmarks and historic districts.[5]

Public designation of a privately owned building as a landmark does not mean that the property must be maintained in good condition by its owner. Plenty of landmarks deteriorate because their owners cannot afford to spend money on repairs or even more expensive restoration. Landmark designation only requires that those repairs meet a standard of appropriateness set by the public agency charged with supervising local landmarks. As the story of Mount Neboh synagogue, discussed later in this chapter, explains, designation does not prevent demolition, either. Some developers even argue against spending more than is absolutely necessary on a building because they want to avoid its later regulation by a government agency that believes fine structures are a uniquely valuable component of an area's heritage. Nevertheless, designation of landmarks and historic districts has become the most prevalent and the most effective method for cultivating the cultural patrimony.

There is now considerable experience with other preservation techniques: public realm investments that increase the value of surrounding properties, thereby encouraging the owners of the surrounding buildings to make similar investments, zoning laws that permit owners to sell development rights, bridge financing from revolving funds with which to purchase and resell threatened landmarks, long-term mortgage financing, local real estate tax benefits, and federal income tax deductions are just a few of the more frequently utilized strategies. As the use of these techniques has become more widespread, there have been increasing challenges from property owners. Sometimes they contend that local statutes impose burdens that inhibit the use of their property. More often, though, they argue that unreasonable regulations keep them from doing so. It has, therefore, become increasingly important to enact legislation that will distribute the costs and benefits of preservation more equitably.

Every preservation technique except government spending on the acquisition and restoration of historic structures depends on the existence of an active real estate market. Development rights, for example, have no value unless somebody wishes to develop a property. Revolving funds will run out of money if nobody wants to buy the buildings that have been purchased. Tax benefits are useless unless there is income from which to pay taxes. Because properties outside active market areas suffer from neglect, we also need programs that will provide property owners in these areas with the money required to maintain and restore historic structures.

Even more serious is the continuing punishment of those who cultivate the cultural patrimony. Virtually every community increases the real estate tax on properties that are renovated. If one is unlucky enough to own a landmark, the price is even higher. Public officials invariably impose extra requirements that add to the cost of the renovation. We need to eliminate this disincentive to the ownership and restoration of historic structures.

Finally, historic preservation should not be dismissed as the enthusiasm of a privileged elite. Historic preservation cannot and should not take place in a vacuum. Cultivating the cultural patrimony is not concerned exclusively with historic preservation; vital to its overall mission is the improvement of the quality of our air and water, the condition of our parks and public spaces, the safety of our streets and highways, the quality of the education we provide our children and the health care we provide the sick, and the character of our entire built and natural environment. With such a wide mandate, there are inevitably tough choices among these competing demands for action. Consequently, historic preservation activities need to be better integrated into the planning, budgeting, and governance of every community.

Preserving the Old and Historic

The obvious way to save a landmark is to purchase it. The earliest example in America occurred when Ann Pamela

Cunningham organized the Mount Vernon Ladies' Association to purchase Mount Vernon, George Washington's home, which had been put up for sale in 1850. The association placed a letter in the *Charleston Mercury* in 1853 seeking contributions, and donations came in from every state in the nation. Five years later it purchased Mount Vernon, and the house has been run as a museum ever since.

Whatever the stated arguments or underlying motives for preservation, the money and effort needed to purchase and restore large sections of cities exceed the financial and administrative resources of most individuals and civic groups. Occasionally, a Rockefeller will contribute the money needed to restore an entire historic site such as Williamsburg, Virginia. However, there are not enough fabulously wealthy philanthropists to bankroll the restoration of every district and town worthy of preservation. Consequently, we look to government to safeguard our heritage.

Government is in a poor position to maintain the cultural patrimony. The Fifth Amendment to the Constitution requires government to compensate property owners whenever it takes any land or buildings for public use. Given the multiplicity of demands for scarce municipal resources, most cities are not about to spend the extraordinary sums needed to buy all the properties suitable for preservation, much less pay for the necessary restoration. During the 1930s, two cities, Charleston, South Carolina, and New Orleans, Louisiana, demonstrated that there were alternatives to massive spending for the acquisition and restoration of historical artifacts like Williamsburg. They found a way to regulate what owners could do with the cultural patrimony.

Williamsburg, Virginia

Williamsburg was settled in 1633. Initially, it differed little from other towns in southwestern Virginia. The first impetus for change came in 1693, when King William and Queen Mary endowed a new college that would bear their names. The second came in 1699, when Williamsburg was designated the new capital of the colony. From then until 1780, when the capital was moved to Richmond, Williamsburg remained the political, intellectual, and social center of Virginia. Thereafter it slowly slid into genteel dilapidation, and by the start of the twentieth century it was little more than a seedy collection of run-down structures.

Williamsburg, 1903. Wythe House prior to restoration. *(Courtesy of Alexander Purves)*

Re-creating colonial Williamsburg, lot by lot, building by building, was the brainchild of the Reverend William Archer Rutherford Goodwin. He came to Williamsburg as rector of Bruton Parish in 1902. One of his first projects was the restoration of his church, the oldest Episcopal structure in constant use in America. Another was the lobbying work he did to have paving installed on Duke of Gloucester Street, Williamsburg's main thoroughfare. His real dream, though, was the complete restoration of colonial Williamsburg. Because he did not have the money, it remained a fantasy for more than two decades.[6]

After taking on a more prosperous parish in Rochester in 1909, Goodwin returned to Williamsburg in 1923 to teach religion at the College of William and Mary, to organize its endowment fund, and to resume his position at Bruton Parish. This time he found a benefactor who paid to bring his fantasy to life.

In 1926, Goodwin persuaded John D. Rockefeller Jr. to pay for some architectural sketches of a restored Williamsburg. Together, they acquired and demolished almost 600 postcolonial structures, rehabilitated nearly 100, and reconstructed another 450. Modern stores were moved to a new business district called Merchants Square. Displaced residents were either moved back as "life tenants" when restoration was completed, provided homes in specially funded low-rent neighborhoods, or paid sufficiently high prices for their houses to allow them to purchase equivalent houses elsewhere. Utilities were buried, paving and street furniture were installed, and trees, shrubs, and flowers were planted. By the time the project was finished, they had spent $68.5 million.

Williamsburg, 1997. Wythe House after restoration. *(Alexander Garvin)*

The re-creation of colonial Williamsburg preserved or rebuilt structures of unusual historic, cultural, and architectural significance. More important, it became a top Virginia tourist destination, attracting 26.5 million visitors to its website and 1.7 million people to its streets in 2010.[7]

"Old and Historic" Charleston, South Carolina

Charleston was founded in 1670 on a narrow peninsula between the Cooper and Ashley Rivers. Its protected, deepwater harbor, 7.5 miles from an ocean sandbar, offers East Coast shipping conditions rivaled only by New York City. During the eighteenth and nineteenth centuries, prosperous Charleston merchants shipped valuable cargoes of indigo, rice, and, later, cotton. They provided its white population with the economic base for a life of luxury and elegance. As described by Constance McLaughlin Green, antebellum Charleston was a very special place:

> *The city was small . . . before 1830, numbering only about three hundred, all knew each other . . . Nowhere else in America was life so truly urbane as in Charleston . . . Charlestonians made gracious living their first business. The concerts of Mozart and Haydn given by the St. Cecilia society, the balls, the theater, the race course, and in 1800 the only golf links in America provided endless diversion.*[8]

The Civil War brought this prosperity to a halt. After the war, other cities continued to grow and change while Charleston faded into somnolent obsolescence. By 1925, DuBose Heyward, in his famous novel, *Porgy,* could describe it as "an ancient, beautiful city that time had forgotten."[9]

It was Charleston's good luck to be a "city that time had forgotten." The wealthy planters that built it had created an environment of unequaled grace and charm. Its residential districts consisted of two- and three-story frame and brick houses lined with *piazzas* (arcaded galleries running the length of the structure) providing shaded outdoor living, surrounded by subtropical gardens, and enclosed in wrought iron and brick walls. Its commercial areas retained the pedestrian scale, pastel colors, and ornamental detail of an antebellum trading center. Elsewhere, such lovely buildings would have been demolished to make way for a "higher and better use." Not only was there little demand for construction sites in Charleston, often there was not even enough money to maintain existing structures.

Neglect was not the only threat to Charleston's heritage. After World War I, the city was discovered by "a steady, but small, flow of discerning visitors. . . . The town found it had to protect itself from collectors of everything from ironwork to complete houses. Some of the buildings were taken down and carried off, from the brickwork of the basement to the timber of the roof."[10]

Charleston, 2011. Residences on Tradd Street in the heart of the Old and Historic District.
(Alexander Garvin)

In 1931, Charlestonians invented a device to deal with this threat: designation of a *historic district* within the zoning ordinance. The Old and Historic Charleston District was a 22-block area at the tip of the peninsula that contained many, but not all, of the city's finest antebellum structures. This special district was added to the zoning ordinance to guarantee "the preservation and protection of the old historic or architecturally worthy structures and quaint neighborhoods which impart a distinct aspect of the city," ensure "the continued construction of buildings in the historic styles . . . form, color, proportion, texture, and material" compatible with the district, "preserve property values," and "attract tourists and residents."[11]

The Board of Architectural Review was responsible for reviewing and approving construction, alteration, demolition, or removal of any structure within the district. Property owners had to apply to the board for a *certificate of appropriateness* or submit a *demolition application* in order to do anything with their property. Such strict regulation caused little hardship. During the Depression, few property owners had the money to do much with their buildings.

After World War II, the exodus to the suburbs further reduced demand for development sites within the historic district. Many property owners could not charge rents high enough to cover repairs. They had little choice but to defer maintenance. Others could not find tenants. Designation as an old and historic district had done little to protect the area from neglect and abandonment. It did, however, prevent demolition, incompatible remodeling, and inappropriate expansion until demand picked up in the 1960s. Thus, when tourists began coming in increasingly large numbers and residents again chose to settle in town rather than in the suburbs, the Old and Historic Charleston District was still there to accommodate them.

By the 1980s, neglect and abandonment were no longer a problem. Most buildings in Charleston's historic district were finally in the hands of owners who had the wherewithal to maintain them properly. Furthermore, the Board of Architectural Review could prevent inappropriate alterations and incompatible construction. By the second decade of the twenty-first century, the district was attracting more than 8.7 million visitors, who spent $5.5 billion annually.[12] Thanks to preservation efforts that began in the 1930s, Charleston is still a remarkably beautiful city, but it is anything but forgotten.

Vieux Carré, New Orleans, Louisiana

In 1937, New Orleans became the second American city to designate for preservation its old and historic district, the Vieux Carré. During the previous year, Louisiana voters had approved a state constitutional amendment giving the city the power to establish a historic district and create a commission that would regulate the actions of its property owners. The city designated a 260-acre area with over 3000 structures, and the mayor appointed an eight-member Vieux Carré Commission. It had power over all exterior work within the district—everything from replacing shutters to major new construction.

The Vieux Carré is a jambalaya of Spanish, Creole, Yankee, African-American, English, and Confederate buildings. Land uses are thrown together in a colorful confusion of jazz joints, stately residences, antique shops, lodging houses, celebrated restaurants, hidden courtyards, and tourist traps. Unlike Williamsburg or Charleston, New Orleans was not faced with economic decline, nor was the Vieux Carré threatened by neglect. New Orleans had been the nation's fifth largest city for most of the nineteenth century. Its busy port had kept it a tourist attraction for nearly two centuries. Cultivating the cultural patrimony in New Orleans therefore required a very different approach.

The Vieux Carré Commission could not create carefully scrubbed fantasies of an imagined past, like Williamsburg, nor conserve "an ancient beautiful city that time had forgotten," like Charleston. It had to maintain the noisy attractions that brought sailors, the quiet little houses that provided refuge for artists, the garish dives that spawned jam sessions, and the elegant mansions that were the envy of elite visitors.

Nurturing the messy vitality of this busy urban center with a distinctive and heterogeneous past requires knowing when to step aside and when to insist on integration with what New Orleans calls the *tout ensemble*. During its first 30 years, the Vieux Carré Commission performed this task with distinction. It supervised everything from the introduction

New Orleans, 1981. Bourbon Sreet in the Vieux Carré. *(Alexander Garvin)*

of wrought-iron gates and magnolia trees to the construction of the new Royal Sonesta Hotel. Then, in the mid-1960s, the commission faced a threat it was powerless to prevent: the Riverfront–Elysian Fields Expressway.

Robert Moses, who had been consulted on the city's transportation problems, proposed a riverfront highway in 1946. He had proposed similar highways for Pittsburgh, Baltimore, Detroit, and other cities that had consulted him in his capacity as the country's leading public works planner. The Riverfront–Elysian Fields Expressway was intended to be an elevated roadway, which, Moses explained:

> *... will lift cars off the street which now interfere with service to the docks. It will facilitate modernization of the piers and give better access to them.... Decatur Street, which bounds Jackson Square, the very heart of the Vieux Carré, and separates it from the market and waterfront, would again be a local service road. Heavy traffic would pulsate along the docks. The Cathedral would still be wedded to the Mississippi, but its precincts would not be choked with needless through traffic.*

Moses intended "no violence to history or tradition." He felt that removing regional traffic from the streets and opening access to the docks would enhance the Vieux Carré as a lively tourist center.[13]

The cost of elevating the 3.4-mile section of the proposed six-lane Riverfront–Elysian Fields Expressway going past the Vieux Carré was estimated at $31 million. If it were to be depressed between Jackson Square and the Mississippi River, another $12.4 million would be required. Despite the fact that the federal government would pay 90 percent of the cost of any version of the expressway, the city council chose an even cheaper alternative: a ground-level highway that would have established a broad barrier of heavy traffic between the Vieux Carré and the Mississippi.[14]

While local designation of a historic district could not stop an interstate highway, designation in 1965 as a historic district on the National Register of Historic Places eventually killed the project. The National Register had been created by the National Parks Service pursuant to the Historic Sites Act of 1935. St. Louis Cathedral, the Cabildo, the Pontalba Buildings, and many other buildings in the Vieux Carré were on the National Register. When Congress passed the comprehensive Historic Preservation Act of 1966, it provided a mechanism which would eventually save the Vieux Carré from the ravages of the proposed highway.

The Historic Preservation Act of 1966 required the newly created Advisory Council for Historic Preservation to examine the impact of actions by federal agencies or federally financed programs on landmarks listed on the National Register. When the Federal Highway Administration recommended the Riverfront–Elysian Fields Expressway, the Advisory Council for Historic Preservation entered the fray. In 1969, after a series of public meetings in Washington and three days of visits to the site, it recommended against construction of the expressway. The secretary of transportation accepted the recommendation because the expressway would have "seriously impaired the historic quality of New Orleans' famed French Quarter."[15]

Impairing the historic quality of the French Quarter would have, by extension, impaired the city's economy. The Vieux Carré has continued to be New Orleans's prime tourist attraction for two centuries. In 2004, prior to Hurricane Katrina, the 10.1 million tourists who visited the city spent nearly $5 billion. Although the number dropped to 3.6 million in 2006, by 2011 the city's more than 8.7 million visitors spent nearly $5.5 billion.[16]

Instant Roots and Ersatz History

For a quarter of a century, Charleston and New Orleans were the only major American cities with statutes that regulated existing privately owned buildings in designated historic districts. Meanwhile, Santa Barbara, California, and Santa Fe, New Mexico, were evolving an entirely different approach to the cultivation of the cultural patrimony—one that concentrated on new construction rather than protection.

America is a country where all but a few families were once newcomers and will soon again be on the move. Wherever they go, they seek to put down roots. Civic leaders in Santa Barbara and Santa Fe understood this need. They wanted to provide instant roots for anybody coming to the area. The ersatz past that they created has a distinctive appearance that is as appealing to tourists as anything by Walt Disney. However, unlike Disneyland, Santa Barbara and Santa Fe are functioning cities filled with traffic, noise, and all the difficulties of urban living. Consequently, they provide a very different model for cities interested in cultivating the cultural patrimony.

Santa Barbara

Santa Barbara was founded in 1782 as a Spanish colonial military outpost. Two years later, settlers started to work on the mission church. Most of the structures that were built while Santa Barbara remained a Spanish and later a Mexican settlement were simple one- and two-story adobe buildings. The Anglo-Americans who emigrated during the second half of the nineteenth century had little respect for this architecture of mud and rough-hewn wood. They were accustomed to milled lumber, manufactured nails, and plate glass. At first they rejected the more "primitive" existing building pattern and built something more like the Western towns depicted in Hollywood movies.

As the nineteenth century drew to a close, Santa Barbara, like all of California, became entranced with its Spanish colonial past. The first manifestation of this was the Mission Revival style (c. 1890–1915) that revived the adobe look by covering brick, hollow tile, stone, or wood walls with stucco. Other elements of this new style included arched openings, scalloped parapets, red tile roofs, low bell towers, shaded porches, rich "Moorish" ornament, and decorative round or quatrefoil windows. Sometimes the buildings seemed more like the red-tiled villas of the French Riviera, while in other instances they were closer to the red-tiled houses of the Italian countryside. Regardless of which specific style they were closer to, one-family houses, bungalow courts, and even small, local railroad stations all over California took on the look of this pseudo- Mediterranean fantasy.[17]

California chose to be represented by buildings of this fantasy architecture at every World's Fair from the Chicago Fair of 1893 to the Panama-California Exposition of 1915 in San Diego. In San Diego, the eastern Beaux Arts architect Bertram Goodhue provided the foundation for an even more elaborate "indigenous" architecture. His California State Building revived the highly ornamented Spanish Churrigueresque style.[18] Combined with elements of the more modest Mission style, it provided the model for the larger, monumental structures that California was then building: hotels, movie palaces, shopping centers, railroad terminals.

The resulting pastiche of Spanish, French, Italian, and Moorish components seemed just right for the state's dramatic coastal landscape and benign climate. Few Californians cared that the busy life of prosperous, urban, twentieth-century Anglo-California would be quite different from the quiet existence in a vast natural landscape interrupted by the missions, presidios, and ranchos of the Spanish colonial past. Nor did they care that life in mass-produced bungalows along congested arteries also would be different from the quiet provincial existence in the red-tiled French and Italian villas that dotted the Mediterranean coast.

Until World War I, only a few private houses and hotels in Santa Barbara were built in this pseudo-Mediterranean style. After the war, fantasy-Mediterranean became the style of choice. In 1919, the city and county of Santa Barbara held a design competition for a combined courthouse, city hall, and veterans' memorial. The winning design, in the new style, proved to be too expensive to be built. Wil-

Santa Barbara, 1923. State Street in 1923, two years prior to the earthquake, when it looked like main streets throughout the city. *(Courtesy of Santa Barbara Historical Society)*

Santa Barbara, 2013. City and County Courthouse, erected in 1929, became the culmination of the Anglo-Californian dream of the city's romantic Hispano-Mediterranean past. *(Alexander Garvin)*

liam Mooser & Company, architects of the second-place design (also fantasy-Mediterranean), together with J. Wilmer Hersey, eventually designed the building that was completed in 1929. The Santa Barbara Courthouse became the culmination of the Anglo-Californian dream of its romantic Hispano-Mediterranean past.

The courthouse is the only building in Santa Barbara that took on the civic scale of Goodhue's San Diego California State Building. The rest of the city was built up in the same Hispano-Mediterranean style, but on a smaller scale. The result—a colorful and intricate combination of courtyards, passages, small shops, outdoor restaurants, and second-story offices—was intended to recreate the narrow shopping streets and squares of the southern Mediterranean, especially Spain.

The effort to cultivate this cultural patrimony began in 1920, when the newly formed Santa Barbara Community Arts Association started to lobby for design regulations. It obtained support for the creation of a Hispano-Mediterranean public square with a new city hall and *Daily News* building in that style. It helped to sponsor a community drafting room that provided drawings illustrating how various parts of the city could be rebuilt in this style. In 1923, it hired Charles Cheney to prepare a general plan, model building and zoning codes, and design guidelines. While they were never adopted, they laid the foundation for the government regulation to come.

The real impact of the Community Arts Association came after a major earthquake struck Santa Barbara in 1925. The association persuaded the city council to establish the Architectural Board of Review, which had to approve the exterior appearance of any structure prior to the issuance of a building permit. Simultaneously, the community drafting room began providing property owners with free plans for any contemplated rebuilding or renovation. Although the ordinance establishing the Architectural Board of Review was revoked the following year, in a few months it had permanently remade the city's architectural image. "At what seemed like the touch from a fairy's wand, a humdrum (it could be anywhere in the U.S.A.) city had become something special."[19]

In 1930, Santa Barbara adopted a comprehensive zoning ordinance that restricted commercial structures to four stories (60 feet) and apartment houses to three stories (45 feet).[20] There was, however, no further regulation of architectural design. In 1947, City Council established the Architectural Review Board; 13 years later, it established the advisory Landmark Committee, which designated a 16-block section of downtown Santa Barbara as *El Pueblo Viejo* District, named after the Spanish presidio that had occupied the site. The irony of declaring the area's charming and colorful fantasy architecture to be "historic" because it occupied the site of a long-demolished military installation seems to have escaped Santa Barbarans.

The Landmark Committee required any construction or remodeling within the district to be compatible with the area's "historical" character. In 1977, El Pueblo Viejo was expanded to include virtually the entire downtown of Santa Barbara. In doing so, the Landmark Committee guaranteed that the most important sections of Santa Barbara would continue to be rebuilt in the image established for it by civic leaders more than half a century earlier.

Like the creators of Williamsburg, Santa Barbarans had discovered the economic value of a distinctive regional appearance. They demonstrated that a community's desire for a "history" could be used to create a unique, artificially constructed environment of considerable beauty and aesthetic significance. By so doing, they also expanded the city's attractiveness to tourists. The centerpiece of this economic engine is a 10-block stretch of State Street, a charming retail corridor with brick and tile paving as well as a variety of decorative trees, flowers, and benches managed by the Downtown Organization, a BID established in 1967. During

Santa Barbara, 2013. State Street has become a colorful combination of small shops and outdoor restaurants. *(Alexander Garvin)*

2008, Santa Barbara attracted more than 10.5 million visitors (28,500 per day), who spent more than $980 million (including $253 million on shopping, $186 million on recreation and culture, and $311 million on dining). Without an attraction like El Pueblo Viejo, there would have been far fewer tourists, lower tax revenues, and fewer jobs.[21]

Santa Fe

The Santa Fe style was created in a manner very similar to that of California's fantasy-Mediterranean style. Santa Fe, like Santa Barbara, was founded as a Spanish colonial outpost in 1610. It was also built up in an adobe architecture rejected by

Santa Barbara, 2013 *(Alexander Garvin)*

San Diego, 1915. The design for the New Mexico State Building at the Panama-California Exposition later became the model for the Museum of Fine Arts in Santa Fe. *(Negative no. 13080: courtesy of Museum of New Mexico)*

the Anglo pioneers who settled there in the nineteenth century. At the beginning of the twentieth century, New Mexico, like California, sought to create a distinctive regional architecture. It also was represented at the Panama-California Exposition of 1915 in San Diego. The exposition's New Mexico State Building, designed by Isaac Hamilton Rapp, set a pattern for what was to become Santa Fe's distinctive fantasy architecture. There the similarity ended.

While the new Santa Fe style was formally inaugurated in San Diego, the Santa Fe Chamber of Commerce had really launched it three years earlier, when it sponsored the New-Old Santa Fe Style Exhibition "to advertise the unique and unrivalled possibilities of the city as 'The Tourist Center of the Southwest' "[22]

While the Chamber of Commerce lobbied for the creation of an "indigenous" architecture as a way of generating tourist revenues, Santa Fe's budding colony of artists and writers lobbied for it as a way of preventing commercial exploitation. They were among the major sponsors of the Old Santa Fe Association, which was founded in 1926 to preserve the area's historic architecture and ensure that new growth would contribute to rather than detract from the area's unique charm.

The new Santa Fe style was closer in appearance to an Indian pueblo than to a Spanish mission. Instead of having slanted red-tiled roofs, the buildings were flat-roofed; instead of deep arched porches and large windows, the buildings had little or no roof overhang and large, flat, horizontal surfaces with few openings; instead of bright pastel or white stucco walls, the buildings were the light, reddish-brown mud color of adobe.

An alliance of artists, writers, and business and civic leaders is an unbeatable combination, especially when it proposes action that will provide additional jobs and tax revenues while nurturing a community's historic identity. Over the next few decades, not only were many new buildings erected in the Santa Fe style, but existing buildings were remodeled to look like "Olde Santa Fe." Thus, when, in 1957, "preservationists" proposed an ordinance defining the Santa Fe style as the only acceptable style for further construction or rehabilitation, it was easily enacted into law. The ordinance requires the city's historic style committee to deny approval to any alteration or new construction that fails to conform to the Old Santa Fe style.[23]

Santa Fe, 1892. The wooden Western town that preceded adoption of the New-Old Santa Fe style. *(Negative no. 1701: courtesy of Museum of New Mexico)*

Santa Fe, 1868. Palace of the Governors prior to remodeling in the New-Old Santa Fe style. *(Alexander Garvin)*

Santa Fe, 1989. Palace of the Governors after remodeling in the New-Old Santa Fe style. *(Alexander Garvin)*

Santa Fe, 1989. Buildings erected along the Old Santa Fe Trail after the adoption of zoning regulations in 1957 that required all buildings to be in the New-Old Santa Fe style. *(Alexander Garvin)*

The fantasy world that emerged resonates with a sense of history and place. It has become just what the Chamber of Commerce wanted when it sponsored the New-Old Santa Fe Style Exhibition in 1912: "the tourist center of the southwest." In 2004–2005, the 1.6 million tourists who visited Santa Fe spent more than $1 billion.[24]

Santa Fe has also become just what the arts colony wanted in 1926 when the Old Santa Fe Association was established—one of the country's major arts centers. Its citizens took the cultural patrimony, produced new buildings with the stamp of New Mexico on them, refined the landscape in a manner respectful of its southwestern surroundings, and produced an enhanced environment of unique character and charm.

Protecting the Cultural Patrimony

During the three decades that followed the creation of Charleston's historic district, fewer than 20 cities established similar districts. Because there had been little development pressure during the Depression, there was little need for preservation. During World War II, construction had come to a standstill, and, thus, there was even less need for preservation. After the war, the rush to satisfy pent-up demand for new buildings decreased the likelihood of any substantial increase in the constituency for preservation.[25]

Postwar new construction changed all that. It eliminated cherished older buildings, replacing them with anonymous and unsightly structures. The preservation movement, which had long been dominated by zealous architecture buffs and colonial dames, started to attract ever-larger numbers of angry citizens who objected to the demolition. By the 1960s, this increasingly vocal constituency was persuading legislative bodies to enact preservation laws.

Pennsylvania Station and New York City's Landmark Legislation

The event that finally transformed this angry citizenry into a powerful lobby for preservation was the demolition of New York City's renowned Pennsylvania Station, designed by McKim, Mead & White. When Penn Station opened in 1910, it occupied nearly 8 acres and handled a quantity of long-haul passengers, commuters, baggage, freight, and mail that still boggles the imagination. Over the years, it had become a New York City icon: "a fitting gateway to the city, symbolically representing the power and position of the railroad."[26]

In 1961, the Pennsylvania Railroad announced that it could not continue to own such a valuable tract of mid-Manhattan property that did not produce substantial revenues. Its solution was to replace the "McKim, Mead & White elephant" with a moneymaking office tower and sports

Manhattan, 1930. Pennsylvania Station. *(Courtesy of Museum of the City of New York)*

arena. Despite newspaper editorials and public demonstrations, the railroad was determined to proceed. Company executives insisted that the railroad was in business to make money, not lose revenues maintaining a public monument. Demolition began in 1963, with construction of the current Penn Station, Madison Square Garden, and office towers above following immediately thereafter. Architectural historian Vincent Scully eloquently captured the difference between the McKim, Mead & White building and the complex that replaced it: where once one entered the city "like a god . . . One scuttles in now like a rat."[27]

In 1965, the year demolition of Penn Station was completed, the New York City Council adopted legislation that civic leaders hoped would prevent further desecration of the cultural patrimony. It established the Landmarks Commission, with responsibility for designating landmark structures and historic districts. The commission was given wide latitude in determining what properties had "a special character or special historical or aesthetic interest or value" or represented "periods or styles of architecture typical of one or more eras in the history of the city." The only limitation was that a landmark had to be at least 30 years old.[28]

Landmarks and properties within historic districts that are designated by the Landmarks Commission cannot be altered without its permission. Property owners must apply for a *permit for minor work, certificate of no effect,* or *certificate of appropriateness.* The commission, in turn, must act within 90 days and specify the reasons for its action. Denia precludes neither further application nor further denial.[29]

At some point, if the commission continues to deny a property owner's application, the decision can be challenged on grounds that it has taken the use of the property from its owner. New York City's landmarks legislation defines this point as having been reached when the property in its existing condition is unable to earn a "reasonable return" of 6 percent per year on its assessed value.[30]

New York City's legislation is similar to that of other cities. The only significant differences are that the number of designated properties is much larger and that property values in New York tend to be so much higher than in other places. It is therefore more likely to be worth litigating landmark issues. As of 2009, the city had designated more than 1181 individual landmarks, 90 historic districts (containing more than 24,000 properties), 10 scenic landmarks, and 114 interior landmarks. If a property owner challenges the Landmarks Commission, the commission has 60 days to develop a plan that will provide the owner with a "reasonable return." The obvious method of achieving this is a partial or full abatement of real estate taxes. Such a plan must be approved by the city council. If the owner rejects this tax relief, the Landmarks Commission can continue denying the permission sought by the owner. The owner then seeks relief in court.[31] In fact, the landmark case in this field was brought to the Supreme Court by a New York City property owner, Penn Central Transportation Company (Penn Central).[32]

Grand Central Terminal

Over the objections of Penn Central, Grand Central Terminal was designated a landmark in 1967. Completed in 1919, the building was the result of a design competition won by Reed & Stem, who were later joined by Warren & Wetmore. The scheme is an ingenious intermingling of separate traffic systems for the pedestrian, automobile, subway, and train.

Like Penn Station, Grand Central Terminal had become an icon. Unlike Penn Station, though, it was located amidst the world's most expensive real estate. The right to build over the vast network of rails leading into the terminal and over the terminal itself is worth billions of dollars. From the beginning, the railroad profited from selling or leasing those rights for construction of buildings like the Yale Club of New York City, the Waldorf-Astoria Hotel, and the Pan Am Building (now the Met Life Building). At the time that Grand Central was designated a landmark, the city's zoning resolution would have permitted additional construction of 1.7 million square feet of floor area on the site.[33]

The Penn Central Railroad opposed landmark designation because it correctly guessed that the Landmarks Commission would not allow it to continue to build over its property. Penn Central had entered into a renewable 50-year lease with Union General Properties, Ltd. (UGP), under which UGP was to build a multistory office building over the terminal. In exchange, it agreed to pay Penn Central $1 million annually during construction and at least $3 million annually thereafter.

The year after landmark designation, UGP sought permission to build over the terminal. Its first proposal was for

Manhattan, 1968. Grand Central Terminal. *(Alexander Garvin)*

a 55-story building designed by Marcel Breuer, which was later revised downward to a 53-story building. The Landmarks Preservation Commission refused both proposals, arguing: "To protect a landmark, one does not tear it down. To perpetuate its architectural features, one does not strip them off. . . . But to balance a 55-story office tower above a flamboyant Beaux-Arts façade seems nothing more than an aesthetic joke."

Since the terminal was exempt from real estate taxes, there was no way of measuring a "reasonable return" or of using real estate tax abatement as a means of creating one. Some preservationists argued that the Penn Central was only entitled to a reasonable return from "transportation purposes." They ignored the fact that it had been earning millions from the non-railroad-related buildings on the site. Others argued that the property's value was derived from the rights-of-way that had been granted by government or from its interrelation with the government-built and -operated transit system. What the state had given, they implied, the state could take away without compensation.[34]

City officials were not sure that Penn Central was not entitled to compensation. More important, they were worried that the entire landmarks statute could be overturned as an unconstitutional taking of property. Accordingly, in 1968 the City Planning Commission amended the zoning resolution to permit transfer of unused development potential.[35] The new provision permitted an owner of a landmark to transfer unused development rights to contiguous lots or to lots across a street or intersection, provided that the development potential of the receiving lot was not increased by more than 20 percent and provided that the City Planning Commission found the transfer would do nothing "to the detriment of the occupants of *buildings* on the *block* or nearby *blocks*."[36]

In 1969, the City Planning Commission and Board of Estimate approved Penn Central's sale of 4 percent of its development rights to Philip Morris for construction of its headquarters across 42nd Street from the terminal. Despite this, Penn Central sued the city of New York, claiming the city had taken (without paying) the 1.7 million square feet of development rights permitted by the zoning ordinance. The railroad argued that the landmark designation represented an unconstitutional taking of private property without just compensation and a deprivation of property without due process of law.[37] It also found that allowing the transfer of development rights was insufficient compensation because only in the unlikely event that one were to demolish the Biltmore Hotel (later remodeled as an office tower occupied by Bank of America), the Commodore Hotel (later remodeled as the Grand Hyatt Hotel), and any of the other profitable large adjacent structures, would there be enough sites to receive Penn Central's development rights. The court of appeals reversed the decision on the basis that the statute did not deprive Penn Central of "all reasonable beneficial use of their property." Finally, in 1978 the Supreme Court upheld the constitutionality of the statute.[38]

The Grand Central Terminal case provided the constitutional foundation for landmark legislation throughout the nation. It also raised some significant issues. Local governments derive their right to regulate (the *police power*) from their responsibility for the protection of health, safety, morals, and general welfare. In the case of zoning, the rationale for regulation is easy to explain (see Chapter 17). Local government is presumed to have examined the capacity of local streets and sidewalks, the capacity of existing infrastructure and community facilities, projected capital expenditures intended to increase that capacity, and the impact of likely development in each part of the city. It is also presumed to have determined standards for providing a suitable level of light and air. Presumably, the zoning resolution balances these factors, determines an optimum level of development for every property, and thereby provides for the health, safety, morals, and general welfare of the community.

Protecting the cultural patrimony is an important additional way of providing for the general welfare. Once that patrimony is destroyed, it is forever gone and can no longer provide citizens with a profound connection to their culture and their history. This is the rationale for such disparate "his-

Manhattan, 1968. The 55-story building proposed for construction of Grand Central Terminal. *(From Exhibit 20A,* Penn Central Transportation Co. v. City of New York, *Supreme Court Appellate Division—First Department)*

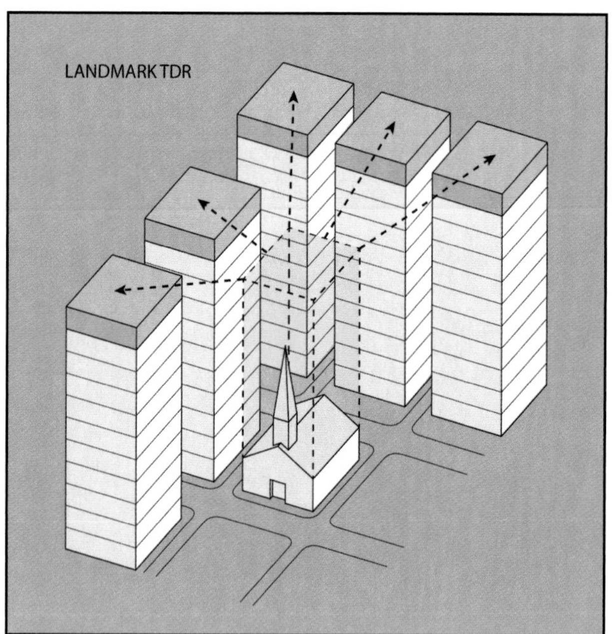

Development rights. Neighboring sites to which the unused bulk over a designated landmark could be transferred. *(Alexander Garvin and Owen Howlett)*

toric" districts as those of Williamsburg, Charleston, New Orleans, Santa Barbara, and Santa Fe.

There is, however, a difference between regulating one specific property and regulating an entire area (whether a historic area or a zoning district). All the properties in the designated area are treated in a similar manner. They suffer the same intrusion into their property rights and derive the same benefits from the area's regulations.

Restricting the rights of the owners of individual properties designated as landmarks may deny them the right to use their property in the same manner as neighboring property owners, especially if landmark regulation forbids any significant further construction. Designated landmarks are probably zoned for development at the same bulk, use, and density as their neighbors. However, if landmark regulation prevents the owner from developing the property in the same manner, that regulation may be unfair.

It was this potentially unequal treatment that the New York City Planning Commission tried to rectify when it amended the zoning ordinance to permit the transfer of unused development rights to neighboring sites. At the same time, it attempted to balance positive benefits with negative impact by restricting the increase in bulk and density on any receiving site to no more than 20 percent and permitting that increase only on contiguous properties or properties across a street or intersection. Had the planning commission permitted development rights to be transferred to an area far from the terminal, the citizens of the receiving area (presumably capable of accommodating only the bulk, density, and use for which it was already zoned) would have been burdened by the impact of additional development (more vehicular traffic, more pedestrians, more noise, less light and air) while property owners in the vicinity of the terminal (which had the capacity to handle the added development) would have benefited (less vehicular traffic, fewer pedestrians, less noise, more light and air).[39]

The Penn Central Railroad was not able to sell further development rights. Most eligible receiving sites were already developed up to, or close to their allowable capacity. By 1988, ten years after the Supreme Court ruling, the railroad had been able to transfer only to the aforementioned Philip Morris building. In 1989, it sought planning commission approval of a transfer of 800,000 square feet of development rights several blocks north of the terminal to 383 Madison Avenue at 46th Street. The owners of this property were prepared to pay the railroad $480 million for these development rights because it would have allowed them to build a 1.4-million-square-foot, 74-story office tower.

The planning commission denied approval because the office-building site would have been developed to a bulk and density 53 percent larger than permitted by the zoning resolution anywhere in New York City.[40] It also denied the contention that a site several blocks away from the terminal was eligible to receive development rights. The developers had argued that the necessary chain of ownership existed in the subsurface lots (used by Penn Central for its railroad tracks) between the terminal and their site. Commission Chair Sylvia Deutsch denied this logic, pointing out that it could conceivably establish "a link or a chain going past Yonkers."[41]

Within days of denying this development rights transfer, the City Planning Commission issued a proposal to create a Grand Central Subdistrict doubling the area that could receive the terminal's development rights. The proposal restricted the permitted bulk to the same maximum level permitted elsewhere in Midtown Manhattan. It was explained that the new district would provide a framework for development rights transfer "based on sound planning, rather than one based on ownership patterns."

After considerable debate and public hearings, the City Planning Commission in 1992 amended the Special Midtown Zoning District to include a Grand Central Subdistrict extending from East 41st to 48th Streets, generally from the midblock west of Madison Avenue to the midblock east of Lexington Avenue, thereby increasing the number of sites eligible for transfer of development rights from landmarks. The commission eliminated the plaza bonus within the subdistrict and established new street-wall, height, setback, and building-entrance requirements. More important, subdistrict regulations limited the transfer of development rights to a maximum of 1 FAR (floor area ratio) to the receiving site, and limited the total FAR on any zoning lot to the maximum permitted anywhere in the city, 21.6. Completed in 2002, the building at 383 Madison Avenue was the first development to make use of the transfer of development rights made possible by the changes to the zoning of the subdistrict.

Paying for Preservation

The story of Grand Central Terminal illustrates the dilemma facing a citizenry that wishes to preserve the cultural patrimony exclusively by regulating what property owners can do with their property. Penn Central, like most property owners, especially those whose property has a high market value, will not willingly give up potential income in order to maintain a landmark that happens to stand on its property. Ordinary homeowners are no different. If they need to deal with leaky windows, walls, and roofs, they will do so. At the same time, they are likely to avoid spending big money to replace original details and materials even though a more cost-effective solution to the leaks may disfigure the building. Property owners are not alone in being unwilling to pay for preservation. Preservationists and local governments also would rather have somebody else pay. But if landmarks and historic districts are to be preserved, the money must come from somewhere.

Several strategies have been proposed that provide funds to forestall demolition, ensure proper maintenance, and still allow owners to benefit from the market value of their property. Each requires money but is within the financial capacity of any community seriously interested in preservation. Moreover, each is a potentially useful tool but must not be interpreted as a panacea that will protect every landmark and work in every situation.

One strategy is to establish an area-wide system for the transfer of development rights. A second involves a revolving fund that provides the money to purchase threatened landmarks and then sells them with appropriate preservation covenants. The most prevalent preservation strategies involve government actions that encourage property owners to cultivate their patrimony. The most obvious of these (strategy three) is for government to eliminate obstacles that make preservation more difficult and expensive. Strategy four is for public agencies to invest in the public realm—the property we all own in common—leaving preservation of individual properties to their owners. Since the money to restore private property may not be easily obtained or the cost of borrowing too high, strategy five is for government action that makes it attractive for institutions to lend the needed funds. The sixth strategy transfers the cost of rehabilitation to local government. Property owners who pay for restoration are reimbursed annually in the form of reduced real estate tax payments. This strategy is particularly attractive to elected officials, because they never have to vote to cut something from the budget or raise the revenues from a new preservation tax. Strategy seven transfers the cost of preservation to the federal government. It also camouflages the price by keeping it out of the budget and spreading it over many years. In this case, the owners of a landmark pay for restoration and are reimbursed by the federal government in the form of deductions from their income tax payments.

Strategy 1: Development Rights Districts

In his book *Space Adrift: Landmark Preservation and the Marketplace,* published in 1974, law professor John Costonis proposes the creation of "development rights transfer districts" to which and from which landmark owners would be entitled to convey excess development rights. The plan assumes that most sites occupied by landmarks are zoned for significantly greater bulk and density, that these sites are "concentrated into reasonably compact areas of the city, usually downtown," or in areas that are about to undergo intensive development, and that the sites which will receive additional development are located where there are plentiful public services and facilities to "absorb large numbers of people with greater efficiency."[42]

Such landmarks are threatened because they do not generate revenues that are commensurate with the property's market value. That market value is based on the development potential of the site under existing zoning. The only way to obtain the potential revenue is to demolish the landmark currently occupying the site and then to erect a building that maximizes the zoning potential.[43]

Costonis argues that demolition can be avoided by identifying and permitting the sale of the unused zoning potential of the site. He also suggests that cities establish a *development rights bank.* The bank would receive donations of development rights from philanthropic property owners and from city-owned landmarks. These rights would be sold to interested developers, and the resulting revenues would be used to fund municipal condemnation of those properties owned by recalcitrant owners who threaten a particular property's condition or long-term survival with either neglect or demolition.

The transfer of development rights (TDR) strategy requires that the benefits of preserving landmarks accrue to the same people and district that will bear the burden of the added noise, traffic, and activity generated at the receiving site. At some point, the added density is transferred to a property that is too far away and the people and businesses in the receiving area will be unfairly carrying a burden that is not placed on those who enjoy the benefits of the landmark. The second pitfall in any TDR strategy is that it depends on the presence of receiving lots in an area where there is great demand for new construction and few sites that are not already built out. That is the reason that TDRs have not yet been successful in Denver, one of the first cities to use this approach.

Denver's Development Rights Transfer District

Denver's Transfer of Development Rights Ordinance was created at a time when millions of square feet of new office space were being built downtown, almost all on sites that had to be cleared of older buildings. When civic leaders realized that existing zoning might be preventing renovation, they decided to amend the city's land use regulations.[44]

The project that initiated this legislation was the Navarre, a landmark listed on the National Register. This four-story brick structure was originally built in 1880 to house the Brinker Collegiate Institute. Nine years later, it was converted into a gambling house and bordello (known as the Richelieu Hotel). In 1914, the property was renamed the Navarre Café.[45]

The Navarre had been vacant for several years when, in 1980, it was purchased for reuse as an office building. The owners of the Navarre tried to obtain a mortgage to finance the rehabilitation of their landmark. Institutional financing was available for what lenders considered the property's "highest and best use," a high-rise office building, but not for renovation. Worse yet, the banks subtracted the cost of demolishing existing buildings from the amount they would lend.

For mortgage appraisal purposes, rehabilitation was deemed to be an expense that increased the eventual cost of site preparation for a skyscraper. Banks were unwilling to lend for an activity that their appraisers believed would decrease the value of the property.

Lisa Purdy, a resourceful preservation specialist at Historic Denver Inc., went to work to reverse this bias against renovation. Taking her cue from Costonis, she proposed legislation that would transform the unused development potential of sites occupied by designated landmarks into a valuable commodity. If such legislation could be enacted, the owners of the Navarre and of other landmarks would, for the first time, have a liquid asset that could be sold without demolishing the historic structures that occupied their land. Moreover, they could use the proceeds from the sale of development rights to finance rehabilitation.[46]

Denver, 1990. Once the Navarre Building, now the American Museum of Western Art. *(Alexander Garvin)*

In 1982, after a year of public review and discussion, the city council amended the zoning ordinance to allow transfer of development rights from designated historic structures. TDR was restricted to two sections of the city: the 93-block central business district (zoned B-5) and the 15-block Lower Downtown (LoDo) section (zoned B-7). In order to ensure that no receiving lot in either zone would overload the existing infrastructure, the permitted increase in floor area was restricted to 25 percent.

The business district, Denver's only B-5 zoning district, permitted commercial use at a FAR of 10. It seemed a perfect location and market for TDRs. Furthermore, the TDR legislation made no change in the area's permitted density, bulk, or land use. It simply allowed the redistribution of some of the B-5 district's bulk and density.[47]

Lower Downtown contained Denver's 1881 railroad station and the bulk of its nineteenth- and early-twentieth-century warehouses, manufacturing lofts, and mercantile structures. No section of the city had a greater concentration of historic structures. Lower Downtown was Denver's only B-7 zoning district, and it permitted commercial land use at a FAR of 4. The legislation only allowed the redistribution of the B-7 zone's authorized bulk and density.[48]

To understand how Denver's TDR works, let's imagine a historic structure containing 20,000 square feet on a 10,000-square-foot lot in a B-5 zone. Since the FAR is 10, the total allowable floor area on the site is 100,000 square feet (i.e., FAR 10 × 10,000 square feet = 100,000 square feet). The existing historic structure uses 20,000 square feet, leaving 80,000 square feet that can be transferred. Let's also imagine that a developer has purchased a 40,000-square-foot vacant lot, 10 blocks away, also in the B-5 zone. The allowable floor area on the site is 400,000 square feet (i.e., FAR 10 × 40,000 square feet = 400,000 square feet); if development rights can be purchased from the owners of designated historic structures, an additional 100,000 square feet can be constructed (i.e., 0.25 × 400,000 square feet = 100,000 square feet). Any developer who has purchased the site for $20 million should be willing to purchase development rights for the same $50 per square foot of floor area (i.e., $20,000,000 ÷ 400,000 square feet = $50 per square foot). The owner of the historic structure who has 80,000 square feet of available development rights should be able to sell them for $4 million (i.e., 80,000 square feet × $50 per square foot = $4,000,000). Thus, the developer will be able to build a 480,000-square-foot building. To max out the development potential at the full 500,000 square feet, though, the developer will be required to purchase of another 20,000 square feet of TDR from some other site with the available capacity to sell such rights.

Although transfer of development rights has been possible in Denver for decades, buildings have not made use of available TDRs. The Denver Athletic Club did transfer 60,000 square feet of development rights to the site of the yet-to-be-built Midland Savings Bank, five blocks away. The D & F Tower entered into a contract to sell its development rights. These rights went unused initially because the energy glut and subsequent recession had brought development to a halt and later because of the generous development opportunities provided by existing downtown zoning meant that TDRs were unnecessary.

TDR legislation did result in the preservation of some historic structures. The owners of the Navarre were able to use the newly created development rights as collateral for a mortgage loan covering the cost of its renovation. When the building was sold for use as the American Museum of Western Art in 1983, the previous owners kept the development rights for future sale. The owners of another building, Odd Fellows Hall, built in 1889, were also able to use their TDRs as part of their collateral for a rehabilitation construction loan. In general, however, this preservation technique has not been of much use in Denver because of the widespread availability of development sites throughout the city.

Strategy 2: The Revolving Fund

Landmarks are often owned by individuals who either do not have the money to maintain them properly, have the money to pay for ordinary upkeep but not for necessary restoration, intend to use their property in a way that would be injurious to its historic character, or are ready to sell to purchasers who would demolish what is regarded as precious. One way to avoid such situations is to acquire the property and then transfer ownership to somebody who will cherish and maintain it. This requires an entity (usually a nonprofit organization) that can act quickly. The most effective means of creating such an entity is to endow it with enough money to acquire and resell a substantial number of threatened landmarks.

The National Trust for Historic Preservation operates the most active revolving fund. Most revolving funds, however, are operated by local entities like the Historic Savannah Foundation or the Pittsburgh History and Landmarks Foundation (see Chapter 13). Once the necessary money has been raised, it is used to purchase threatened properties, which are resold with protective covenants ensuring that the exterior is properly restored and then maintained. Occasionally, a revolving fund is used for necessary restoration prior to resale, for acquisition and demolition of property that is incompatible with a historic district, or for mortgage loans to purchasers who are unable to obtain conventional bank financing. The idea is to keep enlarging the fund, reuse the sale proceeds for additional preservation activity, and always ensure that enough money is on hand to deal with emergencies.

While a revolving fund is applicable to all parts of a city, it can be particularly effective with inexpensive and neglected buildings that do not have development rights to transfer and therefore cannot easily be preserved by the Costonis approach. The money is reused many times, producing many bangs for the same buck. Moreover, when a revolving fund is used as part of a strategy for a whole area, strategic renovation of a few critical buildings will bolster local self-confidence, thereby inducing a private market reaction from neighboring property owners, who will be more likely to make further improvements. This is what happened when a revolving fund was established for the historic wards of Savannah and Charleston.

Savannah's Historic Wards

Savannah's 26 wards provide a unique setting for the lovely eighteenth- and nineteenth-century buildings that were erected around charming public squares (see Chapters 3 and 14). By the middle of the twentieth century, many of these buildings had badly deteriorated. Two squares had been dismembered to provide additional traffic lanes. As a result of these increasingly common compromises to the urban character and fabric of the city, the Historic Savannah Foundation (HSF) was founded in 1954 to limit any further destruction of this extraordinary heritage.[49]

Savannah, 2012. The purchase of the Davenport House (1820) by a group of dedicated citizens in 1955 provided the impetus for further citizen efforts to preserve the city's historic structures. *(Alexander Garvin)*

When the Historic Savannah Foundation was established, there was little public interest in the city's old buildings. With the exception of a citizens group that had successfully fought the insertion of streets into the city's squares, Savannah had neither an organization dedicated to preservation nor laws to preserve any of the landmarks in its historic wards. Few families chose to settle downtown, where property values had been depressed for years. Community leaders became alarmed when the then-dilapidated Davenport House, originally built in 1820, was threatened with demolition. A group of civic leaders raised the money to purchase and eventually restore the building. This effort culminated with the creation of an institution dedicated to the cultivation of the city's cultural patrimony: the Historic Savannah Foundation, a private, nonprofit membership organization.

In 1962, the foundation completed an inventory of historic structures that was first published six years later as a beautifully illustrated book entitled *Historic Savannah.* This survey provided the basis for the historic district zoning enacted in 1972 and helped to generate increased interest in the buildings illustrated in the book. The foundation also promoted awareness of historic properties by publishing a Sunday supplement in Savannah's newspaper and by organizing weekend tours. Together with the Chamber of Commerce, the foundation, in 1965, issued a report entitled "Savannah—A Travel Destination." The thrust of the report was that "a modest investment in the promotion and restoration effort in Savannah could bring tourist revenues of up to $100 million annually."[50] Its initiative, at least initially, was motivated by a desire to preserve the esthetic and historical value of the city and grew to include an effort with the potential to contribute significantly to the local economy.

Historic Savannah's most enterprising activity was its $200,000 revolving fund. The fund was used initially to acquire and renovate threatened landmarks in the 15-acre Pulaski Square–West Jones Street area that surrounds one of Savannah's charming original squares. In 1964, when the Historic Savannah Foundation chose it as a target area, Pulaski Ward was a dilapidated neighborhood with many vacant (albeit historic) structures. Over the next 18 months, Historic Savannah acquired and resold 46 buildings, generating more than $1.5 million in privately financed renova-

PULASKI SQUARE-JONES STREET

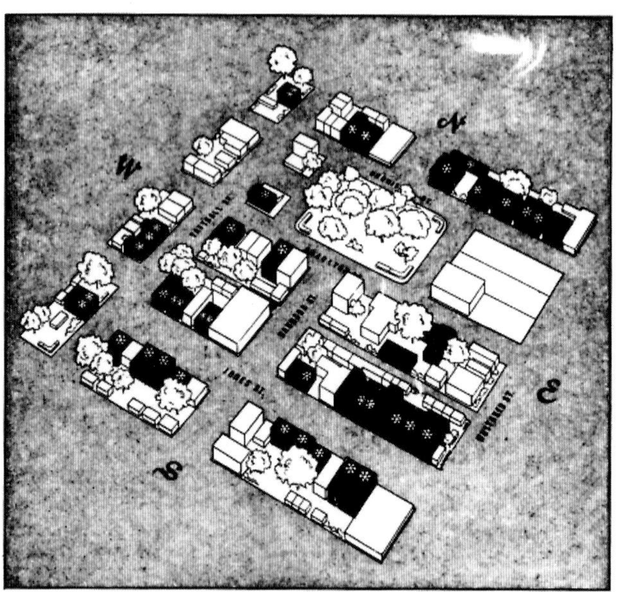

. . . . it was a 13-acre deteriorating, near-slum neighborhood. It is now being reclaimed.

46 buildings have been stabilized . . .

38 were reclaimed by Historic Savannah

28 of the 38 through direct purchase for resale

another 10 influenced by Historic Savannah's efforts since the program's inception.

■ **46 Buildings stabilized (restored or in process)**

* **38 Units reclaimed through Historic Savannah**

THIS WAS DONE IN ONLY 18 MONTHS!

Savannah, 1967. Forty-six buildings were restored or in the process of restoration within 18 months of establishing a revolving fund to purchase and resell deteriorating buildings in Pulaski Ward. *(Courtesy of Historic Savannah Foundation)*

Savannah, 2012. Pulaski Ward historic district four decades after the success of the revolving fund. *(Alexander Garvin)*

tion. Subsequently, another dozen buildings were acquired and rehabilitated privately. Eventually, owners renovated every building in the ward and even began filling in vacant lots with small-scale new construction.

Since its inception more than half a century ago, Historic Savannah has acquired and resold more than 3500 houses that are actively used as residences by their occupants. Moreover, the impacted areas have become tourist destinations for the city's more than 12 million visitors who, in 2011 alone, spent almost $2 billion.[51]

Savannah's Victorian District

Leopold Adler II, who had been a leader in the effort to save Pulaski Square–West Jones Street, decided it was important to extend preservation activity to the 150 blocks of Savannah's Victorian District. Adler was an investment banker who had been deeply involved with the Historic Savannah Foundation in its successful efforts to preserve the city's 26 antebellum wards. Now he wanted to prove that classic structures could be restored while at the same time retaining the neighborhood's racial and economic fabric.

The dozen wards that constitute the Victorian District were built up after the Civil War with opulent one-family houses and more modest row houses. Like countless other inner-city neighborhoods, the larger residences went out of fashion after World War II. The district's lovely housing, however, appealed to the city's African-American population, a portion of which had lived in the modest Beach Institute section of the neighborhood since the Civil War. Because many African-American households could not afford the Victorian District's larger residences, some owners converted their buildings into multiple dwellings. Others just cut back on maintenance.[52]

Rather than depend on HSF, with its citywide constituency and downtown focus, in 1974 Adler organized the Savannah Landmark Rehabilitation Project (SLRP), a nonprofit corporation that was focused exclusively on the Victorian District. Instead of using the revolving fund strategy that had been so successful in Pulaski Square, he created a pool of money for the Victorian District that was intended to completely cover the cost of acquisition *and* restoration. It was to be replenished from the proceeds of the resale of these buildings to responsible owners who would maintain them properly.

Unfortunately, neighborhood residents could not afford to pay more for their housing, so the cost of rehabilitation had to be subsidized. SLRP scrounged for money everywhere: grants from the National Endowment for the Arts, the Ford Foundation, and a variety of public institutions; charitable contributions from interested individuals; and loans from the U.S. Department of Housing and Urban Development (HUD), the National Trust for Historic Preservation, and virtually every bank in Savannah. The primary subsidy however, came from the federal Comprehensive Employment Training Act (CETA), which covered the cost of labor and materials needed to train residents in carpentry, plumbing, wiring, painting, and other aspects of construction. Five years after Adler began the fight to revitalize the neighborhood, HSF joined the fray, establishing its own Victorian District Rehabilitation Project. It was a revolving fund, with $100,000 from the foundation combined with $130,000 in subsidies from Savannah's Community Development Block Grant (CDBG) funds. The city added another $100,000 in

Savannah, 1990. Two decades after preservation efforts in the Victorian District had been started, the revolving fund had failed to generate area revitalization and private developers began to acquire buildings and invest in renovation. *(Alexander Garvin)*

1983. HSF continued its proven policy of acquiring clusters of properties depending on their availability, price, and architectural significance, then reselling them with restrictive covenants requiring prompt, authentic restoration and ongoing maintenance.

In 1983 the CETA program ended. By then, two-thirds of SLRP's 200 CETA workers had obtained jobs in private industry. However, as in New York (see Chapter 11), when Congress terminated the CETA program, subsidized, sweat-equity rehabilitation came to an end. Building occupants could not pay more for their housing. So when CETA ended, further SLRP revolving fund activity ended as well. Similarly, HUD cutbacks in low-interest mortgage programs resulted in the termination of all other subsidized construction. Consequently, neither program could continue. By 1988, HSF had rehabilitated 24 structures containing 51 dwelling units and returned $102,500 to the city of Savannah. Residents were not able to pay higher rents, so there were no property owners who could afford to renovate the buildings that might be acquired by the HSF revolving fund. SLRP had rehabilitated

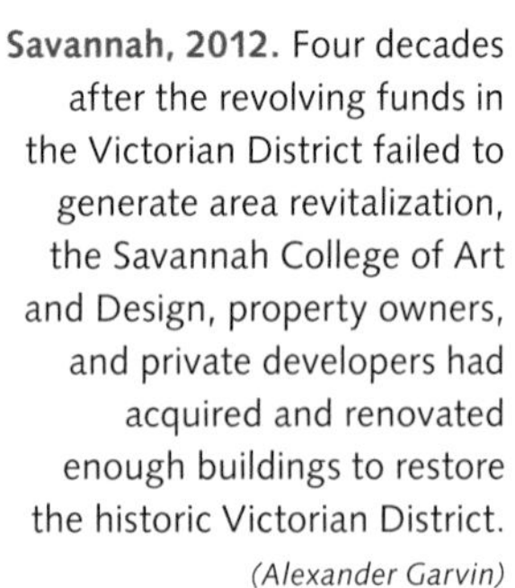

Savannah, 2012. Four decades after the revolving funds in the Victorian District failed to generate area revitalization, the Savannah College of Art and Design, property owners, and private developers had acquired and renovated enough buildings to restore the historic Victorian District. *(Alexander Garvin)*

another 260 housing units and built 44 infill units; it had stopped because subsidy money was no longer available.

After nearly 15 years of hard work by both SLRP and HSF, the Victorian District was still riddled with vacant and dilapidated structures that were a major fire hazard. SLRP and HSF had been unable to generate enough private investment to preserve the bulk of the neighborhood's many beautiful buildings because existing neighborhood residents were too poor to pay more for their housing and, by extension, to cover either additional maintenance expenditures or the additional debt service on rehabilitation loans. With little hope of additional revenue, property owners continued to defer spending on maintenance. The neighbors simply could not follow the pioneers who had renovated buildings in the Victorian District. The situation became sufficiently serious that, in 1989, the Savannah City Council passed an ordinance empowering the city to demolish any privately owned structure considered to be a public nuisance.[53]

People like me, who in 1990 thought the Victorian District would continue to decay, were wrong. The preservation pioneers had opened the area to a new market: art students. In 1978, one year before HSF began its efforts in the district, 4 staff members, 7 faculty members, and 71 students opened the Savannah College of Art and Design (SCAD) in an old building not far from the Victorian District. As of 2012, nearly 10,500 students and 2000 teachers and staff worked in SCAD buildings all over Savannah, including the Victorian District.

Younger households, especially those at SCAD, liked the idea of living in a historic neighborhood whose prices were lower than those in the city's older wards. They began moving into run-down buildings in the Victorian District. Developers responded to the increased demand for space in the district by acquiring and renovating long-vacant structures. In some cases they built entirely new structures that were in character with the surrounding neighborhood. This gentrification did displace some of the lower-income African-American residents, but by the start of the second decade of the twenty-first century, the largely restored Victorian District had blossomed into the well-maintained neighborhood that preservationists envisioned in the 1970s and 1980s—occupied by African-American families that had been there for decades, students who rented rooms in charming nineteenth-century buildings, and middle-income white families that had bought and fixed up some of the older houses.

Savannah, 2012. By the twenty-first century, market conditions had changed sufficiently that entrepreneurs began to rehabilitate Victorian buildings as bed-and-breakfast accommodations for tourists. *(Alexander Garvin)*

Strategy 3: Simplifying Compliance with Government Regulations

Con Howe, Director of Planning for Los Angeles between 1992 and 2005, believes that "when someone is trying to do what the city's plan calls for, government should get out of the way—the faster the better."[54] It, therefore, follows that when a government is trying to retain older buildings and revitalize a downtown, it should eliminate any government obstacles that are in the building owner's way. Unfortunately, those obstacles are frequently zoning, building, seismic, and fire codes that were intended to regulate new construction. These codes usually were enacted based on recommendations of engineers and architects, who assumed that when existing buildings had deteriorated sufficiently they would be discarded, like worn-out clothing. In fact, people often want to retain and reuse cherished older buildings.

When somebody wants to renovate a structure that was erected prior to the regulations that are currently in effect, they frequently find that they cannot comply with the law without tearing out significant sections of the building. A handsome slate roof may have to be replaced because it exceeds the height limit required by the zoning. They may have to rip out a bathroom because it is 6 inches narrower than required to provide access to the disabled and replace a 34-inch-wide wooden stair because the fire code calls for a 36-inch-wide fireproof stair.

Regulations may make it more expensive to rehabilitate the building than to build something from scratch. There may be no way to retain or fix up the building for a new set of occupants and still comply with applicable government regulations. Ironically, however, the unrenovated building may continue to be legally occupied as it is currently used—even though there aren't any users who want to be there.

To deal with these problems, during the last quarter of the twentieth century communities throughout the nation began altering codes to better fit the requirements of building renovation. One of the most effective of these initiatives has been the Los Angeles *Adaptive Reuse Ordinance* (ARO).

"Unzoning" Los Angeles

For years, Los Angeles's plans had called for a "vibrant, mixed-use downtown" with new housing. By the 1990s, the city administration, downtown business interests, property owners, proponents of historic preservation, and many civic organizations had become increasingly concerned with the large number of vacant, old office buildings in downtown Los Angeles. Many of these buildings were significant enough to be listed on the National Register. When vacant, they were susceptible to fire or occupied by squatters. Why not reuse these vacant structures? They left large portions of downtown dead, by day and by night.[55]

A few of these older buildings had been renovated during the previous decade. Con Howe discovered that their owners had had to obtain variances from zoning, fire, and building codes—a process that was sufficiently time-consuming and expensive to discourage most property owners. An interagency task force identified the most frequently needed variances and recommended simple changes to existing codes that make such variances unnecessary when reusing these building for residential purposes.

The Adaptive Reuse Ordinance they proposed was enacted in 1999. Initially, all this zoning applied to downtown business district buildings erected prior to 1974, which met the requirements of a previous building code. Many of the changes involved "unzoned" items that otherwise would have been required for residential buildings: for example, front and side yards. The density requirement, which had previously been calculated based on a minimum amount of floor area per unit, was eliminated. Additional floor area was permitted in the form of mezzanines in apartments with particularly high ceilings. As long as all existing parking was retained, no new parking was required. These changes made conversion an as-of-right procedure, which meant that no review pursuant to the California Environmental Quality Act would be needed. Tom Gilmore, an architect who moved from New York City in the 1990s, was one of the earliest developers to make use of ARO. He began by converting several buildings clustered around Fourth and Main Street, including the former Farmers and Merchants Bank, which resulted in the area becoming known as the Old Bank District. Other developers followed suit. As of 2012, ARO has fostered the creation of more than 9000 downtown apartments in old, previously commercial structures. Not only has 24-hour life returned to the area, the ordinance resulted in political leaders extending its provisions to Hollywood and other areas of Los Angeles. Its provisions have been copied in Oakland, Glendale, Long Beach, and other California communities.

Los Angeles, 2013. The Adaptive Reuse Ordinance resulted in more than 9000 apartments replacing empty or obsolete office space within 14 years. *(Alexander Garvin)*

Strategy 4: Government Investment in the Public Realm

Public agencies can encourage private owners to maintain and improve their properties by investing public money in the public realm. Under this strategy, government creates an environment from which owners can best profit by maintaining their property at the same or a better standard. If property owners do not meet those standards, the shabbiness of their property becomes obvious and they will not be able to obtain rents or sales prices that are as high as those of their neighbors whose properties reflect the quality of the entire neighborhood.

The most common example of this strategy is the red-brick paving that has been slathered, like mercurochrome, along shopping streets throughout the country. It is often accompanied by new lighting fixtures, directional signs, street trees, decorative planters, trash receptacles, bus shelters, and benches. Where there is an untapped market and the new street furniture is intelligently designed, business owners tend to improve the design of their shop fronts and invest in attractive window displays; cafés and restaurants will replace rickety furniture with butcher block tables, designer chairs, and baskets of hanging plants (see Chapter 6).

Between 1949 and 1973, the federal urban renewal program subsidized similar neighborhood improvement investments (see Chapter 12). Local governments found this strategy particularly attractive, because their investments were counted as noncash credits in lieu of the one-third cash contributions that were required to match the two-thirds in federal assistance. Philadelphia's Society Hill Urban Renewal Area is perhaps the very best example of this approach. More than half a century after the city installed the cobblestone streets, granite curbs, brick sidewalks, street trees, and gas lamps, the neighborhood's charming late-eighteenth- and early-nineteenth-century row houses have all been restored. The value of a housing unit in the area increased 78 times between 1950 and 2000, a period during which the values in the city as a whole only increased 16 times.

Philadelphia, 2012. Public investment in brick paving, street trees, gas lamps, and granite curbs enhanced the character of Society Hill and helped to convince property owners and banks to invest in improving the area's historic row houses. *(Alexander Garvin)*

Strategy 5: Government-Assisted Financing

Along with the value of the real estate, so, too, did real estate tax payments from Society Hill increase 78 times from their value prior to the government investment. These additional tax receipts became part of the pool of revenues the city used to pay for garbage collection, police and fire protection, and all other government services. Sometimes local governments choose to direct the tax increment to pay for improvements to the public realm. In this type of *tax increment financing* (TIF) the local government designates an area for redevelopment. The real estate tax revenues that flowed to the city before continue to flow to the city for use in the general fund, but all increases in taxes within the area are turned over to an entity that is responsible for capital spending isolated only to the area designated for redevelopment.

In addition to the investment in public realm infrastructure, the government also helped owners renovate in Society Hill and other urban renewal areas through an urban renewal program that provided mortgage insurance for banks that made loans to properties within the renewal areas. As a result, banks willingly issued mortgage loans that paid for the rehabilitation of buildings in Society Hill, Wooster Square in New Haven (see Chapter 13), and other renewal areas throughout the country.

Strategy 6: Real Estate Tax Benefits

If occupants cannot afford to pay more for their accommodations, if owners are unwilling to pay for "inessential" restoration that will not produce additional earnings, or if purchasers are unwilling to forgo the benefits of a "higher and better use" for their property, the money for historic preservation must come from society as a whole. No community will readily provide funds to a small group of citizens simply because they own or occupy historic structures. It may, however, provide assistance if it is convinced that this money will trigger investment that would otherwise not occur, as well as generate taxes that would otherwise not be collected. Short-term tax breaks are an effective means of encouraging preservation and funding it from society as a whole, provided they result in construction jobs now and additional taxes later.

While forgoing taxes does not provide money up front, any reduction in taxes will, necessarily, increase a property's net cash flow, and this increase may be enough to justify a bank loan to finance renovation. Most states have enacted legislation enabling localities to provide tax incentives for historic preservation. Georgia, Illinois, Oregon, and other states allow cities to freeze real estate tax assessments of designated landmarks. Maryland, Rhode Island, and Wisconsin provide a state income tax credit for the renovation of historic structures. Since passage of the Mills Act in 1976, California provides a one-time state income tax deduction and allows localities to enact legislation that reduces real estate taxes on properties designated locally as landmarks and/or listed on the National Register, provided the owner enters into a contract, which "stipulates that future modifications to the structure, in perpetuity, will meet historic preservation standards."[56]

Some of the most effective historic preservation programs are not targeted exclusively to designated landmarks or to historic districts. New York's J-51 Tax Exemption/Abatement Program, for example, provides assistance to all multiple-family dwellings. By exempting residential buildings from any increase in taxes due to rehabilitation and then abating a large portion of the taxes that would otherwise have had to be paid, J-51 has made possible the preservation of thousands of historic structures (see Chapters 1 and 11).[57]

Initially, J-51 was applicable only to the renovation of multiple-dwelling buildings. In 1975, the benefits were extended to conversion of nonresidential buildings into apartment buildings. In doing so, the city created a powerful instrument for the transformation of districts filled with vacant and underutilized warehouses and loft factories, many of which were significant historical structures that would otherwise have been lost. The transformation of one such area, SoHo, demonstrates how such real estate tax benefits can be used to convert market pressure into a force for historic preservation.

New York City's SoHo

SoHo lofts were not designed to be used as residences. Altering these buildings to meet legal safety requirements was very costly, and restoring a building's historic character added further costs to renovation. As a result, lofts were altered without reference to landmarks regulation.

Financial institutions would not provide a mortgage to pay for conversion of a structure that continued to violate the zoning resolution, the building code, the multiple-dwelling law, and the provisions of the historic district. Even if mortgage financing had been easy to obtain, legal conversion was prohibitively expensive. Moreover, living in legally converted loft buildings was very expensive because the city increased real estate taxes based on the high cost of that conversion. Despite these disincentives, market demand was strong enough for a limited number of buildings to be legally converted to residential use.

The situation changed in 1975, when the J-51 Tax Exemption/Abatement Program was extended to include the conversion of nonresidential structures into multiple dwellings. Instead of discouraging conversion, the government provided enough indirect subsidy to make renovating SoHo buildings very attractive. The new J-51 provisions provided a 12-year exemption from reassessment plus an abatement of annual tax payments equal to 90 percent of the cost of rehabilitation. Banks were willing to provide mortgage commitments based on J-51 benefits. However, they insisted that any renovation be 100 percent legal. In order to obtain the necessary alteration permits, developers had to obtain certificates of appropriateness from the Landmarks Commission.

Demand for retail space on the ground floor and loft residential space upstairs became so intense that many artists could no longer afford to remain in SoHo. Lofts that had been purchased for less than $10,000 in the late 1960s were selling for half a million dollars or more 20 years later. High rents even forced out the innovative retailers who had come to SoHo when there had been few customers. The J-51 Program had been so successful that the city council started cutting back benefits in the early 1980s, then terminated its use for Manhattan loft conversions.

Rezoning and designation as a historic district had not been enough to guarantee the restoration of SoHo's cast-iron architecture. The J-51 Tax Exemption/Abatement Program made available a missing ingredient: mortgage financing. Combined with the already strong market pressure to convert these buildings to residential use, the program accelerated the historic restoration that previously had been financially unattractive.[58]

Similar results were achieved in other neighborhoods in New York (e.g., Tribeca and the West Village) that may not have had as many cast-iron structures but had a similarly obsolete stock of warehouses and multistory manufacturing lofts. During the first six years in which J-51 benefits were available for the conversion of nonresidential structures, 12,000 converted apartments received J-51 benefits, many of them in neighborhoods with a building stock known for its historical character.[59]

Strategy 7: Income Tax Benefits

When preservationists turn to local government for money, they must compete with others who want funds spent on education, police protection, sanitation, and other worthy purposes. The easy way out is to seek the money in Washington, even if most preservationists understand that the federal bureaucracy will not get the job done efficiently, economically, or in a manner that is sensitive to local concerns. Instead,

Manhattan, 1980. Cast-iron buildings in SoHo. *(Alexander Garvin)*

they seek income tax benefits for people who restore designated landmarks and historic districts.[60]

Congress responded to this demand for assistance with a series of changes to the Internal Revenue Code. The process began with the Tax Reform Act of 1976. As a result of this statute, property owners who restored historic buildings could depreciate rehabilitation expenditures over a 60-month period, rather than the longer periods that applied to all other properties. This 60-month depreciation period was, in effect, a subsidy, because the useful life of most rehabilitation extended far beyond five years, but the government was permitting owners to write off their expenses over a much more accelerated time frame.[61]

Just as the 60-month depreciation period had begun to spur interest in historic preservation, Congress provided additional tax benefits in the Revenue Act of 1978. For the first time, owners of historic properties were permitted a tax credit equal to 10 percent of the cost of rehabilitation. Unlike depreciation, which is deducted annually from property income and must be accounted for upon resale, a tax credit is a dollar-for-dollar reduction in income tax that does not have to be returned upon resale. It is a subsidy equal to 10 percent of the cost of rehabilitation, paid by the federal government in the form of uncollected income taxes.[62]

These income tax benefits were minor compared with those provided by the Economic Recovery Tax Act of 1981. This law provided an income tax credit equal to 25 percent of the cost of rehabilitation and also permitted full depreciation of this tax credit. For example (see Table 18.1), developers who acquired landmarks listed on the National Register for $100,000 ($20,000 allocated to land and $80,000 to the building) and spent $1 million on renovation could take a tax credit of $250,000 (25 percent of $1 million) and still depreciate $1,080,000 ($1 million plus $80,000). The example in Table 18.2 demonstrates the probable financial structure of a project of this sort. Although the depreciability of the tax credit was reduced by 50 percent in 1982, it remained an extremely potent incentive.[63]

The impact of these laws on the historic preservation project in the preceding example demonstrates that developers who could sell tax losses often were able to pocket enough money that they needed no cash equity of their own to acquire and renovate a designated landmark. In fact, they could even pocket cash: under the 1976 act, they could net $179,000 ($529,000—$350,000); under the 1978 Act, $228,000 ($578,00—$350,000); and under the 1981 Act, $116,000 ($466,000—$350,000). All this was wiped out by the Tax Reform Act of 1986.

The Tax Reform Act of 1976 generated $140 million in certified rehabilitation. The Revenue Act of 1978 generated $650 million. This was small potatoes compared to the Economic Recovery Tax Act of 1981, which resulted in more than 10,000 restoration projects worth $7.8 billion from 1982 to 1985.[64]

Table 18.1

HISTORIC PRESERVATION PROJECT

Land	$ 20,000
Building	$ 80,000
Rehabilitation	$1,000,000
Soft costs	$ 300,000
Total project cost	$1,400,000
Probable mortgages	$1,050.000
Required equity (25%)	$ 350,000

Congress passed the Tax Reform Act of 1986 as a means of reducing the role of tax loopholes and tax shelters and encouraging developers and lenders to make investment decisions on a market-oriented basis. The act reduced the tax credit for the preservation of designated historic structures from 25 to 20 percent. However, the cut in this tax credit was not the primary cause of the reduction in income tax benefits for historic preservation. The two main causes were lower tax rates and the passive loss rule.

In 1976, when the maximum tax rate had been 70 percent, any individual in the maximum tax bracket who earned $1000 paid $700 in taxes and retained $300. Thus, $1000 in after-tax income was the equivalent of $3333 in pretax cash, because the tax on $3333 would have been $2333 ($3333 × 0.7 = $2333). When the Tax Reform Act of 1986 was enacted 10 years later, that 70 percent tax bracket was reduced to 28 percent. When the tax bracket is only 28 percent, $1000 in after-tax income is worth only $1389 in pretax cash, because the tax on $1389 is $389 ($1389 × 0.28 = $389). By reducing the value of after-tax dollars, the Tax Reform Act of 1986 reduced their impact on investment decisions. As a result, the tax credit became a less attractive incentive for historic preservation.

The impact of the passive loss rule was no less significant. Prior to 1986, property owners and developers could sell tax benefits to passive investors, who used them to shelter other income. Developers used the proceeds of the sale of tax shelter benefits as equity or just pocketed the money. With certain exceptions, the Tax Reform Act of 1986 required that tax credits be applied only to income generated by the property. By eliminating this tax shelter, Congress effectively eliminated investor interest in historic rehabilitation projects that did not generate the cash returns to offset the tax benefits.

Federal tax incentives are difficult to grasp in the abstract. Their impact is easier to understand by examining the evolution of a specific historic district, especially one like the West End in Dallas, which experienced little preservation activity until Congress passed the Economic Recovery Tax Act of 1981.

The West End in Dallas

When Dallas designated 30 blocks of the West End as a historic district in 1975, it was a 67.5-acre area filled with obsolete commercial structures built predominantly during the first two decades of the twentieth century. The three- to eight-story, redbrick warehouses that filled the district had been built for businesses that depended on nearby railroad yards. These buildings were the only dense concentration of late-nineteenth- and early-twentieth-century commercial buildings in Dallas. Unlike the cast-iron lofts in SoHo, however, few of these buildings were significant because of exceptional architectural design.[65]

Dallas did not just designate the West End as a historic district. It developed a preservation plan that included public improvements and amended the city's zoning ordinance to include specific design requirements for all renovation and new construction. The new regulations included a height limit of 100 feet and requirements for façade openings, window setbacks, parking, signs, materials, and color.[66]

Despite a booming economy, the example of adaptive reuse of warehouse districts throughout the country, elaborate development guidelines, and a $4.5 million streetscape improvement program, little development activity occurred prior to 1982. Then, between 1982 and 1985, more than $70 million was spent on rehabilitation in the West End.

The impetus for the adaptive reuse of buildings in the area came from both increased demand for space during the

Table 18.2

HISTORIC PRESERVATION PROJECT

	1976	1978	1981	post-1986
Tax credit	zero	$ 100,000	$ 250,000	$ 200,000
5-year depreciation	$1,080,000	$1,080,000	$1,080,000	27.5 years
Total tax benefit	$1,080,000	$1,180,000	$1,330,000	$ 39,300
Value at 70% tax	$ 756.000	$ 826,000	—	—
Value at 50% tax	—	—	$ 665.000	—
Value at 35% tax	—	—	—	$ 13,755
Probable sale at 70% value	$ 529,000	$ 578,000	$ 466,000	0

Dallas, 2009. Market Street in the West End after several years of renovations and alterations. *(Alexander Garvin)*

early 1980s and the tax benefits provided by the Economic Recovery Tax Act of 1981. As the commercial real estate market became tighter and more expensive in downtown Dallas, smaller firms would increasingly accept secondary locations, such as the West End, for their offices. In addition, the new tax credits and rapid depreciation could be sold to investors who needed tax shelters. Developers who sold these credits used the proceeds of the sale as equity and financed the remaining cost of renovation with institutional mortgages. Thus, the renovation of these historic structures required little or no cash from their developers.

The market for West End office space spilled over to the ground floor. "In 1983, only one retail/restaurant establishment existed in the West End; that restaurant produced less than $100,000 in annual sales and liquor taxes." By 1987, the area had more than 130 retail shops, 30 restaurants, and 10 nightclubs, generating $72 million in retail and restaurant sales, $5.5 million in sales taxes, and $3.3 million in city parking fees and violations. Although the West End was still experiencing a bar, restaurant, and entertainment boom at the end of the decade, development came to a halt. Unfortunately, the situation did not change when the economy took off again at the end of the twentieth century.[67] The rejuvenation of the area was intricately tied to tax structure, which encouraged redevelopment.

Designation as a historic district initially had no effect on the West End. Its preservation was guaranteed only when federal income tax benefits became sufficiently generous to attract developers. If developers are still attracted to the West End when market pressure is restored, the income tax benefits for historic preservation then available will determine whether they choose to demolish or renovate.

An Activist Approach

Owners legitimately trying to repair crumbling façades or leaking roofs have defaced countless landmarks. Anybody who has passed a once-sumptuous Victorian mansion now covered with asphalt tile or imitation flagstone knows the problem. Had the owners been aware of the historic value of their buildings or employed knowledgeable architects, society would now have a restored landmark instead of an unsightly, but watertight, monstrosity. Consequently, preservationists feel justified in demanding that government restrict a property owner's right to alter or demolish historic structures.

Denying permission to alter or demolish old buildings can be taken too far, as it was in the case of New York City's Mount Neboh Synagogue. But waiting for property owners to restore designated landmarks may not be a reasonable alternative, either. While the preservation agency waits for owners to apply for permission to do something, important historical structures may seriously deteriorate. Consequently, more aggressive action, such as the program that brought about the revival of the Art Deco District in Miami Beach, may be necessary.

Mount Neboh Synagogue

In 1982, New York City's Landmarks Preservation Commission officially designated the former Mount Neboh Synagogue as a landmark. Mount Neboh was not listed in the index of architecturally noteworthy buildings of Greater New York issued by the Municipal Arts Society in 1957, nor in its successor volume, *New York Landmarks,* published in 1963.

Bronx, 2012. Two once-similar houses in a historic district: one has been handsomely maintained; other is the victim of ill-conceived "improvements." *(Alexander Garvin)*

It did not appear in Paul Goldberger's 1979 guide to the architecture of Manhattan, *The City Observed.* The American Institute of Architects' *AIA Guide to New York City,* first published in 1967, described it as "another West Side synagogue, which has borrowed heavily from the Byzantine style." Even the chair of the Landmarks Commission recognized that it was "not architecturally significant."[68]

Mount Neboh Synagogue had been acquired by a developer who intended to erect an apartment building. The new structure was to be as tall as neighboring buildings and just as mediocre in design. The abandoned synagogue had been brought to the attention of the Landmarks Commission by a group of West Side residents who had formed the Committee to Save Mount Neboh. They persuaded the commission to hold a public hearing on the designation of this 55-year-old building, which they argued was an example of the "synthesis of Byzantine and other Near Eastern influences." Whether opposition to this privately financed apartment house stemmed from disapproval of tall buildings, mediocre design, high-income tenants, the additional burden to community facilities, or nostalgic affection for the synagogue continues to be debated. What cannot be debated is that neither the committee nor the Landmarks Commission had the money to acquire the property, nor could they find any financially feasible alternative use for the structure.

The owner of Mount Neboh had purchased the site for $2.4 million prior to any discussion of landmark status. He wanted to build market-rate housing on a site zoned for that purpose by the City Planning Commission. Landmark status kept him from proceeding. One year and $1 million later, the Landmarks Commission accepted his pleas of hardship and granted a demolition permit.

By appealing to the state supreme court and court of appeals, the Committee to Save Mount Neboh continued to stall demolition; in addition, they cost the developer still more money, to sustain his own battle in defense. When the owner finally won permission from the courts to develop the site, costs had escalated to the point that he was no longer able to proceed. He sold the site to another developer, who erected an undistinguished apartment house.[69]

The Landmarks Commission did not compensate the developer for the costs incurred as a result of its action. In fact, preservation agencies have not yet had to pay for any infringement of private property rights during the period in which they consider actions proposed by property owners, who, but for landmark status, could lawfully proceed without their permission. They have not had to weigh the value of po-

Manhattan, 2012. The second apartment house from the left was built on the site of Mount Neboh Synagogue. *(Alexander Garvin)*

tential increases in tax revenues and employment from more intensive use of such sites. Nor have they had to consider the impact of the decrease in tax revenues that may result from a drop in market value caused by landmark designation.

South Beach, Miami Beach

In 1976, when Barbara Baer Capitman founded the Miami Design Preservation League, most people found it difficult to understand why a small group of design aficionados cared about Miami Beach's run-down, funny-shaped buildings with neon signs blaring "Rooms $5 a Week." Because cheap air travel had made a short Florida vacation affordable, their persistence paid off. In 1977, the Preservation League received a $10,000 grant from the City Planning Department to conduct a survey and develop a preservation plan for a 20-block section of South Beach.

The League identified 800 distinctive Art Deco buildings erected between the World Wars. *Art Deco* refers to a style that grew out of the 1925 Paris *Exposition Internationale des Arts Decoratifs et Industriels Moderne.* Art Deco buildings in South Beach are characterized by streamlined shapes with curved corners, continuous eyelids shading casement windows, glass block, and porthole windows. Most of the buildings were built of concrete, decorated with ornamental terrazzo floors and coral stone, and painted in pastel colors. Many structures had originally been built as vacation hotels. Others, which had once been modest apartment buildings, later would be converted into hotels. In 1977, however, most people expected these relics to be replaced by high-rise condominiums, similar to those being built on oceanfront property throughout Florida.[70]

In 1979, with the assistance of the report developed following the Planning Department's survey, the city of Miami Beach convinced the National Trust to designate South Beach as a historic district listed on the National Register. While this did not preclude the destruction of some important Art Deco buildings, it did focus national attention on the area, underscored the fact that South Beach contained properties whose value had been underestimated, and offered substantial federal tax incentives for historic preservation. The Preservation League felt that designation by itself was not enough. It advocated an activist approach that combined private investment in an initial nucleus of buildings on Ocean Drive with public investment in the area's infrastructure and community facilities.

One of the league's members, Barbara Capitman's son Andrew, spearheaded the initial investment by concentrating on a cluster of beachfront hotels that the League felt would trigger further investment. To accomplish this, he organized a firm called Art Deco Hotels, whose financial strategy was to capture the income tax benefits available for the rehabilitation of buildings listed on the National Register. In exchange for those benefits, approximately 100 limited partners, many of whom were dedicated preservationists, provided the cash equity needed to acquire seven hotels. Building rehabilitation, which was expected to be financed with bank mortgages, was slow getting started because local banks were skeptical about investing in this seedy section of metropolitan Miami.

Art Deco Hotels accumulated substantial debts, and the partners sold out in 1984. The new owners experienced serious legal and financial difficulties. Eventually, their lender foreclosed and resold the buildings to legitimate real estate developers with experience in South Beach renovation. In spite of all of these problems, the seven-hotel nucleus was discovered by the European photo fashion industry. The attention helped to establish a trendsetting image that set an inspiring example for other developers, who invested in hotels just south of the original seven. Additional investors, particularly gay and Cuban-American entrepreneurs who valued the charming Art Deco architectural features that others had failed to value, concentrated on the area's relatively small,

Miami Beach, 2006. Small Art Deco buildings were among the first structures to be renovated once the area had been designated a historic district. *(Alexandeer Garvin)*

two- and three-story residential apartment houses and seasonal hotels. They purchased buildings on Collins Avenue, one block west of Ocean Drive, and throughout South Beach. Given the limited scale of the buildings that they were acquiring, acquisition and renovation did not require huge investments. Consequently, they were able to raise the relatively modest amounts of cash equity needed to balance bank mortgages. In addition, the more buildings that were rehabilitated, the easier it was to obtain financing from conventional sources that no longer felt the area had the same elevated levels of risk that it had prior to the start of the renovations.

Public action played an equally critical role. Two years after the Art Deco District was listed on the National Register, the Preservation League commissioned a preservation and development plan for the district, which provided the vision that inspired public activity for the next decade and a half. Its recommendations included reconfiguring and reconstructing Ocean Drive, rehabilitating Lummus Park and its Atlantic Ocean beach, and rezoning large sections of the city to permit adaptive reuse of obsolete structures.

Voters approved a bond issue for Ocean Drive and Lummus Park in a 1986 referendum. The work was completed three years later and cost approximately $3 million, half of which was paid by the city and the other half of which was paid by special assessment of the property owners along Ocean Drive. The bonds financed narrowing the street bed of Ocean Drive, widening and rebuilding its western sidewalks, reconstructing its eastern sidewalks, installing new lighting and street furniture, replacing water and sewer lines, and relandscaping. Once completed, the outdoor dining facilities that opened along the widened Ocean Drive sidewalks became one of the district's main tourist attractions.

In 1993, the city purchased air rights and a small adjacent parcel of land for the public Ballet Valet parking garage on Collins Avenue. The building, designed by Arquitectonica, combined ground-floor retail uses with a six-level, 650-car public garage. The garage was so successful that in less than 10 years from its completion it was generating $1 million more than its operating expenses and debt service. More important, it attracted enough retail customers to transform that section of Collins Avenue into a major shopping destination. A year later, the city began working on the revitalization of Lincoln Road. What once had been one of the nation's most successful pedestrianized streets had deteriorated badly and was now pockmarked with vacant and deteriorated structures (see Chapter 6). By creating a capital improvements district that raised more than $15 million and a business improvement district that paid to keep Lincoln Road clean, safe,

Miami Beach, 2009. A $16 million investment in the renovation and relandscaping of Lincoln Road attracted pedestrians, new stores, and restaurants and transformed the street into one of the region's major tourist destinations. *(Alexander Garvin)*

and attractive, the city had made an investment that paid off in hundreds of millions of dollars of tourist spending.

By the second decade of the twenty-first century, South Beach's hotels had become international tourist destinations. Its condominium apartments are vacation trophies for both American and international buyers, and South Beach bars and restaurants attract customers from all over greater Miami. Major developers followed suit. Ian Schrager, an hotelier who had transformed a number of dilapidated New York City buildings into high-fashion hotels, worked the same magic on the Delano, one of the high-rise hotels on Collins Avenue, north of Lincoln Road. What had been a third-rate 1947 hotel, whose only remaining attraction was its beachfront, became a chic, 208-room fantasy resort with a swimming pool, gym, women's spa, and children's playground. The Shelbourne and National Hotels next door were soon upgraded as well.

In order to complete the revitalization of Collins Avenue, the city persuaded Loews to open an 800-room luxury hotel midway between the Ballet Valet Garage and the Delano Hotel. The new 18-story hotel includes 2 ballrooms, 13 meeting rooms, an outdoor pool and grill, 2 bars, 3 restaurants, and a full range of tourist services. The city acquired the site, which it leased back to Loews, and then provided grants for the public facilities that would attract national conferences to Miami Beach.

The city's $51 million expenditure on the Loews Hotel was the last step in transforming the South Beach oceanfront into the thriving Art Deco showpiece about which the Miami Design Preservation League had dared to dream nearly a quarter of a century earlier. By the twenty-first century, Miami Beach had become the U.S. headquarters for most Latin American record, cable TV, and media companies. The European fashion magazines that used it for photo shoots had made the Art Deco District world famous, and few public investments in historic preservation have generated a private market reaction of such magnitude.

Cultivating the Cultural Patrimony

The landscapes of Tuscany and Umbria are the product of a profound approach to cultivating the cultural patrimony. They are marked with the imprint of thousands of years of Italian history. Etruscan masonry provides a foundation for medieval walls. Architectural fragments from Roman basilicas are incorporated into Renaissance churches. Olive trees still grow along contours terraced centuries earlier. Each generation has taken the cultural patrimony, worked to enhance it, and then passed on something greater to its successors.

Etruscan masonry and olive trees may not be great art and the battles that took place among them may not be as significant as the Normandy landing or the siege of Troy, but they are nonetheless treasured components of the Tuscan and Umbrian patrimony. So are the Hungarian folk songs lovingly cataloged by Bela Bartok and Zoltan Kodaly, or the Russian folk melodies included by Tchaikovsky in his symphonies or Rimsky-Korsakov in his operas. In America, however, the cultivation of our cultural patrimony is significantly restricted to buildings of exceptional importance or consistent character, and the approach taken to their preservation rarely gets beyond a requirement that everything be consistent with the architectural style and period in which treasured buildings were erected.

Santa Barbara and Santa Fe are continually building and rebuilding in a style of architecture that each mandates be repeated, if elaborated upon, in the instance of each new building. In Tuscany and Umbria, on the other hand, every change in the landscape is produced by residents imbued with an instinctive understanding that history is a continuous process that never stops. As the United States matures, some communities are developing an instinctive approach to the cultivation of the cultural patrimony that is similar to those in Umbria and Tuscany—albeit one brasher and more heterogeneous than is prevalent in Italy, and one that reflects our very different patrimony. Artists, architects, gays, lesbians, small independent developers, and entrepreneurial nonconformists have pioneered the revival of whole districts that had been abandoned. They have done so by adapting old buildings to entirely new uses and filling in missing teeth where needed with new, inexpensive, easy-to-maintain structures.

Perhaps these innovative preservationists initially settled in districts because they could afford living there; perhaps the art galleries, boutiques, and stylish cafés, bars, and restaurants that followed them forced prices up to levels higher than they could afford; perhaps the shiny new developer-built structures that followed the boutiques and restaurants have brought a very different population. No matter what the rationales or consequences, everybody involved surely selected what was significant, nurtured and enhanced it, made improvements, and are passing on to the next generation an environment that is richer, more fulfilling, and more beautiful. The neighborhoods that are emerging—Third Ward in Milwaukee, the Pearl District in Portland, the Design District in Miami, among others—represent a very American, grassroots approach to our cultural patrimony. One of the liveliest and least conventional of them is evolving in the Northern Liberties district, directly northeast of downtown Philadelphia.

Northern Liberties, Philadelphia

No neighborhood can revive itself, nor do buildings take care of themselves. Cultivation of the cultural patrimony requires people who value the past and want to live in neighborhoods where they can be a part of both the past and the future.

During the 1960s, Edmund Bacon, as explained in Chapters 6 and 12, began the process of reviving the residential life of downtown Philadelphia with the redevelopment of Society Hill and the preservation of its charming row houses. By the 1970s the growing market for that neighborhood's late-eighteenth- and early-nineteenth-century buildings had been established and was spreading south and north to the Queen Village and Old City neighborhoods.

At that time, artists seeking low rents began to move into the Northern Liberties district several blocks north of the Vine Street Expressway, then, as now, the northern boundary of the business district. The district was particularly convenient because of its location on the Market Street–Frankford subway line, a short ride from the downtown business district and the University of Pennsylvania. Many observers expected Northern Liberties to take off in popularity and become "the next neighborhood." But the area's revival was stunted because the area between Northern Liberties and the Expressway did not contain any residences or buildings that were suitable for conversion to residential use. The boom did finally begin, leapfrogging the nonresidential area between Old City and Northern Liberties—only a couple of decades later.[71]

During the 1970s and 1980s pioneers were lured by the area's cheap live-work space, often spending very little money to buy the run-down building into which they moved. They thought of the gritty character of the area with its working-class residents, narrow row houses, small locally owned shops, breweries, lumber and coal yards, gasworks, and tanneries as their patrimony. So, with little money and lots of sweat, they did their best to improve the neighborhood. Their efforts began to take shape in 1975 with the establishment of the Northern Liberties Neighbors Association (NLNA), a nonprofit organization that brought together like-minded residents of the area. It helped to prepare an inventory of the area's buildings that resulted in listing Northern Liberties as a historic district by the National Trust for Historic Preservation in 1985.

Achieving designation by the National Trust, however, did not forestall businesses from closing or moving elsewhere. The growing inventory of vacant properties that remained was susceptible to fires, which many believed were the result of arson. A particularly significant fire occurred in 1990, four years after a complex of six buildings that housed two tanneries closed. The fire left behind a site that had been contaminated by toxic chemicals. Its owner donated this 2-acre disaster to the NLNA. In so doing, it provided the NLNA with the opportunity to make a major investment in improving the neighborhood.

With grants from the Philadelphia Urban Resources Partnership and others, it devised a plan for a community-owned park. The U.S. Environmental Protection Agency (EPA) provided money to have the site decontaminated. The buildings themselves were demolished by the City of Philadelphia, which in 1999 forgave $500,000 in liens on the property. Now known as Liberty Lands, the park quickly made nearby properties attractive to small developers, who built row houses and four-story apartment buildings.

Most of these projects did not appear right after the NLNA opened the park. They followed new buildings erected by Bart Blatstein, a major developer who came to Northern Liberties in 2000. Blatstein, a native Philadelphian who had recently earned substantial sums from the sale of a property in South Philadelphia, had the money to purchase the 16-acre site of the former C. Schmidt & Sons Brewery and other properties in the area at depressed prices that reflected the prevalent vacant lots, empty buildings, and the generally decrepit state of the neighborhood. The ongoing efforts of the NLNA notwithstanding, Northern Liberties at that time was as risky an investment as it had been decades earlier, when many of the buildings were still occupied by functioning businesses, but now it was because many of those buildings were vacant and subject to fire. Consequently, Blatstein began buying them up, eventually purchasing an additional 15 acres scattered throughout the neighborhood. As the danger of fires waned, people began moving in and Northern

Philadelphia, 2012. One of the triggers for the revival of Northern Liberties was Liberty Lands, a community-sponsored park. *(Alexander Garvin)*

Liberties became the hot neighborhood people had predicted decades earlier: a neighborhood with art galleries, boutiques, and distinctive restaurants mixed in with working-class row houses, converted lofts, and small businesses. Blatstein had done market analysis and recognized that a certain density of residents would be required to supply customers to the area's retail stores if the stores were to prosper. As a result, he proposed construction of several seven-story apartment buildings on the sites that he owned, even though neighbors objected because they would have preferred row houses. Blatstein also understood that his market was not Anywhere, U.S.A.; it consisted of people who were already living in Northern Liberties, or were about to move in. In order to tenant buildings, he needed to attract residents who would otherwise move to some other hip neighborhood. That meant enhancing the scene in which they already wanted to participate. He sought out retailers who carried the merchandise they coveted, restaurateurs who offered the cuisine they valued, entertainment venues, fitness centers, and other community gathering places. He offered special rents to trendsetters who would move into retail and office space. The ground-floor nonresidential tenants that he brought to Northern Liberties appealed to future residents of the apartments he was creating above the stores. Finally, he hired Erdy McHenry Architecture and other designers who could devise the environment that both old-time residents and newcomers would enjoy.

Blatstein has been developing his property in stages so that he does not need to wait for months to fill up the buildings, funding interest expense and other carrying costs during the process. Redevelopment began, section by section, with Liberties Walk, a four-block-long stretch of three-story buildings that included local shopping on the ground floor. These pairs of buildings frame a modest walkway connecting the long blocks between Second and Third Streets, which create a pedestrian route to and from the then more populated western section of the neighborhood. Next came the Piazza at Schmidt's, which includes a small, oval office building, an event facility in a remodeled furniture warehouse, and several seven-story apartment buildings that enclose a courtyard. The court has become a community gathering place enclosed by eateries whose tables and chairs spill out into the open space. The high-resolution LED screen mounted on an old warehouse wall displays various sports and entertainment events every day, while the piazza itself is programmed with live events throughout the year. Flanking the piazza are several warehouses that Blatstein renovated and converted for residential use. He removed the cladding from one of them and restored the original industrial windows. In 2011, he opened a branch of Super Fresh, the neighborhood's first supermarket, and additional buildings and retail stores are opening in response to market demand.

Over the past 15 years, newcomers have purchased and restored homes throughout Northern Liberties. Artists have moved into converted lofts, and a bevy of small developers have acquired sites on which they have built modest new buildings. As a result, during the first decade of the twenty-first century the population of Northern Liberties increased by 55 percent. Small businesses have acquired and

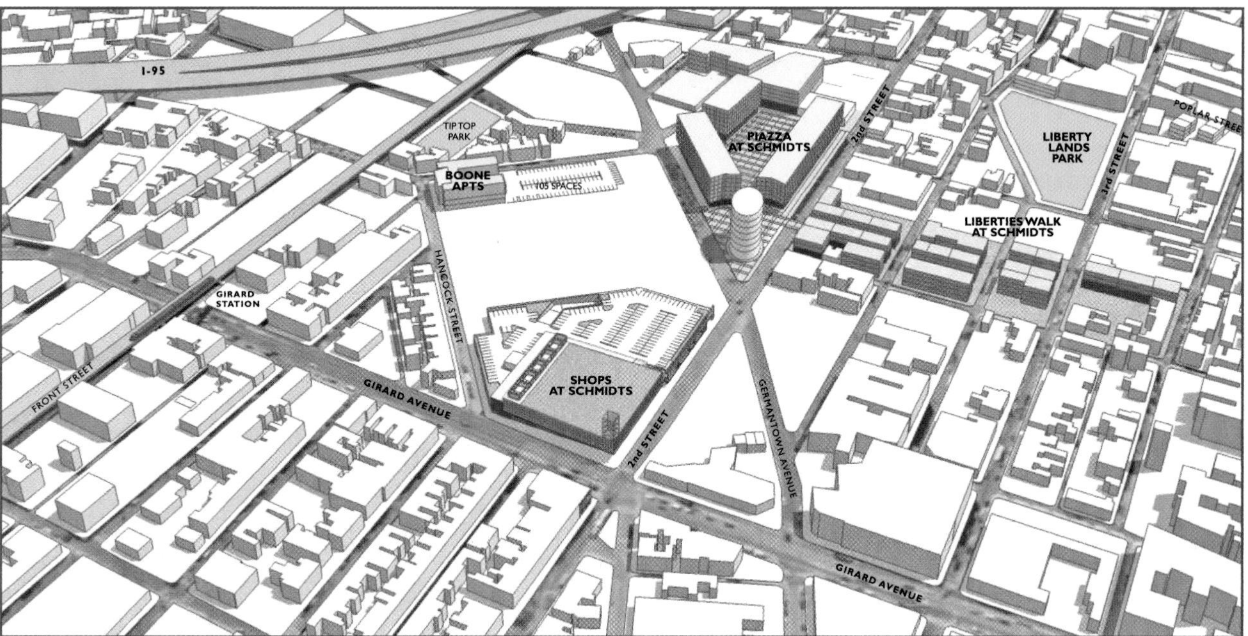

Philadelphia, 2012. Properties that Bart Blatstein developed in Northern Liberties. *(Courtesy of Tower Investments, Inc., Philadelphia)*

Philadelphia, 2012. The initially low property prices were part of the attraction to the gritty combination of building types and periods of construction in the Northern Liberties section of the city. *(Alexander Garvin)*

renovated buildings for their operations, sometimes living there as well. In 2012, KieranTimberlake, one of the city's major architecture firms, acquired a large bottling plant, which it will adapt as the headquarters for its practice.

It is now several decades since the people in Northern Liberties began nurturing and enhancing the neighborhood. They have begun the process of cultivating the cultural patrimony and will be passing on to future residents, businesspeople, and workers a neighborhood that is getting better by the day. It may not be as exquisitely beautiful as the old and historic sections of Charleston, or as consistent in appearance as Santa Barbara, but like the Vieux Carré it is evolving into a peculiarly American jumble of old and new residences, businesses, and entertainment venues.

Ingredients of Success

Preservation is not just a task for people like those who revealed the value of the Victorian District in Savannah or Northern Liberties in Philadelphia. Everybody is responsible for cultivating the cultural patrimony. Too often, government inadvertently hastens the deterioration and destruction of the cultural patrimony by imposing regulations with which it is difficult and expensive to comply. In those instances, property owners either do not make improvements, given the added cost of complying with regulations, or ignore the regulations and make less costly improvements. Such burdens prevent preservation and must be eliminated. Government has a more important role to play in helping property owners cultivate the cultural patrimony through the use of strategic public investment, regulation, and incentives.

Market

Most historic preservation strategies work only when there is market demand. The Victorian District and Northern Liberties were deteriorating neighborhoods through most of the second half of the twentieth century. Only when demand picked up at the end of the century was the preservation of historic structures finally possible. Without buyers for historically significant buildings, development rights, income tax credits, and real estate tax abatement will go unused, and revolving funds will deplete their coffers because they cannot resell the historic properties they have acquired.

Local real estate tax benefits and state and federal income tax incentives can lower the cost of maintaining and restoring historic structures. In some cases, the reduction in tax payments will be large enough to stimulate property owners to renovate old buildings. In Dallas's West End, for example, income tax credits and rapid depreciation schedules allowed developers to offer commercial space at low enough prices to attract users of secondary office space. But tax incentives cannot do much if there are still no consumers at the lower price that these incentives make possible. That is why J-51, which had so significant an effect on the lofts and warehouses of SoHo, had so little impact on many lofts and warehouses in Detroit and St. Louis.

Location

Historic districts begin with a significant locational advantage. The very presence of historically significant structures creates that advantage. Destroy them, and the location loses its attractiveness. Thus, designation as a historic district is in the interest of most property owners. Their only concern will be the added burden of complying with regulations. If designation is accompanied by tax incentives, they may be willing to accept that additional burden.

Proximity also plays a role in determining the success or failure of a preservation strategy. Tax incentives, for example, may lower the price of space in a landmark or historic district enough to attract customers from more expensive nearby properties. The lower prices for apartments in SoHo attracted residents who could not afford such accommodations in Greenwich Village. Similarly, once buildings in the West End had been fixed up, their owners were able to profit from customers who spilled over from downtown Dallas. In Miami Beach, widening sidewalks along Ocean Drive made it easier for the owners of Art Deco hotels to install dining tables that offered views of the beach. The rationale behind both historic and TDR districts is that the costs and benefits of regulation accrue in an equitable fashion to all property owners within the area.

Design

The character of each historic area is perhaps its most important design feature. Eliminate SoHo's cast-iron columns, window walls, and ornament, and you eliminate its charm and attraction to an art-oriented population. Separate and reorganize into neat patterns the jumble of land uses and styles that characterize the Vieux Carré, and you are likely to diminish spending by noisy tourists. Consequently, it is even more important for local preservation agencies to adopt different regulatory strategies for different districts.

The transfer of development rights presents one of the stickiest design problems. For the development rights to retain their value, the receiving property must be of similar value. Similarly, if the additional development is not to be an additional burden to the receiving area, it will have to be fairly close to the property from which these rights have been taken. Thus, the obvious purchaser becomes a neighboring site. If, in contrast, all the additional bulk is concentrated on one receiving site, the result will be a tower that inappropriately dwarfs its neighboring landmark. For this reason, most zoning ordinances restrict the additional bulk to a relatively small percentage (e.g., 20 percent) of the receiving structure.

Preservation agencies are not the only organs of government that help to cultivate the cultural patrimony. Every historic district includes property that is publicly owned and maintained. Manipulation of the public component played a key role in the success of South Beach and the West End. In both places, local agencies resurfaced traffic arteries, replaced street furniture, and planted trees, and in both cities these changes made the ground plane more tourist-friendly. The result was an increase in street-level tourist activity and, by extension, in tourist spending, which justified increased private investment in the preservation of privately owned property.

Rigid conceptions of design are the greatest threat to cultivating the cultural patrimony. Too many preservation agencies adopt a one-solution-only approach to what can be done: restore it to its "proper" appearance. What is the proper appearance for truly significant landmarks such as St. Peter's in Rome? Is it the fourth-century church that was first erected on the site, Bramante's early-sixteenth-century building (with or without the emendations by Raphael, or Peruzzi, or Sangallo, or Michelangelo)? Should Michelangelo's dome be removed? How about Bernini's colonnade? One does not cultivate the cultural patrimony by deciding when to freeze history!

Financing

Historic preservation often requires more money than banks are willing to lend. While there will always be wealthy property owners who will willingly pay extra for the privilege of occupying a landmark, many property owners either will not accept or cannot afford any added burden. The additional money can come from the sale of development rights, real estate tax credits, or state and federal income tax deductions.

Banks tend to base their mortgages on the capitalized value of the net operating income of a property. They usually will agree in advance to increase that mortgage based on the capitalized value of any increased income stream produced by future tax benefits, once those tax benefits are in place. In South Beach, most Art Deco buildings were small enough for their purchasers to raise the needed cash equity. While these property owners were able to pay for renovation before they could be reimbursed from income tax credits, in other situations up-front cash may be difficult to obtain. This is particularly true where real estate tax abatement does not affect the property until renovation has been completed. It is in these situations that a revolving fund can be particularly useful. Unfortunately, few preservationists have tried to exploit this interplay between a revolving fund and tax incentives.

Time

Even if we had the money to acquire and embalm all buildings built prior to whatever point at which preservationists decided history ended, the remaining properties would be unable to sustain a flourishing economy. As the English architectural historian Reyner Banham so cogently explains: "If we let the paranoid preservers manoeuver us into keeping everything, we shall bring the normal life-process of decay and replacement to a halt, we shall straightjacket ourselves in embalmed cities of the past."[72] Societies that genuinely cherish their heritage never separate it from everyday life. For them, cultivating the cultural patrimony means retaining those structures that are genuinely significant, keeping them in good condition, and continually making necessary changes to ensure that they remain a vital part of the environment.

Preservation techniques are as subject to changing conditions as historic structures and therefore need to be written so that they work in good times and bad. The regulations that apply to the buildings in Charleston's Old and Historic District were of little use when there was little development pressure during the 1930s and 1940s. They provided a predictable investment climate during the second half of the twentieth century, when property values increased. Their most important role, however, came in the aftermath of Hurricane Hugo in 1990. Property owners faced with the immediate need to restore damaged property had no choice but to comply with historic district regulations.

Entrepreneurship

The Historic Savannah Foundation played the entrepreneurial role needed to acquire a significant number of historic structures and then convey these buildings to willing buyers. The Miami Design Preservation League provided the strategy that led to the revival of South Beach. In SoHo, relatively poor artists supplied the ingenuity and sweat to adapt industrial lofts to other purposes. Although the Northern Liberties Neighbors Association pioneered the area's preservation effort, the neighborhood's revival finally took off when Bart Blatstein began developing property there. Somebody will always be needed to spark an area's preservation efforts. If we are to preserve the cultural patrimony, government has to create the conditions that will allow the widest variety of these preservationists to flourish.

Building occupants, property owners, and the institutions that finance them invest in improvements when they are sure that the money they invest in restoration will not lose its value. A location within a historic district reduces this risk: investors can rely on the city's landmarks agency to approve only restorations, alterations, additions, and new construction that is consistent with the character of the district. Since they are able to predict the probable future appearance and use of the area, they are assured of relatively stable investment conditions. This is why the developers of the Delano and Loews hotels in South Beach were ready to raise and invest tens of millions of dollars in their rehabilitation.

Property owners also must be certain that any request for permission to make property improvements will be treated fairly and be processed expeditiously. If the application process takes too long, is too expensive, or subjects them to arbitrary demands, they will either abandon the improvements or circumvent the regulatory process. Savannah's zoning ordinance specifies building height, materials, textures, colors, and other design requirements within its historic district. Similar regulations apply in historic districts in Santa Fe, Dallas, and other cities. Owners and developers who comply with these explicit regulations can rely on timely government action and will, therefore, invest in property improvements. In other communities, where their actions are subject to discretionary review, property owners often defer improvements or evade the regulatory process.

The importance of low costs to poor artists renovating their SoHo lofts is easy to understand. The availability of mortgage financing is as critical as the affordability issue. But even when institutional mortgages are available, there will be few takers if lenders require substantial amounts of equity capital. The federal income tax incentives that were available during the late 1970s and early 1980s solved this problem by permitting owners of historic structures to sell future tax benefits to investors. In response to this opportunity, there emerged a group of preservation entrepreneurs who used the proceeds of such tax-shelter sales as equity capital. We need to re-create a similar regulatory and financial climate if we are to attract the entrepreneurial talent needed to restore tens of thousands of historic structures that need attention.

Historic Preservation as a Planning Strategy

The effort to enrich the cultural patrimony must reflect the entire range of community values and interests. This cannot happen as long as historic preservation is treated as a special, privileged government activity. During its infancy, the preservation movement may have required such nurturing. Now that it has come of age, it is time to let the cultivation of the cultural patrimony compete on an equal footing with all the other functions of government. This means changing long-cherished government policies.

Communities must cease punishing the owners of historically significant buildings with increased tax assessments every time they invest in restoration. This would eliminate the disincentive to restoration and remove the threat of demolition from all but those sites that are more valuable when used for other purposes. It would also make communities more selective in designating structures as landmarks because it would mean losing a large portion of the taxes from those properties. In cases where, despite the reduction in taxes, the owner is considering demolition, local government should either appropriate the money to purchase and maintain threatened landmarks or let the organic process of urbanization take its course.

Communities must cease imposing unfunded mandates on owners of historic structures. These owners face a serious burden: paying for desirable but expensive and inessential restoration that may be required by a municipal preservation agency. If it is important enough for government to require an owner to restore a slate roof rather than resurface it with less costly materials, then government should pay for the added burden. The easiest way to cover this cost is to abate real estate tax payments until the owner has taken tax credits equal to the additional cost of the mandated improvement.

The entire nation should cease treating historic preservation as a special activity unrelated to other government functions. Preservation requirements should not be separated from other land use regulations. Charleston made its Old and Historic District a part of the city's zoning ordinance. Similarly, the regulations governing the TDR District in Denver and the West End Historic District in Dallas are set forth in their zoning ordinances. By making historic preservation a significant part of citywide land use policy, subject to review and approval by the local planning agency and the local legislature, public officials will be forced to consider impact on the cultural patrimony of all land use decisions and balance preservation objectives against other community goals.

Too many property owners either avoid improvements that require public review or make them but ignore proper procedures. Since preservation agencies are usually underfunded, they are often unaware of this activity or, if they know about it, do not have the personnel, money, or power to do much about it. Consequently, especially in poorer neighborhoods, designation as a historic district does little to enhance the cultural patrimony. We should admit reality and limit discretionary review to structures of true historic significance. All other designated landmarks and historic districts should be required only to comply with printed regulations, with no further review.

This combination of changes in real estate taxation and land use regulation will place historic preservation on an equal footing with other public objectives. The new real estate tax policies will remove the disincentive to investing in historic structures. Similarly, making historic preservation a part of the zoning ordinance will force public officials to consider historic preservation when they make decisions affecting land use policies. Together, these policies will create conditions in which cultivation of the cultural patrimony will become an easy and ongoing public responsibility.

Notes

1. Christopher Tunnard, "Preserving the Cultural Patrimony," in *Future Environments of North America,* F. Fraser Darling and John P. Milton (editors), Natural History Press, Garden City, NY, 1966, pp. 552–567.
2. David Freeland, *Automats, Taxi Dances, and Vaudeville: Excavating Manhattan's Lost Places of Leisure,* New York University Press, New York, 2009, pp. 85–105.
3. *Penn Central Transportation Company v. City of New York,* 438 U.S. 104 (1978).
4. American Association of Museums, http://www.aam-us.org/.
5. Frank Gilbert, vice president, National Trust for Historic Preservation, personal communication.
6. Historical and statistical information about Williamsburg is derived from Philip Kopper, *Colonial Williamsburg,* Harry N. Abrams, New York, 1986; Colonial Williamsburg Incorporated, *Official Guidebook and Map,* Williamsburg, VA, 1960; and Williamsburg Area Convention & Visitors Bureau.
7. The Colonial Williamsburg Foundation, http://www.history.org/Foundation.
8. Constance McLaughlin Green, *American Cities,* Harper & Row, New York, 1965, pp. 24–25.
9. DuBose Heyward, *Porgy,* George H. Doran, New York, 1925, p. 11.
10. Samuel Gaillard Stoney, *This Is Charleston: A Survey of the Architectural Heritage of a Unique American City,* Carolina Art Association, Historic Charleston Foundations, and the Preservation Society of Charleston, Charleston, SC, 1970 (originally published in 1944), p. 51.
11. City of Charleston, *Zoning Ordinance,* Charleston, SC, 1973, pp. 32–33.
12. Office of Tourism Analysis, College of Charleston, 2012.
13. Robert Moses, *Public Works: A Dangerous Trade,* McGraw-Hill, New York, 1970, pp. 772–776.
14. Christopher Tunnard, *A World with a View,* Yale University Press, New Haven, CT, 1978, pp. 136–141.
15. Ibid., p. 141.
16. New Orleans Tourism Marketing Corporation.
17. Historical and statistical information about Santa Barbara is derived from David Gebhard, "Architectural Imagery, The Mission and California," *Harvard Architectural Review,* vol. 1, Spring 1980, MIT Press, Cambridge, MA, pp. 137–145; Rebecca Conrad and Christopher H. Nelson, *Santa Barbara: A Guide to El Pueblo Viejo,* Capra Press, Santa Barbara, CA, 1986; and the Santa Barbara Conference & Visitors Bureau and Film Commission.
18. This architectural style was based on the work of José de Churriguera (1650–1723), who created a particularly ornate Spanish version of the Baroque style.
19. David Gebhard, "Introduction," Conrad and Nelson, op. cit., p. 17.
20. The first height restrictions were enacted in 1924. They limited commercial structures to six floors and residential buildings to three floors.
21. Santa Barbara Conference & Visitors Bureau and Film Commission.
22. From a 1912 "flyer soliciting financial subscriptions; Carl D. Sheppard, *Creator of the Santa Fe Style—Isaac Hamilton Rapp, Architect,* University of New Mexico Press, Albuquerque, NM, 1988, p. 75.
23. Ellen Beasley, "New Construction in Residential Historic Districts," in *Old and New Architecture—Design Relationship,* National Trust for Historic Preservation, Preservation Press, Washington, DC, 1980, pp. 229–256.
24. Santa Fe Convention and Visitors Bureau; http://nmindustrypartners.org/wp-content/uploads/2011/09/NMTD-Quarterly-Research-Report-January-20121.pdf; http://www.santafechamber.com/community/demographics.aspx; http://www.chron.com/news/article/Buzz-in-Santa-Fe-over-recovering-tourism-industry-3657768.php; and http://www.city-data.com/us-cities/The-West/Santa-Fe-Economy.html.
25. Christopher Tunnard and Boris Pushkarev, *Man-made America: Chaos or Control,* Yale University Press, New Haven, CT, 1963, p. 409.
26. Leland Roth, *McKim, Mead & White, Architects,* Harper & Row, New York, 1983.
27. Vincent Scully, *American Architecture and Urbanism,* Henry Holt, New York, 1988, p. 143.
28. The New York State legislature enacted enabling legislation for the designation of landmarks and historic districts in 1956. It was not

until the controversy over the demolition of Penn Station that the city council began serious consideration of a landmarks preservation law.

29. J. Lee Rankin, "Operation and Interpretation of the New York City Landmarks Preservation Law," *Law and Contemporary Problems,* vol. 36, no. 3 (Summer 1971), pp. 366–372; and Joseph B. Rose, "Landmarks Preservation in New York," *The Public Interest,* no. 74, (Winter 1984), pp. 132–145.
30. New York City's procedure for establishing the assessed value of any property is notoriously imprecise and is successfully challenged by thousands of property owners every year. One can also question the notion that 6 percent of assessed value constitutes a reasonable return. No business would purchase property on the basis of a 6 percent return if it cost more to obtain the money to pay for it. Furthermore, more than three decades have passed since the prevailing rate of interest was lower than 6 percent.
31. A tax abatement plan of this sort has never been submitted for legislative approval.
32. New York City Landmarks Commission, http://www.nyc.gov/html/lpc.
33. New York City Department of City Planning, *Grand Central Area—Proposal for a Special Sub District,* New York, 1989, p. 3.
34. Richard A. Jaffe and Stephen Sherrill, "Grand Central Terminal and the New York Court of Appeals: 'Pure' Due Process, Reasonable Return, and Betterment Recovery," *Columbia Law Review,* vol. 78, no. 1, (January 1978), pp. 134–163.
35. Donald Elliott, chairman of the City Planning Commission 1966–1973, interview, October 17, 1991.
36. New York City, *Zoning Resolution,* Art. VII, Chap. 9, Sec. 74–79.
37. Norman Marcus, "Air Rights Transfers in New York City," *Law and Contemporary Problems,* vol. 36, no. 3, (Summer 1971), pp. 372–379; and Madeleine A. Kleiner, "The Unconstitutionality of Transferable Development Rights," *Yale Law Journal,* vol. 84, no. 5, (April, 1975).
38. *Penn Central Transportation Company v. City of New York,* 438 U.S. 104 (1978).
39. One of the reasons that the New York State Court of Appeals, in the case of *Fred F. French Investing Co. v. City of New York* (1973), held that development rights from Tudor City could not be transferred to a vast area from 38th Street to 60th Street between Third Avenue and Eighth Avenue was that this would have been an unfair allocation of the costs and burdens to uncompensated private owners.
40. The maximum floor area ratio permitted in New York City is 21.6. The transfer would have resulted in a floor area ratio of 33.1.
41. Bret Senft, "Key Players Speak Out—Air rights debate calms long enough for seminar," *Real Estate Weekly,* November 1989, p. 2A.
42. John J. Costonis, *Space Adrift: Landmark Preservation and the Marketplace,* University of Illinois Press, Urbana, IL, 1974, p. 40.
43. Ibid., pp. 28–64.
44. Lisa Purdy and Peter D. Bowes, "Denver's Transferable Development Rights Story," *Real Estate Issues,* American Society of Real Estate Counselors of the National Association of Realtors, Chicago, Spring/Summer 1982, pp. 5–8; and "An Update on Denver's TDR Ordinances," *Real Estate Issues,* American Society of Real Estate Counselors of the National Association of Realtors, Chicago, Spring/Summer 1985, pp. 1–5.
45. Helen Wadsworth, Rachelle Levitt, and Fran von Gerichen, *Denver Metropolitan Area . . . Today 1982,* Urban Land Institute, Washington, DC, 1982, p. 31.
46. Lisa Purdy, interview December 21, 1989.
47. At the time the TDR ordinance was enacted, the B-5 zone included 10 locally designated landmarks with 1,362,000 square feet in available development rights, 898,000 of which were from two government-owned buildings that were unlikely to transfer development rights. There were another seven buildings that were eligible for designation as landmarks. They would generate 1,320,000 square feet in TDRs. However, 810,000 of this is from the U.S. Customs House, which was also unlikely to use its TDRs.
48. The B-7 ordinance was enacted less than a year after the B-5 ordinance and includes incentives for housing as well as historic preservation.
49. Arthur P. Ziegler Jr., Leopold Adler II, and Walter C. Kidney, *Revolving Funds for Historic Preservation: A Manual of Practice,* Ober Park Associates, Pittsburgh, PA, 1975, pp. 26–32, 62–75.
50. Ibid., p. 65.
51. Historic Savannah Foundation, http://www.myhsf.org; Savannah Visitors Bureau, http://savannahvisit.com.
52. Historical and statistical material on the Victorian District is derived from Mary L. Morrison (editor), *Historic Savannah,* Historic Savannah Foundation, Savannah, GA, 1979; Historic Savannah Foundation, "Victorian District Revolving Fund Report 1979–June 1988," unpublished, 1988; Chris Warner, "Attractive Housing For Savannah's Poor," *Urban Land,* Urban Land Institute, Washington, DC, February 1989, pp. 21–23; and Stephanie Churchill (executive director, Historic Savannah Foundation), interview, June 15, 1990.
53. Michael Homans,"Nuisance Code Ready for Council," *Savannah Evening News,* October 4, 1989, and "City Council OK's Ordinance on Crack House Destruction, *Savannah Morning News,* October 20, 1989.
54. Con Howe, interview with the author, January 7, 2013.
55. Ibid.; *The Planning Report,* Los Angeles, vol. XX, no. 2 (December-January 2006–7); and Ken Bernstein, "A Planning Ordinance Injects New Life into Historic Downtown," *Planning Los Angeles* (edited by David C. Sloane), American Planning Association, Washington DC, 2012, pp. 253–262.
56. Katherine Welch Howe, Incentives for Preservation and Rehabilitation of Historic Homes in the City of Los Angeles, The Getty Conservation Institute, Los Angeles, 2004, pp. 2–7.
57. Lesley D. Slavitt, "State and Local Tax Incentives for Historic Preservation," New York City Citizens Housing and Planning Council, unpublished, New York, July 1992; and Richard J. Roddewig, "Preservation Law and Economics," in *A Handbook on Historic Preservation Law,* Christopher J. Duerksen (editor), Conservation Foundation and National Center for Preservation Law, Washington, DC, 1983, pp. 446–456.
58. Special restoration work, even if it is required by the Landmarks Commission, is not currently eligible for J-51 benefits. However, the abatement granted for ordinary rehabilitation is usually enough to make most preservation projects financially feasible.
59. New York City Department of Housing Preservation and Development, Office of Development, Division of Financial Services.
60. John M. Fowler, "The Federal Government As Standard Bearer," in *The American Mosaic: Preserving a Nation's Heritage,* Robert E. Stipe and Antoinette J. Lee (editors), United States Committee, International Council on Monuments and Sites, Washington, DC, 1987, pp. 33–80.
61. The benefits applied to individual landmarks listed in the National Register, buildings within historic districts listed in the National Register and certified by the secretary of the interior as contributing to the historic character of the district, buildings in locally designated historic districts which the secretary of the interior found were based on acceptable criteria that satisfactorily protected the building. See Richard J. Roddewig, "Preservation Law and Economics: Government Incentives to Encourage For-Profit Preservation," in Duerksen (editor), op. cit., pp. 461–464.
62. Richard J. Roddewig, "Preservation Law and Economics: Government Incentives to Encourage For-Profit Preservation," in Duerksen (editor), op. cit., pp. 464–466.
63. Ibid., pp. 467–471.
64. Fowler, op. cit., p. 66.
65. The area's only commercial building of national significance was the Texas School Book Depository, from which it is believed that President John F. Kennedy was assassinated.
66. City of Dallas Ordinance No. 15203, June 1976.
67. Ron Emrich, "West End Historic District, Dallas Texas: Public Sector Revenue Projections," Department of Planning and Development, Dallas, 1988.
68. Alan Burnham (editor), *New York Landmarks,* Wesleyan University Press, Middletown, CT, 1963; Norval White and Elliot Willensky, *AIA Guide to New York City,* Collier Macmillan, New York, 1978; Paul Goldberger, *The City Observed: New York,* Vintage Books, New York, 1979; and Rose, op. cit.

69. Rose, op. cit., pp. 139–140.
70. Historical and statistical information about the Art Deco District in Miami Beach is derived from Barbara Baer Capitman, *Deco Delights—Preserving the Beauty and Joy of Miami Beach Architecture,* E.P. Dutton, New York, 1988; Anderson, Notter, and Finegold, *Preservation and Development Plan,* Miami Design Preservation League, Miami, FL, 1981; Nicholas N. Patricios, *Building Marvelous Miami,* University Press of Florida, Gainesville, FL, 1994; "Miami: A Troubled Paradise," *Metropolis,* New York, December 1995; and interview with Niesan Kasdin, mayor of Miami Beach, July 21, 2001.
71. Historical and statistical information about Northern Liberties in Philadelphia is derived from Harry Kyriakodis, *Northern Liberties: The Story of a Philadelphia River Ward*, The History Press, Charleston SC, 2012; Northern Liberties Neighbors Association and Interface Studio, *Northern Liberties Master Plan*, Philadelphia, November 2005; Daniel Brook, "Bart and the Deal," *Philadelphia Citypaper,* September 20–27, 2001; Inga Saffron, "Let's Not Strip-Maul a Promising Loft District," *Philadelphia Inquirer,* January 26, 2001; Inga Saffron, "Once the Big-Box Bully of Northern Liberties, Developer Bart Blatstein Is Now Its Guru of Good Design," *Philadelphia Inquirer,* March 19, 2004; Inga Saffron, "Zoning Approval Put Off for N. Liberties Projects," *Philadelphia Inquirer,* January 5, 2008; Inga Saffron, "One Man's Oddball Dream," *Philadelphia Inquirer,* May 15, 2009; Inga Saffron, "Massive Project for Site of Schmidt's," *Philadelphia Inquirer,* February 16, 2010; Center City District and Central Philadelphia Development Corporation, *Center City Housing: the Rebound Continues,* Philadelphia, November 2012; interviews with Inga Saffron and Paul Levy, December 13, 2012; and interview with Bart Blatstein, December 12, 2012.
72. Reyner Banham, "Preserve Us from the Paranoid Preservers," *Observer Magazine* (London), October 21, 1973, p. 15.

19

Comprehensive Planning

Philadelphia, 1960. Edmund Bacon working on a model for the future Society Hill. *(Edmund N. Bacon Collection, the Architectural Archives, University of Pennsylvania)*

Comprehensive planning is perhaps the most unfairly discredited of the many prescriptions for fixing the American city. Some have dismissed it as limited by the comprehension of the human mind. Others have ridiculed it as little more than a collection of "civic New Year's resolutions."[1] Despite these wickedly apt observations, comprehensive planning has in some instances proved to be a powerful agent for change. Billions were spent executing proposals put forward in comprehensive plans for Chicago, St. Louis, Philadelphia, and New York. Billions more were spent bringing to cities across the continent the image that Victor Gruen launched in his 1956 plan, *A Greater Fort Worth Tomorrow.*

Daniel Burnham's work in Chicago and Edmund Bacon's work in Philadelphia demonstrate that, like other prescriptions for fixing the city, comprehensive planning can be a powerful instrument for municipal improvement when it is based on an intelligent assessment of location and market plus an effective combination of design, financing, time, and entrepreneurship. The reason for Burnham's and Bacon's success, however, was not merely the persuasiveness of the documents that they called "comprehensive plans." The most important reason for their success was an effective planning process. As Dwight Eisenhower explained, "Plans are worthless, but planning is everything."[2] Indeed, plans that change nothing ***are*** worthless. Plans that result in major improvements to a city, however, are but one element in an ongoing process that Eisenhower, who was referring to the role of a plan in fighting a battle, would have instantly recognized as planning.

Effective comprehensive plans come in three varieties. There are those, frequently prepared by architects, that present compelling visions of the future. They crystallize the collective desires of the polity and set forth major steps that must be taken to achieve them. When comprehensive plans have been successful it is because their authors accepted the advice given by Daniel Burnham in what is the single most famous quotation about city planning:

> *Make no little plans; they have no magic to stir men's blood and will not be realized. Make big plans; aim high in hope and work, remembering that a noble, logical diagram once recorded will never die, but long after we are gone will be a living thing, asserting itself with ever-growing insistency.*[3]

This approach requires so profound an understanding of the topography, economy, social composition, and politics of a community that only a few great artists, like Daniel Burnham, have been successful with it. Such compelling visions of the future also require a level of boldness that is discouraged in our participatory form of government. Accordingly, the approach is usually unsuccessful with the notable exception of some truly extraordinary documents such as the 1909 *Plan of Chicago.*

A second example of successful planning involves comprehensive plans, frequently prepared by appointed public officials, that attempts to strategically deploy already agreed-upon municipal improvements instead of proposing a utopia, as the first type of planning does. This approach was persuasively advocated in 1914 by New York City's Committee on the City Plan, which stated:

> *City planning does not mean the invention of new schemes of public expenditure. It means getting the most out of the expenditures that are bound to be made and saving of future expense in replanning and reconstruction. With or without a comprehensive city plan, the City will probably spend hundreds of millions of dollars on public improvements during the next thirty years. In addition, during this same period property owners will spend some billions of dollars in the improvement of their holdings. To lay down the lines of city development so that these expenditures when made will in the greatest possible measure contribute to the solid and permanent up-building of a great and ever greater city—strong commercially, industrially, and in the comfort and health of its people—furnishes the opportunity and inspiration for city planning.*[4]

This philosophy is easy for public officials to accept and still easier for bureaucrats to implement. That is why the comprehensive plans for St. Louis in the mid-1940s and New York City in the late 1960s successfully generated so many remarkable changes to both cities.

The third approach is process-oriented. It adopts any available "good" ideas, adapting them to meet the demands of the widest variety of constituencies, altering them to fit requirements of legislation, financing, and implementing entities, and adjusting them over and over again until they present a "total vision of the city" that reflects the "collective consciousness of its citizens."[5] It is best illustrated by the ongoing planning efforts in Philadelphia, initiated by Edmund Bacon and his associates during the mid-twentieth century and continued into the twenty-first century by the Center City BID, directed by Paul Levy.

While these three varieties of comprehensive planning could not be more different in philosophy and methodology, they share the same assumptions that "the formless growth of the city is neither economical nor satisfactory" and that "a plan for the well-ordered and convenient city is . . . indispensable."[6] They also share a similar need for gargantuan sums of money, lengthy periods of execution, and continuing acceptance by the city's political establishment.

Communities will not finance such plans unless there is a deep commitment to their recommendations. Until recently, such commitment has occurred only in response to powerful proposals, such as those that emerged from the planning process in Chicago and Philadelphia. It is now possible to base that commitment on widespread familiarity with facts. Computers can tabulate, analyze, and present information on a geographic basis. The maps and charts that they produce can alter the public dialogue in every community. Rather than respond to particular visions of a better tomorrow, citizens are now able to obtain information previously restricted to specialists and decide for themselves what actions to take and what expenditures to make. Some cities are taking steps to make such information more accessible and bringing talented planners to engage the community in a decision-making process that promises to make comprehensive planning an ongoing component of government.

Burnham and Company

America's acceptance of the effectiveness of "a plan for the well-ordered and convenient city" was launched by Daniel Burnham and the other designers of the Chicago Fair of 1893; was elaborated by Burnham and his collaborators on the McMillan Plan for Washington, D.C. (1901–1902), and the plans for Cleveland (1903), Manila (1905), and San Francisco (1904–1906); and reached maturity in 1909 with the *Plan of Chicago* (see Chapters 3 and 6). Burnham came to Chicago with his family at the age of eight and remained there until his death in 1912. He worked as a draftsman until 1873, when he established an architectural firm in partnership with John Wellborn Root. The partnership lasted until Root's death in 1891, when the firm was renamed D. H. Burnham and Company. It was responsible for some of the country's most famous buildings, including the Rookery, Monadnock, and Reliance buildings in Chicago; the Flatiron Building in New York; and Union Station in Washington, D.C. However, the firm's pioneering work in city planning, for which it did not charge a fee in the hope that it would lead to future architectural commissions, made the firm even more famous.[7]

Many talented people were involved in Burnham's city planning practice, but Charles Moore and Edward Bennett were the most important. Moore, whom he met while working on the McMillan Plan, acted largely in the role of an editor. Bennett, an English architect who began working for Burnham not long after graduating from the École des Beaux Arts, became his acknowledged collaborator in developing plans for San Francisco and Chicago. After Burnham's death in 1912, Bennett went on to produce plans for Brooklyn, Minneapolis, Pasadena, and Portland, Oregon.

Chicago, 1893. The World's Columbia Exposition, with its own public services, utility, and transit system, became the model for the well-plannned American City. *(Courtesy of Chicago Historical Society)*

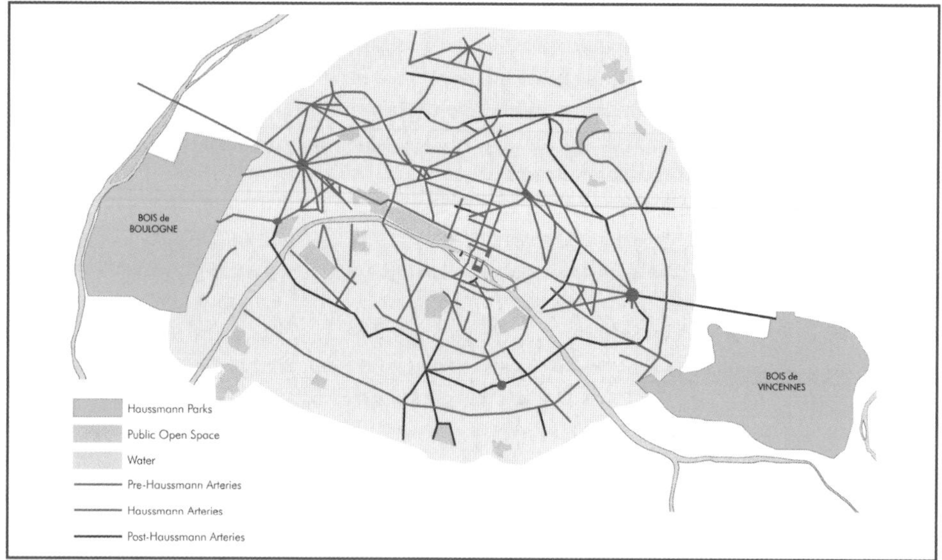

Nineteenth-century Paris. Map of the boulevards Haussmann created and those that pre- and postdate him. *(Alexander Garvin and Joshua Price)*

Impact of the Chicago Fair

The single most important contribution to Burnham's city plans came from a nonparticipant: Baron Georges-Eugène Haussmann. His transformation of Paris (see Chapters 3 and 6) provided the inspiration for the World's Columbian Exposition, commonly referred to as the Chicago Fair of 1893. When Daniel Burnham became chief of construction for the Chicago Fair, he had not seen Haussmann's transformation of Paris. His principal colleagues, Frederick Law Olmsted and Charles McKim, were more than familiar with Haussmann's work. Olmsted's travels to Paris had made him a lifelong admirer of Haussmann's parks and boulevards, while McKim had studied architecture at the École des Beaux Arts in Paris between 1867 and 1870. These three, together with Olmsted's partner Henry Codman, conceived and built the Chicago Fair in a manner that reflected their common admiration of Haussmann's Paris.

The Fair was really a small, "well-ordered and convenient city." It had its own water, sewer, and utility systems; its own fire and police protection; its own sanitation, electric, telephone, and telegraph service; and even its own elevated transit system, which for the first time introduced an electrified third rail. Like Paris, its buildings coalesced into a consistent ensemble. However, this was the result of specific architectural decisions rather than independent private market responses to common development problems. The major buildings around the Court of Honor were built to a common cornice height of 60 feet because the architects all agreed to that, not because developers thought that was as high as their customers would be willing to climb or because it was the maximum height to which they could pipe running water without the added expense of pumping. Major buildings, such as the Administration Building, Terminal Station, and State of Illinois Building, closed axial vistas in conformance with the agreed-upon site plan, not because great boulevards had been cut through the fabric of an existing city to connect major destinations. The buildings were all clad in a mixture

Paris, 2010. Avenue de l'Opera, one of the numerous new boulevards whose buildings were erected by developers who purchased surplus land that was not required for the right-of-way. *(Alexander Garvin)*

of plaster, cement, and jute fiber and painted white because this treatment guaranteed the necessary architectural unity, not because the white cladding was the real estate industry's commonly used building material.[8]

Since the utility systems and public services were not immediately visible, architects, urban planners, and civic leaders came away from the Chicago Fair with the belief that a common building height, axial vistas, and a consistent architectural vocabulary generated "the well-ordered city." These misconstrued characteristics of Haussmann's Paris, along with clustered groups of civic structures, a central railroad station, and a regional park system, became common elements in Burnham's plans for Washington, D.C., Cleveland, Manila, San Francisco, and ultimately for scores of comprehensive plans produced by his followers.

Improvement and Adornment of San Francisco

Burnham came to San Francisco at the invitation of the city's business leaders and political reformers. Ostensibly his visit was for the purpose of speaking to the Committee for the Improvement and Adornment of San Francisco, organized by lawyer, entrepreneur, and former mayor James Phelan. The visit evolved into a pro bono effort to provide the committee with a vision of what Burnham thought San Francisco ought to be.[9]

The project took a year, part of which was devoted to planning in the Philippines. While in San Francisco, Burnham worked from a cabin-studio atop the Twin Peaks of Diamond Heights, which provided a splendid view of the entire metropolitan region. The panoramic drawings made at the site were published in *Report on a Plan for San Francisco,* produced in collaboration with Edward Bennett and Willis Polk.[10]

Report on a Plan for San Francisco offered a series of proposals to transform San Francisco into an American version of Haussmann's Paris, only now reconceived with hills. It proposed creating diagonal boulevards, like those Haussmann had created in Paris, which smashed through the city's street grids, connecting hills, neighborhoods, and new public spaces. At the intersections of the roadways, Burnham envisioned traffic circles, civic squares, and public structures such as Haussmann had created in Paris. Like Paris, the plan included a park system with 9600 acres for large regional parks, 255 acres for 12 smaller parks, and landscaped boulevards/parkways providing the connective tissue for this vast expansion of public open space.

The plan did not simply copy Haussmann's Paris. It also included elements that responded to San Francisco's unique topography and thriving maritime economy. The steep slopes and hilltops that had not yet been developed were to be preserved as public open space. The city's waterfront was set aside for maritime use, but augmented with a perimeter boulevard. A new shipping, warehousing, and manufacturing center was to be created in the Hunter's Point area. Taking a cue from the Chicago Fair, there were to be a central civic center and a rapid transit system. Finally, to ensure that all public and private development would conform to this vision of a well-ordered city, Burnham proposed that an art commission establish requirements for building height, street furniture, signage, statuary, paving, and street trees.

Burnham did not suppose "that all the work indicated [could] or ought to be carried out at once or in the near future." He expected that it was to "be executed by degrees, as the growth of the community demand[ed] and as its finan-

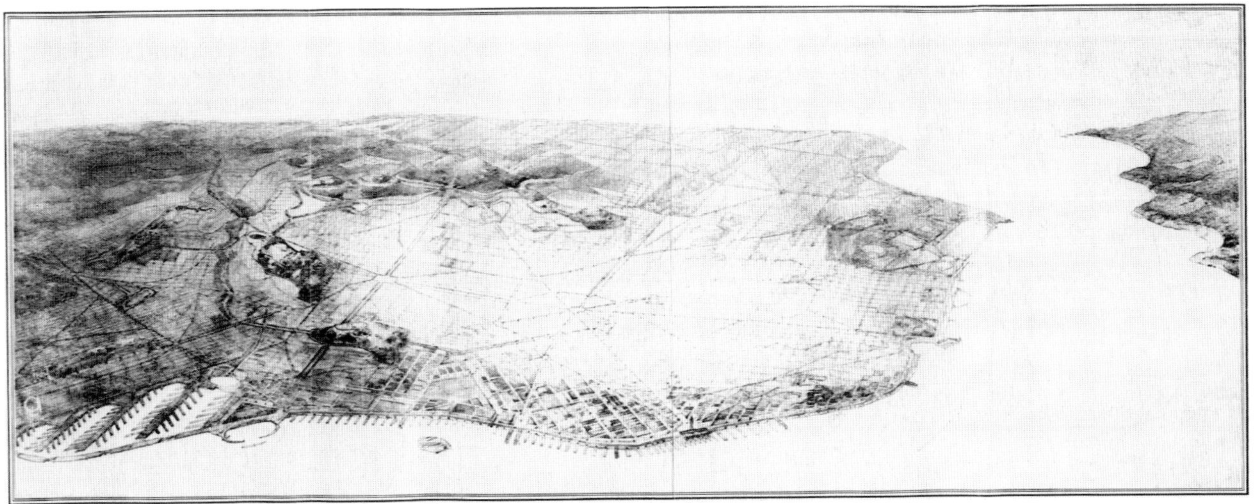

San Francisco, 1905. Bird's-eye perspective of the city from the east showing changes proposed by Daniel Burnham. (*From Burnham and Bennett,* A Report on the Improvement and Adornment of San Francisco, *City of San Francisco, 1905)*

cial ability allow[ed]." He understood that his proposals required both further elaboration and vast expenditures. As he explained, "It is not the province of a report of this kind to indicate the exact details very closely."[11]

Burnham was very wrong, however, in thinking that this plan interfered "as little as possible with the rectangular street system of the city." His diagonal boulevards, public spaces, and civic monuments, like Haussmann's, required massive condemnation, relocation, demolition, and reconstruction. The challenges inherent in a plan of this scope proved to be the primary reason that the *Report on a Plan for San Francisco,* which was presented to the board of supervisors and approved in September 1905, was never implemented. Other reasons included the indirect consequences of the earthquake and fire that, in April 1906, destroyed most of the city.

Burnham and his supporters understood that a comprehensive plan of this magnitude needed a constituency, legislative sanction, a viable implementation mechanism, and lots of money. They planned a major effort to generate public support. The first step was to be distribution of 3000 copies of the report initially authorized by the board of supervisors. However, almost all the copies were destroyed in the earthquake and fire before lobbying could begin. Suddenly, there was no way of marshaling either the necessary support or the money to finance the ideas in the plan. Rather than take the time to consider Burnham's proposals, develop the necessary political consensus, obtain legal sanction for new traffic rights-of-way and block and lot patterns, and then acquire the necessary property, the city quite naturally devoted all its energies to quickly replacing 28,000 buildings lost by an earthquake that left in its wake "a blackened wasteland of more than four square miles—512 blocks."[12]

Plan of Chicago

In his hometown, Daniel Burnham required assistance neither in understanding the city's topography, economic base, demographic composition, and political structure nor in conceiving a strategy to obtain public support. But he did need powerful allies to obtain the legislation, appropriations, and bureaucratic support to implement his ideas. These favorable conditions came together in 1906, when the Chicago Merchant's Club merged with the Commercial Club and established a joint committee to develop a comprehensive plan.[13] Burnham agreed to take charge of drafting their *Plan of Chicago,* donating his services free of charge with the understanding that he would "have an entirely free hand in the choice of . . . associates and assistants."[14]

The *Plan of Chicago* took Burnham, Bennett, Moore (their editor), and a team of architects, drafters, and illustrators three years to produce. It is as much an introduction to "city planning in ancient and modern times" and to the history of Chicago as it is a vision of what they wanted Chicago to become. Almost half of its pages are devoted to photos, prints, and drawings of London, Vienna, and especially Haussmann's Paris, cities that Burnham hoped the citizens of Chicago would be inspired to emulate and surpass. The most important illustrations, however, were Jules Guerin's perspective drawings rendered in watercolor and reproduced in color. They provided the magic that stirred the blood, leading generations of Chicagoans to spend billions of dollars realizing huge sections of this noble diagram.[15]

It was a truly "big plan" that included (1) more than 60,000 acres of parks, parkways, and forest preserves; (2) a regional highway system extending more than 60 miles beyond the business district; (3) a systematic rearrangement of local arteries, including new bridges and through avenues, widened streets, and landscaped boulevards; (4) 23 continuous miles of harbor, pier, lagoon, beach, park, and water-

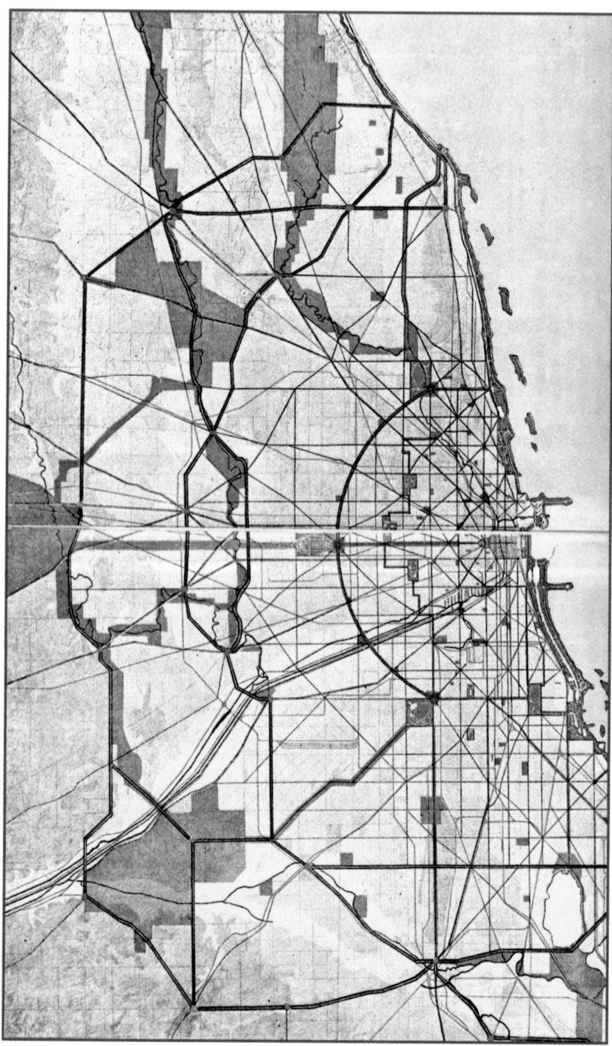

Chicago, 1909. *Plan of Chicago.* Map showing new system of streets, boulevards, parkways, and parks. *(From Burnham and Bennett,* Plan of Chicago, *The Commercial Club of Chicago, 1909)*

Chicago, 1910. View of the Chicago River prior to reconstruction. *(Courtesy of Chicago Historical Society)*

front development along Lake Michigan; (5) a cleaned-up and straightened Chicago River; (6) a consolidated system of freight and passenger railroads, mass transit, and railway terminals; (7) a cluster of existing and new institutional buildings combined into a cultural center in an expanded Grant Park; and (8) a gigantic civic center dominated by a domed city hall soaring to a height of more than 40 stories. Even Haussmann had not conceived of an integrated system of public works on so vast a regional scale.

The authors, like most reformers of the day, believed that government should interfere with the private sector only to provide basic government facilities and services and to ensure the health and safety of the population. Accordingly, the *Plan of Chicago* devoted its attention to infrastructure (roads, bridges, freight and passenger railway improvements, etc.) and community facilities (parks, government buildings, and cultural institutions). In doing so, Burnham and his collaborators expected to remove obstructions to circulation,

Chicago, 1909. Rendering of proposal made in the *Plan of Chicago* for straightening the Chicago River and building a double-decker drive along its edge. *(From Burnham and Bennett,* Plan of Chicago, *The Commercial Club of Chicago, 1909)*

Chicago, 2008. The double-decker Wacker Drive built along the straightened Chicago River was inspired by Burnham's *Plan of Chicago*. *(Alexander Garvin)*

improve the distribution of goods and services, encourage commerce, and provide necessary cultural, educational, and recreational facilities. They did not consider it their role to alter the distribution of income, the patterns of consumption, or the living conditions of any group of citizens. Less than a single page out of the document's 164 pages is devoted to slums, because, as its authors explained, "the slum exists today only because of the failure of the city to protect itself against gross evils and known perils, all of which should be corrected by the enforcement of simple principles of sanitation."[16]

Generations of architects and historians have directed their attention to Burnham's unrealized proposals for majestic public squares and monuments, diagonal boulevards, uniform building heights, and axial vistas. It was not these inflated, academic venerations of Haussmann's Paris, however, to which Chicagoans directed their attention. It was to the inspired proposals for public improvements that Burnham persuaded them would make the city a more "efficient instrument for providing all its people with the best possible conditions of living."[17]

Once the *Plan of Chicago* had been issued, the city council established the 328-member City Plan Commission, published and distributed 165,000 copies of a 93-page booklet entitled *Chicago's Greatest Issue—An Official Plan,* and formally adopted the plan in 1911 after two years of public debate. Since the monumental imitations of Haussmann's Paris involved so much property condemnation, relocation, and community opposition, the city proceeded with those improvements that solved commonly agreed-upon problems and exploited generally accepted opportunities.[18] Within a decade of the plan's publication, the city had spent $237 million on some of the proposals that it put forward, including numerous street, railway, and park improvements and extensions. It also had acquired 14,254 acres of forest preserve. More important, the City Plan Commission proved to be "not a money-spending but a money-earning institution," generating increased property values and city revenues in the areas immediately adjacent to these improvements.[19]

Significant milestones had been achieved by World War II: almost 40,000 acres of forest preserve had been acquired; the double-decker Wacker Drive had been constructed along a straightened Chicago River; Michigan Avenue, Roosevelt Road, and numerous other streets and viaducts had been widened and extended; the produce market had been moved 2 miles from the center of the city; and most dramatically, the city spent more than $1 billion on landfill along Lake Michigan in order to create some of the nation's most extraordinary parkland and provide an enticing setting for new residential and commercial construction.[20]

Burnham's vision for Chicago was accepted because of the persuasive diagnosis he provided of the city's physical, functional, economic, and environmental problems and the convincing proposals he provided to solve those problems. He also explained that the radical changes he proposed in the *Plan of Chicago* "cannot possibly be realized immediately. . . . Therefore it is quite possible that when particular portions of the plan shall be taken up for execution, wider knowledge, longer experience, or change in local conditions may suggest a better solution."[21] That is precisely what happened as the specific design of each project was altered to meet the requirements of the time in which it was brought to completion. Most of the proposals for the reuse of railroad property, for example, proved to be impractical, while the 69-acre complex of Illinois Central railroad properties that Burnham assumed would remain in place forever went out of use and were redeveloped as Central Station (see Chapter 15).

The most important reason that so much of the *Plan of Chicago* was implemented, however, was effective politicking by its supporters. In the 10 years after the plan's publication, slide shows illustrating it were presented to more than 175,000 citizens. During 1912 alone, the City Plan Commission "furnished Plan articles that appeared in 575 magazines, periodicals, trade and club publications . . . [and] permeated the catalogues and magazines of large business concerns."[22] Burnham's supporters even persuaded the board of education to produce 70,000 copies of a simplified version of the

Chicago, 2008. The Burnham Plan created the vision of an entirely different sort of lakeshore that would become the city's major asset. *(Alexander Garvin)*

plan that became the eighth-grade civics textbook in the city's public schools.[23]

Some of Burnham's proposals were commonly desired projects that might have been realized even without a comprehensive plan. The *Plan of Chicago* and the lobbying effort to obtain its implementation provided many others with the push they needed to obtain public support and money. Moreover, the plan's comprehensive scope prevented the city from evolving haphazardly and provided the framework for "a well-ordered, convenient, and unified city" that inspired civic officials in Chicago and elsewhere for almost two generations.

A National Movement

During the year that the *Plan of Chicago* was issued, the first national conference on city planning was held, Wisconsin became the first state to enact legislation authorizing cities to create planning commissions and prepare city plans, and Harvard became the first university to offer a course in city planning. As much as planning had become a part of the national consciousness, no one really knew what "city planning" meant. As Frederick Law Olmsted Jr. explained in 1910 at the second national conference on city planning, "We are dealing here with the play of enormously complex forces which no one clearly understands and few pretend to; and our efforts to control them so often lead to unexpected and deplorable results that sober-minded people are often tempted to give up."[24]

One thing was certain, however: the various early planning documents produced by Charles Mulford Robinson, John Nolen, George Kessler, the Olmsted brothers, and other fledgling city planners could not compare in scope, persuasiveness, or impact with the *Plan for Chicago.* That said, a genuine "city planning" movement, however, would have to be based on more than the artistic genius of a single individual. There had to be a body of competent individuals trained to produce comprehensive plans, plus steady demand for their work.[25]

Despite the junior Olmsted's justifiable worry about the often unexpected and deplorable results of urban planning, there soon were plenty of self-declared professionals promising to fix our cities. Engineers, landscape architects, politicians, businesspeople, social critics, architects, lawyers, and all sorts of other professionals transformed themselves into city planners. In 1915, they banded together to form the American City Planning Association (now the American Planning Association). Harvard took the lead in 1929, creating the first separate School of City Planning to provide training for the new profession. By 2013, there would be 73 master's and 15 bachelor's programs at 77 North American universities.[26]

This growing body of city planning professionals persuaded local, state, and federal agencies to require and even pay for their services. In 1927, Secretary of Commerce Herbert Hoover appointed a nine-member committee of city planners to draft the *Standard City Planning Enabling Act.* As he explained in the foreword to the final report presenting the 1928 act, cities required "a clearly defined permanent planning branch in local government, in the form of a commission which formulates a comprehensive plan and keeps it up to date."[27]

Municipalities everywhere accepted Hoover's recommendation because, as the size, role, and cost of government had mushroomed, public officials needed to collect and present data that provided the basis for decision making, to enumerate and coordinate the rapidly increasing array of government activities, and to compare information on the facilities and services provided to different parts of the city. By 1929, 650 cities had official planning commissions, 200 of which had been the subject of planning reports.

Local governments were happy to employ city planners, particularly when the federal government reimbursed them, which it did in conjunction with the National Industrial Recovery Act of 1933, the Federal Aid Highway Act of 1956, the Demonstration Cities and Metropolitan Development Act of 1966, the Coastal Zone Management Act of 1972, virtually every housing act, and a continuing stream of other legislation. The extent to which this generated stable demand for city planning activities is demonstrated by Section 701 of the Housing Act of 1954, which provided assistance for:

> *(1) preparation, as a guide for long-range development, of general physical plans with respect to the pattern and intensity of land use and the provision of public facilities, together with long-range fiscal plans for such development; (2) programming of capital improvements based on a determination of relative urgency, together with definitive financing plans for the improvements to be constructed in the earlier years of the program; (3) coordination of all related plans of the departments or subdivisions of the government concerned; (4) intergovernmental coordination of all related planned activities among the state and local governmental agencies concerned; and (5) preparation of regulatory and administrative measures in support of the foregoing.*[28]

Despite the vast number of professional planners and the huge sums spent on their work, only a handful of comprehensive plans significantly affected city growth and development. The others failed when their prescriptions proved inappropriate, either because they tried to impose a design concept that had little to do with existing landscape and land use patterns, because the city government lacked the money or the will to spend for those particular proposals, or because they would cause unacceptable disruption and, therefore, political opposition. Plans that succeeded, like those for St. Louis, Philadelphia, and Portland, did so because their authors had produced convincing diagnoses of the problems and opportunities facing those cities, prescriptions the citizenry believed could effectively solve those problems and exploit those opportunities, and implementation strategies whose costs in terms of dollars and dislocation were politically acceptable.

Visions of the Future Metropolis

Most twentieth-century comprehensive plans reflect one of two visions of the future metropolis: Daniel Burnham's amalgam of Haussmann's Paris with the Chicago Fair or Victor Gruen's amalgam of Le Corbusier's City of Tomorrow with the suburban shopping center (see Chapters 4, 6, and 7). Burnham's "noble, logical diagram" reshaped Chicago and provided the model for most city plans produced during the early part of the century. When it was exported to other cities, however, his vision of the City Beautiful provided only planning clichés such as common cornice heights, diagonal boulevards, and monumental civic centers, for example, for published documents that were rarely implemented. With the exception of a handful of civic centers, this planning strategy was rejected by political leaders whose constituents were unwilling to spend gargantuan sums to alter cityscapes that they felt could satisfactorily remain as they were, at no cost to the taxpayer.

Victor Gruen's 1956 plan, *A Greater Fort Worth Tomorrow,* on the other hand, provided more contemporary planning clichés, including circumferential highways providing access to the business district, downtown garages accessible by highway, and pedestrian precincts, which changed the face of virtually every major American city even if the progenitor plan for Forth Worth was ultimately rejected. The philosophy that informed the Fort Worth plan was adopted by governments across the country because it responded to overwhelming local demand for an end to traffic congestion and, more important, because the federal government was ready to pay 90 percent of the bill.

While the Burnham and Gruen visions could not have been more different in appearance, they were similar in other ways. Either could be superimposed on any municipality without paying much attention to its specific characteristics. Both depended on modernization of the city's infrastructure, especially its traffic arteries. Both eliminated the existing, messy mix of land uses, proposing instead to carefully segregate city life into discrete districts. Both required teams of elite professionals to orchestrate their realization. And, finally, both demanded such high levels of spending that they had to excite controversy.

Burnham's vision failed to be implemented outside Chicago because his followers did not have the same profound understanding of the particular city for which they were planning, or lacked the same genius for capturing the collective aspirations of its population, or could not generate the same carefully orchestrated political support. Gruen's vision, on the other hand, required neither profound understanding of a specific city nor artistic genius. It could be applied to any city with a defined central business district surrounded by deteriorating areas that could be acquired in order to accommodate construction of a circumferential highway at relatively low cost without causing much dislocation or generating substantial opposition.

Implementing Burnham's Vision of the City

The standard components of Burnham's "well-ordered city" were proposed wherever planners accepted Burnham's challenge and made "big plans." The 1910 *Report of the New Haven Civic Improvement Commission* by Cass Gilbert and Frederick Law Olmsted Jr. included an extensive regional park system, clearance of the area between the railroad station and the central business district, and a monumental civic center. Virgil Bogue's 1911 *Plan of Seattle* prescribed demolishing whole city blocks and rebuilding entire hills to create a system of diagonal streets and highways, a grandiose civic center, a new harbor, and a vast system of parks and parkways. Carrere & Hastings's 1912 report, *A Plan of the City of Hartford,* proposed a regional park system, a reconstructed waterfront, circumferential boulevards, diagonal avenues, and elaborate outlying districts with "factory sites and workingmen's homes."

Early proponents of comprehensive planning in these and other cities had little difficulty obtaining the civic support needed to commission a comprehensive plan. Their problems began when they tried to get local governments to implement these visions of the "well-ordered, convenient, and unified city." As soon as public officials were faced with budget appropriations, bond issues, and taxes, they avoided the big plans, choosing as usual to implement minor projects

Seattle, 1911. Central Avenue looking north to Central Station as proposed in the *Plan of Seattle* but never executed. *(From V. Bogue,* Plan of Seattle, *Lowman & Hannaford, Seattle, 1911)*

(such as the elimination of grade crossings) or noncontroversial public actions (such as the acquisition of additional parkland in inexpensive outlying areas).

Supporters of comprehensive planning wanted to avoid such piecemeal implementation, because as long as individual projects required political approval, components of a plan that might alienate a section of the city, a large number of voters, or important economic interests would have great difficulty obtaining the necessary support. One way to avoid such parochial voter rejection was to obtain formal approval of an entire plan and make it binding, but even this more democratic method often proves fruitless. This approach was quickly tried and rejected. Virgil Bogue's *Plan of Seattle* was placed on the ballot in 1912. The vote was 14,506 in favor and 24,966 against.[29]

Bogue's attempt to copy Burnham's notion of Paris with hills was even less appropriate to Seattle than to San Francisco. Furthermore, the cost was far beyond the city's pocketbook. A contemporary opposition pamphlet placed its cost at well over $100 million, "a staggering sum in an era when land prices and construction costs were many times less than those of the end of the twentieth century."[30] Bogue, unlike Burnham, had not spent years working with business and political interests to create a constituency for massive spending and, thus, could not withstand opposition to massive government spending.

The voters also rejected this particular approach to the planning process. They were *not* offered what Burnham described as "a noble, logical diagram" that could "be executed by degrees, as the growth of the community demands and its financial ability allows." They were given an all-or-nothing proposition. In 1910, voters had been asked to approve creation of a 21-member commission to formulate a comprehensive plan, a special tax to pay for its preparation, and the requirement that "if a majority of the voters voting thereon shall favor adoption of said City Plan . . . it shall be adopted and shall be the plan to be followed by all City officials in

Fort Worth, 1956. Aerial view of Fort Worth. *(Courtesy of Greater Fort Worth Planning Committee)*

the growth, evolution, and development of . . . Seattle, until modified, or amended at some subsequent election."[31]

They approved all three. However, two years later, when the *Plan of Seattle* was put on the ballot, the voters not only turned down its contents, they also made it very clear that they would not relinquish their city to self-designated experts or give up their role in approving specific actions or let individuals who were not subject to electoral approval spend their hard-earned money on grandiose schemes for municipal improvement.

Implementing Gruen's Vision of the Greater City

As America's leading post–World War II architect of suburban shopping centers and designer of Southdale, the country's first climate-controlled shopping mall, Victor Gruen had an inside view of the competition facing downtown business districts (see Chapter 4). The know-how he gained in developing regional retail centers provided the basis for his "counterattack against urban sprawl and anti-city chaos" and his proposals for "the revitalization of the heart of our cities."[32]

Fort Worth, 1956. Aerial view of Gruen's vision as set forth in *A Greater Fort Worth Tomorrow.* *(Courtesy of Greater Fort Worth Planning Committee)*

Gruen had begun his professional career with great enthusiasm for Le Corbusier's vision of the *City of Tomorrow.* But as he saw its concepts poorly translated into reality, he changed his opinion, explaining that:

> *Though [Le Corbusier] foresaw a great flood of automobiles, [he] underestimated the proportions of the deluge. The amount of traffic necessary to establish any kind of connection between the widely separated towers is indeed of such a scale that it automatically destroys part of the dream.*[33]

The automobile used too much space: 300 square feet to be kept at its point of origin, another 300 square feet when stored at its destination, another 600 square feet in roadways, and 200 square feet where it was sold, repaired, and serviced.[34]

Gruen provided an alternative model in the mid-1950s,

when 20 leading citizens of Fort Worth hired him to devise a comprehensive city plan. At that time the city was suffering from ailments common to cities across the country: declining retail sales, declining building occupancy, and increasing traffic congestion. It also faced lively competition with Dallas.

In *A Greater Fort Worth Tomorrow*, he proposed (1) a series of circumferential and radial highways providing vehicular access to Fort Worth, (2) perimeter parking terminals located along the various loop highways, (3) a central business district concentrated within the inner highway belt and sufficiently compact to ensure no more than 3 minutes' walk to any downtown location, (4) separation of truck and automobile traffic from pedestrians within the inner highway loop, (5) construction of a new outdoor pedestrian level at the second floor of every downtown building, (6) reuse of what had been previously street level (now underground) exclusively for motor vehicles, and (7) a regional mass transit system including downtown shuttle cars for those unable or unwilling to walk a few hundred feet. Gruen proposed establishment of a Central District Development Authority to administer the plan, buy, clear, and sell property, and issue bonds; a Central District Parking Authority to build and operate parking facilities; a Central District Roads Authority to coordinate federal, state, and city arterial programs; and a Central District Transit and Trucking Authority to build and operate transit facilities and basement-level truck roads.[35]

During the 1960s, *A Greater Fort Worth Tomorrow* was the best-publicized and most admired vision of a future metropolis, almost on a par with Le Corbusier's work of the 1920s. Even Jane Jacobs was lavish in her praise:

> *The excellent Gruen plan includes, in its street treatment, sidewalk arcades, poster columns, flags, vending kiosks, display stands, outdoor cafes, bandstands, flower beds, and special lighting effects. . . . It works with existing buildings and this is not just a cost-saving expedient. . . . This mixture is one of downtown's greatest advantages, for downtown streets need high-yield, middling-yield, low-yield, and no-yield enterprises.*[36]

One reason for the plan's appeal was that it required no specialized knowledge of Fort Worth or any other urban region. More important, the proposals could be applied everywhere.

In Fort Worth, however, it was a flop. Parking-lot operators opposed competition from a public authority. Property owners balked at losing ground-level rental income and then spending money to retrofit the second floor to accommodate lobbies and retailing. Taxpayers refused to consider the huge sums needed to create the separate vehicular and pedestrian levels.

Gruen's vision as set forth in *A Greater Fort Worth Tomorrow* may have been stillborn, but it inspired countless similar proposals. His work, *A Revitalization Plan for the City Core of Cincinnati*, was ambitious enough to inspire that city's leadership. The proposed highway loop around the core was rejected, as it had been in Fort Worth. Other components, however, were much-needed changes to the city that were repeated in 1964 in *The Plan for Downtown Cincinnati*, prepared by Rogers, Taliaferro, Kostritsky, and Lamb (RTKL). They included traffic improvements for local streets, reestablishing public gathering places at Fountain Square and Government Square, building downtown apartments to provide street life after office workers had gone for the day, and constructing a convention center at the very location it would eventually be built. Where these improvements occurred, it was because of consumer demand. Those that were rejected were discarded because they were not needed in a city the size of Cincinnati.[37]

The year Gruen published *A Greater Fort Worth Tomorrow*, Congress enacted the National Interstate and Defense Highways Act, which provided 90 percent of the enormous costs of land acquisition, relocation, and demolition for interstate highways penetrating directly into the city (see Chapter 6). In 1962, it approved another highway act that required every urban area of more than 50,000 people to develop "long-range highway plans and programs which are properly coordinated with plans for improvements." It also provided the money to pay for these plans.

During the 1960s, virtually every urban area issued federally required and funded "area transportation studies" proposing interstate highway loops around the central business district and radial interstate highways connecting them with the suburbs. Like Fort Worth, the governments that commissioned these plans chose to ignore Gruen's radical reconstruction of the central business district. However, if granting such preeminence to the automobile encountered opposition, the local government could add a pedestrian street similar to that which Gruen had proposed in 1957 for Kalamazoo. Where there was substantial momentum for housing redevelopment, a convention center, or some other prescription, these were also incorporated into the plan. Whatever the variants, they were all versions of the same noble, logical diagram asserting itself with ever-growing intensity.

For the next quarter of a century, Gruen's diagram remained the common vision of the future metropolis. A growing market of suburban residents needed easy access to downtown jobs, and an increasingly worried group of downtown merchants, property owners, and politicians depended on their consumer spending. There was plenty of federal money to pay for Gruen's diagrammatic solution to their problems and little opposition from occupants of the fringe areas that needed to be acquired for the highway loops. Besides, there was no alternative diagram that better provided for vehicular traffic. Only when most of the 41,000 miles of the Interstate Highway System was reaching completion did Gruen's vision of the future metropolis finally lose its appeal.

Synergistic Government Expenditures

Cities continually change. Consequently, an end-state plan is of limited utility. To be effective, comprehensive planning must be ongoing. Frederick Law Olmsted Jr. explained this in 1911 at the third national conference on city planning:

> *We must disabuse the public mind of the idea that a city plan means a fixed record upon paper of a desire by some group of individuals prescribing, out of their wisdom and authority, where and how the more important changes and improvements in the physical layout of the city are to be made. . . . We must cultivate . . . the conception of a city plan as a device or piece of administrative machinery for preparing, and keeping constantly up to date, a unified forecast and definition of all the important changes, additions and extensions of the physical equipment and arrangement of the city.*[38]

For such planning even to take place, it must be an established function of municipal government, like street maintenance or fire protection. During the first decades of the twentieth century, when Burnham, Bennett, Nolen, and the Olmsted brothers began preparing city plans, that notion seemed far-fetched. In the twenty-first century, when municipal governments everywhere include large numbers of "planners," civic leaders still question the need for planning. Thus, whenever budget cuts are required, city planning is among the first to suffer. Nevertheless, the profession has become an important, ongoing government activity because, as New York's Committee on the City Plan explained in 1914, "It means getting the most out of the expenditures that are bound to be made." More important, the strategic deployment of these expenditures can obtain effects that are greater than would be produced if they were made independently.

Ongoing strategic deployment of municipal expenditures requires a standardized planning methodology and budget procedures that reflect that planning. Lots of cities engage in such planning without having printed city plans. Among cities that have printed comprehensive plans, two stand out because of their strategic use of government expenditures: the 1947 *Comprehensive City Plan* for St. Louis and the 1969 *Plan for New York City*. Both are based on extensive (almost atlas-like) analyses of their city's topography, infrastructure, community facilities, demography, and economy. Both demonstrate understanding and respect for past and present municipal expenditures. Yet neither presents any vision of the future metropolis. They are snapshots of city planning at that particular time in that particular city, and their effectiveness is simply a reflection of the effectiveness of local planning activity.

Harland Bartholomew and the St. Louis Comprehensive City Plan

The 1947 *Comprehensive City Plan* for St. Louis was the work of Harland Bartholomew, who had moved to St. Louis in 1916 to become its first city planning "engineer." He remained in that position until his retirement in 1950, devoting half his time to this municipal job and the rest to establishing and administering what was for decades the country's preeminent private consulting firm for city planning.[39]

Bartholomew, who was born in 1889 and died 100 years later, started his career in 1912 as a civil engineer in the New York City office of E. P. Goodrich. The just-established Newark City Plan Commission had hired Goodrich and architect George B. Ford to prepare a comprehensive city plan. Bartholomew, who like Goodrich and Ford had never worked on such a plan, was assigned to the job. Despite his inexperience and relative youth at 23, Bartholomew's role was so important that the resulting report, *City Planning for Newark,* lists him as "assistant engineer" directly beneath "expert advisors" Goodrich and Ford.[40]

Bartholomew was different from many of his pioneering colleagues because he believed effective city planning required total immersion in the locality and direct involvement with municipal government. This approach grew out of his work in Newark, where he had experienced the suspicion with which visiting "experts" are viewed by full-time civil servants. As a result, he jumped at the opportunity of becoming the City Plan Commission's engineer in 1914, when its contract with Goodrich and Ford expired. The position made him the nation's first full-time, salaried, professional planner on any municipal payroll. More important, it underscored for him the importance of being part of the bureaucracy entrusted with deciding on and then implementing any recommendations. Bartholomew moved to St. Louis two years later, before much of his work could be brought to fruition. However, his experience in Newark provided the basis for his work in St. Louis and hundreds of other cities.

Three years after moving to St. Louis, he established Harland Bartholomew and Associates, a city planning consulting firm, which he directed for 42 years. Because he lived in and worked out of St. Louis, he could not provide other cities that contracted for his services with the same intimate involvement. Consequently, key personnel moved to any city that retained his firm. They had to remain there for the term of the contract, which was usually three years. Once relocated, they worked with citizen advisory committees that both reviewed their work and organized political support.

Each plan was based on a standardized approach that had been devised by Bartholomew, who also was responsible for

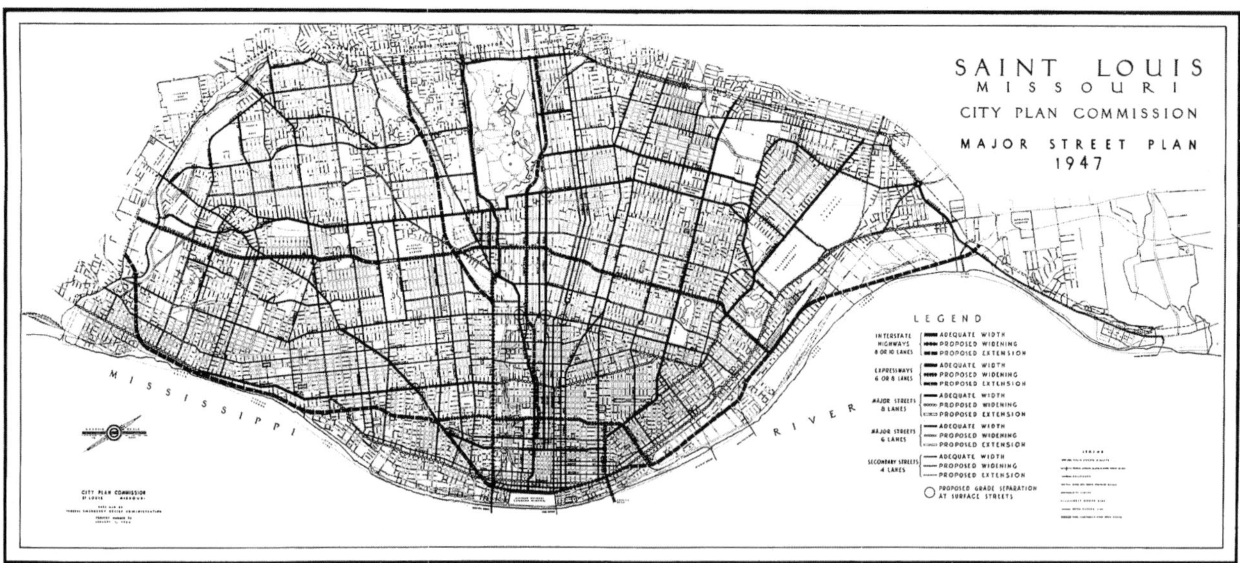

St. Louis, 1947. Highway Plan proposed by Harland Bartholomew in the *Comprehensive City Plan* indicates proposed widening and extensions of interstate highways, expressways, eight-lane major streets, six-lane major streets, and four-lane secondary streets. *(Courtesy of Art and Architecture Library, Yale University)*

its basic strategy and contents. It usually included sections on park and recreation facilities, streets and highways, transit, water and rail transportation, sewers and water supply, city appearance, zoning, the legal and financial aspects of the city plan, and (sometimes) housing. Each of these components was also standardized. For example, the section on streets and highways included typical cross sections illustrating the proper dimensions and layout for a range of arteries. Between 1920 and 1926 the firm produced 20 of the nation's 87 comprehensive city plans. Its foremost competitor, John Nolen, was responsible for only 12.[41]

The firm was no less successful when it came to implementing recommendations. Bartholomew was not interested in merely producing documents. He wanted to institutionalize city planning as an effective component of municipal government. Thus, "if it was the wish of the contract city to retain its Bartholomew representative on a permanent basis (as it often was) and the latter was agreeable, Bartholomew acted as a kind of sponsor for the union."[42] As a result, 408 cities, especially in the Middle West, employed planning officials who had started their careers with Harland Bartholomew and Associates.[43]

The approach to planning espoused by Bartholomew and his followers became known as the *City Efficient* or *City Practical,* in contrast to the *City Beautiful* approach ostensibly practiced by Burnham, Bennett, and Bogue. The term may have evolved because of the many engineers attracted to the profession, or because the comprehensive plan proposals that were implemented most frequently tended to include elimination of grade crossings, installation of water and sewer systems, and street widenings, or because practicality was easier to sell politically than esthetics. Whatever the reason, it is a misnomer. All professional planners of that era were committed to clear presentation of carefully collected data and scientific analysis of this information, restructuring of city government to improve its efficiency, and expansion of the bureaucracy to ensure better delivery of municipal services. The city plans they produced (whether called City Beautiful or City Efficient) included similar proposals for park and parkway expansion, civic center development, and street improvement.

By participating in municipal government, Bartholomew ensured the entrepreneurial element needed for plan implementation. His success in St. Louis, however, came from concentrating on the money his recommendations required. Between 1916, when he prepared the first report on the city plan, and 1947, when he produced his last major plan for St. Louis, Bartholomew guided the development and expenditure of $157 million in bond issues. As a result, virtually all the projects he proposed came to fruition, including the establishment, paving, and widening of streets, bridges, and viaducts; the construction and reconstruction of water and sewer systems; the creation of major public plazas, parks, and playgrounds; the removal of countless railroad grade crossings; and the modernization, enlargement, and construction of public hospitals, firehouses, schools, garages, and even airports.[44]

By the time the 1947 *Comprehensive City Plan* was issued, Bartholomew had developed the same deep understanding of St. Louis that Burnham had of Chicago. Rather than inspire readers with European planning paradigms and gorgeous renderings of his vision of the future metropolis,

Bartholomew presented quantifiable information describing St. Louis. This included graphs showing current and projected population, by decade, between 1800 and 1970, tables presenting the amount of land devoted to various uses (also by decade) between 1915 and 1945, maps of existing and proposed facilities, and charts indicating existing dimensions and proposed changes to local streets, major arteries, and interstate highways. It was all intended to convince civic leaders, government employees, and voters that the recommendations made in the plan were unavoidable given the probable course of city development.

The *Comprehensive City Plan* assumed that by 1970 the city would have grown by 10 percent and then listed the improvements needed both to accommodate this growth and to catch up with the unavoidable neglect that occurred during the Depression and World War II. Major recommendations included revision of the zoning ordinance; construction of express highways; development of extensive off-street parking facilities to eliminate downtown congestion; provision of a citywide system of neighborhood parks, playfields, and playgrounds; clearance and redevelopment of slum districts to provide decent housing for low- and moderate-income families; and creation of 35 airfields (including three major airports).

In fact, the population of St. Louis dropped 24 percent, from 816,000 in 1940 to 622,000 in 1970, and was just 319,000 in 2010.[45] Nevertheless, most of Bartholomew's recommendations were implemented. The city continued with site assemblage and development of the Jefferson National Expansion Memorial to be created along the waterfront. It continued work on the Memorial Plaza Civic Center (see Chapter 5). It condemned and cleared a series of "blighted areas" that became public housing and redevelopment projects. It built limited-access highways that had been recommended both along the waterfront and into the suburbs. The only major element in the plan that was ignored was its proposed 35 airfields. Bartholomew had completely misjudged the future of commercial and private aviation.

Bartholomew possessed neither Daniel Burnham's artistic genius nor his inspirational vision. However, by institutionalizing the planning function within city government, he was able to develop an equally deep involvement with his city and to persuade its citizens to spend the large sums needed to implement his proposals.

The Plan for New York City

Comparing the 1947 St. Louis *Comprehensive City Plan* with the 1969 *Plan for New York City* is like comparing a pocket dictionary with an encyclopedia. The *Plan for New York City* is a 25-pound boxed set of six oversized volumes. Unlike Bartholomew's work, it did not reflect a 30-year commitment to urban planning as an integral function of municipal government. The *Plan for New York City* was produced because the newly elected Lindsay administration decided to use the City Planning Commission to spearhead fundamental reform of city policies and programs.[46]

Although the 1938 city charter had established a City Planning Commission and required it to prepare a comprehensive city plan, none had ever been published. More important, the federal government was threatening to cut off housing and urban renewal subsidies that, since 1959, had required such a document. The absence of a comprehensive city plan thus became the excuse for producing a document that presented the administration's policies and programs.[47]

Mayor John V. Lindsay appointed one of his closest associates, Donald H. Elliott, to be chairman of the City Planning Commission. During the first 11 months of the administration, Elliott had been counsel to the mayor. Before that, he had been director of research for the Lindsay campaign and a member of the mayor's law firm. For the first time in its history, the City Planning Commission was to be headed by an individual who had unusual rapport with the mayor and broad sanction to develop administration policy. Upon taking office, Elliott rashly promised to release the much-demanded comprehensive plan. When he was given a copy of the existing draft, he junked it and exploited the situation to reorganize the Department of City Planning, to refocus its activities, and to set forth publicly the planning initiatives of the new administration.[48]

Up to that time the Department of City Planning had been a small, insular agency located in one of Lower Manhattan's faceless office buildings. Elliott quickly opened offices in each of the city's five boroughs, established an urban-design group to propose major planning initiatives, and added expert professional staff who worked with operating agencies on such citywide issues as housing, mass transit, economic development, and the environment. The borough offices were responsible for helping to divide the city into community planning districts, provide the forthcoming *Plan for New York City* with individual chapters on each district, and maintain ongoing liaisons with these districts and their appointed community planning boards.[49]

Produced under the direction of Edward Robin, the administration's new director of comprehensive planning, the *Plan for New York City* included one volume devoted to critical issues facing the city and individual volumes on each of the city's five boroughs. The borough volumes are mini atlases, with sections on the borough as a whole, on each planning district, and on areas of special concern, such as the Model Cities neighborhoods (see Chapter 12). The district sections (which were also published separately so they could be distributed to residents) included (1) an aerial photograph; (2) photographs of characteristic neighborhood scenes; (3) color maps presenting the city's land use policy based on the zoning; (4) maps locating *all* community facilities, public and

New York, 1969. The *Plan for New York City* proposed to direct 60 percent of the city's resources to major action areas and preventive renewal areas. *(Courtesy of New York City Department of City Planning)*

publicly assisted housing, transit stations, urban renewal and historic districts, and scheduled capital budget expenditures; (5) charts detailing public school utilization and enrollment, hospital and nursing-care facilities and capacity, size and apartment distribution of all public and publicly aided housing projects, scheduled capital construction projects, subway travel times, and listing virtually all other public facilities; (6) tables presenting population characteristics by age, racial composition, household size, and income; and (7) a text describing the area, its history, its population, and the planning issues it was facing. No American city government had ever before printed a document that presented such comprehensive information on all its neighborhoods. Not until 1988 and 1991, when the Cleveland City Planning Commission issued the two volumes of its *Cleveland Civic Vision 2000,* would any city government produce a document of similar scope.

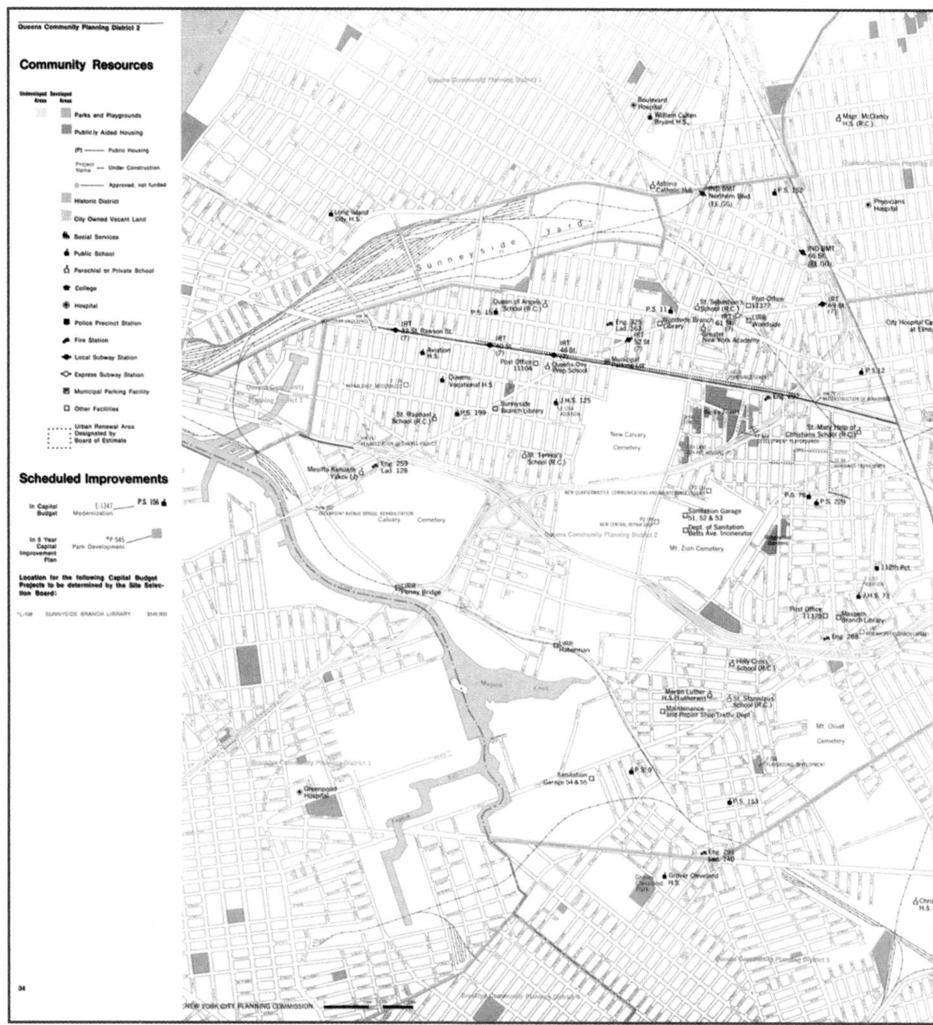

Queens, 1969. Map of community resources and scheduled capital improvements for Community Planning District 2. *(Courtesy of New York City Department of City Planning)*

The Critical Issues volume also included dozens of additional maps, graphs, charts, and photographs. It set forth the Lindsay administration's strategy for reinforcing the city's role as the national center, increasing opportunities for self-realization, improving the city as a living environment, and making city government more responsive. William H. Whyte, doing essentially what Moore had done for Burnham, was its unacknowledged editor.

Comprehensive city plans had always concentrated on improving the environment. The *Plan for New York City* was different in reflecting local concern for the three other critical issues: the national center, opportunity, and government. The plan concentrated on the national center because its authors rightly recognized that it was the engine that would enable the rest of the plan to be implemented—generating the money that made everything else possible. By advocating programs and policies that encouraged expansion of national-center activities, the Lindsay administration hoped to generate economic expansion and therefore the taxes to pay for additional services in poverty areas. The focus on increasing opportunity grew out of the city's historical role in providing immigrants, minorities, and especially the poor with a chance to make better lives for themselves. Here again, by emphasizing business expansion in Manhattan, the administration hoped to create additional employment and thereby additional opportunity for its citizens. The attention to government reflected the administration's profound commitment to citizen participation and decentralization in the delivery of municipal services. This was not an attempt to balance the administration's commitment to the national center. It was an attempt to alter government by an entrenched civil service, which Lindsay believed was unaccountable to its elected officials, geographically separated from the neighborhoods that it served, and, as a result, unresponsive to its residents.

Comprehensive city plans usually presented public expenditures without offering priorities. The *Plan for New York City* proposed a strategy for selecting priorities. It separated the city into three general categories: *major action areas* were essentially concentrations of poverty where the plan proposed to direct 60 percent of the city's resources; *preventive renewal areas* were those neighborhoods that could go downhill very fast without additional investment and where it proposed to concentrate 30 percent of the city's resources; and *sound areas,* which the plan assessed to be in relatively good shape and would need only 10 percent of the city's resources. (See Tables 19.1 and 19.2). Other comprehensive city plans promised a better city once their proposals were implemented. The authors of the *Plan for New York City* made no such promise. Instead, they wrote:

> *We are . . . optimistic. But we are also New Yorkers. We cannot see Utopia. Even if all of [our] recommendations were carried out, if all the money were somehow raised, ten years from now all sorts of new problems will have arisen, and New Yorkers will be talking of the crisis of the City, what a hopeless place it is, and why does not somebody do something. Our hope for this Plan is that it will help give them good choices to make.*[50]

In order to avoid controversy, the *Plan for New York City* was not released until after Mayor Lindsay's reelection in 1969. The delay did not prevent a storm of criticism. The document became a lightning rod for middle-class resentment of the administration's focus on the poor, especially African Americans, Puerto Ricans, and other minorities. When the commission scheduled public hearings in each of the 62 community planning boards, it was followed from place to place by irate citizens who opposed not only siting public housing projects in "decent neighborhoods," but also civilian review of police practices, community control of ghetto schools, and a variety of other Lindsay policies presented in the document.

The plan was also opposed by the city bureaucracy, which was not at all interested in shifting its priorities or opening itself to community review. Furthermore, elected officials refused to reallocate funds ostensibly being allocated to provide all citizens with municipal services in order to compensate for the greater needs of poverty areas. Given these obstacles, there never was any chance of concentrating 60 percent of the city's resources in those areas that the City Planning Commission designated for "major action."

Civic groups and professional organizations, impatient with the voluminous data on neighborhoods that did not concern them, complained that there was no vision for the city's future. They wanted elaborate renderings of their favorite projects. Without these projects, there was no way to obtain their support and therefore no way to develop the necessary citywide constituency. While the document was a treasure trove of data, it did not include the necessary graphic representation or vision—no matter how trite or contrived they might be—to make it excite the public the way that Burnham's plan had generations earlier in Chicago.

The *Plan for New York City* was quietly abandoned during the last year of the Lindsay administration. Nevertheless, it has continued to have substantial, though largely unnoticed,

Table 19.1

SCHEDULED CAPITAL IMPROVEMENTS[51]

Number	Description	Estimated cost, $
E-789	New West Queens High School	14,700,000
E-1265	Addition to Intermediate School 126	550,000
E-1327	Renovation to Intermediate School 126	1,289,000
F-16	New firehouse for engine company 262	567,000
HW-147	Rehabilitation of Queens tunnel viaduct	6,125,000
HW-163	Paving of Hazen Street	650,000
HW-182	Paving and repaving of 31st Avenue	715,000
L-159	New East River Branch Library	731,420
P-127	New playground at Public School 17	(lump sum item)
P-127	New playground at Public School 76	(lump sum item)
PO-106	New 114th Precinct police station	2,440,000
PW-18	Bower Bay pollution control plant, 1st and 2d stages	49,093,000
PW-133	Storehouse	49,093,000
PW-186	Modernization of old county courthouse	197,000
PW-231	Rehabilitation of Queensboro Bridge	2,800,000
PW-237	Bower Bay pollution control plant, 3d stage	32,000,000
T-65	East 63rd Street—41st Street subway route	37,663,360
TF-446	New parking facility (Queensbridge Plaza)	6,125,000
TF-452	Parking field (Steinway Street and 31st Avenue)	90,000
HS-32*	New ambulatory health services station	1,289,000

*Queens Community District No. 2.

TABLE 19.2
PUBLIC LIBRARIES*[52]

Name and address	Number of Books			Hours open weekly
	Total	Adult	Child	
Astoria: 14-01 Astoria Blvd.	26,862	14,602	12,260	40
Broadway: 20-40 Broadway	63,641	43,178	20,463	50
Queensbridge:10-43 41st Street	13,344	6,922	6,422	30
Ravenswood: 35-32 21st Street	24,392	14,689	9,703	40
Steinway: 21-45 31st Street	34,986	22,952	12,304	47

*Queens Community District No. 2.

impact. Billions of dollars have been spent on the development of Roosevelt Island, the Javits Convention Center, the North River Pollution Control Plant, the redevelopment of downtown Brooklyn, and countless housing projects, all initiated during Elliott's chairmanship of the Planning Commission and all advocated by the plan. The Beame administration, which took control of City Hall in 1974, made "preventive renewal" part of the city's housing strategy when it implemented the Neighborhood Preservation Program (see Chapter 13). In 1979, when the Koch administration incorporated the presentation of planning data by the community district into the annual budget process, the City Planning Department simply institutionalized techniques used in the *Plan for New York City.* Although city agencies now provide substantial amounts of local and citywide data on their websites, more than four decades after publication, the *Plan for New York City* is still the only comprehensive atlas of the city of New York.

Like Bartholomew's plan for St. Louis, the *Plan for New York City* affected its city's future because it provided a snapshot of ongoing planning activity that had been initiated years prior to publication and because it was the product of its authors' determined effort to institutionalize city planning as an ongoing municipal function. Also like the *Comprehensive City Plan* for St. Louis, however, it influenced only the city for which it was prepared and had no impact on the course of American city planning. A mass of data without a noble, logical diagram just does not stir the blood.

Planning as a Continuing Process

Once the World's Colombian Exposition and the *Plan of Chicago* had demonstrated the possibility of successfully planning an entire city and once virtually every municipal government had institutionalized city planning as a core municipal function, civic leaders faced new and continuing demands for participation in the planning process. These demands came from developers who were unhappy with existing zoning, from neighborhood groups who wanted a role in determining the future of their communities, from architects who demanded to remake the landscape to fit their vision of a better city, from civic organizations unhappy with existing government priorities and budget allocations, from political leaders who felt they should participate in any municipal decision making, and from city residents who just wanted to improve their living conditions.

Professional planners were initially suspicious of any interference. Outside consultants hired to prepare comprehensive city plans questioned intrusion by any party who lacked the necessary "expertise." Planners working inside government agencies were not at all eager to jeopardize their role in determining what property owners were permitted to do with their land and buildings or how government's own projects would be shaped. Neither group could hold out for long. People who want to solve their own problems and spend their own money usually find ways of doing so.

Every day in every city people launch proposals that do not originate within the planning profession. They sidestep strategies for municipal expenditure and visions of the good city in order to proceed with those proposals that have sufficient political support to engender action. This widespread citizen participation in the planning process need not produce sad results. The trick is to involve all the participants in an ongoing city planning process, but one that is nonetheless guided by a strong vision. This is what happened in Philadelphia in the 1940s, 1950s, and 1960s under the guidance of its planning director, Edmund Bacon (see Chapters 2, 6, and 12) and takes places wherever the planning process is successful.

Reinventing Downtown Philadelphia

Demand for a comprehensive reconstruction of Philadelphia originated during the Great Depression. Having agreed to refinance the city's debt during a period of plummeting tax collections, banks had imposed drastic spending restraints. Meanwhile, the corrupt Republican machine that ran the city continued its outmoded, patronage-ridden, but politically profitable system for deciding where and how municipal services would be allocated. This combination of limited funds and inefficient government resulted in continuing deterioration of the city's capital plant and demands for greater levels of public investment.[53]

Planning, slum clearance, and housing reform had been dominated by the Philadelphia Housing Association since its creation in 1909. During the late 1930s, a group of reformers, self-described "young Turks," entered the fray. This informal collection of young business leaders and professionals concentrated on charter reform and opposition to the Republican machine.

They were defeated at the polls in 1939, only to regroup

as the City Policy Committee, this time concentrating on urban planning. After considerable lobbying, they obtained the mayor's approval for the creation of a powerful planning commission. The commission was formally established by a unanimous vote of the City Council in 1942. At their suggestion, the mayor appointed Robert B. Mitchell as its first director and agreed to focus the commission's efforts on demographic and topographic analysis, the condition of the city's capital plant, and mass transit.

In order to generate additional support and monitor the new commission, reformers organized the Citizens' Council on City Planning in 1943. One year later, the Fairmount Park Association proposed clearing the slum just north of Independence Hall and replacing it with a formal, landscaped mall. The Center City Residents' Association, formed in 1946, demanded a major effort to fight downtown housing deterioration and to prevent residential displacement by commercial land uses that were incompatible with neighboring residences.

All these demands for action coalesced in 1947 at the Better Philadelphia Exhibition. The show was sponsored by a blue-chip committee organized by Edward Hopkinson and Walter Philips, respectively chairmen of the Planning Commission and the Citizens' Council. They raised $400,000 to mount the show and persuaded Gimbels Department Store to house it at no charge. The idea, as articulated in the official exhibition brochure, was "to dramatize city planning—to gain the confidence of a public made cynical by utopian futuramas and the inertia of local politicians."

During the two months that the show as open, it attracted more than 385,000 visitors, who were presented with aerial maps, drawings, cartoons, a diorama, lights that went on and off, and all manner of bells and whistles. The show's appearance had been conceived by architect Oscar Stonorov, and it's content was a true expression of the ideas, concerns, and interests of all the proponents of city planning in Philadelphia. The most impressive of the displays was a 30-foot-long model of downtown Philadelphia with sections that flipped over to illustrate how the Philadelphia of 1947 could be transformed into a much better Philadelphia by implementing the proposals outlined at the exhibition. It included Independence Mall, Penn Center, Society Hill, and many of the other planning projects that were to be executed over the next few decades.

One year after the exhibition closed, yet another civic organization, the Greater Philadelphia Movement, was established. Its purpose was to bring together the city's 35 most powerful business executives as a force for municipal improvement, to fight for charter reform, and to implement the projects launched at the Better Philadelphia Exhibition.

The man who would synthesize these disparate citizens groups and ideas for a better Philadelphia was Edmund Bacon. Bacon received an architecture degree from Cornell and settled in Philadelphia, where he quickly became one of the young Turks. He left to study design and planning with Eliel Saarinen at Cranbrook Academy in suburban Detroit. Following completion of his studies, he worked for three years as a planner and housing expert for the city of Flint, Michigan, before returning to Philadelphia. In 1941, Bacon

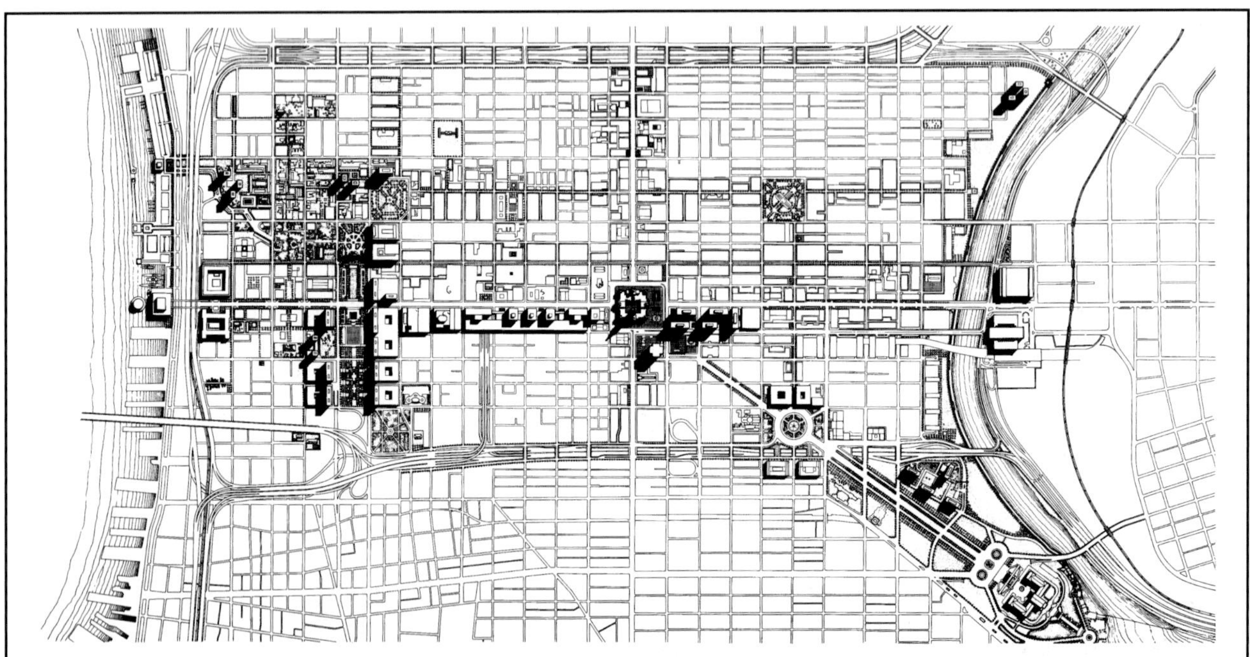

Philadelphia, 1960. Redevelopment of Center City as proposed in the comprehensive plan. *(Courtesy of Department of City Planning, Philadelphia)*

became managing director of the Philadelphia Housing Association. It was a position that enabled him to bridge the gulf between traditional housing reformers and the rapidly growing number of advocates for other planning activity. It was also perfect preparation for his appointment in 1949 as executive director of the Planning Commission.

The year 1949 proved critical for planning in Philadelphia. The reformers and their agenda were endorsed at the ballot box. Joseph S. Clark Jr. was elected city controller, and Richardson Dilworth was elected city treasurer. A Charter Commission was appointed. In 1951, it proposed and obtained voter approval for a strengthened mayoralty and a powerful planning commission that would tie together physical and fiscal planning. The new planning commission would be responsible for enacting zoning regulations, approving housing and redevelopment projects, coordinating transportation and park development, and preparing an annual capital construction budget and a six-year capital improvement program.

When Clark was elected mayor in 1951, he had a ready-made planning constituency, a generally agreed-upon planning strategy, a planning commission with all the powers needed to initiate it, and, as time would prove, a planning director who was uniquely qualified to coordinate its implementation. In an act pregnant with symbolism, Clark had city hall scrubbed. He also replaced gas lamps with electric lights, regularized street cleaning, modernized garbage collection, established an $80 million sewage-treatment system, and inaugurated the $150 million capital program proposed by the planning commission.

Bacon, who was executive director of the City Planning Commission until 1970, has described his approach to urban planning as "the painful search for form." In fact, it was a continuation of a participatory process that had long been under way. Each of the projects on which he worked underwent what he called "democratic feedback." Each was forced to change, and change often, until the project satisfied the requirements of the consumer, the institutions providing the financing, the developer, neighborhood groups, citywide civic organizations, local politicians, and (if the money were coming from Washington) federal regulations.

Independence Mall illustrates the complexity of any

Philadelphia, 1950. Aerial view of the approach to the Old Broad Street Station (popularly known as the "Chinese Wall") that became the site of Penn Center. *(Courtesy of The Free Library of Philadelphia)*

Philadelphia, 2010. The absence of pedestrian traffic to and from the basement-level retail concourse at Penn Center has guaranteed that the open space between its buildings remains empty and uninviting. *(Alexander Garvin)*

participatory planning process. The project was initially conceived in 1932 by architect Roy Larson and adjusted by a group of civic leaders, who created the Independence Hall Association in 1942, to advocate federal acquisition and development of "a suitable approach to Independence Hall." It was further altered to meet the requirements of the National Park Service, which in 1952 agreed to finance the project. After it opened in 1965, it was augmented by buildings that were part of a contiguous federally assisted urban renewal project. The mall itself was relandscaped in 2001. Since then, the mall has been augmented by the Independence Visitors Center, a new home for the Liberty Bell, and the National Constitution Center and Museum.[54]

In contrast to Independence Mall, changes to Penn Center were more of a reflection of real estate practices, while the evolution of Society Hill was a reflection of virtually all the participants in the planning process (see Chapters 2, 6, and 12). In the case of Penn Center, Clark and Bacon began by persuading the Pennsylvania Railroad to vacate its Broad Street Station, demolish its viaduct, and make the site available for construction of the commercial center that had been envisioned at the Better Philadelphia Exhibition. Bacon's 1952 design proposed three similar city blocks organized around a continuous, 1400-foot-long sunken concourse that was open to the sky along its 150-foot width, except where it was bridged over by the city's street grid. The concourse was intended to provide an attractive, unobstructed, pedestrian connection between a suburban railroad station and the center of town. Each block was to be bounded on the north and south by retail shops that opened both to the concourse level below and to the street level above, while the eastern edge of each block was to be anchored by large office slabs.[55]

The scheme could not generate sufficient revenue to justify the high price that the Pennsylvania Railroad sought for the station. The project's developer, Robert Dowling, believed the project could be made feasible by increasing rental revenues and decreasing development costs. He proposed to eliminate the underground concourse, which he believed would be concentrated with money-losing shops, and to increase revenues by substantially enlarging the floor area devoted to office space by increasing the number of office buildings and relocating them to the longer north and south portions of each block. These revisions required pedestrians coming from public transit facilities to cross city streets.

Mayor Clark was willing to increase the office space, but he forced Dowling to restore the basement-level concourse. Without the continuous opening to the sky, the midblock space between office slabs became a shadowed alley that could attract only minimal pedestrian activity. The concourse ultimately became a maze of tunnels, interrupted only by those limited number of spots where small courtyards with staircases to street level provided pedestrian access and a bit of natural light. Even though the compromise was poor urban design, it enabled the project to proceed. More important, it triggered private construction of dozens of office buildings on neighboring blocks.

Society Hill required similar compromises. The original plan did not anticipate that the Federal Aid Highway Act would finance the city's construction of an inner expressway loop along the edge of the neighborhood by the Delaware River. In 1957, the Planning Commission hired architects Larson, Stonorov, and Kling to prepare a redevelopment plan that adjusted to these new realities. The following year the plan was again revised because it did not meet the requirements for greater rehabilitation assistance being requested by downtown homeowners. When sites were finally offered for development, the design was changed again to meet the marketing objectives of the winning team, developer William Zeckendorf Sr. and architect I. M. Pei.[56]

The nation's most sophisticated and complex planning process had been under way for two decades when, in May 1960, the City Planning Commission, under Bacon's leadership, issued a formal planning document entitled *The Comprehensive Plan of the City of Philadelphia.* It presented an

Philadelphia, 2010. Market Street was completely rebuilt by individual property owners once Penn Center replaced the "Chinese Wall." *(Alexander Garvin)*

analysis of the existing conditions and desired goals in 103 pages of maps, charts, and text. There were concise discussions of the city's history, population, economy, industry, commerce, recreation, community facilities, residence, transportation, and land use. The presentation of transportation, retailing, parks, and everything else reflected a logical hierarchy stepping upward in size from the smallest neighborhood phenomenon, past the community level, the district level, the regional level, till the hierarchy came to its peak at the center-city level. The plan tried to provide every citizen with at least a minimum level of service. It identified what was missing at each level and recommended action. It also proposed channeling additional growth to the center city and to major high-density corridors. Any similarity to the complex, participatory planning process that produced Penn Center or Society Hill, however, was purely coincidental. It was a snapshot of what was at that time under discussion as part of the planning process. As Bacon later confessed, the real reason it was produced was to satisfy HUD requirements for federal urban renewal funding.[57]

Over the next decade, the commission issued dozens of amplifying documents that stated that they were "in conformity with the Comprehensive Plan." There were major district plans (for Northwest and West Philadelphia), small but detailed redevelopment area plans, and functional plans, such as the 1968 *Comprehensive Plan for Swimming Pools* (a ludicrous contradiction in terms). Perhaps these documents were needed as part of the capital programming process or to meet specific federal requirements. However, with the exception of the redevelopment area plans, they bore little relation to the tremendous changes Edmund Bacon was effecting on the cityscape. The area around Penn Center was becoming the city's premier office district, and the area around city hall was once again a municipal asset. In an attempt to balance Penn Center, a second, combination commuter-railroad, transit, retail, office complex was taking shape along East Market Street. Additional redevelopment projects (perhaps not as exquisite as Society Hill) were under way, and interstate highways were spreading across town.

Only during the late 1960s, when community opposition to highway construction had become a national phenomenon, did any part of Bacon's "total vision of the city" face defeat. In 1970, the 2.8-mile South Street link in the center-city loop highway was stopped by a coalition of environmentalists who hated highways, African Americans who feared displacement, and liberals who opposed creating a "Mason-Dixon line" that would separate upper-middle-class whites from the African Americans living south of the proposed loop.[58]

Edmund Bacon had greater impact on the planning and development of his hometown than any individual except Robert Moses in New York and Daniel Burnham in Chicago. Like Burnham, he had created a noble, logical diagram that had the power to stir the blood. Like Bartholomew, he had institutionalized the planning function in government and then exploited the power it brought. *The Comprehensive Plan of the City of Philadelphia* had little to do with that impact. Bacon's extraordinary effectiveness was the product of the planning process he helped to create.

By remaining an integral part of Philadelphia's planning process for three decades, just as Bartholomew had in St. Louis, Bacon himself became the focus of the city's "collective consciousness." By constantly adjusting to "democratic feedback," he met the needs of the other participants in the planning process. By seeking and employing resources from wherever they might be available, he implemented the collective "total vision of the city." This "patient search for form" had the same dramatic impact on the course of development in Philadelphia that Burnham's *Plan of Chicago* had on that city in the preceding 50 years.[59]

The transformation of downtown Philadelphia has continued into the twenty-first century, though now it is spearheaded by Paul Levy, who has been the founding CEO of Center City District (CCD) since 1990, when the business improvement district was created (see Chapter 8). Levy may not play as dominant a role as Bacon did, but there are many similarities in their approach to changing Philadelphia. Like Bacon, he works hard to learn everything he can about this changing city and makes sure that the information is shared with everybody. The CCD's publications and websites, like the city planning documents that Bacon produced, provide regularly updated data on the health of downtown Philadelphia. Providing this common base of information is the first requirement of building a consensus around any planning initiative. Like Bacon, he has developed a close working relationship with the city's real estate industry, business community, and political leaders. Also like Bacon, he adopts good ideas wherever they originate, promotes a persuasive vision that can rally support, and then adjusts and recalibrates it until he has created a constituency for important planning initiatives. Unlike Bacon, but like Haussmann, he manages an institution responsible for delivering services. In 2013, the CCD will spend $20 million cleaning streets, preventing crime, maintaining the streetscape, and supporting a wide variety of downtown activities.[60]

Philadelphia, 2005. Society Hill towers announced and began the rebirth of an entire downtown residential neighborhood. *(Alexander Garvin)*

The improvements to downtown during his tenure include the construction of the Pennsylvania Convention Center (1993–1994) and its expansion (2011); evolution of South Broad Street into an "Avenue of the Arts" (begun in 1992), and the creation of a park along the Schuylkill River. The principal project for which he personally can take credit is Dilworth Plaza in front of City Hall (2013), part of his remarkable effort to improve the pedestrian environment in downtown Philadelphia. As with Bacon, these initiatives all involved a myriad of other actors and the preparation of countless plans. But as Eisenhower explained, the essential ingredient has been the ongoing planning process.

Planning Portland, Oregon

Portland has a long history of city plans. In 1903, the Board of Park Commissioners asked John Olmsted to prepare a proposal for a major park and parkway system. After failing to attract Daniel Burnham to develop a comprehensive plan, the Civic Improvement League hired Edward Bennett, who produced the 1911 *Greater Portland Plan*. Seven years later, Charles Cheney issued a series of reports on housing and planning that provided the basis for the establishment of a planning commission in 1918, a housing code in 1919, and a zoning ordinance in 1924. This was followed in 1930–1932 by studies of transportation and land use by Harland Bar-

Portland, 1974. Harbor Drive along the Willamette River was replaced with Tom McCall Park. *(Courtesy of Oregon Historical Society)*

tholomew and in 1943 by Robert Moses's 85-page plan, *Portland Improvement.*

As in most cities, these documents provided support for some projects that were implemented. Most of this planning history, however, was junked during the 1970s and 1980s. By 1970, opposition to highways and urban renewal resulted in the election of a younger, more reform-minded city council. This group included Neil Goldschmidt, who, as mayor between 1972 and 1979, spearheaded the creation of the suburban light-rail system, pedestrian/transit mall, and waterfront park. The growing power of citizen action was amply demonstrated by the role played by "The Friends of Pioneer Courthouse Square" in forcing the city to proceed with this project and then raising money to finance it.

The commitment to city plans continued, but with a very different approach. It was embodied in a series of new documents: *Planning Guidelines—Portland Downtown Plan* (1972), *Comprehensive Plan* (1980), and *Portland Center City Plan* (1988). Unlike earlier plans, these documents were prepared by municipal civil servants, not by outside experts. They were policy statements issued by the Portland Planning Commission and adopted by the city council, developed over several years, and inclusive of public consultation. Consequently, they reflected the opinions and desires of the city's population more accurately than earlier documents prepared by outside consultants; for the same reasons, they included many more proposals that were likely to be implemented.

By the 1970s, virtually every agency in city and state governments included "city planners." There was no longer separation between functional agencies concerned with traffic or water supply, on one hand, and the city planning department on the other. More important, the thinking of the planners themselves had changed radically. Artistically determined visions of the City Beautiful and carefully engineered proposals for the City Efficient had been replaced by relativistic pluralism. Professional planners now saw themselves as part of an adversarial system in which they could responsibly make different proposals depending on which government agency, civic organization, or private developer they represented.

In 1967, Paul Davidoff, the most articulate spokesperson for this approach, wrote a widely read article proposing to empower citizens who were left out of the decision-making process. Government money would be appropriated to hire "advocate planners" for neighborhoods that could not afford to pay for professional services. While Davidoff's specific proposals were not implemented in Portland or anywhere else, they played a major role in shaping the character of further

Portland, 2007. Tom McCall Park. *(Alexander Garvin)*

Portland, 2007. The Tri-Met light-rail system. *(Alexander Garvin)*

planning in Portland, New York, and other cities.[61] A growing number of Portland's younger civil servants believed in widespread citizen participation, and not just because it reflected their planning philosophies. Citizen participation had become an integral part of the federal urban renewal programs that funded their salaries. Consequently, they had every reason to include residents in every aspect of their work.

Mayor Goldschmidt institutionalized community planning itself in 1974, when he obtained city council approval for the creation of the Office of Neighborhood Associations (ONA). Beginning in 1975, ONA coordinated the production of "neighborhood needs reports." It became as much a part of the city budget process as the community district budgets initiated by the *Plan for New York City.* As a result of this reshaped planning process, the number of active neighborhood groups in Portland increased twofold between 1974 and 1979, from approximately 30 to nearly 60.[62]

Portland provides a vivid demonstration of the continuing impact of planning. The two redbrick pedestrian transitways and the 27-stop light-rail system that the city installed during the 1970s and 1980s transformed the area around Pioneer Courthouse Square into the most convenient spot in downtown Portland—perfect for retail shopping. The response from the private sector was both predictable and impressive. The Rouse Company acquired the nearby Olds & King department store and converted it into "The Galleria," a 75-foot-high atrium surrounded by a variety of restaurants, cafés, and retail stores. Nordstrom built a new store facing the square, and Saks Fifth Avenue added a store as well. The block adjacent to Saks was rebuilt as Pioneer Place, a multistory, air-conditioned atrium with shops, restaurants, and tourist-oriented retail outlets.

The evolution of the civic center, the cultural center, McCall Park, and RiverPlace in the 1980s and 1990s, as well as the opening of the Westside Max light-rail, furthered the city's transformation. Four decades have passed since the publication of the *Portland Downtown Plan* of 1972. During that time Portland has been the site of more than 80 residential developments (new construction and rehabilitation). The city's downtown office inventory increased from 5.7 million square feet in 1972 to 20.5 million square feet in 2010, while city employment grew from 156,000 in 1970 to 292,000 in 2010. An estimated 7.9 million tourists came to Portland in 2011, generating more than $3.8 billion in direct spending. These statistics demonstrate the effectiveness of Portland's planning process.[63]

Portland, 2007. In 1979, the city acquired and demolished a garage in order to create Pioneer Courthouse Square. The following year, it held a design competition which attracted 162 submissions. The result was a major new civic space right in the center of downtown. *(Alexander Garvin)*

Planning in Baton Rouge, Louisiana

The commitment to planning in Baton Rouge is more recent than in Philadelphia or Portland. It began when local citizens commissioned *Plan Baton Rouge*, prepared in 1998 under the guidance of Duany/Plater-Zyberk & Company (see Chapters 14, 16, and 17). It was followed by one waterfront plan prepared in 2003 by Michael Van Valkenburgh Associates with Dana Nunez Brown, and another drafted in 2006 by Hargreaves Associates. They were followed by *Plan Baton Rouge II*, prepared in 2009 by the Cambridge planning firm of Chan Krieger Sieniewicz. All these documents set forth general goals and presented desirable projects that were advocated by city leaders determined to improve Baton Rouge.

When *Plan Baton Rouge* was initiated, this city's downtown should have been thriving. As the capital of Louisiana and the site of the nation's largest oil refinery, Baton Rouge had been a growing city with a healthy economy for the entire twentieth century. Between 1950 and 1990, when many older cities were in decline, the population of Baton Rouge had increased 75 percent, outside the center city. The city had invested in the usual prescriptions for maintaining a healthy downtown, including everything from a semipedestrian shopping street and a convention center to casino gambling. None of these measures had been enough to reverse the steady loss of customers. Consequently, at the end of the twentieth century, downtown Baton Rouge was filled with empty stores and vacant buildings.[64]

Since the usual recommendations for urban revitalization had failed, the Baton Rouge Area Foundation, the state of Louisiana, and the parish of East Baton Rouge decided to try an old remedy: comprehensive planning. In late 1997, they issued a request for proposals. The winners were Duany Plater-Zyberk (architects and town planners), Robert Gibbs (retail consultant), Walter Kulash (traffic consultant), and me (implementation consultant).

DPZ agreed to coordinate a state-of-the-art New Urbanist planning charrette modeled on the 157 similar events it had previously managed (see Chapter 16). Its objective was to provide material for a comprehensive downtown plan that would be submitted to the city council for approval. Thereafter, its progress would be monitored by a small organization established to promote the plan.

During the four months prior to the charrette, Duany, Kulash, Gibbs, and I traveled separately to Baton Rouge to study the situation. We delivered public lectures that were covered by local newspapers and television. We met with many of the people who would be participating in the planning process. By the time we returned for the charrette, in the last week of June 1998, there had been so many meetings and news stories that the entire city was ready to participate in the planning process.

The team of professionals that came to the charrette included 14 members of the DPZ staff, 5 local designers, 2 transportation consultants, 3 retail consultants, and me. The charrette began with a public lecture delivered by Andrés Duany to an overflow crowd in the Old State Capitol, a national landmark that was first occupied in 1849. The next day the team returned there with copies of old maps, drawings, photos, and previous studies of Baton Rouge. DPZ brought computers already loaded with useful planning materials, including prototypical building designs for everything from parking garages and town houses to multiplex cinemas and farmer's markets.

Everybody was invited to propose ideas. As soon as an idea was presented, the team prepared computer and freehand drawings that examined its physical feasibility and what it might look like. Participating consultants took these schemes and examined their functional and financial feasibility. Then they were tested with property owners, interested developers, lending institutions, relevant city agencies, and interested local elected officials. Later on the same day Duany presented the results at a scheduled public review session.

By the time DPZ began its work in Baton Rouge, the firm had evolved into a well-honed town planning organization with a palette of downtown development techniques and a strong commitment to regional variation. Duany and Plater-Zyberk had developed a common urban design philosophy that included construction along the property line, "liner" buildings that eliminated the gaps in pedestrian shopping, multiplex cinemas that attracted customers at night when office workers returned home, narrower vehicular rights-of-way, and a variety of other devices intended to enhance the pedestrian environment. Many of these are techniques that have been employed in urban environments for centuries. As Duany repeated throughout the charrette, there is every reason to repeat and refine successful examples of traditional urbanism.

At the end of six days, we knew what might work and what definitively would not. The next day was spent preparing the materials we needed to present our plan to the public. We delivered our proposals to more than 700 people over a

Baton Rouge, 1999. *Plan Baton Rouge* identified the Auto Hotel, an empty garage, as an ideal site for an art center that would attract people from the entire metropolitan region to the city's then-somnolent downtown. *(Alexander Garvin)*

period of four hours. These included administrative actions, changes to existing projects, changes in government procedure, legislation, and proposals for public action and private development.

A large number of downtown projects had been approved but were not yet fully designed. For example, the state was scheduled to build two garages. We proposed that other public facilities be included at ground level: shops, a health club with day care facilities to be operated by the YMCA, and even a farmer's market. The most innovative public action called for converting a derelict garage into a cultural center. Proposals for additional private development included a multiplex cinema, renovation of existing hotels, new convenience retail, and new downtown housing.

Plan Baton Rouge was submitted to the city council in early 1999. By 2001, the city of Baton Rouge had adopted the plan's recommended traffic-calming proposal for River Road, planted 300 street trees at recommended locations, and commissioned a series of additional traffic studies to examine the impact of various other proposals. Based on the plan's suggested changes to its parking garages, the state of Louisiana entered into an agreement in which the YMCA would occupy the ground floor of the state-sponsored LaSalle Parking Garage that opened in 2001. At the state's other project, the Galvez Building and Garage, a public market was included on the ground floor. It also appropriated $4.2 million to pay for the transformation of the city's long-vacant Auto Hotel (garage) into the art center that was supported by the city's leadership and specifically included in *Plan Baton Rouge*. The $28 million building reopened in 2005 as the 125,000-square-foot Shaw Center for the Arts and includes the Louisiana State University (LSU) Museum of Art, the LSU School of Art Gallery, a 325-seat theater, classrooms, a restaurant, and a rooftop garden.

Property owners have steadily invested in improving historic residences in two downtown neighborhoods. As a result, there are now over 1200 residential units in the heart of the central business district and another 588 units currently under construction or in planning. In 2006, the long-abandoned Heidelberg Hotel was renovated at a cost of $72 million and reopened as the 290-room Hilton Baton Rouge Capitol Center. Five years later, $25 million was invested to transform historic Hotel King into an upscale boutique hotel. Construction of a new Hampton Inn and Suites Hotel on Lafayette Street will be completed in early 2013.

The initial public investments supported by *Plan Baton Rouge* continue to encourage business to return downtown. As of 2013, downtown Baton Rouge contains over 60 restaurants, 21 bars and lounges, 8 art galleries, and 5 indoor and outdoor concert venues. It is now home to close to 40,000 daily office workers and attracts 3 million visitors annually. During 2012 alone, Ameritas Technologies, the NOLA Media Group, and 16 other businesses moved downtown.

Andrés Duany and his team produced a plan that had crystallized deep-rooted ideas in this remarkable community. Once they had been advocated by the team that produced *Plan Baton Rouge*, projects like the Shaw Center gained credibility with public officials, property owners, and developers. Consequently, once *Plan Baton Rouge* was formally adopted, its implementation was not particularly controversial because the community had been invested from the outset of the planning process. The executive director of the plan, Elizabeth "Boo" Thomas, shepherded its individual proposals through a public review process involving every relevant person in the city. The most important *Plan Baton Rouge* supporter proved to be Louisiana Commissioner of Administration Mark Drennen, who went out of his way to have the state of Louisiana adopt all the proposals that applied to the state government.

The charrette organized by DPZ for *Plan Baton Rouge* was fundamental to creating momentum for the planning process. As in other DPZ charrettes, *Plan Baton Rouge* was covered daily on television and in the newspapers. Unlike other charrettes, however, this one involved virtually everybody who had any role in deciding what would happen downtown. This process produced a wealth of research on community preferences and ideas that otherwise would have required considerably more time to obtain, if it even could have been collected. It also reduced the chances of making mistakes or suggesting something that would have produced too much local opposition. More important, it generated the public support without which *Plan Baton Rouge* could not have been implemented. Once that support was in place, the city's unusually committed civic leaders, particularly Thomas, Drennan, and Davis Rhorer (Director of the Downtown Development District, its BID), were able to bring many of its proposals to fruition.

While the renovation and reopening of the Shaw Center and the Hilton Baton Rouge Capitol Center, along with the reorientation of River Road, brought real change to the city, few of the other proposals advocated in the DPZ Plan came to fruition. Its most important impact, however, was initiating the city's commitment to a planning process, which included the publishing of three important plans. The Van Valkenburgh Associates/Nunez *Master Plan for the Riverfront District* proposed creating a Riverfront District whose identity would emerge from an entirely rebuilt system of riverfront streets, sidewalks, streetlights, street trees, and landscaping, all of which would terminate at a waterfront esplanade along River Road. *Reconnecting with the River,* the Hargreaves "master plan concept" for the riverfront, went even further, proposing $245 million in capital expenditures just one year after Baton Rouge had begun dealing with the aftermath of Hurricane Katrina. It advocated a vision that was beyond the city, state, or federal government's pocketbook. Nevertheless, some of its ideas are still under consideration, and, as Burn-

Baton Rouge, 2012. When the Auto Hotel reopened as the Shaw Center for the Arts and the city invested in relandscaping the open spaces across the street, tourists and residents flocked to downtown Baton Rouge to attend the public events there. *(Courtesy of the Baton Rouge Downtown Development District)*

ham's plan proved a century earlier, many planning proposals require significant time before they are either affordable or palatable. *Plan Baton Rouge II* included general recommendations that by that time had become generally accepted by the nation's city planners: greening the city, encouraging mixed land use, improving connectivity among the various parts of the city, and supporting the downtown BID's efforts to make Baton Rouge cleaner, safer, and more attractive. *Plan Baton Rouge II* also proposed creating an entertainment corridor and a revolving housing loan fund, and establishing a city parking policy.

It is clear that documents did not produce desired changes to the city. Rather, it was public involvement in an ongoing, serious planning process led by nonprofit organizations dedicated to that process: the Baton Rouge Area Foundation, the Center for Planning Excellence, and the Downtown Development District (its BID). Paraphrasing Edmund Bacon, the planning process in Baton Rouge gave birth to a collective "total vision of the city" that provides a framework through which its citizens are pursuing their "patient search for form."

Ingredients of Success

Effective comprehensive planning, like that practiced in Chicago at the beginning of the twentieth century, Philadelphia in the middle of that century, and Baton Rouge at the start of the twenty-first century, provides an attractive alternative to formless growth. Few people today believe in such planning. The reason was identified in the 1947 brochure for the Better Philadelphia Exhibition: "a public made cynical by utopian futuramas and the inertia of local politicians." We must overcome that cynicism. But that is only the first step. It must be followed by activities that generate a constituency for comprehensive planning.

Daniel Burnham clearly understood the fundamental role of a constituency for comprehensive planning when he called for big plans that stir the blood. From time to time, great artists do appear. Like Burnham, they can provide the necessary "noble diagram." In those instances, the lucky city should grab this unique opportunity. As Burnham also understood, that noble diagram needs to be marketed. He worked with civic leaders to mount the single most effective lobbying effort anywhere for the implementation of a comprehensive plan.

Artistic genius is a scarce commodity that cannot be provided to hundreds of cities at the same time. Since third-rate talent just will not provide the necessary inspiration, proponents of comprehensive planning are more likely to succeed if they generate a constituency either within municipal government or from widespread participation in a public decision-making process. The latter does not, however, mean that planners defer all judgment to the public; the best publicly advocated plans always begin with a vision of some sort and then amend it based on participation from the public.

An institutionalized city planning process worked for Harland Bartholomew in St. Louis in the 1940s, just as it worked in Portland in the 1970s, because it simplifies and standardizes proposals so they can be understood by all the participants. It also worked when it was applied to the budget process in New York City in the 1960s. As long as a Harland Bartholomew or a Donald Elliott is there to provide leadership, the results are excellent. But when that process is leaderless, the results inevitably will be as chaotic.

Market

Perhaps the least understood and most important ingredient in comprehensive planning is its market. As so many authors of neglected and discarded plans discovered, asserting a "need" does not mean that people will desire it enough to pay the price. Nor does portraying a "better" city mean that the electorate will support it.

Proposing public action to satisfy a need or obtain a better city is futile if there is no demand for it. Too often, planners recommend public action without truly understanding its market context. Even as sophisticated a planner as Harland Bartholomew can misjudge the market, as he did in 1947 when he proposed 35 neighborhood airfields for St. Louis. All the more reason to avoid planning that is not market-oriented.

In St. Louis and Baton Rouge, a substantial portion of the market had shifted to the suburbs long before the city residents engaged in comprehensive planning. Planning for St. Louis, which lost half its population after publication of the plan, failed to reverse that trend. Planning for Baton Rouge, on the other hand, has begun the slow process of reviving its downtown business district. If a market exists, intelligent planning can shape its effect on the city. For example, Edmund Bacon accurately identified demand for office space in downtown Philadelphia. He also identified some of the impediments preventing businesses from remaining there: inconvenient access from suburban areas, the blighting influence of the "Chinese Wall," and inadequate residential opportunities for a middle-class labor force that wanted to walk to work. Penn Center and Society Hill successfully eliminated some of these impediments. The marketplace did the rest.

Location

It is not enough to identify proposed locations for transportation, recreation, housing, commerce, industry, and public facilities. Hundreds of cities have published elaborately colored maps that prescribe the correct location for every land use. Few if any of these maps have transformed the cities for which they were prepared. Those that have succeeded did so because they exploited or compensated for the characteristics of specific locations or increased proximity to desired facilities.

The *Plan of Chicago*, for example, identified Lake Michigan as a locational asset and proposed actions that would allow the city to benefit from that asset. Based on those recommendations, the city spent hundreds of millions transforming the lakefront into the location of choice for office and apartment buildings. Similarly, the plan identified the Chicago River as an obstacle to downtown commercial development and proposed straightening its course and building what became Wacker Drive. By following those suggestions, the city eliminated impediments to development along the river.

The *Comprehensive City Plan* for St. Louis proposed projects that would increase the desirability of locations throughout the city by placing public facilities near them. In the business district, it concentrated on highway access and parking; for residential neighborhoods, it proposed schools and recreational facilities; for the city as a whole, it advocated airports. Decentralizing airport locations was clearly a mistake, but the locational analysis of parking, schools, and recreation was of real benefit when it came time to make capital budget expenditures.

To be effective, a comprehensive plan must identify a few critical components and locate them so that they can become an interdependent combination that generates more than the sum of its parts. Portland achieved this result by strategically locating improvements to its circulation system, adding open space, and building new public facilities. The interdependent

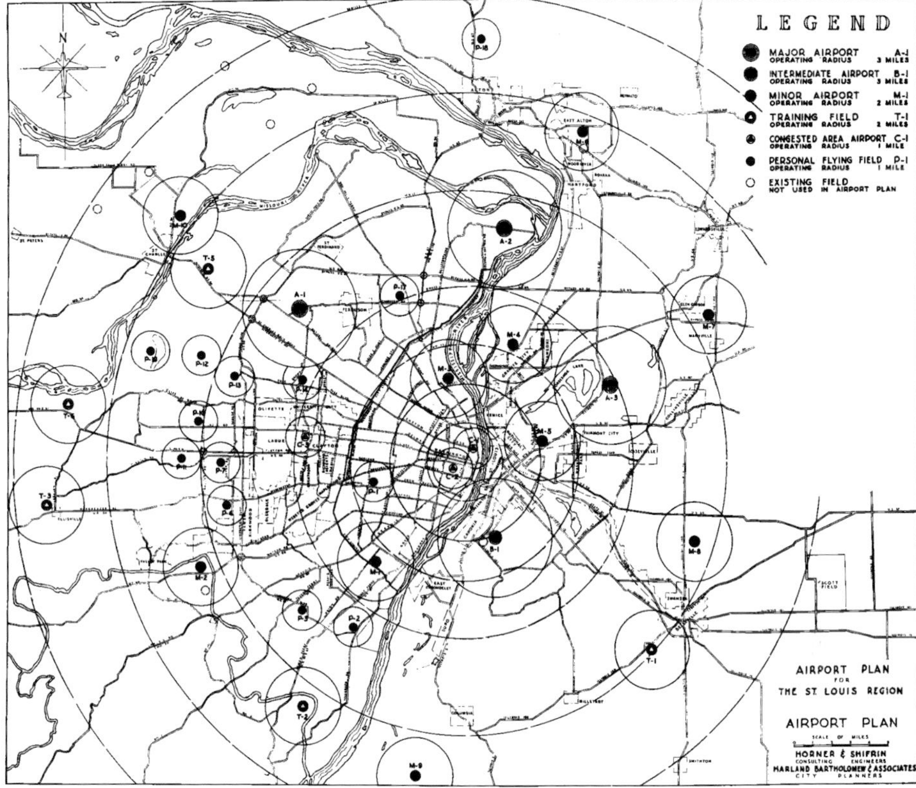

St. Louis, 1947. The airport locations proposed by Harland Bartholomew in the *Comprehensive City Plan* were based on the mistaken notion that air travel was a neighborhood rather than a regional service. *(Courtesy of Art and Architecture Library, Yale University)*

combination that was created attracted a large portion of the regional market to Portland's downtown office, retail, and tourist businesses.

Design

The power of a "noble, logical diagram" lies in its clarity, which, as Burnham so persuasively argues, "will be a living thing, asserting itself with ever growing insistency." This is also its liability. The entire planning profession was initially beguiled by Daniel Burnham's combination of Haussmann's Paris with the Chicago Fair. But whether it was applied to hilly San Francisco and or rainy Seattle, this scheme proved to be stillborn. Only when Burnham himself applied it to Chicago was any part of it successful, and then only because inappropriate components were ignored.

Another approach is to tailor comprehensive planning to the topography, history, and physical development of an existing city. Edmund Bacon took this approach in planning for Philadelphia. But repeating the Society Hill approach in another place is difficult if the place does not contain significant eighteenth-century architecture, just as projects like Penn Center would be impossible in cities without suburban train stations and subway systems.

Dimensions, arrangement, and character of the components of the plan play important roles. Portland began with blocks that were 200 feet square. This frequency of streets and sidewalks encouraged greater than usual reliance on pedestrian activity. The blocks created by the 1683 plan for Philadelphia were 425 feet by 500 to 675 feet. By introducing greenways, Bacon was able to make these oversized blocks more pedestrian-friendly.

Bacon's scheme also rearranged the components of the city. Moving the produce market from Society Hill to South Philadelphia improved the efficiency of food distribution and removed an impediment to downtown residential development. Such rearrangement, though, is possible only when it reduces the cost of doing business. Rearranging the components of a city is futile if that rearrangement is not cost-effective. Gruen's proposal for creating a new pedestrian level for Fort Worth and banishing service vehicles to underground streets would have improved the delivery of goods and services and increased pedestrian and vehicular safety. The cost of retrofitting downtown Fort Worth, however, would have been borne by property owners. They would have had to pay to move ground-floor uses to a rebuilt second level, while losing revenue from at least one story. Understandably, they scuttled Gruen's plan.

Financing

There is no way to implement comprehensive planning without money. As both Burnham's *Plan of Chicago* and Bartholomew's *Comprehensive City Plan* for St. Louis demonstrate, this is not just a matter of dollars. Cities will appropriate billions when they believe they will be getting their money's worth. Burnham's proposals for the Chicago lakeshore proved easy to finance when it became clear that private developers would follow suit by building along the lakeshore and that new construction would generate more than enough additional tax revenues to retire the bond issues that paid for the park and parkway development. Gruen's diagram was applied to cities across the country because the federal government was ready to pay 90 cents for every 10 cents contributed locally. Bacon's plan for Society Hill became a reality only when it became eligible for federal urban renewal assistance.

Money is available for all sorts of projects. As New York City's Committee on the City Plan wrote in 1914, comprehensive planning "means getting the most out of the expenditures that are bound to be made and saving future expense in replanning and reconstruction." Both Bartholomew in St. Louis and Bacon in Philadelphia tried to develop equitable methods for prioritizing these expenditures. Consequently, both cities made expenditures in communities that did not have their fair share of public facilities. More important, Bacon and Bartholomew also identified major projects that would have citywide impact and thereby helped to create constituencies for their implementation.

If comprehensive plans are to be effective, they must, like the *Plan for New York*, specify all recommended projects, their cost, and the source of the money to pay for them. Not only is this responsible planning, but it also ensures that any plan that results from the planning process will be marketable to the constituencies with the power to approve it.

Time

Few comprehensive plans consider time in terms of movement through different parts of the city. Society Hill's greenways are among the rare exceptions. Fewer still examine the quality of life, 24 hours a day, throughout the year. The combination of Portland's riverfront park, concentration of cultural facilities, light-rail system, and downtown residential buildings have achieved what became long-discarded hopes for St. Louis. However, if cities thought in these terms, far fewer places would empty out in the morning when people go to work or become deserted at night after they have gone home.

The period of time least considered in comprehensive planning is the time it takes to implement recommendations. One reason for the effectiveness of the 1947 Better Philadelphia exhibition was that Edmund Bacon, the man largely responsible for the show, in 1949 became the executive director of the City Planning Commission and remained in office for 21 years. So much time is required for implementation that comprehensive plans can become obsolete quite quickly. Once downtown revival was under way in Philadelphia, the constituency for further spending waned and opponents were able to defeat the downtown highway loop. In New York City, once President Nixon terminated federal housing assistance programs, the ambitious urban renewal strategy proposed in the *Plan for New York City* was no longer feasible.

Successful comprehensive plans must be both politically and financially feasible. As Burnham explained, a comprehensive plan can only "be executed by degrees, as the growth of the community demands and as its financial ability allows."[65] For this reason, he recommended not indicating the exact details too closely. The more important reason so many proposals in his *Plan of Chicago* were executed is that they could be altered to meet the demands of the political and financial context of the period in which each implementation became possible. Authors of future plans ought to follow his example.

Entrepreneurship

The entrepreneurial problem in comprehensive planning is that planners do not implement their own proposals. Burnham solved the problem by helping to organize a marketing effort that inspired citizens to vote for bond issues, legislators to appropriate funds, civil servants to spend this money, and private developers to build his vision for the city. Bartholomew solved the problem by placing planners within the municipal government for which he was planning, thereby generating implementation from within the bureaucracy. Bacon's genius in getting his plans realized was in adjusting his proposals until they satisfied the requirements of both private developers and government officials. Those planners who forget that they do not operate in a vacuum, that they have neither the power nor the ability to make anything they propose happen by itself, are doomed to be forever irrelevant.

If comprehensive plans are to be implemented, they must specify the actions that are necessary and who is responsible for those actions. But even when the actions and the actors have been specified, effective public officials can accelerate implementation. By using the formal activities of the City Planning Commission, Donald Elliott was able to use the planning commission's role in drafting the city's capital budget (eliminated in the charter reform of 1975) to advance proposals listed in the *Plan for New York City*.

Philadelphia, 2009. Urban renewal activities restored Society Hill to its traditional role as one of the city's most desirable residential areas. *(Alexander Garvin)*

Likewise, Boo Thomas was able to keep proposals from *Plan Baton Rouge* in the forefront of that city's agenda by encouraging press coverage and promoting public discussion of the plan's contents.

Comprehensive Planning as a City Strategy

There is no way to force municipal governments to produce imaginative comprehensive plans like the *Plan for Chicago*. Nor is there any way to guarantee implementation. At best, one can legislate procedures that make planning issues a part of the public dialogue, provide the public with the information needed to avoid single-function decision making, and ensure that necessary personnel are in place should a municipal government decide to engage in comprehensive planning.

The significant word is *planning*, not *plan*. Virtually every city has produced at least one "comprehensive city plan." In most cases, these plans have lacked the vision to inspire public support, and their authors have lacked the political skill and drive needed to guarantee implementation. As a result, they are only comprehensively ignored.

When citizens have ready access to information about their neighborhoods and the city as a whole, there is a better chance for a political consensus to emerge and, along with it, the demand for comprehensive planning. Inexpensive computer technology now permits both the quantitative and graphic presentation of virtually any information on an area-by-area basis. There is, therefore, every reason to provide the electorate with the basis for making intelligent planning decisions about their neighborhoods. Citizens in the cities with websites that provide that information are increasingly able to participate in the planning process.

Information by itself does not change cities, however. Change requires a knowing interpretation of that informa-

Chicago, 2008 *(Alexander Garvin)*

tion, a constituency that seeks change, a specific set of actions that will implement those changes, money to pay for them, and an entity responsible for implementing them. The reason planning was so effective in early-twentieth-century Chicago, mid-twentieth-century Philadelphia, and early-twenty-first-century Baton Rouge is that the leadership of those cities worked so diligently to obtain relevant information about their city, hired people who could provide them with a knowing interpretation of that information, devised effective public action, developed the necessary constituency, raised the money to pay for proposed action, and ensured there was an entity ready, willing, and able to execute recommendations.

The plans they published provided visions of a future to which they aspired, but one that is subject to continuous adjustment that allows for regular disposal of proposals in response to the realities on the ground. Successful planning never rigidly adheres to initial conceptions. As Daniel Burnham explained, even the best-laid plans will have to be adjusted when "wider knowledge, longer experience, or change in local conditions may suggest a better solution."[66]

Chicago, Philadelphia, and Baton Rouge all proposed initiatives that would never work. Some, like the diagonal boulevards in the 1909 *Plan of Chicago,* failed because they were unnecessary and required unreasonable amounts of money and dislocation. Others, like Philadelphia's proposed South Street Expressway, succumbed to fierce local opposition. Planning succeeds only when there is a real partnership among the planners, the implementing entities, and the electorate, thereby ensuring that needed public action will be supported long enough to bring the desired projects to fruition. Only then will public action generate the widespread, sustainable private market reaction that is the product of successful planning.

Notes

1. Norton E. Long, *The Polity,* Rand McNally, Chicago, 1962, p. 192.
2. President Dwight D. Eisenhower, Speech to the National Executive Reserve Conference, Washington DC, November 14, 1957.
3. Despite the fact that there is no known source for this quotation, it is always attributed to Daniel Burnham. It is quoted in a 1918 Christmas card from Willis Polk to Edward Bennett as a statement by Burnham made in 1907.
4. Committee on the City Plan, *Development and Present Status of City Planning in New York City,* City of New York, Board of Estimate and Apportionment, Committee on the City Plan, New York, 1914, p. 12.
5. Edmund Bacon, public comments made at a conference sponsored by the Institute for Urban Design held in New York City, April 22, 1988.
6. Daniel H. Burnham and Edward H. Bennett, *Plan of Chicago* (first published in Chicago in 1909), Da Capo Press, New York, 1970, p. 1.
7. Historical material on the life of Daniel Burnham is derived from Thomas S. Hines, *Burnham of Chicago,* Oxford University Press, New York, 1974.
8. See Chapter 5, note 2.
9. Hines, op. cit., pp. 174–196.
10. While Edward Bennett was formally acknowledged as assisting in the preparation of the *Report on a Plan for San Francisco,* Willis Polk was not. Polk was an American architect who had spent two years with D. H. Burnham in Chicago before starting his own architectural practice in San Francisco. He closed his firm in order to take

charge of the San Francisco office of D. H. Burnham and Company. Polk's role in developing the *Report* is presented in Richard Longstreth, *On the Edge of the World,* MIT Press, Cambridge, MA, 1989, pp. 298–304.
11. Daniel H. Burnham, assisted by Edward H. Bennett, *A Report on a Plan for San Francisco, A Facsimile Reprint of the 1906 Plan,* Urban Books, Berkeley, CA, 1971, p. 35.
12. Mel Scott, *The San Francisco Bay Area: A Metropolis in Perspective,* University of California Press, Berkeley, CA, 1985, p. 109.
13. Charles Norton and Charles Wacker, chairman and vice chairman of the Chicago Plan Committee, were to become Burnham's major allies in this endeavor. In 1909, the Chicago City Council appointed Wacker the first chairman of the City Plan Commission, a position that gave him a primary role in getting the *Plan of Chicago* adopted and then implemented. Norton, who left Chicago in 1909, became a major advocate for planning in New York City, where in 1914 he chaired the Advisory Commission that produced *Development and Present Status of City Planning in New York City,* later spearheading the creation of the Committee on Regional Plan, which produced the 1929 *Regional Plan and Survey of New York and Its Environs.*
14. Hines, op. cit., pp. 318–322.
15. Robert Bruegmann, "Burnham, Guerin, and the City as Image," in *The Plan of Chicago: 1909–1979,* Catalogue to the Exhibition of the Burnham Library of Architecture, The Art Institute, Chicago, 1979, pp. 16–28.
16. Burnham and Bennett, *Plan of Chicago,* op. cit., p. 108.
17. Ibid., p. 1.
18. For a more detailed account of the Chicago planning process, see Alexander Garvin, *The Planning Game: Lessons from Great Cities,* W. W. Norton & Company, New York, 2013, pp. 97–122.
19. Chicago Plan Commission, *Ten Years' Work of the Chicago Plan Commission 1909–1919,* Chicago Plan Commission, Chicago, 1920, pp. 2–7.
20. Harold M. Meyer and Richard C. Wade, *Chicago: Growth of a Metropolis,* University of Chicago Press, Chicago, 1969.
21. Burnham and Bennett, *Plan of Chicago,* op. cit., p. 2.
22. Chicago Plan Commission, op. cit., p. 13.
23. Walter D. Moody, *Wacker's Manual of the Plan of Chicago,* Chicago Plan Commission, Chicago, 1916.
24. Frederick Law Olmsted Jr., "Introductory Address on City Planning," Rochester, New York, May 2–4, 1910, reproduced in Roy Lubove, *The Urban Community,* Prentice-Hall, Englewood Cliffs, NJ, 1967, pp. 81–94. For a review of his city plans, see Susan L. Klaus, "Efficiency, Economy, Beauty: The City Planning Reports of Frederick Law Olmsted, Jr., 1905–1915," *Journal of the American Planning Association,* vol. 57, no. 4 (Autumn 1991), American Planning Association, Chicago, pp. 456–470.
25. Historical and statistical information on American city planning prior to 1929 is derived from Henry Vincent Hubbard, *Our Cities Today and To-morrow: A Survey of Planning and Zoning Progress in the United States,* Harvard University Press, Cambridge, MA, 1929. Additional material going through to 1965 can be found in Mel Scott, *American City Planning Since 1890: A History Commemorating the Fiftieth Anniversary of the American Institute of Planners,* University of California Press, Berkeley, CA, 1969, and through 1980, in M. Christine Boyer, *Dreaming the Rational City: The Myth of American City Planning,* MIT Press, Cambridge, MA, 1983.
26. http://www.planningaccreditationboard.org.
27. Secretary of Commerce Herbert Hoover, *Standard City Planning Enabling Act,* 1928.
28. Sec. 701(d), Housing Act of 1954.
29. William H. Wilson, *The City Beautiful Movement,* Johns Hopkins University Press, Baltimore, 1989, pp. 213–233.
30. Wilson, op. cit., pp. 213–233.
31. Virgil G. Bogue, *Plan of Seattle,* Lowman & Hanford Co., Seattle, WA, 1911, p. 11.
32. Victor Gruen, *The Heart of Our Cities,* Simon & Schuster, New York, 1964, p. 198.
33. Ibid.
34. Victor Gruen, "No More Offstreet Parking in Congested Areas," originally published in *The American City,* September 1959, and reprinted in *Readings in Urban Transportation,* George M. Smerk (editor), Indiana University Press, Bloomington, IN, 1968, pp. 78–81.
35. Victor Gruen & Associates, *A Greater Fort Worth Tomorrow,* Greater Fort Worth Planning Committee, Fort Worth, TX, 1956.
36. Jane Jacobs, "Downtown Is for People," in *The Exploding Metropolis,* The Editors of *Fortune,* Doubleday Anchor Books, Garden City, NY, 1957, p. 146.
37. Alexander Garvin, "Make No Little Plans," in *Unbuilt Cincinnati,* David R. Scheer and J. A. Chewning (editors), The Cincinnati Forum for Architecture and Urbanism, Cincinnati, OH, 1998, pp. 20–25.
38. Frederick Law Olmsted Jr., "President's Address of Welcome," *Proceedings of the Third National Conference on City Planning,* New York, June 3, 5, and 6, 1911, p. 12.
39. Historical and statistical information on Harland Bartholomew is derived from Bartholomew's city plans, from Eldridge Lovelace, *Harland Bartholomew: His Contributions to American City Planning,* University of Illinois Office of Printing Services, Urbana, IL, 1993, and from Norman J. Johnson, "Harland Bartholomew: Precedent for the Profession," in *The American Planner: Biographies and Recollections,* Donald A. Krueckeberg (editor), Methuen, New York, 1983, pp. 279–300.
40. E. P. Goodrich and George B. Ford, *City Planning for Newark,* L.J. Hardham Printing Co., Newark, NJ, 1913, p. iv.
41. Johnson, op. cit., p. 284.
42. Ibid.
43. Lovelace, op. cit., pp. A-15–A-19.
44. Harland Bartholomew, *Comprehensive City Plan, St. Louis, Missouri,* City Plan Commission, St. Louis, MO, 1947, pp. 70–74.
45. U.S. Bureau of the Census.
46. The 1961 City Charter provided for a City Planning Commission with seven members, six appointed by the mayor for staggered eight-year terms and a chairman who served at the pleasure of the mayor. The charter also designated the chairman as director of the Department of City Planning, which was responsible for a variety of ordinary government functions (maintaining the city map, administering the zoning ordinance, preparing a five-year capital improvement program, etc.) and for providing the mayor, the planning commission, and the city's elected officials with information and advice regarding all matters related to the development of the city.
47. The Housing Act of 1954 required every community receiving urban renewal assistance to have a *workable program* for community improvement. The workable program was required to include seven components: (1) codes and ordinances, (2) a comprehensive plan, (3) neighborhood analyses, (4) an effective administrative organization, (5) a financing plan, (6) a relocation plan, and (7) a citizen participation program. At first, most communities didn't qualify. To continue receiving assistance, therefore, they had to demonstrate steady progress toward completing each component. The comprehensive plan itself was required to include six components: (1) a land use plan, (2) a thoroughfare plan, (3) a community facilities plan, (4) a public improvements program, (5) a zoning ordinance, and (6) a subdivision ordinance. While New York City had long argued that all six already existed, local officials of the Department of Housing and Urban Development kept pressing for a single document that tied them together into a coherent planning strategy.
48. Donald H. Elliott, interview, April 18, 1991.
49. The 1961 City Charter required the City Planning Commission to designate community planning districts, each with an appointed planning board that advised elected officials and city agencies on planning issues. In 1968, the City Planning Commission designated 62 districts. This number was reduced to 59 pursuant to the charter changes of 1975.
50. New York City Planning Commission, *Plan for New York City,* vol. 1, Department of City Planning, 1969, p. 5.
51. New York City Planning Commission, *Plan for New York City,* vol. 5, Department of City Planning, 1969, p. 25.
52. Ibid.
53. Historical material on city planning in Philadelphia is derived from

John F. Bauman, *Public Housing Race and Renewal—Urban Planning in Philadelphia 1920–1974,* Temple University Press, Philadelphia, 1987; Jeanne R. Lowe, *Cities in a Race for Time,* Random House, New York, 1967, pp. 313–404; Edmund Bacon, *Design of Cities,* Viking Press, New York, 1967, pp. 243–271; Robert B. Mitchell (editor), "Special Issue: Planning and Development in Philadelphia," *Journal of the American Institute of Planners,* vol. 26, no. 3 (August 1960), American Institute of Planners, Baltimore, pp. 155–241; "Philadelphia Story," *Progressive Architecture,* Reinhold Publishing, Stamford, CT, April 1976, pp. 45–83; Michelle Osborne, "A History of the Ups and Downs That Finally Resulted in the Defeat of The Expressway," *Architectural Forum,* Whitney Publications, New York, October 1971, pp. 39–41; and Gregory L. Heller, "Biography of Edmund N. Bacon" (working title), Unpublished MS, January 2010.

54. Bacon, op. cit., pp. 243–271; and Garvin, op. cit., p. 170.
55. Ibid., pp. 175–182.
56. Ibid., pp. 182–188.
57. Edmund Bacon, interview with the author, November 22, 1998.
58. Garvin, op. cit., pp. 191–193.
59. Bacon, op. cit., pp. 243–271.
60. Garvin, op. cit., pp.194–195.
61. Paul Davidoff, "Advocacy and Pluralism in Planning," *Journal of the American Institute of Planners,* vol. 31, American Institute of Planners, Chicago, November 1965, pp. 186–197; Richard Bolan, "Emerging Views of Planning," *Journal of the American Institute of Planners,* vol. 33, American Institute of Planners, Chicago, July 1967, pp. 233–245; Edmund M. Burke, "Citizen Participation Strategies," *Journal of the American Institute of Planners,* vol. 34, American Institute of Planners, Chicago, September 1965, pp. 186–197.
62. Carl Abbott, Portland: Planning, Politics, and Growth in a Twentieth-Century City, University of Nebraska Press, Lincoln, NE, 1983, pp. 199–202.
63. Statistical and historical material on Portland is derived from http://www.travelportland.com and Bay Area Economics, http://www.portlandoregon.gov/bps/article/322772.
64. For a more detailed account of *Plan Baton Rouge,* see Alexander Garvin, "A Mighty Turnout in Baton Rouge," *Planning,* vol. 64, no. 10 (October 1998), American Planning Association, Chicago, pp. 18–20; and Alexander Garvin, "Baton Rouge Presses Ahead to Implement Downtown Plan," *Planning,* vol. 65, no. 9 (September 1999), American Planning Association, Chicago; http://cpex.org/work/plan-baton-rouge; http://www.downtownbatonrouge.org; and http://www.shawcenter.org.
65. Burnham and Bennett, *A Report on a Plan for San Francisco,* op. cit., p. 35.
66. Burnham and Bennett, *Plan of Chicago,* op. cit., p. 2

Index

Numbers in italics indicate pages with illustrations